Abiotic Stress Responses in Plants

Parvaiz Ahmad • M.N.V. Prasad
Editors

Abiotic Stress Responses in Plants

Metabolism, Productivity and Sustainability

Springer

Editors
Parvaiz Ahmad
Department of Botany
Amar Singh College
University of Kashmir
Srinagar, Jammu and Kashmir
India
parvaizbot@rediffmail.com

M.N.V. Prasad
Department of Plant Sciences
University of Hyderabad
Andhra Pradesh, Hyderabad
India
prasad_mnv@yahoo.com
mnvsl@uohyd.ernet.in

ISBN 978-1-4614-0633-4 e-ISBN 978-1-4614-0634-1
DOI 10.1007/978-1-4614-0634-1
Springer New York Dordrecht Heidelberg London

Library of Congress Control Number: 2011940823

Printed on acid-free paper

Springer is part of Springer Science+Business Media (www.springer.com)

Preface

Climate constrained world represents an ideal scenario of abiotic stresses in which there has been a change in the statistical distribution of weather (temperature, soil moisture, salinity, ecohydrology, soil fertility, emission of greenhouse gases, etc.) over periods of time that range from decades to centuries to millions of years. Plants do respond to these changes in the process of acclimation and acquiring tolerance – morphologically, structurally, physiologically, biochemical and molecular mechanisms.

Abiotic stress cause changes in soil–plant–atmosphere continuum which is responsible for reduced yield in several of the major crops in different parts of the world. Therefore, the subject of abiotic stress response in plants – metabolism, productivity and sustainability is gaining considerable significance in the contemporary world.

This is a collective and companion volume to our previous edition *Environmental Adaptations and Stress Tolerance of Plants in the Era of Climate Change*. This volume deals with an array topics in the broad area of abiotic stress responses in plants focusing "*metabolism, productivity and sustainability*" by selecting some of the widely investigated themes. Chapter 1: Abiotic stress responses in plants – present and future. Chapter 2: Abiotic stress-induced morphological and anatomical changes in plants. Chapter 3: Abiotic stress responses in plants – metabolism to productivity. Chapter 4: Approaches to increasing salt tolerance in crop plants. Chapter 5: Understanding and exploiting the impact of drought stress on plant physiology. Chapter 6: Sustainable fruit production in Mediterranean orchards subjected to drought stress. Chapter 7: Drought stress-induced reactive oxygen species and antioxidants in plants. Chapter 8: Role of glutathione reductase in plant abiotic stress. Chapter 9: Flavonoids as antioxidants in plants under abiotic stresses. Chapter 10: Proteomic markers for oxidative stress – new tools for reactive oxygen species and photosynthesis research. Chapter 11: Environmental stress and role of arbuscular mycorrhizal symbiosis. Chapter 12: Effects of exogenous application of 5-aminolevulinic acid (ALA) in crop plants. Chapter 13: Abiotic stress and role of salicylic acid in plants. Chapter 14: Trehalose and abiotic stress tolerance. Chapter 15: Uptake of mineral elements during abiotic stress. Chapter 16: Effect of micronutrient deficiencies on plants stress responses. Chapter 17: Stress-induced flowering. Chapter 18: Postharvest stress treatments in fruits and vegetables. Chapter 19: Abscisic acid signalling in plants. Chapter 20: Plant tolerance and fatty acid profile in responses to heavy metals. Chapter 21: Cadmium accumulation and subcellular distribution in plants and their relevance to the trophic transfer of Cd. Chapter 22: The role of soil organic matter in trace element bioavailability and toxicity. Chapter 23: Oxidative stress and phytoremediation. Chapter 24: Phytoremediation of low levels of heavy metals using duckweed (*Lemna minor*). We fervently believe that this volume will provide good information and understanding of abiotic stress tolerance in plants.

We are extremely thankful to all the contributors for comprehensive and cogent reviews which ultimately resulted in the present form. We are pleased to place on record the superb and skillful job of Amna Ahmad, Andy Kwan and the rest of the technical team at the production unit for publishing this work in record time.

Srinagar, Jammu & Kashmir, India — Parvaiz Ahmad
Hyderabad, Andhra Pradesh, India — M.N.V. Prasad

Contents

Contributors

Parvaiz Ahmad
Department of Botany, A.S. College, University of Kashmir, Srinagar, Jammu & Kashmir, India

Adolfo David Arenas
Dpto. de Química y Suelos, Decanato de Agronomía, Universidad Centroccidental Lisandro Alvarado, Cabudare, Venezuela

Graziella Berta
Dipartimento di Scienze dell'Ambiente e della Vita, Università del Piemonte Orientale "Amedeo Avogadro", Alessandria, Italy

Cecilia Brunetti
Department of Plant, Soil and Environmental Science, University of Florence, Sesto Fiorentino, Firenze, Italy

Teresa Casacchia
CRA, Centro di Ricerca per l'Olivicoltura e l'Industria Olearia, Rende, Cosenza, Italy

Ruby Chandna
Molecular Ecology Lab, Department of Botany, Jamia Hamdard, Hamdard Nagar, New Delhi, India
National Institute for Plant Genomics and Research, New Delhi, India

Bartolomeo Dichio
Dipartimento di Scienze dei Sistemi Colturali, Forestali e dell'Ambiente, Università degli Studi della Basilicata, Potenza, Italy

Kinga Drzewiecka
Department of Chemistry, University of Life Sciences, Poznań, Poland

Fatih Duman
Department of Biology, Faculty of Science, Erciyes University, Kayseri, Turkey

Martina Di Ferdinando
Department of Plant, Soil and Environmental Science, University of Florence, Sesto Fiorentino, Firenze, Italy

Alessio Fini
Department of Plant, Soil and Environmental Science,
University of Florence, Sesto Fiorentino, Firenze, Italy

Rebecca Ford
Department of Agriculture and Food Systems, Melbourne School of Land and Environment, The University of Melbourne, Parkville, VIC, Australia

Jun Furukawa
Graduate School of Life and Environmental Sciences,
University of Tsukuba, Tsukuba, Japan

Anna Fusconi
Dipartimento di Biologia Vegetale, Università degli Studi di Torino,
Torino, Italy

Piotr Goliński
Department of Chemistry, University of Life Sciences, Poznań, Poland

Olga M. Grant
Department of Biology, National University of Ireland, Maynooth,
Co. Kildare, Ireland

R. Hajiboland
Plant Science Department, University of Tabriz, Tabriz, Iran

Khalid Ul Rehman Hakeem
Molecular Ecology Lab, Department of Botany, Jamia Hamdard,
Hamdard Nagar, New Delhi, India

Asiya Hameed
Department of Botany, Jamia Hamdard, Hamdard Nagar, New Delhi, India

Miyuki Hara
Graduate School of Life and Environmental Sciences,
University of Tsukuba, Tsukuba, Japan

Yoshihiro Imahori
Graduate School of Life and Environmental Sciences,
Osaka Prefecture University, Osaka, Japan

S.M. Impa
Crop and Environmental Sciences Division, International Rice
Research Institute, Metro Manila, Philippines

S.V. K. Jagadish
Crop and Environmental Sciences Division, International Rice
Research Institute, Metro Manila, Philippines

Ratna Karan
School of Plant, Environmental, and Soil Sciences, Louisiana State
University Agricultural Center, Baton Rouge, LA, USA

Ahmet Korkmaz
Department of Horticulture, Faculty of Agriculture,
Kahramanmaras Sutcu Imam University, Kahramanmaras, Turkey

Carmen Lluch
Departamento de Fisiología Vegetal, Facultad de Ciencias,
Universidad de Granada, Granada, Spain

Miguel López-Gómez
Departamento de Fisiología Vegetal, Facultad de Ciencias,
Universidad de Granada, Granada, Spain

Andrea Furtado Macedo
Laboratório Integrado de Biologia Vegetal, Departamento de Botânica,
Instituto de Biociências, CCBS, Universidade Federal do Estado do Rio
de Janeiro, Rio de Janeiro, RJ, Brazil

Mahmooduzzafar
Department of Botany, Jamia Hamdard, Hamdard Nagar, New Delhi, India

Nitin Mantri
School of Applied Sciences, Health Innovations Research Institute,
RMIT University, Melbourne, VIC, Australia

Kenji Miura
Graduate School of Life and Environmental Sciences,
University of Tsukuba, Tsukuba, Japan

Tsuyoshi Mizoguchi
Graduate School of Life and Environmental Sciences,
University of Tsukuba, Tsukuba, Japan

Mirosław Mleczek
Department of Chemistry, University of Life Sciences, Poznań, Poland

M.S. Monteiro
CESAM and Department of Biology, University of Aveiro, Aveiro, Portugal

S. Nadaradjan
Crop Physiology Unit, Department of Plant Breeding and Genetics,
Pandit Jawaharlal Nehru College of Agriculture and Research Institute,
Karaikal, Puducherry, India

Gabrijel Ondrasek
Faculty of Agriculture, University of Zagreb, Zagreb, Croatia

Assunta Maria Palese
Dipartimento di Scienze dei Sistemi Colturali, Forestali e dell'Ambiente,
Università degli Studi della Basilicata, Potenza, Italy

Edwin Pang
School of Applied Sciences, Health Innovations Research Institute,
RMIT University, Melbourne, VIC, Australia

Lué-Merú Marcó Parra
Dpto. de Química y Suelos, Decanato de Agronomía,
Universidad Centroccidental Lisandro Alvarado, Cabudare, Venezuela

Vikas Patade
Defence Research and Development Organisation,
Defence Institute of Bio-Energy Research, Goraparao, Uttarakhand, India

Angelos Patakas
Laboratory of Plant Production, University of Ioannina, Agrinio, Greece

Suprasanna Penna
Functional Plant Biology Section, Nuclear Agriculture and Biotechnology Division, Bhabha Atomic Research Centre, Trombay, Mumbai, India

Tabasum N. Qadri
Department of Botany, Jamia Hamdard, Hamdard Nagar, New Delhi, India

Zed Rengel
School of Earth and Environment, University of Western Australia, Crawley, WA, Australia

Korina Rodríguez
Dpto. de Química y Suelos, Decanato de Agronomía,
Universidad Centroccidental Lisandro Alvarado, Cabudare, Venezuela

Erick Sánchez
Dpto. de Química y Suelos, Decanato de Agronomía,
Universidad Centroccidental Lisandro Alvarado, Cabudare, Venezuela

Aiko Sato
Graduate School of Life and Environmental Sciences,
University of Tsukuba, Tsukuba, Japan

T.O. Siddiqi
Department of Botany, Jamia Hamdard, Hamdard Nagar, New Delhi, India

A.M.V.M. Soares
CESAM and Department of Biology, University of Aveiro, Aveiro, Portugal

Adriano Sofo
Dipartimento di Scienze dei Sistemi Colturali, Forestali e dell'Ambiente,
Università degli Studi della Basilicata, Potenza, Italy

Prasanta K. Subudhi
School of Plant, Environmental, and Soil Sciences, Louisiana State University Agricultural Center, Baton Rouge, LA, USA

Kiyotoshi Takeno
Department of Biology, Faculty of Science, Niigata University, Ikarashi, Niigata, Japan

Massimiliano Tattini
Consiglio Nazionale delle Ricerche, Istituto per la Protezione delle Piante,
Sesto Fiorentino, Firenze, Italy

Gosmyr Torres
Dpto. de Química y Suelos, Decanato de Agronomía,
Universidad Centroccidental Lisandro Alvarado, Cabudare, Venezuela

Radomira Vankova
Laboratory of Hormonal Regulations in Plants, Institute of Experimental Botany AS CR, Prague, The Czech Republic

Agnieszka Waśkiewicz
Department of Chemistry, University of Life Sciences, Poznań, Poland

Cristos Xiloyanni
Dipartimento di Scienze dei Sistemi Colturali, Forestali e dell'Ambiente, Università degli Studi della Basilicata, Potenza, Italy

Peerzada Yasir Yousuf
Molecular Ecology Lab, Department of Botany, Jamia Hamdard, Hamdard Nagar, New Delhi, India

Abiotic Stress Responses in Plants: Present and Future

1

Nitin Mantri, Vikas Patade, Suprasanna Penna, Rebecca Ford, and Edwin Pang

Abstract

Drought, cold, high-salinity and heat are major abiotic stresses that severely reduce the yield of food crops worldwide. Traditional plant breeding approaches to improve abiotic stress tolerance of crops had limited success due to multigenic nature of stress tolerance. In the last decade, molecular techniques have been used to understand the mechanisms by which plants perceive environmental signals and further their transmission to cellular machinery to activate adaptive responses. This knowledge is critical for the development of rational breeding and transgenic strategies to impart stress tolerance in crops. Studies on physiological and molecular mechanisms of abiotic stress tolerance have led to characterisation of a number of genes associated with stress adaptation. Techniques like microarrays have proven to be invaluable in generating a list of stress-related genes. Some of these genes are specific for a particular stress while others are shared between various stresses. Interestingly, a number of

N. Mantri (✉) • E. Pang
School of Applied Sciences, Health Innovations Research Institute, RMIT University, Melbourne 3000, VIC, Australia
e-mail: nitin.mantri@rmit.edu.au

V. Patade
Defence Research and Development Organisation, Defence Institute of Bio-Energy Research, Goraparao, Dist-Nainital 263139, Uttarakhand, India

S. Penna
Functional Plant Biology Section, Nuclear Agriculture and Biotechnology Division, Bhabha Atomic Research Centre, Trombay, Mumbai 400 085, India

R. Ford
Department of Agriculture and Food Systems, Melbourne School of Land and Environment, The University of Melbourne, Parkville 3010, VIC, Australia

P. Ahmad and M.N.V. Prasad (eds.), *Abiotic Stress Responses in Plants: Metabolism, Productivity and Sustainability*, DOI 10.1007/978-1-4614-0634-1_1, © Springer Science+Business Media, LLC 2012

genes are shared in abiotic and biotic stress responses. This highlights the complexity of stress response and adaptation in plants. There is a whole cascade of genes involved in abiotic stress tolerance; starting from stress perception to transcriptional activation of downstream genes leading to stress adaptation and tolerance. A number of these genes have been discovered but we still do not have the complete list with all interactions. There is also significant number of genes with unknown functions found to be regulated by abiotic stresses. Understanding the function of these genes and their interaction with other known genes to effect stress adaptation is required.

The recent discovery that microRNAs regulate gene expression adds another layer of complexity to our understanding of abiotic stress tolerance. Significant amount of work will be needed to identify microRNAs associated with abiotic stress response, and understand their interaction with each other and their mechanism of regulating abiotic stress response. The promising side is the development of next-generation sequencing techniques that has allowed deep sequencing of mRNAs and microRNAs associated with abiotic stress response. A complete understanding on physiological and molecular mechanisms especially signalling cascades in response to abiotic stresses in tolerant plants will help to manipulate susceptible crop plants and increase agricultural productivity in the near future.

Keywords

Abiotic stress • Antioxidants • Ion homeostasis • MicroRNA • Osmotic adjustments • Signal transduction • Transgenic approaches

1 Introduction

The major abiotic stresses (drought, high salinity, cold, and heat) negatively influence the survival, biomass production and yields of staple food crops up to 70% (Vorasoot et al. 2003; Kaur et al. 2008; Thakur et al. 2010) hence, threaten the food security worldwide. Dehydration stress imparted by drought, salinity and temperature severity is the most prevalent abiotic stress that limits plant growth and productivity (Vorasoot et al. 2003; Jaleel et al. 2009; Thakur et al. 2010). Since tolerance to this stress is multigenic and quantitative in nature (Collins et al. 2008), a massive challenge exists to understand the key molecular mechanisms for advanced selective breeding purposes. Traditional plant-breeding approaches have been marginally successful in improving the tolerances to these stresses (Flowers and Yeo 1995; Flowers et al. 2000). The molecular mechanisms underlying abiotic stress tolerances in plants are being unravelled with various high throughput sequencing and functional genomics tools in particular to advance the understanding of stress signal perception and transduction of the associated molecular regulatory networks (Heidarvand and Amiri 2010; Ray et al. 2010; Sanchez et al. 2011). Understanding the mechanisms by which plants perceive environmental signals and further their transmission to cellular machinery to activate adaptive responses is of critical importance for the development of rational breeding and transgenic strategies to impart stress tolerance in crops. Ultimately, marrying the physiological, biochemical and gene regulatory network knowledge will be essential to develop or select for stress-tolerant and high-yielding food crop cultivars.

2 Physiological and Molecular Mechanisms of Abiotic Stress Tolerance

2.1 Ion Transport and Homeostasis

The effect of salinity on plant growth limitation is proposed to be due to the osmotic effect from the ion imbalance in the earlier phase and a direct effect of the ions themselves in the latter phase of low to moderate stress (Munns and Tester 2008). At high salinity levels, salt-sensitive species lack the ability to control Na^+ transport, where ionic effects dominate the osmotic effect. For normal metabolic reactions, plant cells need to maintain high K^+ (100–200 mM) and low Na^+ (less than 10–20 mM) levels (Flowers and Dalmond 1992; Carden et al. 2003). Therefore, tolerance to salinity stress must involve maintaining or quickly re-establishing both osmotic and ionic homeostasis (Munns and Tester 2008).

In general, plants employ one or both of the following to survive high salinity environments to ensure internal osmotic and ionic homeostasis; (1) avoidance – to keep sensitive plant tissues away from regions of concentrated salt ions and (2) tolerance – to exclude ions from roots or compartmentalise ions away from the cytoplasm of physiologically active cells (Silva et al. 2010). Indeed, efficient exclusion of excess Na^+ ions from the cytoplasm and accumulation of Na^+ ions within vacuoles are the main adaptive tolerance mechanisms to salinity stress (Munns and Tester 2008). Exclusion is typically carried out by transmembrane transport proteins that exclude Na^+ from the cytosol in exchange for H^+, a secondary transport process which is energy-dependent and driven by the proton motive force generated by the plasma membrane H^+-ATPase. Likewise, compartmentalisation is generally carried out by vacuolar membrane H^+-ATPase and H^+-pyrophosphatase proteins (Rodríguez-Rosales et al. 2009; Ye et al. 2009; Leidi et al. 2010; Pasapula et al. 2011). By increasing the cellular levels of proteins (such as vacuolar antiporter proteins), number of abiotic stress-tolerant transgenic plants have been produced to control the transport functions. *AtSOS* from *Arabidopsis* has been shown to encode plasma membrane Na^+/H^+ antiporter (*NHX*) with significant sequence similarity to the respective antiporter from bacteria and fungi (Shi et al. 2000). Overexpression of the vacuolar Na^+/H^+ antiporter (*NHX1*) or the *Arabidopsis thaliana vacuolar H^+-translocating pyrophosphatase* (*AVP1*) gene energized the pumping of Na^+ into the vacuole and increased both accumulation and tolerance to Na^+ (Gaxiola et al. 2001; Pasapula et al. 2011). More efficient sequestration of these ions to the vacuole may improve tissue tolerance to salinity by reducing the cytosolic Na^+ concentrations. The importance of Na^+ sequestration in salt tolerance has been further demonstrated in transgenic plants overexpressing *AtNHX1* (Leidi et al. 2010; Silva et al. 2010).

2.2 Osmotic Adjustments and Controlling Factors

Intracellular water lost from the cell due to salt, drought and cold, leads to cellular dehydration. To prevent this and protect the cellular proteins, plants accumulate many organic compounds such as amino acids (proline), quaternary and other amines (glycine betaine and polyamines), a variety of sugars (mainly fructose and sucrose), sugar alcohols, complex sugars (like trehalose and fructans) and organic acids (oxalate, malate) (Valliyodan and Nguyen 2006). These metabolites with osmolytic function are also known as compatible solutes or osmoprotectants and may accumulate to high levels without disturbing the intracellular biochemistry (Ford 1984). By reducing the water potential within the cell, water loss is prevented and osmotic adjustment is facilitated (Delauney and Verma 1993).

Transgenic studies have been carried out for developing tolerant genotypes through manipulation of enzymes that synthesize specific osmolytes (Chen and Murata 2008; Szabados and Savoure 2010). The success of these studies on imparting stress tolerance has varied since the function of the targeted osmolytes is not restricted to osmotic adjustment, but also confers osmoprotection (Krishnan et al. 2008). In several studies, accumulation of osmolytes provided protection through scavenging of reactive oxygen species (ROS) and chaperone-like activities in maintaining protein structures and functions (Bohnert and Shen 1999; Krishnan et al. 2008). Pleiotropic effects such as necrosis and growth retardation were also observed due to disturbances in endogenous pathways of primary metabolisms. Patade et al. (in press) reported differential osmotic adjustment in sugarcane where salt-stressed plants appeared to use salt as an osmoticum and PEG stressed plants relied on accumulation of sugars.

A substantial number of transgenic studies have been performed to overexpress genes encoding osmoprotectants such as glycine-betaine (Bensen et al. 2008; Chen and Murata 2008) and proline (Delauney and Verma 1993; Verdoy et al. 2006; Szabados and Savoure 2010). Also, a number of sugars and sugar alcohols (mannitol, trehalose, myo-inositol and sorbitol) have been targeted for the engineering of compatible-solute overproduction, thereby protecting the membrane and protein complexes during stress (Gao et al. 2000; Suprasanna et al. 2005). In particular, there is a growing body of research on trehalose metabolism as a means of engineering stress tolerance in crops (Suprasanna 2003). In transgenic tomato, trehalose overproduction using the yeast *trehalose-6-phosphate synthase gene* led to significant tolerance to salinity, drought and oxidative stress (Cortina and Culianez-Macia 2005). Similarly, transgenic plants engineered for the overexpression of polyamines exhibited increased tolerance to multiple abiotic stresses such as heavy metal, salinity, drought, low and high temperature and fungal disease resistance (Capell et al. 2004; Prabhavathi and Rajam 2007).

Aside from osmotolerance, osmolyte accumulation also plays a vital role in the maintenance of cellular activities. For example, proline accumulation through overexpression of the *P5CS* gene in *Medicago truncatula* resulted in enhanced osmotolerance and also aided in maintaining nitrogen-fixing activity under osmotic stress (Verdoy et al. 2006). In order to minimize possible negative pleiotropic effects such as those previously mentioned, engineering of pathways for overproduction of compatible solutes should be through stress-inducible and/or tissue specific regulation (Su and Wu 2004).

2.3 Cold Acclimation

Plants survive freezing temperatures either through avoidance, primarily by super cooling of tissue water, or through freezing tolerance. Several plant species have the ability to increase freezing tolerance (FT) in response to low non-freezing temperatures (below 10°C) within a short photoperiod, a phenomenon known as cold acclimation (Thomashow 2010). The level of FT obtained through cold acclimation is not static but can vary seasonally and is rapidly lost upon return to a warm non-acclimating temperature. FT can be induced by osmotic stresses (Li et al. 2002) as well as treatment with abscisic acid (Li et al. 2003). Programmed dehydration is characteristic of overwintering tissues and at least partly contributes to FT (Welling et al. 2004). Further, cellular changes related to accumulation of storage proteins, sugars and starch are triggered by alteration of source–sink relationships after growth cessation in response to low temperature exposure (Zhu and Coleman 2001). The essential accumulation of sugars for cold acclimation was demonstrated by the inability of an *Arabidopsis* sucrose synthase mutant to cold acclimate (Uemura et al. 2003). The high abundance of sugars in cold acclimated plants suggests a role in osmoregulation, whereas less abundant sugars might have a role in cryoprotection or as signalling molecules (Stitt and Hurry 2002).

Recent progress has been made in elucidating the physiological and molecular mechanisms underpinning freezing tolerance. FT is a genetically complex trait, reflected by large number of genes that are affected by low temperature, thus estimated to be up to 25% of the entire transcriptome (Krebs et al. 2002). Altered expression of specific cold responsive-*COR* genes results in various physiological and biochemical changes during the process of cold acclimation, and the combined effect of the gene products is manifested in the level of FT obtained (Chinnusamy et al. 2006; Novillo et al. 2007; Thomashow 2010). The activation of *COR* genes is controlled by a set of signalling pathways triggered by exposure to the LT stimulus (Chinnusamy et al. 2006). The *A. thaliana* CBF (C-repeat binding factor) cold response pathway is most likely the best understood regulatory pathway involved in cold acclimation. This occurs through a rapid cold induction of CBF transcription factors, followed by expression of the regulon genes, which imparts freezing tolerance (Thomashow 2010). Specifically, in *Arabidopsis*, three CBF genes, *CBF1* (*DREB1b*), *CBF2* (*DREB1c*) and *CBF3* (*DREB1a*), were induced within 15 min of low temperature exposure (Gilmour et al. 1998; Liu et al. 1998). These CBFs encodes closely related members of the AP2/ERF (Apetala2/Ethylene-responsive element binding factor), a family of transcription factors (Riechmann et al. 2000) that binds to CRT/DRE (C-repeat/dehydration responsive element) DNA regulatory elements found in the promoters of CBF-targeted genes (Stockinger et al. 1997; Liu et al. 1998). The CBF proteins induce the expression of many CBF regulon genes (Maruyama et al. 2004; Vogel et al. 2005). This leads to an increase in freezing tolerance (Jaglo-Ottosen et al. 1998; Liu et al. 1998) through the accumulation of low molecular weight cryoprotective metabolites, such as raffinose, sucrose and proline (Cook et al. 2004; Kaplan et al. 2004), along with the production of cryoprotective polypeptides, such as COR15a (Steponkus et al. 1998).

Cytoskeletal reorganization serves as a link between membrane rigidification and Ca^{2+} influx in the early stages of cold acclimation, and is needed for the development of maximum FT (Orvar et al. 2000; Sangwan et al. 2002). Low-temperature-induced changes in cytosolic calcium correlate with the expression of cold-responsive genes and the development of FT. In *Arabidopsis*, the increase in cytosolic calcium comes from rapid cold-induced release of calcium from both extracellular and vacuolar stores (Knight et al. 1996). Following the cold stimulus, the Ca^{2+} homeostasis in cells is restored to resting levels by active Ca^{2+} transporters. A connection between the calcium spikes and cold-regulated gene expression has been demonstrated to involve induction of DREB genes (Shinozaki and Yamaguchi-Shinozaki 1996). Overexpression of *CBF1*, a *DREB1A* homolog, enhanced freezing-stress tolerance and increased the expression of cold regulated genes (*cor15a*, *cor6.6*, and *cor47*) (Jaglo-Ottosen et al. 1998). Overexpression of *DREB1A* also enhanced drought and salt tolerance in transgenic plants, demonstrating the cross-stress protective function of this gene family (Kasuga et al. 1999).

In *Arabidopsis*, the rapid influx of calcium into the cytosol is required for normal cold induction of the CBF target genes *KIN1* and *KIN2* (Knight et al. 1996; Tahtiharju et al. 1997). Accumulation of dehydrins, proteins which accumulate in vegetative tissues during dehydration stresses, was linked to the development of FT both in herbaceous and woody plants (Peng et al. 2008; Xu et al. 2008). Recently, a close link between the up-regulation of low temperature-associated proteins and vernalization fulfilment in wheat (*Triticum aestivum*) was reported (Sarhadi et al. 2010).

2.4 Antioxidant Defence for Abiotic Stress Tolerance

Reactive Oxygen Species (ROS) such as singlet oxygen, hydrogen peroxide molecules, superoxide and hydroxyl radicals are constantly produced in chloroplasts, mitochondria and peroxisomes by aerobic processes (Apel and Hirt 2004). Thought to be integral to downstream defense/tolerance responses, the elevated levels of ROS are often associated with exposure to biotic (e.g. pathogens or pests) and abiotic (e.g. high light, UV radiation, temperature extremes, heavy metals, air pollutants, drought stress, salt stress, mechanical/physical stress) factors (Neill et al. 2002; Imlay 2003; Einset et al. 2007). Overproduction of ROS leads to oxidative damage such as lipid peroxidation of cell membranes (Imlay 2003) or even cell death (Jones 2000). In order to control ROS levels and protect cells from oxidative injury, plants possess both enzymes and non-enzymatic metabolites that may play a significant role in ROS signalling in plants (Vranova et al. 2002).

The harmful effects of ROS are prevented by the presence of lipid soluble antioxidants (α-tocopherol and carotenoids), water-soluble reductants (glutathione and ascorbate) and antioxidant enzymes such as catalase (CAT, EC 1.11.1.6), ascorbate peroxidase (APX, EC 1.11.1.11) and superoxide dismutase (SOD, EC 1.15.1.1) present in plant cells (Desikan et al. 2004). In response to stress, some of the osmolytes accumulate in plant cells, besides a role in scavenging of free radicals and protecting enzymes (Krishnan et al. 2008). The ability to activate protective mechanisms, such as an increase in the activity of scavenging enzymes, is vital for oxidative stress tolerance. Transgenic improvements for abiotic stress tolerances have been achieved through detoxification strategies by overexpressing the enzymes involved in oxidative protection. For example, salt or thermal stress treatment inhibited the growth of wild tobacco and caused increased lipid peroxidation, while overexpression of tobacco glutathione-S-transferase (GST) and glutathione peroxidase (GPX) reduced oxidative damage in the stressed transgenic seedlings (Roxas et al. 2000). Furthermore, overexpression of *CuZn* superoxide dismutase (SOD) and ascorbate peroxidase (APX) in transgenic sweet potato enhanced tolerance and recovery from drought stress. This was due to a considerable increase in expression of antioxidant enzymes that reduced the levels of malondialdehyde and electrolyte leakage (Lu et al. 2010). Likewise, *Arabidopsis* transformed with antisense barley 2-cystein peroxiredoxin sequence resulted in high expression of APX and monodehydroascorbate reductase (MDHAR; Baier et al. 2000).

Overexpression of an alternative oxidase (AOX) gene reduced oxidative stress in transgenic *Arabidopsis* under cold exposure (Sugie et al. 2006). Vitamin E was shown to be another participant in the protective mechanism against oxidative stress since Vitamin-E deficient *Arabidopsis* mutants were chilling sensitive. This was proposed to be because of a defective export of photoassimilate (Zhu et al. 2007).

2.5 Signal Transduction in Response to Abiotic Stresses: Specificity and Cross-Talk

Abiotic stresses are complex stimuli (ionic imbalance and osmotic stress), perceived by multiple primary sensors that cause alteration in the expression of many genes. The cascade of molecular responses ranges from stress perception, to signal transduction to cytoplasm and nucleus, to gene expression and finally metabolic changes leading to stress tolerance. Rapid increase in cytosolic Ca^{2+} levels in response to the various environmental stress stimuli are controlled by four major families of calcium-binding proteins; calmodulins, calmodulin-like proteins, calcineurin B-like proteins and calcium-dependent protein kinases (CDPKs) (Snedden and Fromm 2001; Luan et al. 2002; Sanders et al. 2002). Following the Ca^{2+} influx, signals are proposed to be mediated by combinations of phosphorylation/dephosphorylation cascades and is thought to be controlled by members of the Ca^{2+}-dependent protein kinase (CDPK) gene family (Zhang et al. 2005). Members of the CDPK family are also reported to activate ABA/stress responsive gene expression. Altered expression of *Oryza sativa CDPK* (*OsCDPK*) has been correlated with tolerance to cold, salt and drought stress (Saijo et al. 2000).

Plants demonstrate both, stress-specific as well as shared responses that protect them from several environmental stresses (Mantri et al. 2010b). Plants respond to stress by regulating gene expression leading to both common and distinctive changes in transcript levels of stress responsive genes (Shinozaki and Yamaguchi-Shinozaki 2000). Indeed, overlap has been reported in gene expression induced by different stresses (Chen et al. 2002; Mantri et al. 2007; Seki et al. 2009). Plants universally appear to suffer from osmotic and oxidative stress under salt, drought and cold stress (Beck et al. 2007; Munns and Tester 2008). However, prevention of the osmotic stress is performed by stress-specific and general tolerance mechanisms. For example, in salt stress, osmotic adjustment maintains osmotic homeostasis while endurance through the period of freezing-induced osmotic stress relies on avoidance or interruption of ice nuclei formation (Pearce 2001).

Chen et al. (2002) identified groups of transcription factors regulated either singly i.e., abiotic stress (class I) or by both, biotic and abiotic stresses (class II) in *Arabidopsis*. Among the class I group, ~20 genes were preferentially induced by abiotic stresses such as salinity, osmotic, cold and jasmonic acid treatments. These transcription factors include DRE/CRT binding factors activated by cold stress, CCA1 and Athb-8 (regulated by hormones, Baima et al.

2001), Myb proteins as well as bZIP/HD-ZIPs and AP2/EREBP domain proteins (Kizis et al. 2001). Further, Seki et al. (2002) employed a full-length cDNA microarray, containing 7,000 independent *Arabidopsis* cDNAs to identify cold, drought and salinity-induced target genes and stress-related transcription factor family members such as DREB, ERF, WRKY, MYB, bZIP, helix-loop-helix and NAC. ABA is not only involved in drought-specific responses but also there is a cross-talk in cold and salinity stress responses (Seki et al. 2002).

2.5.1 Cross Talk Between Biotic and Abiotic Stress Signalling

Plants have developed various methods to deal with biotic and abiotic stresses. Traditionally, the molecular mechanisms associated with tolerance to each stress have been studied independently. Therefore, the knowledge of signalling pathways that are shared during biotic and abiotic stress responses remain rudimentary. In a recent study in chickpea, plant responses to fungal infection (*Ascochyta blight*) were found to be more similar to high-salinity stress than drought and cold stresses (Mantri et al. 2010a). Supporting this, abscisic acid-induced myb1 (*SlAIM1*) gene from tomato (*Solanum lycopersicum*) encoding an R2R3MYB transcription factor was induced by pathogens, plant hormones, salinity and oxidative stress (Abuqamar et al. 2009). Further, silencing the *SlAIM1* by RNA interference led to an increased susceptibility to the necrotrophic fungus *Botrytis cinerea*, and increased sensitivity to salt and oxidative stress. Also an ectopic expression of *SlAIM1* led to high salinity and oxidative stress tolerance (Abuqamar et al. 2009). This suggested that *SlAIM1* regulates a transmembrane ion flux, an indication of an early response to abiotic stress and pathogen infection, perhaps preceding hypersensitive cell death and necrosis.

Misregulation of ion fluxes can result in impaired plant tolerance to necrotrophic infection or abiotic stress (Abuqamar et al. 2009). Emerging evidence suggests that hormone signalling pathways like those controlled by jasmonic acid, abscisic acid, ethylene, and salicylic acid are central to the crosstalk between abiotic and biotic stress responses (Fujita et al. 2006). Recent studies have indicated several transcription factors and kinases are important candidates leading to crosstalk in stress signalling pathways. Mitogen-activated protein kinases (MAPKs) have been shown to be involved in developmental, hormonal, abiotic, and biotic stress signalling (Colcombet and Hirt 2008; Rodriguez et al. 2010). The activation of components of MAPK cascades by more than one type of stress, suggests that MAPK cascades serve as crossroads for numerous abiotic and biotic stress signalling pathways. Furthermore, as the *Arabidopsis* genome is reported to have around 20 MAPKs, 10 MAPKKs and 60 MAPKKKs, the signals recognized by the 60 MAPKKKs have to be transferred via 10 MAPKKs to 20 MAPKs, offering great chances for crosstalk between different stress signals.

Spatial and temporal expression patterns based on cell biological analysis combined with biochemical characterization of the signalling components, mainly identification of signalling complexes, is necessary to establish specificity or crosstalk of the signalling pathways (Chinnusamy et al. 2004). In the coming years, with the further development and incorporation of "omics" tools and computational approaches, deeper understanding of the signalling pathways, specificity and cross talk should be targeted. Currently, only a limited number of pathways and their components have been unravelled. In nature, however, plants face and respond to an overabundance of stimuli (including biotic as well as abiotic) simultaneously.

3 Involvement of Other Novel Genes Like MicroRNA in Plant Stress Tolerance

Discovery and functional association of microRNAs (miRNAs) have led to a large new research area in the previously unsuspected world of non-coding RNAs (Lee et al. 1993; Reinhart et al. 2002). The miRNAs are endogenous, small 21–24 nucleotide, single stranded, non-protein coding RNAs that have recently emerged as important regulators of gene expression (Bartel 2004).

These regulate target gene expression by catalyzing posttranscriptional gene silencing (Palatnik et al. 2003) or translation repression (Chen 2004). Targets of miRNA comprises transcription factors or other regulatory proteins that function in plant development or signal transduction. Recently, research on micro-RNAs (miRNAs) have suggested an association between miRNAs and plant stress responses (Patade and Suprasanna 2010). However, the relationship between micro-RNAs and stress response is just beginning to be explored. Several miRNAs are either up- or down-regulated by abiotic stresses, suggesting to be involved in stress-responsive gene expression and stress adaptation affecting a variety of cellular and physiological processes (Sunkar and Zhu 2004; Shukla et al. 2008).

Sunkar and Zhu (2004) identified novel and abiotic stress-regulated miRNAs and reported differential expression of some of the identified miRNAs in *Arabidopsis* seedlings exposed to dehydration, salinity, or cold stress. In order to unravel function of microRNA, Zhao et al. (2007) studied transcript expression profiles of miRNAs in rice (*O. sativa*) under drought stress. The drought-induced expression of *miR*-169g and *miR*393 was validated by microarray expression profiling and confirmed greater expression of *miR*-169g in roots rather than shoots. Sequence analysis revealed occurrence of two proximate DREs (dehydration-responsive element) in the upstream of the *MiR*-169g, suggesting transcript expression regulation of *miR*-169g by CBF/DREBs.

Sunkar et al. (2006) provided evidence on involvement of miRNA in oxidative stress responses by targeting cytosolic and chloroplastic superoxide dismutases that detoxify superoxide radicals. Transcript expression of *miR*398 in response to oxidative stress was down-regulated, leading to posttranscriptional accumulation of the SOD mRNA and thus oxidative stress tolerance. Moreover, transgenic *Arabidopsis* plants overexpressing a *miR*398-resistant form of SOD accumulated more mRNA than plants overexpressing a regular form and were consequently much more tolerant to high light, heavy metals and other oxidative stresses. *Arabidopsis* have been shown to trigger the accumulation of *miR159* in response to ABA, drought stress, and gibberellic acid (GA) treatment and the miRNA was predicted to target four MYB transcription factors (Reyes and Chua 2007). Recently, Patade and Suprasanna (2010) characterized transcript expression of mature *miR159* in response to short- and long-term salt and PEG-induced osmotic stress in sugarcane. A change in mature transcript levels of *miR159* was not detected in response to long-term (15 days) NaCl or iso-osmotic (–0.7 MPa) PEG stress. However, short-term (up to 24 h) salt or PEG stresses increased transcript level of the mature miRNA as compared to the control. The early induction of the gene under the short treatments supports its involvement in the regulation of genes involved in stress perception and/or signalling.

Zhou et al. (2008) developed a computational transcriptome-based approach to annotate stress-inducible miRNAs in plants. Interestingly, the promoter analysis of the miRNA genes revealed the presence of many known stress-responsive cis-regulatory elements. Continued efforts are needed to identify the complete set of miRNAs and other small RNAs that are fundamental to the stress regulation pathways. The identification and functional validation of stress-regulated small RNAs including miRNAs will help in designing new strategies for improving stress tolerance (Sunkar et al. 2006; Katiyar-Agarwal et al. 2007).

4 Strategies for Improving Abiotic Stress Tolerance

Many strategies undertaken for improving abiotic stress tolerance in a particular genetic background have included screening of diverse genetic resources, wide crossing and subsequent recurrent backcrossing; identification and selection of the major conditioning genes through linkage mapping and quantitative trait loci (QTL) analysis; the production and screening of mutant populations and the transgenic introduction of novel genes (Fig. 1.1). Although some success has been achieved in introducing tolerance traits into crop varieties from wild relatives (i.e. barley; Forster et al. 2000 and tomato; Foolad et al. 2001), in

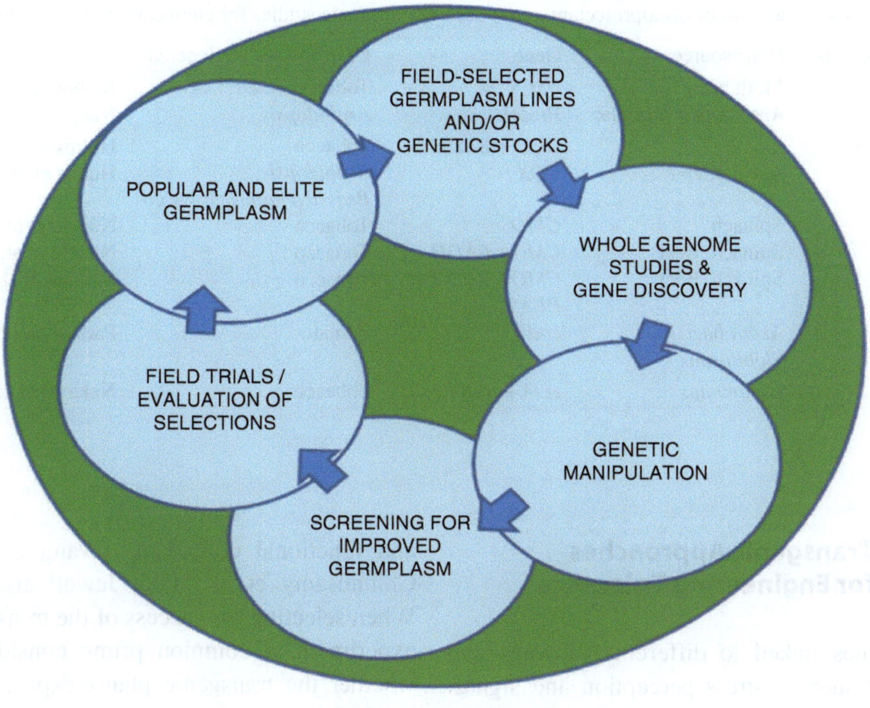

Fig. 1.1 Integrated components in the development of improved germplasm for abiotic stress tolerance

general there has been very little success reported in achieving high abiotic tolerance into elite germplasm (Flowers 2004).

As previously mentioned, breeding for, or induction of, abiotic stress tolerance traits is almost always limited by the genetic complexity of the underpinning mechanisms as well as potential interaction among genetic determinants. Also, differential selection of a particular stress may be affected by additional environmental factors, plant development stage, poor or irreproducible selection techniques, and the logistical constraints of physiological screening of large breeding populations on a field scale (Flowers et al. 2000). In this regard, the identification of discrete chromosomal regions that have a major effect on the specific tolerance trait through quantitative trait loci (QTL) mapping and marker-assisted selection remains a valuable option for many breeding programs (Cushman 2009; Cuartero et al. 2010). This is particularly so when whole genome knowledge is lacking and no candidate tolerance genes are known.

For accurate selection of the related phenotype, reliable and realistic screening techniques are required. However, uniformity and reliability of field-based screening may suffer from heterogeneity in the stress across the site (i.e. boron or salinity level) as well as the potential compounding environmental factors (i.e. disease, rainfall, temperature). Also, when the starting material is genetically wide, heterogeneity among the genetic backgrounds may also impact on the ability to accurately select the most superior or different tolerances. As an alternative, cellular-based mutant induction and subsequent selection initially under controlled in vitro conditions offers a method to quickly screen large populations with homogeneous backgrounds for novel fortuitous changes related to tolerance. Subsequent field screening then ensures adequate performance of the tolerance trait under the external potentially mitigating factors previously mentioned. Unsurprisingly, this method has generated great interest in selecting for abiotic stress tolerances in several crop species (Suprasanna et al. 2008).

Table 1.1 Some examples of osmoprotectant genes used in transgenic studies for engineering abiotic tolerance

Osmoprotectants	Gene source	Gene	Crop species engineered	References
Proline	Moth bean	*P5CS*	Tobacco	Kishor et al. (1995)
	Arabidopsis thaliana	*ProDH*	*Arabidopsis*	Nanjo et al. (1999)
Glycine betaine	*E. coli*	*CDH + BADH*	Tobacco	Holmstrom et al. (2000)
	Arthrobacter	*COX*	*Arabidopsis*, *Brassica napus*, Tobacco	Huang et al. (2000)
	Spinach	*CMO*	Tobacco	Nuccio et al. (1998)
	Spinach, Beet	*CMO+BADH*	Tobacco	Nuccio et al. (2000)
	Spinach, Beet	*CMO+BADH+ PEAMT*	Tobacco	McNeil et al. (2001)
	Arthrobacter globiformis	*codA*	Tomato	Park et al. (2004)
Ectoine	*Halomonas*	*ectA+ectB+ectC*	Tobacco	Nakayama et al. (2000)

4.1 Transgenic Approaches for Engineering Tolerance

Many genes linked to different pathways and processes such as stress perception and signalling, contributing to molecular, biochemical, cellular, physiological and morphological adaptations are differentially regulated in response to plant stress (Munns and Tester 2008). Stress responsive genes include those that alleviate the effect of the stress and lead to adjustment of the cellular environment and plant tolerance. The gene products are classified into three major groups: those encoding products that directly protect plant cells against stress, those that are involved in signalling cascades and in transcriptional control and those that are involved in water and ion uptake and transport.

Engineering metabolic and stress-signalling pathways to produce stress-tolerant crops is one of the major interests of agricultural research. Genetic transformation with stress-inducible genes has been employed to gain an understanding of their functional role in the tolerance response and ultimately to improve the tolerance trait in the target genotype (Zhang et al. 2004, Cuartero et al. 2010). To date, by far, majority of these studies have been limited to single-gene transfers within known multigenic pathways and mostly those involved in signalling and regulatory pathways, or effector genes that code for enzymes catalysing the synthesis of structural and functional defendants (Wang et al. 2003; Chinnusamy et al. 2005; Jewell et al. 2010). When selecting for success of the transformation experiment, a common prime consideration is whether the transgenic plants express a higher level of the transgene (i.e. an osmoprotectant or a protein) only under the stress conditions (Zhu 2001). In general, specific inducible promoters are used rather than constitutive promoters since the tolerance/stress-induced mechanisms may be energy and nucleic acid greedy and divert essential resources away from normal growth processes (Su et al. 1998).

As examples, transgenic rice plants developed with choline oxidase (*codA*), d-pyrroline-5-corboxylate synthase (*P5CS*), LEA protein group 3 (*HVA1*), alcohol dehydrogenase (*ADH*) and pyruvate decarboxylase (*PDC*) genes exhibited drought tolerance (Datta 2002; Soren et al. 2010). Potato and rice (Yeo et al. 2000 and Garg et al. 2002, respectively) transformed with trehalose synthesis genes displayed tolerance to drought (in case of potato), and salt, drought, and low-temperature stress (in case of rice). Tobacco plants transformed with ectoine biosynthesis genes from the halophilic bacterium *Halomonas elongate* showed enhanced salt tolerance. Also transformation with genes for sorbitol (Sheveleva et al. 1997) or mannitol (Shen et al. 1997) resulted in an increased accumulation of these osmolytes and tolerance to high salinity (Table 1.1). Overexpression of genes encoding the enzymes pyrroline-5-carboxylate

(P5C) synthetase (P5CS) and P5C reductase (P5CR) resulted in proline overproduction and enhanced abiotic stress tolerance (Szabados and Savoure 2010). *P5CS* overexpression in transgenic tobacco dramatically elevated free proline (Kishor et al. 1995) with improved germination and growth of seedlings under salt stress. Transgenic petunia plants transformed with *Arabidopsis P5CS* gene showed resistance to drought conditions for longer duration than control plants (Yamada et al. 2005).

The enhancement of glycine betaine (GB) synthesis in transgenic plants using genes that encode for enzymes (choline monooxygenase, betaine aldehyde dehydrogenase and choline oxidase) in GB biosynthesis is another strategy to achieve enhanced tolerance to drought, salt and chilling stress (Rontein et al. 2002; Chen and Murata 2008). Transgenic rice plants expressing the *codA* (*choline oxidase*) gene recovered from an initial growth inhibition under salt and low-temperature stress, and grew normally than the wild type (Sakamoto et al. 1998). Several other plants that have been genetically engineered for obtaining salt, drought, freezing and heat tolerance through GBS accumulation include; *A. thaliana, Brassica napus, Brassica juncea, Gossypium hirsutum, Lycopersicon esculentum, Nicotiana tabacum, Solanum tuberosum* and *Zea mays* (Chen and Murata 2008).

Trehalose is a non-reducing disaccharide and an effective osmoprotectant (Goddijn and van Dunn 1999). Transgenic plants overexpressing trehalose biosynthetic genes showed increased tolerance to different abiotic stress conditions (Penna 2003; Almeida et al. 2007). A stress-inducible promoter has been utilised to overexpress *Escherichia coli* trehalose biosynthesis genes (*otsA* and *otsB)* as a fusion gene (TPSP) in rice, to confer tolerance to different abiotic stresses (Garg et al. 2002). The TPSP fusion gene is dually advantageous as both the genes can be simultaneously introduced into the rice genome leading to increased catalytic efficiency for trehalose synthesis (Jang et al. 2003; Almeida et al. 2007).

Research on genetic engineering efforts with other osmolytes such as mannitol, fructans, ononitol, proline, glycinebetaine and ectoine have also shown promise for generating tolerant genotypes (Suprasanna et al. 2005). To avoid overproduction of compatible solutes burdening the plant's metabolic machinery and potentially diminishing pleiotropic effects, engineering for overproduction should be done under stress-inducible and/or tissue specific regulation. In addition, production of the osmolytes should be targeted to the chloroplast by placing a signal sequence in front of the engineered enzymes (Shen et al. 1997).

As previously stated, abiotic stress generates an increase in reactive oxygen species that may be deleterious to normal cellular functions. Therefore, several oxidative-stress-related genes have been employed in developing transgenic plants tolerant to various stresses (Hussain et al. 2008). For example, transgenic tobacco plants overexpressing chloroplastic Cu/Zn-SOD showed increased resistance to oxidative stress caused by salt exposure (Tanaka et al. 1999; Bartel 2001). Transgenic alfalfa (*Medicago sativa*) plants expressing Mn-SOD had reduced injury from water-deficit stress, as determined by chlorophyll fluorescence, electrolyte leakage and regrowth (McKersie et al. 1996). Simultaneous expression of genes encoding three antioxidant enzymes: copper zinc superoxide dismutase, ascorbate peroxidase and dehydroascorbate reductase in the chloroplasts of tobacco plants conferred enhanced tolerance to oxidative stresses caused by paraquat and salt (Lee et al. 2007). Similarly, overexpression of *AtNDPK2* efficiently modulated oxidative stress caused by various environmental stresses in sweet potato through enhanced antioxidant enzyme activities such as peroxidase, ascorbate peroxidase and catalase (Kim et al. 2010). Thus it seems promising to target detoxification pathways as an approach for obtaining plants with multiple stress-tolerance traits.

Transgenic manipulation of detoxification pathways through overexpressing genes involved in oxidative protection, such as glutathione peroxidase, superoxide dismutase, ascorbate peroxidases and glutathione reductases is an area of current interest. Constitutive expression of the *Nicotiana PK1* gene (regulatory protein NPK1) enhanced freezing, heat and salinity tolerance in transgenic maize plants (Shou et al. 2004b). In a further study, Shou et al. (2004a) expressed a tobacco MAPKKK (NPK1) constitutively in maize resulting in enhanced drought tolerance. The transgenic maize plants maintained significantly higher photosynthesis rates, suggesting, NPK1 induced a mechanism that protected photosynthesis machinery from dehydration damage.

Under salt stress, tolerant plant cells must maintain high K^+ (100–200 mM) and lower Na^+ (less than 1 mM) levels for normal metabolic function. An important strategy for achieving greater tolerance to salinity stress is to help plants to re-establish homeostasis under stressful environments, restoring both ionic and osmotic homeostasis. This strategy continues to be a major approach to improve salt tolerance in plants through genetic engineering, where the target is to achieve Na^+ excretion, or vacuolar storage. A number of abiotic stress-tolerant transgenic plants have been produced by increasing the cellular levels of proteins (such as vacuolar antiporter proteins) that control the transport functions. For example, *AtSOS* from *Arabidopsis* has been shown to encode a plasma membrane Na^+/H^+ antiporter (NHX) with significant sequence similarity to the respective antiporter from bacteria and fungi (Shi et al. 2000). Constitutive expression of vacuolar Na^+/H^+ antiporter (*NHX1*) or *AVP1* (*A. thaliana* vacuolar H^+-translocating pyrophosphatase) gene energized the pumping of Na^+ into the vacuole, and increased both accumulation and Na^+ tolerance in *Arabidopsis* (Gaxiola et al. 2001). Thus more efficient sequestration of these ions to the vacuole could improve tissue tolerance to salinity by reducing the cytosolic Na^+ concentrations. The importance of Na^+ sequestration in salt tolerance has been further demonstrated in transgenic tomato plants overexpressing the *AtNHX1* gene (Zhang and Blumwald 2001). Also, a vacuolar chloride channel gene, *AtCLCd*, involved in cation detoxification has been cloned as well as overexpressed in *Arabidopsis* and shown to confer salt tolerance. Up-regulation of the *Salt Overly Sensitive 1* (*SOS1*) gene in *Arabidopsis* resulted in a greater proton motive force necessary for elevated Na^+/H^+ antiporter activities (Shi et al. 2000).

Apart from the single gene approach, tolerance towards multiple stresses may be achieved by targeting a stress inducible signal transduction molecule and/or transcription factor (Chinnusamy et al. 2005). The transcription factors play an important role in the acquisition of stress tolerance, which ultimately contribute to agricultural and environmental practices (Century et al. 2008). A large number of transcription factors are involved in the plant response to abiotic stress (Vincour and Altman 2005). Most of these falls into several large transcription factor families, such as AP2/ERF, bZIP, NAC, MYB, MYC, Cys2His2 zinc-finger and WRKY. Accordingly, overexpression of the functionally conserved *At-DBF2* gene resulted in wide and high levels of multiple stress tolerances in *Arabidopsis* (Lee et al. 1999). Salt stress-tolerant tobacco plants were produced by overexpressing the calcineurin, a Ca^{2+}/calmodulin-dependent protein phosphatase gene, formally identified as being involved in salt-stress signal transduction in yeast (Pardo et al. 1998; Grover et al. 1999).

Some stress responsive genes may share the same transcription factors, as indicated by the significant overlap of the gene expression profiles that are induced in response to drought and cold stress (Seki et al. 2001; Chen and Murata 2002; Mantri et al. 2007). The activation of stress-induced genes has been possible in transgenic plants by overexpressing one or more transcription factors that recognize regulatory elements of these genes. In *Arabidopsis*, the transcription factor DREB1A specifically interacts with the DRE and induces expression of stress tolerance genes (Shinozaki and Yamaguchi-Shinozaki 1997). CaMV 35S promoter-driven overexpression of DREB1A cDNA in transgenic *Arabidopsis* plants provided tolerance to salt, freezing and drought stress through strong constitutive expression of the stress inducible genes (Liu et al. 1998).

The transcription factors involved in the ABA-dependent (such as *NAC, AREB/ABF*, and *MYB*) and –independent (*AP2/ERF* gene) stress response pathways regulate cascade of downstream genes and events that enhance tolerance to drought stress. Transforming crops with such transcription factor genes should be more meaningful in the development of drought tolerance (Zhang et al. 2004; Ashraf 2010). Overexpressing *Arabidopsis CBF1* (*CRT/DRE*) cDNA in tomato improved tolerance to salt, chilling and drought stress; however, the plants exhibited a dwarf phenotype as well as reduced fruit set and seed number (Hsieh et al. 2002). Overexpression of Alfin1 (transcriptional regulator) in alfalfa plants exhibited salinity tolerance through regulated endogenous *MsPRP2* (NaCl-inducible gene) mRNA levels (Winicov and Bastola 1999).

4.2 The Future of Transgenic Approaches

The current plant genetic engineering approach for developing salt stress-tolerant transgenic plants includes altering the expression levels of native genes or incorporating alien genes for osmolytes, ion transporters, transcription factors and other signalling molecules. The advent of global transcription profiling has demonstrated that large numbers of other genes are also up- and down-regulated simultaneously in response to salt stress. This second category of genes encode proteins related to the regulation of transcriptional and translational machineries with distinct roles in mediating the salt stress response. Particularly, coordinated induction and action of the transcript of several RNA binding proteins, ribosomal genes, helicases, cyclophilins, F-box proteins, dynamin-like proteins, translation initiation and elongation factors seems to be important in salt stress tolerance. The functionality of these genes at the cellular level should also be investigated to assess aptness for targeted transgenic approaches (Sahi et al. 2006).

The evaluation of genetically engineered salt-tolerant transgenic lines needs critical, careful, and thorough experimentation (Flowers 2004). The fourth or fifth generation genotypes should be evaluated along with parental (wild-type) lines under controlled saline and non-saline treatment conditions. Validation should not stop at the laboratory or green house level, since quantitative measures of growth are required throughout the plant life cycle in field conditions.

5 Conclusions and Future Perspective

In the last decade, significant progress has been made in our understanding of the complex mechanisms governing abiotic stress tolerance in crop plants. However, we are still far from pinning the exact battery of gene activation responsible for tolerance to a particular abiotic stress condition. This situation is complicated when one considers plants have to simultaneously cope with numerous biotic stresses along with various abiotic stresses. Our struggle to understand these complex mechanisms is ongoing and recent development of new tools for high-throughput genotyping and phenotyping gives us a new ray of hope. A complete understanding on physiological and molecular mechanisms especially signalling cascades in response to abiotic stresses in tolerant plants will help to manipulate susceptible crop plants and increase agricultural productivity in the near future.

References

Abuqamar S, Luo H, Laluk K, Mickelbart MV, Mengiste T (2009) Crosstalk between biotic and abiotic stress responses in tomato is mediated by the AIM1 transcription factor. Plant J 58(2):347–360

Almeida A, Cardoso L, Santos D, Torné J, Fevereiro P (2007) Trehalose and its applications in plant biotechnology. In Vitro Cell Dev Biol Plant 43(3):167–177

Apel K, Hirt H (2004) Reactive oxygen species: metabolism, oxidative stress, and signal transduction. Annu Rev Plant Biol 55:373–399

Ashraf M (2010) Inducing drought tolerance in plants: recent advances. Biotechnol Adv 28:169–183

Baier M, Noctor G, Foyer CH, Dietz KJ (2000) Antisense suppression of 2-Cysteine peroxiredoxin in Arabidopsis specifically enhances the activities and expression of enzymes associated with ascorbate metabolism but not glutathione metabolism. Plant Physiol 124(2): 823–832

Baima S, Possenti M, Matteucci A, Wisman E, Altamura M, Ruberti I, Morelli G (2001) The Arabidopsis ATHB-8 HD-Zip protein acts as a differentiation-promoting transcription factor of the vascular meristems. Plant Physiol 126(2):643–655

Bartel D (2001) Targeting detoxification pathways: an efficient approach to obtain plants with multiple stress tolerance. Trends Plant Sci 6:284–286

Bartel D (2004) MicroRNAs: genomics, biogenesis, mechanism, and function. Cell 116(2):281–297

Beck E, Fettig S, Knake C, Hartig K, Bhattarai T (2007) Specific and unspecific responses of plants to cold and drought stress. J Biosci 32(3):501–510

Bensen R, Castiglioni P, Korte J, Bell E, Hinchey B, Loida P, Ahrens J (2008) Transgenic plants with increased glycine-betaine. United States Patent 7410800. Application Number:10/839092

Bohnert HJ, Shen B (1999) Transformation and compatible solutes. Sci Hortic 78:237–260

Capell T, Bassie L, Christou P (2004) Modulation of the polyamine biosynthetic pathway in transgenic rice confers tolerance to drought stress. Proc Natl Acad Sci USA 101:9909–9914

Carden DE, Walker DJ, Flowers TJ, Miller AJ (2003) Single-cell measurements of the contributions of cytosolic Na+ and K+ to salt tolerance. Plant Physiol 131:676–683

Century K, Reuber TL, Ratcliffe OJ (2008) Regulating the regulators: the future prospects for transcription-factor-based agricultural biotechnology products. Plant Physiol 147:20–29

Chen X (2004) A microRNA as translational repressor of APETALA2 in Arabidopsis flower development. Science 303:2022–2025

Chen TH, Murata N (2002) Enhancement of tolerance of abiotic stress by metabolic engineering of betaines and other compatible solutes. Curr Opin Plant Biol 5:250–257

Chen THH, Murata N (2008) Glycinebetaine: an effective protectant against abiotic stress in plants. Trends Plant Sci 13(9):499–505

Chen W, Provart NJ, Glazebrook J, Katagiri F, Chang HS, Eulgem T, Mauch F, Luan S, Zou G, Whitham SA, Budworth PR, Toa Y, Xie Z, Chen X, Lam S, Kreps JA, Harper JF, Si-Ammour A, Mauch-Mani B, Heinlein M, Kobayashi K, Hohn T, Dangl JL, Wang X, Zhu T (2002) Expression profile matrix of *Arabidopsis* transcription factor genes suggests their putative functions in response to environmental stresses. Plant Cell 14:559–574

Chinnusamy V, Schumaker K, Zhu JK (2004) Molecular genetic perspectives on cross-talk and specificity in abiotic stress signaling in plants. J Exp Bot 55(395): 225–236

Chinnusamy V, Jagendorf A, Zhu JK (2005) Understanding and improving salt tolerance in plants. Crop Sci 45:437–448

Chinnusamy V, Zhu J, Zhu JK (2006) Gene regulation during cold acclimation in plants. Physiol Plant 126:52–61

Colcombet J, Hirt H (2008) Arabidopsis MAPKs: a complex signalling network involved in multiple biological processes. Biochem J 413(2):217–226

Collins NC, Tardieu F, Tuberosa R (2008) Quantitative trait loci and crop performance under abiotic stress: where do we stand? Plant Physiol 147:469–486

Cook D, Fowler S, Fiehn O, Thomashow MF (2004) A prominent role for the CBF cold response pathway in configuring the low-temperature metabolome of Arabidopsis. Proc Natl Acad Sci USA 101: 15243–15248

Cortina C, Culianez-Macia FA (2005) Tomato abiotic stress enhanced tolerance by trehalose biosynthesis. Plant Sci 169:75–82

Cuartero J, Bolarin MC, Moreno V, Pineda B (2010) Molecular tools for enhancing salinity tolerance in plants. In: Jain SM, Brar DS (eds) Molecular techniques in crop improvement. Springer, New York, pp 373–405

Cushman JC (2009) Functional genomics of plant abiotic stress tolerance. In: Genomics of plants and fungi (Book Chapter). Marcel Dekker, Inc., New York

Datta SK (2002) Recent developments in transgenics for abiotic stress tolerance in rice. JIRCAS Work Rep: 43–53

Delauney AJ, Verma DP (1993) Proline biosynthesis and osmoregulation in plants. Plant J 4:215–223

Desikan R, Cheung MK, Bright J, Henson D, Hancock JT, Neill SJ (2004) ABA, hydrogen peroxide and nitric oxide signaling in stomatal guard cells. J Exp Bot 55:205–212

Einset J, Winge P, Bones A (2007) ROS signaling pathways in chilling stress. Plant Signal Behav 2(5):365–367

Flowers TJ (2004) Improving crop salt tolerance. J Exp Bot 55:307–319

Flowers TJ, Dalmond D (1992) Protein synthesis in halophytes: the influence of potassium, sodium and magnesium in vitro. Plant Soil 146:153–161

Flowers TJ, Yeo AR (1995) Breeding for salinity tolerance in crop plants: where next? Aust J Plant Physiol 22:875–884

Flowers TJ, Koyama ML, Flowers SA, Sudhakar C, Singh KP, Yeo AR (2000) QTL: their place in engineering tolerance of rice to salinity. J Exp Bot 51:99–106

Foolad MR, Zhang LP, Lin GY (2001) Identification and validation of QTLs for salt tolerance during vegetative growth in tomato by selective genotyping. Genome 44:444–454

Ford CW (1984) Accumulation of low molecular weight solutes in water stress tropical legumes. Phytochemistry 22:875–884

Forster BP, Ellis RP, Thomas WT, Newton AC, Tuberosa R, This D, el-Enein RA, Bahri MH, Ben Salem M (2000) The development and application of molecular markers for abiotic stress tolerance in barley. J Exp Bot 51:19–27

Fujita M, Fujita Y, Noutoshi Y, Takahashi F, Narusaka Y, Yamaguchi-Shinozaki K, Shinozaki K (2006) Crosstalk between abiotic and biotic stress responses: a current view from the points of convergence in the stress signaling networks. Curr Opin Plant Biol 9:436–442

Gao M, Sakamoto A, Miura K, Murata N, Sugiura A, Tao R (2000) Transformation of Japanese persimmon (*Diospyros kaki* Thunb.) with a bacterial gene for choline oxidase. Mol Breed 6:501–510

Garg AK, Kim JK, Owens TG, Ranwala AP, Choi YC, Kochian LV, Wu RJ (2002) Trehalose accumulation in rice plants confers high tolerance levels to different abiotic stresses. Proc Natl Acad Sci USA 99: 15898–15903

Gaxiola RA, Li J, Unurraga S, Dang LM, Allen GJ, Alper SL, Fink GR (2001) Drought- and salt-tolerant plants result from over expression of the AVP1 H^+-pump. Proc Natl Acad Sci USA 98:11444–11449

Gilmour SJ, Zarka DG, Stockinger EJ, Salazar MP, Houghton JM, Thomashow MF (1998) Low temperature regulation of the Arabidopsis CBF family of AP2 transcriptional activators as an early step in cold-induced COR gene expression. Plant J 16:433–442

Goddijn OJM, van Dunn K (1999) Trehalose metabolism in plants. Trends Plant Sci 4:315–319

Grover A, Sahi C, Sanan N, Grover A (1999) Taming abiotic stresses in plants through genetic engineering: current strategies and perspective. Plant Sci 143:101–111

Heidarvand L, Amiri RM (2010) What happens in plant molecular responses to cold stress? Acta Physiol Plant 32:419–431

Holmström KO, Somersalo S, Mandal A, Palva TE, Welin B (2000) Improved tolerance to salinity and low temperature in transgenic tobacco producing glycine betaine. Journal of Experimental Botany 51(343): 177–185

Hsieh TH, Lee JT, Charng YY, Chan MT (2002) Tomato plants ectopically expressing Arabidopsis CBF1 show enhanced resistance to water deficit stress. Plant Physiol 130:618–626

Huang J, Hirji R, Adam L, Rozwadowski KL, Hammerlindl JK, Keller WA, Selvaraj G (2000) Genetic Engineering of Glycinebetaine Production toward Enhancing Stress Tolerance in Plants: Metabolic Limitations. Plant Physiology 122:747–756

Hussain TM, Thummala C, Mahamed H, Zafar S, Saleh BK, Rama GG (2008) Recent advances in salt stress biology – a review. Biotechnol Mol Biol Rev 3(1):008–013

Imlay JA (2003) Pathways of oxidative damage. Annu Rev Microbiol 57:395–418

Jaglo-Ottosen KR, Gilmour SJ, Zarka DG, Schabenberger O, Thomashow MF (1998) Arabidopsis CBF1 over expression induces COR genes and enhances freezing tolerance. Science 280:104–106

Jaleel CA, Manivannan P, Wahid A, Farooq M, Somasundaram R, Panneerselvam R (2009) Drought stress in plants: a review on morphological characteristics and pigments composition. Int J Agric Biol 11:100–105

Jang IC, Oh SJ, Seo JS, Choi WB, Song SI, Kim CH, Kim YS, Seo HS, Choi YD, Nahm BH, Kim JK (2003) Expression of a bifunctional fusion of the *Escherichia coli* genes for trehalose-6-phosphate synthase and trehalose-6-phosphate. Phosphatase in transgenic rice plants increases trehalose accumulation and abiotic stress tolerance without stunting growth. Plant Physiol 131:516–524

Jewell MC, Campbell BC, Ian D (2010) Transgenic plants for abiotic stress resistance. In: Kole C et al (eds) Transgenic crop plants. Springer, Berlin, pp 67–132

Jones A (2000) Does the plant mitochondrion integrate cellular stress and regulate programmed cell death? Trends Plant Sci 5:273–278

Kaplan F, Kopka J, Haskell DW, Zhao W, Schiller KC, Gatzke N, Sung DY, Guy CL (2004) Exploring the temperature stress metabolome of Arabidopsis. Plant Physiol 136:4159–4168

Kasuga M, Liu Q, Miura S, Yamaguchi-Shinozaki K, Shinozaki K (1999) Improving plant drought, salt, and freezing tolerance by gene transfer of a single stress-inducible transcription factor. Nat Biotechnol 17: 287–291

Katiyar-Agarwal S, Gao A, Vivian-Smith J (2007) A novel class of bacteria-induced small RNAs in Arabidopsis. Genes Dev 21:3123–3134

Kaur G, Kumar S, Nayyar H, Upadhyaya HD (2008) Cold stress injury during the pod-filling phase in chickpea (*Cicer arietinum* L.): effects on quantitative and qualitative components of seeds. J Agron Crop Sci 194(6):457–464

Kim Y-H, Kim MD, Choi YI, Park S-C, Yun D-J, Noh EW, Lee H-S, Kwak S-S (2010) Transgenic poplar expressing *Arabidopsis NDPK2* enhances growth as well as oxidative stress tolerance. Plant Biotechnol J 9:334–347

Kishor PB, Hong Z, Miao G-H, Hu C-AA, Verma DP (1995) Overexpression of delta-pyrroline-5-carboxylate synthetase increases proline production and confers osmotolerance in transgenic plants. Plant Physiol 108: 1387–1394

Kizis D, Lumbreras V, Pagès M (2001) Role of AP2/EREBP transcription factors in gene regulation during abiotic stress. FEBS Lett 498(2–3):187–189

Knight H, Trewavas AJ, Knight MR (1996) Cold calcium signaling in Arabidopsis involves two cellular pools and a change in calcium signature after acclimation. Plant Cell 8:489–503

Krebs JA, Wu Y, Chang H-S, Zhu T, Wang X, Harper JF (2002) Transcriptome changes for Arabidopsis in response to salt, osmotic, and cold stress. Plant Physiol 130:2129–2141

Krishnan N, Dickman MB, Becker DF (2008) Proline modulates the intracellular redox environment and protects mammalian cells against oxidative stress. Free Radic Biol Med 44(4):671–681

Lee RC, Feinbaum RL, Ambros V (1993) The *C. elegans* heterochronic gene lin-4 encodes small RNAs with antisense complementarity to lin-14. Cell 75:843–854

Lee JH, van Montagu M, Verbruggen N (1999) A highly conserved kinase is an essential component for stress tolerance in yeast and plant cells. Proc Natl Acad Sci USA 96:5873–5877

Lee YP, Kim SH, Bang JW (2007) Enhanced tolerance to oxidative stress in transgenic tobacco plants expressing three antioxidant enzymes in chloroplasts. Plant Cell Rep 26:591–598

Leidi EO, Barragán V, Rubio L, El-Hamdaoui A, Ruiz MT, Cubero B, Fernández JA, Bressan RA, Hasegawa PM, Quintero FJ, Pardo JM (2010) The AtNHX1 exchanger mediates potassium compartmentation in vacuoles of transgenic tomato. Plant J 61(3):495–506

Li C, Puhakainen T, Welling A, Viherä-Aarnio A, Ernstsen A, Junttila O, Heino P, Palva ET (2002) Cold acclimation in silver birch (*Betula pendula*). Development of freezing tolerance in different tissues and climatic ecotypes. Physiol Plant 116:478–488

Li C, Junttila O, Heino P, Palva ET (2003) Different responses of northern and southern ecotypes of *Betula pendula* to exogenous ABA application. Tree Physiol 23:481–487

Liu Q, Kasuga M, Sakuma Y, Abe H, Miura S, Yamaguchi-Shinozaki K, Shinozaki K (1998) Two transcription factors, DREB1 and DREB2, with an EREBP/AP2 DNA binding domain separate two cellular signal transduction pathways in drought- and low temperature-responsive gene expression, respectively, in Arabidopsis. Plant Cell 10:1391–1406

Lu YY, Deng XP, Kwak SS (2010) Over expression of CuZn superoxide dismutase (CuZn SOD) and ascorbate peroxidase (APX) in transgenic sweet potato enhances tolerance and recovery from drought stress. Afr J Biotechnol 9(49):8378–8391

Luan S, Kudla J, Rodriguez-Concepcion M, Yalovsky S, Gruissem W (2002) Calmodulins and calcineurin B-like proteins: calcium sensors for specific signal response coupling in plants. Plant Cell 14(Suppl): S389–S400

Mantri NL, Ford R, Coram TE, Pang ECK (2007) Transcriptional profiling of chickpea genes differentially regulated in response to high-salinity, cold and drought. BMC Genomics 8:303

Mantri NL, Ford R, Coram TE, Pang ECK (2010a) Evidence of unique and shared responses to major biotic and abiotic stresses in chickpea. Environ Exp Bot 69(3):286–292

Mantri NL, Pang ECK, Ford R (2010b) Molecular biology for stress management. In: Yadav SS, McNeil DN, Weeden N, Patil SS (eds) Climate change and management of cool season grain legume crops. Springer, Heidelberg, pp 377–408. ISBN 978-90-481-3708-4

Maruyama K, Sakuma Y, Kasuga M, Ito Y, Seki M, Goda H, Shimada Y, Yoshida S, Shinozaki K, Yamaguchi-Shinozaki K (2004) Identification of cold-inducible downstream genes of the Arabidopsis DREB1A/CBF3 transcriptional factor using two microarray systems. Plant J 38:982–993

McKersie BD, Bowley SR, Harjanto E, Leprince O (1996) Water-deficit tolerance and field performance of transgenic alfalfa overexpressing superoxide dismutase. Plant Physiol 111(4):1177–1181

McNeil SD, Nuccio ML, Ziemak MJ, Hanson AD (2001) Enhanced synthesis of choline and glycine betaine in transgenic tobacco plants that overexpress phosphoethanolamine N-methyltransferase. PNAS 98: 10001–10005

Munns R, Tester M (2008) Mechanisms of salinity tolerance. Annu Rev Plant Biol 59:651–681

Nakayama H, Yoshida K, Ono H, Murooka Y, Shinmyo A (2000) Ectoine, the Compatible Solute of Halomonas elongata, Confers Hyperosmotic Tolerance in Cultured Tobacco Cells. Plant Physiology 122:1239–1248

Nanjo T, Kobayashi M, Yoshiba Y, Kakubari Y, Yamaguchi-Shinozaki K, Shinozaki K (1999) Antisense suppression of proline degradation improves tolerance to freezing and salinity in Arabidopsis thaliana. FEBS Letters 461:205–210

Neill S, Desikan R, Hancock J (2002) Hydrogen peroxide signaling. Curr Opin Plant Biol 5:388–395

Novillo F, Medina J, Salinas J (2007) Arabidopsis CBF1 and CBF3 have a different function than CBF2 in cold acclimation and define different gene classes in the CBF regulon. Proc Natl Acad Sci USA 104:21002–21007

Nuccio ML, Russell BL, Nolte KD, Rathinasabapathi B, Gage DA, Hanson AD (1998) The endogenous choline supply limits glycine betaine synthesis in transgenic tobacco expressing choline monooxygenase. The Plant Journal 16(4):487–496

Nuccio ML, McNeil SD, Ziemak MJ, Hanson AD, Jain RK, Selvaraj G (2000) Choline Import into Chloroplasts Limits Glycine Betaine Synthesis in Tobacco: Analysis of Plants Engineered with a Chloroplastic or a Cytosolic Pathway. Metabolic Engineering 2:300–311

Orvar BL, Sangwan V, Omann F, Dhindsa RS (2000) Early steps in cold sensing by plant cells: the role of actin cytoskeleton and membrane fluidity. Plant J 23:785–794

Palatnik JF, Allen E, Wu X, Schommer C, Schwab R, Carrington JC, Weigel D (2003) Control of leaf morphogenesis by microRNAs. Nature 425:257–263

Pardo JM, Reddy MP, Yang S (1998) Stress signaling through Ca^{2+}/Calmodulin dependent protein phosphatase calcineurin mediates salt adaptation in plants. Proc Natl Acad Sci USA 95:9681–9683

Park EJ, Jeknic Z, Sakamoto A, DeNoma J, Yuwansiri R, Murata N, Chen TH (2004) Genetic engineering of glycinebetaine synthesis in tomato protects seeds, plants, and flowers from chilling damage. The Plant Journal 40(4):474–487

Pasapula V, Shen G, Kuppu S, Paez-Valencia J, Mendoza M, Hou P, Chen J, Qiu X, Zhu L, Zhang X, Auld D, Blumwald E, Zhang H, Gaxiola R, Payton P (2011) Expression of an Arabidopsis vacuolar H^+-pyrophosphatase gene (AVP1) in cotton improves drought- and salt tolerance and increases fibre yield in the field conditions. Plant Biotechnol J 9(1):88–99

Patade VY, Suprasanna P (2010) Short-term salt and PEG stresses regulate expression of MicroRNA, *miR159* in sugarcane leaves. J Crop Sci Biotechnol 13(3): 177–182

Patade VY, Bhargava S, Suprasanna P. Salt and drought tolerance of sugarcane under iso-osmotic salt and water stress: growth, osmolytes accumulation and antioxidant defense. J Plant Interact, in press

Pearce R (2001) Plant freezing and damage. Ann Bot 87:417–424

Peng Y, Reyes JL, Wei H, Yang Y, Karlson D, Covarrubias AA, Krebs SL, Fessehaie A, Arora R (2008) RcDhn5, a cold acclimation-responsive dehydrin from Rhododendron catawbiense rescues enzyme activity from dehydration effects in vitro and enhances freezing tolerance in RcDhn5-overexpressing Arabidopsis plants. Physiol Plant 134(4):583–597

Penna S (2003) Building stress tolerance through overproducing trehalose in transgenic plants. Trends Plant Sci 8(8):355–357

Prabhavathi VR, Rajam MV (2007) Polyamine accumulation in transgenic eggplant enhances tolerance to multiple abiotic stresses and fungal resistance. Plant Biotechnol 24:273–282

Ray S, Dansana PK, Bhaskar A, Giri J, Kapoor S, Khurana JP, Tyagi AK (2010) Emerging trends in functional genomics for stress tolerance in crop plants. In: Hirt H (ed) Plant stress biology: from genomics to systems biology. Wiley-VCH Verlag GmbH and Co. KGaA, Weinheim, Germany

Reinhart BJ, Weinstein EG, Rhoades MW, Bartel B, Bartel DP (2002) microRNAs in plants. Genes Dev 16: 1616–1626

Reyes JL, Chua NH (2007) ABA induction of *miR159* controls transcript levels of two *MYB* factors during *Arabidopsis* seed germination. Plant J 49:592–606

Riechmann JL, Heard J, Martin G, Reuber L, Jiang CZ, Keddie J, Adam L, Pineda O, Ratcliffe OJ, Samaha RR, Creelman R, Pilgrim M, Broun P, Zhang JZ, Ghandehari D, Sherman BK, Yu GL (2000) Arabidopsis transcription factors: genome-wide comparative analysis among eukaryotes. Science 290:2105–2110

Rodriguez MC, Petersen M, Mundy J (2010) Mitogen-activated protein kinase signaling in plants. Annu Rev Plant Biol 61:621–649

Rodríguez-Rosales MP, Gálvez FJ, Huertas R, Aranda MN, Baghour M, Cagnac O, Venema K (2009) Plant NHX cation/proton antiporters. Plant Signal Behav 4(4):265–276

Rontein D, Basset G, Hanson AD (2002) Metabolic engineering of osmoprotectant accumulation in plants. Metab Eng 4:49–56

Roxas VP, Lodhi SA, Garrett DK, Mahan JR, Allen RD (2000) Stress tolerance in transgenic tobacco seedlings that overexpress Glutathione S-Transferase/Glutathione Peroxidase. Plant Cell Physiol 41(11): 1229–1234

Sahi S, Singh A, Blumwald E, Grover A (2006) Beyond osmolytes and transporters: novel plant stress tolerance related genes from transcriptional profiling data. Physiol Plant 127:1–9

Saijo Y, Hata S, Kyozuka J, Shimamoto K, Izui K (2000) Over-expression of a single Ca^{2+}-dependent protein kinase confers both cold and salt/drought tolerance on rice plants. Plant J 23:319–327

Sakamoto A, Murata A, Murata N (1998) Metabolic engineering of rice leading to biosynthesis of glycinebetaine and tolerance to salt and cold. Plant Mol Biol 38:1011–1019

Sanchez DH, Pieckenstain FL, Szymanski J, Erban A, Bromke M, Hannah MA, Kraemer U, Kopka J, Udvardi MK (2011) Comparative functional genomics of salt stress in related model and cultivated plants identifies and overcomes limitations to translational genomics. PLoS One 6(2):e17094

Sanders D, Pelloux J, Brownlee C, Harper JF (2002) Calcium at the crossroads of signalling. Plant Cell 14(Suppl):S401–S417

Sangwan V, Orvar BL, Beyerly J, Hirt H, Dhindsa RS (2002) Opposite changes in membrane fluidity mimic cold and heat stress activation of distinct plant MAP kinase pathways. Plant J 31:629–638

Sarhadi E, Mahfoozi S, Hosseini SA, Salekdeh GH (2010) Cold acclimation proteome analysis reveals close link between the up-regulation of low-temperature associated proteins and vernalization fulfillment. J Proteome Res 9(11):5658–5667

Seki M, Narusaka M, Abe H, Kasuga M, Yamaguchi-Shinozaki K, Carninci P, Hayashizaki Y, Shinozaki K (2001) Monitoring the expression pattern of 1300 *Arabidopsis* genes under drought and cold stresses by using a full-length cDNA microarray. Plant Cell 13:61–72

Seki M, Narusaka M, Ishida J, Nanjo T, Fujita M, Oono Y, Kamiya A, Nakajima M, Enju A, Sakurai T, Satou M, Akiyama K, Taji T, Yamaguchi-Shinozaki K, Carninci P, Kawai J, Hayashizaki Y, Shinozaki K (2002) Monitoring expression profile of 7000 *Arabidopsis* genes under drought, cold-and high-salinity stresses using a full-length cDNA microarray. Plant J 31:279–292

Seki M, Okamoto M, Matsui A, Kim JM, Kurihara Y, Ishida J, Morosawa T, Kawashima M, Kim T, Shinozaki K (2009) Microarray analysis for studying the abiotic stress responses in plants. Mol Tech Crop Improve 3:333–355

Shen B, Jensen RG, Bohnert HJ (1997) Increased resistance to oxidative stress in transgenic plants by targeting mannitol biosynthesis to chloroplasts. Plant Physiol 113:1177–1183

Sheveleva E, Chmara W, Bohnert HJ, Jensen RG (1997) Increased salt and drought tolerance by d-ononitol production in transgenic *Nicotiana tabacum* L. Plant Physiol 115:1211–1219

Shi H, Ishitani M, Kim C, Zhu JK (2000) The *Arabidopsis thaliana* salt tolerance gene *SOS 1* encodes a putative Na^-/H^+ antiporter. Proc Natl Acad Sci USA 97:6896–6901

Shinozaki K, Yamaguchi-Shinozaki K (1996) Molecular responses to drought and cold stress. Curr Opin Biotechnol 7:161–167

Shinozaki K, Yamaguchi-Shinozaki K (1997) Gene expression and signal transduction in water-stress response. Plant Physiol 115:327–334

Shinozaki K, Yamaguchi-Shinozaki K (2000) Molecular response to dehydration and low temperature: differences and cross-talk between two stress signalling pathways. Curr Opin Plant Biol 3:217–223

Shou H, Bordallo P, Fan JB, Yeakley JM, Bibikova M, Sheen J, Wang K (2004a) Expression of an active tobacco mitogen-activated protein kinase kinase kinase enhances freezing tolerance in transgenic maize. Proc Natl Acad Sci USA 101:3298–3303

Shou H, Bordallo P, Wang K (2004b) Expression of the *Nicotiana* protein kinase (NPK1) enhanced drought tolerance in transgenic maize. J Exp Bot 55: 1013–1019

Shukla LI, Chinnusamy V, Sunkar R (2008) The role of microRNAs and other endogenous small RNAs in plant stress responses. Biochim Biophys Acta 1779:743–748

Silva P, Facanha AR, Tavares RM, Geros H (2010) Role of tonoplast proton pumps and Na^+/H^+ antiport system in salt tolerance of *Populus euphratica* oliv. J Plant Growth Regul 29:23–34

Snedden WA, Fromm H (2001) Calmodulin as a versatile calcium signal transducer in plants. New Phytol 151:35–66

Soren KR, Ali K, Tyagi A, Tyagi V (2010) Recent developments in transgenics for abiotic stress tolerance in rice. Indian J Biotechnol 9:233–251

Steponkus PL, Uemura M, Joseph RA, Gilmour SJ, Thomashow MF (1998) Mode of action of the COR15a gene on the freezing tolerance of *Arabidopsis thaliana*. Proc Natl Acad Sci USA 95:14570–14575

Stitt M, Hurry V (2002) A plant for all seasons: alterations in photosynthetic carbon metabolism during cold acclimation in Arabidopsis. Curr Opin Plant Biol 5:199–206

Stockinger EJ, Gilmour SJ, Thomashow MF (1997) *Arabidopsis thaliana* CBF1 encodes an AP2 domain-containing transcriptional activator that binds to the C-repeat/DRE, a cis-acting DNA regulatory element that stimulates transcription in response to low temperature and water deficit. Proc Natl Acad Sci USA 94:1035–1040

Su J, Wu R (2004) Stress-inducible synthesis of proline in transgenic rice confers faster growth under stress conditions than that with constitutive synthesis. Plant Sci 166(4):941–948

Su J, Shen Q, Ho T-H, Wu R (1998) Dehydration-stress-regulated transgene expression in stably transformed rice plants. Plant Physiol 117:913–922

Sugie A, Naydenov N, Mizuno N, Nakamura C, Takumi S (2006) Overexpression of wheat alternative oxidase gene Waox1a alters respiration capacity and response to reactive oxygen species under low temperature in transgenic Arabidopsis. Genes Genet Syst 81: 349–354

Sunkar R, Zhu JK (2004) Novel and stress-regulated microRNAs and other small RNA from Arabidopsis. Plant Cell 16:2001–2019

Sunkar R, Kapoor A, Zhu JK (2006) Posttranscriptional induction of two Cu/Zn superoxide dismutase genes in Arabidopsis is mediated by down regulation of miR398 and important for oxidative stress tolerance. Plant Cell 18:2051–2065

Suprasanna P (2003) Building stress tolerance through over-producing trehalose in transgenic plants. Trends Plant Sci 8(8):355–357

Suprasanna P, Teixeira da Silva JA, Bapat VA (2005) Plant abiotic stress, sugars and transgenics: a perspective. In: Teixeira da Silva JA (ed) Floriculture, ornamental and plant biotechnology: advances and topical issues, 1st edn. Global Science Publishers, London, UK, pp 86–93

Suprasanna P, Sidha M, Bapat VA (2008) Integrated approaches of mutagenesis and *in vitro* selection for crop improvement. In: Kumar A, Shekhawat NS (eds) Plant tissue culture, molecular markers and their role in crop productivity. IK International Publishers, New Delhi, pp 73–92

Szabados L, Savoure A (2010) Proline: a multifunctional amino acid. Trends Plant Sci 15(2):89–97

Tahtiharju S, Sangwan V, Monroy AF, Dhindsa RS, Borg M (1997) The induction of kin genes in cold-acclimating *Arabidopsis thaliana*: evidence of a role for calcium. Planta 203:442–447

Tanaka Y, Hibino T, Hayashi Y, Tanaka A, Kishitani S, Takabe T, Yokota S, Takabe T (1999) Salt tolerance of transgenic rice overexpressing yeast mitochondrial Mn-SOD in chloroplasts. Plant Sci 148:131–138

Thakur P, Kumar S, Malik JA, Berger JD, Nayyar H (2010) Cold stress effects on reproductive development in grain crops: an overview. Environ Exp Bot 67(3):429–443

Thomashow MF (2010) Molecular basis of plant cold acclimation: insights gained from studying the CBF cold response pathway. Plant Physiol 154:571–577

Uemura M, Warren G, Steponkus PL (2003) Freezing sensitivity in the sfr4 mutant of Arabidopsis is due to low sugar content and is manifested by loss of osmotic responsiveness. Plant Physiol 131:1800–1807

Valliyodan B, Nguyen HT (2006) Understanding regulatory networks and engineering for enhanced drought tolerance in plants. Curr Opin Plant Biol 9:1–7

Verdoy D, Coba de la peña T, Redondo FJ, Lucas MM, Pueyo JJ (2006) Transgenic *Medicago truncatula* plants that accumulate proline display nitrogen-fixing activity with enhanced tolerance to osmotic stress. Plant Cell Environ 29:1913–1923

Vincour B, Altman A (2005) Recent advances in engineering plant tolerance to abiotic stress: achievements and limitations. Curr Opin Plant Biol 16:123–132

Vogel JT, Zarka DG, van Buskirk HA, Fowler SG, Thomashow MF (2005) Roles of the CBF2 and ZAT12 transcription factors in configuring the low temperature transcriptome of Arabidopsis. Plant J 41:195–211

Vorasoot N, Songsri P, Akkasaeng C, Jogloy S, Patanothai A (2003) Effect of water stress on yield and agronomic characters of peanut. Songklanakarin J Sci Technol 25(3):283–288

Vranova E, Inze D, Breusegem F (2002) Signal-transduction during oxidative stress. J Exp Bot 53: 1227–1236

Wang W, Vinocur B, Altman A (2003) Plant responses to drought, salinity and extreme temperatures: towards genetic engineering for stress tolerance. Planta 218: 1–14

Welling A, Rinne P, Vihera-Aarnio A, Kontunen-Soppela S, Heino P, Palva ET (2004) Photoperiod and temperature differentially regulate the expression of two dehydrin genes during overwintering of birch (*Betula pubescens* Ehrh.). J Exp Bot 55:507–516

Winicov I, Bastola DR (1999) Transgenic over expression of the transcription factor Alfin1 enhances expression of the endogenous MsPRP2 gene in alfalfa and improves salinity tolerance of the plants. Plant Physiol 120:473–480

Xu J, Zhang Y, Guan Z, Wei W, Han L, Chai T (2008) Expression and function of two dehydrins under environmental stresses in *Brassica juncea* L. Mol Breed 21(4):431–438

Yamada M, Morishita H, Urano K, Shiozaki N, Yamaguchi-Shinozaki K, Shinozaki K, Yoshiba Y (2005) Effects of free proline accumulation in petunias under drought stress. J Exp Bot 56:1975–1981

Ye CY, Zhang HC, Chen JH, Xia XL, Yin WL (2009) Molecular characterization of putative vacuolar NHX-type Na(+)/H(+) exchanger genes from the salt-resistant tree *Populus euphratica*. Physiol Plant 137(2): 166–174

Yeo ET, Kwon HB, Han SE, Lee JT, Ryu JC, Byu MO (2000) Genetic engineering of drought resistant potato plants by introduction of the trehalose-6-phosphate synthase (TPS1) gene from *Saccharomyces cerevisiae*. Mol Cells 10:263–268

Zhang HX, Blumwald E (2001) Transgenic salt-tolerant tomato plants accumulate salt in foliage but not in fruit. Nat Biotechnol 19:765–768

Zhang JZ, Creelman RA, Zhu JK (2004) From laboratory to field. Using information from Arabidopsis to engineer salt, cold, and drought tolerance in crops. Plant Physiol 135:615–621

Zhang T, Wang Q, Chen X, Tian C, Wang X, Xing T, Li Y, Wang Y (2005) Cloning and biochemical properties of CDPK gene *OsCDPK14* from rice. J Plant Physiol 162(10):1149–1159

Zhao BT, Liang RQ, Ge LF, Li W, Xiao HS, Lin HX, Ruan KC, Jin YX (2007) Identification of drought-induced microRNAs in rice. Biochem Biophys Res Commun 354:585–590

Zhou X, Wang G, Sutoh K, Zhu JK, Zhang W (2008) Identification of cold-inducible microRNAs in plants by transcriptome analysis. Biochim Biophys Acta 1779:780–788

Zhu JK (2001) Plant salt tolerance. Trends Plant Sci 6:66–71

Zhu B, Coleman GD (2001) Phytochrome-mediated photoperiod perception, shoot growth, glutamine, calcium, and protein phosphorylation influence the activity of the poplar bark storage protein gene promoter (bspA). Plant Physiol 126:342–351

Zhu J, Dong CH, Zhu JK (2007) Interplay between cold-responsive gene regulation, metabolism and RNA processing during plant cold acclimation. Curr Opin Plant Biol 10:290–295

2 Abiotic Stress-Induced Morphological and Anatomical Changes in Plants

Angelos Patakas

Abstract

Plant abiotic stress responses include changes in both physiological and biochemical processes as well as in anatomical and developmental patterns. Despite the diversity in plant species and abiotic stresses, a generic "stress-induced" response at the plant anatomical level is reported which is mainly comprises three components: inhibition of cell elongation, localized stimulation of cell division, and alterations in cell differentiation status. This result in changes in anatomical characteristics of basic plants organs mainly roots, xylem, and leaves which contribute in adaptation to unfavorable environmental conditions. Taking into consideration that drought consist the most important environmental constraint to plant growth and production, this chapter refers to the holistic approach of anatomical changes at both organ and organism level under limiting soil water availability. The agronomical significance and perspectives of this stress-induced anatomical alterations are discussed.

Keywords

Abiotic Stress • Morphogenic responses • Auxin • Ethylene • ROS • Anatomical changes • Xylem anatomy • Leaf anatomy

1 Introduction

It is well known that abiotic stresses such as extremes in temperature, drought, salinity, heavy metals, and radiation represent the most limiting factors for agricultural productivity worldwide (Doupis et al. 2011). Plant responses to abiotic stresses comprises morphological, physiological, and biochemical changes that either decrease plant's stress exposure and/or limit damage and facilitate recovery of impaired systems (Potters et al. 2007). However, understanding abiotic stress responses in plants is difficult due to the complexity, interrelationship, and variability of mechanisms and molecules involved a fact that consist their evaluation an important and challenging topic in plant research.

A. Patakas (✉)
Laboratory of Plant Production, University of Ioannina, 30100 Agrinio, Greece
e-mail: apatakas@cc.uoi.gr

P. Ahmad and M.N.V. Prasad (eds.), *Abiotic Stress Responses in Plants: Metabolism, Productivity and Sustainability*, DOI 10.1007/978-1-4614-0634-1_2, © Springer Science+Business Media, LLC 2012

2 Abiotic Stress-Induced Morphogenic Responses in Plants

As far as the morphological adaptations in response to abiotic stress is concerned, recent studies revealed the development of distinct morphological responses in plants (stress-induced morphogenesis responses – SIMRs) irrespective the abiotic stresses applied. In fact, different stressors was found to induce similar morphogenic responses which mainly comprised of an inhibition of elongation, localized stimulation of cell division, and complex changes in cell differentiation (Fig. 2.1). This is completely surprising

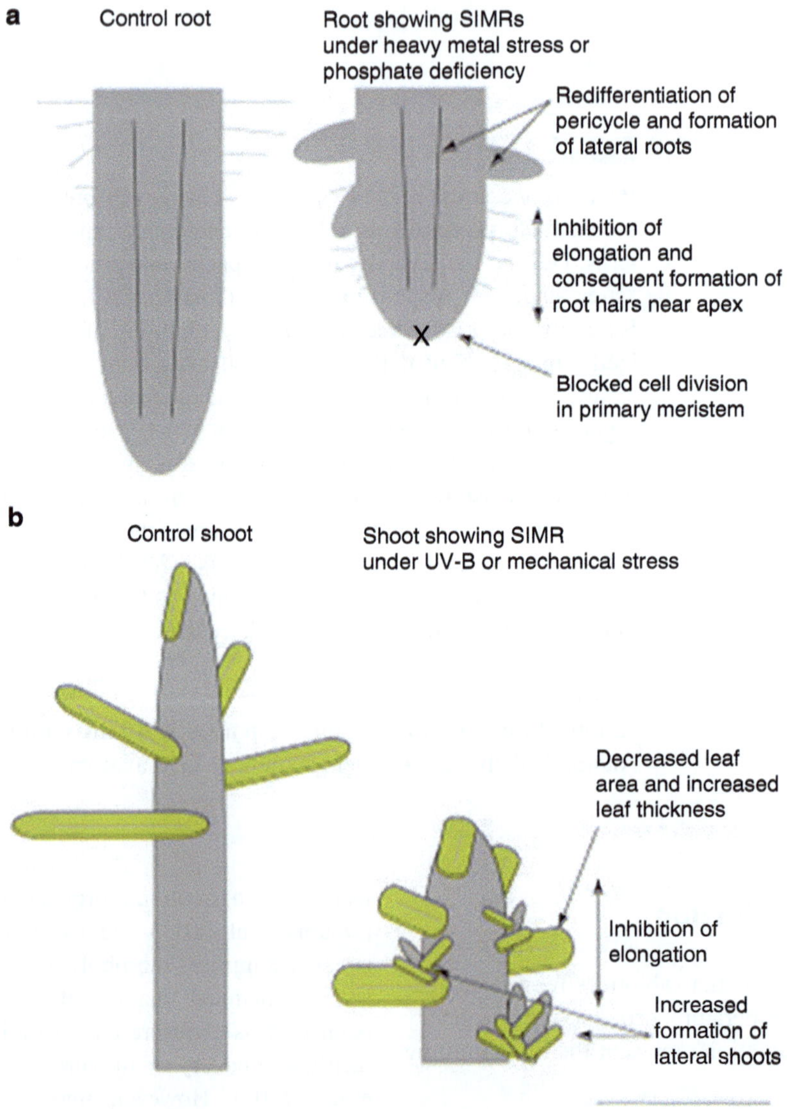

Fig. 2.1 Schematic presentation of the effects of abiotic stresses on plant morphology. (**a**) Overview of root stress-induced morphogenic responses, including an inhibition of root elongation, blocked cell division in the primary meristem, and increased formation of lateral roots. (**b**) Overview of shoot stress-induced morphogenic responses, consisting the inhibition of shoot elongation, increased formation of lateral shoots, and increased leaf thickness (reproduced from: Potters et al. 2007). Stress-induced morphogenic responses: growing out of trouble? Trends in Plant Science 12: 98–105; permission from Elsevier

taking into consideration the differences in types of stress and plants species (Potters et al. 2007, 2009), as well as in stress perception mechanisms, target tissues, and effects on cellular metabolism. For instance, increased concentration of cadmium in plants which affects calcium equilibrium (Perfus-Barbeoch et al. 2002), drought which affects cell wall elasticity and cell redox balances (Sgherri et al. 2007), and elevated UV-B which impacts on DNA transcription and replication, all induce similar morphogenesis responses which contribute to plant adaptation. These common plant morphogenic responses under abiotic stress imply the possible existence of a distinct mechanism operating at either the cell or the organism level. This mechanism should be responsible for sensing, perception, and transduction of the environmental stresses signals leading to specific changes in plants morphology. Till now there are several theories concerning the structure and functioning of the above mentioned mechanism. The first one (reviewed by Prusinkiewicz and Rolland-Lagan 2006) suggests that morphogenesis is controlled by organ and organism-wide gradients of signaling molecules, the so-called "morphogenes." Abiotic stressors that caused morphogenesis are translated into specific patterns of "morphogene" distribution at the organism, organ, or tissue scale, leading to cellular regulation of the processes that are linked to single cells. This theory was expanded recently by incorporating the cellular signaling network (Rauch and Millonas 2004). An alternative interpretation considered morphogenesis as a cellular process that is controlled by the rate of division and expansion of individual cells (Sugimoto-Shirasu and Roberts 2003). Taking into consideration that both cell division and elongation are the major cellular processes by which a plant expands, any morphogenetic process should be based on these two processes. Thus, changes in plant morphology in response to different abiotic stressors are the ultimate consequence of changes in the division and/or expansion rate of the cells. Recently an integrated approach of morphogenesis, whereby growth is being controlled at both the cellular and the organismal levels, has also been proposed (Beemster et al. 2003). Following this approach, the growth and organs formation responses under different abiotic stress are coordinated via growth substance gradients and signaling molecules; all operating at the organ and organismal levels. This theory clearly suggests the occurrence of certain cellular and organismal components that control SIMR phenotype. Whatever interpretation is true the similarities in the morphogenic responses induced by distinct stresses, implies common processes at least at molecular level. At this level, increasing data tend to confirm that stress-induced changes in absolute values as well as interactions between the three morphogenes – auxin, Reactive Oxygen Species (ROS), and ethylene – mainly mediate and control SIM responses in plants (Fig. 2.2).

2.1 The Role of Auxin, Ethylene, and ROS

Recent studies indicate that morphogenesis is tightly linked to hormonal homeostasis, with several hormones controlling cell elongation, cell division, and reorientation of growth. Among them, auxin [indole-3-acetic acid (IAA)] and its gradients are found to be closely associated with lateral root formation and axillary branching, two key components of the SIM responses of plants exposed to abiotic stress (Potters et al. 2007).

Several mechanisms have been proposed to explain stress-induced changes in auxin metabolism with those referring to auxin transport and catabolism to be considered as the preponderant ones. In particular, various abiotic stressors are found to affect expression of auxin efflux carrier genes (Schrader et al. 2003) resulting in modification of auxin distribution throughout the plant organism (Leyser 2005; Paponov et al. 2005). An alternative interpretation suggests that the accumulation of distinct phenolic compounds in response to stress exposure (Winkel-Shirley 2002) contribute in inhibition of polar auxin transport. Indeed, Peer et al. (2004) demonstrate an inverse relation between auxin transport and flavonoid content in a series of *Arabidopsis* flavonoid mutants. As far as the auxin catabolism is concerned, recent

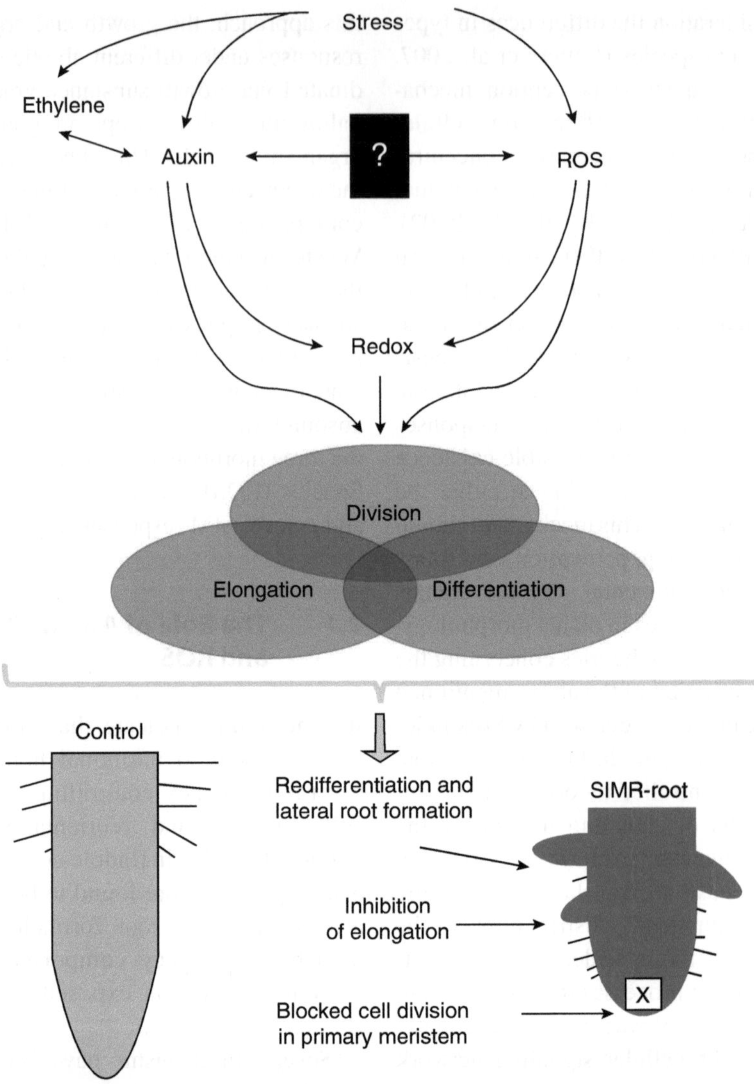

Fig. 2.2 Interaction analysis between environmental parameters, reactive oxygen species (ROS), auxins, and ethylene leading to the stress-induced morphogenic response (SIMR) phenotype (reproduced from: Potters et al. 2009). Different stresses, similar morphogenic responses: integrating a plethora of pathways. Plant, Cell and Environment 32: 158–169; permission from John Wiley and Sons

data conclude that IAA concentration is also subjected to modulation under stress conditions (Jansen et al. 2001). IAA could be degraded via a peroxidative mechanism coupled to a very efficient branched-chain process in which organic peroxide is formed and/or via an oxidative mechanism, using molecular oxygen as electron sink (Savitsky et al. 1999). Additionally, stress-induced changes in auxin conjugation as well as in auxin sensitivity can alter IAA activity and hence impact on morphogenesis (Jiang et al. 2007).

Ethylene involved in the responses of several plant species to heat, drought, or ozone stress (Rao et al. 2002). However, most data suggest ethylene minor contribution in the SIMRs process under abiotic stress conditions mainly indirectly through alteration in the auxin sensitivity of plants (Takahashi et al. 2003).

Enhanced ROS production is associated with a broad range of abiotic stresses including drought (Beis and Patakas 2010), heat stress, enhanced UV-B radiation stress (Doupis et al. 2011), heavy metal stress and anoxia. This increase in ROS production induce an upregulation of ROS scavenging systems, involving enzymatic components such as superoxide dismutase, catalase, and ascorbate peroxidase, as well as antioxidants such as ascorbate and glutathione (Noctor and Foyer 1998). Generally, this integrated system prevents oxidative damage constituting a common component of the response of plants to many distinct stresses. Despite the already known role of ROS as signaling molecules there are also increasing evidences that may also contribute in controlling developmental processes (Gapper and Dolan 2006). In particular, ROS is reported to influence cell development (e.g., xylem vessel formation; Ros-Barcelo et al. 2002), cell division (i.e., temporarily inhibiting cell cycle activity; Reichheld et al. 1999), cell elongation (Schopfer 2001), somatic embryogenesis (Pasternak et al. 2007), and adventitious root formation (Li et al. 2007). Taking into consideration that both ROS and SIMRs are common components of many distinct stresses, it could be assumed that ROS are intermediates between the stress and the development of the SIM responses. Indeed, a direct correlation between ROS and the SIM responses has been demonstrated by Pasternak et al. (2005) who found that *Arabidopsis thaliana* plantlets subjected to either ROS-generating compounds (e.g., paraquat), or a hydrogen peroxide derivative (Pasternak et al. 2005) had an SIMR-like phenotype, similar to that induced by, for example, under enhanced copper concentration.

However, the above mentioned signaling pathways involving either auxins, ethylene, or ROS should not be considered functioning independently one to other. For instance, ROS may modulate auxin sensitivity, by downregulation of auxin-inducible gene expression, a process that involves changes in MAPK activity (Kovtun et al. 2000). Moreover, the auxin-induced elongation of root cells is easily mimicked using either superoxide or hydrogen peroxide (Schopfer et al. 2002). Conversely, auxins, and cytokinins might modulate H_2O_2 effects on stomatal closure, by regulating H_2O_2 scavenging (cytokinins) or production (auxins) (Song et al. 2006). Furthermore, the possible role of nitric oxide and other reactive nitrogen species (RNS) on plants morphogenesis control (Kolbert et al. 2008) as well as an interaction between nitric oxide and auxin in the control of cell division should be reconsidered (Ötvös et al. 2005). All these data confirms that the assumption that SIMRs are based solely on the three morphogenes – auxin, ROS, and ethylene – changes might be an oversimplification. It seems possible that these changes consist only a small part of a more complex mechanism involving interrelationship and interaction between a broader range of molecules whose activation depends on the kind and intensity of abiotic stress as well as on the plant species. In this frame, different stressors are likely to activate a plethora of specific sensors that operates in parallel to each other. Each sensor could regulate a complex signaling cascade, with several interactions resulting in acclimation responses to mild stress. Thus, more detailed analysis concerning interactions analysis between various environmental stress parameters and signals that control SIMRs is needed. Additionally, the specific stress-induced anatomical changes at plant organ level, that is, root, xylem, and leaf should be examined and their relevant significance and potential role to plant adaptation must be carefully evaluated. Considering that drought consist the most important and limiting environmental factor for plant production worldwide, this chapter mainly refers to the holistic approach of plant anatomical changes under limited soil water availability.

3 Drought-Induced Anatomical Changes in Roots

Roots are the primary sites of water and nutrient uptake by plants. Roots also have a remarkable capacity to sense and respond to most of the physicochemical parameters of the soil by adjusting their growth and water transport properties accordingly; these functions being tightly linked to shoot physiology (Bengough et al. 2011). Therefore, under limiting soil water availability conditions, roots may play an important role in maintaining the water status of the whole plant. In particular, morphological and anatomical alterations induced in roots in response to drought result in modification of their soil water extrapolation ability through changes in either allometric parameters – in terms of root branching (lateral root formation) and rate and direction of growth of individual roots and/or in roots water conductivity (Bengough et al. 2011). The importance and relative contribution of these two mechanisms in evaluating the different ability of plants to withstand drought conditions is becoming of increasing interest to plant scientists (Gewin 2010).

3.1 Root Growth Pattern Under Drought Conditions

Generally plants exposed to drought condition exhibit an inhibition of root growth (Westgate and Boyer 1985; Robertson et al. 1990). A significant reduction in root elongation caused by the application of osmotic stress has been found in maize and bean (Robertson et al. 1990). In cotton and in seedlings of three boreal conifers, a reduction in lateral roots number was also observed (Croser et al. 2001; Zolla et al. 2010). Furthermore, using *Arabidopsis* plants, osmotic stress severely represses the formation of lateral roots immediately after their initiation (Deak and Malamy 2005; Xiong et al. 2006). On the contrary, there are also several reports indicating an increase in root-to-shoot biomass ratio (*R*:*S*) in droughted plants (Robertson et al. 1990). A possible explanation for this contradiction might be based on the fact that the observed increase in root-to-shoot ratio might be attributed to a larger decrease in plant growth rather than an increase in root growth rate per se. Furthermore, under mild water stress conditions some species was found to promote absolute increases in root elongation rates resulting in significant increase in total root biomass (Robertson et al. 1990). This differential behavior of plants root system in response to drought could be related not only to different genotypes examined but also to differences in the rate and the intensity of drought stress applied (Zolla et al. 2010). Another possible explanation might be based on indirect effects of drought to root growth. For example, mechanical impedance (soil that is too hard for roots to penetrate rapidly) may significantly affect root growth and development under field conditions (Whalley et al. 2005). As soils dry, capillary forces make matric potential more negative, often causing soil strength to increases rapidly (Whitmore and Whalley 2009). Thus, mechanical impedance could be a major limitation to root growth especially in compacted dry soils with soil water potential less than −100 kPa (Whalley et al. 2005). These effects are exacerbated by further decreasing soil water content, limiting root growth to a relatively greater extent than water stress per se. Penetrometer resistance is a common empirical method used for soil strength quantification. It is equal to the force required to push a metal cone into the soil divided by its cross-sectional area. Penetrometer resistance of 2 MPa is often taken as an indicator of a soil where mechanical impedance will be a major limitation to root elongation, unless a network of channels or fissures exists for roots to exploit (Bengough et al. 2006). Even where such channels exist, root growth may become very clustered in these channels that by-pass the hard soil, restricting water, and nutrient uptake from any impenetrable areas in between (Bengough et al. 2011). Penetrometer resistances in excess of 2 MPa could occur even in relatively moist soils (e.g., matric potentials of −100 kPa to −200 kPa), resulting in sufficient reduction of root elongation rate to less than half of its unimpeded rate. However, even these data concerning root elongation rate decreases in response to both increasing penetrometer resistance and decreasing matric potential appeared controversy (Fig. 2.3). A possible explanation could be attributed to genotypes differences as well as to the methodological difficulties related to accurate control and estimation of soil matric potential at the root surface.

Overall roots seem to elongate more slowly in drying soils due to a combination of drought and

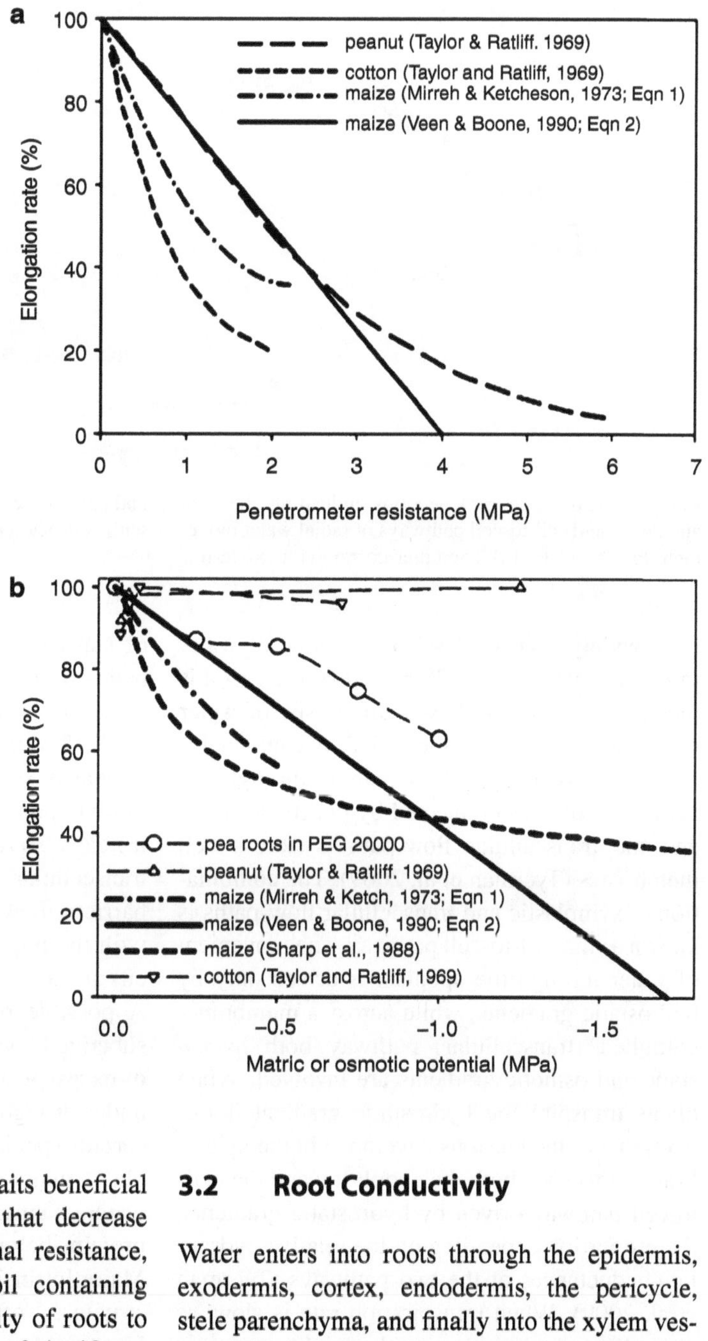

Fig. 2.3 Changes in root elongation rate in response to (**a**) penetrometer resistance and (**b**) matric or osmotic potential (reproduced from: Bengough et al. 2011). Root elongation, water stress, and mechanical impedance: a review of limiting stresses and beneficial root tip traits. Journal of Experimental Botany 62: 59–68; permission from Oxford University Press

mechanical impedance. Root tip traits beneficial to root penetration include traits that decrease cavity expansion pressure, frictional resistance, or axial cell wall tension. In soil containing macropores and channels, the ability of roots to exploit such channels may also be of significant importance, and root hairs are probably sufficiently strong to aid root tip anchorage significantly. Thus, it is essential to consider root responses to soil strength when developing strategies to breed drought-resistant crops and, in order to address this adequately, it may require to develop some novel screening approaches.

3.2 Root Conductivity

Water enters into roots through the epidermis, exodermis, cortex, endodermis, the pericycle, stele parenchyma, and finally into the xylem vessels (Fig. 2.4). The radial conductance of roots is about two orders of magnitude lower than the axial conductance which is largely determined by the dimensions, and the number of xylem vessels (Bramley 2006; Bramley et al. 2007; Tyerman et al. 2009). The composite transport model which comprises apoplastic, symplastic, and transcellular flow-paths operating in parallel has

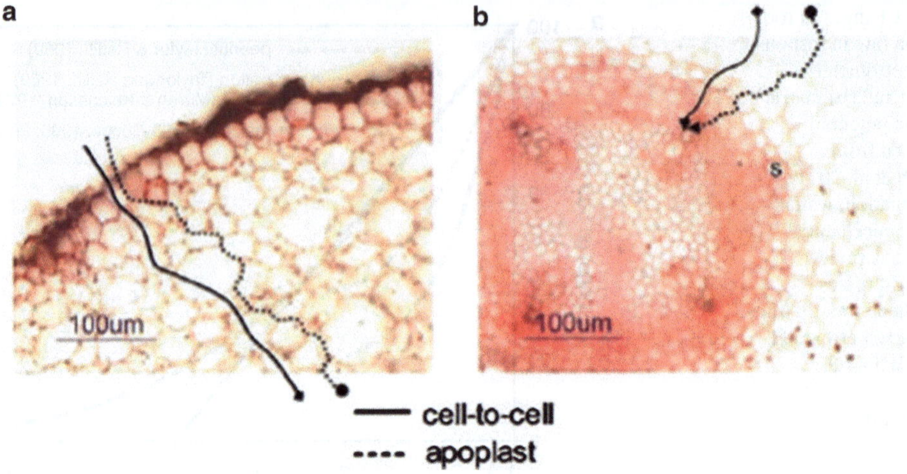

Fig. 2.4 Transverse root sections indicating the root apoplastic and cell-to-cell pathways of radial water movement to the xylem. (**a**) The outer cortex with exodermis and passage cells and (**b**) stele and endodermis consisting with cells showing suberin (s) lamellae are shown

been widely used to describe the flow of water through roots (Steudle 2000). According to this model the apoplastic flow-path consist of water movement outside of the cells' plasma membrane, the symplastic flow-path is through the cytoplasm of cells connected by plasmodesmata, and the transcellular flow-path is across cell membranes (Tyerman et al. 2009). The combination of symplastic and transcellular flow-paths is known as the cell-to-cell pathway. The movement of water through the apoplast is driven only by hydrostatic gradients, while across a membrane-delimited (transcellular) pathway both hydrostatic and osmotic gradients are involved. When plants transpire the hydrostatic gradient dominates due to the tensions developed in the xylem. Water moves via both the apoplastic and the cell-to-cell pathway driven by hydrostatic gradients, the proportion depending on the relative hydraulic conductances of the two pathways (Tyerman et al. 2009). When transpiration rate is slow, as normally occurred during the night or under drought conditions, osmotic flow may dominate, because without large hydrostatic-driven water flows ions in the stele are not diluted, creating an osmotic gradient. However, under normal transpiration conditions the water flow-path taken is mainly influenced by root anatomy. In particular, the apoplastic pathway can be inhibited by the presence of Casparian bands, which are deposits of suberin or lignin in the cell wall. Casparian bands occur in radial and transverse walls of the endodermis and exodermis (Steudle and Peterson 1998; Tyerman et al. 2009). Suberin lamellae may also occur on the tangential walls to further inhibit apoplastic flow (Fig. 2.4). Suberin lamellae can also restrict movement of water along the transcellular pathway. The formation of these barriers to water movement is often associated with the imposition of stress such as water deficits and aging of the plant (Vandeleur et al. 2008). A possible role of this enhanced formation of suberized layers might be correlated to reduction of excess water losses to soil which might occur under drought conditions. On the other hand, in certain species, the transcellular path seems to play a major role as it is efficiently facilitated by water channel proteins named aquaporins. These proteins belong to the ubiquitous super family of Major Intrinsic Proteins (Maurel et al. 2008). The structure of several aquaporins (Tornroth-Horsefield et al. 2006) enables them to insert as tetramers in the membrane forming four individual pores which allow the passage of water or of small neutral molecules (Maurel et al. 2008, 2010). In plants, aquaporins fall into four or five homology subfamilies, among which the Plasma membrane Intrinsic Proteins (PIPs) represent the most abundant aquaporins at the plasma membrane. Because this membrane is a potential obstacle to transcellular water flow, PIPs can control a large part of the root water permeability or hydraulic conductivity (Lpr; Tournaire-Roux et al. 2003). A large array of environmental and hormonal stimuli are known to trigger short-term (minutes to hours) adjustments of Lpr. Among them drought and salinity stress, usually induces a significant decrease in hydraulic conductivity, whereas Absciscic acid (ABA) can exert either an up- or a downregulating effect, depending on time, dose or species (Parent et al. 2009). Soil compaction or flooding which restrict oxygen diffusion in the soil, result in root anoxia which, in turn, downregulates hydraulic conductivity in certain plant species (Tournaire-Roux et al. 2003; Bramley et al. 2010). There is now substantial pharmacological and genetic evidence that most of the short-term changes in root hydraulics caused by abiotic environmental stress are mediated through the regulation of root aquaporin expression and activity. A variety of mechanisms involving transcriptional control (Alexandersson et al. 2005), stimulus-induced internalization of PIPs (Boursiac et al. 2008), or regulated channel opening and closing (gating) by cytosolic calcium, cytosolic protons, or aquaporin phosphorylation has been revealed (Boursiac et al. 2008; Verdoucq et al. 2008). It is important to note that most of the soil stress conditions, including water, nutrient, or oxygen deprivation, all influence hydraulic conductivity and induce an accumulation of ROS in root tissues. A conserved signaling chain involving ROS and acting downstream on aquaporin phosphorylation and subcellular relocalization mediates, in part, the downregulating effects of these stresses on hydraulic conductivity (Boursiac et al. 2008). A ROS-induced stimulation of hydraulic conductivity has also been reported in certain plant species (Benabdellah et al. 2009; Maurel et al. 2010).

3.3 Optimization of Soil Water Uptake Under Drought

Under drought conditions, the rapid adjustment of root hydraulic conductivity to soil water content could be considered as an important mechanism conferring on plants the capability for optimizing soil water uptake. In particular, it was suggested that a transient increase in hydraulic conductivity during the onset of drought might be of great importance facilitating water uptake from the drying soil (Hose et al. 2000; Maurel et al. 2010). Conversely, the long-term downregulation of hydraulic conductivity in droughted plants might be interpreted as a survival reaction to postpone soil water shortage under prolonged stress. It is proposed that a reduction in hydraulic conductivity may primarily impact on water potential gradients along the soil–root–shoot continuum, thereby inducing stomatal closure as well as other water-saving mechanisms operating at the leaf level. Furthermore, drought-induced inhibition of hydraulic conductivity might be beneficial regarding water uptake from soils with a nonuniform water distribution. In such conditions, a reduction of hydraulic conductivity in roots exposed to low water availability would lead to a lowering of soil water uptake by these roots while those exposed to other horizons would compensate by increasing their water uptake due to a decrease in xylem water potential (Ehlert et al. 2009; Maurel et al. 2010). One possible advantage of this mechanism is that it would help the sub fraction of the roots exposed to the driest soil zone to survive, without affecting the overall water uptake capacity of the plant. Downregulation of hydraulic conductivity can also serve as a plant protective reaction restricting a possible backflow of water from the plant into the most desiccated zones of the soil, which could especially occur in the absence of any transpirational driving force at night (Doussan et al. 2006).

4 Functional and Ecological Xylem Anatomy

4.1 The Cavitation Formation and Recovery Mechanism

In xylem tissues of plants, water flows under negative pressure along the pathway from root tips to stomata (Pockman and Sperry 2000). The so-called tension forces, originated by water evaporation in the substomatal chambers during transpiration, permit water uptake from the soil, and water transport along the plant's conduction system (Steudle 2001) which is mainly consisted of two components; tracheids and vessel. Both components lack a protoplast when fully mature and generally lack the end walls between adjacent cells to reduce the resistance to water flow. This model of water transport suggests that water columns could be interrupted by air formation (cavitation) (Tyree and Zimmermann 2002) when tensions increase under drought conditions. This results in the blockage of vascular tissues due to embolism, causing loss of hydraulic conductance of the plant (Sperry et al. 2006). There are large differences between species in vulnerability to embolism and differences have been also observed even between cultivars of the same species (Lovisolo and Schubert 2006; Lovisolo et al. 2008, 2010). In plants with an anisohydric type response to drought (Hukin et al. 2005), a higher cavitation vulnerability of shoots was found compared to isohydric ones. In the former species stomatal closure occurred relatively late, well after shoot hydraulic conductance was significantly affected by embolisms, suggesting a possible role of cavitation to the plants survival strategy.

In order to reintegrate vessel functionality, plants have developed different repair mechanisms, which, in some cases, are associated with positive root pressure, and in other cases involve active and energy-consuming processes in shoot conductive tissues (Salleo et al. 2004). In grapevines, experiments conducted with the metabolic and water transport inhibitor mercuric chloride, indicated the occurrence of an active mechanism consisting both living cells and aquaporins (Lovisolo and Schubert 2006). A positive role of aquaporins in embolism repair was also reported for other species in which aquaporins, localized in xylem vessel parenchyma cells, were activated during embolism recovery (Martre et al. 2002; Sakr et al. 2003). Similar result were obtained using tobacco (*Nicotiana tabacum* L.) RNAi plants (Kaldenhoff et al. 2008; Lovisolo et al. 2010). Furthermore, an ABA/aquaporin interaction in embolism repair and in aquaporin activation during drought is reported (Kaldenhoff et al. 2008). In particular, ABA was found to be involved in gating mechanisms of water channels by facilitating their structural restoration, possibly acting from the cytoplasmatic side of aquaporins (Wan et al. 2004). Although downregulation of aquaporins after ABA treatment has also been reported (Suga et al. 2002), the majority of results suggest that aquaporins, if responsive, were upregulated by this hormone (Jang et al. 2004; Parent et al. 2009; Lovisolo et al. 2010).

4.2 Hydraulic Architecture and Drought

The hydraulic architecture of water conducting elements plays an important role in plants water economy. Recent studies indicated significant plasticity in xylem anatomical characteristics in response to environmental changes which influence not only the hydraulic characteristics through changes in water conductivity but also the ability of plants to adapt at different abiotic stresses (Arend and Fromm 2007). For example, the observed decrease in vessels diameter in response to plants exposure to drought might contribute to drought tolerance through the already known increasing resistance to drought-induced xylem embolism (Lo Gullo et al. 1995; Hacke et al. 2006). However, in most studies, drought stress results in smaller vessels but increased vessel frequencies compared with nonstressed trees (Junghans et al. 2006; Arend and Fromm 2007). Thus, the sum of vessel lumen area remained unchanged under mild stress compared with nonstressed plants because an increased vessel frequency compensated the decreased size of individual vessels. Cell wall thickness was also increased leading to higher resilience of the conductive system to decreasing turgor pressure (Junghans et al. 2006). Such structural changes have reported to occur in many plants exposed to drought a fact that underlie the ability of trees to adapt their wood anatomy characteristics to environmental requirements (Searson et al. 2004).

Despite the progress in drought-induced changes in xylem anatomical characteristics there is still little information concerning possible molecular events and signals which probably mediate wood structure adjustment to environmental conditions. Phytohormones such as ABA and auxin are considered as the key components for structural flexibility of wood formation (Mellerowicz et al. 2001). Arend and Fromm (2007) speculated that the apparent differences in seasonal wood characteristics are closely related to seasonal variation in sensitivity of cambial cells to ABA. Furthermore several observations suggest that dynamic changes in auxin levels and changes in auxin responsiveness and not the amount of auxin per se are part of rather complex molecular mechanism involved in adjusting wood anatomy to environmental changes. It is obvious that future studies will require systems approaches at cell, tissue, and plant level to understand the integration of environmental input signals and their transduction into adjustment of wood formation to various abiotic stresses. This knowledge might be useful in selection programs for drought tolerant plant species.

4.3 Abscission as a Drought Avoidance Strategy

Trees show a consistent decline in their leaf-specific xylem conductivity as water moves from trunk to minor branches and leaves (Zimmermann 1978). These results in ever steeper pressure gradients and lower xylem pressure (Px) toward the distal ends of the flow-path. According to the "segmentation hypothesis" (Zimmermann 1983) this could be an adaptive response ensuring that cavitation would be confined to "cheap" distal organs that could be sacrificed during a drought. Especially in trees, avoidance of xylem cavitation is considered very important since water has to be transported to greater heights, and any damage could not easily be overcome by root sap pressure or capillary forces as in annual plants. Furthermore by separating lateral organs form the hydraulic continuum, the overall transpiration of the plant is greatly reduced relieving xylem tension in more basal parts of the plant, and therefore contributing to plants survival under water stress. Leaf shedding as a drought adaptive measure is well known to occur in tropical trees but has also been observed in temperate and boreal tree species (Ogaya and Penuelas 2006; Castro-Diez and Navarro 2007; Fischer and Polle 2010). Chen et al. (2002) showed that drought stress induces rapidly leaf abscission in *P. euramericana* (cv. Italica) but not in *P. popularis*. After leaf abscission the water status of plants rapidly recovered. In addition to single leaves abscission, several species exhibit premature senescence of leaves and leaf abscission on whole branches and subsequently necrosis and loss of these branches; a process, which is referred to as "branch sacrifice" (Fischer and Polle 2010). In contrast to simple leaf abscission, growth of the proximal buds can not be resumed by entering more favorable conditions. Thus, branch dieback can be considered as a more extreme form of stress avoidance than leaf abscission. Dead branches will be retained in the crown or eventually be separated from the tree body by mechanical forces. Interestingly, this phenomenon of "branch sacrifice" can be mainly observed in species, which are native to semiarid areas and therefore well-adapted to drought stress; whereas in genotypes adapted to more humid areas, branch dieback only rarely was occurred during drought periods (Rood et al. 2000; Fischer and Polle 2010). Branch dieback often correlates with higher xylem vulnerability a fact that indicates the importance of this mechanism for survival under drought (Rood et al. 2000). It should also be taken into consideration that branch sacrifice might be energy consuming as it includes a significant loss of tree tissues. However, dieback in most cases proceeds from younger, more apical to older, more basal parts of a branch; a fact that alleviates the energy cost (Rood et al. 2000; Fischer and Polle 2010). In addition to stomatal closure, which reduces the transpiration rate per unit leaf area, branch sacrifice seem to be an efficient mechanism conferring plants the capability to maintain water balance by reducing the total transpirational area of the canopy and hence the total plant water losses.

4.4 Xylem Capacitance

The hydraulic capacitance (*C*) of plants represents the amount of water that could be storage in various plant tissues. Among different tissues that can contribute to the total *C* the sapwood is considered as the major source of stored water that could be withdrawn and recharged on a seasonal (Waring et al. 1979) and daily basis (Cermák et al. 2007; Scholz et al. 2007); the intrinsic value varying widely among species. During the day, xylem water flux and, consequently, tension are rarely at steady state due to the continual fluctuations in stomatal conductance and atmospheric evaporative demand. Transient, transpiration-induced increases in xylem tension result in the capacitive discharge of water into the transpiration stream, effectively bypassing a portion of the soil-to-leaf hydraulic resistance and lengthening the time required for tension and flow to attain steady-state values throughout the plant (Phillips et al. 2004; Sperry et al. 2008). Consistent with an Ohm's law analogue for xylem water transport, the *C* of a tissue is *t* defined as the ratio of change in its water content to change in its water potential (d*W*/d*Y*). However, for comparisons among species and plants of different sizes, it is often more informative to express *C* in terms of the mass of water released per tissue volume per change in water potential (Scholz et al. 2007; Sperry et al. 2008), or the total mass of water withdrawn daily from internal storage per change in water potential between two points in the soil–plant system. Although absolute amounts of water derived from *C* may constitute only 10–30% of the total daily transpiration (Phillips et al. 2004; Sperry et al. 2008), the buffering impact of *C* on the daily dynamics of plant–water relations and maximum tensions generated in the terminal portions of the water transport pathway can be substantial, even in relatively small plants with a limited total water storage capacity. Recent studies of tropical trees suggest that the stomata regulate transpiration in a manner that optimizes the capacitive discharge of water from stem tissue, while at the same time avoiding excessive embolism. Species-specific set points for the daily minimum water potential of terminal branches appear to represent a compromise that maximizes the reliance on stored water over the range where *C* is nearly constant as sapwood water deficit increases (Fig. 2.5a), but minimizes the risk of embolism as both *C* and its buffering effect diminish beyond this point and embolism begins to increase exponentially (Fig. 2.5b).

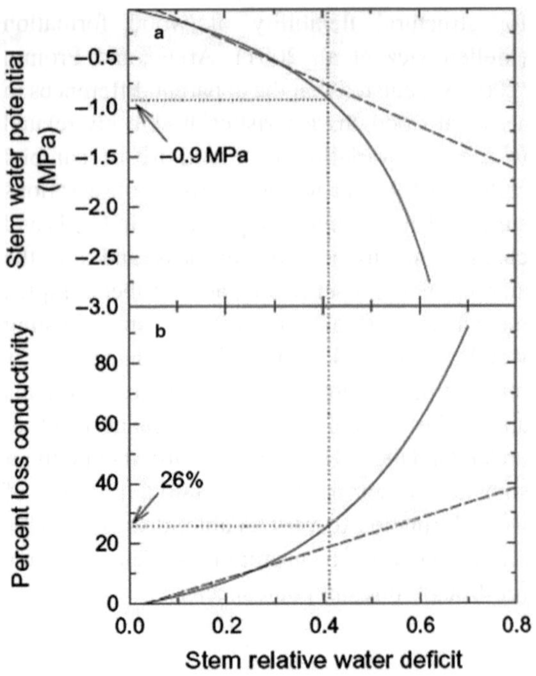

Fig. 2.5 Changes in (**a**) stem water potential and (**b**) stem hydraulic conductivity loss in relation to stem relative water deficit. Dashed lines represent linear regressions fitted to the initial nearly portions of the curves (reproduced from: Sperry et al. 2008). Safety and efficiency conflicts in hydraulic architecture: scaling from tissues to trees. Plant, Cell and Environment 31:632–645; permission from John Wiley and Sons

In this frame, capacitance might be effective in increasing plants xylem conductance while maintaining xylem safety (Sperry et al. 2008).

5 Leaf Anatomy and Drought

Leaves are complex structures mainly consisting of two dissimilar layers (spongy and palisade mesophyll) of photosynthesizing cells, with different packing and cell orientation, interspersed by vascular tissues, all between two epidermis, which are perforated by stomatal pores. This structure results in an appreciable volume, so that the photosynthesizing cells and chloroplasts are located at some distance from the points of entry of CO_2. Once CO_2 reaches the leaf surface, diffusion into the leaf depends on the stomatal resistance as the

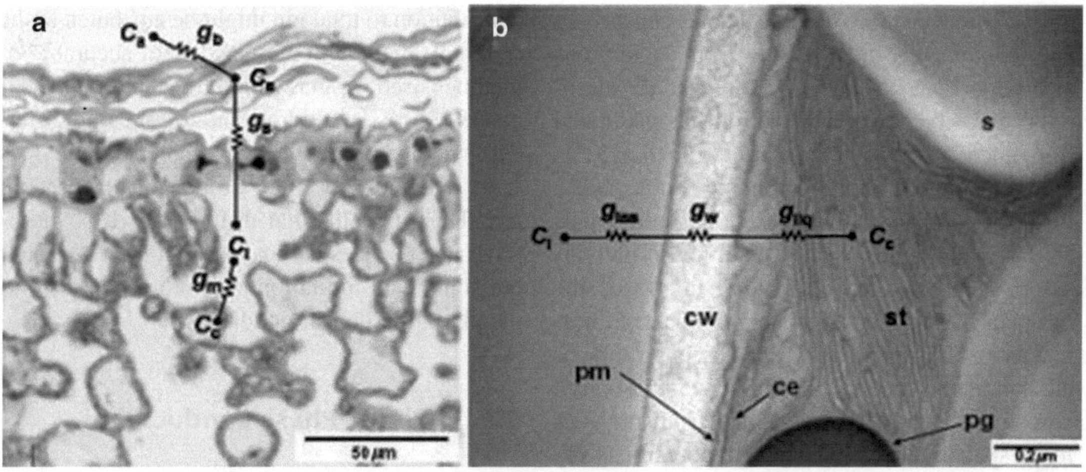

Fig. 2.6 (**a**) Detailed micrograph of the abaxial surface of an olive leaf (*bottom side up*), presenting CO_2 pathway of from ambient (Ca) through leaf surface (Cs) and intercellular air spaces (Ci) to the chloroplast (Cc). Boundary layer conductance (gb), stomatal conductance (gs), and mesophyll conductance (gm) are indicated. (**b**) Electron micrograph of a grapevine leaf where cell wall (cw), plasma membrane (pm), the chloroplast envelope (ce), and stroma thylakoid (st) can be observed. The differences in CO_2 concentration between Ci and Cc depends on intercellular air space (gias), cell wall (gw), and liquid (gliq) CO_2 conductances. A grain of starch (s) and a plastoglobule (pg) can be also seen (reproduced from Flexas et al. 2008). Mesophyll conductance to CO_2: current knowledge and future prospects. Plant, Cell and Environment 31: 602–612; permission from John Wiley and Sons. Photos by A. Diaz-Espejo

cuticle is usually regarded as effectively CO_2 impermeable (Kerstiens 2006; Morison and Lawson 2007). During photosynthesis, the CO_2 entering the leaves through stomata has to diffuse from substomatal internal cavities to the sites of carboxylation inside the stroma through the leaf mesophyll. Therefore, understanding CO_2 diffusion in leaves is considered very important because the characteristics of the overall diffusion pathway are one of the determinants of the photosynthetic rate (Flexas et al. 2004). It is generally assumed that the overall leaf internal diffusion conductance in the photosynthetic pathway (often refereed as mesophyll conductance gm) can be divided in at least two main components; a gaseous component through the intercellular air spaces (gias) and a liquid component referring to CO_2 diffusion from the cell walls to the chloroplasts (gliq) (Fig. 2.6). Taking into consideration that both components can be significantly affected by the structure of the mesophyll as well as that abiotic stress considerably affects leaf anatomy it is obvious that abiotic stresses can influence photosynthetic performance through alterations in leaf structure.

5.1 The Gias Component

Clearly the first anatomical feature that influences gias is the stomatal distribution between leaf surfaces. Stomata can occur on either both surfaces (amphistomatous) or on one surface (usually lower only, hypostomatous, but also upper only, hyperstomatous), with different stomatal sizes and densities on the two surfaces (Morison and Lawson 2007). Because of these differences there is potentially a substantial difference in gias between amphistomatous leaves and those with stomata on only one surface, as the apparent CO_2 path length is halved. However, most studies tend to conclude that the real path length for CO_2 diffusing in the leaf depends on leaf anatomy and mesophyll organization (Morison and Lawson 2007) with gias estimations for hypostomatous leaves to range from 1/3 to 1/2 that of amphistomatous leaves (Terashima et al. 2006).

The proportion of the leaf that occupied by air spaces varies substantially, although part of this variation must be attributed to different measurement methods used. In studies concerning 56 predominately dicot species from 21 families in a wide range of habitats, the average mesophyll porosity (air space fraction *fias* estimated from microscopic sections) varied from 4 to 51%, with a median of 15% (Slaton and Smith 2002). In species with succulent leaves, mesophyll cells are large and tightly packed resulting in low *fias* (Morison and Lawson 2007) while in grasses, the *fias* values are generally lower compared to normal dicots. Except different species, environmental parameters could also influence mesophyll porosity. Shaded leaves obtained from the four forest species tended to have high porosities (34–48%) compared to those fully exposed to sun (Slaton and Smith 2002). Sack et al. (2003) also reported significantly lower *fias* in sun than in shade leaves of various tree species in agreement with similar results concerning herbaceous dicot species (Ivanova et al. 2006; Morison and Lawson 2007). On the other hand, several studies have reported smaller or even no differences between sun and shade leaves (Morison and Lawson 2007). Except different environmental conditions, plants exposure to various abiotic stresses could also result in alterations in leaf anatomy characteristics impairing CO_2 diffusion. For example, growth under saline conditions reduced *fias* in *Spinacia oleracea* from 32 to 24% (Delfine et al. 1998), and high and low altitude-grown *Polygonum cuspidatum* had *fias* values of 27 and 49%, respectively (Morison and Lawson 2007). In addition, comparative studies of paradermal leaf sections of two olive varieties subjected to different irrigation regimes indicate a significant reduction in intercellular air spaces in drought stressed compared to well irrigated plants (Chartzoulakis et al. 1999). Furthermore, *fias* values as well as estimated values of gas phase conductance (gias) were significantly higher in a more drought adapted olive variety "koroneiki", suggesting an important role of these leaf anatomical alterations to the drought adaptation (Chartzoulakis et al. 2002).

Parkhurst and Mott (1990) reported that gias accounted for 10–60% of total mesophyll conductance (gm), being more important in hypostomatous leaves. This great variability in gias contribution to total gm might be attributed to the different methodologies and the accuracy of methods used for gias estimations. Other investigators (Morison and Lawson 2007) using different methodologies have shown that gias contributes only a small proportion to the total gm. Thus it seems that diffusion of CO_2 mesophyll in air spaces in most cases consist a smaller but significant limitation to photosynthesis compared to liquid phase conductance.

5.2 Liquid Phase Conductance

Liquid phase component of mesophyll conductance is quite complex depending on many leaf anatomical traits such as mesophyll cell surface area, chloroplast surface area exposed to intercellular air spaces (Evans and Loreto 2000), chloroplast rearrangements, and cell wall thickness (Terashima et al. 2006). However, the CO_2 diffusion coefficient in the liquid phase is suggested to be much slower compared to that in intercellular air spaces, indicating significant lower gliq values (Morison and Lawson 2007). Actually, a reevaluation of gias using the comparison of diffusivities in air and helox, revealed that gliq was much smaller than gias, and therefore it consists the most limiting factor for photosynthesis (Piel et al. 2002). Furthermore, Gorton et al. (2003) using a photoacoustic technique, also concluded that low gliq as the most limiting factor for CO_2 diffusion in the mesophyll. Gillon and Yakir (2000) combining ^{13}C and ^{18}O discrimination by leaves were able to partition gm into the two components, gias and a cellular component that was termed chloroplast conductance (gchl) instead of gliq (Fig. 2.6). As chloroplasts are usually tightly coupled to cell membranes facing intercellular air spaces, it was assumed that CO_2 would not have to cross the cytosol, entering the chloroplasts directly after crossing the cell wall and plasma and chloroplast membranes. These authors showed that gias was lower than gchl in thick leaves of oaks, but gchl was lower than gias in the mesophytic leaves of soybean and tobacco. In contrast, Piel et al. (2002) showed that even in the sclerophyll oak *Quercus ilex* rliq (the inverse of gliq) contributed to 70% of the total internal resistance, while rias (the inverse of gias) contributed the remaining 30%. Therefore, gliq (or gchl) seems to be the most important component of gm not only in mesophytic species but probably also in sclerophylls.

5.3 Ecophysiological Significance of Mesophyll Conductance

Many studies revealed gm variability not only among different species and cultivars, but also in response to environmental variables. In particular, it was found that gm decrease in response to water stress, low nitrogen availability, salinity, high altitude, water logging, leaf temperature, and leaf aging (Flexas et al. 2008). In addition to these long-term and developmental changes in gm recent data indicate also short-term gm responses to several environmental variables suggesting that gm is not only finite, but it also acclimates and responds both in the long (days, weeks) and short (minutes, hours) terms to many environmental variables. For example, leaf desiccation, rapid changes in leaf temperature, changes in CO_2 concentration, and changes in light intensity all result in significant changes in gm. In particular, cutting the leaf petiole resulted in immediate reduction of both stomatal conductance gs and gm of about 30% after only 10 min. Similarly changes in leaf temperature and in external CO_2 concentration also results in large responses of gm within 20–30 min, in plants acclimated to both low and high temperatures (Gorton et al. 2003; Warren and Dreyer 2006). All these data confirm similar responses of both gm and stomata to changes in almost all environmental variables indicating that gm and gs are tightly coregulated in order to efficient adjust CO_2 availability in the chloroplasts in response to environmental changes. These changes in gm might consider very important, conferring on plants the capability to efficiently and rapidly regulate photosynthetic rate in response to the environment. If this is true then it would be expected that gm changes in plants would result in maintaining photosynthetic performance under adverse environmental conditions. Indeed, under drought conditions an increase in photosynthesis by increasing stomatal conductance is impossible due to consequent increase in transpiration plant water losses. On the contrary, an increase in photosynthetic performance by means of increasing gm may result in increasing Pn without additional water losses; thus increasing plants water use efficiency (WUE; Aranda et al. 2007). Lauteri et al. (1997) showed that provenances of *Castanea sativa* from low rainfall areas had a higher WUE and a higher gm than provenances from high rainfall areas. Plants originated from areas characterized by high mean annual precipitation exhibited significantly higher gm/gs ratio compared to those from more arid areas. A similar relationship between WUE and gm/gs was observed in provenances of the annual herb *Crepis tirasii*, an ancient endemism from the Balearic Islands with a very fragmentary distribution in isolated populations (Flexas et al. 2008). When growing in a common environment, plants from different populations from sites with high precipitation exhibited gm/gs ratios between 0.3 and 0.4 mol mol^{-1}. However, a separated population originated from ore arid environment showed a significant higher WUE, in correspondence with increased gm/gs. Thus, it seems that increasing the ratio gm/gs enhances photosynthesis and WUE in plants evolved under arid environments (Flexas et al. 2008). While further studies including more species and larger numbers of cultivars are needed, these results suggest that gm could be a good target for the improvement of crop WUE through biotechnology.

6 Conclusion and Future Perspective

Plants have evolved a large variety of distinct anatomical alterations operating in both organs and organism level in order to adjust to unfavorable environmental conditions. At organs level significant changes in certain anatomical characteristics in plants exposed to abiotic stresses have been observed. In the last 20 years, significant progress has been made in understanding the relationship between changes in root, xylem, and leaf anatomy as well as in evaluation of possible links between theses anatomical alterations and plants adaptation. These changes seem to be coordinated under stress conditions resulting in similar anatomic responses irrespective the plant species and abiotic stress applied. Despite the fact that the role of specific molecules has been revealed more analysis is needed in order to fully evaluate the molecular and physiological basis of this orchestrated adjustment of anatomical changes under stress conditions. In order to achieve this, a holistic approach is needed consisting from two different but complimentary research directions. The first one refers to a system-based approach relating changes in environmental parameters and molecular signals to anatomical changes. The second focus on the detailed analysis of certain anatomical changes induced in different organs by abiotic stress giving emphasis on their contribution in plant adaptation. Results obtained will be used in order to elucidate different adaptation potential of various plant species as well as to reveal possible novel stress tolerance mechanisms to abiotic stress.

References

Alexandersson E, Fraysse L, Sjovall-Larsen S, Gustavsson S, Fellert M, Karlsson M, Johanson U, Kjellbom P (2005) Whole gene family expression and drought stress regulation of aquaporins. Plant Mol Biol 59:469–484

Aranda I, Pardos M, Puértolas J, Jiménez MD, Pardos JA (2007) Water-use efficiency in cork oak (*Quercus suber*) is modified by the interaction of water and light availabilities. Tree Physiol 27:671–677

Arend M, Fromm J (2007) Seasonal change in the drought response of wood cell development in poplar. Tree Physiol 27:985–992

Beemster G, Fiorani F, Inzé D (2003) Cell cycle: the key to plant growth control? Trends Plant Sci 8:154–158

Beis A, Patakas A (2010) Differences in stomatal responses and root: shoot signalling between two grapevine varieties subjected to drought. Funct Plant Biol 37:1–8

Benabdellah K, Ruiz-Lozano JM, Aroca R (2009) Hydrogen peroxide effects on root hydraulic properties and plasma membrane aquaporin regulation in *Phaseolus vulgaris*. Plant Mol Biol 70:647–661

Bengough AG, Bransby MF, Hans J, McKenna SJ, Roberts TJ, Valentine TA (2006) Root responses to soil physical conditions: growth dynamics from field to cell. J Exp Bot 57:437–447

Bengough AG, McKenzie BM, Hallett PD, Valentine TA (2011) Root elongation, water stress, and mechanical impedance: a review of limiting stresses and beneficial root tip traits. J Exp Bot 62:59–68

Boursiac Y, Boudet J, Postaire O, Luu DT, Tournaire-Roux C, Maurel C (2008) Stimulus-induced down-regulation of root water transport involves reactive oxygen species-activated cell signalling and plasma membrane intrinsic protein internalization. Plant J 56:207–218

Bramley H (2006) Water flow in the roots of three crop species: the influence of root structure, aquaporin activity and water logging. The University of Western Australia, Perth

Bramley H, Turner NC, Turner DW, Tyerman SD (2007) Comparison between gradient dependent hydraulic conductivities of roots using the root pressure probe: the role of pressure propagations and implications for the relative roles of parallel radial pathways. Plant Cell Environ 30:861–874

Bramley H, Turner NC, Turner DW, Tyerman SD (2010) The contrasting influence of short-term hypoxia on the hydraulic properties of cells and roots of wheat and lupin. Funct Plant Biol 37:183–193

Castro-Diez P, Navarro J (2007) Water relations of seedlings of three *Quercus* species: variations across and within species grown in contrasting light and water regimes. Tree Physiol 27:1011–1018

Cermák J, Kucera J, Bauerle WL, Phillips N, Hinckley TM (2007) Tree water storage and its diurnal dynamics related to sap flow and changes in stem volume in old-growth Douglas-fir trees. Tree Physiol 27:181–198

Chartzoulakis K, Patakas A, Bosabalidis A (1999) Influence of leaf anatomical structure on the photosynthetic rate of two olive cultivars grown under drought conditions. Environ Exp Bot 42:113–120

Chartzoulakis K, Patakas A, Kofidis G, Bosabalidis A (2002) Water stress affects leaf anatomy, gas exchange and growth of two avocado cultivars. Sci Hortic 1778:1–13

Chen S, Li J, Wang T (2002) Osmotic stress and ion-specific effects on xylem abscisic acid and the relevance to salinity tolerance in poplar. J Plant Growth Regul 21:224–233

Croser C, Renault S, Franklin J, Zwiazek J (2001) The effect of salinity on the emergence and seedling growth of *Picea mariana*, *Picea glauca*, and *Pinus banksiana*. Environ Pollut 115:9–16

Deak KI, Malamy J (2005) Osmotic regulation of root system architecture. Plant J 43:17–28

Delfine S, Alvino A, Zacchini M, Loreto F (1998) Consequences of salt stress on conductance to CO_2 diffusion, Rubisco characteristics and anatomy of spinach leaves. Aust J Plant Physiol 25:395–402

Doupis G, Chartzoulakis K, Beis A, Patakas A (2011) Allometric and biochemical responses of grapevines subjected to drought and enhanced ultraviolet-B radiation. Aust J Grape Wine Res 17:36–42

Doussan C, Pierret A, Garrigues E, Pages L (2006) Water uptake by plant roots. II. Modelling of water transfer in the soil–root system with explicit account of flow within the root system – comparison with experiments. Plant Soil 283:99–117

Ehlert C, Maurel C, Tardieu F, Simonneau T (2009) Aquaporin mediated reduction in maize root hydraulic conductivity impacts cell turgor and leaf elongation even without changing transpiration. Plant Physiol 150:1093–1104

Evans JR, Loreto F (2000) Acquisition and diffusion of CO_2 in higher plant leaves. In: Leegood RC, Sharkey TD, von Caemmerer S (eds) Photosynthesis: physiology and metabolism. Kluwer Academic Publishers, Dordrecht, The Netherlands, pp 321–351

Fischer U, Polle A (2010) Populus responses to abiotic stress. In: Jansson S et al (eds) Plant genetics and genomics: crops and models. Springer Science+Business Media, New York, pp 225–245

Flexas J, Bota J, Loreto F, Cornic G, Sharkey TD (2004) Diffusive and metabolic limitations to photosynthesis and salinity in C3 plants. Plant Biol 6:269–279

Flexas J, Ribas-Carbo M, Diaz-Espejo A, Galmes J, Medrano H (2008) Mesophyll conductance to CO_2: current knowledge and future prospects. Plant Cell Environ 31:602–612

Gapper C, Dolan L (2006) Control of plant development by reactive oxygen species. Plant Physiol 141:341–345

Gewin V (2010) Food: an underground revolution. Nature 466:552–553

Gillon JS, Yakir D (2000) Internal conductance to CO_2 diffusion and C18OO discrimination in C3 leaves. Plant Physiol 123:201–213

Gorton HL, Herbert SK, Vogelmann TC (2003) Photoacoustic analysis indicates that chloroplast movement does not alter liquid-phase CO_2 diffusion in leaves of *Alocasia brisbanensis*. Plant Physiol 132:1529–1539

Hacke UG, Sperry JS, Wheeler JK, Castro L (2006) Scaling of angiosperm xylem structure with safety and efficiency. Tree Physiol 26:689–701

Hose E, Steudle E, Hartung W (2000) Abscisic acid and hydraulic conductivity of maize roots: a study using cell- and root-pressure probes. Planta 211:874–882

Hukin D, Cochard H, Dreyer E, Thiec DL, Bogeat-Triboulot MB (2005) Cavitation vulnerability in roots and shoots: does *Populus euphratica* Oliv., a poplar from arid areas of Central Asia, differ from other poplar species? J Exp Bot 56:2003–2010

Ivanova LA, Petrov MS, Kadushnikov RM (2006) Determination of mesophyll diffusion resistance in *Chamaerion angustifolium* by the method of three-dimensional reconstruction of the leaf cell packing. Russ J Plant Physiol 53:316–324

Jang JY, Kim DG, Kim YO, Kim JS, Kang H (2004) An expression analysis of a gene family encoding plasma membrane aquaporins in response to abiotic stresses in *Arabidopsis thaliana*. Plant Mol Biol 54:713–725

Jansen MAK, Van den Noort RE, Tan A, Prinsen E, Lagrimini M, Thorneley RNF (2001) Phenol-oxidizing peroxidises contribute to the protection of plants from ultraviolet radiation stress. Plant Physiol 126:1012–1023

Jiang C, Gao X, Liao L, Harberd NP, Fu X (2007) Phosphate starvation root architecture and anthocyanin accumulation responses are modulated by the gibberellin-DELLA signalling pathway in Arabidopsis. Plant Physiol 145:1460–1470

Junghans U, Polle A, Düchting P et al (2006) Effects of salt stress on the anatomy and auxin physiology of poplar xylem. Plant Cell Environ 29:1519–1531

Kaldenhoff R, Ribas-Carbo M, Flexas J, Lovisolo C, Heckwolf M, Uehlein N (2008) Aquaporins and plant water balance. Plant Cell Environ 31:658–666

Kerstiens G (2006) Water transport in plant cuticles: an update. J Exp Bot 57:2493–2499

Kolbert Z, Bartha B, Erdei L (2008) Osmotic stress- and indole- 3-butyric acid-induced NO generation are partially distinct processes in root growth and development in *Pisum sativum*. Physiol Plant 133:406–416

Kovtun Y, Chiu WL, Tena G, Sheen J (2000) Functional analysis of oxidative stress-activated mitogen-activated protein kinase cascade in plants. Proc Natl Acad Sci USA 97:2940–2945

Lauteri M, Scartazza A, Guido MC, Brugnoli E (1997) Genetic variation in photosynthetic capacity, carbon isotope discrimination and mesophyll conductance in provenances of Castanea sativa adapted to different environments. Funct Ecol 11:675–683

Leyser O (2005) Auxin distribution and plant pattern formation: how many angels can dance on the point of PIN? Cell 121:819–822

Li SW, Xue LG, Xu SJ, Feng HY, An LZ (2007) Hydrogen peroxide involvement in formation and development of adventitious roots in cucumber. Plant Growth Regul 52:173–180

Lo Gullo MA, Salleo S, Piaceri EC, Rosso R (1995) Relations between vulnerability to xylem embolism and xylem conduit dimensions in young trees of *Quercus cerris*. Plant Cell Environ 8:661–669

Lovisolo C, Schubert A (2006) Mercury hinders recovery of shoot hydraulic conductivity during grapevine rehydration: evidence from a whole plant approach. New Phytol 172:469–478

Lovisolo C, Perrone I, Hartung W, Schubert A (2008) An abscisic acid related reduced transpiration promotes gradual embolism repair when grapevines are rehydrated after drought. New Phytol 180:642–651

Lovisolo C, Perrone I, Carra A, Ferrandino A, Flexas J, Medrano H, Schubert A (2010) Drought-induced changes in development and function of grapevine (*Vitis* spp.) organs and in their hydraulic and non-hydraulic interactions at the whole-plant level: a physiological and molecular update. Funct Plant Biol 37: 98–116

Martre P, Morillon R, Barrieu F, North GB, Nobel PS, Chrispeels MJ (2002) Plasma membrane aquaporins play a significant role during recovery from water deficit. Plant Physiol 130:2101–2110

Maurel C, Verdoucq L, Luu DT, Santoni V (2008) Plant aquaporins: membrane channels with multiple integrated functions. Annu Rev Plant Biol 59:595–624

Maurel C, Simonneau T, Sutka M (2010) The significance of roots as hydraulic rheostats. J Exp Bot 61: 3191–3198

Mellerowicz EJ, Baucher M, Sundberg B, Boerjan W (2001) Unravelling cell wall formation in the woody dicot stem. Plant Mol Biol 47:239–274

Morison JIL, Lawson T (2007) Does lateral gas diffusion in leaves matter? Plant Cell Environ 30: 1072–1085

Noctor G, Foyer CH (1998) Ascorbate and glutathione: keeping active oxygen under control. Annu Rev Plant Physiol Plant Mol Biol 49:249–279

Ogaya R, Penuelas J (2006) Contrasting foliar responses to drought in *Quercus ilex* and *Phillyrea latifolia*. Biol Plantarum 50:373–382

Ötvös K, Pasternak TP, Miskolczi P, Domoki M, Dorjgotov D, Szucs A, Bottka S, Dudits D, Fehér A (2005) Nitric oxide is required for, and promotes auxin-mediated activation of, cell division and embryogenic cell formation but does not influence cell cycle progression in alfalfa cell cultures. Plant J 43:849–860

Paponov IA, Teale WD, Trebar M, Blilou I, Palme K (2005) The PIN auxin efflux facilitators: evolutionary and functional perspectives. Trends Plant Sci 10: 170–177

Parent B, Hachez C, Redondo E, Simonneau T, Chaumont F, Tardieu F (2009) Drought and abscisic acid effects on aquaporin content translate into changes in hydraulic conductivity and leaf growth rate: a trans-scale approach. Plant Physiol 149:2000–2012

Parkhurst DF, Mott KA (1990) Intercellular diffusion limits to CO_2 uptake by leaves. Plant Physiol 94: 1024–1032

Pasternak T, Rudas V, Potters G, Jansen MAK (2005) Morphogenic effects of abiotic stress: reorientation of growth in *Arabidopsis thaliana* seedlings. Environ Exp Bot 53:299–314

Pasternak TP, Ötvös K, Domoki M, Fehér A (2007) Linked activation of cell division and oxidative stress defense in alfalfa leaf protoplast-derived cells is dependent on exogenous auxin. Plant Growth Regul 51:109–117

Peer WA, Bandyopadhyay A, Blakeslee JJ, Makam SI, Che RJ, Masson PH, Murphy AS (2004) Variation in expression and protein localization of the PIN family of auxin efflux facilitator proteins in flavonoid mutants with altered auxin transport in *Arabidopsis thaliana*. Plant Cell 16:1898–1911

Perfus-Barbeoch L, Leonhardt N, Vavasseur A, Forestier C (2002) Heavy metal toxicity: cadmium permeates through calcium channels and disturbs the plant water status. Plant J 32:539–548

Phillips NG, Oren R, Licata J, Linder S (2004) Time series diagnosis of tree hydraulic characteristics. Tree Physiol 24:879–890

Piel C, Frak E, Le Roux X, Genty B (2002) Effect of local irradiance on CO_2 transfer conductance of mesophyll in walnut. J Exp Bot 53:2423–2430

Pockman WT, Sperry JS (2000) Vulnerability to cavitation and the distribution of Sonoran Desert vegetation. Am J Bot 87:1287–1299

Potters G, Pasternak TP, Guisez Y, Palme KJ, Jansen MAK (2007) Stress-induced morphogenic responses: growing out of trouble? Trends Plant Sci 12:98–105

Potters G, Pasternak TP, Guisez Y, Jansen MAK (2009) Different stresses, similar morphogenic responses: integrating a plethora of pathways. Plant Cell Environ 32:158–169

Prusinkiewicz P, Rolland-Lagan AG (2006) Modelling plant morphogenesis. Curr Opin Plant Biol 9:83–88

Rao MV, Lee H, Davis KR (2002) Ozone-induced ethylene production is dependent on salicylic acid, and both salicylic acid and ethylene act in concert to regulate ozone-induced cell death. Plant J 32:447–456

Rauch EM, Millonas MM (2004) The role of trans-membrane signal transduction in turing-type cellular pattern formation. J Theor Biol 226:401–407

Reichheld JP, Vernoux T, Lardon F, Van Montagu M, Inzé D (1999) Specific checkpoints regulate plant cell cycle progression in response to oxidative stress. Plant J 17: 647–656

Robertson JM, Hubick KT, Yeung EC, Reid DM (1990) Developmental responses to drought and abscisic acid in sunflower roots. J Exp Bot 41:325–337

Rood SB, Patino S, Coombs K, Tyree MT (2000) Branch sacrifice: cavitation-associated drought adaptation of riparian cottonwoods. Trees 14:248–257

Ros-Barcelo A, Pomar F, Lopez-Serrano M, Martinez P, Pedreno MA (2002) Developmental regulation of the H_2O_2-producing system and of a basic peroxidase isoenzyme in the *Zinnia elegans* lignifying xylem. Plant Physiol Biochem 40:325–332

Sack L, Cowan PD, Jaikumar N, Holbrook NM (2003) The 'hydrology' of leaves: co-ordination of structure and function in temperate woody species. Plant Cell Environ 26:1343–1356

Sakr S, Alves G, Morillon R, Maurel K, Decourteix M, Guilliot A, Fleurat-Lessard P, Julien JL, Chrispeels MJ (2003) Plasma membrane aquaporins are involved in winter embolism recovery in walnut tree. Plant Physiol 133:630–641

Salleo S, Lo Gullo MA, Trifilo P, Nardini A (2004) New evidence for a role of vessel-associated cells and phloem in the rapid xylem refilling of cavitated stems of *Laurus nobilis* L. Plant Cell Environ 27: 1065–1076

Savitsky PA, Gazaryan IG, Tishkov VI, Lagrimini LM, Ruzgas T, Gorton L (1999) Oxidation of indole-3-acetic acid by dioxygen catalysed by plant peroxidases: specificity for the enzyme structure. Biochem J 340: 579–583

Scholz FG, Bucci SJ, Goldstein G, Meinzer FC, Franco AC, Miralles-Wilhelm F (2007) Biophysical properties and functional significance of stem water storage tissues in neotropical savanna trees. Plant Cell Environ 30:236–248

Schopfer P (2001) Hydroxyl radical-induced cell-wall loosening in vitro and in vivo: implications for the control of elongation growth. Plant J 28:679–688

Schopfer P, Liszkay A, Bechtold M, Frahry G, Wagner A (2002) Evidence that hydroxyl radicals mediate auxin-induced extension growth. Planta 214:821–828

Schrader J, Baba K, May ST, Palme K, Bennett M, Bhalerao RP, Sandberg G (2003) Polar auxin transport in the wood-forming tissues of hybrid aspen is under simultaneous control of developmental and environmental signals. Proc Natl Acad Sci USA 100: 10096–10101

Searson MJ, Thomas DS, Montagu KD, Conroy JP (2004) Wood density and anatomy of water limited *Eucalyptus*. Tree Physiol 24:1295–1302

Sgherri C, Quartacci MF, Navarri-Izzo F (2007) Early production of activated oxygen species in root apoplast of wheat following copper excess. J Plant Physiol 164:1152–1160

Slaton MR, Smith WK (2002) Mesophyll architecture and cell exposure to intercellular air space in alpine, desert, and forest species. Int J Plant Sci 163:937–948

Song XG, She XP, He JM, Huang C, Song TS (2006) Cytokinin- and auxin-induced stomatal opening involves a decrease in levels of hydrogen peroxide in guard cells of *Vicia faba*. Funct Plant Biol 33:573–583

Sperry JS, Hacke UG, Pittermann J (2006) Size and function in conifer tracheids and angiosperm vessels. Am J Bot 93:1490–1500

Sperry JS, Meinzer F, McCulloh K (2008) Safety and efficiency conflicts in hydraulic architecture: scaling from tissues to trees. Plant Cell Environ 31:632–645

Steudle E (2000) Water uptake by roots: effects of water deficit. J Exp Bot 51:1531–1542

Steudle E (2001) The cohesion-tension mechanism and the acquisition of water by plant roots. Annu Rev Plant Physiol Plant Mol Biol 52:847–875

Steudle E, Peterson CA (1998) How does water get through roots? J Exp Bot 49:775–788

Suga S, Komatsu S, Maeshima M (2002) Aquaporin isoforms responsive to salt and water stresses and phytohormones in radish seedlings. Plant Cell Physiol 43:1229–1237

Sugimoto-Shirasu K, Roberts K (2003) 'Big it up': end reduplication and cell-size control in plants. Curr Opin Plant Biol 6:544–553

Takahashi H, Kawahara A, Inoue Y (2003) Ethylene promotes the induction by auxin of the corticalmicrotubule randomization required for low-pH-induced root hair initiation in lettuce (*Lactuca sativa* L.) seedlings. Plant Cell Physiol 44:932–940

Terashima I, Hanba YT, Tazoe Y, Vyas P, Yano S (2006) Irradiance and phenotype: comparative eco-development of sun and shade leaves in relation to photosynthetic CO_2 diffusion. J Exp Bot 57:343–354

Tornroth-Horsefield S, Wang Y, Hedfalk K, Johanson U, Karlsson M, Tajkhorshid E, Neutze R, Kjellbom P (2006) Structural mechanism of plant aquaporin gating. Nature 439:688–694

Tournaire-Roux C, Sutka M, Javot H, Gout E, Gerbeau P, Luu DT, Bligny R, Maurel C (2003) Cytosolic pH regulates root water transport during anoxic stress through gating of aquaporins. Nature 425:393–397

Tyerman SD, Vandeleur RK, Shelden MC, Tilbrook J, Mayo G, Gilliham M, Kaiser BN (2009) Water transport and aquaporines in grapevine. In: Roubelakis-Angelakis KA (ed) Grapevine molecular physiology & biotechnology, 2nd edn. Springer, New York, pp 73–96

Tyree M, Zimmermann MH (2002) Xylem structure and the ascent of sap. Springer, New York

Vandeleur RK, Mayo G, Shelden MC, Gilliham M, Kaiser BN, Tyerman SD (2008) The role of PIP aquaporins in water transport through roots: diurnal and drought stress responses reveal different strategies between isohydric and anisohydric cultivars of grapevine. Plant Physiol 149:445–460

Verdoucq L, Grondin A, Maurel C (2008) Structure–function analysis of plant aquaporin AtPIP2;1 gating by divalent cations and protons. Biochem J 415: 409–416

Wan XC, Steudle E, Hartung W (2004) Gating of water channels (aquaporins) in cortical cells of young corn roots by mechanical stimuli (pressure pulses): effects of ABA and of $HgCl_2$. J Exp Bot 55:411–422

Waring RH, Whitehead D, Jarvis PG (1979) The contribution of stored water to transpiration in Scots pine. Plant Cell Environ 2:309–317

Warren CR, Dreyer E (2006) Temperature response of photosynthesis and internal conductance to CO2: results from two independent approaches. J Exp Bot 57:3057–3067

Westgate ME, Boyer JS (1985) Osmotic adjustment and the inhibition of leaf, root, stem and silk growth at low water potentials in maize. Planta 164:540–549

Whalley WR, Leeds-Harrison PB, Clark LJ, Gowing DJG (2005) Use of effective stress to predict the penetrometer resistance of unsaturated agricultural soils. Soil Tillage Res 84:18–27

Whitmore AP, Whalley WR (2009) Physical effects of soil drying on roots and crop growth. J Exp Bot 60:2845–2857

Winkel-Shirley B (2002) Biosynthesis of flavonoids and effects of stress. Curr Opin Plant Biol 5:218–223

Xiong L, Wang RG, Mao G, Koczan JM (2006) Identification of drought tolerance determinants by genetic analysis of root response to drought stress and abscisic acid. Plant Physiol 142:1065–1074

Zimmermann MH (1978) Hydraulic architecture of some diffuse porous trees. Can J Bot 56:2286–2295

Zimmermann MH (1983) Xylem structure and the ascent of sap. Springer, Berlin

Zolla G, Heimer Y, Barak S (2010) Mild salinity stimulates a stress-induced morphogenic response in *Arabidopsis thaliana* roots. J Exp Bot 61:211–224

Abiotic Stress Responses in Plants: Metabolism to Productivity

3

Andrea Furtado Macedo

Abstract

Plants are sessile beings, so the lack of mechanisms to escape from adverse conditions has fostered, through evolution, the development of unique and sophisticated responses to environmental stress. Depending on the degree of plasticity that a plant possesses to deal with a new environmental situation, in response to abiotic stress, morphological, anatomical, and physiological changes may occur. These changes can affect plant growth, productivity in agriculture, metabolic profile, and plant nutritional potential, for example. Therefore, plant abiotic stress has been a matter of concern for the maintenance of human life on earth and especially for the world economy. To meet these challenges, genes, transcripts, proteins, and metabolites that control the architecture and/or stress resistance of crop plants in a wide range of environments will need to be identified, in order to facilitate the biotechnological improvement of crop productivity. The combination of different "omics" tools, which rather than investigating a limited number of substances, enable the large-scale scanning of various substances, offers great potential for postgenomics to elucidate the genotype–phenotype relationships. This chapter is intended to be a synopsis of current knowledge on this regard. It focuses on plant proteome and metabolome affected by abiotic factors. It will include informations on recent advances in methods of omics like proteomics and metabolomics, which should be considered as a new opportunity to understand abiotic responses and identify genes responsible for important crop traits.

Keywords

Abiotic stress • Plant responses • Productivity • Proteome • Metabolome • Tolerance

A.F. Macedo (✉)
Laboratório Integrado de Biologia Vegetal,
Departamento de Botânica, Instituto de Biociências,
CCBS, Universidade Federal do Estado do
Rio de Janeiro, Rio de Janeiro, RJ, Brazil
e-mail: andreafm@unirio.br; andreafm22@yahoo.com.br

P. Ahmad and M.N.V. Prasad (eds.), *Abiotic Stress Responses in Plants: Metabolism, Productivity and Sustainability*, DOI 10.1007/978-1-4614-0634-1_3,

1 Introduction: Abiotic Stress Responses – Importance

Stress is the negative effect that an organism may suffer, and can be classified as internal or external. Internal stress is that derived from mutations or abnormal cell divisions that can lead to metabolic changes. External stress can have a biotic or abiotic origin. Biotic stress may be caused by the attack of herbivores and pathogens. Biotic stress refers to the physical or chemical changes in the environment of the individual (Madlung and Comai 2004). Among the most common abiotic stresses are those related to drought, excess salt in the soil, extremes of temperature, and the presence of toxins contaminating the environment (Bhatnagar-Mathur et al. 2008).

Plants are sessile beings, so the lack of mechanisms to escape from adverse conditions has fostered, through evolution, the development of unique and sophisticated responses to environmental stress. The chain of events that culminate in a response begins with the perception of a specific signal. This signal will generate a specific set of internal responses that lead from the changes in gene expression to changes in metabolism (Hazen et al. 2003; Shao et al. 2007; Agrawal et al. 2010). All these sets of changes represent nothing less than the effort of this sessile organism to overcome the stress situation, maintain homeostasis and adapat (Altman 2003; Hazen et al. 2003).

Survival in hostile environments involves developing mechanisms of tolerance, resistance, or avoidance. Plants that develop tolerance to a given factor can, over time, overcome the effects of this factor without injury. For instance, *Anastatica hierochuntica* and several species in the genus *Selaginella* are called "resurrection plants" because of their ability to withstand and recover from extended periods of internal water deficit. Another tolerance mechanism to avoid dehydration is the accumulation of osmolytes and changes in metabolism (Bouchabke et al. 2008).

To develop resistance means to submit to a given environment by means of counter measures. Acceleration of the plant life cycle to allow flowering before a drought period is a good example of this strategy. Many arid-land grain crops have been improved through breeding programs that allow the crop to avoid seasonal dry periods (Des Marais and Juenger 2010).

Avoidance prevents exposure to the stress (Madlung and Comai 2004). A good example of this strategy is what happens to plants subjected to osmotic stress (drought). Plants can adjust their absorption and water loss by regulating the physiological function of the roots and transpiration, respectively. Stomatal regulation is a strategy to avoid dehydration (Buckley et al. 2003). However, despite this conservative strategy, reductions of photosynthesis and growth can occur.

Plants are often unable to adjust to a certain condition and become sensitive to it (Wang et al. 2003). Depending on the degree of plasticity that a plant possesses to deal with a new environmental situation, in response to abiotic stress, morphological, anatomical, and physiological changes may occur. These changes can affect plant growth, productivity in agriculture, metabolic profile, and plant nutritional potential, for example (Altman 2003). Therefore, plant abiotic stress has been a matter of concern for the maintenance of human life on earth and especially for the world economy.

The main aim of improvements in agricultural production is the eradication of hunger for the ever-increasing human population. It is worrying that 70% of the extremely poor who suffer from hunger live in rural areas (Sanchez and Swaminathan 2005). It is important to improve the nutritional quality of food for 854 million people (about 14% of our population) worldwide who are chronically or acutely malnourished (FAO 2004). It is urgent to increase agricultural productivity with the most nutritious plants; however, the predominant concern is the maintenance of a healthy environment and conservation of local biodiversity, both of which are becoming progressively degraded and subject to accelerated global climate change (Hu et al. 2010).

Specifically in agriculture and the eco-environment, abiotic stresses such as extreme temperatures, salinity, and drought have decreased productivity as much as 50% (Boyer 1982; Bray 1997). Osmotic stress may reduce crop yields to less than half of their potential (Boyer 1982). Forecasts for the year 2050 indicate that up to 50% of farmland may become saline (Wang et al. 2003). Salinization is a problem today, and presently affects 22% of arable areas (FAO 2004). All these data are tightly linked to plant biology, because plants offer the globe its only renewable resource, not only of food, but also of building material and energy. Knowledge of plant biology is also a powerful tool to use natural resources reasonably (Agrawal et al. 2003; Beer and Tavazoie 2004).

In view of this general situation, a major question arises: how to overcome all these adverse factors? One simple answer is to study the responses to different stresses. A key challenge for plant breeding is to investigate these responses at the genetic level, identifying genes responsible for important crop traits. Studying plant responses to different abiotic stresses can reveal how some plant species overcome these adverse conditions by developing resistance or tolerance, the nature of environmental changes can be explored, and finally, tolerant and/or resistant plants can be developed (Meyerowitz 2002; Gesch et al. 2003).

Many studies in this direction have been implemented in attempts to identify stress-regulated genes. Some have shown that plants that are exposed to different stresses, have genes that are regulated in singular ways, but that nevertheless induce similar defense responses (Ozturk et al. 2002; Altman 2003). Probably this is because drought, salinity, extreme temperatures, and oxidative stress are interconnected, and may induce similar effects on plants. Salinity and drought, for example, cause similar responses in plant cells: membrane and protein damage and disruption in the distribution of ions (Vinocur and Altman 2005; Rácz et al. 2008; Hu et al. 2010). Stress inducers from abiotic as well as biotic factors also have some common signal and response pathways in plants (Hodge 2004; Bray 2004; Chinnusamy et al. 2004; Hinsinger et al. 2005; Hongbo et al. 2005; Liu and Li 2005; Munns 2005; Leakey et al. 2006; Humphreys et al. 2006) and thereby have the potential to moderate each other's effects through cross-talking (Shigeoka et al. 2002; Shinozaki et al. 2003; Soltis and Soltis 2003; Hongbo et al. 2005). Investigation of those responses that follow a similar pattern can be useful in developing sustainable agriculture by reducing the need for chemicals (e.g., fertilizers, herbicides, insecticides, fungicides) and preserving/optimizing natural resources (e.g., water, reclaiming wasteland for intensive agriculture) (Wang et al. 2003; Agrawal et al. 2010).

Responses to abiotic stress at the gene level fall into one of three types: (a) genes coding proteins that play an important role in signaling cascades and in transcriptional control (Zhu 2001), (b) genes whose products immediately confer protection on membranes and proteins (Bray 1997), and (c) those that are involved in water and ion uptake and transport, such as aquaporins and ion transporters (Blumwald 2000). Examples of the first option are MyC, MAP kinases and SOS kinase (Zhu 2001), phospholipases (Frank et al. 2000), and many transcription factors (TFs) that regulate transcription by binding, and belong to several gene families including AP2/EREBPs (APETALA2 and ethylene-responsive element-binding proteins), HSF, CBF/DREB (dehydration-responsive element/C-repeat-binding), ABF/ABAE families, bZIP (basic-domain leucine zipper), NAC, MYB/MYC, Cys2/His2 zinc-finger, and WRKY (Umezawa et al. 2006; Hongbo et al. 2005). Transcriptional elements can activate or suppress the transcriptional effect of corresponding genes (Beer and Tavazoie 2004; Bray 2004; Liu and Baird 2004).

Genes that code for products that directly confer protection on membranes and proteins and therefore the function of plant cells to resist environmental stress, are those that synthesize proteins related to the support of the integrity of cellular structures, or destruction of structures damaged by osmotic stress. Late embryogenesis abundant proteins (LEA67), heat shock proteins (HSP68) (Bray 1997), antifreezing proteins, osmotic regulatory proteins, free-radical scavengers (Wang et al. 2003), and various proteinase inhibitors are examples of the latter type of proteins (Des Marais and Juenger 2010). LEA and chaperones often have conservative sequences and polar amino acids, so they are stable and can cooperate in stabilizing the structures of proteins and cell membranes (Fu et al. 2007; Jyothsnakumari et al. 2009).

From investigations on genetic identification and/or molecular responses of stress-related plant responses, modern molecular techniques were developed to breed better crops. The principal objective in plant breeding is to obtain plants that combine higher yields, reliable yield stability, better quality, and obvious stress-resistant characters (abiotic and biotic) over different years and locations (Bray 2004; Chaves and Oliveira 2004). The identification and use of molecular markers and introgression of genomic portions (QTLs) involved in stress tolerance is one good alternative, although undesirable agronomic characteristics from the donor plants may be introduced into the target plant (Roessner and Pettolino 2007).

Techniques that are more accurate than conventional or molecular breeding, such as genetic engineering, allow the selection of genes and their overexpression and/or introduction into the genome. By these methods, new cultivars can be produced more rapidly and efficiently with less chance of failure. Genetic engineering techniques can also overcome barriers to sexual crossing, so that genes of interest arising from taxonomically distant organisms can be selected and introduced (Bhatnagar-Mathur et al. 2008).

According to data collected on the productivity of rice, wheat, and corn during the last three decades, the observed increase is related to breeding and selection of high-yielding genotypes (Wang et al. 2003). However, this improvement in productivity is not followed by an increase in the potential yield of crops. That is, even in optimum environmental conditions, without infection by pathogens and without limitation of resources, both old and new cultivars give the same yield. Therefore, better understanding of the responses of cultivars to abiotic stress, in plant breeding, can implement plant improvement at a very practical level (Wang et al. 2003).

Some molecular responses to abiotic stress, or levels of stress, have been well established and can be used to optimize the production of more resistant individuals. Many of these studies were carried out with *Arabidopsis*. However, apart from specific stress responses at the gene or metabolic level, there is a common signal transduction pathway model for stress, which is shared by many higher plants. This model proceeds through the perception of the environmental signal, and subsequently the production of a secondary messenger (such as inositol phosphates and reactive oxygen species – ROS), which will regulate the endogenous levels of Ca^{2+}. From this point, a chain of events occurs that affects protein phosphorylation, reaching proteins linked to the protection of cellular structures or transcription factors controlling specific sets of stress-regulated genes. These genes are related to the production of regulatory molecules such as the plant hormones abscisic acid (ABA), ethylene, and salicylic acid (SA). Some of these regulatory molecules can, in turn, initiate a second round of circulation (Shao et al. 2007). From the analysis of all the data that can be generated from this model of response, from the standpoint of molecular biology as well as physiological, metabolic, and environmental stress, several questions arise. How to integrate all this available information? How to analyze the data completely? How to establish a relationship among different data sets at different levels and obtain accuracy? These questions can be answered by using techniques of the postgenomic era such as proteomics and metabolomics, which will be discussed in the next section.

2 Metabolic Profile of Plant Reponses to Abiotic Stresses: Proteomic and Metabolomic Approaches

2.1 Context in the Postgenomic Era

To cope with food shortages, classical plant breeding methods alone, such as were intensively employed during the Green Revolution in the 1960s, will not achieve the expected result. To satisfy the expanding food requirements of the rapidly growing world population, production of grain crops needs to increase a further 50% by 2025 (Khush 2003). This increase in production must be accompanied by optimization of growing conditions, which nowadays are suboptimal for plant growth. About 70% of the potential yield is estimated to be lost because of unfavorable physical and chemical factors, even on farms in developed countries (Boyer 1982). To meet these challenges, genes, transcripts, proteins, and metabolites that control the architecture and/or stress resistance of crop plants in a wide range of environments will need to be identified, in order to facilitate the biotechnological improvement of crop productivity. Furthermore, several questions must be elucidated, such as: Which genes and proteins are up- or downregulated by the different types of abiotic stresses? What are the functions of these stress-responsive genes, proteins, and metabolites? What are the characteristics of the following events: stress perception ⇨ signal transduction ⇨ gene activation ⇨ protein expression ⇨ metabolite production ⇨ whole plant response (Altman 2003)?

Another important topic for understanding the relevance of modern technologies is breeding experiments for environmental stresses, specifically abiotic stresses. These experiments are slow and inefficient because of certain limitations, such as: (1) the types of stresses are highly variable in terms of timing during the plant growth cycle, (2) a type of stress that occurs in a specific period affects various tissues and involves multiple responses, making it very difficult to investigate the genetic control of crop tolerance, (3) the phenotyping of plant materials is very sensitive to environmental conditions (e.g., soil chemistry, soil texture, and weather) (Salekdeh and Komatsu 2007).

The identification of genes, transcripts, proteins, and metabolites involves the use of molecular tools. The molecular tools cover genome-wide genetic and physical maps of the chromosome for mapping, isolating, and sequencing of important genes, microarray, proteomics, and metabolomics for high throughput analysis of gene expression, and transformation and marker-aided selection (MAS) for validating candidate genes and utilizing them in molecular breeding. RNA and DNA microarrays can be used to detect gene expression in organisms. Nevertheless, to understand biological systems we must to go further. The set of proteins and mainly metabolites, for example, which are the final products from genes, directly reflecting the surrounding environment, need to be understood through their interactions and modifications (Salekdeh and Komatsu 2007; Ryan and Robards 2006; Bundy et al. 2009).

Thus, in the postgenomic era, the molecular tools to be used are those related to functional genomics such as transcriptomics, proteomics, and also metabolomics, which together, under a holistic view of the plant, called systems biology, will enable a better understanding of the complex regulatory networks associated with stress adaptation and tolerance (Urano et al. 2010). Systems biology is the integration of data from physiology, genomics, transcriptomics, proteomics, and metabolomics into models that might, eventually, represent and simulate the physiology of the organism. All these platforms combined in systems biology represent a new approach to discovering the genes and pathways that are crucial for stress responsiveness and tolerance. The integration of different data sets derived from a single sample will increase our understanding of data through a more holistic overview (May et al. 2011).

Proteomics and metabolomics, specifically, rely on label-free quantitative MS techniques, enabling absolutely essential high throughput analyses. To better exemplify, two phenotypes of *Arabidopsis* were compared in their responses to abiotic temperature stress, in order to better distinguish them. More refined results were obtained when differentiation integrated the results metabolite/protein dataset, than when only protein or only metabolites were examined (Morgenthal et al. 2007). To integrate these postgenomic platforms, bioinformatics, and computational tools are necessary (Kitano 2002).

The combination of different "omics" tools, which rather than investigating a limited number of substances (e.g., Van Dam and Poppy 2008), enable the large-scale scanning of various substances, offers great potential for postgenomics to elucidate the genotype-phenotype relationships (Wienkoop et al. 2010)

2.2 Proteomic Approach

But what is proteomics? Proteomics is the global study of proteins that are expressed in a given organ, tissue, or cell line. This approach provides unique insights into biological systems that cannot be provided by genomic or transcriptomic approaches, simply because there are many more proteins than protein-coding genes (Wienkoop et al. 2010). Proteomics has been used for systematic purposes, qualitative and quantitative profiling, and evaluation of the functions of proteins that are present in plant cells, tissues, or organelles. Therefore, proteomics is also a good tool to elucidate the elements that are involved in stress perception and transduction, and some reviews covering this area have already been published (Thurston et al. 2005; Jorrín et al. 2006).

The process of proteomics research in plant breeding follows a path that begins with the identification of stress-response proteins through comparison between stressed and control plants. Studies of proteome responses to stress generally compare protein profiles among resistant or tolerant organisms such as wild plants, mutants from genetic model species such as *Arabidopsis*, or crop plants, especially rice, wheat, and maize, or transgenics with susceptible or nontolerant individuals (Cooper and Farrant 2002). Following the numerous attempts to improve cultivars through classical crossover, several lines are available that have different degrees of tolerance (Salekdeh and Komatsu 2007). Different proteins, selected from contrasts between resistant/tolerant versus susceptible/nontolerant, or between optimal growth conditions versus stressed growth conditions, are taken as candidates involved in the stress response. The detection of these candidate proteins may allow correlations with the stress and tolerance trait. Plant growth, the level and duration of stress, and plant phenotyping are relevant topics for stress proteome study (Salekdeh and Komatsu 2007). Irrespective of which stress is applied and what plant species is utilized, most of the different proteins identified appear to be either constitutively present (preformed defenses) or are specifically induced in the resistant/tolerant plants (Cooper and Farrant 2002).

The course of a standard proteomics experiment often includes the following procedures: experimental design, sampling, tissue/cell or organelle preparation, protein extraction/fractionation/purification, labeling/modification, separation, Mass spectrometry (MS) analysis, protein identification, and statistical analysis of data and validation (Jorrín-Novo et al. 2009). The extraction of proteins is a crucial step in reaching the later stages of protein detection and identification. At this stage it is necessary to extract and solubilize proteins. Several extraction protocols are available, but two types of protocols are mostly used for plant material: tissue homogenization in buffer-based media, or in organic-solvent media (TCA-acetone, phenol, precipitation protocols). In order to achieve maximum efficiency in the extraction stage, capturing the greatest possible diversity of proteins is necessary, and this is often accomplished by combining different procedures. To be considered an ideal method, the extraction protocol should be reproducible, while at the same time it should reduce the level of contaminants and minimize artifactual protein degradation and modification (Carpentier et al. 2005; Rossignol et al. 2006).

Separation techniques may involve either gel-based or gel-free approaches. For gel-based studies, 1-DE and 2-DE are the preferred techniques used in combination with MS (Lilley and Dupree 2006; Jorrín et al. 2007; Görg et al. 2009). One of the major criticisms of 2-DE is its low precision, with relative standard deviations reported to fall in the range of 15–70%. Major sources of variability for this technique may include the transfer between the first and the second dimension, the analyst's expertise and the detection of separated proteins (Schröder et al. 2008).

Gel-free liquid chromatography tandem mass spectrometry (LC–MS/MS) analysis, called shotgun proteomics (Leitner and Lindner 2009), can increase the number of different proteins that can be identified from complex samples, compared to more traditional gel-based approaches. Shotgun proteomics has become the method of choice for the analysis of complex protein mixtures (Wolters et al. 2001; Gerster et al. 2010). However, the combination of SDS-PAGE, band cutting, trypsin digestion, and LC separation of the resulting peptides is the most powerful proteomics tool to cover the majority of proteins (de Godoy et al. 2006; Tribl et al. 2008).

The so-called "second generation" MS technologies for Quantitative Proteomics include difference gel electrophoresis (DIGE), isotope-coded affinity tags (ICAT) (Shiio and Aebersold 2006), isobaric tags for relative and absolute quantitation (iTRAQ) (Wiese et al. 2007; Gan et al. 2007), and stable isotope labeling by amino acids in cell culture (SILAC) (Nelson et al. 2007; Palmblad et al. 2008) are now beginning to be successfully applied to plants for quantitative and large-scale proteomics studies. The gel-free multidimensional protein identification technology (MudPIT) is particularly well suited for the identification of hydrophobic proteins (Tjalsma et al. 2004; Görg et al. 2009) and allows the detection of a much larger number of proteins compared to gel-based methods, its drawback being the lack of quantitative data (Bayer et al. 2006). The gel-based 2-D DIGE technique is adequate for quantitative proteomics, and requires only a small amount of protein (0.025–0.050 mg) compared with 2-DE (ca. 0.7–1.0 mg) and therefore avoids the limitation of the existence of highly abundant proteins in the protein samples (Majeran et al. 2005; Ndimba et al. 2005; Casati et al. 2005; Dunkley et al. 2006).

To investigate highly complicated proteomics, label-free approaches by means of LC–MS, an IT or Fourier transform mass spectrometer have been used (Wang et al. 2006). The simplicity and cost-effectiveness of this technique make its validation with plant extracts desirable (Jorrín et al. 2007).

Although Bottom-up Proteomics (analysis of proteolytic peptide mixtures) remains the predominant platform, top-down strategies (analysis of intact proteins) should allow a more complete characterization of the proteome, including protein isoforms and posttranslational modifications (PTM). All these aspects have been discussed in detail in recent reviews (Aebersold and Mann 2003; Cravatt et al. 2007; Zubarev and Mann 2007; Molina et al. 2007; Good et al. 2007).

Using classical quadrupole and ion trap mass analyzers, intact protein masses can be determined with standard deviations in the range of 2–5 kDa. The use of Fourier transform mass spectrometry ion cyclotron resonance (Meng et al. 2007) can avoid the problems relating to complex mixtures of protein isoforms, which may complicate the determination of protein mass (Katz et al. 2007; Bräutigam et al. 2008). Surprisingly, it has been reported that a set of proteins can only be detected by a specific technology (Komatsu et al. 2006; McDonald et al. 2006; Wu et al. 2006), which is in agreement with the idea that a combination of different methodologies is still needed to characterize entire proteomes.

Some innovations in the field of proteomics have allowed leveraging of resources to better detect and identify proteins. In the past few years, the development of new Orbitrap and dissociation methods such as electron-transfer dissociation, have opened up new possibilities in proteome analysis. The mass spectrometer, despite constant improvement in terms of machines, software, and protocols, has reached the limit of its capacity (Jorrín et al. 2007).

Proteomics platforms have a number of restrictions, such as sensitivity, resolution, and speed of data capture. They also face a number of challenges, such as deeper proteome coverage, proteomics of unsequenced "orphan" organisms (Carpentier et al. 2005), top-down proteomics (Han et al. 2006) and protein quantitation (Cox and Mann 2007). These restrictions and challenges arise from the huge diversity of proteins, with widely differing physical and chemical characteristics, that are present in organisms.

Finally, in silico proteomics, although it is as yet only applicable where the full genomic sequence is available (i.e., *Arabidopsis* and rice), is useful in both predicting and validating experimental data (Heazlewood et al. 2007). Because of the large amounts of data generated by proteomics analyses over the past year, there have been efforts to form a database where proteomics information can be deposited and made available to the scientific community: the PPDB, http://ppdb.tc.cornell.edu (Sun et al. 2009); the PODB, http://proteome.dc.affrc.go.jp/Soybean/; the Organellome, http://podb.nibb.ac.jp/Organellome (Mano et al. 2007); and the knowledge-based UniProt (Jorrín-Novo et al. 2009).

Efforts to form a searchable database of MS/MS reference spectra have been implemented by committees such as the Subcommittee of the

Multinational Arabidopsis Steering Committee, through projects such as the "Green Proteome" (Weckwerth et al. 2008; Hummel et al. 2007), Plant Proteomics in Europe (COST Action FA0603). The database permits authentic protein identification through a genome-independent approach, since newly generated MS/MS spectra can be matched against previous experimental MS/MS spectra. This approach allows semiquantitative analysis at the same time as spectrum matching.

Initiatives have also begun to create a guide for conducting proteomics experiments to achieve more consistent results, because many papers contain errors in the experimental design, the analysis, and the interpretation of the data (Nesvizhskii et al. 2007). More consistent data cannot rely upon speculation, especially when the genome or transcriptome of the species being studied is still unknown. Analysis of the greatest possible number of proteins, rather than only a fraction, also improves the consistency of results. Therefore, the HUPO's Proteomic Standard Initiative has developed guidance modules (Orchard and Hermjakob 2008) that have been translated into Minimal Information about a Proteomic Experiment (MIAPE) documents. The MIAPE documents recommend proteomics techniques that should be considered and followed when conducting a proteomics experiment. Proteomics journals should be, and in fact are, extremely strict in recommending that investigators follow the MIAPE standards for publishing a proteomics experiment (Jorrín-Novo et al. 2009).

What are the protein profiles that are found in plants under abiotic stress?

According to individual studies and reviews, few proteins are specific for the type of stress applied (Bolwell et al. 2001; Cooper and Farrant 2002; Skylas et al. 2002; Hajheidari et al. 2007). For some differential proteins, multiple isoforms or specific PTMs may be detected, each responding differently according to the stress applied (Hammond-Kosack et al. 1998). Proteins that are expressed by the same stressors, clearly confer a physiological advantage under stress conditions, and thus are simultaneously potential targets for marker-assisted selection and rational candidate genes for the identification of quantitative trait loci.

Drought stress, metal toxicity, and salt-osmotic stress are the types of abiotic stress that are most often investigated. In contrast to the intensive study of the influence of water and nutrient status on plant proteomes, studies of plant responses to light and temperature stress are rare. Various sources of plant material were examined in proteome experiments: leaves and cotyledons, roots, fruits, phloem and xylem saps, apoplastic fluid, entire seedlings, shoots, stem segments, seeds, nuclear fractions, gametophores, and meristem tissue.

Drought conditions may induce proteins related to detoxifying reactive oxygen species (ROS) (Hajheidari et al. 2007), but many other abiotic stresses can enhance production of ROS resulting from photosynthesis, respiration, and NADPH oxidase (Hammond-Kosack et al. 1998). This observation makes sense, since most stressors increase production of reactive oxygen species ROS in plants. Cells exposed to high amounts of (ROS) may be damaged. ROS act as secondary messengers involved in the stress-response signal transduction pathway. Therefore, to detoxify the cell, that is, remove excess ROS, plants have two mechanisms. The most important ways to combat ROS are those that involve SOD (Hajheidari et al. 2005), the water–water cycle, the ascorbate–glutathione cycle, glutathione peroxidase, and catalase (del Río et al. 2006). In the early stages of drought stress, many proteins associated with root morphogenesis and carbon/nitrogen metabolism, which may contribute to drought avoidance by enhancing root growth are stimulated (Yoshimura et al. 2008). 2-Cysteine peroxiredoxin is a protein that can be synthesized from drought stress, and belongs to the group that reduces H_2O_2 and alkyl hydroperoxide (Dietz et al. 2002). This protein constitutes an important alternative to detoxification under oxidative stress conditions. Small heat shock proteins (sHSPs) are also induced by heat and drought stresses. HSPs function as chaperones and play an integral role in protein folding and assembly (Sun et al. 2002). Therefore, sHSPs are promising protein markers for marker-assisted breeding programs to increase stress tolerance. The response to drought stress (Hajheidari et al. 2005) also involves the expression of cytosolic Cu–Zn SOD, cyclophilin, nucleoside-diphosphate kinase, a nascent polypeptide-associated complex a-chain, and the large subunit of Rubisco. Nucleoside diphosphate kinase (NDPKs) is also more strongly expressed after heat and drought stress (Escobar Galvis et al. 2001; Moon et al. 2010). NDPK uses ATP to maintain the cellular levels of CTP, GTP, and UTP (Moon et al. 2010) and cooperates in cellular redox regulation. The overexpression of AtNDPK2 leads to decreased constitutive ROS levels and increased tolerance to multiple environmental stresses.

Actin depolymerizing factor 4 (ADF) is also correlated with responses to drought and salt stress (Salekdeh et al. 2002; Ali and Komatsu 2006; Yan et al. 2010). ADF is related to osmoregulation under osmotic stress. This group of proteins is involved in the regulation of different cellular processes including cytokinesis, remodeling of actin filaments, cytoplasmic streaming, and signal transduction events (Dong et al. 2001). The upregulation of ADF under drought and salt stress indicates that this protein might be associated with dynamic reorganization of the cytoskeleton during drought stress. Redox proteins such as glutathione dehydrogenase (At1g19570) are affected in stress regulation (Morgenthal et al. 2007; Wienkoop et al. 2008).

Mitogen-activated protein kinases (MAPKs) are upstream regulators of many aspects of plant cell signaling. MAPK cascades usually require three components: MAPK kinase kinases (MPKKKs), which phosphorylate MAPK kinases (MPKKs), which phosphorylate MAPKs, which phosphorylate diverse proteins (Chinnusamy et al. 2004; Ren et al. 2008). After MAPK is activated, it further activates transcription factors in the nucleus, or phospholipid-cleaving enzymes in the cytoplasm. This set of enzymes is related to stress response (Cheong et al. 2002; Xu et al. 2003; Chinnusamy et al. 2004; Hu et al. 2006). MPK4 and MPK6 have received considerable attention for their role in abiotic stress signaling. Posttranslational activation of these two kinases is stimulated by cold, low humidity, salt, wounding, reactive oxygen species, and touch (Ichimura et al. 2000; Yuasa et al. 2001).

Salt stress responses involve the substrate-binding proteins of ABC transporters. Products including H1 transporting ATPases, signal transduction-related proteins, transcription/translation-related proteins, detoxifying enzymes, amino acid, and purine biosynthesis-related proteins, proteolytic enzymes, HSPs, and carbohydrate metabolism-associated proteins are also involved in salt stress (Des Marais and Juenger 2010).

Excessive light enhances production of proteins involved in photosynthesis, as well as some known light stress-related proteins, such as HSP, dehydroascorbate reductase, and SOD (Cushman and Bohnert 2000). The cold stress response leads to accumulation of dehydrins and low-temperature-induced protein (Uno et al. 2000).

2.3 Metabolomic Approach

What is metabolomics? Metabolomics is the untargeted analysis of a set of metabolites that are produced by an organism, so the metabolome is the set of metabolites, specifically low-molecular-weight molecules (typically 3,000 m/z), present in a cell, tissue, or organ in a particular physiological or developmental state (Oliver et al. 1998). It is the layer downstream from large-scale analysis of RNA (transcriptomics) and proteins (proteomics) (Weckwerth 2003; Bino et al. 2004).

Understanding the metabolome is important to elucidate the complex network related to abiotic stress. The idea that metabolites are only the final product of gene expression is outmoded (Hollywood et al. 2006). It is increasingly understood that metabolites themselves regulate macromolecular operations through, for example, feedback inhibition and as signaling molecules. The cellular processes are in reality intimately networked, with many feedback loops, and thus should be represented as dynamic protein complexes interacting with neighborhoods of metabolites (Caspi 2006). Metabolomics analyses are therefore destined to provide an integrated perspective of the functional status of an organism (Dixon et al. 2006). More than this perspective, metabolome investigation is complementary to transcriptomics and proteomics, and may have special advantages. While changes in the levels of individual enzymes may be expected to have little effect on metabolic fluxes, they can and do have significant effects on the concentrations of a variety of individual metabolites. In addition, as the "downstream" result of gene expression, changes in the metabolome are amplified relative to changes in the transcriptome and the proteome, which is likely to allow for increased sensitivity (Dixon et al. 2006). Finally, it is known that metabolic fluxes are regulated not only by gene expression but also by posttranscriptional and posttranslational events, and as such, the metabolome can be considered to be closer to the phenotype (Siritunga and Sayre 2003).

Metabolomics is not intended to identify a particular metabolite or set of metabolites, as is done in traditional phytochemical studies. The broader purpose of this technique allows the evaluation not of only a very small fraction of the metabolism, but of the maximum possible number of metabolites. This is because none of the existing techniques allows the evaluation of all the metabolites that are present in an organism (Ryan and Robards 2006). To capture all of them, different analytical platforms must be combined, considering that plant metabolites have different chemical properties (Fernie et al. 2004; Moco et al. 2007). Their differences are based on the degree of volatility, polarity, and concentration in a given tissue (Weckwerth 2003). Because of this wide variability of physicochemical characteristics, metabolomics studies are usually based on substances with certain chemical affinities.

The most widely used model for studying this platform is *Arabidopsis thaliana*, but other species including food plants such as tomato and potato have been investigated by means of this approach (Catchpole et al. 2005; Kristensen et al. 2005; Keurentjes et al. 2006; Moco et al. 2006; Leiss et al. 2009; Kunin et al. 2009).

Metabolomics investigations are based on techniques that include nuclear magnetic resonance (NMR), Fourier transform ion cyclotron resonance coupled with mass spectrometry (FT-ICR-MS), and separation-based techniques such as gas chromatography and liquid chromatography coupled with mass spectrometry (GC–MS and LC–MS). These analytical tools can profile the impact of time, stress, nutritional status, and environmental perturbation on hundreds of metabolites simultaneously, resulting in massive, complex data sets. This information, in association with transcriptomics and proteomics, has the capacity to produce a more holistic view of the composition of food and feed products, to optimize crop trait development, and to enhance diet and health (Dixon et al. 2006).

Samples intended for this approach are prepared using rapid freezing that stops enzyme activity. Subsequently, metabolites are extracted by different methods, for example, with methanol to extract semipolar metabolites. The extract can then be analyzed by many different methods and approaches (Hollywood et al. 2006).

Because of the unique structural composition and three-dimensional configuration of each compound, NMR yields a specific spectrum for each substance. The advantage of this method is that it is highly reproducible and nondestructive, and can also quantify the metabolites (Verpoorte et al. 2007). Despite this, metabolites that are present in smaller quantities will not be detected. While NMR uses magnetic resonance, all the other metabolomics platforms use mass spectrometry (MS) for identification (Macel et al. 2010).

The most widely used metabolomics platforms are MS combined with chromatographic separations, because of the availability and relatively low cost of these techniques. With MS, metabolites are ionized (charged) and their mass-to-charge ratios (*m/z*) are measured using electric and magnetic fields in a mass analyzer. These mass-to-charge ratios are specific for each metabolite. The disadvantage of the MS platform is that quantification of the substances is difficult and can generally only be measured in relative terms. The reproducibility is lower compared to NMR, although the MS method is much more sensitive. "Hyphenated" techniques of LC–MS (Yamazaki et al. 2003a, b) combine retention times (the time needed to pass through the column that separates the compounds) with MS for identification, normally, of the nonvolatile metabolites, particularly the semipolar secondary metabolites such as flavonoids, alkaloids, and glucosinolates, but also sugars and amino acids (Macel et al. 2010). GC–MS is widely used to analyze low-molecular-weight volatiles

(Fiehn et al. 2000; Roessner et al. 2001). Analytical methods for metabolic fingerprinting analyses of crude extracts with no previous separation steps, involve Fourier-transform ion cyclotron resonance (FTICR) mass spectrometry and time-of-flight (TOF) mass spectrometry (Aharoni et al. 2002; Brown et al. 2005), which cover a broad range of substances (Hagel and Facchini 2007). The mass-to-charge ratios are established from the cyclotronic frequency of the ions in a fixed magnetic field (Macel et al. 2010). Although FT-ICR-MS has a higher sensitivity and resolution compared to NMR, it mainly gives the elemental composition of a metabolite (through MS) without providing much extra information about the chemical structures of the molecules (Macel et al. 2010).

In contrast to transcriptome studies (but in common with protein analysis), no tools are available for amplification of metabolites, and consequently sensitivity is a major issue. Metabolites have huge chemical differences, and are often present in a wide dynamic range. All of these challenges need to be adequately addressed by the analysis strategy employed.

Because of the huge amount of data that can be generated by the techniques mentioned above, as also occurs in proteomics, data processing is required (Lommen 2009). For data analysis, knowledge of bioinformatics is required (Smilde et al. 2005; Sumner et al. 2007). Data can be examined by multivariate statistics such as principal components analysis (PCA), nonmetric multidimensional scaling (NMDS), and partial least squares discriminant analysis (PLS-DA) (Westerhuis et al. 2008). These multivariate methods will show whether the metabolome, and to a certain extent also which metabolites, differ between treatments or species. To investigate the behavior of individual metabolites, Student's *t*-tests or univariate analyses of variance (ANOVA) can be used in combination with correction for false discovery rates (FDR) (Macel et al. 2010).

With the intention of gathering the largest possible amount of metabolomics data, a World Wide Web-access system was created. The PlantMetabolomics.org (PM) website allows public consultation of the MS-based plant metabolomics experimental results from multiple analytical and separation techniques. PM has extensive annotation links between the identified metabolites and metabolic pathways in AraCyc (Mueller et al. 2003) at The Arabidopsis Information Resource (Rhee et al. 2003) and the Plant Metabolic Network (www.plantcyc.org), the Kyoto Encyclopedia of Genes and Genomes (KEGG) (Kanehisa 2004), and MetNetDB (Wurtele et al. 2007). The rationale for the development of PM as an information portal is to provide free public access to experimental data, along with cross-references to related genetic, chemical, and pathway information. The portal also serves as an information resource for the field of metabolomics by providing tutorials on how to conduct metabolomics experiments. It describes minimum reporting standards (Fiehn et al. 2007; Sumner et al. 2007) for plant metabolomics experiments, based on the recommendations of the MSI. In addition, PM contains background information about the experimental design and tools that can be used to analyze the collected data (Bais et al. 2010).

Using the appropriate technology, there are different strategies to investigate the metabolome. The strategy most often used to study abiotic stress is metabolite target analysis, which is an approach that is restricted to metabolites of, for example, a particular enzyme system that would be directly affected by abiotic or biotic perturbation (Hollywood et al. 2006).

What are the metabolic profiles found in plants under abiotic stress?

Dehydration–stress response metabolome studies have shown that both ABA-dependent and ABA-independent pathways are involved in this kind of stress (Yamaguchi-Shinozaki and Shinozaki 2006). The endogenous ABA level rises in response to water-deficit stress, to modulate physiological stress responses and gene expression. ABA produced during dehydration affects the accumulation of various amino acids and sugars such as glucose and fructose. In particular, the dehydration-inducible accumulations of BCAAs (branch-chain amino acids), saccharopine, proline, and agmatine are correlated with the dehydration-inducible expression of their key biosynthetic genes (BCAT2, LKR/SDH, P5CS1, and ADC2, respectively), which are regulated by endogenous ABA (Urano et al. 2009). On the other hand, the levels of raffinose and galactinol are not regulated by ABA during dehydration stress. Thus, it seems that ABA has an important role in regulating metabolism during water stress (Urano et al. 2010).

Some studies indicate that more glucose, malate, and proline tend to be produced in plants under dehydration stress than under salt stress. Probably this difference results from the need for plants subjected to salt stress to make a greater osmotic adjustment, detoxify ROS, and ameliorate photoinhibition (Cramer et al. 2006). When the plant treatment occurs under more severe conditions such as dehydration and heat-stress treatment, sucrose replaces proline as the major osmoprotectant (Rizhsky et al. 2004).

The temperature stress response such as to cold and other stresses, involves the DREB1/CBF (dehydration-responsive element-binding factor/C-repeat) transcriptional network. A correlation between the metabolome (monosaccharides, disaccharides, oligosaccharides, and sugar alcohols) and the DREB1A/CBF3 transcription factor under low temperature was also observed. The low-temperature-inducible accumulation of galactinol and raffinose is correlated with the expression of the GolS3 gene, which is a direct target of DREB1A/CBF3. Some studies indicate that the expression of DREB1A affects the accumulation of low-temperature regulated metabolites, especially sucrose, raffinose, galactinol, and myoinositol (Maruyama et al. 2009).

According to some metabolome studies the majority of metabolites produced in response to heat shock overlapped with those produced in response to cold shock. Furthermore, these results indicate that a metabolic network of compatible solutes including proline, monosaccharides (glucose and fructose), galactinol and raffinose has an important function in tolerance to temperature stress (Kaplan et al. 2004).

Salt stress responses are correlated with higher levels of various osmolytes, such as fructose, sucrose, complex sugars, malate, and proline (Gong et al. 2005). Short-term response to salt stress seems to involve the simultaneous induction of several pathways: the methylation cycle for the supply of methyl groups, the phenylpropanoid pathway for lignin production, and glycine betaine (GB) biosynthesis (Kim et al. 2007). In the long-term response to salt stress, however, glycolysis and sucrose metabolism were coinduced, and then the methylation cycle was coreduced. As observed in experiments with drought stress, under salt stress ABA was also shown to have an important role in establishing metabolite profiles (Kempa et al. 2008). Complex readjustment of carbohydrate metabolism occurs throughout the period of salt stress, and ABA triggers the initial stages of carbon mobilization.

To integrate multiple datasets from metabolite, transcript, and protein information, some initiatives have been developed, such as the creation of DOME (database for OMEs). This platform allows the storage of DNA microarray data, protein fragment mass spectral data from two-dimensional gel separations, and metabolite MS data after separation by GC, LC, or CE. Additional databases and programs that allow integration of metabolite with transcript data are AraCyc (http://arabidopsis.org/tools/aracyc/) as mentioned above, MAPMAN, and KaPPA-View (http://kpv.kazusa.or.jp/kappa-view/) (Dixon et al. 2006).

3 Plant Stress Tolerance or Resistance: Productivity and Prospects

A single cross between two plants, as used in conventional breeding, joins two sets of about 15,000–25,000 genes. This is the assumption that guides classical agriculture. Plant breeding has been performed since the start of agricultural practices thousands of years ago. Classical breeding relies on the process of homologous recombination between two genomes, creating novel genetic diversity. In contrast, modern biotechnological methods allow only a few genes to be modified at the same time, leaving the rest of the genome unaltered (Akhond and Machray 2008). Moreover, genes from any source can be introduced into a crop plant, genes, and their products can be tested to evaluate their safety, and genes can be altered and assessed under laboratory conditions to change their properties before being introduced into a plant.

Combining the resources of classical breeding with modern biotechnology, a novel variety of genotypes and phenotypes can be created, agricultural productivity can be increased, and human survival in the face of population growth and climate change can be achieved (Altman 2003). This is an important subject for agricultural research, as increased competition with other land uses pushes farms into harsher environments, fresh water becomes scarcer, and the climate change anticipated by some scientists increases environmental stress. Therefore, to increase productivity by engineering plants that are more resistant or tolerant to abiotic stress, genes, and their products are the target of this initiative. However, this task seems more daunting than engineering plants that are resistant to pests and herbicides. Biotic stress is largely dependent on monogenic traits, while abiotic stresses are multigenic and thus more difficult to control and engineer (Vinocur and Altman 2005).

According to James, "In 2007, the global area of biotech crops increased for the twelfth consecutive year at an annual growth rate of 12%. While the technology was initially applied in developed countries, 12 million farmers in farmers were in 2007 biotech crops were grown by 23 countries covering 114.3 million hectares and over 90% of the beneficiary resource-poor in developing countries, with increased incomes from biotech crops contributing to the alleviation of poverty" (Akhond and Machray 2008).

The sequence of breeding for plant tolerance to abiotic stress consists of several stages: (1) conventional breeding and germ plasm selection; (2) clarification of the specific molecular control process in tolerant and sensitive genotypes; (3) biotechnology-oriented improvement of selection and breeding operations by functional genomics investigations, use of molecular probes and markers for selection among natural and bred populations, and transformation with specific genes; (4) large-scale propagation (seed or vegetative) of the engineered and selected genotypes; and (5) improvement and adaptation of current agricultural practices (Altman 2003).

Stress-induced gene expression can be broadly categorized into three groups: (1) genes encoding proteins with known enzymatic or structural functions, (2) proteins with as yet unknown functions, and (3) regulatory proteins. Transgenic plants tolerant to certain types of abiotic stresses were developed based on the manipulation of genes that protect and maintain the function and structure of cellular components. Initially, some studies have focused on identifying genes responsible for the synthesis of a single metabolite. In the case of abiotic stress related to salinity and drought, studies have evaluated proteins that are involved in water channels, the synthesis of osmolytes (proline, betaine, sugars such as trehalose), or transport to the work of transformation. Metabolic traits, especially pathways with relatively few enzymes, have been characterized genetically and appear more amenable to manipulation than do structural and developmental traits (Bhatnagar-Mathur et al. 2008). However, this perspective neglects the likelihood that abiotic stress tolerance involves many genes at once, and that single-gene tolerance is unlikely to be sustainable. Given this limitation, new prospects have arisen for the development of transformed plants that have resistance or tolerance to abiotic stress. One possibility involves the manipulation of genes that belong to the third category mentioned above, genes that express regulatory proteins. Through these proteins, many genes associated with stress responses can be simultaneously regulated by a single-gene encoding the stress-inducible transcription factor (Kasuga et al. 1999), thus providing conditions to enhance tolerance to multiple stresses including drought, salinity, and freezing. This new ability to engineer more resistant or tolerant plants, recognized as the "second wave", coincides with a better combination of genetic engineering with plant physiology. Gene cassettes driven by stress-induced promoters are being used to generate transgenics, since stress-induced promoters (particularly those induced by anaerobic conditions, low or high temperatures, and salt stresses) have now been characterized (Bhatnagar-Mathur et al. 2008).

Some transformation experiments showing promising results for abiotic tolerance have been implemented (Wang et al. 2003). Many reviews elucidating the process of abiotic stress tolerance and on engineering tolerance to stress have been published in the last few years (Akhond and Machray 2008).

As mentioned before, plants subjected to water stress usually tend to produce sugars and similar compounds that act as osmoprotectants. One such substance is trehalose. Trehalose levels have been increased in GM rice by overexpressing genes encoding trehalose biosynthetic enzymes from the bacterium *Escherichia coli* (Garg et al. 2002). Rice plants subjected to this type of experiment showed better performance under salt, drought, and low-temperature stress conditions. On the other hand, some results showed that transformed plants that produced high levels of osmoprotectants suffer from deleterious pleiotropic effects, such as dwarfing (Hazen et al. 2003).

Some genes have been found to be multifunctional. Examples include *BADH*, *P5CS*, and *HAV*, which are involved in preventing drought, salt, osmotic, and heat stress (Hu et al. 2010). Specifically, *BADH* is responsible for the production of betain aldehyde dehydrogenase, and is involved in the biosynthesis of GB, which acts as a compatible solute in plants. Transformed plants with *BADH* have osmoregulation ability, but also improved salt and heat tolerance (Hu et al. 2010).

Accumulation of osmotically active compounds can prevent osmotic damage to cell structures, protein destabilization, and the negative effects of ROS (Wang et al. 2003). The accumulation of proline, betaine, free amino acids, sugars, sugar alcohols, alkaloids, etc. can be achieved by overexpression of enzymes associated with their biosynthesis, or by suppression of enzymes that induce their destruction (Chen and Murata 2002). Transgenic rice, soybean, tobacco, and wheat overexpressing pyrroline-5-carboxylate synthetase (P5CS), which induces the biosynthesis of the above-mentioned enzymes, showed, in specific cases, salinity, drought, salt, and heat resistance, and increased biomass under water stress (Sokhansanj et al. 2006; Vendruscolo et al. 2007).

Other alternatives have been explored to solve the problem of excess salt in the soil. This problem, usually caused by irrigation, now affects vast cultivable areas. An important strategy for achieving greater tolerance to abiotic stress is to help plants to reestablish homeostasis under stressful environments, restoring both ionic and osmotic homeostasis. The target is to achieve Na+ excretion from the root, or its storage in the vacuole, so overexpression of a gene that encodes a vacuolar Na+/H+ antiport pump could be one solution (Apse and Blumwald 2002). Such a pump would allow for more effective removal of salt from the cytoplasm and its transfer to the vacuole. These results have been obtained in GM tomato plants, which showed a higher tolerance to salt concentrations than did nontransformed individuals, and survived better in areas that were previously considered useless for agriculture. Furthermore, the fruit does not accumulate salt, and is edible. This effort to produce food in large areas, that are presently impractical for farming, has also been extended to problems of soil contamination by heavy metals (Shewry et al. 2008).

The transcription factors activate cascades of genes that act together in enhancing tolerance towards multiple stresses. Transcriptional activation of stress-induced genes has been possible in transgenic plants with overexpression of TFs, which also belong to the multifunctional gene family, recognize promoter regulatory elements of these genes, and can induce stress-responsive gene expression and increase tolerance to abiotic stress. DREB can increase the drought, salt, and cold tolerance of many species, as confirmed by transgenic researches (Ito et al. 2006; Sakuma et al. 2006).

Overexpression of 9-*cis*-epoxycarotenoid dioxygenase (NCED) is connected to ABA biosynthesis, results in a relaxation of stress symptoms related to cold, drought, and salt (Jung et al. 2008).

Abiotic stress signaling in plants involves receptor-coupled phospho-relay, phosphoionositol-induced Ca^{2+} changes, the MAPK cascade, and transcriptional activation of stress-response genes. Plant acclimatization to environmental stress involves activation of various kinases, and in turn, a single kinase gene can affect various kinds of stress resistance. From this principle, it has been shown that maize transformed with the tobacco MAPKKK/NPK has an oxidative signal cascade activated, improving cold, heat, and salt tolerance (Shou et al. 2004).

Overexpression of sensors that can perceive stress signals through the combination reactions of signals (Wang et al. 2007), such as calcium-dependent protein kinases (CDPKs), was implemented in barley. CDPKs are unique Ca^{2+} sensors in plants, and transgenic barley with this improvement responds better to cold and salt stress. Other kinds of sensors are salt sensors (Qiu et al. 2004) and osmosensors (Hu et al. 2010).

To avoid stress caused by oxygen radicals, transgenic plants designed to overexpress enzymes involved in oxidative protection, such as glutathione peroxidase, superoxide dismutase, ascorbate peroxidases, and glutathione reductases (Hazen et al. 2003). Overexpression of antioxidative enzymes (such as SOD) in transgenic alfalfa, rice, *Arabidopsis,* and cabbage (Serrot et al. 2008) induces higher cold, drought, and salinity resistance than in wild plants (Samis et al. 2002; Tseng et al. 2007).

Transgenic plants overexpressing genes encoding LEA proteins and molecular chaperones such as HAV1, a group 3 LEA protein gene, can enhance resistance to drought, salt, cold, and other stresses (Jyothsnakumari et al. 2009). For example, transgenic wheat, oats, and rice with overexpression of the HAV1 gene have drought and salt resistance, and show improved growth (Fu et al. 2007). Genetic engineering for increased thermotolerance by enhancing heat shock protein synthesis in plants has been achieved in a number of plant species (Katiyar-Agarwal et al. 2003).

4 Conclusion and Future Perspective

In conclusion, huge efforts have been made in identifying plant abiotic stress responses and proteomics and metabolomics has been identified as the tools of choice for comprehensive analysis of genes functions, protein and metabolites interactions, and modifications. As the gene sequences of new species are being elucidated, comparative studies with *Arabidopsis* may contribute to the elucidation of certain answers.

Plants grown on the stress conditions can be compared to plants with tolerance or resistance to a particular factor. In addition, this study can be conducted in field conditions or in tissue culture. Although potentially expensive, plant tissue culture can offer some advantages over traditional field growing practices. These advantages include manipulation of culture system to occlude the influence of variables that are not desirable such as some environmental factors: climate, nutrient availability, and disease. Functional genomic approaches, including metabolomics, will certainly allow description of biosynthetic and regulatory pathways. These approaches will enable rational plant engineering to produce transgenics of interest on demand.

References

Aebersold R, Mann M (2003) Mass spectrometry-based proteomics. Nature 422:198–207

Agrawal N, Dasaradhi PVN, Mohmmed A, Malhotra P, Bhatnagar RK, Mukherjee SK (2003) RNA Interference: biology, mechanism, and applications. Microbiol Mol Biol Rev 67:657–685

Agrawal GK, Jawa N-S, Lebrun M-H, Job D, Rakwal R (2010) Plant secretome: unlocking secrets of the secreted proteins. Proteomics 10:799–827

Aharoni A, de Vos CH, Verhoeven HA, Mailiepaard CA, Kruppa G, Bino RJ, Goodenowe DB (2002) Nontargeted metabolome analysis by use of Fourier transform ion cyclotron mass spectrometry. Omics 6: 217–234

Akhond MAY, Machray GC (2008) Biotech crops: technologies, achievements and prospects. Euphytica 166: 47–59

Ali GM, Komatsu S (2006) Proteomic analysis of rice leaf sheath during drought stress. J Proteome Res 5:396–403

Altman A (2003) From plant tissue culture to biotechnology: scientific revolutions, abiotic stress tolerance, and forestry. In Vitro Cell Dev Biol Plant 39:75–84

Apse MP, Blumwald E (2002) Engineering salt tolerance in plants. Curr Opin Biotechnol 13:146–150

Bais P, Moon SM, He K, Leitão R, Dreher K, Walk T, Sucaet Y, Barkan L, Wohlgemuth G, Roth MR, Wurtele ES, Dixon P, Fiehn O, Lange BM, Shulaev V, Sumner LW, Welti R, Nikolau BJ, Rhee SY, Dickerson JA (2010) Plant metabolomics.org: a web portal for plant metabolomics experiments. Plant Physiol 152:1807–1816

Bayer EM, Bottrill AR, Walshaw J, Vigouroux M, Naldrett MJ, Thomas CL, Maule AJ (2006) *Arabidopsis* cell wall proteome defined using multidimensional protein identification technology. Proteomics 6:301–311

Beer MA, Tavazoie S (2004) Predicting gene expression from sequence. Cell 117:185–198

Bhatnagar-Mathur P, Vadez V, Sharma KK (2008) Transgenic approaches for abiotic stress tolerance in plants: retrospect and prospects. Plant Cell Rep 27: 411–424
Bino RJ, Hall RD, Fiehn O, Kopka J, Saito K, Draper J, Nikolau BJ, Mendes P, Roessner-Tunali U, Beale MH, Trethewey RN, Lange BM, Wurtele ES, Sumner LW (2004) Potential of metabolomics as a functional genomics tool. Trends Plant Sci 9:418–425
Blumwald E (2000) Sodium transport and salt tolerance in plants. Curr Opin Cell Biol 12:431–434
Bolwell PP, Page A, Pi lewska M, Wojtaszek P (2001) Pathogenic infection and the oxidative defences in plant apoplast. Protoplasma 217:20–32
Bouchabke O, Chang F, Simon M, Voisin R, Pelletier G, Durand-Tardif M (2008) Natural variation in *Arabidopsis thaliana* as a tool for highlighting differential drought responses. PLoS One 3:e1705
Boyer JS (1982) Plant productivity and environment. Science 218:443–448
Bräutigam A, Hoffmann-Benning S, Weber APM (2008) Comparative proteomics of chloroplast envelopes from C3 and C4 plants reveals specific sdaptations of the plastid envelope to C4 photosynthesis and candidate proteins required for maintaining C4 metabolite fluxes. Plant Physiol 148:568–579
Bray EA (1997) Plant responses to water deficit. Trends Plant Sci 2:48–54
Bray EA (2004) Genes commonly regulated by water-deficit stress in *Arabidopsis thaliana.* J Exp Bot 55:2331–2341
Brown SC, Kruppa G, Dasseux JL (2005) Metabolomics applications of FT-ICR mass spectrometry. Mass Spectrom Rev 24:223–231
Buckley TN, Mott KA, Farquhar GD (2003) A hydromechanical and biochemical model of stomatal conductance. Plant Cell Environ 26:1767–1785
Bundy J, Davey M, Viant M (2009) Environmental metabolomics: a critical review and future perspectives. Metabolomics 5:3–21
Carpentier SC, Witters E, Laukens K, Deckers P, Swennen R, Panis B (2005) Preparation of protein extracts from recalcitrant plant tissues: an evaluation of different methods for two-dimensional gel electrophoresis analysis. Proteomics 5:2497–2507
Casati P, Zhang X, Burlingame AL, Walbot V (2005) Analysis of leaf proteome after UV-B irradiation in maize lines differing in sensitivity. Mol Cell Proteomics 4:1673–1685
Caspi R (2006) MetaCyc: a multiorganism database of metabolic pathways and enzymes. Nucleic Acids Res 34:D511–D516
Catchpole GS, Beckmann M, Enot DP, Mondhe M, Zywicki B, Taylor J, Hardy N, Smith A, King RD, Kell DB, Fiehn O, Draper J (2005) Hierarchical metabolomics demonstrates substantial compositional similarity between genetically modified and conventional potato crops. Proc Natl Acad Sci USA 102:14458–14462
Chaves MM, Oliveira MM (2004) Mechanisms underlying plant resilience to water deficits: prospects for water-saving agriculture. J Exp Bot 55:2365–2384
Chen THH, Murata N (2002) Enhancement of tolerance of abiotic stress by metabolic engineering of betaines and other compatible solutes. Curr Opin Plant Biol 5:250–257
Cheong YH, Chang H-S, Gupta R, Wang X, Zhu T, Luan S (2002) Transcriptional profiling reveals novel interactions between wounding, pathogen, abiotic stress, and hormonal responses in *Arabidopsis*. Plant Physiol 129:661–677
Chinnusamy V, Schumaker K, Zhu J (2004) Molecular genetic perspectives on cross - talk and specificity in abiotic stress signalling in plants. J Exp Bot 55: 225–236
Cooper K, Farrant JM (2002) Recovery of the resurrection plant *Craterostigma wilmsii* from desiccation: protection versus repair. J Exp Bot 53:1805–1813
Cox J, Mann M (2007) Is proteomics the new genomics? Cell 130:395–398
Cramer GR, Ergül A, Grimplet J, Tillett RL, Tattersall EAR, Bohlman MC, Vincent D, Sonderegger J, Evans J, Osborne C, Quilici D, Schlauch KA, Schooley DA, Cushman JC (2006) Water and salinity stress in grapevines: early and late changes in transcript and metabolite profiles. Funct Integr Genomics 7:111–134
Cravatt BF, Simon GM, Yates JR III (2007) The biological impact of mass-spectrometry-based proteomics. Nature 450:991–1000
Cushman JC, Bohnert HJ (2000) Genomic approaches to plant stress tolerance. Curr Opin Plant Biol 3:117–124
de Godoy L, Olsen J, de Souza G, Li G, Mortensen P, Mann M (2006) Status of complete proteome analysis by mass spectrometry: SILAC labeled yeast as a model system. Genome Biol 7:R50
del Río LA, Sandalio LM, Corpas FJ, Palma JM, Barroso JB (2006) Reactive oxygen species and reactive nitrogen species in peroxisomes. Production, scavenging, and role in cell signaling. Plant Physiol 141:330–335
Des Marais DL, Juenger TE (2010) Pleiotropy, plasticity, and the evolution of plant abiotic stress tolerance. Ann N Y Acad Sci 1206:56–79
Dietz KJ, Horling F, König J, Baier M (2002) The function of the chloroplast 2-cysteine peroxiredoxin in peroxide detoxification and its regulation. J Exp Bot 53:1321–1329
Dixon RA, Gang DR, Charlton AJ, Fiehn O, Kuiper HA, Reynolds TL, Tjeerdema RS, Jeffery EH, German JB, Ridley WP, Seiber JN (2006) Applications of metabolomics in agriculture. J Agric Food Chem 54:8984–8994
Dong C-H, Xia G-X, Hong Y, Ramachandran S, Kost B, Chua N-H (2001) ADF proteins are involved in the control of flowering and regulate f-actin organization, cell expansion, and organ growth in arabidopsis. Plant Cell 13:1333–1346
Dunkley TPJ, Hester S, Shadforth IP, Runions J, Weimar T, Hanton SL, Griffin JL, Bessant C, Brandizzi F, Hawes C, Watson RB, Dupree P, Lilley KS (2006) Mapping the *Arabidopsis* organelle proteome. Proc Natl Acad Sci USA 103:6518–6523
Escobar Galvis ML, Marttila S, Håkansson G, Forsberg J, Knorpp C (2001) Heat stress response in pea involves

interaction of mitochondrial nucleoside diphosphate kinase with a novel 86-kilodalton protein. Plant Physiol 126:69–77

FAO (Food, Agriculture Organization of the United Nations) (2004) FAO production yearbook. FAO, Rome

Fernie AR, Trethewey RN, Krotzky AJ, Willmitzer L (2004) Metabolite profiling: from diagnostics to systems biology. Nat Rev Mol Cell Biol 5:763–769

Fiehn O, Kopka J, Dormann P, Altmann T, Trethewey RN, Willmitzer L (2000) Metabolite profiling for plant functional genomics. Nat Biotechnol 18:1157–1161

Fiehn O, Sumner LW, Rhee SY, Ward J, Dickerson J, Lange BM, Lane G, Roessner U, Last R, Nikolau B (2007) Minimum reporting standards for plant biology context information in metabolomic studies. Metabolomics 3:195–201

Frank W, Munnik T, Kerkmann K, Salamini F, Bartels D (2000) Water deficit triggers phospholipase d activity in the resurrection plant craterostigma plantagineum. Plant Cell 12:111–124

Fu D, Huang B, Xiao Y, Muthukrishnan S, Liang G (2007) Overexpression of barley hva1 gene in creeping bentgrass for improving drought tolerance. Plant Cell Rep 26:467–477

Gan CS, Chong PK, Pham TK, Wright PC (2007) Technical, experimental, and biological variations in Isobaric Tags for Relative and Absolute Quantitation (iTRAQ). J Proteome Res 6:821–827

Garg AK, Kim J-K, Owens TG, Ranwala AP, Choi YD, Kochian LV, Wu RJ (2002) Trehalose accumulation in rice plants confers high tolerance levels to different abiotic stresses. Proc Natl Acad Sci USA 99:15898–15903

Gerster S, Qeli E, Ahrens CH, Bühlmann P (2010) Protein and gene model inference based on statistical modeling in k-partite graphs. Proc Natl Acad Sci USA 107:12101–12106

Gesch RW, Kang I-H, Gallo-Meagher M, Vu JCV, Boote KJ, Allen L, Bowes G (2003) Rubisco expression in rice leaves is related to genotypic variation of photosynthesis under elevated growth CO_2 and temperature. Plant Cell Environ 26:1941–1950

Gong Q, Li P, Ma S, Indu Rupassara S, Bohnert HJ (2005) Salinity stress adaptation competence in the extremophile *Thellungiella halophila* in comparison with its relative *Arabidopsis thaliana*. Plant J 44:826–839

Good DM, Wirtala M, McAlister GC, Coon JJ (2007) Performance characteristics of electron transfer dissociation mass spectrometry. Mol Cell Proteomics 6:1942–1951

Görg A, Drews O, Lück C, Weiland F, Weiss W (2009) 2-DE with IPGs. Electrophoresis 30:S122–S132

Hagel JM, Facchini PJ (2007) Plant metabolomics: analytical platforms and integration with functional genomics. Phytochem Rev 7:479–497

Hajheidari M, Abdollahian-Noghabi M, Askari H, Heidari M, Sadeghian SY, Ober ES, Salekdeh GH (2005) Proteome analysis of sugar beet leaves under drought stress. Proteomics 5:950–960

Hajheidari M, Eivazi A, Buchanan BB, Wong JH, Majidi I, Salekdeh GH (2007) Proteomics uncovers a role for redox in drought tolerance in wheat. J Proteome Res 6:1451–1460

Hammond-Kosack KE, Tang S, Harrison K, Jones JDG (1998) The tomato Cf-9 disease resistance gene functions in tobacco and potato to confer responsiveness to the fungal avirulence gene product Avr9. Plant Cell 10:1251–1266

Han X, Jin M, Breuker K, McLafferty FW (2006) Extending Top-Down Mass Spectrometry to proteins with masses greater than 200 kilodaltons. Science 314:109–112

Hazen SP, Wu Y, Kreps JA (2003) Gene expression profiling of plant responses to abiotic stress. Funct Integr Genomics 3:105–111

Heazlewood JL, Verboom RE, Tonti-Filippini J, Small I, Millar AH (2007) SUBA: the *Arabidopsis* subcellular database. Nucleic Acids Res 35:D213–D218

Hinsinger P, Gobran GR, Gregory PJ, Wenzel WW (2005) Rhizosphere geometry and heterogeneity arising from root-mediated physical and chemical processes. New Phytol 168:293–303

Hodge A (2004) The plastic plant: root responses to heterogeneous supplies of nutrients. New Phytol 162:9–24

Hollywood K, Brison DR, Goodacre R (2006) Metabolomics: current technologies and future trends. Proteomics 6:4716–4723

Hongbo S, Zongsuo L, Mingan S, Bochu W (2005) Impacts of PEG-6000 pretreatment for barley (*Hordeum vulgare* L.) seeds on the effect of their mature embryo in vitro culture and primary investigation on its physiological mechanism. Colloids Surf B Biointerfaces 41:73–77

Hu X, Song F, Zheng Z (2006) Molecular characterization and expression analysis of a rice protein phosphatase 2C gene, OsBIPP2C1, and overexpression in transgenic tobacco conferred enhanced disease resistance and abiotic tolerance. Physiol Plant 127:225–236

Hu XJ, Zhang ZB, Xu P, Fu ZY, Hu SB, Song WY (2010) Multifunctional genes: the cross-talk among the regulation networks of abiotic stress responses. Biol Plantarum 54:213–223

Hummel J, Niemann M, Wienkoop S, Schulze W, Steinhauser D, Selbig J, Walther D, Weckwerth W (2007) ProMEX: a mass spectral reference database for proteins and protein phosphorylation sites. BMC Bioinformatics 8:216

Humphreys MW, Yadav RS, Cairns AJ, Turner LB, Humphreys J, Skøt L (2006) A changing climate for grassland research. New Phytol 169:9–26

Ichimura K, Mizoguchi T, Yoshida R, Yuasa T, Shinozaki K (2000) Various abiotic stresses rapidly activate *Arabidopsis* MAP kinases ATMPK4 and ATMPK6. Plant J 24:655–665

Ito Y, Katsura K, Maruyama K, Taji T, Kobayashi M, Seki M, Shinozaki K, Yamaguchi-Shinozaki K (2006) Functional analysis of rice DREB1/CBF-type transcription factors involved in cold-responsive gene expression in transgenic rice. Plant Cell Physiol 47:141–153

Jorrín JV, Rubiales D, Dumas-Gaudot E, Recorbet G, Maldonado A, Castillejo MA, Curto M (2006) Proteomics: a promising approach to study biotic interaction in legumes. A review. Euphytica 147:37–47

Jorrín JV, Maldonado AM, Castillejo MA (2007) Plant proteome analysis: a 2006 update. Proteomics 7:2947–2962

Jorrín-Novo JV, Maldonado AM, Echevarría-Zomeño S, Valledor L, Castillejo MA, Curto M, Valero J, Sghaier B, Donoso G, Redondo I (2009) Plant proteomics update (2007–2008): second-generation proteomic techniques, an appropriate experimental design, and data analysis to fulfill MIAPE standards, increase plant proteome coverage and expand biological knowledge. J Proteomics 72:285–314

Jung C, Seo JS, Han SW, Koo YJ, Kim CH, Song SI, Nahm BH, Choi YD, Cheong J-J (2008) Overexpression of AtMYB44 enhances stomatal closure to confer abiotic stress tolerance in transgenic *Arabidopsis*. Plant Physiol 146:623–635

Jyothsnakumari G, Thippeswamy M, Veeranagamallaiah G, Sudhakar C (2009) Differential expression of LEA proteins in two genotypes of mulberry under salinity. Biol Plantarum 53:145–150

Kanehisa M (2004) The KEGG resource for deciphering the genome. Nucleic Acids Res 32:277D–280D

Kaplan F, Kopka J, Haskell DW, Zhao W, Schiller KC, Gatzke N, Sung DY, Guy CL (2004) Exploring the temperature-stress metabolome of *Arabidopsis*. Plant Physiol 136:4159–4168

Kasuga M, Liu Q, Miura S, Yamaguchi-Shinozaki K, Shinozaki K (1999) Improving plant drought, salt, and freezing tolerance by gene transfer of a single stress inducible transcription factor. Nat Biotechnol 17:287–291

Katiyar-Agarwal S, Agarwal M, Grover A (2003) Heat-tolerant basmati rice engineered by over-expression of *hsp 101*. Plant Mol Biol 51:677–686

Katz E, Fon M, Lee YJ, Phinney BS, Sadka A, Blumwald E (2007) The citrus fruit proteome: insights into citrus fruit metabolism. Planta 226:989–1005

Kempa S, Krasensky J, Dal Santo S, Kopka J, Jonak C (2008) A central role of abscisic acid in stress-regulated carbohydrate metabolism. PLoS One 3:3935

Keurentjes JJB, Fu J, de Vos CHR, Lommen A, Hall RD, Bino RJ, van der Plas LHW, Jansen RC, Vreugdenhil D, Koornneef M (2006) The genetics of plant metabolism. Nat Genet 38:842–849

Khush G (2003) Productivity improvements in rice. Nutr Rev 61:S114–S116

Kim JK, Bamba T, Harada K, Fukusaki E, Kobayashi A (2007) Time-course metabolic profiling in *Arabidopsis thaliana* cell cultures after salt stress treatment. J Exp Bot 58:415–424

Kitano H (2002) Computational systems biology. Nature 420:206–210

Komatsu S, Zang X, Tanaka N (2006) Comparison of two proteomics techniques used to identify proteins regulated by gibberellin in rice. J Proteome Res 5: 270–276

Kristensen C, Morant M, Olsen CE, Ekstrøm CT, Galbraith DW, Lindberg Møller B, Bak S (2005) Metabolic engineering of dhurrin in transgenic *Arabidopsis* plants with marginal inadvertent effects on the metabolome and transcriptome. Proc Natl Acad Sci USA 102:1779–1784

Kunin WE, Vergeer P, Kenta T, Davey MP, Burke T, Woodward FI, Quick P, Mannarelli M-E, Watson-Haigh NS, Butlin R (2009) Variation at range margins across multiple spatial scales: environmental temperature, population genetics and metabolomic phenotype. Proc R Soc Lond B Biol Sci 276:1495–1506

Leakey ADB, Uribelarrea M, Ainsworth EA, Naidu SL, Rogers A, Ort DR, Long SP (2006) Photosynthesis, productivity, and yield of maize are not affected by open-air elevation of CO_2 concentration in the absence of drought. Plant Physiol 140:779–790

Leiss KA, Choi YH, Abdel-Farid IB, Verpoorte R, Klinkhamer PGL (2009) NMR metabolomics of thrips (*Frankliniella occidentalis*) resistance in *Senecio* hybrids. J Chem Ecol 35:219–229

Leitner A, Lindner W (2009) Chemical tagging strategies for mass spectrometry-based phospho-proteomics. Methods Mol Biol 527:229–243

Lilley KS, Dupree P (2006) Methods of quantitative proteomics and their application to plant organelle characterization. J Exp Bot 57:1493–1499

Liu X, Baird VW (2004) Identification of a novel gene, HAABRC5, from *Helianthus annuus* (Asteraceae) that is upregulated in response to drought, salinity, and abscisic acid. Am J Bot 91:184–191

Liu H-S, Li F-M (2005) Photosynthesis, root respiration, and grain yield of spring wheat in response to surface soil drying. Plant Growth Regul 45:149–154

Lommen A (2009) MetAlign: interface-driven, versatile metabolomics tool for hyphenated full-scan mass spectrometry data preprocessing. Anal Chem 81: 3079–3086

Macel M, Van Dam NM, Keurentjes JJB (2010) Metabolomics: the chemistry between ecology and genetics. Mol Ecol Res 10:583–593

Madlung A, Comai L (2004) The effect of stress on genome regulation and structure. Ann Bot 94: 481–495

Majeran W, Cai Y, Sun Q, Van Wijk KJ (2005) Functional differentiation of bundle sheath and mesophyll maize chloroplasts determined by comparative proteomics. Plant Cell 17:3111–3140

Mano S, Miwa T, S-I N, Mimura T, Nishimura M (2007) The plant organelles database (PODB): a collection of visualized plant organelles and protocols for plant organelle research. Nucleic Acids Res 36:D929–D937

Maruyama K, Takeda M, Kidokoro S, Yamada K, Sakuma Y, Urano K, Fujita M, Yoshiwara K, Matsukura S, Morishita Y, Sasaki R, Suzuki H, Saito K, Shibata D, Shinozaki K, Yamaguchi-Shinozaki K (2009) Metabolic pathways involved in cold acclimation identified by integrated analysis of metabolites and transcripts regulated by DREB1A and DREB2A. Plant Physiol 150:1972–1980

May P, Christian N, Ebenhöh O, Weckwerth W, Walther D (2011) Integration of proteomic and metabolomic profiling as well as metabolic modeling for the functional analysis of metabolic networks. Methods Mol Biol 694:341–363

McDonald T, Sheng S, Stanley B, Chen D, Ko Y, Cole RN, Pedersen P, Van Eyk JE (2006) Expanding the subproteome of the inner mitochondria using protein separation technologies. Mol Cell Proteomics 5:2392–2411

Meng F, Wiener MC, Sachs JR, Burns C, Verma P, Paweletz CP, Mazur MT, Deyanova EG, Yates NA, Hendrickson RC (2007) Quantitative analysis of complex peptide mixtures using FTMS and Differential Mass Spectrometry. J Am Soc Mass Spectrom 18:226–233

Meyerowitz EM (2002) Plants compared to animals: the broadest comparative study of development. Science 295:1482–1485

Moco S, Bino RJ, Vorst O, Verhoeven HA, de Groot J, van Beek TA, Vervoort J, de Vos CHR (2006) A liquid chromatography-mass spectrometry-based metabolome database for tomato. Plant Physiol 141:1205–1218

Moco S, Vervoort J, Bino RJ, De Vos RCH, Bino R (2007) Metabolomics technologies and metabolite identification. TrAC Trends Analyt Chem 26:855–866

Molina H, Horn DM, Tang N, Mathivanan S, Pandey A (2007) Global proteomic profiling of phosphopeptides using electron transfer dissociation tandem mass spectrometry. Proc Natl Acad Sci USA 104:2199–2204

Moon JH, Kim SN, Kang BW, Chae YS, Kim JG, Ahn JS, Kim YK, Yang DH, Lee JJ, Kim HJ, Choi YJ, Shin HJ, Chung JS, Cho GJ, Sohn SK (2010) Early onset of acute GVHD indicates worse outcome in terms of severity of chronic GVHD compared with late onset. Bone Marrow Transplant 45:1540–1545

Morgenthal K, Wienkoop S, Wolschin F, Weckwerth W (2007) Integrative profiling of metabolites and proteins: improving pattern recognition and biomarker selection for systems level approaches. Methods Mol Biol 358:57–75

Mueller LA, Zhang P, Rhee SY (2003) AraCyc: a biochemical pathway database for *Arabidopsis*. Plant Physiol 132:453–460

Munns R (2005) Genes and salt tolerance: bringing them together. New Phytol 167:645–663

Ndimba BK, Chivasa S, Simon WJ, Slabas AR (2005) Identification of *Arabidopsis* salt and osmotic stress responsive proteins using two-dimensional difference gel electrophoresis and mass spectrometry. Proteomics 5:4185–4196

Nelson CJ, Huttlin EL, Hegeman AD, Harms AC, Sussman MR (2007) Implications of 15N - metabolic labeling for automated peptide identification in *Arabidopsis thaliana*. Proteomics 7:1279–1292

Nesvizhskii AI, Vitek O, Aebersold R (2007) Analysis and validation of proteomic data generated by tandem mass spectrometry. Nat Methods 4:787–797

Oliver SG, Winson MK, Kell DB, Baganz F (1998) Systematic functional analysis of the yeast genome. Trends Biotechnol 16:373–378

Orchard S, Hermjakob H (2008) The HUPO proteomics standards initiative – easing communication and minimizing data loss in a changing world. Brief Bioinform 9:166–173

Ozturk ZN, Talamé V, Deyholos M, Michalowski CB, Galbraith DW, Gozukirmizi N, Tuberosa R, Bohnert HJ (2002) Monitoring large-scale changes in transcript abundance in drought- and salt-stressed barley. Plant Mol Biol 48:551–573

Palmblad M, Mills DJ, Bindschedler LV (2008) Heat-shock response in *Arabidopsis thaliana* explored by multiplexed quantitative proteomics using differential metabolic labeling. J Proteome Res 7:780–785

Qiu Q-S, Guo Y, Quintero FJ, Pardo JM, Schumaker KS, Zhu J-K (2004) Regulation of vacuolar Na+/H+ exchange in *Arabidopsis thaliana* by the salt-overly-sensitive (SOS) pathway. J Biol Chem 279: 207–215

Rácz I, Páldi E, Szalai G, Janda T, Pál M, Lásztity D (2008) S-methylmethionine reduces cell membrane damage in higher plants exposed to low-temperature stress. J Plant Physiol 165:1483–1490

Ren D, Liu Y, Yang K-Y, Han L, Mao G, Glazebrook J, Zhang S (2008) A fungal-responsive MAPK cascade regulates phytoalexin biosynthesis in *Arabidopsis*. Proc Natl Acad Sci USA 105:5638–5643

Rhee SY, Beavis W, Berardini TZ, Chen G, Dixon D, Doyle A, Garcia-Hernandez M, Huala E, Lander G, Montoya M, Miller N, Mueller LA, Mundodi S, Reiser L, Tacklind J, Weems DC, Wu Y, Xu I, Yoo D, Yoon J, Zhang P (2003) The *Arabidopsis* Information Resource (TAIR): a model organism database providing a centralized, curated gateway to *Arabidopsis* biology, research materials and community. Nucleic Acids Res 31:224–228

Rizhsky L, Liang H, Shuman J, Shulaev V, Davletova S, Mittler R (2004) When defense pathways collide. The response of Arabidopsis to a combination of drought and heat stress. Plant Physiol 134:1683–1696

Roessner U, Pettolino F (2007) The importance of anatomy and physiology in plant metabolomics. Metabolomics 18:253–278

Roessner U, Luedemann A, Brust D, Fiehn O, Linke T, Willmitzer L, Fernie AR (2001) Metabolic profiling allows comprehensive phenotyping of genetically or environmentally modified plant systems. Plant Cell 13:11–29

Rossignol M, Peltier J-B, Mock H-P, Matros A, Maldonado AM, Jorrín JV (2006) Plant proteome analysis: a 2004–2006 update. Proteomics 6:5529–5548

Ryan D, Robards K (2006) Metabolomics: the greatest omics of them all? Anal Chem 78:7954–7958

Sakuma Y, Maruyama K, Osakabe Y, Qin F, Seki M, Shinozaki K, Yamaguchi-Shinozaki K (2006) Functional analysis of an *Arabidopsis* transcription factor, DREB2A, involved in drought-responsive gene expression. Plant Cell 18:1292–1309

Salekdeh GH, Komatsu S (2007) Crop proteomics: aim sustainable agriculture of tomorrow. Proteomics 7: 2976–2996

Salekdeh GH, Siopongco J, Wade LJ, Ghareyazie B, Bennett J (2002) Proteomic analysis of rice leaves during drought stress and recovery. Proteomics 2:1131–1145

Samis K, Bowley S, McKersie B (2002) Pyramiding Mn - superoxide dismutase transgenes to improve persistence and biomass production in alfalfa. J Exp Bot 53:1343–1350

Sanchez PA, Swaminathan MS (2005) Cutting world hunger in half. Science 307:357–359

Schröder S, Zhang H, Yeung ES, Jänsch L, Zabel C, Wätzig H (2008) Quantitative gel electrophoresis: sources of variation. J Proteome Res 7:1226–1234

Serrot PH, Sabater B, Martín M (2008) Expression of the ndhCKJ operon of barley and editing at the 13th base of the mRNA of the ndhC gene. Biol Plantarum 52:347–350

Shao H-B, Guo Q-J, Chu L-Y, Zhao X-N, Su Z-L, Hu Y-C, Cheng J-F (2007) Understanding molecular mechanism of higher plant plasticity under abiotic stress. Colloids Surf B Biointerfaces 54:37–45

Shewry PR, Jones HD, Halford NG (2008) Plant biotechnology: transgenic crops. Adv Biochem Eng Biotechnol 11:149–186

Shigeoka S, Ishikawa T, Tamoi M, Miyagawa Y, Takeda T, Yabuta Y, Yoshimura K (2002) Regulation and function of ascorbate peroxidase isoenzymes. J Exp Bot 53:1305–1319

Shiio Y, Aebersold R (2006) Quantitative proteome analysis using isotope-coded affinity tags and mass spectrometry. Nat Protoc 1:139–145

Shinozaki K, Yamaguchi-Shinozaki K, Seki M (2003) Regulatory network of gene expression in the drought and cold stress responses. Curr Opin Plant Biol 6:410–417

Shou H, Bordallo P, Wang K (2004) Expression of the *Nicotiana* protein kinase (NPK1) enhanced drought tolerance in transgenic maize. J Exp Bot 55:1013–1019

Siritunga D, Sayre RT (2003) Generation of cyanogen-free transgenic cassava. Planta 217:367–373

Skylas DJ, Cordwell SJ, Hains PG, Larsen MR, Basseal DJ, Walsh BJ, Blumenthal C, Rathmell W, Copeland L, Wrigley CW (2002) Heat shock of wheat during grain filling: proteins associated with heat-tolerance. J Cereal Sci 35:175–188

Sokhansanj A, Sadat Noori SA, Niknam V (2006) Comparison of bacterial and plant genes participating in proline biosynthesis with Osmotin gene, with respect to enhancing salinity tolerance of transgenic tobacco plants. Russ J Plant Physiol 53:110–115

Soltis DE, Soltis PS (2003) The role of phylogenetics in comparative genetics. Plant Physiol 132:1790–1800

Sumner LW, Amberg A, Barrett D, Beale MH, Beger R, Daykin CA, Fan TW-M, Fiehn O, Goodacre R, Griffin JL, Hankemeier T, Hardy N, Harnly J, Higashi R, Kopka J, Lane AN, Lindon JC, Marriott P, Nicholls AW, Reily MD, Thaden JJ, Viant MR (2007) Proposed minimum reporting standards for chemical analysis. Metabolomics 3:211–221

Sun W, Van Montagu M, Verbruggen N (2002) Small heat shock proteins and stress tolerance in plants. Biochim Biophys Acta 1577:1–9

Sun Q, Zybailov B, Majeran W, Friso G, Olinares PDB, van Wijk KJ (2009) PPDB, the plant proteomics database at Cornell. Nucleic Acids Res 37:D969–D974

Thurston G, Regan S, Rampitsch C, Xing T (2005) Proteomic and phosphoproteomic approaches to understand plant-pathogen interactions. Physiol Mol Plant Pathol 66:3–11

Tjalsma H, Antelmann H, Jongbloed JDH, Braun PG, Darmon E, Dorenbos R, Dubois J-YF, Westers H, Zanen G, Quax WJ, Kuipers OP, Bron S, Hecker M, van Dijl JM (2004) Proteomics of protein secretion by *Bacillus subtilis*: separating the "secrets" of the secretome. Microbiol Mol Biol Rev 68:207–233

Tribl F, Lohaus C, Dombert T, Langenfeld E, Piechura H, Warscheid B, Meyer HE, Marcus K (2008) Towards multidimensional liquid chromatography separation of proteins using fluorescence and isotope-coded protein labelling for quantitative proteomics. Proteomics 8:1204–1211

Tseng MJ, Liu C-W, Yiu J-C (2007) Enhanced tolerance to sulfur dioxide and salt stress of transgenic Chinese cabbage plants expressing both superoxide dismutase and catalase in chloroplasts. Plant Physiol Biochem 45:822–833

Umezawa T, Fujita M, Fujita Y, Yamaguchi-Shinozaki K, Shinozaki K (2006) Engineering drought tolerance in plants: discovering and tailoring genes to unlock the future. Curr Opin Biotechnol 17:113–122

Uno Y, Furihata T, Abe H, Yoshida R, Shinozaki K, Yamaguchi-Shinozaki K (2000) *Arabidopsis* basic leucine zipper transcription factors involved in an abscisic acid-dependent signal transduction pathway under drought and high-salinity conditions. Proc Natl Acad Sci USA 97:11632–11637

Urano K, Maruyama K, Ogata Y, Morishita Y, Takeda M, Sakurai N, Suzuki H, Saito K, Shibata D, Kobayashi M, Yamaguchi - Shinozaki K, Shinozaki K (2009) Characterization of the ABA - regulated global responses to dehydration in *Arabidopsis* by metabolomics. Plant J 57:1065–1078

Urano K, Kurihara Y, Seki M, Shinozaki K (2010) "Omics" analyses of regulatory networks in plant abiotic stress responses. Curr Opin Plant Biol 13:132–138

Van Dam NM, Poppy GM (2008) Why plant volatile analysis needs bioinformatics – detecting signal from noise in increasingly complex profiles. Plant Biol 10: 29–37

Vendruscolo ECG, Schuster I, Pileggi M, Scapim CA, Molinari HBC, Marur CJ, Vieira LGE (2007) Stress-induced synthesis of proline confers tolerance to water deficit in transgenic wheat. J Plant Physiol 164:1367–1376

Verpoorte R, Choi YH, Kim HK (2007) NMR-based metabolomics at work in phytochemistry. Phytochem Rev 6:3–14

Vinocur B, Altman A (2005) Recent advances in engineering plant tolerance to abiotic stress: achievements and limitations. Curr Opin Biotechnol 16:123–132

Wang W, Vinocur B, Altman A (2003) Plant responses to drought, salinity and extreme temperatures: towards genetic engineering for stress tolerance. Planta 218:1–14

Wang W, Vignani R, Scali M, Cresti M (2006) A universal and rapid protocol for protein extraction from recalcitrant plant tissues for proteomic analysis. Electrophoresis 27:2782–2786

Wang Y, Liu C, Li K, Sun F, Hu H, Li X, Zhao Y, Han C, Zhang W, Duan Y, Liu M (2007) *Arabidopsis* EIN2 modulates stress response through abscisic acid response pathway. Plant Mol Biol 64:633–644

Weckwerth W (2003) Metabolomics in systems biology. Annu Rev Plant Biol 54:669–689

Weckwerth W, Baginsky S, van Wijk K, Heazlewood JL, Millar H (2008) The multinational *Arabidopsis* steering subcommittee for proteomics assembles the largest proteome database resource for plant systems biology. J Proteome Res 7:4209–4210

Westerhuis JA, Hoefsloot HCJ, Smit S, Vis DJ, Smilde AK, Velzen EJJ, Duijnhoven JPM, Dorsten FA (2008) Assessment of PLSDA cross validation. Metabolomics 4:81–89

Wienkoop S, Morgenthal K, Wolschin F, Scholz M, Selbig J, Weckwerth W (2008) Integration of metabolomic and proteomic phenotypes. Mol Cell Proteomics 7:1725–1736

Wienkoop S, Baginsky S, Weckwerth W (2010) *Arabidopsis thaliana* as a model organism for plant proteome research. J Proteomics 73:2239–2248

Wiese S, Reidegeld KA, Meyer HE, Warscheid B (2007) Protein labeling by iTRAQ: a new tool for quantitative mass spectrometry in proteome research. Proteomics 7:340–350

Wolters DA, Washburn MP, Yates JR (2001) An automated multidimensional protein identification technology for shotgun proteomics. Anal Chem 73:5683–5690

Wu WW, Wang G, Baek SJ, Shen R-F (2006) Comparative study of three proteomic quantitative methods, DIGE, CICAT, and iTRAQ, using 2D Gel- or LC-MALDI TOF/TOF. J Proteome Res 5:651–658

Wurtele ES, Li L, Berleant D, Cook D, Dickerson JA, Ding J, Hofmann H, Lawrence M, E-k L, Li J, Mentzen W, Miller L, Nikolau BJ, Ransom N, Wang Y (2007) Concepts in plant metabolomics, vol 10. Springer, Dordrecht, The Netherlands, pp 145–157

Xu L, Xu Y, Dong A, Sun Y, Pi L, Xu Y, Huang H (2003) Novel as1 and as2 defects in leaf adaxial-abaxial polarity reveal the requirement for ASYMMETRIC LEAVES1 and 2 and ERECTA functions in specifying leaf adaxial identity. Development 130:4097–4107

Yamaguchi-Shinozaki K, Shinozaki K (2006) Transcriptional regulatory networks in cellular responses and tolerance to dehydration and cold stresses. Annu Rev Plant Biol 57:781–803

Yamazaki M, Nakajima J, Yamanashi M, Sugiyama M, Makita Y, Springob K, Awazuhara M, Saito K (2003a) Metabolomics and differential gene expression in anthocyanin chemo-varietal froms of *Perilla frutescens*. Phytochemistry 62:987–995

Yamazaki Y, Urano A, Sudo H, Kitajima M, Takayama H, Yamazaki M, Aimi N, Saito K (2003b) Metabolite profiling of alkaloids and strictosidine synthase activity in camptothecin producing plants. Phytochemistry 62:461–470

Yan J, Tang Y, Sun C, Su Y, Mao B (2010) STM study on nonionic fluorosurfactant zonyl FSN self-assembly on Au(100): (3/1–1/1) Molecular lattice, corrugations, and adsorbate-enhanced mobility. Langmuir 26: 3829–3834

Yoshimura K, Masuda A, Kuwano M, Yokota A, Akashi K (2008) Programmed proteome response for drought avoidance/tolerance in the root of a C3 xerophyte (wild watermelon) under water deficits. Plant Cell Physiol 49:226–241

Yuasa T, Ichimura K, Mizoguchi T, Shinozaki K (2001) Oxidative stress activates ATMPK6, an *Arabidopsis* homologue of map kinase. Plant Cell Physiol 42:1012–1016

Zhu J-K (2001) Cell signaling under salt, water and cold stresses. Curr Opin Plant Biol 4:401–406

Zubarev R, Mann M (2007) On the proper use of mass accuracy in proteomics. Mol Cell Proteomics 6: 377–381

Approaches to Increasing Salt Tolerance in Crop Plants

4

Ratna Karan and Prasanta K. Subudhi

Abstract

Soil salinity is widely recognized as a major threat to global food security. Salinity and other abiotic stresses, which are expected to be more frequent in future due to disturbances in global climate, pose a serious challenge for plant scientists to ensure food supply for the growing world population. Several approaches have been advocated to address the salinity problem, but the most logical solution to maintain crop productivity in salt-affected areas is to enhance salt tolerance of crop plants. Due to the genetic and physiological complexity associated with salt tolerance, efforts to breed salt-tolerant plants have met with limited success. Although progress has been made in deciphering the genetic basis of salt tolerance, sustained efforts are needed to systematically dissect and utilize the natural variability in the available germplasm for improving crop adaptation in saline environments using modern genomics tools. Wide range of variability for salt tolerance in wild relatives, cultivars of major field crops, and halophytes offers bright prospect for discovery of superior salt-tolerant alleles for crop improvement. With an enhanced understanding of molecular mechanisms and the associated genes for component traits of salt tolerance, it would be possible to breed salt-tolerant plants using an integrated approach involving conventional breeding, physiological analysis, marker-assisted selection, and transgenic technology.

Keywords

Abiotic stress tolerance • Conventional breeding • Genomics • Marker-assisted selection • Mutagenesis • Quantitative trait loci • Transgenics

R. Karan • P.K. Subudhi (✉)
School of Plant, Environmental, and Soil Sciences,
Louisiana State University Agricultural Center,
Baton Rouge, LA 70803, USA
e-mail: psubudhi@agctr.lsu.edu

P. Ahmad and M.N.V. Prasad (eds.), *Abiotic Stress Responses in Plants: Metabolism, Productivity and Sustainability*, DOI 10.1007/978-1-4614-0634-1_4, © Springer Science+Business Media, LLC 2012

1 Introduction

Abiotic stresses such as drought, salinity, submergence, extreme temperatures, mineral toxicities, and deficiencies impair crop growth and productivity and threaten global food security (Gao et al. 2007; Witcombe et al. 2008). Among these stresses, salinity is impacting more farm lands worldwide due to poor quality irrigation water, inadequate drainage, salt water flooding of coastal land, and salt accumulation in dry areas (Kijne 2006). Around 21% of world's irrigated land is estimated to be affected by salinity (Ghassemi et al. 1995) and it continues to be a major problem in the arid and semi-arid regions. The negative impact of climate change on food crops is well recognized. Global temperatures are estimated to rise between 1.1°C and 6.4°C during the next century (IPCC 2007). The increased temperatures will disrupt weather patterns, leading to regular occurrence of floods, drought, and salinity. Melting ice caps and glaciers are expected to cause a rise in sea level (Wassmann et al. 2004; Melloul and Collin 2006), which may seriously affect crop productivity in coastal areas due to increased soil salinity.

World population is increasing at an alarming rate and it is expected to grow from 6 billion today to nearly 8.3 billion by 2030 (FAO 2010). With no prospect of expanding arable land due to urbanization, rapid industrialization, and water scarcity in many populous developing countries of the world (Rengasamy 2010), providing food security for the world population will require at least 57% increase in food grain production by 2050 (Wild 2003). Although most major crops have witnessed increased productivity in the past, productivity has been stagnant in recent years and replicating the success of the past to increase food production further may not be easy. Increasing salinity tolerance of the world's major food crops is an important goal of plant scientists as the world's population is increasing more quickly than the area of agricultural land to support it (FAO 2010). To alleviate the negative impact of salinity on food production, use of halophytic species, improved water management, and enhancing salinity tolerance in major field crops have been advocated. But improving adaptation of crop plants in saline environments remains a challenging task due to complex genetic basis of salt tolerance mechanisms. In this chapter, we discuss about different approaches for the development of salinity-tolerant crop plants to boost food grain production.

2 Effect of Salinity on Crop Plants

Salinity inhibits seed germination and alters the physiology and anatomy of the plants resulting in reduced crop yield. Salt accumulates in soil due to movement of salty water from adjacent areas during flooding or poor quality irrigation water. When water recedes after flooding, salt from lower soil profiles comes to the soil surface due to the capillary movement. Salinity is a soil condition characterized by a high concentration of soluble salts, mostly chloride and sulfates of sodium in the soil. Soils are classified as saline when the electrical conductivity (EC) is 4 dS/m or more, which is equivalent to approximately 40 mM NaCl and generates an osmotic pressure of approximately 0.2 MPa (USDA-ARS 2008). Both chloride and sodic salts cause damage to the root system of crops. The chloride-triggered injury is identifiable by the extensive leaf blade scorching symptoms whereas the accumulation of sodic salts results in leaf mottling and leaf necrosis. Soil salinity creates both osmotic and ionic stresses in plants. Presence of salts in the soil solution reduces the ability of the plant to take up water, and this leads to reduction in the growth rate, referred as the osmotic or water-deficit effect of salinity. If an excessive amount of salt enters the plant in the transpiration stream, it causes injury to cells in the transpiring leaves, resulting in further reduction in plant growth. This is called the salt-specific or ionic effect of salinity (Greenway and Munns 1980). Salinity stress is finally quantified as a decrease in water potential. Plants resist low water potential in different ways by accumulating osmolytes and modifying the properties of cell walls through production of protective proteins (Verslues et al. 2006). Plants differ greatly in their tolerance to salinity, as reflected in their different growth responses. Most of the commonly used food crops are sensitive to salinity (Flowers and Colmer 2008). Among cereals, rice (*Oryza sativa*) is the most sensitive and barley (*Hordeum vulgare*) is the most tolerant. Bread wheat (*Triticum aestivum*) is moderately tolerant than durum wheat (*Triticum turgidum* ssp. durum). The variation in salinity tolerance in dicotyledonous species is even greater than in monocotyledonous species. Rice, wheat, and barley are the most extensively studied cereal crop plants for understanding the physiological and molecular basis of salt tolerance. The comprehensive survey of salt tolerance for crops and pasture species published by the US Salinity Laboratory (Maas and Hoffman 1977; USDA-ARS 2005), presents a threshold salinity below which there is no reduction in yield, and then a linear reduction in yield with increasing salinity. Salinity reduces the rate of leaf expansion, and closes stomata and thereby reduces photosynthesis, through the soil water deficit caused by the osmotic stress (Rahnama et al. 2010). Upon exposure to soil salinity, plants accumulate toxic concentrations of Na^+ in leaves, which impose an additional limitation to growth by reducing the longevity of photosynthetic tissues (Munns 2002). The control of Na^+ transport and its effective exclusion from the mesophyll cells of leaves is therefore an important requirement for salinity tolerance. Na^+ exclusion from leaves is associated with salt tolerance in cereal crops including rice (Asch et al. 2000; Haq et al. 2010), durum wheat (Munns and James 2003), bread wheat (Cuin et al. 2009, 2010), barley (Shavrukov et al. 2010), and its wild relatives (Garthwaite et al. 2005), tall wheatgrass (Colmer et al. 2006), and *Triticum tauschii* (Schachtman et al. 1991). The major components that govern salt tolerance are reduced salt uptake or salt exclusion, enhanced K^+/Na^+ ratio, tissue tolerance, closure of stomata, upregulation of antioxidant system for protection against reactive oxygen species (ROS), synthesis of osmolytes, water use efficiency (WUE), early flowering, and vigorous growth to dilute the salt concentration in plant tissue (Colmer et al. 2005; Ismail et al. 2007).

3 Approaches for Developing Salt-Tolerant Crop Plants

3.1 Conventional Approach

3.1.1 Germplasm Screening and Classical Breeding

The success of the crop-breeding program largely depends on the availability of natural genetic variation among the germplasm resources. Large number of cultivated and wild germplasm in major crops, preserved in the Consultative Group on International Agricultural Research (CGIAR) institutions and national centers, provide unique resources for systematic screening for discovery of novel variability to improve adaptation of crop plants in saline environments. Particularly, the wild relatives, land races, and traditional cultivars are the potential reservoirs of novel alleles to improve abiotic stress tolerance. Accurate phenotyping procedures are critical for identifying useful germplasm for crop improvement program as well as for deciphering the genetic basis of the mechanisms associated with salinity tolerance. Several parameters for salinity tolerance are studied by growing the germplasm in a variety of culture techniques such as hydroponics, pot culture, and field screening. Plant materials are evaluated from germinating seeds through seedlings up to mature plants. Salinity causes not only ion toxicity and imbalance, but also reduces photosynthesis in plants. Classical screening methods are based on assessment of yield responses to salt stress. Although screening based on yield represents the combined genetic and environmental effects on plant growth and includes integration of the physiological mechanisms conferring salinity tolerance at the whole plant level, it is more convenient and practical if indirect indicators of salt tolerance can be employed at the whole plant, tissue, or cellular levels (Ashraf and Harris 2004). Faster screening methods can be employed for identification of potential parents in a breeding program through selection for high leaf K^+/Na^+ ratios in the presence of salinity, and high K^+/Na^+ discrimination that has been described as a physiological index for salinity tolerance in bread wheat (Dvorak et al. 1994), durum wheat (Munns et al. 2000), and rice (Asch et al. 2000).

Gregorio et al. (1997) used Standard Evaluation System (SES) for rice seedling salinity screening based on the percent of leaf damage, which in turn is used to assign the SES scores. These SES scores measure overall survival and/or vigor of the plant and are therefore good indicators of performance of the plant under stress. SES scores as well as low Na^+ uptake, K^+ uptake and low Na^+/K^+ ratio have been reported to be tightly linked to seedling salinity tolerance (Lee et al. 2003a, Lisa et al. 2004). Analysis of Na^+ uptake, K^+ uptake and Na^+/K^+ ratio of rice seedlings under salt stress, however, are difficult to measure in the large populations associated with breeding programs. Therefore, screening for salinity tolerance at seedling level using SES method is ideal.

Artificial salinized soil in pots or irrigation with saline water under field condition has been used in rice (Aslam et al. 1993). Absolute shoot fresh and dry weights along with percent mortality at seedling stage in rice under salinity stress revealed an efficient, reproducible, reliable, and simple method for assessing relative salinity tolerance in breeding program (Aslam et al. 1993). Physiological characters such as leaf area index (LAI), measurement of Na^+, K^+, and Ca^+ along with amount of photosynthetically active radiation (PAR) absorbed by plants under salinity stress have been used to evaluate salt tolerance in rice (Zeng et al. 2003b). Leaf area index (LAI) was found to be significantly contributing to the yield of grain than other physiological parameters under salt stress. Salt tolerance as defined by the grain yield and amount of PAR absorbed by a plant was found to be strongly related to LAI (Zeng et al. 2003a). Nutrient solution popularly known as Yoshida medium (Yoshida et al. 1976), supplemented with different concentrations of NaCl, is commonly used for salt tolerance screening in rice. A number of cultivars, landraces, and advanced breeding lines such as Pokkali, Nonabokra, SR26B, Damodar, Cheriviruppu, CSR11, Getu, FL378, FL 478, IR 51500-AC17, IR 51500-AC11-1, IR 4595-4-1-13, IR 51491-AC10 have been identified as useful sources for salt tolerance in rice (Dwivedi et al. 2010). Using these lines as donors, few salt-tolerant lines have been released (Gregorio et al. 2002; Ismail et al. 2007). Large number of introgression lines with enhanced abiotic stress tolerance have been developed in a massive backcross breeding program using three recurrent parents and 203 donor lines with tolerance to several abiotic stresses in rice (Ali et al. 2006).

Genetic variation in two key physiological traits, leaf Na^+ and K^+/Na^+ discrimination ratio, among the genotypes of barley and wheat, indicates the possibility of genetic improvement of salt tolerance. Bread wheat, which maintains a lower shoot Na^+ concentration than durum wheat, restricts Na^+ transport to leaf tissues through Na^+ exclusion and maintains high selectivity of K^+ over Na^+, while barley is not so efficient with respect to these traits. However, the adverse effects of Na^+ within leaves of barley are minimized by its compartmentalization into vacuoles (with Cl^-) in leaves by a mechanism known as tissue tolerance (Munns and James 2003; Colmer et al. 2005) and the production of organic solutes to osmotically balance the cytosol (Garthwaite et al. 2005). Khan et al. (2006) compared performance of 16 wheat genotypes under saline condition using gravel culture technique in lysimeters with four salinity levels, i.e., control (1.5 dS m^{-1}), low saline (6.0 dS m^{-1}), medium saline (9.0 dS m^{-1}), and highly saline (12.0 dS m^{-1}). On the basis of less than 50% reduction in yield and yield components, they found five genotypes viz. LU-26 s, HT-45, ESW-9525, V-8319, Sarsabz were tolerant, whereas Bhittai, Marvi, Chakwal-86, DS-17, Sussi (SD-66), Zardana were medium tolerant, SD1200/51, Khirman, V-7012 medium sensitive, and RWM-9313, SH-43 sensitive. Tolerant wheat genotypes were successful in maintaining low Na^+ and high K^+ uptake and high K^+/Na^+ ratio. A durum wheat line 149 has low Na^+ concentrations and high K^+/Na^+ ratios in the leaf blade (Munns et al. 2000) due to presence of two genes, *Nax1* and *Nax2*, which are responsible for exclusion of salts from leaves and roots, respectively (Munns and James 2003). However, the concentration of Na^+ in shoots of line 149 as a whole was not as low as bread wheat (Husain et al. 2004), suggesting retention of Na^+

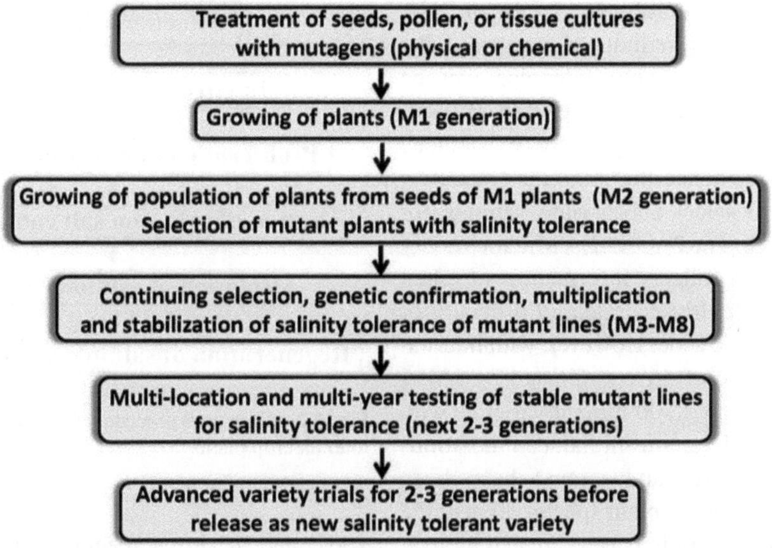

Fig. 4.1 A generalized approach for developing salinity-tolerant crop plants using mutation breeding

in the leaf sheath. The physiological traits associated with Na^+ accumulation include the rate of transfer of Na^+ from the root to the shoot (net root xylem loading) and this was much lower in Line 149 than the durum cv Tamaroi (Davenport et al. 2005). The genotypes did not differ in unidirectional root uptake of Na^+. The major differences in Na^+ transport between the genotypes were in the rate of transfer to the shoot and the preferential accumulation of Na^+ in the leaf sheath versus the leaf blade (Davenport et al. 2005). The K^+/Na^+ ratio at early seedling stage has been shown to be a convenient and reliable indicator of salt tolerance in wheat and barley (Tajbakhsh et al. 2006; Thalji and Shalaldeh 2007). Stomatal conductance and chlorophyll content in leaves can be measured by a non-destructive, rapid, and simple technique using a porometer and SPAD meter, respectively. El-Hendawy et al. (2007) used net photosynthetic rate, stomatal conductance, and SPAD values to screen wheat genotypes for salinity tolerance and reported genetic variation for these traits. Based on this study, SPAD reading could be used as a potential tool for large-scale screening for salt tolerance.

Wild relatives of wheat and barley exhibit genetic variation for Na^+ and Cl^- exclusion (Colmer et al. 2005) which provides opportunity to improve salt tolerance through interspecific hybridization. In case of wheat, the diploid progenitors, *Aegilops tauschii (*DD) and *Triticum urartu* (AA), and synthetic hexaploid wheat involving *Triticum urartu*, were superior in K^+/Na^+ discrimination. Similarly, wild *Hordeum* species including *Hordeum vulgare* subsp. *spontaneum* showed high leaf K^+ concentration and enhanced ability to exclude Na^+ and Cl^-. Wild relatives of wheat, Tall wheatgrass (*Agropyron elongatum*) and *Lophopyrum elongatum* provide a source of novel genes for improvement of salt tolerance of bread wheat. A comprehensive list of salt-tolerant germplasm in important cereal crops is reported by Dwivedi et al. (2010).

3.1.2 Mutation Breeding

Creation of novel and useful genetic variation in important agronomic traits is the most important prerequisite for a crop-breeding program. Mutagenic agents, such as X-rays, gamma rays, fast neutrons, and chemical mutagens such as ethyl methane sulfonate have been used to induce mutations in seeds to generate genetic variation for crop improvement (Fig. 4.1). It offers the possibility of inducing desired characters that either cannot be found in nature or have been lost during evolution. The mutagen treatment breaks the nuclear DNA and during the process of DNA repair, new mutations are induced randomly. The mutants with abiotic stress tolerance can be selected by plant breeders to develop salinity-tolerant crop plants. The purpose of induced mutations is to enhance the mutation frequency rate in order to select appropriate variants for salinity tolerance. The FAO/IAEA Mutant Variety Database (MVD) collects information on plant mutant varieties (cultivars) released officially or commercially worldwide. However, with international collaborative effort coordinated by IAEA and FAO, more than 2,700 mutant varieties with one or more useful traits from induced mutations (mainly from X-rays and γ-rays) have been released in 170 different plant species all over the world. Diamant variety in barley, created by irradiation of dormant seeds with X-rays, has very high yield, short stem, very good grain, and malting quality as well as lodging resistance. Calrose 76 was the first semi-dwarf table rice variety released in the US produced by irradiation with γ-rays. Some outstanding examples of mutant rice cultivars are VND 95-20 and VND 99-3 which have long grains with excellent grain quality and wide adaptation to acid sulphate soil and salinity. These mutants not only increased biodiversity, but also provided valuable breeding material for crop improvement. The database of mutant varieties can be found at the web http://mvgs.iaea.org/AboutMutantVarieties.aspx, maintained by Plant Breeding and Genetics section of joint IAEA/FAO program.

3.2 Tissue Culture Approach

Plant tissue culture techniques provide a promising and feasible approach to develop salt-tolerant crop plants. Haploid culture, double haploidy, somaclonal variation, and in vitro-induced mutagenesis has been used to create variability to improve salinity tolerance in crop plants. Cell and tissue culture techniques have been used to obtain salt-tolerant plants through in vitro culture approaches: selection of mutant cell lines from cultured cells and subsequent plant regeneration (Zair et al. 2003; Gandonou et al. 2006; Lu et al. 2007; Queirs et al. 2007) and in vitro screening of plant germplasm for salt tolerance (Arzani and Mirodjagh 1999; Dziadczyk et al. 2003; Lee et al. 2003b; Wheatley et al. 2003; Dasgupta et al. 2008). Cereal tissue culture is of great economic importance for selection of improved cell lines under in vitro condition (Fig. 4.2). Tissue culture has been used as a breeding tool for rapid screening of genetic materials for salt tolerance in wheat (Arzani and Mirodjagh 1999). Houshmand et al. (2005) evaluated salt-tolerant genotypes of durum wheat derived from the in vitro culture in field experiments under both saline and non-saline field conditions as well as under greenhouse condition using salinized solution culture. In spite of the smaller range of genotypes used from the in vitro method, tolerant genotypes performed comparably with those of the field-derived tolerant genotypes for grain yield under saline field conditions. Overall, in vitro selected tolerant genotypes showed significantly better performance for biomass production under high salinity condition than field-derived tolerant genotypes.

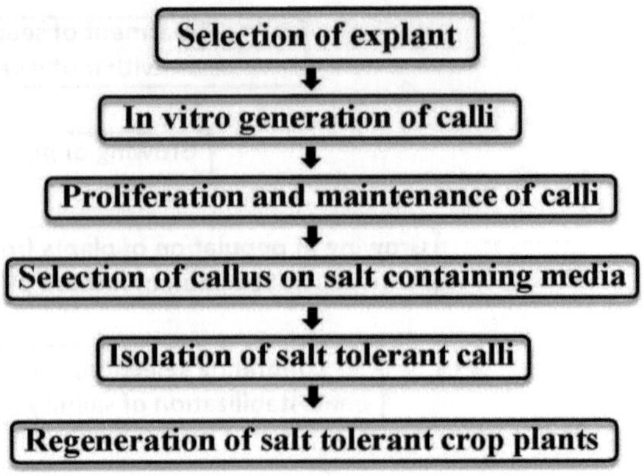

Fig. 4.2 An in vitro procedure for regeneration of salinity tolerant crop plants

3.2.1 Somaclonal Variants

Somaclonal variation refers to the variation seen in plants that have been originated through plant tissue culture. It is particularly common in plants regenerated from callus. The amount of variation that can be expected under in vitro condition may vary with the clone, age of the clone, and use of selection pressure applied to single cells for

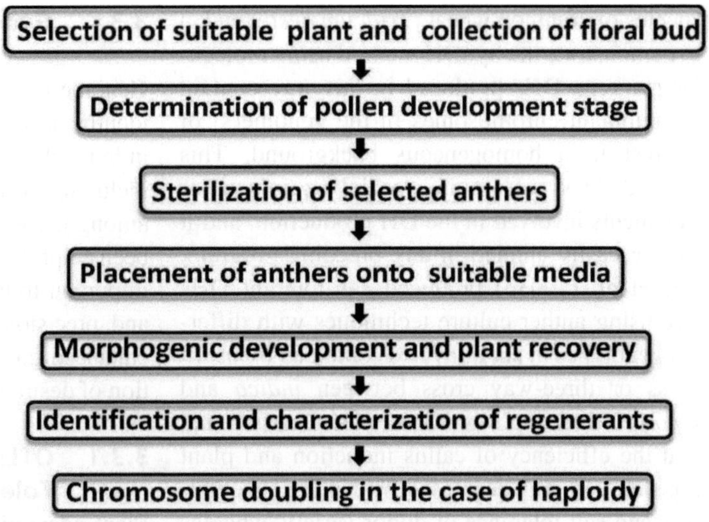

Fig. 4.3 A common procedure for haploid and dihaploid production in crop plants

salinity stress. Somaclonal variations are stable and occur at high frequencies. Novel gene mutations may result during the tissue culture process. It can be performed in vegetatively or sexually or asexually propagated plants. This approach reduces the time required for the release of new variety compared to mutation breeding and has been useful in breeding programs (Zhu et al. 2000).

3.2.2 Double Haploids

A major limitation of the conventional breeding is the long-time frame in developing varieties. Double haploid breeding enables breeders to produce genetically uniform lines within one generation. This effectively bypasses the lengthy process of self-pollination and selection normally required to produce true breeding genotypes. Double Haploids (DH) are plants that have undergone chromosome duplication from haploid plants. The production of haploids and DHs through gametic embryogenesis is the most effective way for the development of complete homozygous lines from heterozygous parents in comparison with the conventional breeding methods that employ several generations of selfing for getting homozygous plants. DH technique is well established in a range of economically important crop species, including major cereals (Wedzony et al. 2009). Various methods such as chromosome elimination subsequent to wide hybridization, pollination with irradiated pollen, selection of twin seedlings, in vivo or in vitro pollination with pollen from a triploid plant, gynogenesis and pollen embryogenesis through in vitro anther or isolated microspore culture were used to obtain DHs (Forster and Thomas 2005). Anther culture and DH production is influenced by various internal and external factors. A generalized method for haploid and DH production is shown in Fig. 4.3. A list of haploid-derived varieties of asparagus, barley, brassica, eggplant, melon, pepper, rapeseed, rice, tobacco, triticale, and wheat are available at http://www.scri.sari.ac.uk/assoc/COST851/Default.htm. The development of protocols to produce haploid and DH plants has significant impact on agricultural systems. Mostly in vitro-derived anther or isolated microspore culture method are preferably used to obtain haploids and DHs from diploid plants (Germanà 2010). Recently, a simple method for synthesizing DHs (SynDH) especially for allopolyploid species has been reported by utilizing meiotic restitution genes (Zhang et al. 2011a). This method involved three steps: hybridization to induce recombination, interspecific hybridization to extract haploids, and spontaneous chromosome doubling by selfing the interspecific F_1s. Zhang et al. (2011a) used *Triticum turgidum* L. and *Aegilops tauschii* Coss, the two ancestral species of common wheat (*Triticum aestivum* L.) to demonstrate the SynDH method using molecular markers. DHs produced in this way contain recombinant chromosomes in the genome(s) of interest in a homogeneous background. This method does not require special equipment or treatments involved in the DH production, and it can be easily applied in any breeding program. Lee et al. (2003b) produced salt-tolerant DHs rice using anther culture techniques with different genotypes in six F_1 hybrids obtained by backcross or three-way cross between *indica* and *japonica* differing in salt tolerance. It was found that the efficiency of callus induction and plant regeneration was decreased by NaCl concentration and salt tolerance of donor variety, whereas induction in *japonicas* was higher than those in indicas. The percentages of callus induction in Gyehwa 5 (*japonica*, tolerant) and IR61633-B-2-2-1 (*japonica*, sensitive) were 21.1% and 13.5% on agar medium containing 0.3% NaCl, respectively. In four F_1 hybrids, the frequencies of high salt-tolerant DHs were 21.4% and 8.9% in 0.3% NaCl medium and the control, respectively. Therefore, the high frequency of salt-tolerant DHs could be selected in the callus induction medium (0.3% NaCl) and in the combinations crossed with salt-tolerant *japonica* as the third parent. F_1 anther culture has become an effective tool to attain homozygosity of recombinants within the shortest possible time. The technique also offers the opportunity to screen haploid materials at the early stage of tissue culture. This allows recessive mutants to be identified under a variety of selection pressures. This approach was used to develop salt-tolerant homozygous recombinants from diverse cross combinations, which led to the identification of the promising rice varieties IR51500-AC-17, IR51485-AC-1, IR51500-AC11-1, and AC6534-4 for salinity, AC6533-3 for sodicity, and AC6534-1 for dual tolerance (Singh et al. 1992; Singh and Mishra 1995; Senadhira et al. 2002). Rahman et al. (2010) developed DH lines from the crosses involving salt-tolerant IRRI-derived lines using anther culture and in a field study one line AC-1 was promising for cultivation in saline areas of Bangladesh.

3.3 Genomics Approach

Genome-mapping techniques are accelerating identification of exact position and function of individual genes controlling agronomic traits including tolerance to salinity. Striking similarities among the genomes of different crop species has been helpful in expanding the genetic variability of important traits for crop improvement. The scope and precision of current breeding programs are enhanced due to use of linked markers for selection of desirable alleles for the target traits.

3.3.1 QTLs for Components of Salinity Tolerance

Plant adaptation to unfavorable environments is governed by morphological, physiological, and unique genetic architecture. By integrating physiological and genetic strategies, we can obtain a better understanding of the molecular basis of crop adaptation thus paving the way toward a more targeted breeding approach for enhancing abiotic stress tolerance in crop plants. QTL mapping is revealing genetic components of salt tolerance for genetic improvement of existing varieties. QTLs controlling salinity tolerance related traits have been mapped in several mapping populations in major field crops: rice (Table 4.1), wheat (Table 4.2), and barley (Table 4.3). To date, there are over 10,000 mapped QTLs reported for rice and maize in the Gramene database (http://www.gramene.org).

Bonilla et al. (2002) identified a major QTL, designated *Saltol*, on chromosome 1, using an RIL population between the highly tolerant landrace Pokkali and sensitive IR29. The QTL accounts for about 45% of the variation for seedling and shoot Na^+/K^+ ratio. Ismail et al. (2007) mapped the *Saltol* QTL within 1.2 Mb, which is currently being introgressed into several popular salt-sensitive rice cultivars. But multiple alleles were identified at the *Saltol* locus when several Pokkali accessions and near-isogenic lines were analyzed (Thomson et al. 2010). Using an RIL population developed from the cross Co39 X Moroberekan, Haq et al. (2010) identified a major effect on QTL for leaf Na^+ concentration and K^+:Na^+ ratio on chromosome 1 which may harbor

Table 4.1 Molecular markers associated with quantitative trait loci for salt tolerance in rice

Molecular markers	QTLs	Cross	References
	Chr 1		
RM8094, RM10793, SKC1, RM493	Standard Evaluation System (SES) score	Pokkali X IR29	Alam et al. (2011)
C813-C86	Survival days of seedling	Nona Bokra X Koshihikari	Lin et al. (2004)
C1211-S2139	Shoot K^+ concentration	Nona Bokra X Koshihikari	
RM562-RM543, RM8086-RM8231	Na^+ uptake	Taromnahelli X Khazar	Sabouri and Sabouri (2008)
RM562-RM9	Shoot K^+ concentration	IR64 X Binam	Zang et al. (2008)
E12M51_1 to E12M37_1	Na^+ uptake, K^+ Conc., Na^+:K^+ ratio	IR4630 X IR15324	Koyama et al. (2001)
Est_2-RZ569A	Seedling salinity tolerance	Milyang 23 X Gihobyeo	Lee et al. (2007)
R117–R2417	Shoot length	Niponbare X Kasalath	Takehisa et al. (2004)
	Chr 2		
OSR17-RM211, RM530-RM250	Survival days of seedling	IR64 X Binam	Zang et al. (2008)
R418-R3393	Tiller number	Niponbare X Kasalath	Takehisa et al. (2004)
	Chr 3		
RM5639,RM5626, RM3867	Standard Evaluation System (SES) score Survival days of seedling	Pokkali X IR29	Alam et al. (2011)
RM231-RM175	Shoot K^+ concentration	IR64 X Binam	Zang et al. (2008)
RM81B-RM22	Shoot Na^+ concentration	IR64 X Binam	
RM231-RM175	Na^+ uptake	IR64 X Binam	
RM416-RM5626	K^+ uptake	Taromnahelli X Khazar	Sabouri and Sabouri (2008)
RM1022-RM6283	Na^+:K^+ uptake ratio	Taromnahelli X Khazar	
RM6832 RM7389	Shoot length	Taromnahelli X Khazar	
R1927-R250	Seedling salinity tolerance	Niponbare X Kasalath	Takehisa et al. (2004)
RG179-RZ596		Milyang 23 X Gihobyeo	Lee et al. (2007)
	Chr 4		
RM6659	Standard Evaluation System score (SES)	Pokkali X IR29	Alam et al. (2011)
RM261-E12M79_6	Na^+ conc., K^+ uptake, Na^+:K^+ ratio	IR4630 X IR15324	Koyama et al. (2001)
C891-C513	Root K^+ concentration	Nona Bokra X Koshihikari	Lin et al. (2004)
C1016	Shoot length	Niponbare X Kasalath	Takehisa et al. (2004)
	Chr5		
RM163	Standard Evaluation System (SES) score	Pokkali X IR29	Alam et al. (2011)
RM421-RM440	Seedling root dry weight	Taromnahelli X Khazar	Sabouri and Sabouri (2008)
C624-C1268	Shoot fresh weight	Niponbare X Kasalath	Takehisa et al. (2004)
	Chr 6		
RM20224	Standard Evaluation System (SES) score	Pokkali X IR29	Alam et al. (2011)
RM50-RM539	Shoot K^+ concentration	IR64 X Binam	Zang et al. (2008)
RM527-RM3	Shoot Na^+ concentration	IR64 X Binam	
RM3827-RM340	Na^+:K^+ uptake ratio	Taromnahelli X Khazar	Sabouri and Sabouri (2008)
C214-R2549	Survival days of seedling	Nona Bokra X Koshihikari	Lin et al. (2004)
R1167-R1608	Shoot fresh weight	Niponbare X Kasalath	Takehisa et al. (2004)
G1314-Rz413b	Na^+ conc., K^+ uptake	IR4630 X IR15324	Koyama et al. (2001)
	Chr 8		
RM4955-RM152	K^+ uptake	Taromnahelli X Khazar	Sabouri and Sabouri (2008)

(continued)

Table 4.1 (continued)

Molecular markers	QTLs	Cross	References
RM38-RM25	Survival days of seedling	IR64 X Binam	Zang et al. (2008)
	Chr 9		
RM1553-RM5702	Na^+ uptake	Taromnahelli X Khazar	Sabouri and Sabouri (2008)
R1751-R2638	Root Na^+ concentration	Nona Bokra X Koshihikari	Lin et al. (2004)
C506-C1263	Tiller number	Niponbare X Kasalath	Takehisa et al. (2004)
RM205, E12M31_1	K^+ uptake	IR4630 X IR15324	Koyama et al. (2001)
	Chr 10		
RM222	Standard Evaluation System (SES)score	Pokkali X IR29	Alam et al. (2011)
RM7545-RM4455	Na^+ uptake	Taromnahelli X Khazar	Sabouri and Sabouri (2008)
	Chr 11		
RM26063, RM224	Standard Evaluation System (SES) score	Pokkali X IR29	Alam et al. (2011)
RM120-RM181	Shoot K^+ concentration	IR64 X Binam	Zang et al. (2008)

Table 4.2 Molecular markers associated with quantitative trait loci for salt tolerance in wheat

Molecular markers	QTLs	Cross	References
	Chr 1		
Xm71p78.5	Grain yield	Chinese spring X SQ1	Quarrie et al. (2005)
XksuD1.2-Xcdo426	Seedling shoot dry weight	Opata85 X W7984	Ma et al. (2007)
Gbm1153-barc028a	Leaf symptoms	Berkut X Krichauff	Genc et al. (2010)
Barc028a-gwm164	Tiller number	Berkut X Krichauff	
wPt-4647-wmc147	Shoot K^+ conc.	Berkut X Krichauff	
	Chr 2		
XksuD22, XksuE16,Xgwm312	Shoot Na^+ conc.	Tamaroi X Line 149	Lindsay et al. (2004)
Xm86p65.1	Grain yield	Chinese spring X SQ1	Quarrie et al. (2005)
Xfba70.1-Xcdo447	Seedling dry weight	Opata85 X W7984	Ma et al. (2007)
Xcdo1281-Xfba106	Seedling shoot fresh weight	Opata85 X W7984	
Gwm102-wmc027	Leaf symptoms	Berkut X Krichauff	Genc et al. (2010)
Gwm095-cfa2263	Seedling biomass	Berkut X Krichauff	
wPt-3114-wmc170	Shoot Na^+ conc.	Berkut X Krichauff	
wmc272-barc349	Shoot Na^+ conc.	Berkut X Krichauff	
	Chr 3		
Xbarc042-Xgwm383	Grain yield	Chinese spring X SQ1	Quarrie et al. (2005)
Xglk683-Xtam61	Germination salt tolerance	Opata85 X W7984	Ma et al. (2007)
Xglk683-Xtam61	Germination radical dry weight	Opata85 X W7984	
Xfbb168-Xbcd147	Germination radical dry weight	Opata85 X W7984	
Xtam47-Xcdo460	Seedling salt injury	Opata85 X W7984	
Xfbb168-Xbcd147	Seedling shoot dry weight	Opata85 X W7984	
Xfbb168-Xbcd147	Seedling shoot fresh weight	Opata85 X W7984	
Xfbb117-Xfbb156	Seedling dry weight	Opata85 X W7984	
Xfbb156-Xfba220	Seedling fresh weight	Opata85 X W7984	
Xcdo1406-Xbcd288226	Chlorophyll content	Opata85 X W7984	
Gwm299-gwm247	Shoot K^+ conc.	Berkut X Krichauff	Genc et al. (2010)
Cfd223-cfd152	Shoot K^+ conc.	Berkut X Krichauff	
	Chr 4		
Xm92p78.8-Xpsr490.2Ss1	Grain yield	Chinese spring X SQ1	Quarrie et al. (2005)
Xgwm165.2	Grain yield	Chinese spring X SQ1	

(continued)

Table 4.2 (continued)

Molecular markers	QTLs	Cross	References
Xcdo1081-Xfbb226	Germination salt tolerance	Opata85 X W7984	Ma et al. (2007)
Xcdo1081-Xfbb226	Germination radical fresh weight	Opata85 X W7984	
Xbcd588-Xbcd129	Seedling fresh weight	Opata85 X W7984	
Xbcd588-Xbcd129	Seedling shoot fresh weight	Opata85 X W7984	
Xbcd588-Xbcd129	Seedling root fresh weight	Opata85 X W7984	
wPt-7062-gwm6	Tiller number	Berkut X Krichauff	Genc et al. (2010)
wPt-7062-gwm6	Seedling biomass	Berkut X Krichauff	
wPt-7919-wPt-0150	Shoot K^+ conc.	Berkut X Krichauff	
	Chr 5		
Xwg232.2-Xbarc074	Grain yield	Chinese spring X SQ1	Quarrie et al. (2005)
Xbarc044	Grain yield	Chinese spring X SQ1	
Xbcd1871-Xcdo749	Germination stage salt tolerance	Opata85 X W7984	Ma et al. (2007)
Xfbb12.2-Xfba127	Seedling salt injury	Opata85 X W7984	
Gwm304-gwm186	Leaf symptoms	Berkut X Krichauff	Genc et al. (2010)
wPt-1370-Vrn1A	Tiller number	Berkut X Krichauff	
wPt-1370-Vrn1A	Seedling biomass	Berkut X Krichauff	
wPt-1370-Vrn1A	Chlorophyll content	Berkut X Krichauff	
wPt-1370-Vrn1A	Shoot K^+ conc.	Berkut X Krichauff	
	Chr 6		
Xm87p78.5a	Grain yield	Chinese spring X SQ1	Quarrie et al. (2005)
Xfba345-Xglk479	Seedling salt injury	Opata85 X W7984	Ma et al. (2007)
Xfbb231.1-Xpsr106	Seedling root fresh weight	Opata85 X W7984	
Xfbb231.1-Xpsr106	Seedling root dry weight	Opata85 X W7984	
cfd287-cfd076a	Leaf symptoms	Berkut X Krichauff	Genc et al. (2010)
cfd287-cfd076a	Seedling biomass	Berkut X Krichauff	
cfd080-barc171	Shoot Na^+ conc.	Berkut X Krichauff	
	Chr 7		
Xpsp3094.1Xm68p78.6	Grain yield	Chinese spring X SQ1	Quarrie et al. (2005)
Xfba72-Xfba127	Germination stage radical dry weight	Opata85 X W7984	Ma et al. (2007)
Xfba311-Xbcd178	Seedling fresh weight	Opata85 X W7984	
Xwg380-XksuD2	Chlorophyll content	Opata85 X W7984	
wPt-5153-ksm019	Leaf symptoms	Berkut X Krichauff	Genc et al. (2010)
wPt-5153-ksm019	Shoot K^+ conc.	Berkut X Krichauff	
gwm282-wPt-0961	Seedling biomass	Berkut X Krichauff	
wPt-4744-gwm282	Shoot Na^+ conc.	Berkut X Krichauff	

multiple genes for salt tolerance whose impact varies with stress duration, leaf age, and leaf type. Lang et al. (2001) reported a micro-satellite marker, RM223 on rice chromosome 8, associated with salt tolerance at both vegetative and reproductive stages. Recently, Alam et al. (2011) identified salinity-related QTLs at seedling stage using SES scores and identified *Saltol* as well as non-*Saltol* related QTLs at seedling-stage in Pokkali, which may assist in QTL pyramiding and marker-assisted breeding programs. A study by Lin et al. (2004) employing the tolerant *indica* landrace Nonabokra with the susceptible *japonica* Koshihikari, identified several large-effect QTLs, including the *SKC1* QTL and a QTL for shoot Na^+ concentration. While the salt-tolerant landraces Pokkali and Nonabokra were routinely used in the past for breeding, the level of tolerance attained by new lines is always lower than traditional donors (Gregorio et al. 2002), and the existing tolerant varieties seem to be superior in only few traits associated with tolerance. The QTL *SKC1*, originally detected by its effect on K^+ concentration, was cloned by map-based cloning and was

Table 4.3 Molecular markers associated with quantitative trait loci for salt tolerance in barley

Molecular markers	QTLs	Cross	References
	Chr 1H		
Glb1-ABC160	Seedling salt tolerance	Steptoe X Morex	Mano and Takeda (1997)
ABC160-His4A	Seedling salt tolerance	Steptoe X Morex	
WG789B-ABR337	Seedling salt tolerance	Steptoe X Morex	
Drun8-ABC261	Germination salt tolerance	Harrington X TR306	
bPb-2240-bPb-0631	Spikes per plant	CM72 X Gairdner	Xue et al. (2009)
	Chr 2H		
ABG459-Pox	Seedling salt tolerance	Steptoe X Morex	Mano and Takeda (1997)
ABC152D-Rrn5S1	Seedling salt tolerance	Steptoe X Morex	
His3C-ABC152D	Seedling salt tolerance	Steptoe X Morex	
P21M12d	Shoot dry weight	Derkado X B83-12/21/5	
bPb-6088-bPb-4377	Dry weight per plant	CM72 X Gairdner	Ellis et al. (2002)
Bmag0381-bPb-0827	Grain number per plant	CM72 X Gairdner	Xue et al. (2009)
bPb-3536-bPb-1103	Shoot Na^+ Conc.	CM72 X Gairdner	
	Chr 3H		
bPb-0049-bPb-4564	Plant height	CM72 X Gairdner	Xue et al. (2009)
bPb-7989-bPb-4660	Spikes per plant	CM72 X Gairdner	
	Chr 4H		
MWG634-WG622	Germination salt tolerance	Steptoe X Morex	Mano and Takeda (1997)
P17M62f	Shoot dry weight	Derkado X B83-12/21/5	
bPb-1278-bPb-3512	Tiller number	CM72 X Gairdner	Ellis et al. (2002)
bPb-1278-bPb-3512	Spikes per line	CM72 X Gairdner	Xue et al. (2009)
bPb-0130-bPb-8437	Spikes per plant	CM72 X Gairdner	
	Chr 5H		
WG889-ABC324	Germination salt tolerance	Steptoe X Morex	Mano and Takeda (1997)
ABC309-MWG632	Germination salt tolerance	Harrington X TR306	
iEst9-WG908	Seedling salt tolerance	Steptoe X Morex	
WG364-MWG514B	Seedling salt tolerance	Steptoe X Morex	
CDO504-ABG712	Seedling salt tolerance	Harrington X TR306	
TubA3-MWG740	Seedling salt tolerance	Harrington X TR306	
Bmag337	Shoot dry weight	Derkado X B83-12/21/5	Ellis et al. (2002)
	Chr 6H		
ABG387B-ABG458	Germination salt tolerance	Steptoe X Morex	Mano and Takeda (1997)
BCD340E-ksuD17	Seedling salt tolerance	Steptoe X Morex	
bPb-6421-bPb-3921	Spikes per line	CM72 X Gairdner	
bPb-7323-bPb-2751	Grain yield	CM72 X Gairdner	Xue et al. (2009)
bPb-8889-bPb-7323	Na^+: K^+ ratio	CM72 X Gairdner	
	Chr 7H		
bPb-1209-bPb-6821	Spikes per line	CM72 X Gairdner	Xue et al. (2009)
P40M38b	Tiller number	Derkado X B83-12/21/5	Ellis et al. (2002)

identified as the sodium transporter *OsHKT8* (Ren et al. 2005). Combining superior alleles underlying the component traits could potentially result in higher levels of tolerance. Based on QTL-linked marker profile, Manneh et al. (2007) identified superior salt-tolerant genotypes of rice to improve selection efficiency while selecting for yield in stress environments. Wang et al. (2011) mapped 16 QTLs for salt tolerance of rice at the seed germination stage and four QTLs with major effect could be useful to improve stand establishment under saline areas. QTLs associated with tolerance at various developmental stages will be needed for more stable performance in salt affected areas.

Improving salt tolerance and productivity wheat is a major challenge in breeding program. In studies involving hexaploid, tetraploid, and diploid types, it was suggested that the D genome of wheat carries gene *Kna1* that controls the relative concentration of K^+ and Na^+ in the shoots of plants grown in saline hydroponic culture (Shah et al. 1987; Gorham et al. 1987, 1997; Gorham 1994). This gene was mapped on to the chromosome 4DL (Dvorak et al. 1994) and then fine mapped as a single gene (Dubcovsky et al. 1996). Ma et al. (2007) used a wheat RIL population derived from the cross Opata85 × W7984 (international mapping population) and identified 47 QTLs based on salt tolerance index, salt injury index, biomass, shoot length, root length, chlorophyll content, and proline content on all wheat chromosomes except 1B, 1D, 4B, 5D, and 7D. Ten of these QTLs were effective during germination stage and 37 QTLs were important at the seedling stage. In another study, genetic analysis of wheat Line 149 identified two major genes *Nax1* and *Nax2* for Na^+ exclusion (Munns et al. 2003). *Nax1* was located on chromosome 2A by QTL analysis (Lindsay et al. 2004) and has been fine mapped as an Na^+ transporter of the HKT gene family, *HKT7* (Huang et al. 2006). *Nax2* located on chromosome 5A was identified as *HKT8* (Byrt et al. 2007). *Nax1* removes Na^+ from the xylem in roots and the lower parts of leaves, the leaf sheaths, while *Nax2* removes Na^+ from the xylem only in the roots (James et al. 2006). *Nax2* has the same phenotype as *Kna1*, the QTL for Na^+ exclusion and enhanced K^+/Na^+ selectivity in bread wheat, *T. aestivum* (Dvorak et al. 1994). *Nax2* was shown to be homologous to *Kna1* in *T. aestivum*, namely *TaHKT8* (Byrt et al. 2007). The *HKT* gene family encodes transporters in the plasma membrane that mediate the uptake of Na^+ or K^+ from the apoplast (Hauser and Horie 2010). They are important for cellular Na^+ and K^+ homeostasis, and their expression in the stele, particularly in the xylem parenchyma cells lining the xylem vessels, helps in retrieving Na^+ from the transpiration stream and thus contributing to Na^+ exclusion from leaves (Munns and Tester 2008; Hauser and Horie 2010). James et al. (2011) analyzed a population derived from crossing of a tetraploid durum wheat (*Triticum turgidum* ssp. durum) and hexaploid bread wheat (*Triticum aestivum*) by marker-assisted selection (MAS) for hexaploid plants containing either *Nax1*, *Nax2*, or both *Nax1* and *Nax2*. *Nax1* line decreased the leaf blade Na^+ concentration by 50%, whereas *Nax2* line decreased it by 30%, and both genes together decreased it by 60%. High Na^+ sheath:blade ratio in *Nax1* lines conferred extra advantage under waterlogged and saline conditions. The effect of *Nax2* on lowering the Na^+ concentration in bread wheat was surprising as this gene is already present in bread wheat, putatively at the *Kna1* locus. The results indicate that both *Nax* genes have the potential to improve the salt tolerance of bread wheat. A list of salinity-related QTLs for wheat and barley is given in Tables 4.2 and 4.3, respectively.

3.3.2 Marker-Assisted Selection

The application of QTL mapping provided the means to genetically dissect tolerance traits into discrete QTLs that can then be pyramided into high-yielding varieties using marker-assisted selection (MAS). Marker-assisted selection is the use of molecular markers linked to useful traits to select individuals with desirable genetic makeup during the variety development process. It provides a dramatic improvement in the efficiency with which breeders can select plants with desirable combination of genes. DNA markers should enhance the recovery rate of the isogenic recurrent genome after hybridization and facilitate the introgression of quantitative trait loci necessary to increase stress tolerance (Fig. 4.4). Molecular marker techniques were used successfully to transfer alleles of interest from wild relatives into commercial cultivars (Tanksley and McCouch 1997). Use of permanent mapping populations, such as recombinant inbred lines (RILs) or chromosomal segment substitution lines (CSSLs) enables testing stress-tolerant traits in replicated experiments across different environments, which can help in differentiating the QTLs based on their effectiveness at different stress levels. Once important QTL targets are identified, it can be used as single introgressions in a set of near-isogenic lines (NILs), which will help in identifying the complexity of different traits by limiting the variation between lines to focus only on the locus of interest.

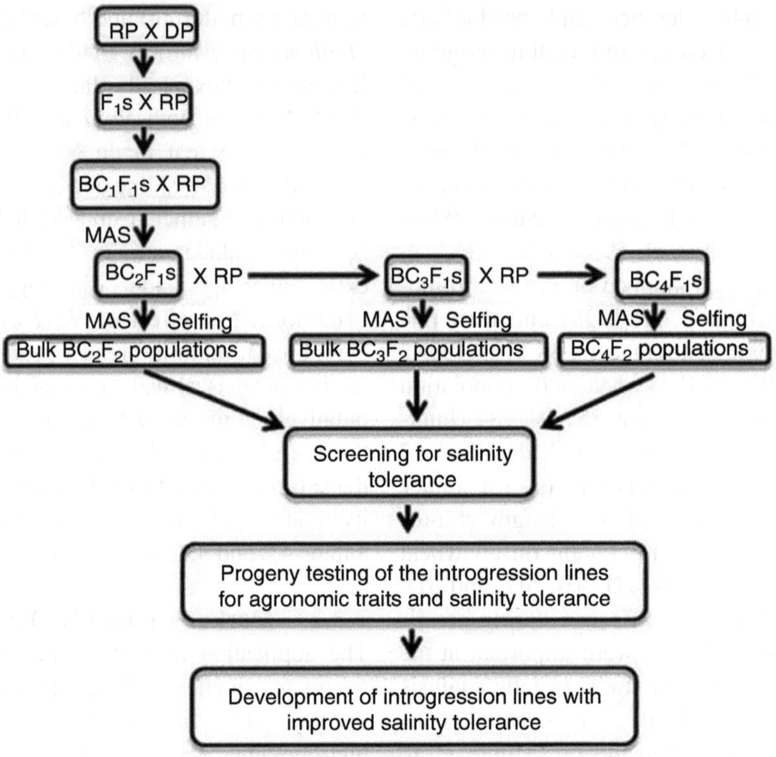

Fig. 4.4 A schematic representation of marker-assisted backcross breeding procedure for development of introgression lines with enhanced salt tolerance. *RP* recurrent parent, *DP* donor parent, *MAS* Marker assisted selection

The recent advances in genomics have paved the way for clear and reliable methods for MAS in plants starting from QTL identification, NIL development, and fine-mapping to transferring the QTL into popular varieties using a precise marker-assisted backcrossing (MABC) strategy (Collard et al. 2005, 2008 Collard and Mackill 2008;). MABC involves the manipulation of genomic regions involved in the expression of particular traits of interest through DNA markers, and combines the power of a conventional backcrossing program with the ability to differentiate parental chromosomal segments (Fig. 4.4).

3.4 Transgenic Approach

Progress in genomics is instrumental in discovery and characterization of large number of salt stress-related candidate genes offering unique opportunity for exploiting transgenic technology to enhance salinity tolerance in crop plants. Development of transgenic plants with improved tolerance to abiotic stresses has brought some hope for sustainable agriculture under harsh environmental conditions. Genetic engineering is an attractive option when genes of interest are present in cross barrier species, distant relatives, or non-plant sources (Bhatnagar-Mathur et al. 2008). It is also faster to introduce beneficial genes than the conventional or molecular breeding. Due to the complexity associated with salt tolerance mechanism, effort should be made to introduce and fine tune multiple genes with synergistic effect under suitable stress inducible promoters for controlling their expression at a specific time, in a specific organ, or under specific conditions of stress.

3.4.1 Discovery of Candidate Genes

When a plant is subjected to abiotic stresses, expressions of a number of genes are changed, resulting in altered levels of several proteins and metabolites inside the cell. Altered expression of these genes may be responsible for conferring protection or susceptibility to abiotic stresses. Candidate genes for salinity stress tolerance have been identified from prokaryotic extremophiles, lower eukaryotes, tolerant wild relatives, landraces, cultivars, and T-DNA mutant plants using high throughput techniques such as differential display polymerase chain reaction (DD-PCR), cDNA-amplified fragment length polymorphism (cDNA-AFLP), suppression subtractive hybridization (SSH), serial analysis of gene expression (SAGE), DNA microarray, and two-dimensional gel electrophoresis.

Differentially expressed genes under osmotic stress have been identified by DDRT-PCR in barley, sunflower, and pak-choi (Wei et al. 2001; Liu and Baird 2003; Qiu et al. 2009). Wei et al. (2001) identified a gene encoding the barley vacuolar ATPase subunit B (BSVAP) from salt-sensitive barley cultivar Maythorpe. In barley roots, this enzyme may be involved in the sequestration of Na^+ and Ca^{2+} ions into the vacuole, since the proton gradient produced by the ATPase is used by Na^+/H^+ and Ca^{2+}/H^+ antiporters to drive the uptake of Na^+ and of Ca^{2+} (Taiz 1992). Liu and Baird (2003) isolated 17 cDNA clones that are differentially expressed in sunflower under drought or salinity and 13 of these cDNAs were confirmed by quantitative RT-PCR to be expressed differentially in response to osmotic stress. Qiu et al. (2009) generated 101 cDNA fragments differentially expressed under salt stress in Pak-choi (*Brassica campestris* L. ssp. chinensis (L.) Makino var. communis Tsen et Lee), and found seven cDNA sequences highly homologous to some known expression genes or the genes related to the signaling pathways in plants under different abiotic stress.

Kumari et al. (2009) applied SSH technique in a salt-sensitive rice cultivar IR64 and a salt-tolerant landrace Pokkali and isolated 1,194 salinity-regulated cDNAs. Gene expression analysis of selected genes using macroarrays and Northern blots indicated that salinity tolerance of Pokkali may be due to constitutive overexpression of many salt responsive genes which are stress inducible in IR64. Few clones mapped near *Saltol* locus on chromosome 1. Using global gene expression analyses, Walia et al. (2005, 2007) observed gene expression changes in large number of genes under salinity stress in salt-sensitive genotypes compared with the tolerant lines. A total of 164 ESTs were upregulated in the tolerant line (FL478) under salinity stress during vegetative stage (Walia et al. 2005), whereas during panicle initiation stage, 292 genes were upregulated and 346 genes were downregulated in salt-sensitive *japonica* M103. Interestingly, in the salt-tolerant *japonica* only 54 were upregulated and 54 downregulated (Walia et al. 2007). Cotsaftis et al. (2011) carried out root-specific transcriptome studies in contrasting genotypes of rice (FL478, IR29, Pokkali and IR63731) under salinity stress and identified several gene families with known links to salinity tolerance.

Chen et al. (2003) used cDNA-AFLP to analyze differentially expressed genes in wheat under salinity stress and found a large number of gene fragments related to salt stress. One of the cDNA encoding glycogen synthase kinase-shaggy kinase (*TaGSK1*) was induced by NaCl stress as a part of the signal transduction component. This technique has also been used to isolate differentially expressed ESTs under salinity stress in soybean (Aoki et al. 2005) and *Spartina alterniflora* (Baisakh et al. 2008). An expressed sequence tag (EST) analysis in a grass halophyte *Spartina alterniflora* at early stages of salt stress (Baisakh et al. 2008) produced 1,227 quality ESTs of which 27% of ESTs represented genes for stress response. Transcript abundance analysis of eight known genes of various metabolic pathways and nine transcription factor genes showed temporal and tissue-dependent variation in expression under salinity stress.

A combination of genetic mapping and bulked transcriptome profiling was used by Pandit et al. (2010) resulting in identification of two genes, an integral transmembrane protein DUF6 and a chloride cotransporter, which colocalized within the QTL interval. Walia et al. (2006) performed expression analysis of genes in barley (*Hordeum vulgare* L.) during salinity stress at the seedling stage of barley using microarray. Genes, differentially regulated by salinity, were also associated with various abiotic stresses such as low temperature, heat stress, and drought stress. Expression level of genes related to jasmonic acid (JA) biosynthesis and jasmonic acid-responsive genes (JRGs) was found to be differentially regulated. Jasmonic acid may function as a "master switch" for stress-induced signaling pathway leading to changes in gene expression (Wasternack et al. 1998). It has been reported that JA and ABA affect gene expression in a synergistic manner through more or less independent signaling pathways (Ortel et al. 1999). This indicates that ABA signaling pathway is also activated along with JA pathway in response to salinity stress in barley. Ozturk et al. (2002) analyzed drought and salinity-induced responses in barley (*Hordeum vulgare* L. cv. Tokak) transcriptome using cDNA microarray and reported alteration in 5% of the genes under salinity stress compared to 15% under drought stress. Upregulation under both drought and salt stress was restricted to ESTs for metallothionein-like and LEA proteins, while increases in ubiquitin-related transcripts characterized salt stress.

Yan et al. (2005) investigated the salt stress-responsive proteins in the root of rice (*Oryza sativa* L. cv. Nipponbare) in proteomics approach using 3-week-old seedlings treated with 150 mM NaCl for 24, 48 and 72 h. Two-dimensional gel electrophoresis showed 34 upregulated and 20 downregulated proteins. Mass spectrometry analysis could identify 12 spots representing 10 different proteins involved in regulation of carbohydrate, nitrogen, and energy metabolism, reactive oxygen species scavenging, mRNA and protein processing, and cytoskeleton stability. Comparative proteomic analysis in two contrasting hybrid rice variety by Ruan et al. (2011) led to identification of new components involved in salt-stress signaling. One protein that was upregulated during salt stress was homologous to cyclophilin 2 (*OsCYP2*). Chitteti and Peng (2007) reported the differential expression of phosphoproteome under salinity stress in the root of rice. They identified 17 differentially upregulated and 11 differentially downregulated putative phosphoproteins. Witzel et al. (2009) conducted two-dimensional gel electrophoresis using a series of hydroponics-based salinity stress experiments in contrasting genetic mapping parents of barley cvs Steptoe and Morex. The proteome analysis of roots from both genotypes revealed cultivar-specific and salt stress-responsive protein expression. Twenty-six proteins could be identified by mass spectrometry. Among those, two proteins involved in the glutathione-based detoxification of reactive oxygen species (ROS) were more abundant in the tolerant genotype.

The identification of various abiotic stress-specific changes in gene expression has been achieved by comparing gene expression in non-induced and salinity stress-induced tissues or by comparing contrasting cultivars (Sahi et al. 2003; Baisakh et al. 2008; Karan et al. 2009; Kumari et al. 2009). In one such study, it was noted that less number of genes were induced under salt stress in *Thellungiella halophila* (a salt-tolerant relative of *Arabidopsis*) in comparison to *Arabidopsis* (Inan et al. 2004). This indicated that the stress tolerance of *Thellungiella halophila* may be due to constitutive overexpression of a few salt tolerance related genes which are stress inducible in *Arabidopsis* (Taji et al. 2004). Sengupta and Majumder (2009) analyzed changes in leaf protein expression under salt stress in the wild halophytic rice *Porteresia coarctata* and salt-sensitive *Oryza sativa* and identified 16 proteins involved in osmolyte synthesis, photosystem functioning, RubisCO activation, cell wall synthesis, and chaperone functions. These differentially regulated genes in different plant species may serve as candidates to improve salinity tolerance in crop plants using transgenic approach.

3.4.2 Transformation in Crop Plants

Most common genes used for genetic engineering of stress-tolerant plants include transcription factors, signal transduction genes, water channel proteins, ion transporters, detoxifying genes, molecular chaperones, dehydrins, and osmoprotectants (Table 4.4).

Calcium acts as a secondary messenger in various signal transduction pathways in plants. Xu et al. (2011) isolated a calcium-binding multi-stress-responsive gene *OsMSR2* from rice which

Table 4.4 Examples of transgenic intervention in crop plants to improve salt tolerance

Gene	Gene product	Gene source	Target plant	Effect of the transgene	Reference
HvCBF4	CBF transcription factor	*Hordeum vulgare*	Rice	Enhanced salinity, drought, and cold tolerance	Oh et al. (2007)
OSNAC5	Transcription factor	*Oryza sativa*	Rice	Enhanced salt tolerance	Takasaki et al. (2010)
ABP9	Transcription factor	*Zea mays*	Arabidopsis	Enhanced drought, high salt, freezing temperature, and oxidative stresses tolerance	Zhang et al. (2011b)
TaMYB2A	Transcription factor	*Triticum aestivum*	*Arabidopsis*	Enhanced salt, drought, and freezing tolerance	Mao et al. (2011)
TaSRG	Transcription factor	*Triticum aestivum*	Arabidopsis	Improved salt tolerance	He et al. (2011)
CgNHX1	Vacuolar Na^+/H^+ exchanger,	*Chenopodium glaucum*	Rice	Improved salt tolerance	Li et al. (2008)
PgNHX1	Vacuolar Na^+/H^+ antiporter	*Pennisetum glaucum*	Rice	Enhanced salt tolerance	Verma et al. (2007)
AtNHX1	Na^+/H^+ antiporter	*Arabidopsis thaliana*	Maize	Improved salt tolerance	Li et al. (2010b)
ThNHX1	Na^+/H^+ antiporter	*Thellungiella halophila*	Arabidopsis	Improved salt tolerance	Wu et al. (2009)
SsNHX2	Na^+/H^+ antiporter	*Suaeda salsa*	Arabidopsis	Improved salt tolerance	Li et al. (2009)
AtNHX1	Vacuolar Na^+/H^+ antiporter	*Arabidopsis*	Groundnut	Improved salt and drought tolerance	Asif et al. (2011)
OsKAT1	Shaker family K^+ channel	*Oryza sativa*	Rice	Enhanced salt tolerance	Obata et al. (2007)
PutHKT2;1	K^+ transporter	*Puccinellia tenuiflora*	Arabidopsis	Improved salt tolerance	Ardie et al. (2009)
ThIPK2	Inositol poly-phosphate kinase	*Thellungiella halophila*	*Brassica napus*	Improved abiotic stress tolerance	Zhu et al. (2009)
ZmSIMK1	Mitogen activated protein kinase	*Zea Mays*	Arabidopsis	Increased tolerance to salt stress	Gu et al. (2010)
TaSnRK2.4	Sucrose non-fermenting 1 type serine/threonine protein kinase	*Triticum aestivum*	Arabidopsis	Enhanced tolerance to drought, salt, and freezing stresses	Mao et al. (2010)
ZmMKK4	Mitogen activated protein kinase kinase kinase	*Zea mays*	*Arabidopsis*	Salt and cold tolerance	Kong et al. (2011)
ZmALDH2.2A1	Aldehyde dehydrogenase	*Zea mays*	Tobacco	Improved tolerance to salt and drought	Huang et al. (2008)
OsBADH	Betaine aldehyde dehydrogenase	*Oryza sativa* (*Indica*)	Rice (*Japonica*)	Improved salt tolerance	Hasthanasombut et al. (2011)
BADH	Betaine aldehyde dehydrogenase	*Suaeda liaotungensis*	Maize	Improved salt tolerance	Wu et al. (2008)
H^+–PPase	H(+)-pyrophosphatase	*Thellungiella halophila*	Cotton	Improved salt tolerance	Lv et al. (2008)
PcINO1	Myo-inositol 1-phosphate synthase	*Porteresia coarctata*	Brassica, Rice	Improved salt tolerance	Das-Chatterjee et al. (2006)
MIPS	Myo-inositol 1-phosphate synthase	*Spartina alterniflora*	Tobacco, Rice	Improved salt tolerance	Baisakh et al. (2009)
SOD	Cu/Zn superoxide dismutase	*Avicennia marina*	Rice	Improved salt tolerance	Prashanth et al. (2008)
ThCBL9	Calcineurin B-like protein	*Thellungiella halophila*	Arabidopsis	Improved salt and osmotic stress tolerance	Sun et al. (2008)
GST	Glutathione S-transferase	*Suaeda salsa*	Rice	Abiotic stress resistance	Zhao and Zhang (2006)
SbGSTU	Glutathione S-transferase	*Salicornia brachiata*	Tobacco	Enhanced salt tolerance	Jha, Sharma and Misra (2010)
GLYII	Glyoxalase II	*Oryza sativa*	Rice	Higher salt tolerance	Singla-Pareek et al. (2008)
PcSrp	Serine-rich-protein	*Porteresia coarctata*	Finger millet	Improved salt tolerance	Mahalakshmi et al. (2006)

is strongly upregulated by a wide spectrum of stresses, including cold, drought, and heat in different tissues at different developmental stages of rice. Expression of *OsMSR2* gene into *Arabidopsis* conferred enhanced tolerance to high salt and drought through ABA mediated pathway. Calcium-dependent protein kinases (CDPKs) regulate downstream components in calcium signaling pathway. Rice has 29 genes for CDPKs and constitutes a large multigene family. Asano et al. (2011) used mini-scale full-length CDNA overexpressor (FOX) gene hunting system and generated 250 independent transgenic rice plants overexpressing individual rice CDPKs. Transgenic rice plants over expressing *OsCDPK21-FOX* had more survival rate under salinity stress than wild type plants and also found to be involved in the positive regulation of ABA and salt stress signaling pathways.

The members of NAM, ATAF, and CUC (NAC) transcription factor family has been used to improve abiotic stress tolerance through genetic engineering. Transgenic rice plants constitutively overexpressing *OsNAC6* showed tolerance to dehydration, salinity, and blast disease (Nakashima et al. 2007). However, the plants exhibited growth retardation and low reproductive yields. By utilizing stress-inducible promoters, such as the native *OsNAC6* promoter, it may be possible to reduce negative impact on plant growth and reproduction. *SNAC1* (Stress responsive NAC1), which is expressed in guard cells under drought stress, improved both salt and drought tolerance in transgenic rice plants with upregulation of many stress-responsive genes (Hu et al. 2006). *OsNAC5* is induced by abiotic stresses such as salinity, drought, cold, abscisic acid, and methyl jasmonic acid and interacts with *OsNAC5, OsNAC6,* and *SNAC1.* Transgenic plants overexpressing *OsNAC5* had improved tolerance to high salinity compared to control plants (Takasaki et al. 2010).

Li et al. (2010a) isolated and characterized a basic helix-loop-helix (bHLH) protein gene *OrbHLH001* from wild rice (*Oryza rufipogon*), which encodes an ICE1-like protein containing multiple homopeptide repeats and is induced by salt stress in the shoots. Transgenic *Arabidopsis* expressing *OrbHLH001* enhanced salt and freezing tolerance independent of a CBF/DREB1 cold-response pathway. Oh et al. (2005) reported that *DREB1A* and *ABF3* rice transgenic lines did not show any stunted growth despite constitutive expression. However, transgenic plants exhibited drought and salinity stress tolerance but low level of freezing stress tolerance. DREB homologs have been isolated from rice: *OsDREB1A* and *OsDREB1B* are cold inducible, *OsDREB2A* is induced by dehydration and salinity stress (Dubouzet et al. 2003). Over-expression of the *OsDREB1* gene in rice and *Arabidopsis* correlated with improved salt, drought, and low temperature tolerance (Ito et al. 2006). *OsDREB1F* is one of the most upregulated genes isolated from upland rice. The transgenic plants harboring *OsDREB1F* gene led to enhanced tolerance to salt, drought, and low temperature in both rice and *Arabidopsis* (Wang et al. 2008). The HARDY (HRD) gene, an AP2/ERF-like transcription factor from *Arabidopsis* enhanced drought resistance and salt tolerance, accompanied by an enhancement in the expression of abiotic stress associated genes (Karaba et al. 2007).

The Na^+/H^+ antiporters play an important role in the maintenance of cellular ion homoeostasis, cytoplasmic pH regulation, and cell turgor leading to salt tolerance in plants (Horie and Schroeder 2004). Ohta et al. (2002) used a *NHX1* gene from a halophyte *Atriplex gmelini* to generate transgenic rice lines which showed eight-fold higher activity of this gene compared to the wild rice. These transgenic lines survived in 300 mM NaCl for 3 days but the wild types failed to survive. Over-expression of vacuolar Na^+/H^+ antiporter gene *OsNHX1* enhanced salt tolerance in rice (Fukuda et al. 2004), whereas over-expression of the same gene from wheat together with H^+-pyrophosphatase (TVP1) improved both salt and drought tolerance in *Arabidopsis* (Brini et al. 2007). A field trial of wheat transgenic lines expressing *AtNHX1* gene exhibited higher grain yield, heavier and larger grains in saline conditions and there was less Na^+ and more K^+ accumulation in leaves in transgenic lines compared with the non-transgenic lines (Xue et al. 2004).

Garg et al. (2002) overexpressed *E. coli* trehalose bisynthetic genes (*otsA* and *otsB*) under the control of either tissue-specific or stress responsive promoter in transgenic rice plants which accumulated 3–10 times more trehalose and showed better growth, less photo-oxidative damage and favorable mineral balance under salt, drought, and low temperature stress conditions. Transgenic rice overexpressing *OsCYP2* encoding a cyclophilin 2 protein showed increased salinity tolerance with lower levels of lipid peroxidation products and higher activities of antioxidant enzymes than wild type seedlings (Ruan et al. 2011).

Halophytes represent an ideal target for understanding the genetic and molecular basis of their adaptation in saline environments (Subudhi and Baisakh 2011). Several types of genes belonging to different metabolic functions have been identified and used for overexpression into glycophytic plants to enhance salinity stress tolerance. Ohta et al. (2002) reported enhanced salinity tolerance in transgenic rice using the vacuolar Na^+/H^+ antiporter (*AgNHX1*) from *Atriplex gmelini*. Similarly, transgene expression of Na^+/H^+ antiporter (*SsNHX1*), *SsNHX2* (an alternative splicing variant of *SsNHX1*) from *Suaeda salsa* in *Arabidopsis* resulted into higher salt tolerance as well as vigorous growth of transgenics along with higher fresh, dry weights, and Na^+ and K^+ accumulation compared with the wild type plants (Li et al. 2009). *Tau* class glutathione transferases genes are specific to plant and important for protecting plants against oxidative damage. Expression of *SbGSTU* from a halophyte *Salicornia brachiata* in tobacco led to suppressed growth of transgenic seedlings at higher salt, but these transgenic seedlings had continuous growth significantly better than WT seedlings below 300 mM NaCl (Jha et al. 2010). Jithesh et al. (2006) isolated a cDNA (*Sod1*) encoding a cytosolic Cu/ZnSOD that accumulated during oxidative stress in the mangrove species, *Avicennia marina*. Transgenic Pusa Basmati-1 rice plants with *Sod1* were more tolerant to oxidative, salinity (150 mM NaCl) and drought stresses indicating mangrove as a potential source for other abiotic stress tolerance candidate genes (Prashanth et al. 2008). Majee et al. (2004) isolated and characterized a novel salt-tolerant gene L-myo-inositol 1-phosphate synthase from halophytic rice, *Porteresia coarctata*. Incorporating this gene into rice and other plant species increased synthesis of inositol under salinity stress and enhanced tolerance to salinity stress (Das-Chatterjee et al. 2006). Jacobs et al. (2011) identified a sodium pumping ATPase gene *PpENA1* from the moss *Physcomitrella patens*, which is not found in higher plants. Expression of *PpENA1* gene increased the salinity tolerance of transgenic rice.

4 Conclusions and Future Perspective

Salinity stress causes extensive crop losses in many parts of the world due to lack of salt tolerance in major field crops. Enhancing tolerance to salinity in crops will be an important goal of plant breeders in future to ensure food supply for the growing world population (Flowers 2004). Wide range of variation in level of salt tolerance found in halophytes, wild species, and crop germplasm clearly demonstrates the genetic basis of salt tolerance. Although it is widely recognized that the genetic and physiological basis of salt tolerance in plants is inherently complex due to involvement of multigene controlled traits or mechanisms, lack of thorough understanding of these mechanisms and their contribution toward salt tolerance is a major limitation to develop salt-tolerant plants. Additionally, lack of proper screening methods, low heritability, and low selection efficiency of component traits, and limited knowledge of the interactions among the genotype, plant growth stage, and the timing of stress also contributed to slow progress in breeding salinity-tolerant crops (Flowers and Flowers 2005; Ismail et al. 2007).

Conventional breeding technologies coupled with molecular genetic analysis particularly the QTL mapping studies are revealing important genetic components to enhance salt tolerance in field crops. Salt-tolerant germplasm have been identified and developed by breeding, mutation, and tissue culture techniques. QTL mapping and marker-assisted breeding will be increasingly employed with the effort for systematic dissection and utilization of natural variation in the available germplasm for improving crop performance in saline environments (Collins et al. 2008). Particularly, the wild relatives of crops and landraces hold enormous promise to mine the superior alleles for enhancing crop adaptation under salinity (Tanksley and McCouch 1997; Feuillet et al. 2008).

A large number of candidate genes involved in different salt tolerance mechanisms from diverse sources have been identified. It is expected that modern genomics tools will continue to be employed in model organisms such as Arabidopsis and rice, salt-tolerant crop accessions and wild relatives, and halophytes to provide genetic clues for plant adaptation to salinity through identification and verification of candidate genes (Collins et al. 2008; Subudhi and Baisakh 2011). Genomic technology is expected to contribute significantly toward discovery of useful candidate genes for various component traits which can be targeted to improve elite cultivars using transgenic pyramiding (Takeda and Matsuoka 2008). The utility of transgenic technology can be further enhanced through discovery and exploitation of stress inducible promoters which could enhance salt tolerance with minimum undesirable pleiotropic effect on crop growth and yield. It is conceivable that designing crop genotypes with improved salinity tolerance may be difficult but not impossible. However, collaboration of geneticists, molecular breeders, physiologists, and genomicists will be needed to implement an integrated approach to discover, test, and introgress the superior alleles to enhance salt tolerance in major food crops.

Acknowledgments The financial support for this study from the United States Department of Agriculture-CSREES is gratefully acknowledged. This manuscript is approved for publication by the Director of Louisiana Agricultural Experiment Station, USA as manuscript number 2011-306-5892.

References

Akoi A, Kanegami A, Mihara M, Kojima T, Shiraiwa M, Takahara H (2005) Molecular cloning and characterization of a novel soybean gene encoding a leucine-zipper-like protein induced to salt stress. Gene 356:135–145

Alam R, Sazzadur Rahman M, Seraj ZI, Thomson MJ, Ismail AM, Tumimbang-Raiz E, Gregorio GB (2011) Investigation of seedling-stage salinity tolerance QTLs using backcross lines derived from *Oryza sativa* L.Pokkali.PlantBreed.doi:10.1111/j.1439-0523.2010.01837.x

Ali AJ, Xu JL, Ismail AM, Fu BY, Vijayakumar CHM, Gao YM, Domingo J, Maghirang R, Yu SB, Gregorio G, Yanaghihara S, Cohen M, Carmen B, Mackill D, Li ZK (2006) Hidden diversity for abiotic and biotic stress tolerances in the primary gene pool of rice revealed by a large backcross breeding program. Field Crops Res 97:66–76

Ardie SW, Xie LN, Takahashi R, Liu SK, Takano T (2009) Cloning of a high-affinity K^+ transporter gene *PutHKT2, 1* from *Puccinellia tenuiflora* and its functional comparison with *OsHKT2, 1* from rice in yeast and Arabidopsis. J Exp Bot 60:3491–3502

Arzani A, Mirodjagh SS (1999) Response of durum wheat cultivars to immature embryo culture, callus induction and in vitro salt stress. Plant Cell Tiss Org Cult 58:67–72

Asano T, Hakata M, Nakamura H, Aoki N, Komatsu S, Ichikawa H, Hirochika H, Ohsugi R (2011) Functional characterisation of *OsCPK21*, a calcium-dependent protein kinase that confers salt tolerance in rice. Plant Mol Biol 75:179–191

Asch F, Dingkuhn M, Dörffling K, Miezan K (2000) Leaf K^+/Na^+ ratio predicts salinity induced yield loss in irrigated rice. Euphytica 113:109–118

Ashraf M, Harris PJC (2004) Potential biochemical indicators of salinity tolerance in plants. Plant Sci 166:3–16

Asif MA, Zafar Y, Iqbal J, Iqbal MM, Rashid U, Ali GM, Arif A, Nazir F (2011) Enhanced expression of *AtNHX1* in transgenic groundnut (*Arachis hypogaea* L.) improves salt and drought tolerance. Mol Biotech. DOI: 10.1007/s12033-011-9399-1

Aslam M, Qureshi RH, Ahmed N (1993) A rapid screening technique for salt tolerance in rice (*Oryza sativa* L.). Plant Soil 150:99–107

Baisakh N, Subudhi PK, Bhardwaj P (2008) Primary responses to salt stress in a halophyte, smooth cordgrass (*Spartina alterniflora* Loisel). Funct Integr Genomics 8:287–300

Baisakh N, Rajasekharan K, Deleon T, Biradar H, Parco A, Singh P, Subudhi PK (2009) Overexpression of myo-inositol phosphate synthase gene from a halophyte *Spartina alterniflora* confers salt tolerance in transgenic tobacco and rice. Plant and Animal Genome

XVII, San Diego, CA, Jan 10–14 2009, Poster No. 616, Final abstract guide pp 117

Bhatnagar-Mathur P, Vadez V, Sharma KK (2008) Transgenic approaches for abiotic stress tolerance in plants, retrospect and prospects. Plant Cell Rep 27: 411–424

Bonilla P, Dvorak J, Mackill DJ, Deal K, Gregorio GB (2002) RFLP and SSLP mapping of salinity tolerance genes in chromosome 1 of rice (*Oryza sativa* L.) using recombinant inbred lines. Philipp Agric Sci 85:68–76

Brini F, Hanin M, Mezghani I, Berkowitz GA, Masmoudi K (2007) Overexpression of wheat Na^{+}/H^{+} antiporter *TNHX1* and H^{+}-pyrophosphatase *TVP1* improve salt- and drought-stress tolerance in *Arabidopsis thaliana* plants. J Exp Bot 58:301–308

Byrt CS, Platten JD, Spielmeyer W, James RA, Lagudah ES, Dennis ES, Tester T, Munns R (2007) *HKT1,5*-like cation transporters linked to Na^{+} exclusion loci in wheat, *Nax2* and *Kna1*. Plant Physiol 143:1918–1928

Chen GP, Ma WS, Huang ZJ, Xu T, Xue YB, Shen YZ (2003) Isolation and characterization of *TaGSK1* involved in wheat salt tolerance. Plant Sci 165: 1369–1375

Chitteti BR, Peng Z (2007) Proteome and phosphoproteome differential expression under salinity stress in rice (*Oryza sativa*) roots. J Proteome Res 6:1718–1727

Collard BCY, Mackill DJ (2008) Marker assisted selection: an approach for precision plant breeding in the twenty-first century. Phil Trans Roy Soc Lond B Biol Sci 363:557–572

Collard BCY, Jahufer MZZ, Brouwer JB, Pang ECK (2005) An introduction to markers, quantitative trait loci (QTL) mapping and marker-assisted selection for crop improvement: the basic concepts. Euphytica 142:169–196

Collard BCY, Veracruz CM, McNally KL, Virk PS, Mackill DJ (2008) Rice molecular breeding laboratories in the genomics era: current status and future considerations. Int J Plant Genomics 2008:524847. doi:10: 1155/2008/524847

Collins NC, Tardieu F, Tuberosa R (2008) Quantitative trait loci and crop performance under abiotic stress: where do we stand? Plant Phys 147:469–486

Colmer TD, Munns R, Flowers TJ (2005) Improving salt tolerance of wheat and barley: future prospects. Aust J Exp Agric 45:1425–1443

Colmer TD, Flowers TJ, Munns R (2006) Use of wild relatives to improve salt tolerance in wheat. J Exp Bot 57:1059–1078

Cotsaftis O, Plett D, Johnson AAT, Walia H, Wilson C, Ismail AM, Close T, Tester M, Baumann U (2011) Root-specific transcript profiling of contrasting rice genotypes in response to salinity stress. Mol Plant 4:25–41

Cuin TA, Tian Y, Betts SA, Chalmandrier R, Shabala S (2009) Ionic relations and osmotic adjustment in durum and bread wheat under saline conditions. Funct Plant Biol 36:110–119

Cuin TA, Parsons D, Shabala S (2010) Wheat cultivars can be screened for salinity tolerance by measuring leaf chlorophyll content and shoot sap potassium. Funct Plant Biol 37:255–263

Das-Chatterjee A, Goswami L, Maitra S, Ghosh-Dastidar K, Ray S, Majumder AL (2006) Introgression of a novel salt-tolerant L-*myo* inositol 1-phosphate synthase from *Porteresia coarctata* (Roxb.) Tateoka (*PcINO1*) confers salt tolerance to evolutionary diverse organisms. FEBS Lett 580:3980–3988

Davenport R, James RA, Zakrisson-Plogander A, Tester M, Munns R (2005) Control of sodium transport in durum wheat. Plant Physiol 137:807–818

Dubcovsky J, Santa-Maria G, Epstein E, Luo MC, Dvo ák J (1996) Mapping of the K/Na discrimination locus *Kna1* in wheat. Theor Appl Genet 92:448–454

Dubouzet JG, Sakuma Y, Ito Y, Kasuga M, Dubouzet EG, Miura S, Seki M, Shinozaki K, Yamaguchi-Shinozaki K (2003) OsDREB genes in rice, *Oryza sativa* L., encode transcription activators that function in drought, high salt and cold-responsive gene expression. Plant J 33:751–763

Dasgupta M, Sahoo MR, Kole PE, Mukherjce A (2008) Evaluation of orange-fleshed sweet potato (I pomea betatas L) genotypes for salt tolerance through shoot apex culture under in vitro Nace mediated salinity stress condition. Plant Cell Tiss. Organ Culture 94:161–170

Dvorak J, Noaman MM, Goyal S, Gorham J (1994) Enhancement of the salt tolerance of *Triticum turgidum* L by the *Kna1* locus transferred from *Triticum aestivum* L. chromosome 4D by homoeologous recombination. Theor Appl Genet 87:872–877

Dwivedi S, Upadhyaya H, Subudhi PK, Gehring C, Bajic V, Ortiz R (2010) Enhancing abiotic stress tolerance in cereals through breeding and transgenic interventions. Plant Breed Rev 33:31–114

Dziadczyk P, Bolibok H, Tyrka M, Hortynski JA (2003) In vitro selection of strawberry (*Fragaria × ananassa* Duch.) clones tolerant to salt stress. Euphytica 132:49–55

El-Hendawy SE, Hu Y, Schmidhalter U (2007) Assessing the suitability of various physiological traits to screen wheat genotypes for salt tolerance. J Integr Plant Biol 49:1352–1360

Ellis RP, Forster BPP, Gordon DC, Handley LL, Keith RP, Lawrence P, Meyer R, Powell W, Robinson D, Scrimgeour CM, Young G, Thomas WTB (2002) Phenotype/genotype associations of yield and salt tolerance in a barley mapping population segregating for two dwarfing genes. J Exp Bot 53:1163–1176

FAO (2010) FAO Land and plant nutrition management service. http://www.fao.org

Feuillet C, Langridge P, Waugh R (2008) Cereal breeding takes a walk on the wild side. Trends Genet 24:24–32

Flowers TJ (2004) Improving crop salt tolerance. J Exp Bot 55:307–319

Flowers TJ, Colmer TD (2008) Salinity tolerance in halophytes. New Phytol 179:945–963

Flowers TJ, Flowers SA (2005) Why does salinity pose such a difficult problem for plant breeders? Agric Water Management 78:15–24

Forster BP, Thomas WTB (2005) Doubled haploids in genetics and plant breeding. Plant Breed Rev 25: 57–88

Fukuda A, Nakamura A, Tagiri A, Tanaka H, Miyao A, Hirochika H, Tanaka Y (2004) Function, intracellular localization and the importance in salt tolerance of a vacuolar Na^+/H^+ antiporter from rice. Plant Cell Physiol 45:146–159

Gandonou C, Errabii T, Abrini J, Idaomar M, Senhaji N (2006) Selection of callus cultures of sugarcane (*Saccharum* sp.) tolerant to NaCl and their response to salt stress. Plant Cell Tiss Org Cult 87:9–16

Gao JP, Chao DY, Lin HX (2007) Understanding abiotic stress tolerance mechanisms: recent studies on stress response in rice. J Integr Plant Biol 49:742–750

Garg AK, Kim JK, Owens TG, Ranwala AP, Choi YD, Kochian LV, Wu R (2002) Trehalose accumulation in rice plants confers high tolerance levels to different abiotic stresses. Proc Natl Acad Sci USA 99: 15898–15903

Garthwaite AJ, von Bothmer R, Colmer TD (2005) Salt tolerance in wild Hordeum species is associated with restricted entry of Na^+ and Cl^- into the shoots. J Exp Bot 56:2365–2378

Genc Y, Oldach K, Verbyla AP, Lott G, Hassan M, Tester M, Wallwork H, McDonald GK (2010) Sodium exclusion QTL associated with improved seedling growth in bread wheat under salinity stress. Theor Appl Genet 121:877–894

Germanà MA (2010) Anther culture for haploid and doubled haploid production. Plant Cell Tiss Organ Cult 104:283–300

Ghassemi F, Jakeman A, Nix HA (1995) Salinisation of land and water resources: human causes, extent, management and case studies. UNSW Press, Sydney, Australia

Gorham J (1994) Salt tolerance in the Triticeae: K/Na discrimination in some perennial wheat grasses and their amphiploids with wheat. J Exp Bot 45:441–447

Gorham J, Hardy C, WynJones RG, Joppa LR, Law CN (1987) Chromosomal location of a K/Na discrimination character in the D genome of wheat. Theor Appl Genet 74:584–588

Gorham J, Bridges J, Dubcovsky J, Dvo ák J, Hollington PA, Luo MC, Khan JA (1997) Genetic analysis and physiology of a trait for enhanced K^+/Na^+ discrimination in wheat. New Phytol 137:109–116

Greenway H, Munns R (1980) Mechanisms of salt tolerance in non-halophytes. Ann Rev Plant Physiol 31:149–190

Gregorio GB, Senadhira D, Mendoza RD (1997) Screening rice for salinity tolerance. IRRI Discussion Paper Series 22. International Rice Research Institute, P.O. Box 933, Manila 1099, Philippines

Gregorio GB, Senadhira D, Mendoza RD, Manigbas NL, Roxas JP, Guerta CQ (2002) Progress in breeding for associated salinity tolerance and associated traits in rice. Field Crops Res 76:91–101

Gu L, Liu Y, Zong X, Liu L, Li DP, Li DQ (2010) Overexpression of maize mitogen-activated protein kinase gene, *ZmSIMK1* in Arabidopsis increases tolerance to salt stress. Mol Biol Rep 37:4067–4073

Haq TU, Gorham J, Akhtar J, Akhtar N, Steele KA (2010) Dynamic quantitative trait loci for salt stress components on chromosome 1 of rice. Funct Plant Biol 37:634–645

Hasthanasombut S, Supaibulwatana K, Mii M, Nakamura I (2011) Genetic manipulation of *Japonica* rice using the *OsBADH1* gene from *Indica* rice to improve salinity tolerance. Plant Cell Tiss Organ Cult 104:79–89

Hauser F, Horie T (2010) A conserved primary salt tolerance mechanism mediated by HKT transporters: a mechanism for sodium exclusion and maintenance of high K^+/Na^+ ratio in leaves during salinity stress. Plant Cell Environ 33:552–565

He X, Hou X, Shen Y, Huang Z (2011) *TaSRG*, a wheat transcription factor, significantly affects salt tolerance in transgenic rice and Arabidopsis. FEBS Lett 585:1231–1237

Horie T, Schroeder JI (2004) Sodium transporters in plants. Diverse genes and physiological functions. Plant Physiol 136:2457–62

Houshmand S, Arzani A, Maibody SAM, Feizi M (2005) Evaluation of salt-tolerant genotypes of durum wheat derived from in vitro and field experiments. Field Crops Res 91:345–354

Hu H, Dai M, Yao J, Xiao B, Li X, Zhang Q, Xiong L (2006) Overexpressing a NAM, ATAF, and CUC (NAC) transcription factor enhances drought resistance and salt tolerance in rice. Proc Natl Acad Sci USA 35:12987–12992

Huang S, Spielmeyer W, Lagudah ES, James RA, Platten RD, Dennis ES, Munns R (2006) A sodium transporter (*HKT7*) is a candidate for *Nax1*, a gene for salt tolerance in durum wheat. Plant Physiol 142:1718–1727

Huang W, Ma X, Wang Q, Gao Y, Xue Y, Niu X, Yu G, Liu Y (2008) Significant improvement of stress tolerance in tobacco plants by overexpressing a stress-responsive aldehyde dehydrogenase gene from maize (*Zea mays*). Plant Mol Biol 68:451–463

Husain S, von Caemmerer S, Munns R (2004) Control of salt transport from roots to shoots of wheat in saline soil. Funct Plant Biol 31:1115–1126

Inan G, Zhang Q, Li P, Wang Z, Cao Z, Zhang H, Zhang C, Quist TM, Goodwin SM, Zhu J, Shi H, Damsz B, Charbaji T, Gong Q, Ma S, Fredricksen M, Galbraith DW, Jenks MA, Rhodes D, Hasegawa PM, Bohnert HJ, Joly RJ, Bressan RA, Zhu JK (2004) Salt Cress. A halophyte and cryophyte Arabidopsis relative model system and its applicability to molecular genetic analyses of growth and development of extremophiles. Plant Physiol 135:1–20

IPCC (Intergovernmental Panel on Climate Change) (2007) Climate change 2007: the physical science basis (Summary for policy makers). IPPC Secretariat, WMO, Geneva, Switzerland, p 21

Ismail AM, Heuer S, Thomson MJ, Wissuwa M (2007) Genetic and genomic approaches to develop rice germplasm for problem soils. Plant Mol Biol 65:547–570

Ito Y, Katsura K, Maruyama K, Taji T, Kobayashi M, Seki S, Shinozaki K, Yamaguchi-Shinozaki K (2006) Functional analysis of rice DREB1/CBF type transcription factors involved in cold responsive gene expression in transgenic rice. Plant Cell Physiol 47:141–153

Jacobs A, Ford K, Kretschmer J, Tester M (2011) Rice plants expressing the moss sodium pumping ATPase *PpENA1* maintain greater biomass production under salt stress. Plant Biotech J. doi:10.1111/j.1467-7652.2011.00594.x

James RA, Davenport RJ, Munns R (2006) Physiological characterization of two genes for Na+exclusion in durum wheat, Nax1 and Nax2. Plant Physiol 142:1537–1547

James RA, Blake C, Byrt CS, Munns R (2011) Major genes for Na^+ exclusion, *Nax1* and *Nax2* (wheat *HKT1,4* and *HKT1,5*), decrease Na^+ accumulation in bread wheat leaves under saline and waterlogged conditions. J Exp Bot. doi:10.1093/jxb/err003

Jha B, Sharma A, Mishra A (2010) Expression of *SbGSTU* (*tau* class glutathione S-transferase) gene isolated from *Salicornia brachiata* in tobacco for salt tolerance. Mol Biol Rep. 38:4823–4832

Jithesh MN, Prashanth SR, Sivaprakash KR, Parida AK (2006) Monitoring expression profiles of antioxidant genes to salinity, iron, oxidative, light and hyperosmotic stresses in highly salt tolerant grey mangrove, *Avicennia marina* (Forsk.) Vierh by mRNA analysis. Plant Cell Rep 25:865–876

Karaba A, Dixit S, Greco R, Aharoni A, Trijatmiko KR, Marsch-Martinez N, Krishnan A, Nataraja KB, Udayakumar M, Pereira A (2007) Improvement of water use efficiency in rice by expression of *HARDY*, an *Arabidopsis* drought and salt tolerance gene. Proc Natl Acad Sci USA 104:15270–15275

Karan R, Singla-Pareek SL, Pareek A (2009) Histidine kinase and response regulator genes as they relate to salinity tolerance in rice. Funct Integr Genomics 9:411–417

Khan MA, Shirazi MU, Ali M, Mumtaz S, Sherin A, Ashraf MY (2006) Comparative performance of some wheat genotypes growing under saline water. Pak J Bot 38:1633–1639

Kijne JW (2006) Abiotic stress and water scarcity: identifying and resolving conflicts from plant level to global level. Field Crops Res 97:3–18

Kong X, Pan J, Zhang M, Xing X, Zhou Y, Liu Y, Li D, Li D (2011) *ZmMKK4*, a novel group C MAPK kinase in maize (*Zea mays*), confers salt and cold tolerance in transgenic Arabidopsis. Plant Cell Environ. doi:10.1111/j.1365-3040.2011.02329.x

Koyama ML, Levesley A, Koebner RMD, Flowers TJ, Yeo AR (2001) Quantitative trait loci for component physiological traits determining salt tolerance in rice. Plant Physiol 125:406–422

Kumari S, Sabharwal VP, Kushwaha HR, Sopory SK, Singla-Pareek SL, Pareek A (2009) Transcriptome map for seedling stage specific salinity stress response indicates a specific set of genes as candidate for saline tolerance in *Oryza sativa* L. Funct Integr Genomics 9:109–123

Lang NT, Yanagihara S, Buu BC (2001) A microsatellite marker for a gene conferring salt tolerance on rice at the vegetative and reproductive stages. SABRAO J Breed Genet 33:1–10

Lee KS, Choi WR, Ko JC, Kim TS, Gregoria GP (2003a) Salimity tolerance of *japonica* and *indica* (*Bryza sativa*) at the seedling stage. Planta 216:1043–1046

Lee SY, Lee JH, Kwon TO (2003b) Selection of salt-tolerant doubled haploids in rice anther culture. Plant Cell Tiss Organ Cult 74:143–149

Lee SY, Ahn JH, Cha YS, Yun DW, Lee MC, Ko JC, Lee KS, Eun MY (2007) Mapping QTLs related to salinity tolerance of rice at the young seedling stage. Plant Breed 126:43–46

Li JY, He XW, Xu L, Zhou J, Wu P, Shou HX, Zhang FC (2008) Molecular and functional comparisons of the vacuolar Na^+/H^+ exchangers originated from glycophytic and halophytic species. J Zhejiang Univ Sci B 9:132–140

Li WH, Zhang Q, Kong XQ, Wu CX, Ma XL, Zhang H, Zhao YX (2009) Salt tolerance is conferred in Arabidopsis by over-expression of the vacuolar Na^+/H^+ antiporter gene *SsNHX2*, an alternative splicing variant of *SsNHX1*, from *Suaeda salsa*. J Plant Biol 52:147–153

Li F, Guo S, Zhao Y, Chen D, Chong K, Xu Y (2010a) Overexpression of a homopeptide repeat-containing bHLH protein gene (*OrbHLH001*) from Dongxiang wild rice confers freezing and salt tolerance in transgenic Arabidopsis. Plant Cell Rep 29:977–986

Li TX, Zhang Y, Liu H, Wu YT, Li WB, Zhang HX (2010b) Stable expression of Arabidopsis vacuolar Na^+/H^+ antiporter gene *AtNHX1*, and salt tolerance in transgenic soyabean for over six generations. Chinese Sci Bull 55:1127–1134

Lin HX, Zhu MZ, Yano M, Gao JP, Liang ZW, Su WA, Hu XH, Ren ZH, Chao DY (2004) QTLs for Na^+ and K^+ uptake of the shoots and roots controlling rice salt tolerance. Theor Appl Genet 108:253–260

Lindsay MP, Lagudah ES, Hare RA, Munns R (2004) A locus for sodium exclusion (*Nax1*), a trait for salt tolerance, mapped in durum wheat. Funct Plant Biol 31:1105–1114

Lisa LA, Seraj ZI, Elahi CMF, Das KC, Biswas K, Islam MR, Salam MA, Gomosta AR (2004) Genetic variation in microsatellite DNA, physiology and morphology of coastal saline rice (*Oryza sativa* L.) landraces of Bangladesh. Plant Soil 263:213–228

Liu X, Baird WV (2003) Differential expression of genes regulated in response to drought or salinity stress in sunflower. Crop Sci 43:678–687

Lu SY, Peng XX, Guo ZF, Zhang GY, Wang ZC, Wang CY, Pang CS, Fan Z, Wang JH (2007) In vitro selection of salinity tolerant variants from triploid bermudagrass (*Cynodon transvaalensis* × *C-dactylon*) and their physiological responses to salt and drought stress. Plant Cell Rep 26:1413–1420

Lv S, Zhang KW, Gao Q, Lian LJ, Song YJ, Zhang JR (2008) Overexpression of an H^+-PPase gene from *Thellungiella halophila* in cotton enhances salt tolerance and improves growth and photosynthetic performance. Plant Cell Physiol 49:1150–1164

Ma L, Zhou E, Huo N, Zhou R, Wang G, Jia J (2007) Genetic analysis of salt tolerance in a recombinant inbred population of wheat (*Triticum aestivum* L.). Euphytica 153:109–117

Maas EV, Hoffman GJ (1977) Crop salt tolerance-current assessment. J Irrig Drain Div ASCE 103:115–134

Mahalakshmi S, Christopher GSB, Reddy TP, Rao KV, Reddy VD (2006) Isolation of a cDNA clone (PcSrp) encoding serine-rich-protein from *Porteresia coarctata* T. and its expression in yeast and finger millet (*Eleusine coracana* L.) affording salt tolerance. Planta 224: 347–359

Majee M, Maitra S, Ghosh Dastidar K, Pattnail S, Chatterjee A, Hait NC, Das KP, Majumdar AL (2004) A novel salt-tolerant l-myo-inositol-1-phosphate synthase from *Porteresia coarctata* (Roxb.) Tateoka, a halophytic wild rice: molecular cloning, bacterial overexpression, characterization, and functional introgression into tobacco-conferring salt tolerance phenotype. J Biol Chem 279:28539–28552

Manneh B, Stam P, Struik PC, Bruce-Oliver S, van Eeuwijk FA (2007) QTL-based analysis of genotype-by-environment interaction for grain yield of rice in stress and non-stress environments. Euphytica 156:213–226

Mano Y, Takeda K (1997) Mapping quantitative trait loci for salt tolerance at germination and the seedling stage in barley (*Hordeum vulgare* L.). Euphytica 94: 263–272

Mao X, Zhang H, Tian S, Chang X, Jing R (2010) TaSnRK2.4, an SNF1-type serine/threonine protein kinase of wheat (*Triticum aestivum* L.), confers enhanced multistress tolerance in Arabidopsis. J Exp Bot 61:683–696

Mao X, Jia D, Li A, Zhang H, Tian S, Zhang X, Jia J, Jing R (2011) Transgenic expression of *TaMYB2A* confers enhanced tolerance to multiple abiotic stresses in Arabidopsis. Funct Integr Genomics. doi:10.1007/s10142-011-0218-3

Melloul A, Collin M (2006) Hydrogeological changes in coastal aquifers due to sea level rise. Ocean Coastal Management 49:281–297

Munns R (2002) Comparative physiology of salt and water stress. Plant Cell Environ 25:239–250

Munns R, James RA (2003) Screening methods for salinity tolerance: a case study with tetraploid wheat. Plant Soil 253:201–218

Munns R, Tester M (2008) Mechanisms of salinity tolerance. Annu Rev Plant Biol 59:651–681

Munns R, Hare RA, James RA, Rebetzke GJ (2000) Genetic variation for improving the salt tolerance of durum wheat. Aust J Agric Res 51:69–74

Munns R, Rebetzke GJ, Husain S, James RA, Hare RA (2003) Genetic control of sodium exclusion in durum wheat. Aust J Agr Res 54:627–635

Nakashima K, Tran LS, Van Nguyen D, Fujita M, Maruyama K, Todaka D, Ito Y, Hayashi N, Shinozaki K, Yamaguchi-Shinozaki K (2007) Functional analysis of a NAC-type transcription factor *OsNAC6* involved in abiotic and biotic stress-responsive gene expression in rice. Plant J 51:617–630

Obata T, Kitamoto HK, Nakamura A, Fukuda A, Tanaka Y (2007) Rice shaker potassium channel *OsKAT1* confers tolerance to salinity stress on yeast and rice cells. Plant Physiol 144:1978–1985

Oh SJ, Song SI, Kim YS, Jang HJ, Kim M, Kim YK (2005) *Arabidopsis* CBF3/DREB1A and ABF3 in transgenic rice increased tolerance to abiotic stress without stunting growth. Plant Physiol 138: 341–351

Oh SJ, Chang-Woo Kwon CW, Dong-Woog Choi DW, Sang Ik Song SI, Kim JK (2007) Expression of barley HvCBF4 enhances tolerance to abiotic stress in transgenic rice. Plant Biotech J 5:646–656

Ohta M, Hayashi Y, Nakashima A, Hamada A, Tanaka A, Nakamura T, Hayakawa T (2002) Introduction of a Na^+/H^+ antiporter gene from *Atriplex gmelini* confers salt tolerance to rice. FEBS Lett 532:279–82

Ortel B, Atzorn R, Hause B, Feussner I, Miersch O, Wasternack C (1999) Jasmonate-induced gene expression of barley (*Hordeum vulgare*) leaves the link between jasmonate and abscisic acid. Plant Growth Regul 29:113–122

Ozturk ZN, Talame V, Deyhoyos M, Michalowski CB, Galbraith DW, Gozukirmizi N, Tuberosa R, Bohnert HJ (2002) Monitoring large-scale changes in transcript abundance in drought-and salt-stressed barley. Plant Mol Biol 48:551–573

Pandit A, Rai V, Bal S, Sinha S, Kumar V, Chauhan M, Gautam RK, Singh R, Sharma PC, Singh AK, Gaikwad K, Sharma TR, Mohapatra T, Singh NK (2010) Combining QTL mapping and transcriptome profiling of bulked RILs for identification of functional polymorphism for salt tolerance genes in rice (*Oryza sativa* L.). Mol Genet Genom 284:121–136

Prashanth SR, Sadhasivam V, Parida A (2008) Over expression of cytosolic copper/zinc superoxide dismutase from a mangrove plant *Avicennia marina* in *indica* Rice var Pusa Basmati-1 confers abiotic stress tolerance. Transgenic Res 17:281–291

Qiu Y, Li X, Zhi H, Shen D, Lu P (2009) Differential expression of salt tolerance related genes in *Brassica campestris* L. ssp. *chinensis* (L.) Makino var. *communis* Tsen et Lee. J Zhejiang Univ Sci B 10:847–851

Quarrie SA, Steed A, Calestani C, Semikhodskii A, Lebreton C, Chinoy C, Steele N, Pljevljakusi D, Waterman E, Weyen J, Schondelmaier J, Habash DZ, Farmer P, Saker L, Clarkson DT, Abugalieva A, Yessimbekova M, Turuspekov Y, Abugalieva S, Tuberosa R, Sanguineti M-C, Hollington PA, Aragués R, Royo A, Dodig D (2005) A high-density genetic map of hexaploid wheat (*Triticum aestivum* L.) from the cross Chinese Spring x SQ1 and its use to compare QTLs for grain yield across a range of environments. Theor Appl Genet 110:865–880

Queirs F, Fidalgo F, Santos I, Salema R (2007) In vitro selection of salt tolerant cell lines in *Solanum tuberosum* L. Biol Plant 51:728–734

Rahman MS, Das KC, Das DK, Biswas K, Chowdhury MBH, Karim NH, Salam MA, Seraj ZI (2010) Breeding and anther derived lines of rice (*Oryza sativa* L.) for saline coastal areas of Bangladesh. Bangladesh J Bot 39:71–78

Rahnama A, Poustini K, Munns R, James RA (2010) Stomatal conductance as a screen for osmotic stress tolerance in durum wheat growing in saline soil. Funct Plant Biol 37:255–263

Ren Z, Gao J, Li L, Cai X, Huang W, Chao D, Zhu M, Wang Z, Luan S, Lin H (2005) A rice quantitative trait locus for salt tolerance encodes a sodium transporter. Nat Genet 37:1141–1146

Rengasamy P (2010) Soil processes affecting crop production in salt affected soils. Funct Plant Biol 37: 613–620

Ruan SL, Ma HS, Wang SH, Fu YP, Xin Y, Liu WZ, Wang F, Tong JX, Wang SZ, Chen HZ (2011) Proteomic identification of *OsCYP2*, a rice cyclophilin that confers salt tolerance in rice (*Oryza sativa* L.) seedlings when overexpressed. BMC Plant Biol 11:34. doi:10.1186/1471-2229-11-34

Sabouri H, Sabouri A (2008) New evidence of QTLs attributed to salinity rice. Afr J Biotech 7:4376–4383

Sahi C, Agarwal M, Reddy MK, Sopory SK, Grover A (2003) Isolation and expression analysis of salt stress-associated ESTs from contrasting rice cultivars using a PCR-based subtraction method. Theor Appl Genet 106:620–628

Schachtman DP, Munns R, Whitecross MI (1991) Variation of sodium exclusion and salt tolerance in *Triticum tauschii*. Crop Sci 31:992–997

Senadhira D, Zapata-Arias FJ, Gregorio GB, Alejar MS, de la Cruz HC, Padolina TF, Galvez AM (2002) Development of the first salt-tolerant rice cultivar through *indica/indica* anther culture. Field Crops Res 76:103–110

Sengupta S, Majumder AL (2009) Insight into the salt tolerance factors of a wild halophytic rice, *Porteresia coarctata*: a physiological and proteomic approach. Planta 229:911–929

Shah SH, Gorham J, Forster BP, Wyn Jones RG (1987) Salt tolerance in the Triticeae: the contribution of the D-genome to cation selectivity in hexaploid wheat. J Exp Bot 38:254–269

Shavrukov Y, Gupta NK, Miyazaki J, Baho MN, Chalmers KJ, Tester M, Langridge P, Collins NC (2010) *HvNax3*-a locus controlling shoot sodium exclusion derived from wild barley (*Hordeum vulgare ssp. spontaneum*). Funct Integr Genomics 10:277–291

Singh RK, Mishra B (1995) Screening F_1 anther culture derivatives of rice for salt tolerance. In: Sharma B et al. (eds) Genetic research and education: current trends and the next fifty years. Indian Society of Genetics and Plant Breeding New Delhi. Ind J Genet Plant Breed (Special Issue): 509–513

Singh RK, Mishra B, Senadhira D (1992) Promising salt tolerant F_1 anther culture derivatives (ACDs). Int Rice Res Newslett 17:17

Singla-Pareek SL, Yadav SK, Pareek A, Reddy MK, Sopory SK (2008) Enhancing salt tolerance in a crop plant by overexpression of glyoxalase II. Transgenic Res 17:171–180

Subudhi PK, Baisakh N (2011) *Spartina alterniflora* Loisel., a halophyte grass model to dissect salt stress tolerance. In Vitro Cell Dev Biol-Plant 47:441–457

Sun ZB, Qi XY, Li PH, Wu CX, Zhao YX, Zhang H, Wang ZL (2008) Overexpression of a *Thellungiella halophila* CBL9 homolog, *ThCBL9*, confers salt and osmotic tolerances in transgenic *Arabidopsis thaliana*. J Plant Biol 51:25–34

Taiz L (1992) The plant vacuole. J Exp Biol 172: 113–122

Tajbakhsh M, Zhou M, Chen Z, Mendham NJ (2006) Physiological and cytological response of salt-tolerant and non-tolerant barley to salinity during germination and early growth. Aust J Exp Agr 46:555–562

Taji T, Seki M, Satou M, Sakurai T, Kobayashi M, Ishiyama K, Narusaka Y, Narusaka M, Zhu JK, Shinozaki K (2004) Comparative genomics in salt tolerance between Arabidopsis and Arabidopsis-related halophyte salt cress using Arabidopsis microarray. Plant Physiol 135:1697–1709

Takasaki H, Maruyama K, Kidokoro S, Ito Y, Fujita Y, Shinozaki K, Yamaguchi-Shinozaki K, Nakashima K (2010) The abiotic stress-responsive NAC-type transcription factor OsNAC5 regulates stress-inducible genes and stress tolerance in rice. Mol Genet Genomics 284:173–183

Takeda S, Matsuoka M (2008) Genetic approaches to crop improvement: responding to environmental and population changes. Nat Rev Genet 9:444–457

Takehisa H, Shimodate T, Fukuta Y, Ueda T, Yano M, Yamaya T, Kameya T, Sato T (2004) Identification of quantitative trait loci for plant growth of rice in paddy field flooded with salt water. Field Crops Res 89:85–95

Tanksley SD, McCouch SR (1997) Seed banks and molecular maps: unlocking genetic potential from the wild. Science 277:1063–1066

Thalji T, Shalaldeh G (2007) Screening wheat and barley genotypes for salinity tolerance. J Agron 6:75–80

Thomson MJ, de Ocampo M, Egdane J, Rahman MA, Sajise AG, Adorada DL, Tumimbang-Raiz E, Blumwald E, Seraj ZI, Singh RK, Gregorio GB, Ismail AM (2010) Characterizing the *Saltol* quantitative trait locus for salinity tolerance in rice. Rice 3: 148–160

USDA-ARS (2005) Agricultural weather facility, Global crop production review. 1–5

USDA-ARS (2008) Research Databases. (http://www.ars.usda.gov/Services/docs.htm?docid=8908)

Verma D, Singla-Pareek SL, Rajagopal D, Reddy MK, Sopory SK (2007) Functional validation of a novel isoform of Na^+/H^+ antiporter from *Pennisetum glaucum* for enhancing salinity tolerance in rice. J Biosci 32:621–628

Verslues PE, Agarwal M, Katiyar-Agarwal S, Zhu J, Zhu JK (2006) Methods and concepts in quantifying resistance to drought, salt and freezing, abiotic stresses that affect plant water status. Plant J 45:523–539

Walia H, Wilson C, Condamine P, Liu X, Ismail AM, Zeng L, Wanamaker SI, Mandal J, Xu J, Cui X, Close TJ (2005) Comparative transcriptional profiling of two contrasting rice genotypes under salinity stress during the vegetative growth stage. Plant Physiol 139: 822–835

Walia H, Wilson C, Wahid A, Condamine P, Cui X, Close TJ (2006) Expression analysis of barley (*Hordeum vulgare* L.) during salinity stress. Funct Integr Genomics 6:143–156

Walia H, Wilson C, Zeng L, Ismail AM, Condamine P, Close TJ (2007) Genome-wide transcriptional analysis of salinity stressed *japonica* and *indica* rice genotypes during panicle initiation stage. Plant Mol Biol 63:609–623

Wang QY, Guan YC, Wu YR, Chen HL, Chen F, Chu CC (2008) Overexpression of a rice *OsDREB1F* gene increases salt, drought and low temperature tolerance in both *Arabidopsis* and rice. Plant Mol Biol 67:589–602

Wang ZF, Wang JF, Bao YM, Wu YY, Zhang HS (2011) Quantitative trait loci controlling rice seed germination under salt stress. Euphytica 178:297–307

Wassmann F, Hien NX, Hoanh CT, Tuong TP (2004) Sea level rise affecting the Vietnamese Mekong Delta: water elevation in the flood season and implications for rice production. Climate Change 66:89–107

Wasternack C, Miersch O, Kramell R, Hause B, Ward J, Beale M, Boland W, Parthier B, Feussner I (1998) Jasmonic acid:biosynthesis, signal transduction, gene expression. Fett-Lipid 100:139–146

Wedzony M, Forster BP, Zur I, Golemiec E, Szechynska-Hebda M, Dubas E, Gotebiowska G, Wedzony M (2009) Progress in doubled haploid technology in higher plants. In: Touraev A, Forster BP, Jain SM (eds) Advances in haploid production in higher plants. Springer, Berlin, pp 1–34

Wei W, Bilsborrow P, Hooley P, Fincham D, Forster B (2001) Variation between two near isogenic barley *(Hordeum uulgare)* cultivars in expression of the B subunit of the vacuolar ATPase in response to salinity. Hereditas 135:227–231

Wheatley AO, Ahmad MH, Asemota HN (2003) Development of salt adaptation in in vitro greater yam (*Dioscorea alata*) plantlets. In Vitro Cell Dev Biol-Plant 39:346–353

Wild A (2003) Soils, land and food: managing the land during the twenty-first century. Cambridge University Press, Cambridge, UK

Witcombe JR, Hollington PA, Howarth CJ, Reader S, Steele KA (2008) Breeding for abiotic stresses for sustainable agriculture. Philos Trans R Soc B Biol Sci 363:703–716

Witzel K, Weidner A, Surabhi GK, Borner A, Mock HP (2009) Salt stress-induced alterations in the root proteome of barley genotypes with contrasting response towards salinity. J Exp Bot 60:3545–3557

Wu W, Su Q, Xia XY, Wang Y, Luan YS, An LJ (2008) The *Suaeda liaotungensis* kitag betaine aldehyde dehydrogenase gene improves salt tolerance of transgenic maize mediated with minimum linear length of DNA fragment. Euphytica 159:17–25

Wu C, Gao X, Kong X, Zhao Y, Zhang H (2009) Molecular cloning and functional analysis of a Na^+/H^+ antiporter gene *ThNHX1* from a halophytic plant *Thellungiella halophila*. Plant Mol Biol Rep 27:1–12

Xu GY, Rocha PSCF, Wang ML, Xu ML, Cui YC, Li LY, Zhu YX, Xia X (2011) A novel rice calmodulin-like gene, OsMSR2, enhances drought and salt tolerance and increases ABA sensitivity in Arabidopsis. Planta. doi:10.1007/s00425-011-1386-z

Xue ZY, Zhi DY, Xue GP, Zhang H, Zhao YX, Xia GM (2004) Enhanced salt tolerance of transgenic wheat (*Triticum aestivum* L.) expressing a vacuolar Na^+/H^+ antiporter gene with improved yields in saline soils in the field and a reduced level of leaf Na^+. Plant Sci 167:849–859

Xue D, Huang Y, Zhang X, Wei K, Westcott S, Li C, Chen M, Zhang G, Lance R (2009) Identification of QTLs associated with salinity tolerance at late growth stage in barley. Euphytica 169:187–196

Yan S, Tang Z, Su W, Sun W (2005) Proteomic analysis of salt stress-responsive proteins in rice root. Proteomics 5:235–244

Yoshida S, Forno DA, Cock JH, Gomez KA (1976) Laboratory manual for physiological studies of rice 3rd edition. International Rice Research Institutes, Manila, Philippines, p 61

Zair I, Chlyah A, Sabounji K, Tittahsen M, Chlyah H (2003) Salt tolerance improvement in some wheat cultivars after application of in vitro selection pressure. Plant Cell Tiss Org Cult 73:237–244

Zang J, Sun Y, Wang Y, Yang J, Li F, Zhou YL, Zhu LH, Reys J, Mohammadhosein F, Xu JL, Li Z (2008) Dissection of genetic overlap of salt tolerance QTLs at the seedling and tillering stages using backcross introgression lines in rice. Sci China C Life Sci 51:583–591

Zeng L, Lesch SM, Grieve CM (2003a) Rice growth and yield respond to changes in water depth and salinity stress. Agric Water Manage 59:67–75

Zeng L, Poss JA, Wilson C, Draz ASE, Gregorio GB, Grieve CM (2003b) Evaluation of salt tolerance in rice genotypes by physiological characters. Euphytica 129:281–292

Zhang L, Zhang L, Luo J, Chen W, Hao M, Liu B, Yan Z, Zhang B, Zhang H, Zheng Y, Liu D, Yen Y (2011a) Synthesizing double haploid hexaploid wheat populations based on a spontaneous alloploidization process. J Genet Genomics 38:89–94

Zhang X, Wang L, Meng H, Wen H, Fan Y, Zhao J (2011b) Maize *ABP9* enhances tolerance to multiple stresses in transgenic Arabidopsis by modulating ABA signaling and cellular levels of reactive oxygen species. Plant Mol Biol 75:365–378

Zhao FY, Zhang H (2006) Salt and paraquat stress tolerance results from co-expression of the *Suaeda salsa* glutathione S-transferase and catalase in transgenic rice. Plant Cell Tiss Org Cult 86:349–358

Zhu GY, Kinet JM, Bertin P, Bouharmont J, Lutts S (2000) Crosses between cultivars and tissue culture-selected plants for salt resistance improvement in rice, *Oryza sativa*. Plant Breed 119:497–504

Zhu JQ, Zhang JT, Tang RJ, Lv QD, Wang QQ, Yang L, Zhang HX (2009) Molecular characterization of *ThIPK2*, an inositol poly- phosphate kinase gene homolog from *Thellungiella halophila*, and its heterologous expression to improve abiotic stress tolerance in *Brassica napus*. Physiol Plantarum 136:407–425

Understanding and Exploiting the Impact of Drought Stress on Plant Physiology

5

Olga M. Grant

Abstract

Despite the enormous volume of literature relating to plant responses to drought, there is still much work to be done to fully unravel the impact of a range of possibly interacting processes on plant physiology during drought. This review highlights some of the key processes. Some aspects of plant physiological response to drought can actually be beneficial in agronomy, if exploited correctly. It is also clear that substantial genetic variation in response to drought exists within many species. Exploiting such variation in conventional breeding has led to the release of drought-tolerant varieties. Increasing demand for food and other plant-based products coupled with increasing frequency and distribution of drought means that more rapid development of suitable varieties is now required. Understanding the genetic basis of drought tolerance is therefore essential, and the explosion in genomic data for a wide range of plant species is currently being harnessed in enhancing genetic improvement programmes.

Keywords

Aquaporins • Deficit irrigation • Gas exchange • Gene expression • Osmotic adjustment • Photosynthesis • Quantitative trait loci • Reactive oxygen species • Stomatal conductance • Water use efficiency

Abbreviations used

A	Assimilation rate
ABA	Abscisic acid
AQP	Aquaporin
APX	Ascorbate peroxidase
AsA	Ascorbic acid
CAT	Catalase
$\delta^{13}C$	Carbon isotope composition
DHA	Dehydroascorbate
DHAR	Dehydroascorbate reductase
E	Evapotranspiration rate
EST	Expressed sequence tag
GPX	Glutathione peroxidase
GR	Glutathione reductase
g_s	Stomatal conductance

O.M. Grant (✉)
Department of Biology, National University of Ireland, Maynooth, Co. Kildare, Ireland
e-mail: olga.grant@nuim.ie

P. Ahmad and M.N.V. Prasad (eds.), *Abiotic Stress Responses in Plants: Metabolism, Productivity and Sustainability*, DOI 10.1007/978-1-4614-0634-1_5, © Springer Science+Business Media, LLC 2012

GSH	Glutathione
MDA	Monohydroascorbate
MDAR	Monohydroascorbate reductase
MTX	Methotrexate
p_a	Partial pressure of CO_2 in the air
PEG	Polyethylene glycol
p_i	Intercellular partial pressure of CO_2
PIP	Plasma membrane intrinsic protein
PLD	Phospholipase D
POD	Guaicol peroxidase
PRI	Photochemical reflectance index
PSI	Photosystem I
PSII	Photosystem II
QTL	Quantitative trait loci
RFLP	Restriction fragment length polymorphism
ROS	Reactive oxygen species
SOD	Superoxide dismutase
TIP	Tonoplast intrinsic protein
WUE	Water use efficiency
WUE_i	Instantaneous water use efficiency.

1 Introduction

Worldwide, crop production is limited by drought more than by any other environmental stress (Cattivelli et al. 2008). Climate change, which will result in increased crop demand for water, will exacerbate this problem. Dwindling natural resources coupled with consumer concern have led to an increasing requirement for crops to be produced with limited impact on the environment; nonetheless the demand for crop products is growing with the increasing global population, and consumers demand high quality and uniformity. Only about 15% of the world's agricultural land is irrigated, but irrigated land accounts for almost half of global food production (Fereres and Connor 2004). Fresh water scarcity is regarded as the single largest water problem worldwide (Shiklomanov 1997, Jury and Vaux 2005) with irrigated agriculture being the single-largest component of fresh water withdrawal, accounting for 70% of diverted water and 90% of consumed water globally (Shiklomanov 1997).

The literature on drought is overwhelming. A search for articles on "drought" in the Science Citation Index returns over 7,000 entries under the "Plant Sciences" category – with a further 1,000 under "Agronomy", and a further 600 under "Biochemistry and Molecular Biology". Articles published in 2010 give an idea of where the current interests in this topic lie. Abscisic acid (or ABA), drought tolerance, and stomata are the most frequently occurring key-words (Fig. 5.1), followed closely by photosynthesis/CO_2 assimilation/photosynthetic rate and root traits/root development, or more specific key-words relating to root traits. Growth, yield, stomatal conductance, water use efficiency, and hydraulics also featured relatively frequently. The focus of the articles, however, is very diverse, with the most-frequent key-word, abscisic acid (or ABA), only appearing in 7% of articles. About 5% of the articles relate to the model plant *Arabidopsis*, with similar numbers of articles on wheat and on rice, and approximately 4% of articles deal with maize.

The objective of this article is to attempt to integrate current understanding of diverse aspects of plant responses to drought, and highlight how understanding of these responses can be exploited.

2 Understanding the Impact of Drought Stress on Plant Physiology

2.1 Gas Exchange

Under mild to moderate water deficits, stomatal closure is one of the earliest plant responses, concomitant with the reduced water potential and turgor associated with even a small decrease in relative water content (Chaves et al. 2003, Lawlor and Tezara 2009). Reduced stomatal conductance limits water loss and CO_2 diffusion, and hence photosynthetic assimilation. Intercellular partial pressure of CO_2 (p_i) is determined both by the supply of CO_2 to the chloroplasts and the demand for CO_2 for photosynthesis by the chloroplasts. If photosynthetic capacity is uninhibited, then the demand for CO_2 by the chloroplasts remains unchanged, but the supply of CO_2 to the chloroplasts will be reduced in parallel with stomatal closure and reduced transpiration. Ultimately, reduced photosynthetic assimilation rates result in

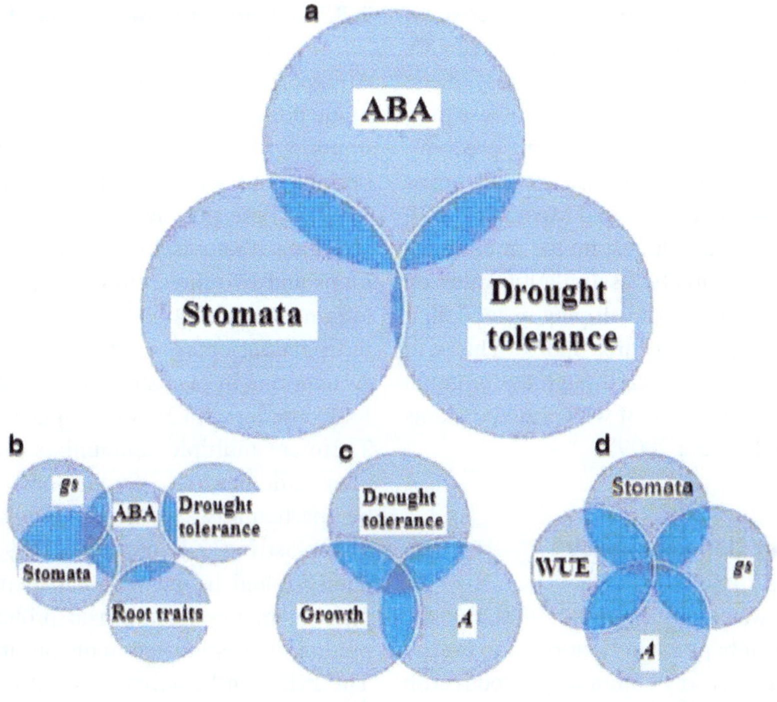

Fig. 5.1 The three main topics covered by articles published on "drought" in 2010 (**a**), and the main topics which overlap with each of those three topics (**b–d**), with overlapping *circles* indicating that both topics are combined in some of the articles. Topics were determined by assessing the number of articles containing relevant keywords in the 745 articles in the "Plant Sciences" category of the Science Citation Index for that year. *ABA* abscisic acid, g_s stomatal conductance, *A* photosynthesis

reduced vegetative growth, and for many crops even mild drought stress results in reduced yield. Nonetheless, in some crops both stomatal conductance and carbon assimilation can be maintained until water potential falls to relatively low levels (Flexas and Medrano 2002). On the other hand, in species which are adapted to dry conditions, stomatal closure can be seen even where tissue water content is still high – in response to very mild soil moisture reduction or at midday on hot days (Pinheiro and Chaves 2011). Such stomatal closure helps to prevent xylem pressure exceeding cavitation thresholds (David et al. 2007).

Both hydraulic and chemical signals sent from drying roots to the shoot are involved in the regulation of stomatal closure and decreased growth during soil drying (Tardieu et al. 2010). Whether hydraulic limitation or chemical signalling dominates, appears to depend on the species and growing conditions. The importance of abscisic acid (ABA) as a root-sourced signal transported via the xylem and involved in stomatal regulation during drought has been highlighted in several studies (e.g., Dodd et al. 2006), but other compounds such as the precursors of ABA or cytokinins also play a role, as do changes in mineral composition or pH of the xylem (e.g., Wilkinson and Davies 2008). Where xylem sap is acidic, ABAH is partitioned into alkaline components in the symplast of the leaf cells, away from the sites of action of ABA on the stomata. As pH increases, the proportion of ionised ABA transported in the xylem sap increases, and has a greater impact on stomatal behaviour (Hartung et al. 1998). Sharp and Davies (2009) found that changes in xylem sap pH induced by drought are more common in herbaceous species than in woody perennials.

The main cause of reduced photosynthetic rate under mild to moderate water deficits is a reduction in the diffusion of atmospheric CO_2 to the site of carboxylation (Chaves et al. 2009). This is as a result of both stomatal closure and a reduction in mesophyll conductance – although the extent of the influence of mesophyll conductance is still debated (Pinheiro and Chaves 2011). Water stress also directly impacts on internal transport of CO_2 and on enzyme activity and hence photosynthetic capacity (Lawlor and Tezara 2009), and these metabolic and diffusive limitations become predominant relative to stomatal limitation as water stress becomes more severe (Flexas and Medrano 2002).

It should be noted that despite the importance of reduced photosynthetic assimilation rates on growth and yield, slowly developing water deficit can result in a small leaf area index. This will impact on productivity even though assimilation rates may be close to those of well-watered plants (see Lawlor and Tezara 2009).

2.2 Water Use Efficiency

Instantaneous water use efficiency (WUE_i), the ratio of photosynthetic assimilation (A) to evapotranspiration (E), is a key component of both crop water use efficiency (WUE) and yield potential. WUE_i is often increased when water availability is limited, as a result of stomatal closure and hence reduced E (Farquhar et al. 1982). This can ultimately lead to an increase in WUE at the whole plant scale (biomass/water transpired) or even crop scale (yield of harvestable product/water used or applied).

2.3 Transport of Water

The naturally occurring gradient in water potential between the environment of plant roots and the environment of the shoots drives the uptake of water (Knipfer and Fricke 2011). Hydraulic resistances both in the root and shoot can limit the flow of water through the plant. The main hydraulic barrier to water uptake by roots is the radial transport pathway between root epidermis and xylem, rather than the axial path along xylem conduits (Steudle and Peterson 1998). The radial resistance to water flow can be divided into an apoplastic component (cell wall, middle lamella, and intercellular air space) and a cell-to-cell component (through plasmodesmata and across membranes) (Knipfer and Fricke 2011).

2.3.1 Role of Aquaporins in Water Transport

Aquaporin (AQP) proteins are a class of membrane proteins that are now understood to play an important role in water transport, both under optimal and stress conditions (Kjellbom et al. 1999). These proteins consist of six membrane-spanning domains linked by three extracellular loops and two intercellular loops. They occur as tetramers, each monomer forming a functionally independent pore (Luu and Maurel 2005). Aquaporins in plants occur as multiple isoforms, for example, with 35 homologues in *Arabidopsis*. There are multiple subfamilies of aquaporins in plant genomes, two of which have been shown to be particularly involved in water transport. Tonoplast intrinsic proteins (TIPs) are predominantly found in the vacuolar membranes, while plasma membrane intrinsic proteins (PIPs) are predominantly located in the plasma membranes. The PIPs can be subdivided into PIP1 and PIP2 homology subgroups and are considered to be exclusive transporters of water as shown by the highly conserved topology of the aromatic/arginine selectivity filter (Wallace and Roberts 2004). They are particularly important in controlling transcellular water transport, and are especially abundant in roots, where they play a substantial role in the uptake of water from soil (Javot and Maurel 2002).

2.3.2 Impact of Drought on Aquaporin Expression

Regulation of AQP genes can be positively correlated with hydraulic conductance (Secchi et al. 2007, Secchi and Zwieniecki 2010) and transpiration rate (Aroca et al. 2006). Thus increased aquaporin expression during drought is often seen as a mechanism of maintaining hydraulic conductance and transpiration (Sade et al. 2010). Considerable variation, however, exists between species and even between cultivars in expression of AQP isoforms in response to drought. Cocozza et al. (2010) found that a *Populus nigra* clone which reduced stomatal conductance (as indicated by enrichment in carbon isotope composition in leaves) under drought also showed enhanced expression in the roots of two aquaporin genes, during drought. Another clone, which showed proline accumulation under drought in old leaves (this may allow osmotic adjustment – see later section – and hence maintenance of stomatal conductance), showed down-regulation of the same genes. The authors deduced that the clone that reduced stomatal conductance under drought increased the permeability of vascular tissue by overexpressing aquaporin genes, in order to facilitate water transport, whereas the proline-accumulating clone increased water conservation in root cells by down-regulating aquaporins. Drought was found to increase the expression of one of three PIPs examined in the leaves of *Phaseolus vulgaris*, and of all three PIPs in the roots (Aroca et al. 2006). *PvPIP2;1* gene expression and PIP1 protein abundance were increased in the leaves also when ABA or methotrexate (MTX), an inhibitor of stomatal opening, were applied. None of the treatments in that study changed the leaf water status, suggesting that rapid stomatal closure allowed leaf water status to be maintained. Almeida-Rodriguez et al. (2010) also found up-regulation of six out of 11 studied aquaporin genes in response to drought in the poplar *Populus simonii* × *balsamifera*, which they found to be drought avoiding, rapidly reducing stomatal conductance in response to stress. In contrast, no up-regulation of these genes was found in response to stress in the less drought-avoiding *P. balsamifera*. Secchi et al. (2006) consider that the trunk diameter fluctuations found in drought-sensitive *Olea europaea* require rapid water transport, involving aquaporin expression and gating. They found that expression of a PIP1, a PIP2, and a TIP1 was drastically decreased in response to 3–4 weeks of drought stress. Twig water potential and hydraulic conductivity were also decreased by the stress. Up-regulation of a PIP isoform occurred in the anisohydric grapevine "Chardonnay" during drought stress, but not in "Grenache", which shows far stronger stomatal control of water loss during drought (Vandeleur et al. 2009), indicating that there is no consistent correlation between the relative stomatal control in different species and aquaporin expression during drought. PIP isoforms in leaves of the model plant *Arabidopsis thaliana* were generally down-regulated upon gradual drought stress (which was reflected at the protein level) (Alexandersson et al. 2005).

Up-regulation and down-regulation of genes during drought stress has been found to be dependent on the method of water withdrawal employed (Bray 2004). In leaves of Richter-110 (a *Vitis* hybrid), expression levels were strongly decreased under short-term moderate water stress but maintained or increased under short-term severe water stress. After maintenance for 7 days under water stress, however, transcript abundance was approximately 50% of that in well-watered plants, regardless of whether stress was mild or severe (Galmés et al. 2007).

Some of the contrasting behaviour of different aquaporins (up- vs. down-regulation) may be explained by the hypothesis that two "classes" of AQP exist: (1) AQPs involved in the maintenance of cellular water (osmotic) status, which are not involved in controlling large stress-related variations in water potential, but which buffer local variations at the cell level or allow the movement of some gases of small non-electrolytes through membranes, and (2) AQPs that are specifically expressed and/or regulated following stresses in appropriate organs to compensate for the altered water potential (Hachez et al. 2006). In the stem parenchyma of *Populus trichocarpa*, PIP1 aquaporins were up-regulated in response to drought stress, but PIP2 aquaporins were not (Secchi and Zwieniecki 2010). Interestingly, in that study, two PIP1 aquaporin genes which were *not* found to be up-regulated in response to drought *were* found to be up-regulated in response to xylem vessel embolism artificially induced without the presence of water stress. In addition, those two genes were down-regulated after embolism removal, suggesting a local role of these particular channels in refilling of embolisms. Expression profiles can also differ for the same aquaporin gene in different organs (Aroca et al. 2006; Guo et al. 2006; Galmés et al. 2007). Zhang et al. (2008) compared the expression of a TIP in seedlings of a cultivar of wheat growing in water and growing in polyethylene glycol (PEG). Expression in shoots was up-regulated while that in roots was down-regulated in the PEG treatment.

Expression in shoots was also up-regulated when salt was added to the water. When ABA was added to the water, no effect on transcript abundance was detected. This may suggest a direct effect of the osmotic potential of the solution on the aquaporin expression, as opposed to the expression being regulated by the transpirational demand. Plants were also grown with split roots, where half the roots were in PEG and the expression in these roots was down-regulated, while that in the other half of the roots was up-regulated. In this case, there was no change in transcript abundance in shoots compared to control – and osmotic potential of the shoots of the split-root plants was very similar to that of control plants. The authors suggest that this aquaporin must be involved in distribution of water from where there is enough water to where there is less.

Comprehensive studies using macro or microarrays are required to determine with certainty whether AQP genes are expressed in a coordinated fashion (Javot and Maurel 2002). Such a study would ideally be undertaken in combination with assessment of hydraulic conductivity, to determine whether expression parallels water transport. It should be noted that there have been several reports showing that the respective abundance in aquaporin transcripts and in the encoded proteins are not necessarily correlated (see discussion in Luu and Maurel 2005). While much of the assessment of the role of aquaporins in water transport during drought stress has been based on studies of levels of aquaporin transcription. Water stress also acts on aquaporin protein relocalisation and on gating via reversible phosphorylation or via direct effects of osmotic or hydrostatic gradients (Luu and Maurel 2005).

2.4 Reactive Oxygen Species

Stomatal closure as a result of drought coincides with exposure to high photosynthetically active radiation. When leaves are subjected to excess incident radiation relative to the available intracellular CO_2, the rate of electron production exceeds the rate of electron use in the Calvin cycle. Reactive oxygen species (ROS), such as the superoxide anion ($O_2^{\bullet-}$), hydrogen peroxide (H_2O_2), the hydroxyl radical ($HO^{\bullet}$), and singlet oxygen ($^1O_2$), are therefore produced, particularly in the chloroplasts, which are both the main producers as well as targets of ROS (Sofo et al. 2005). The ROS react with proteins and lipids, causing damage to cellular structures and metabolism, particularly those associated with photosynthesis (Lawlor and Tezara 2009). This situation will ultimately damage the photosynthetic apparatus, unless either photoprotective mechanisms are available to down-regulate photosynthesis, or the decline in CO_2 assimilation coincides with an increase in the strength of another sink for the absorbed radiation. Photoprotective mechanisms include thermal dissipation in the xanthophylls or lutein cycles (Demmig-Adams and Adams III 1996, Garcia-Plazaola et al. 2003), while alternative sinks include photorespiration (Harbinson et al. 1990) or the Mehler peroxidase reaction, in which electrons are transferred from reduced ferredoxin to $O_2^{\bullet-}$. ROS accumulation under such conditions depends on the balance between ROS synthesis and ROS dissipation (discussed below).

Unfortunately, ROS synthesis, dissipation, and damage associated with ROS accumulation has not been quantified under clearly defined levels/duration of irradiance, and levels of water deficit (Lawlor and Tezara 2009), leaving the impact of this system poorly understood as yet. Also, ROS are generally considered to lead to photodamage, but Nishiyama et al. (2011) argue that the impact of ROS is more related to inactivation of repair of photodamaged PSII than to the photodamage itself. Another complication in understanding the role of ROS in drought-stressed plants is that in addition to causing damage, ROS also act as signal molecules that activate multiple defence responses: Increased ROS production and the high redox state of the electron transport chain during water deficit induce expression of genes coding for components of energy-dissipating and regulation systems in chloroplasts, which assists in acclimation (Pfannschmidt et al. 2009).

2.4.1 Scavenging of Reactive Oxygen Species

Plants use both enzymatic and non-enzymatic antioxidant defence mechanisms to scavenge ROS. The enzymatic system includes superoxide dismutases (SOD), which act as the first line of defence against superoxide radicals as they catalyse the dismutation of superoxide radicals to H_2O_2 and O_2 (Fridovich 1995). The subsequent defences are mostly concerned with depleting the resulting H_2O_2 before it can be converted to the highly reactive (and extremely damaging) hydroxyl radical ($HO^{\bullet}$) by the Fenton reaction, in the presence of ferrous (Fe^{2+}) ions (Mittler 2002). The enzymes involved include catalase (CAT), guaiacol-type peroxidases (POD), and enzymes of the ascorbate-glutathione cycle (Mittler 2002), such as ascorbate peroxidase (APX). In the process of converting $O_2^{\bullet-}$ to H_2O_2 and O_2, ascorbic acid (AsA) is oxidised to form the monohydroascorbate radical (MDA) that is reduced back to AsA by either reduced ferredoxin or by NADPH, catalysed by MDA reductase (MDAR). Dehydroascorbate (DHA) is produced when MDAR fails to reduce MDA to AsA, and is reduced to AsA by DHA reductase (DHAR). Alternatively, in the glutathione peroxidase (GPX) cycle, glutathione (GSH) is required to restore AsA as the electron donor (Pfannschmidt et al. 2009); glutathione reductase (GR) catalyses the NADPH-dependent reduction of oxidised glutathione to its reduced form (Ahmad et al. 2010). Polyphenol oxidase isoenzymes, located mainly in the thylakoid lumen, oxidise *o*-diphenolic substrates to *o*-quinones, and are therefore involved in the metabolism of phenols, which have a non-enzymatic antioxidant action. In another cycle, the catalase cycle (Mittler 2002), catalases – heme-containing enzymes particularly abundant in the glyoxysomes – destroy the H_2O_2 generated by oxidases involved in the β-oxidation of fatty acids, and in the peroxisomes of green leaves, where they scavenge the H_2O_2 arising from the oxidation of the photorespiratory-produced glycolate.

Changes of expression and activities of antioxidant enzymes have been detected in many species of plants in response to adverse environmental conditions, such as water deficit and other abiotic, biotic, and developmental stimuli (Ahmad et al. 2010). Sofo et al. (2005) found that the activities of SOD, APX, CAT, and POD increased in relation to the severity of drought stress in both leaves and roots of olive (grown under high temperature and irradiance). In particular, a marked increase in APX activity was found in leaves of plants during severe drought stress. The authors suggested that up-regulation of the antioxidant system might be an important attribute linked to drought tolerance, which could limit cellular damage caused by ROS during water deficit. APX in the roots, in contrast, showed reduced levels of activity, possibly indicating that APX activity could be attributed mainly to the chloroplast-located enzyme (chlAPX) of leaf tissues. The huge increase in APX activity in the leaves under drought could explain how the chloroplasts were sufficiently protected against reactive oxygen species to maintain high electron transport rates.

Over-expression of one or more ROS-scavenging enzymes in various compartments has been shown to relieve oxidative stress (Miller et al. 2010). Eltayeb et al. (2007) found that over-expression of a MDAR gene in tobacco resulted in enhanced tolerance of PEG-induced water stress; the authors suggested this may be due to increased levels of AsA which mainly resulted from the enhanced activity of MDAR.

Accurate characterisation of the complex stress tolerance phenotypes of transgenic plants (over-)expressing a variety of antioxidant enzymes has been identified as a significant challenge in understanding antioxidant defences (Allen et al. 1997). To date, assessment of the behaviour of mutants with altered ROS-scavenging capacity has focussed on stress factors other than drought. Thus much work is still needed to better understand the role of ROS and ROS-scavenging in drought tolerance.

2.5 Osmotic Adjustment

Osmotic adjustment relates to the lowering of osmotic potential due to the net accumulation of solutes in response to water deficits (Zhang et al. 1999).

Osmotic adjustment is often induced during drought (Chaves et al. 2009), with solutes accumulating, resulting in the maintenance of a higher turgor potential at a given leaf water potential (Zhang et al. 1999). Different types of compatible solutes can be responsible e.g. various sugars, organic acids, amino acids, sugar alcohols, and ions. Concentrations of soluble sugars (sucrose, glucose, and fructose) are altered by drought – in general concentrations increase (Chaves and Oliveira 2004) – although under severe dehydration they may decrease (Pinheiro et al. 2001). Soluble sugars act as signalling molecules under stress (Chaves and Oliveira 2004), interact with hormones, and modify the expression of genes involved in photosynthetic metabolism – generally resulting in a reduction in source activity such as photoassimilate export and an increase in sink activity such as production of lipids and proteins (Chaves et al. 2009).

Osmotic adjustment in plant cells can aid the maintenance of water uptake and cell turgor during stress, and therefore can allow a plant to continue growth during water deficit, since zero turgor occurs at a lower water potential in osmotically adjusted leaf tissue. However, where water supply is not replenished, continued abstraction of water will ultimately be detrimental – and thus osmotic adjustment is not always advantageous (Sinclair and Purcell 2005). Engineering osmoprotectant synthesis pathways into model plant species has led to significant (albeit modest) improvements in stress tolerance; adding multiple genes to increase osmoprotectant flux in response to stress may be more beneficial (Rathinasabapathi 2000).

2.6 Other Adjustments

Developments in molecular biology have opened up the possibility of exploring the role of diverse molecules in drought tolerance. Many molecular adjustments have been found during drought stress, and comparison of drought-tolerant and non-drought tolerant lines has been used to indicate whether or not the extent of such adjustments may in some way be related to drought tolerance. It has been suggested, for example, that microRNAs may play a role in drought tolerance in maize (Kong et al. 2010). MicroRNAs are small RNA molecules that are important regulators of gene expression at the post-transcriptional level by repressing mRNA expression (Covarrubias and Reyes 2010). The expression of a wide range of genes is altered during drought. Dehydrins are among the most frequently observed proteins induced by dehydration and may help in stabilizing membranes or proteins during stress (Egerton-Warburton et al. 1997). A relationship has been suggested between both water-soluble inositol-polyphophates and membrane lipid polyphosphoinositides and drought stress (Munnik and Vermeer 2010). The activity of phospholipase D (PLD), which regulates the production of phosphatidic acid – a key class of lipid mediators in plant response to environmental stress, increases under drought (Hong et al. 2010). PLDα1 is particularly interesting with respect to drought tolerance, since it promotes stomatal closure and reduces water loss.

The method of imposing drought in many molecular papers, however, limits their application to "real" drought situations. For example, in the above-mentioned microRNA publication (Kong et al. 2010), "drought" actually involved dehydration by removing plants from soil and leaving them on filter paper – and Pinheiro and Chaves (2011) found that results from such experiments are very different to those where plants are subjected to drought conditions in soil/growing media. The lack of measurement of plant water relations in many molecular studies also means comparisons cannot be made across studies. A specific disadvantage of transcriptomic analysis is that in most comparisons protein abundance correlated very poorly with gene expression (Deyholos 2010), which can be particularly problematic in stress physiology, where sometimes only a small portion of the transcripts representing a specific subset of genes are actively translated. Deyholos (2010) pointed out another problem with many transcriptomic studies: they tend to focus on young tissue, which may not be the most relevant tissue in "real" crops in the field. Nonetheless, collaboration between ecophysiologists, agronomists, and molecular biologists in improving these investigations should be encouraged and is essential to optimise our understanding of plant responses to drought. In particular, the generation of mapping populations of contrasting cultivars or ecotypes has provided powerful new resources for dissecting the genetic basis of differences in drought tolerance and/or water use efficiency (e.g. Juenger et al. 2010). "Model" plants, however, are still not necessarily well understood in relation to physiological responses to water stress (Secchi and Zwieniecki 2010) – this needs rapid correction in order to fully exploit the wealth of genetic and genomic data available for such plants.

As highlighted in the introduction, the response of plants to drought is a huge topic. Water stress has an impact on many processes e.g. inflorescence development (Setter et al. 2011) that are outside the scope of this chapter. Several different processes interact. To give some examples, Kadioglu and Terzi (2007) highlight links between ROS-scavenging, osmolyte accumulation, and leaf rolling in dehydration avoidance; AQP down-regulation during drought stress may be a response to a cascade of events triggered initially by ROS accumulation (Luu and Maurel 2005); while ABA induces transcription factors that regulate the expression of PIP AQPs (Kaldenhoff et al. 1996).

3 Exploiting the Impact of Drought Stress on Plant Physiology

3.1 Exploiting Stomatal Closure: Deficit Irrigation and Partial Rootzone Drying

It has been known for some time that chemical compounds synthesised in drying roots can act as long-distance signals, which induce stomatal closure in the leaf or restrict leaf growth via arrest of meristematic development. As a result, in some cases, stomatal closure can occur without significant changes in shoot water status. This occurs where plant water potential is buffered by controlling stomatal aperture via feed-forward mechanisms. Plants that show this response are said to be "isohydric". Even where the leaf water potential is similar, plants exposed to a water deficit will have a lower xylem water potential – and it is this that controls leaf growth – on account of a reduced gradient of water potential from the roots through the leaves due to reduced water flux (Tardieu et al. 2010); this difference between xylem and leaf water potential may result in reduced leaf growth in the plants experiencing a water deficit. Careful manipulation of soil water availability to induce a mild water deficit allows minimisation of the impact of the deficit on shoot water status (Davies et al. 2000). This has been exploited in "Deficit Irrigation" strategies, where less than 100% of crop evapotranspiration is replaced by irrigation, and in variations on Deficit Irrigation, one of which is known as Partial Rootzone Drying. In Partial Rootzone Drying, water is applied only to one side of the roots, so that the other side is exposed to drying soil, with the side being irrigated switched at intervals. Such techniques have allowed water savings without reducing yield (e.g., dos Santos et al. 2003, Grant et al. 2004).

The potential of exploiting plant responses to dry soil/substrate is not confined to food production. In landscaping under semi-arid conditions, transplantation is more likely to result in successful growth if the plants have been pre-conditioned to dry conditions. Thus deficit irrigation is increasingly being used in the production of ornamental nursery stock with reduced shoot height and/or leaf area, increased root-collar diameter, root growth potential, and root:shoot ratio, increased osmotic adjustment and water use efficiency, and low stomatal conductance, leaf water and turgor potentials, and relative water content (Franco et al. 2006). Variations on the idea of deficit irrigation can be applied to this end – for example Bañón et al. (2006) exposed *Nerium oleander* seedlings to both deficit irrigation and low air humidity on the nursery, prior to transplant, with the result that mortality after transplant was reduced from 92 to 32% compared to control plants; Franco et al. (2006) highlight that microclimate management during the nursery phase can be an effective means of producing high-quality seedlings capable of withstanding transplant shock and capable of rapid establishment in arid landscapes. Even where water is not limiting, deficit irrigation can be used to control the size and quality of hardy ornamental nursery stock (Cameron et al. 2006), particularly where the application of deficit irrigation is combined with an efficient method of sensing plant water requirements that is suited to application on the nursery (Grant et al. 2011).

In addition to the impact of long-distance signalling, mild water deficits also exert direct or indirect impacts on yield and the quality of harvested products: for example, reduced leaf growth may improve the light environment around fruit while the fruit are developing.

3.2 Exploiting Genotypic Variation in Breeding Programmes

Long-term sustainability requires not only more resource-efficient production systems but the development of crops with improved resource use efficiency, and crops which are tolerant to specific stresses. Clearly there is substantial variation between cultivars in tolerance of water deficits, which could be exploited in breeding programmes. This has been clear for many years in the major food crops (e.g., Fischer and Wood 1979, Quarrie 1983, Farquhar and Richards 1984), but such variation is also now being revealed in a wide range of other crops – for example, fruit crops (Klamkowski and Treder 2008, Grant et al. 2010), energy crops (Zub and Brancourt-Hulmel 2010), and forest trees (Brendel et al. 2008).

Despite such genetic variation, there has been limited progress to date in developing drought tolerance (Tuberosa and Salvi 2006). This is partly due to a shortage of research exploring the physiological impact of altering crop genetics. The genetic resources for investigating drought tolerance have increased dramatically in recent years, but phenotypic research has not kept pace (Miflin 2000, Verslues et al. 2006). This in turn is largely a consequence of the difficulty of measuring plant physiological processes with the required spatial and temporal resolution to match molecular investigations.

Stomatal conductance (g_s) largely determines transpirational water loss, and also influences photosynthetic assimilation and hence yield. It is therefore useful to screen genotypes for g_s in programmes to improve yield or water use efficiency (e.g., Gutierrez-Rodriguez et al. 2000). Cuvette-methods for measuring g_s, however, are not suited to large-scale screening. Infrared thermometry (and more particularly infrared thermal imaging) has opened up the possibility of remotely assessing the temperature of individual leaves or whole crop canopies (Jones 2004) and may be a useful tool for early-generation selection of physiologically superior lines in breeding programmes aiming to increase yield or reduce the impact of stressful environments (Blum et al. 1982, Amani et al. 1996, Reynolds et al. 1999, Olivares-Villegas et al. 2007, Lopes and Reynolds 2010). Chlorophyll fluorescence provides information on photochemistry, including photochemical capacity and electron transport rate. Chlorophyll fluorescence imaging in the field is still in its infancy, but the photochemical reflectance index (PRI) derived from images taken at the 530 nm and 570 nm wavelengths may provide a good indication of the photosynthetic functioning of plant leaves (Inoue and Peñuelas 2006), and has also been suggested as a good indicator of water stress (Súarez et al. 2010). Combined together, thermal and chlorophyll fluorescence images taken in quick succession on the same area of leaves could be used to rapidly determine the ratio of photosynthetic assimilation to stomatal conductance i.e., WUE_i (see review by Chaerle et al. 2007), but to date imaging has not been used for this purpose.

From the above, a picture emerges of imaging techniques potentially providing a solution to the problem of needing to screen physiological responses over large numbers of plants and at different time-points during the imposition of stress or during the crop's development (Munns et al. 2010), as they allow real-time continuous assessment of physiological processes across organs, whole plants, or even whole plant populations or crops. More established phenotyping techniques include assessment of stable isotope composition. During photosynthesis, the extent of discrimination against the naturally abundant and heavier isotope of carbon, ^{13}C, is related to the ratio of internal to external partial pressure of CO_2 (p_i/p_a), which is controlled by both stomatal conductance and photosynthetic capacity, and therefore is indirectly related to water use efficiency (Farquhar and Richards 1984), with greater enrichment being associated with greater photosynthetic water use efficiency. The level of enrichment can be measured by determining the ratio of ^{13}C and ^{12}C and comparing the ratio to that of a standard, to determine the carbon isotope composition, $\delta^{13}C$. $\delta^{13}C$ integrates the ratio of assimilation to transpiration over the duration in which dry matter is assimilated. The measurement is suited to screening in crop breeding programmes, since only small samples of material are required, harvesting plant material is rapid compared to measurement of, for example, photosynthetic or transpiration rates, and once dried the plant material can be stored until used for isotope analysis. Additionally, $\delta^{13}C$ shows high heritability (Condon and Richards 1992, Richards 1996). Use of $\delta^{13}C$ to screen for variation in WUE_i has been successfully applied in breeding programmes for water-limited environments (Condon et al. 2004). In the case of bread wheat under rainfed conditions, selection for high $\delta^{13}C$ was more efficient than direct selection for either high biomass or high yield, and has led to increases in grain yield (Rebetzke et al. 2002). Genotypes with higher photosynthetic water use efficiency should be more productive where water availability is limited. However, under irrigation, genotypes with lower photosynthetic water use efficiency (relating to higher transpiration) have been found in some cases to perform better (Araus et al. 2003). Low water use efficiency in such situations is associated with faster growth, and consequently, at least in cereals, greater total biomass and higher yields (Condon et al. 2004). In such cases, higher yield is associated with poorer enrichment in ^{13}C. For irrigated crops, this needs to be borne in mind in prediction of genotypic advantages on the basis of $\delta^{13}C$. $\delta^{13}C$ analysis may need to be combined with other techniques to select for genotypes with both relatively high instantaneous water use efficiency and relatively high transpiration under the desired environmental conditions. Assessment of oxygen isotope composition ($\delta^{18}O$, which reflects evaporative enrichment in leaves due to transpiration, and has been shown in some crops to correlate negatively with transpiration rate (Cabrera-Bosquet et al. 2009) in parallel with $\delta^{13}C$ could indicate the extent to which WUE_i is influenced by transpiration. While isotope composition is undoubtedly an effective measure in many situations, isotope composition analysis (particularly with respect to $\delta^{18}O$) is relatively costly and preparation of the plant material is labour-intensive.

3.2.1 QTL Mapping

Although conventional breeding has been successful to some extent in developing drought-tolerant cultivars, such selection programmes are expensive and slow, and it has proven impossible to incorporate certain traits, for example, osmotic adjustment, via conventional breeding (Zhang et al. 1999). More rapid development of varieties with drought resistance should be possible through exploitation of mapped molecular markers (Price and Courtois 1999). Restriction fragment length polymorphism (RFLP) and other molecular markers allow loci-controlling traits related to drought tolerance to be identified and located in the genome. Mapping of quantitative trait loci (QTL) allows dissection of complex traits into their components, each of which is controlled by one or more QTL. Genetic markers at the identified QTL can then be used in rapid selection of plants with the desired alleles. As explained by Price and Courtois (1999), the process is as follows: A trait with potential value in breeding for drought tolerance is chosen, parental lines displaying extreme phenotypes for this trait are identified, these lines are crossed to produce progenies that segregate for the trait of interest, the trait in question is determined in each of the progeny, the parents are screened for genetic polymorphism in many molecular markers, the genotype of each progeny is determined for the selected markers, a genetic map is constructed from the marker data, and finally markers associated with the trait are identified using analysis of variance or interval mapping. Near-isogenic lines, produced through repeated back-crossing, allow fine-scale mapping of QTL (Yadav et al. 2011). If QTL for traits expected to relate to drought tolerance are developed, then near-isogenic lines can be used to characterise the impact of specific QTL on yield (or other commercially-desirable characteristics) in the presence or absence of drought (Price and Courtois 1999). Alternatively, recombinant inbred lines can be used – Sanchez et al. (2002), for example, identified four QTL associated with the stay-green trait in sorghum – a trait which confers resistance to premature senescence under water stress post-flowering – using a recombinant inbred line population; these QTL accounted for more than 50% of the phenotypic variance in field trials. Another example of the power of QTL mapping is the development of near-isogenic lines from two elite cotton cultivars (Levi et al. 2011); each near-isogenic line expressed advantageous osmotic adjustment in comparison to the parents. While there was a tendency for the near-isogenic lines to show higher concentrations of certain metabolites than the parents, the authors acknowledge that further analysis is necessary in order to establish a direct impact of these metabolite concentrations on drought tolerance, but highlight the value of developing genomics information on different crops for dissecting the causes of enhanced drought tolerance. Work is also underway to determine QTL associated with such traits as root morphological characters (Price and Courtois 1999).

Although QTL mapping holds great promise, mapping drought tolerance remains complicated. One reason is that naturally occurring drought is often unpredictable and therefore researchers must artificially create drought – but while this allows control of the timing and duration of drought, it may not be representative of the "real" droughts to which the crops are normally exposed (Price and Courtois 1999). As with conventional breeding, the choice of trial sites and the way in which drought is imposed remain hugely important.

Stress-induced transcript expression profiles have been used to identify candidate genes associated with QTLs (Diab et al. 2004, Gorantla et al. 2005, Street et al. 2006). Where both a genome sequence and microarrays are available for a species, expressed sequence tags (ESTs) spotted on microarrays can be located on the genome sequence; if QTL regions have already been identified, then all ESTs on a microarray and within a QTL region can be examined for differential expression (Street et al. 2006). The expression levels when quantified within a mapping population are used to map "expression QTL".

3.3 Exploiting Gene Expression During Drought: Genetic Engineering

It is now appreciated that an understanding of photosynthetic metabolism under drought is essential if plants are to be engineered that maintain high yield under water deficit (Lawlor and Tezara 2009). A couple of examples provide an indication of the potential, but also the complications, involved in genetic engineering for drought tolerance.

As the first example, over-expression of SODs that function in the water–water cycle in chloroplasts in theory would improve photosynthesis during environmental stress, allowing the maintenance of close to normal levels of PSII (photosystem II) and PSI (photosystem I) activity during stress, decreasing the inhibition of photosynthesis and reducing ROS levels. As discussed earlier, while ROS have the potential to cause oxidative damage to cells during environmental stress, they also appear to play a key role as signal transduction molecules involved in mediating responses to pathogen infection, environmental stress, and different developmental stimuli (Miller et al. 2010). The steady-state level of ROS in cells therefore needs to be carefully controlled – both under normal metabolism and under stress. Elucidating the mechanisms that control ROS signalling in cells during environmental stress could provide a powerful means to enhance the tolerance of crops to such stress.

As a second example, a number of genes encoding osmolyte biosynthesis have been transformed into model plant species (Zhang et al. 1999). The transformants generally show enhanced drought tolerance, yet these studies have not actually shown that the increase in accumulated solutes in the transformants caused osmotic adjustment. Zhang et al. (1999) suggest that osmolytes may play a more complex role in conferring drought resistance than simply contributing to osmotic adjustment, but the emergence of the final expected result (increased drought tolerance) without the expected intermediate result (enhanced osmotic adjustment) is a salutary example of the complicated interaction of different processes in drought response. This also highlights how much work is still required before genetically engineered drought-tolerant varieties are likely to be available for crop production.

4 Conclusions and Future Perspective

In addition to the complications already mentioned in this chapter, it needs to be borne in mind that drought often co-occurs with other stress factors, such as high temperature and irradiance, increased soil salinity, reduced nutrient availability, and hard, compacted soil that impedes root growth (Wilkinson and Davies 2010). Indeed, more than 15% of the 2010 articles on drought relate also to another stress such as salinity, high or low temperature stress, or biotic stress. Such interactions make the already difficult task of understanding plant responses to drought even more difficult, but the very existence of multiple stresses makes the exploitation of "positive" responses to drought (reduced transpiration, enhanced fruit quality, natural tolerance in wild relatives) highly desirable. While huge advances have been made in understanding plant responses to drought and mechanisms of drought tolerance, a large-scale integrated approach is now required to increase the speed and relevance of research in this area. Despite the seriousness of increasing global water shortage, it is clear that plants have many mechanisms to deal with drought stress. The agricultural management and choice of cultivars to target these mechanisms can in some cases enhance the quality of plant products while simultaneously reducing the exploitation of an essential and dwindling natural resource.

Acknowledgements The author is funded by Marie Curie IEF grant 252196, "IMaging of Plant Responses to Environmental StresS" (IMPRESS). Advice from Nada Šurbanovski and Phil Dix on the sections on aquaporins and reactive oxygen species, respectively, is gratefully acknowledged.

References

Ahmad P, Umar S, Sharma S (2010) Mechanism of free radical scavenging and role of phytohormones in plants under abiotic stresses. In: Ashraf M, Ozturk M, Ahmad MSA (eds) Plant adaptation and phytoremediation. Springer Science+Business Media, Heidelberg, pp 99–118

Alexandersson E, Fraysse L, Sjövall-Larsen S, Gustavsson S, Fellert M, Karlsson M, Johanson U, Kjellbom P (2005) Whole gene family expression and drought stress regulation of aquaporins. Plant Mol Biol 59:469–484

Allen RD, Webb RP, Schake SA (1997) Use of transgenic plants to study antioxidant defences. Free Radic Biol Med 23:473–479

Almeida-Rodriguez AM, Cooke JEK, Yeh F, Zwiazek JJ (2010) Functional characterization of drought-responsive aquaporins in *Populus balsamifera* and *Populus simonii×balsamifera* clones with different drought resistance strategies. Physiol Plant 140:321–333

Amani I, Fischer RA, Reynolds MP (1996) Canopy temperature depression association with yield of irrigated spring wheat cultivars in a hot climate. J Agron Crop Sci 176:119–129

Araus JL, Bort J, Steduto P, Villegas D, Royo C (2003) Breeding cereals for Mediterranean conditions: ecophysiological clues for biotechnology application. Ann Appl Biol 14:129–141

Aroca R, Ferrante A, Vernieri P, Chrispeels MJ (2006) Drought, abscisic acid and transpiration rate effects on the regulation of PIP aquaporin gene expression and abundance in *Phaseolus vulgaris* plants. Ann Bot 98:1301–1310

Bañón S, Ochoa J, Franco JA, Alarcón JJ, Sánchez-Blanco MJ (2006) Hardening of oldeander seedlings by deficit irrigation and low air humidity. Environ Exp Bot 56:36–43

Blum A, Mayer J, Golan G (1982) Infrared thermal sensing of plant canopies as a screening technique for dehydration avoidance in wheat. Field Crops Res 5:137–146

Bray EA (2004) Genes commonly regulated by water-deficit stress in *Arabidopsis thaliana*. J Exp Bot 55:2331–2341

Brendel O, Le Thiec D, Scotti-Saintagne C, Bodenes C, Kremer A, Guehl JM (2008) Quantitative trait loci controlling water use efficiency and related traits in Quercus robur L. Tree Genet Genom 4:263–278

Cabrera-Bosquet L, Molero G, Nogues S, Araus JL (2009) Water and nitrogen conditions affect the relationships

of $\Delta^{13}C$ and $\Delta^{18}O$ to gas exchange and growth in durum wheat. J Exp Bot 60:1633–1644
Cameron RWF, Harrison-Murray RS, Atkinson CJ, Judd HL (2006) Regulated deficit irrigation – a means to control growth in woody ornamentals. J Hort Sci Biotechnol 81:435–443
Cattivelli L, Rizza F, Badeck FW, Mazzucotelli E, Mastrangelo EM, Francia E, Marè C, Tondelli A, Stanca AM (2008) Drought tolerance improvement in crop plants: an integrated view from breeding to genomics. Field Crops Res 105:1–14
Chaerle L, Leinonen I, Jones HG, Van Der Straeten D (2007) Monitoring and screening plant populations with combined thermal and chlorophyll fluorescence imaging. J Exp Bot 58:773–784
Chaves MM, Flexas J, Pinheiro C (2009) Photosynthesis under drought and salt stress: regulation mechanisms from whole plant to cell. Ann Bot 103:551–560
Chaves MM, Maroco JP, Pereira JS (2003) Understanding plant responses to drought – from genes to the whole plant. Funct Plant Biol 30:239–264
Chaves MM, Oliveira MM (2004) Mechanisms underlying plant resilience to water deficits: prospects for water-saving agriculture. J Exp Bot 55:2365–2384
Cocozza C, Cherubini P, Regier N, Saurer M, Frey B, Tognetti R (2010) Early effects of water deficit on two parental clones of *Populus nigra* grown under different environmental conditions. Funct Plant Biol 37:244–254
Condon AG, Richards RA (1992) Broad sense heritability and genotype×environment interaction for carbon isotope discrimination in field-grown wheat. Aust J Agric Res 43:921–934
Condon AG, Richards RA, Rebetzke GJ, Farquhar GD (2004) Breeding for high water-use efficiency. J Exp Bot 55:2447–2460
Covarrubias AA, Reyes JL (2010) Post-transcriptional gene regulation of salinity and drought responses by plant microRNAs. Plant Cell Environ 33:481–489
David TS, Henriques MO, Kurz-Besson C, Nunes J, Valente F, Vaz M, Pereira JS, Siegwolf R, Chaves MM, Gazarini LC, David JS (2007) Water-use strategies in two co-occurring Mediterranean evergreen oaks: surviving the summer drought. Tree Physiol 27:793–803
Davies WJ, Bacon MA, Thompson DS, Sobeih W, Rodríguez LG (2000) Regulation of leaf and fruit growth in plants growing in drying soil: exploitation of the plants' chemical signalling system and hydraulic architecture to increase the efficiency of water use in agriculture. J Exp Bot 51:1617–1626
Demmig-Adams B, Adams WW III (1996) The role of xanthophyll cycle carotenoids in the protection of photosynthesis. Trends Plant Sci 1:21–26
Deyholos MK (2010) Making the most of drought and salinity transcriptomics. Plant Cell Environ 33: 648–654
Diab AA, Teulat-Merah B, This D, Ozturk NZ, Benscher D, Sorrells ME (2004) Identification of drought-inducible genes and differentially expressed sequence tags in barley. Theor Appl Genet 109:14147–11425
Dodd IC, Theobald JC, Bacon MA, Davies WJ (2006) Alternation of wet and dry sides during partial rootzone drying irrigation alters root-to-shoot signalling of abscisic acid. Funct Plant Biol 33:1081–1089
dos Santos TP, Lopes CM, Rodrigues ML, de Souza CR, Maroco JP, Pereira JS, Silva JR, Chaves MM (2003) Partial rootzone drying: effects on growth and fruit quality of field-grown grapevines (*Vitis vinifera*). Funct Plant Biol 30:663–671
Egerton-Warburton LM, Balsamo RA, Close TJ (1997) Temporal accumulation and ultrastructural localization of dehydrins in Zea mays. Physiol Plant 101: 545–555
Eltayeb AE, Kawano N, Badawi GH, Kaminaka H, Sanekata T, Shibahara T, Inanaga S, Tanaka K (2007) Overexpression of monodehydroascorbate reductase in transgenic tobacco confers enhanced tolerance to ozone, salt and polyethylene glycol stresses. Planta 225:1255–1264
Farquhar GD, O'Leary MH, Berry JA (1982) On the relationship between carbon isotope discrimination and the intercellular carbon dioxide concentration in leaves. Aust J Plant Physiol 9:121–137
Farquhar GD, Richards RA (1984) Isotopic composition of plant carbon correlates with water-use-efficiency of wheat genotypes. Aust J Plant Physiol 11:539–552
Fereres E, Connor DJ (2004) Sustainable water management in agriculture. In: Cabrera E, Cobacho R (eds) Challenges of the new water policies for the XXI century. A.A. Balkema, Lisse, pp 157–170
Fischer RA, Wood JT (1979) Drought resistance in spring wheat cultivars. III. Yield association with morphophysiological traits. Aust J Agric Res 30:1001–1020
Flexas J, Medrano H (2002) Drought-inhibition of photosynthesis in C_3 plants: stomatal and non-stomatal limitations revisited. Ann Bot 89:183–189
Franco JA, Martínez-Sánchez JJ, Fernández JA, Bañón S (2006) Selection and nursery production of ornamental plants for landscaping and xerogardening in semiarid environments. J Hort Sci Biotechnol 81:3–17
Fridovich I (1995) Superoxide radical and superoxide dismutases. Annu Rev Biochem 64:97–112
Galmés J, Pou A, Alsina MM, Tomàs M, Medrano H, Flexas J (2007) Aquaporin expression in response to different water stress intensities and recovery in Richter-110 (*Vitis* sp.): relationship with ecophysiological status. Planta 226:671–681
Garcia-Plazaola JI, Hernández A, Olano JM, Becerril JM (2003) The operation of the lutein epoxide cycle correlates with energy dissipation. Funct Plant Biol 30: 319–324
Gorantla M, Babu PR, Lachagari VBR, Feltus FA, Paterson AH, Reddy AR (2005) Functional genomics of drought stress response in rice: Transcript mapping of annotated unigenes of an indica rice (*Oryza sativa* L. cv. Nagina 22). Curr Sci 89:496–514
Grant OM, Davies MJ, Longbottom H, Herrero A, Harrison-Murray R (2011) Application of deficit irrigation to controlling growth of hardy nursery stock. Acta Horticulturae 889:409–416

Grant OM, Johnson AW, Davies MJ, James CM, Simpson DW (2010) Physiological and morphological diversity of cultivated strawberry (*Fragaria × ananassa*) in response to water deficit. Environ Exp Bot 68:264–272

Grant OM, Stoll M, Jones HG (2004) Partial rootzone drying does not affect fruit yields of raspberries. J Hort Sci Biotechnol 79:125–130

Gutierrez-Rodriguez M, Reynolds MP, Larque-Saavedra A (2000) Photosynthesis of wheat in a warm, irrigated environment – II. Traits associated with genetic gains in yield. Field Crops Res 66:51–62

Guo L, Wang ZY, Lin H, Cui WE, Chen J, Liu M, Chen ZL, Qu LJ, Gu H (2006) Expression and functional analysis of the rice plasma-membrane intrinsic protein gene family. Cell Research 16:277–286

Hachez C, Zelazny E, Chaumont F (2006) Modulating the expression of aquaporin genes in planta: a key to understand their physiological functions? Biochim Biophys Acta 1758:1142–1156

Harbinson J, Genty B, Foyer CH (1990) Relationship between photosynthetic electron transport and stromal enzyme activity in pea leaves. Plant Physiol 94: 545–553

Hartung W, Wilkinson S, Davies W (1998) Factors that regulate abscisic acid concentrations at the primary site of action at the guard cell. J Exp Bot 51:361–367

Hong YY, Zhang WH, Wang XM (2010) Phospholipase D and phosphatidic acid signalling in plant response to drought and salinity. Plant Cell Environ 33:627–635

Inoue Y, Peñuelas J (2006) Relationship between light use efficiency and photochemical reflectance index in soybean leaves as affected by soil water content. Int J Remote Sens 27:5109–5114

Javot H, Maurel C (2002) The role of aquaporins in root water uptake. Ann Bot 90:301–313

Jones HG (2004) Application of thermal imaging and infrared sensing in plant physiology and ecophysiology. Adv Bot Res 41:107–163

Juenger TE, Sen S, Bray E, Stahl E, Wayne T, McKay J, Richards JH (2010) Exploring genetic and expression differences between physiologically extreme ecotypes: comparative genomic hybridization and gene expression studies of Kas-1 and Tsu-1 accessions of *Arabidopsis thaliana*. Plant Cell Environ 33:1268–1284

Jury WA, Vaux H (2005) The role of science in solving the world's emerging water problems. Proc Natl Acad Sci USA 102:15715–15720

Kadioglu A, Terzi R (2007) A dehydration avoidance mechanism: leaf rolling. Bot Rev 73:290–302

Kaldenhoff R, Kolling A, Richter G (1996) Regulation of the *Arabidopsis thaliana* aquaporin gene AthH2 (PIPlb). J Photochem Photobiol 36:351–354

Kjellbom P, Larsson C, Johansson I, Karlsson M, Johanson U (1999) Aquaporins and water homeostasis in plants. Trends Plant Sci 4:1360–1385

Klamkowski K, Treder W (2008) Response to drought stress of three strawberry cultivars grown under greenhouse conditions. J Fruit Ornament Plant Res 16:179–188

Knipfer T, Fricke W (2011) Water uptake by seminal and adventitious roots in relation to whole-plant water flow in barley (*Hordeum vulgare* L.). J Exp Bot 62: 717–733

Kong YM, Elling AA, Chen B, Wang Deng X (2010) Differential expression of microRNAs in maize inbred and hybrid lines during salt and drought stress. Am J Plant Sci 1:69–76

Lawlor DW, Tezara W (2009) Causes of decreased photosynthetic rate and metabolic capacity in water-deficient leaf cells: a critical evaluation of mechanisms and integration of processes. Ann Bot 103:561–579

Levi A, Paterson AH, Cakmk I, Saranga Y (2011) Metabolite and mineral analyses of cotton near-isogenic lines introgressed with QTLs for productivity and drought-related traits. Physiol Plant 141: 265–275

Lopes MS, Reynolds MP (2010) Partitioning of assimilates to deeper roots is associated with cooler canopies and increased yield under drought in wheat. Funct Plant Biol 37:147–156

Luu D-T, Maurel C (2005) Aquaporins in a challenging environment: molecular gears for adjusting plant water status. Plant Cell Environ 28:85–96

Miflin B (2000) Crop improvement in the 21st century. J Exp Bot 51:1–8

Miller G, Suzuki N, Ciftci-Yilmaz S, Mittler R (2010) Reactive oxygen species homeostasis and signalling during drought and salinity stresses. Plant Cell Environ 33:453–467

Mittler R (2002) Oxidative stress, antioxidants and stress tolerance. Trends Plant Sci 11:15–19

Munnik T, Vermeer JEM (2010) Osmotic stress-induced phosphoinositide and inositol phosphate signalling in plants. Plant Cell Environ 33:655–669

Munns R, James RA, Sirault XRR, Furbank RT, Jones HG (2010) New phenotyping methods for screening wheat and barley for beneficial responses to water deficit. J Exp Bot 61:3499–3507

Nishiyama Y, Allakhverdiev SI, Murata N (2011) Protein synthesis is the primary target of reactive oxygen species in the photoinhibition of photosytem II. PhysiologiaPlantarum.doi:10.1111/j.1399-3054.2011.01457.x

Olivares-Villegas JJ, Reynolds MP, McDonald GK (2007) Drought-adaptive attributes in the Seri/Babax hexaploid wheat population. Funct Plant Biol 34:189–203

Pfannschmidt T, Bräutigam K, Wagner R, Dietzel L, Schröter Y, Steiner S, Nykytenko A (2009) Potential regulation of gene expression in photoysnthetic cells by redox and energy state: approaches towards better understanding. Ann Bot 103:599–607

Pinheiro C, Chaves MM (2011) Photosynthesis and drought: can we make metabolic connections from available data? J Exp Bot 62:869–882

Pinheiro C, Chaves MM, Ricardo CP (2001) Alterations in carbon and nitrogen metabolism induced by water deficit in the stems and leaves of *Lupinus albus*. J Exp Bot 52:1063–1070

Price A, Courtois B (1999) Mapping QTLs associated with drought resistance in rice: progress, problems and prospects. Plant Growth Regul 29:123–133

Quarrie SA (1983) Characterization of spring wheat genotypes differing in drought-induced abscisic acid accumulation: II. Leaf Water Relat 34:1528–1540

Rathinasabapathi B (2000) Metabolic engineering for stress tolerance: installing osmoprotectant synthesis pathways. Ann Bot 86:709–716

Rebetzke GJ, Condon AG, Richards RA, Farquhar GD (2002) Selection for reduced carbon isotope discrimination increases aerial biomass and crop yield of rainfed bread wheat. Crop Sci 42:739–745

Reynolds MP, Rajaram S, Sayre KD (1999) Physiological and genetic changes of irrigated wheat in the post-green revolution period and approaches for meeting projected global demand. Crop Sci 39:1611–1621

Richards RA (1996) Defining selection criteria to improve yield under drought. Plant Growth Regul 20:157–166

Sade N, Gebretsadik G, Seligmann RA, Schwartz A, Wallach R, Moshelion M (2010) The role of tobacco aquaporin in improving water use efficiency, hydraulic conductivity, and yield production under salt stress. Plant Physiol 152:245–254

Sanchez AC, Subudhi PK, Rosenow DT, Nguyen HT (2002) Mapping QTLs associated with drought resistance in sorghum (*Sorghum bicolor* L. Moench). Plant Mol Biol 48:713–726

Secchi F, Lovisolo C, Schubert A (2007) Expression of OePIP2.1 aquaporin gene and water relations of *Olea europaea* twigs during drought stress and recovery. Ann Appl Biol 150:163–167

Secchi F, Lovisolo C, Uehlein N, Kaldenhoff R, Schubert A (2006) Isolation and functional characterization of three aquaporins from olive (*Olea europaea* L.). Planta 225:381–392

Secchi F, Zwieniecki MA (2010) Patterns of PIP gene expression in *Populus trichocarpa* during recovery from xylem embolism suggest a major role for the PIP1 aquaporin subfamily as moderators of refilling process. Plant Cell Environ 33:1285–1297

Setter TL, Yan J, Warburton M, Ribaut J-M, Xu Y, Sawkins M, Buckler ES, Zhang Z, Gore MA (2011) Genetic association mapping identifies single nucleotide polymorphisms in genes that affect abscisic acid levels in maize floral tissues during drought. J Exp Bot 62:701–716

Sharp RG, Davies WJ (2009) Variability among species in the apoplastic pH signalling response to drying soils. J Exp Bot 60:4363–4370

Shiklomanov IA (1997) Comprehensive assessment of the freshwater resources of the world: assessment of water resources and water availability in the world. World Metereorological Organisatization/Stockholm Environment Institute, Geneva, Switzerland

Sinclair TR, Purcell LC (2005) Is a physiological perspective relevant in a 'genocentric' age? J Exp Bot 56: 2777–2782

Sofo A, Dichio B, Xiloyannis C, Masia A (2005) Antioxidant defences in olive trees during drought stress: changes in activity of some antioxidant enzymes. Funct Plant Biol 32:45–53

Steudle E, Peterson CA (1998) How does water get through roots? J Exp Bot 49:775–788

Street NR, Skogström O, Sjödin A, Tucker J, Rodríguez-Acosta M, Nilsson P, Jansson S, Taylor G (2006) The genetics and genomics of the drought response in *Populus*. Plant J 48:321–341

Súarez L, Zarco-Tejada PJ, González-Dugo V, Berni JAJ, Sagardoy R, Morales F, Fereres E (2010) Detecting water stress effects on fruit quality in orchards with time-series PRI airborne imagery. Remote Sens Environ 114:286–298

Tardieu F, Parent B, Simonneau T (2010) Control of leaf growth by abscisic acid: hydraulic or non-hydraulic processes? Plant Cell Environ 33:636–647

Tuberosa R, Salvi S (2006) Genomics-based approaches to improve drought tolerance of crops. Trends Plant Sci 11:405–411

Vandeleur RK, Mayo G, Shelden MG, Gilliham M, Kaiser BM, Tyerman SD (2009) The role of plasma membrane intrinsic protein aquaporins in water transport through roots: diurnal and drought stress responses reveal different strategies between isohydric and anisohydric cultivars of grapevine. Plant Physiol 149: 445–460

Verslues PE, Agarwal M, Katiyar-Agarwal S, Zhu J, Zhu J-K (2006) Methods and concepts in quantifying resistance to drought, salt and freezing, abiotic stresses that affect plant water status. Plant J 45:523–539

Wallace IS, Roberts DM (2004) Homology modeling of representative subfamilies of *Arabidopsis* major intrinsic proteins. Classification based on the aromatic/arginine selectivity filter. Plant Physiol 135:1059–1068

Wilkinson S, Davies WJ (2008) Manipulation of the apoplastic pH of intact plants mimics stomatal and growth responses to water availability and microclimatic variation. J Exp Bot 59:619–631

Wilkinson S, Davies WJ (2010) Drought, ozone, ABA and ethylene: new insights from cell to plant community. Plant Cell Environ 33:510–525

Yadav RS, Sehgal D, Vadez V (2011) Using genetic mapping and genomics approaches in understanding and improving drought tolerance in pearl millet. J Exp Bot 62:397–408

Zhang J, Deng Z, Cao S, Wang X, Zhang A, Zhang X (2008) Isolation of six novel aquaporin genes from *Triticum aestivum* L. and functional analysis of *TaAQP6* in water redistribution. Plant Mol Biol Rep 26:32–45

Zhang J, Nguyen HT, Blum A (1999) Genetic analysis of osmotic adjustment in crop plants. J Exp Bot 50: 291–302

Zub HW, Brancourt-Hulmel M (2010) Agronomic and physiological performances of different species of *Miscanthus*, a major energy crop. A review. Agron for Sustain Dev 30:201–214

Sustainable Fruit Production in Mediterranean Orchards Subjected to Drought Stress

6

Adriano Sofo, Assunta Maria Palese, Teresa Casacchia, Bartolomeo Dichio, and Cristos Xiloyannis

Abstract

Drought stress is the main cause of reduced fruit tree growth and productivity in Mediterranean semi-arid regions and causes a complex of responses at molecular, cellular, physiological and developmental level. In particular, the response of fruit trees to water scarcity is a species- and cultivar-dependent and a series of studies have been carried out to clarify and deepen the mechanisms of their adaptation, avoidance, resistance or tolerance against drought. Considering that 16% of the total cultivable land of the Mediterranean area is occupied by fruit crops, the choice of an appropriate and rational irrigation management is of key importance. Furthermore, plant water status in an orchard is related to many biotic and abiotic factors, such as the amount of light intercepted, plant densities and canopy architecture, which play a key role in determining orchard productivity and fruit quality. The recent research on the physiology of fruit trees and on soil chemical and biological fertility in fruit orchards have revealed that sustainable and innovative soil management systems, with a particular emphasis on water management (e.g., sustained deficit irrigation, regulated deficit irrigation and partial root-zone drying), can determine an optimal plant nutritional equilibrium, avoid nutrients accumulation and leaching risks, improve irrigation efficiency and prevent soil erosion and root asphyxia. The application, optimization and innovation of sustainable agricultural techniques with a low negative environmental impact allow to recover or increase the normal levels of total fertility in fruit agro-ecosystems, so

A. Sofo (✉) • A.M. Palese • B. Dichio • C. Xiloyannis
Dipartimento di Scienze dei Sistemi Colturali, Forestali e dell'Ambiente, Università degli Studi della Basilicata, Viale dell'Ateneo Lucano 10, 85100 Potenza, Italy
e-mail: adriano.sofo@unibas.it

T. Casacchia
CRA, Centro di Ricerca per l'Olivicoltura e l'Industria Olearia, c. da Li Rocchi-Vermicelli, 87036 Rende, Cosenza, Italy

P. Ahmad and M.N.V. Prasad (eds.), *Abiotic Stress Responses in Plants: Metabolism, Productivity and Sustainability*, DOI 10.1007/978-1-4614-0634-1_6, © Springer Science+Business Media, LLC 2012

maintaining top yields of high quality. On this basis, the aim of this work is to give an detailed information on drought stress in fruit trees.

Keywords
Drought stress • *Prunus* species • Olive tree • *Citrus* species • Mediterranean fruit species

1 Introduction

Plants are sessile organisms and their only alternative to a rapidly changing environment is a fast adaptation to abiotic and biotic stresses. This concept is particularly valid for the productive, physiological and biochemical responses against water deficit (Chaves et al. 2003; Foyer et al. 2005; Morales et al. 2006). In particular, trees carry on the same processes as other seed plants, but their larger size, slower maturation and much longer life accentuate their susceptibility to drought stress in comparison to smaller plants having a shorter life span (Pallardy 2008).

Drought stress is the main cause of reduced fruit tree growth and productivity in Mediterranean semi-arid regions and causes a complex of responses at molecular, cellular, physiological and developmental level. In Mediterranean ecosystems, in which the summer months are characterized by lack of precipitation, elevated temperatures and high irradiance levels, fruit tree species are subjected to a continuous and severe water deficit. Under these adverse environmental conditions, photoinhibition, photooxidation and photorespiration occur, which in turn can negatively affect fruit yield and quality.

Fruit orchards constitute an integral and significant part of the Mediterranean environment and culture, and their ecological importance has only recently been acknowledged. Considering that 16% of the total cultivable land of the Mediterranean area is occupied by fruit orchards (Olesen and Bindi 2002), it is easy to understand that the study of the response of fruit trees to drought stress is of key importance for the agriculture and the economy of the Mediterranean countries. The plant water status in an orchard is related to many biotic and abiotic factors, such as the amount of light intercepted, plant densities and canopy architecture, which play a key role in determining orchard productivity and fruit quality (Grossman and DeJong 1998).

In the Mediterranean basin, fruit trees are often grown in marginal and unfertile lands with low levels of soil organic matter, and orchards are often subjected to soil degradation (Celano et al. 2002; Xiloyannis et al. 2005). Adoption of agricultural systems utilizing non-sustainable techniques, such as frequent and intensive cultivation, zero organic matter inputs (e.g., from cover crops and pruning residues), and use of excessive amounts of water and chemical fertilizers, can further worsen the situation. While such non-sustainable practices are still common among growers, there is evidence that suitable agricultural management practices, such as minimum tillage or no tillage, recycling of the carbon sources internal to the fruit grove (pruning material, spontaneous or/and seeded cover crops, compost amendments) and adequate irrigation, fertilization and pruning, are recommended to maintain an optimal water status in plants, save and use rationally irrigation water, restore soil organic matter, reduce erosion process and environmental pollution and increase the CO_2 sequestration processes from the atmosphere into the soil (Lal 2004; Dichio et al. 2007; Sofo et al. 2010a, b). Furthermore, the adoption of environmentally sustainable agricultural techniques in productive orchards has positive effects on soil microbiota, enhancing soil fertility, plant growth and fruit yields and quality by increasing nutrients availability and turnover in the soil (Kushwaha et al. 2000; Gruhn et al. 2000; Jagadamma et al. 2008; Sofo et al. 2010a, b). Finally, a sustainable orchard management is of particular importance in Mediterranean climates, where soil mineralisation rate is high (Kimmins 1997).

The aim of this work is to give an up-to-date overview of these studies, that will be herein discussed in detail for each tree species. We decided to include the most important and typical productive fruit species of the Mediterranean basin.

2 The Genus *Prunus*

The genus *Prunus* (family Rosaceae) comprises more than 400 species adapted to temperate areas and cultivated in Europe. In particular, stone fruit crops, such as peach (*Prunus persica* L.), plum (*P. cerasifera* L. and *P. domestica* L.), almond (*P. dulcis* L.), apricot (*P. armeniaca* L.) and cherry tree (*P. avium* L.), are typical and economically important and mainly localized in Mediterranean regions. Productive stone fruit trees are usually grafted plants with a lower part, the rootstock and an upper grafted part, which is the genotype of the commercial variety. Rootstocks are important for agronomic purposes, as they have a different genetic background compared to the commercial varieties and can be used to confer various traits such as drought stress resistance.

A better understanding of the effects of water deficit on *Prunus* species has a primary importance for improved management practices (Girona et al. 2005b), breeding programmes (Rieger et al. 2003) and for predicting fruit growth and quality (Torrecillas et al. 1996; Besset et al. 2001; Esparza et al. 2001; Girona et al. 2002). During periods of water deficit, species of the genus *Prunus* show significant decrease in gas exchange (Ruiz-Sánchez et al. 2000a; Besset et al. 2001; Klein et al. 2001). The decrease of soil humidity, together with high values of vapour pressure deficit (VPD), causes reduction in leaf water potential (LWP), carbon assimilation and transpiration in different species of *Prunus* (Rieger and Duemmel 1992; Rieger 1995; Berman and DeJong 1996; Alar on et al. 2000; Besset et al. 2001; Esparza et al. 2001; Klein et al. 2001; Rieger et al. 2003; Matos et al. 2004; Romero et al. 2004c; Gomes-Laranjo et al. 2006; Intrigliolo and Castel 2006; Dichio et al. 2007; Godini et al. 2008; Egea et al. 2010). Some studies also highlighted the activation of antioxidant defenses as a strategy to face drought-dependent oxidative stress in this genus (Scebba et al. 2001; Sofo et al. 2005; Sorkheh et al. 2011).

2.1 Peach and Apricot

The peach tree (*Prunus persica* L.) and the apricot tree (*Prunus armeniaca* L.) are two of the most common and economically important species of the Mediterranean basin (Grossman and DeJong 1998; Alar on et al. 2000; Besset et al. 2001; Girona et al. 2002). The drought tolerance of peach and apricot is mainly based on stomatal control (Arndt et al. 2000) and morphological characteristics (Rieger et al. 2003), together with some degree of osmotic adjustment (Alar on et al. 2000). Work in these two species covered subjects from the physiological processes adopted to regulate water status under drought conditions (Ruiz-Sánchez et al. 2000a; Rieger et al. 2003) to the biochemistry underlying plant response to water deficits and oxidative stress (Arndt and Wanek 2002; Sofo et al. 2005). Peach and apricot trees are highly sensitive to drought stress at particular phenological stages, such as flowering and fruiting, and during stem extension and fruit growth (Berman and DeJong 1997; Xiloyannis et al. 2005). Considering the sensitivity of these two species to water deficit, several authors highlighted the importance of considering wetting patterns, soil depth and root exploration in peach and apricot irrigation management (Ruiz-Sánchez et al. 2000a; Girona et al. 2002). Keeping in view an efficient use of water in peach and apricot orchards, it is of key importance to consider the type of training system and plant architecture, and particularly the distribution of light in the various parts of the canopy, as well as the system of irrigation and its management (Xiloyannis et al. 2010).

Among the indicators used for monitoring water status of peach and apricot trees, two of the most reliable were indicated by Alar on et al. (2000) and Arndt and Wanek (2002). The formers used foliar carbon isotope composition ($\delta^{13}C$)

measured in leaves of peach under water deficit as a tracer to study whole plant carbon allocation patterns. In fact, it is known that foliar carbon isotope discrimination decreases in water-deficit situations as discrimination by the photosynthetic primary carboxylation reaction decreases. On the other side, Arndt and Wanek (2002) used sap flow measured with a heat-pulse technique as an indicator of transpiration and the water status of young apricot plants. They observed that when apricot trees are drought-stressed, measures of sap flow slightly underestimate actual transpiration, confirming an increasing hydraulic resistance under drought conditions. In apricot (cv. "Búlida"), a preconditioning treatment using a substantial reduction in the irrigation water (25% of crop evapotranspiration, ET_c) promotes a better drought-hardening of the plants due to a greater osmotic adjustment (0.77 MPa) that prevents severe plant dehydration and leaf abscission (Ruiz-Sánchez et al. 2000a). This treatment may be valuable for young apricot plants in the nursery stage in order to improve their subsequent resistance to drought.

Regarding the relationship between drought and fruit yield, Berman and DeJong (1996) demonstrated that in well-watered peach plants (cv. "Elegant Lady"), tree water status is independent of crop load, whereas in trees receiving reduced irrigation, the degree of drought stress increased with increasing crop load. The same authors found that drought stress induces fruit fresh weight reductions at all crop loads, whereas fruit dry weight is not reduced by drought stress in trees having light to moderate crop loads. These results suggest that the degree of drought stress imposed did not affect the dry weight sink strength of peach fruit. On the other hand, drought-stressed trees with heavy crop loads had significantly reduced fruit dry weights, which were likely due to carbohydrate source limitations resulting from large crop carbon demands and drought stress limitations on photosynthesis. Crisosto et al. (1994) observed a higher density of trichomes and a continuous and much thicker cuticle on peaches (cv. "O'Henry") from the deficit and optimum irrigation treatments than from the excess irrigation treatment, indicating that a well-designed irrigation management can improve fruit quality and extend shelf life. In peach trees, there is a direct correlation between water availability and carbohydrate synthesis (Girona et al. 2002), and between photosynthetic rate and types of carbohydrates synthesised (Escobar-Gutiérrez et al. 1998). During fruit growth, high photosynthetic rates are necessary for growth requirements of peach (Besset et al. 2001). Sorbitol and sucrose are the two main photosynthetic carbohydrates of peach plants and their function depends on the organ of utilization and its developmental stage (Lo Bianco et al. 2000). In well-watered peach plants, these sugars are translocated from their sources, mainly mature leaves, and then absorbed by sink organs, such as shoot apices (Lo Bianco et al. 2000), developing fruits (Grossman and DeJong 1998) and buds during dormancy release (Marquat et al. 1998). Under drought stress, sucrose metabolism is only marginally reduced, whereas sorbitol accumulates in sinks and sources, contributing up to 80% to osmotic adjustment (Lo Bianco et al. 2000).

Regulated deficit irrigation (RDI), the practice of reducing applied water at selected phenological stages less sensitive to water deficit, was successfully applied to both peach and apricot (Ruiz-Sánchez et al. 2000b; Girona et al. 2005a; Dichio et al. 2007). The application of RDI, based on imposing plant drought stress in a controlled manner, is a feasible water-saving practice for Mediterranean arid areas. Moreover, RDI extended over a long period lead to adaptation of peach tree to dry conditions due to a better extraction of water from deeper soil. The success of RDI strongly depends on the appropriate use of localized irrigation techniques, which allows the control of soil water content (SWC) and plant water status. Moreover, an efficient use of irrigation water is particularly important for improving water uptake by root system. In peach, the application of RDI during the early stages of fruit growth until the end of shoot growth slightly influences fruit size and number (Boland et al. 2000) and a water deficit treatment during the final stage of rapid fruit growth causes decreases in fruit size but also significant increases in total fruit soluble solids (Crisosto et al. 1994; Besset et al. 2001; Naor et al. 2001). Thus, peach quality and taste can be considered as being improved by a water deficit applied in this phonological phase.

On the other hand, withholding irrigation applied after harvest reduces vegetative growth of early-maturing peach trees (Johnson et al. 1992; Ghrab et al. 1998; Girona et al. 2005a) and can improve fruit quality (Gelly et al. 2004; Dichio et al. 2007). It is important to note that RDI, though applied during the post-harvest stage, has to be performed avoiding high levels of drought stress, which could negatively influence the accumulation of reserve carbohydrates, flower development and thus, indirectly, crop yield of the following year (Xiloyannis et al. 2005). Dichio et al. (2007) evaluated the effects of RDI applied in the post-harvest stage of mature peach plants (cv. "Springcrest") trained to transverse Y in an experimental field located in Southern Italy. These authors confirmed the possibility to reduce the irrigation water by applying RDI during phenological stages less sensitive to water deficit without negatively affecting peach growth and yield. In their experiment, from bud break to harvest, irrigation was carried out by applying 100% ET_c, while from harvest to early autumn, plants were separated into three groups and subjected to different irrigation treatments (100%, 57% and 34% ET_c). RDI determined the reduction in the growth of waterspouts and lateral shoots but did not influence the growth of fruiting shoots. No significant reductions in crop yield and quality were observed in the 57% ET_c treatment, whereas about 1,100, 1,800 and 2,500 m^3 ha^{-1} of water were saved in the first, the second and the third year, respectively. In the second year of the trial, the use of RDI in the post-harvest stage determined carbohydrate and nitrogen (N) accumulation in roots, branches, shoots and floral buds. Therefore, the results of Dichio et al. (2007) demonstrate that, under scarce water supply conditions, a clear benefit for both vegetative growth, and carbon and N allocation of peach trees can be obtained through the use of RDI during the post-harvest stage. In another study, Ruiz-Sánchez et al. (2000b) evaluated the response of apricot trees (cv. "Búlida") to RDI under Mediterranean climate. The authors applied an RDI treatment irrigated at 100% ET_c during the critical periods (second rapid fruit growth period and 2 months after harvest) and with a reduction of 40% ET_c during the other periods. An average water saving of 34% was achieved in the fourth year RDI treatment and apricot quality was not modified by RDI treatment. Furthermore, when irrigation water saving was around 25%, the yield obtained was similar to that of the control treatment. It is noteworthy that apricot fruit growth showed few differences between the control and the RDI treatment during the deficit irrigation period, but an accelerated rate of growth was noted when irrigation was increased to 100% ET_c. In conclusion, the satisfactory yield obtained with RDI in Mediterranean peach and apricot orchards suggests to adopt it for early ripening cultivars grown in semi-arid areas with limited water resources in order to improve irrigation efficiency and save water while maintaining top yields of high quality.

With the adoption of a sustainable management under semi-arid climatic conditions (an example is reported in Fig. 6.1), peach and apricot yield can be enhanced up to 25–30%, and the amounts of water and of N, P, K and soil carbon inputs annually incorporated into the soil increase significantly if compared to a non-sustainable orchard (Xiloyannis et al. 2010). Among the sustainable practices, cover crops are of key importance, as in Mediterranean apricot groves the use of mixture of herbaceous species with a high biomass production, such as *Vicia faba/Avena sativa*, can produce approximately 1.0 ton ha^{-1} of humus, with clear benefits for soil fertility (Celano et al. 2002). Moreover, in rain-fed conditions, it is beneficial to sow cover crops in autumn and sow them just before spring in order to avoid water and nutrient competition (Xiloyannis et al. 2005). Sustainable practices also have positive effects on soil microbiota, that influences soil fertility and plant growth by regulating nutrient availability and increasing their turnover (Kushwaha et al. 2000; Borken et al. 2002; Widmer et al. 2006; Govaerts et al. 2008). A molecular approach was often used to reveal qualitative changes in the structure of soil bacterial and fungal communities in various Mediterranean agro-ecosystems

SUSTAINABLE MANAGEMENT	CONVENTIONAL MANAGEMENT
• Minimum tillage and cover crops (30 kg ha^{-1} of Trifolium subterraneumseeds and spontaneous grass)	• Conventional tillage (strong and deep soil plowing)
• Guided fertilization (fertigation based on plant nutrient demand evaluated by leaf mineral analyses and on soil measured nitrogen levels) • Compost amendment (15 t ha^{-1} fresh weight) • Incorporation of cover crop and pruning residues into the soil (light harrowing at a depth of 10 cm carried out in Autumn)	• Chemical fertilization (100 kg N, 10 kg P, 20 kg K ha^{-1}) without considering soil nutrient levels and plant nutrient requirements • Removal of pruning residues from the field
• Guided drip irrigation based on crop evapotranspiration (3 drip emitters per plant along the tree lines with a capacity of 4 L h^{-1} each)	• Empirical irrigation (using excessive amounts of water, without considering soil moisture and crop evapotranspiration)
• Pruning aimed to vegetative-productive equilibrium of plants (winter pruning based on the selection of shoots with a high number of floral buds and on a better light interception in the canopy)	• Empirical pruning

Fig. 6.1 Comparison between a sustainable and a conventional management of a peach orchard (*Prunus persica* (L.) Batsch Nectarine, cv. "Supercrimson" grafted on GF677) located in Southern Italy, under a semi-arid climate with an average annual rainfall of 525 mm. Peach trees were trained to vase (500 plants ha^{-1}) with a north-south row orientation (data from Sofo et al. 2010a)

(Bending et al. 2002; Marschner et al. 2003) but little is known on the molecular and metabolic aspects of soil microbial community at orchard level. One of the few researches on this subject was carried out by Sofo et al. (2010a), that examined the short-time effects (after 4 years) of two different management (sustainable and non-sustainable) systems on microbial genetic, functional and metabolic diversity of a Mediterranean peach orchard (cv. "Supercrimson"), evaluated by a combination of culture-dependent and culture-independent techniques. They revealed qualitative and quantitative changes of soil microbial communities (different electrophoretic patterns of bacterial 16S ribosomal and fungal 18S ribosomal RNA genes and higher indexes of microbiological diversity) in response to a sustainable soil management.

2.2 Almond

Almond (*Prunus dulcis* L.) is the most important tree nut produced on a global basis, and its limited gene pool limits the cultivation to specific areas with Mediterranean climate (Sorkheh et al. 2011). This species is one of the oldest tree nut crops, and today represents the largest production of any commercial tree nut product. The response to water deficit of almond trees is a well-documented process (Esparza et al. 2001, 2010; Klein et al. 2001; Gomes-Laranjo et al.; 2006, Rouhi et al. 2007; Egea et al. 2010). The results obtained on the agronomic response of almond trees to different deficit irrigation strategies demonstrate the prevalence of direct and strong links between the intensity of the water restriction and the response of several parameters related to tree growth, yield and water status (Rouhi et al. 2007; Egea et al. 2010). Besides predawn or midday LWP, midday stem water potential (SWP) and midday leaf stomatal conductance, a series of indicators of plant water status were applied on drought-stressed almond trees. The results obtained by Nortes et al. (2005) indicate that both maximum daily trunk shrinkage and trunk growth rate in almond are sensitive to drought stress and that the second is the most useful parameter for quantifying water deficit intensity and duration.

As regard to fruit production, in almond trees (cv. "Nonpareil") an average loss in yield of 7.7 kg $tree^{-1}$ occurs in response to each 1 MPa decrease in stem water potential (SWP) below −1.2 MPa, if a severe irrigation deprivation is carried out during the harvest period of the previous year (Esparza et al. 2001). This yield loss is likely due to the decrease in the number of fruiting positions per tree, even though the authors did not observe effects of irrigation deficit on the percentage of spurs that flowered or set fruit during subsequent years. In another study, Esparza et al. 2001 confirmed that a severe drought stress during the harvest period in almond causes a reduction in non-structural carbohydrates content but not in N content per tree, so limiting vegetative growth in the following year and impacting subsequent fruit-bearing capacity rather than directly affecting flowering, fruit set or fruit growth. Differences in N-allocation patterns between fruiting and non-fruiting shoots were recently observed by Nortes et al. (2009) in drought-stressed almond plant (50% ET_c during the entire growing season), if compared to fully watered plants (100% ET_c). They found that in the 50% ET_c treatment, a high N status is maintained in the leaves of fruit-bearing shoots, to the detriment of N resources allocated to vegetative shoots.

The studies on RDI applied in Mediterranean almond orchards and aimed to improve fruit yield and quality are numerous (Romero et al. 2004a, b, c; Girona et al. 2005b; Egea et al. 2010). It was found that an RDI of 20% ET_c applied during the pre-harvest period (kernel-filling stage) does not cause reduction in kernel yield and size in almond (cv. "Cartagenera") and improves water-use efficiency, but only if predawn LWP is maintained above a threshold value of −2.0 MPa (Romero et al. 2004c). In the same cultivar, Romero et al. (2004a, b) indicated that a severe RDI (20% ET_c) during the kernel-filling stage, and a recovery at 75% ET_c during the post-harvest phase allows to save 220–273 mm yr^{-1} irrigation water without negatively affecting plant growth and fruiting.

Related almond species, interspecific crosses and spontaneous interspecific hybrids demonstrate a greater resistance to abiotic and biotic stresses and so represent valuable germplasm sources for rootstock breeding, especially under non-irrigated conditions (Browicz and Zohary 1996). The wide adaptation of the related wild almond species indicate their potential as sources for resistance to drought stress as well as modified tree and nut traits. On this basis, the selection for drought resistance in almond rootstock material and for the increase of quality of cultivated almond production under stress conditions is particularly important. Among the different wild almond varieties, Rouhi et al. (2007) studied the intrinsic water use efficiency, defined as the ratio of assimilation rate over stomatal conductance, in cultivated and wild almond species, finding that *P. dulcis* is the species most tolerant to drought, *P. scoparia* tries to avoid drought, and *P. lycioides* has an intermediate behavior and this latter can have potential for use as rootstock for commercial almond production. As wild almond species can be very important for rootstock selection, in a recent paper, Sorkheh et al. (2011) examined the changes of antioxidant enzyme activities and the level of some antioxidant compounds involved in the ascorbate-glutathione cycle in drought-stressed plants of eight wild almond species from different geographical points of Iran. The authors found that after 70 days without irrigation, mean pre-dawn LWP in all the species fell from 0.32 to –2.30 MPa and marked decreases in CO_2 uptake and transpiration occurred. The activities of the antioxidant enzymes involved in the ascorbate-glutathione cycle increased in relation to the severity of drought stress in all the wild species studied. Furthermore, the levels in total ascorbate and glutathione and H_2O_2 were directly related to the increase of drought stress. The up-regulation of the activities of some antioxidant compounds during drought stress is an immediate and efficacious response to scavenge the excess of activated oxygen species (AOS), such as superoxide anion ($O_2^{\cdot-}$), hydrogen peroxide (H_2O_2), hydroxyl radical ($HO^{\cdot}$) and singlet oxygen ($^1O_2$), and it was also observed in other fruit tree species, such as apricot (Scebba et al. 2001), olive (Sofo et al. 2004a) and plum rootstocks (Sofo et al. 2005).

As seen for olive (Sofo et al. 2004b), the role of proline during drought stress is particularly important for the osmotic homeostasis of the plants. The results of a forthcoming paper on wild almond (Sorkheh et al. 2011) highlight that the cell membrane damage is a direct consequence of oxidative stress by H_2O_2 and that the application of exogenous proline can alleviate these detrimental effects. Thus, it is possible to recommend exogenous proline treatment of wild species of almond in order to increase their antioxidant defenses when subjected to drought.

2.3 Plum and Cherry

Many plum genotypes are used as rootstock for almost all other *Prunus* species and, among them, Myrobalan plum (*Prunus cerasifera* L.) clones often show positive agronomic features for resistance to pathogens and abiotic stresses (Lecouls et al. 2004; Intrigliolo and Castel 2006). In the Mediterranean regions, drought is the main limiting factor for plum growth (Rato et al. 2008). Sofo et al. (2005) studied the effects of water deficit on photosynthetic performance and on the components of the ascorbate-glutathione cycle in four interspecific plum hybrids, used as rootstocks, hypothesizing that an excess of reducing power, with the consequent increase in H_2O_2 and other AOS concentration, causes the up-regulation of some antioxidant enzymes during a drought period. Their results showed that the activities of antioxidant enzymes and the levels of the molecules involved in the ascorbate-glutathione cycle (antioxidant enzymes, total ascorbate and glutathione and H_2O_2) increased in all the hybrids examined in parallel to the severity of drought stress. After 70 days of water shortage, mean pre-dawn LWP of all the hybrids fell from –0.34 to –3.30 MPa and marked decreases in net photosynthesis and transpiration occurred. All these physiological and biochemical responses could limit cellular damage caused by AOS during periods of water deficit. On the basis of these results, it appears that the ability of *Prunus* hybrids to regulate the enzymatic antioxidant system during drought stress can be an important attribute linked to drought tolerance. As regard to yield susceptibility to reduced irrigation in plum (*P. salicina*, cv. "Black Gold"), Intrigliolo and Castel (2006) pointed out that an RDI applied from pit hardening to harvest reduces fruit weight by 10–21%, indicating that phase III of fruit growth is a phenological period highly sensitive to water deficits in this species.

As in plum, water relations and photosynthesis of sweet cherry (*P. avium* L.) grown in Mediterranean environments are mainly influenced by the rootstock genotype, and the regulation of fruit quality is mainly dependent on the cultivar genotype (Gonçalbes et al. 2005; Godini et al. 2008). An interesting and wide comparison among cherry rootstocks subjected to non-irrigated conditions in Southern Italy highlighted the satisfactory performance of "SL 64" and the promising performance by dwarfing "Weiroot® 158" and semi-dwarfing "MaxMa 14" under water-limited growing conditions. An optimal water and nutrient management is of primary importance for cherry trees grown under semi-arid conditions, and drip fertigation is a valid irrigation system for this species (Neilsen et al. 2005). Furthermore, water scarcity could determine an higher percentage of double fruits as well as a phytohormonal disequilibrium, with consequent losses of commercial product (Engin and Ünal 2008).

3 The Olive Tree

The genus *Olea* encloses more than 30 species belonging to the family *Oleaceae*. The cultivated species, *Olea europaea* L., is a very important crop in the world and occupies an area of about 9 Mha (FAOSTAT, 2010) with more than 2,600 cultivars, many of which may be ecotypes (Therios 2009). Olive is an evergreen and long-lived species cultivated between 30° and 45° of latitude but it finds the optimal conditions for its growth and productive expression in Mediterranean countries. In such area, climate is characterized by winter rains, which fall in a short cold season, and a dry and hot summer period.

Olive is cultivated for its edible fruits subjected to different processing technologies to obtain table olives or oil, both characterized by high nutritional and organoleptic values. Such products are among the basic aliments of the so-called Mediterranean diet and show interesting properties for human health preservation (Visioli et al. 1999, 2006). The olive tree products (oil, leaves) can be usefully employed in the herboristic and cosmetic sectors. A promising perspective could be the recovery from olive mill wastewater (OMW) of phenolic compounds, such as hydroxytyrosol, with high added value due to their powerful antioxidant and potential beneficial properties for human health and effective antimicrobial activity. These compounds could be used as integrators in food, pharmaceutical and cosmetic products or as a natural pesticide against a variety of seed infections (Allouche et al. 2004; Yangui et al. 2009). In addition, the recovery process could partially solve the long-standing problem of OMW disposal. OMW, whose worldwide production is around 30 Mm^3 yr^{-1}, has high polluting effects for the environment, especially when its disposal is carried out using unsuitable procedures (Allouche et al. 2004; Celano et al. 2010).

3.1 Olive Responses to Drought

Olive tree shows a great capacity to tolerate the long summer water shortage by means of numerous strategic devices aimed to control water losses and increase water uptake from soil. Besides the anatomical and morphological features of leaves (small size, high specific leaf weight, thick and waxy cuticle, hairy leaf surfaces, high stomatal density), typical adaptations of drought-tolerant plants, olive presents specialized physiological and biochemical mechanisms.

Under severe drought stress, olive tree significantly lowers water content and water potentials of its tissue establishing a high potential gradient between leaves and roots (predawn LWP values of −7.0 MPa and −3.5 MPa, respectively) which allows the root system to utilize water up to soil water potential of −2.5 MPa. Such value is well below the permanent wilting point, measured at −1.5 MPa for most of the fruit species. Under such conditions, and especially in soil characterized by a good water storage capacity, olive plants have access to a greater and readily available soil water (between field capacity and –2.5 MPa), so withstanding long drought period (Xiloyannis et al. 2003). In olive tree, stomata progressively reduce their activity starting from predawn LWP below –0.9 MPa, and they can remain open up to –7.0 MPa (Xiloyannis et al. 1999). A progressive closure of stomata as predawn LWP decreased was observed in other fruit tree species but their stomatal closure was reached at values of predawn LWP ranging from –1.5 to –2.5 MPa (Lakso 1979; Castel and Fereres 1982). Under stressful conditions, olive tissues are able to transpire large amounts of water, accumulated during the afternoon and night, ensuring a certain level of leaf functionality. As a matter of fact, olive leaves can give up to transpiration about 60% of the water stored in their tissues contributing to the demands of transpiration as stress increases up to extreme values (Xiloyannis et al. 1999). At predawn LWP of –6.0 MPa, olive maintains a certain transpirative and photosynthetic activity (around 10% and 20%, respectively, of that of well-watered plants), that allows the plants to produce assimilates and accumulate them in the various organs. Particularly, long-term soil water deficit reduces in young olive trees the development of the above-ground organs with respect to the under-ground part (roots and stump), so raising the under/above-ground ratio. The effect is particularly marked in leaf area, that is significantly reduced under rain-fed conditions (47% lesser than irrigated plants at the seventh year from planting) (Dichio et al. 2002). Such reduction of canopy size limits the water demand for transpiration.

Another strategy adopted by the olive tree to overcome water deficit is osmotic adjustment which consists in either active synthesis and accumulation of osmotically active compounds (carbohydrates, some aminoacids, organic and inorganic acids, cations and anions) within cells (active osmotic adjustment) or loss of water from plant cells, with the consequent increase in osmolyte concentration (passive osmotic adjustment) (Xiloyannis et al. 1999; Cataldi et al. 2000; Sofo et al. 2004b; Dichio et al. 2009). This physiological process is measured by the variation in osmotic potential within plant tissues (Dichio et al. 2007). A higher concentration of osmolytes (particularly mannitol, glucose and proline) facilitates water diffusion in cells and maintains the turgor of plant tissues essential for plant physiological activity. The maintenance of cell turgor in roots also avoids or delays the separation of these organs from the soil. Under drought conditions, olive trees activate metabolic processes to produce substances that increase cell tissue rigidity, likely by regulating some enzymes involved in lignin biosynthesis such as peroxidases (Sofo et al. 2004b). This mechanism results in an increase in elastic modulus (ε) as cell walls become more rigid or thicker. Higher ε values produces a faster turgor loss of cells for a given percentage of dehydration. An increase of cell tissue rigidity together with low values of Ψ_p, due to active and passive osmotic adjustment, can be responsible for the observed high gradients of water potential between leaves and soil, and thus can facilitate water extraction from the soil.

In olive trees, the activities of some antioxidant enzymes significantly increase in leaves and roots of drought-stressed plants (Sofo et al. 2004a). These enzymes limit the cellular damages caused by AOS, so allowing the plant to maintain a photosynthetic efficiency also under severe drought conditions (Xiloyannis et al. 2003). Significant increases of lipoxygenase (LOX) activity and malondialdehyde (MDA) content, two markers of oxidative stress, were also found during the progressive increment of drought stress in both leaf and root tissues of olive plants (Sofo et al. 2004a, b), so suggesting that water deficit is associated with the oxidation of membrane lipids. In olive plants under drought stress, the damage of photosynthetic apparatus, and the resulting decrease in photosynthetic efficiency, occurs particularly by means of the light-dependent inactivation of the photosystem II (photoinhibition) and the oxidation of chloroplastic pigments (photo-oxidation) (Angelopoulos et al. 1996; Sofo et al. 2009). Despite these damages, olive tree is able to recover its water status faster (5 days) than other fruit tree species even if it shows a slow recovery of photosynthesis and transpiration (Angelopoulos et al. 1996).

Finally, olive tree can respond to short period stress by regulating the activity and the expression of its root water channels (aquaporins)

(Tataranni 2009). As the adverse conditions continue, root suberification occurs, so avoiding dehydration. In fact, an increase of suberification process was observed in root cell walls at exodermis and endodermis level. Under such conditions, root activity recovery is preceded by the emergence of root primordia (Tataranni 2009).

3.2 Effects of Irrigation Management on Productivity, and Fruit and Oil Quality

Generally, irrigation raises significantly the vegetative growth of olive tree and its productive response. This leads to early bearing, steady and satisfactory yields, and improvement of fruit features. In addition, as the productive tree performances are not influenced by moderate levels of drought stress, a reduced irrigation is recommended in arid and semi-arid areas to save water. Deficit irrigation strategies in olive orchards can be applied following different approaches (Fereres and Soriano 2007).

Sustained deficit irrigation (SDI) distributes a reduced water volume, as percentage of ET_c, throughout the whole irrigation season. Many studies, carried out under diverse pedo-climatic conditions, compared irrigation regimes based on different levels of ET_c restitution and their influence on fruit and oil quality of different olive cultivars. Patumi et al. (2002), Magliulo et al. (2003), d'Andria et al. (2004), Grattan et al. (2006), Berenguer et al. (2006) and Dabbou et al. (2010) found that a restitution ranging from 66 to 75% of ET_c is enough to obtain good yields similar to those harvested from fully irrigated trees. However, phenolic compounds in oils significantly decreased passing from the lowest to the highest irrigation levels. Although reduction in polyphenol content modified slightly sensory properties of oils decreasing their bitterness and pungency, it did not compromise oil storage capacity. Stefanoudaki et al. (2009) referred about a contradictory effect of irrigation which decreased contents of both undesirable (pungent and bitter attributes) and favourable sensory qualities (intense green notes). As a matter of fact, irrigation could be managed to meet consumer's particular needs. Patumi et al. (1999), Tovar et al. (2001) and Tovar et al. (2002) studied the effect of several irrigation treatments on L-phenylalanine ammonia-lyase activity (PAL) in developing fruits. PAL is the key enzyme in phenolic biosynthesis and a high PAL activity is associated with the accumulation of anthocyanins and other phenolic compounds in tissues of several fruit species (Weaver and Herrmann 1997; Ryan et al. 2002). PAL activity and phenolic level decreased during fruit development and were influenced by irrigation, being lowered as the water supplied increased.

Regulated deficit irrigation (RDI), firstly proposed by Chalmers et al. (1981), reduces water supplies during specific periods characterized by a less plant sensibility to water stress with minimal effects on yield. While water deficit can reduce fruit and oil yields due to the effect on flowering, fruit set and oil accumulation phases, many researchers agree in identifying pit hardening, generally occurring in midsummer, as the less sensitive phenological stage of olive tree (Lavee and Wodner 1991; Goldhamer 1999; "Moriana et al. 2003; Orgaz and Fereres 2004; Iniesta et al. 2009). On the other hand, in environments characterized by good spring rainfall and deep soil profiles, irrigation applied from the beginning of pit hardening to early fruit veraison could control tree vigour while maintaining crop yield and oil quality (Gómez-Rico et al. 2006; Tognetti et al. 2006, 2007; d'Andria et al. 2009).

Partial root-zone drying (PRD) is an irrigation strategy aimed to maintain in a drying state at least half of the tree root system while the other half is kept under wet soil conditions. Such technique is based on the existence of a chemical signal between root and shoot which determines plant responses to soil drought stress limiting shoot and leaf growth. Particularly, under mild soil drought stress, abscisic acid (ABA), moving in the xylem from the roots, reaches the epigean parts of the tree, where it regulates stomatal movement and shoot meristem activity. The alternation of wet and dry conditions in the soil is a requirement to allow roots to produce ABA. Generally, a PRD cycle lasts 10–15 days, depending on soil type and other factors such as rainfall and temperature (Davies et al. 2000; Stoll et al. 2000; Stikic et al. 2003;

Sepaskhah and Ahmadi 2010). Wahbi et al. (2005) reported that PRD strategies slightly reduced yield (15–20%) and increased plant water use efficiency of 60–70%. Aganchich et al. (2008) showed that PRD irrigation of "Picholine marocaine" plants, besides water saving (50%), positively affects both fruit biometric parameters and oil production (highest oil content, precocious fruit ripeness), and causes increases in total polyphenol. Instead, Fernández et al. (2006), comparing PRD and RDI treatments (50% ET_c), did not find significant improvement of the physiological parameters measured. Such findings led the authors to advise against the use of PRD because of its high cost and difficulty in management.

Moriana and Orgaz (2003) proposed an irrigation scheduling adapted to the typical alternate bearing habit of the olive which supplies water only in "on" years. Although this approach was successfully tested in pistachio plants (Stevenson and Shackel 1998), the authors expressed some doubts on the viability of such program for olive. As a matter of fact, an exceptional severe drought during the rainfed "off" year, able to completely deplete water in the soil profile, could have an important impact in flowering and fruit set of the following "on" year resulting in very low yields. On the other hand, Palese et al. (2010) reported that after a rainfed "off" year, olive trees continuously non-irrigated showed a great capacity of recovery, which led to a vegetative activity and productive response similar to those of the irrigated plants. This is due to a complete replenishment of soil water reserve following autumn–winter rains. As reported by Martín-Vertedor et al. (2011), the application of SDI during "off" year could be advisable when a lower water consumption occurred. Therefore, the optimal irrigation amount could be determined each year, according to crop load levels.

3.3 Strategies for Rainwater Capture and Storage Under Rainfed Conditions

In traditional olive cultivation areas of Mediterranean Basin, rainfall is the only source of water for the olive tree. Therefore, under rainfed conditions, strategies aimed to improve the recharge of rainwater in soils by using specific soil management techniques, or to capture rainwater in collection systems (i.e. Tunisian "jessour", hand-made stone terraces, basin at farm and hydrographic level), are recommended (de Graaff and Eppink 1999; Fleskens et al. 2005; Tubeileh et al. 2009).

Among the soil management techniques, mechanical tillage is still the most common in Mediterranean olive orchards, where it is performed also as a dry farming technique with the aim of reducing soil evaporation by interrupting water capillary rise and increasing soil surface roughness (Ozpinar and Cay 2006). Furthermore, tillage should improve infiltration and percolation into soil of rainfall water but such effects often occur only for a short period of time immediately after the machine passage (Pastor et al. 2000). Unfortunately, continuous tillage may result in the degradation of soil structure which can significantly reduce water infiltration rate causing runoff, erosion processes and fertility loss (Abid and Lal 2008). These degradation mechanisms are quickened by the high air temperatures that induce an intense microbial biomass activity and the mineralization of the labile fraction of organic matter, the most active in the soil. A significant loss of organic matter leads to a further deterioration of soil hydraulic properties directly involved in the recharge and storage of rainfall into the soil (Lipecki and Berbeć 1997; Strudley et al. 2008).

Autumn–winter cover crops, spontaneous or sown, can represent an alternative to tillage in rainfed olive orchards showing a beneficial effect in intercepting raindrops, reducing runoff, facilitating and speeding infiltration of excess surface water into the deepest soil layers even thanks to the channels left by their dense death root network (Pastor et al. 2000; Pardini et al. 2002; Hernández et al. 2005; Durán-Zuazo et al. 2009; Palese et al. 2009a). A study carried out by means of a non-invasive geophysical techniques (electrical resistivity imaging, ERI) revealed that a cover cropped mature olive orchard was more efficient to intercept and store rainwater than tilled grove, resulting in a significant water reserve at the deepest soil layers (>1.0 m), convenient for the root system of rainfed olive trees in the driest months (Celano et al. 2011). On the other hand, cover crops show very high hydric consumptions from the soil (from 200 up to 350 mm per year) and so they could compete with olive trees for water, especially when annual rainfall is less than 500 mm (Bellini 1983; Pardini et al. 2002). Therefore, it is fundamental to choose the most opportune date for cover crops suppression (by mechanical or chemical means), avoiding the overlapping between weed growth and some critical phases for the olive productive performance such as flowering and fruit set (Orgaz and Fereres 2004).

The improvement of soil water holding capacity can be reached also by means of techniques aimed to increase and/or preserve soil carbon content. Pruned material represents an important source of dry matter internal to the olive orchard and characterised by high content of lignin, low nitrogen level (C/N > 25) and slow decomposition process (Celano et al. 2003). Once cut and buried in the soil, pruning material is able, in the long period, to build up soil organic matter which, in turn, improve soil hydraulic features (Pastor et al. 2000; Hernández et al. 2005). The recycle of polygenic organic material inside the olive orchard (spontaneous cover crops + pruned material), offering mixed organic substrates, strongly affects the activity of soil microbial communities which show a higher complexity and diversity at genetic, functional and metabolic levels (Sofo et al. 2010b).

3.4 Use of Non-conventional Water Sources for Irrigation

Olive trees are widely diffused in arid and semiarid environments where water shortage and competition among the different water consumption sectors are relevant problems. For this reason, the use of low quality water for irrigation (e.g., saline water or municipal wastewater) could represent a realistic way to overcome the scarcity of "conventional" water assigning it especially for human consumption. In addition, an increase of the irrigated olive-grown area could lead to improved farmers' income, with a general benefit to the local rural economy.

In the Mediterranean regions, large amounts of saline water (with an electrical conductivity, EC > 2.0 dS m^{-1}) are available for irrigation. Among Mediterranean fruit tree species, olive tree is moderately salt tolerant (Ayers and Westcot 1976; FAO 1985; Rugini and Fedeli 1990), and it shows a different tolerance behaviour depending on cultivars, salt concentration (EC from 5.0 to 13.7 dS m^{-1}, the latter identified as the tolerance limit) and salt type dissolved in the irrigation water (Rugini and Fedeli 1990; Chartzoulakis 2005). Salt tolerance in olive cultivars is basically related to salt-exclusion mechanisms occurring within roots, which prevent salt translocation rather than salt absorption by keeping Na^+ and Cl^- at root level and limit the accumulation of such ions into actively growing shoots. Furthermore, Ca^{2+} has a main role in regulating the selectivity of the ionic absorption, decreasing Na^+ uptake and its transport to the shoot and reducing toxic effects of Na^+ on integrity of the plasmatic membrane in root cells (Benlloch et al. 1991; Tattini et al. 1995; Melgar et al. 2009). As a matter of fact, the increase of Ca/Na ratio by adding Ca^{2+} to irrigation water has been recommended to mitigate the detrimental effects of salinity stress (Rinaldelli and Mancuso 1996; Melgar et al. 2009). The correction of water irrigation, together with the use of drip irrigation and the choice of a tolerant cultivar, can be useful tools for an appropriate employment of saline water. As reported by Melgar et al. (2009), the long-term irrigation of mature olive trees of cv. "Picual", a salt-tolerant cultivar, with saline water (EC up to 10.0 dS m^{-1}) did not affect growth and yield, and no salt accumulation was found in the upper 30 cm soil layer thanks to the ion leaching linked to the rain season (annual precipitation of 702 mm). On the other hand, irrigation with saline water could be harmful in low rainfall areas (less than 250 mm). Under such conditions, it is essential to plan a proper soil leaching management (Wiesman et al. 2004).

Another alternative water resource is reclaimed urban wastewater. Olive trees can lend themselves to irrigation with this low quality water because their fruits are usually harvested 1 month, or more, after the last water application (according to the variety and its maturation time), and they are eaten after processing (to obtain oil or table olives). Such conditions decrease risk of fruit microbial contamination. Furthermore, the use of microirrigation system avoids the contact among wastewater, fruits and leaves allowing the production of safe high-value olive yields and avoiding health risk for the farm workers and the consumers (Palese et al. 2006, 2009b; Bedbabis et al. 2009). A sustainable orchard management (Fig. 6.2) coupled with an intense water absorption by the roots of olive trees and cover crops active in the wetted soil volume, excluded water logging by runoff and percolation to deeper soil layers avoiding aquifer pollution by faecal bacteria (Palese et al. 2009b). From an agronomic point of view, wastewater is rich of mineral elements (particularly P, N and K) and organic matter, both important for yield and vegetative development of olive trees and soil fertility, and often eliminated during the sewage treatment (Ramirez-Fuentes et al. 2002; Yadav et al. 2002; Tarchouna Gharbi et al. 2010a). The reduction of the treatment level decreases fertilization costs and pollution and the price of the treated water allowing, in economic terms, its sustainable reuse (Lopez et al. 2006; Palese et al. 2009b). Nutrients supplied by wastewater should be taken into account in preparing the annual fertilization plan (Palese et al. 2008). On the other hand, reclaimed urban wastewater can be an important source of both salts and potentially toxic metals. Although urban wastewater usually shows a low concentration of heavy metals, long-term irrigation can increase their concentration into soils even if not at critical values (Ramirez-Fuentes et al. 2002; Yadav et al. 2002; Tarchouna Gharbi et al. 2010b; Klay et al. 2010). Therefore, a systematic monitoring of metal content in wastewater, soil and plant is recommended to avoid hazardous situations for populations and environment.

4 The Genus *Citrus*

An efficient water management for *Citrus* spp. trees in any cropping situation requires accurate quantitative information on water use. Interpretation of the water relations of most *Citrus* spp. cultivars is difficult due to the occurrence of stomatal oscillations whose origin is not well known and which cause sampling problems in irrigation management (Dzikiti et al. 2006, 2008; Wright 2008).

Vegetative growth, and particularly leaf development and stem diameter, of orange trees (*Citrus sinensis* (L.) Osbeck) is particularly susceptible to water scarcity (Dzikiti et al. 2006; Aiyelaagbe and Orodele 2007), and plants respond to drought by changes in gas exchange, phytohormonal balance and polyamine contents (Wang and Liu 2009). Moreover, internal water storage contributed significantly to the daily total leaf transpiration in the species (Dzikiti et al. 2006, 2008). García Petillo et al. (2004) compared the effects of different irrigation volumes on "Washington Navel" orange yields during a 5-year period (0, 50%, 100% and 150% ET_c). To apply these treatments, one irrigation drip line per tree row, with drippers, of 2, 4 and 6 $L h^{-1}$ capacity, separated 1 m apart were used for the 50%, 100% and 150% ET_c treatments. Another treatment received the same amount of water as 100% ET_c, but with two drip lines spaced 1 m apart per tree row and 2 $L h^{-1}$ drippers and showed significant increases in total fruit yield and fruit size if compared to 100% ET_c. The application of the PRD irrigation method (50 and 100% ET_c) to orange trees was evaluated over two growing seasons by Dzikiti et al. (2008b). The authors found that stomatal conductance in the PRD treatments was lower than in the control fully-watered treatment but no significant changes in average fruit yield were found between the two PRD treatments and the control plants. Regarding RDI, it has been demonstrated that the irrigation cut-off during the final fruit growth and maturity process (phase III) in orange (cv. "Lane late" grafted on "Carrizo" citrange) reduces midday SWP, does not reduce fruit yield and increases total soluble solids and titrable acidity, without altering fruit quality and the final maturity index (Pérez-Pérez et al. 2009). On the contrary, García-Tejero et al. (2010) found that RDI applied during the flowering and early fruit-growth phases (cv. "Navelina" grafted onto "Carrizo" citrange), both yield and fruit quality (in terms of total soluble solids and titrable acidity) were negatively affected.

SUSTAINABLE MANAGEMENT	CONVENTIONAL MANAGEMENT
• No tillage - Spontaneous weeds and grasses mowed at least twice a year • Pruning material cut and left on the ground as mulch	• Conventional tillage (milling at 10 cm soil depth) performed 2-3 times per year in order to keep the soil bare
• Guided fertilization: fertigation based on a nutrient balance approach which takes into account nutrient input (by wastewater), output (by yield), and recycling/immobilisation in the grove system (by pruned material, senescent leaves, cover crops)	• Mineral fertilization carried out empirically once per year by using granular product applied to the soil
• Guided drip irrigation with treated municipal wastewater based on crop evapotranspiration calculated according to FAO equation: $ET_c = K_r \times K_c \times ET_0$ (K_r = reduction coefficient; K_c = crop coefficient; ET_0 = potential evapotranspiration) - (6 self-compensating drippers per tree delivering 8 L h^{-1})	• No irrigation
• Light winter pruning performed each year in order to reach vegetative-reproductive balance of trees	• Heavy pruning carried out every two years - Pruned residues burned out of the olive grove

Fig. 6.2 Comparison between a sustainable and a conventional management of a mature olive orchard (*Olea europaea* L., cv. "Maiatica") located in Southern Italy, under a semi-arid climate with an average annual rainfall of 561 mm. Olive trees were vase trained (156 plants ha^{-1}) (data from Palese et al. 2008, 2009b)

Other *Citrus* species showed responses against drought stress similar to those found in orange. Huang et al. (2000) examined the growth changes generated by mild drought stress on potted tangerine trees (*Citrus reticulata* Blanco, cv. "Zhuju") during early juice sac expansion stage. They observed that fruit growth was inhibited by drought stress but a greater water uptake was caused by a lower water potential in fruits of stressed plants, likely due to a higher water loss from fruit to transpiring leaves during water shortage and some active adaptive physiological responses (osmotic adjustment and cell wall loosening) of fruit to this stress. In satsuma mandarin trees (*Citrus unshiu* Marc.), a positive relationship between flower-bud induction and the level of endogenous plant hormones was found as a result of the application of mild (predawn LWP = from −0.5 to −1.0 MPa) and moderate drought stress (predawn LWP = from −1.5 to −2.0 MPa) (Yoshita and Takahara 2004). The results indicate that gibberellin levels were enhanced by severe drought stress, higher in the leaves from the branches that produce fewer flowers during flower-bud induction periods, whereas the levels of indole-3-acetic acid were higher in the leaves from the branches that produced more flowers during the season when flower-buds develop. As in *C. sinensis*, the measurement of maximum daily trunk shrinkage is a suitable and reliable indicator of the level water deficit reached by plants of other *Citrus* species, such as lemon (*C. limon* (L.) Burm. fil.) and sour orange (*C. aurantium* L.) (Ortuno et al. 2009).

5 Other Mediterranean Fruit Species

5.1 Pomegranate

Pomegranate trees (*Punica granata* L.) are considered as a crop with a high level of tolerance to soil water deficit. Pomegranate cultivation is mainly confined to the tropics and subtropics and it grows well in arid and semi-arid climates, but it is now widely cultivated in Mediterranean (Hepaksoy et al. 2009). In Spain, for example, its culture is concentrated in the south east, where fresh water available for agriculture is very scarce. The water relations of field grown pomegranate trees grown under different drip irrigation regimes were recently investigated by Intrigliolo et al. (2011). These authors observed that during spring and autumn, midday SWP was not significantly different between irrigation treatments while there were considerable differences in leaf photosynthesis and stomatal conductance, suggesting a near-isohydric behaviour of pomegranate trees. This means that plants control gas exchange such that daytime water content is almost unaffected by soil water deficits, and that other mechanisms (e.g., ABA production and signaling) can be responsible for the regulation of plant water status. There is little knowledge about the response of pomegranate to drought, and in general to abiotic stresses. In one of the few researches, Bhantana and Lazarovitch (2010) studied the evapotranspiration, crop coefficient and growth of two young pomegranate varieties under salt stress, confirming that this species exhibits a high tolerance under adverse environmental conditions. If compared to other irrigation techniques, drip irrigation is the best way to increase fruit yield and plant growth of pomegranate, as its root system is particularly inhibited by water stagnation, whereas fruit yield is not significantly influenced by the level of irrigation (Sulochanamma et al. 2005). Furthermore, irrigation of pomegranate trees is very important, as fruit splitting and cracking can occur, unless they are regularly irrigated. Excess watering or excessive rain during the maturation period may also cause similar damage to the fruits (Hepaksoy et al. 2009). Finally, vitamin C, reducing sugar and total sugar content were observed in fruit of drought-stressed plants (Lawand et al. 1992).

5.2 Pistachio

Pistachio is a crop indigenous to western and central Asia but its cultivation has spread to the Mediterranean region, which has become its second most important centre of diversity after Iran. The importance of *Pistacia* spp. is not limited to this product alone: the tree's great tolerance to drought stress and their ability to thrive in poor soil conditions make them particularly suitable for forestry programmes on marginal lands, where they can also represent a source of additional income for local farmers (Padulosi et al. 1998; Sedaghat 2008). Pistachio cultivation requires the use of rootstock because grafting is the only form of vegetative propagation, thus the choice of the most effective rootstocks plays a key role, as they determine the physiological and biochemical responses of the plants to drought (Ranjbarfordoei et al. 2000, 2002; Gijón et al. 2010). The extreme drought resistance of *Pistacia* spp. enables farmers in arid and semi arid lands to grow this nut without irrigation (Kaska 2002). Despite the economic importance of edible pistachio (*Pistacia vera* L.), very little information is available on its nutrient requirements and water needs. Potassium (K) fertilization is found to be effective in increasing leaf K status, nut yield and quality in this species, and K uptake occurs mainly during the nut fill period (Zeng et al. 1998). Tajabadipour et al. (2006), studied the effects of three irrigation frequency and five K levels on the plant water relations and growth of three pistachio cultivars ("Badami", "Ghazvini" and "Sarakhs"), founding that the dry weights of leaves, stems and roots significantly decreased with increasing irrigation intervals, whereas K application had no significant effect on LWP, osmotic potential and turgor potential. From a molecular point of view, Yakubov et al. (2005) observed the accumulation of dehydrin-like proteins both in the inflorescence bud and in the bark of young pistachio stems, suggesting that they may have a role in drought and cold tolerances, as well as serving as storage proteins. An irrigation experiment involving pistachio (*P. vera*, cv. "Kerman", on *P. terebinthus* rootstocks) was performed by Gijón et al. (2009) over a 4-year period to determine the effect of RDI (at 65% and 50% of control irrigation) on nut yield and quality. The growth season was divided into three phenological stages: stage I – from sprouting until the end of rapid nut growth; stage II – from maximum nut size until the beginning of kernel growth; and stage III – from the beginning of kernel growth until harvest. The plants subjected to RDI were only significantly stressed during stage II, showing midday LWP of around −1.4 MPa. The application of RDI resulted in smaller nut diameter and lower total yield. Moreover, trees subjected to RDI had a total yield and percentage of split nuts similar to those of the controls, and did not show the normal alternate bearing pattern of this tree crop. The authors concluded that this rootstock-scion combination presents a high degree of drought-resistance and could be efficiently applied in pistachio cultivation.

5.3 Prickly Pear

Opuntia, also known as "nopales" or "paddle cactus", is a genus in the family Cactaceae. The most common culinary species belonging to this genus is the Indian fig *Opuntia ficus-indica* (L.) Miller, commonly known as "prickly pear". This species is native to Mexico but it is also found in southern Europe and northern Africa, where it contributes, like olive tree, to the typical Mediterranean landscape. The prickly pear tree is able to store high water amounts in its succulent organs and it has a very wide (even though not deep) root system with dense and rapidly regenerating root hairs, that allow plants to efficiently use extremely low rainfall (Mulas and Mulas 2004). Prickly pear has a CAM photosynthesis and thus maintains the stomata of mature cladodes open only in the night, but, under extremely severe water deficits, the stomata remain closed all day long and so the plants use to photosynthesize only the CO_2 deriving from the respiration (Nieddu et al. 1997). Pimienta-Barrios et al. (2000) evaluated the effects of seasonal variation in temperature, irradiation and soil moisture content on the photosynthetic rates of prickly pear. They demonstrated that this species is strongly adapted to arid climates and that stem photosynthesis by cladodes (stem modified for photosynthesis that looks like leaves) allows plants to fix carbon to be used during the periods when soil water content is very low. Drought significantly affects cladode morphology and inhibits new cladode production, as these latter have a C_4-photosynthesis and open the stomata during the day, with consequent water losses (Nieddu et al. 1997). Furthermore, cladodes can reach temperatures 15°C higher than the environmental ones, maintaining their enzymatic activities up to 60°C.

The fruit yield of prickly pear is quite low, likely due to limiting environmental factor (low water amounts, soils with low levels of organic matter), but a fertilization up to 160 kg ha^{-1} determines a yield increase and a high fruit quality (Mulas and Mulas 2004). Mulas and D'hallewin (1997) estimated that fruit yield in irrigated plants is at least two folds higher than that of un-watered plants, due to higher fruit number per cladode and not to increases in fruit weight. On the other hand, irrigated plants presented an increase in fruit peel thickness, which reduced the juice percentage, and in seed weight (Mulas and D'hallewin 1997). Snyman (2006) aimed at quantifying the effects of drought stress on the growth of tap roots, side roots and rain roots of the species *Opuntia ficus-indica* (L.) (cv. "Morado", with green cladodes) and *O. robusta* Wendl. (cv. "Monterey", with blue cladodes), both having edible fruits. They planted 1-year-old cladodes in root boxes and pots in a greenhouse. Placing the cladodes flat on the soil, more areoles came in contact with the soil, and each areole complex formed on average three roots. From the analysis of the growth of tap roots, side roots and rain roots and from the data on root size and density, *O. robusta* appeared to be less sensitive to drought than *O. ficus-indica*.

5.4 Loquat

Loquat (*Eriobotrya japonica* Lindl.), also called Japanese medlar or Japanese plum, is a subtropical evergreen tree crop indigenous to southeastern China but very well adapted to mild-winter areas of the Mediterranean basin (Hueso and Cuevas 2008). Drought stresses causes significant decreases in leaf expansion rate, area and photosynthetic pigments and in stomata size, and increases in stomata density (Luo et al. 2007). Both deficit irrigation during the entire season and post-harvest RDI from mid-May through the end of August (reduction of 20% water needs, with water savings established around 1,450 m^3 ha^{-1} yr^{-1}) were successfully applied in this species (Hueso and Cuevas 2008). It was observed that post-harvest RDI usually advances full bloom 10–20 days, allowing to obtain a more precocious and valuable yield, whereas the effects of continuous deficit irrigation is less noticeable (Hueso and Cuevas 2008; Fernández et al. 2010). On the contrary, yield and fruit quality are not affected for the different deficit irrigation strategies. The optimal month for the application of post-harvest RDI in loquat seem to be July, due to the positive effects on the advancement of bloom and harvest date and its harmlessness for flower development, even though every RDI applied in the period June–August (with a water reduction up to 75%) does not influence negative fruit set, size and yield (Cuevas et al. 2007).

6 Conclusions and Future Perspectives

A new approach in fruit orchard management is imposed by the environmental emergencies that are marking this recent period (e.g., soil degradation as a result of erosion and desertification, water shortage, greenhouse effect). In semi-arid Mediterranean lands, the adoption of agricultural systems by means of conventional, non-sustainable techniques causes the reduction of soil organic matter, groundwater contamination, soil deficiency of mineral elements (in particular phosphorus and nitrogen), alkalinization/salinization and nutritional imbalances in plants. On the other hand, the recent researches on the physiology of fruit trees and on soil chemical and biological fertility in fruit orchards have revealed that sustainable and innovative soil management systems, with a particular emphasis on irrigation, allow to obtain an optimal plant nutritional equilibrium, avoid nutrients accumulation and leaching risks, improve irrigation efficiency and prevent soil erosion and root asphyxia. As highlighted in this chapter, the definition of appropriate irrigation techniques (e.g., SDI, RDI, PRD) and soil management in Mediterranean fruit orchards are indispensable requisites for preserving soil quality, positively affecting soil microbial activity and fertility and maintaining top yields of high quality. Considering the scientific, practical and socio-economical importance of these topics, and the increasing environmental emergencies related to water scarcity, a conspicuous number of studies is expected in the next years. At the moment, it is clear that the application, optimization, innovation of sustainable agricultural techniques with a low negative environmental impact can allow to recover or increase the normal levels of total fertility in agro-ecosystems, with positive effects on both soil and yield quality.

References

Abid N, Lal R (2008) Tillage and drainage impact on soil quality: I. Aggregate stability, carbon and nitrogen pools. Soil Till Res 100:89–98

Aganchich B, El Antari A, Wahbi S, Tahi H, Wakrim R, Serraj R (2008) Fruit and oil quality of mature olive trees under partial rootzone drying in field conditions. Grasas Aceites 59(3):225–233

Aiyelaagbe I, Orodele O (2007) Leaf gas exchange and growth response of juvenile 'Valencia' orange trees to dry season irrigation in south western Nigeria. Proceedings conference on international agricultural research for development tropentag 2007. University of Kassel-Witzenhausen and University of Göttingen, Germany, October 9–11, 2007

Alar on JJ, Domingo R, Green SR, Sánchez-Blanco MJ, Rodrfiguez P, Torrecillas A (2000) Sap flow as an indicator of transpiration and the water status of young apricot trees. Plant Soil 227:77–85

Allouche N, Fki I, Sayadi S (2004) Toward a high yield recovery of antioxidants and purified hydroxytyrosol from olive mill wastewaters. J Agric Food Chem 52:267–273

Angelopoulos K, Dichio B, Xiloyannis C (1996) Inhibition of photosynthesis in olive trees (*Olea europaea* L.) during water stress and rewatering. J Exp Bot 47: 1093–1100

Arndt SK, Wanek W (2002) Use of decreasing foliar carbon isotope discrimination during water limitation as a carbon tracer to study whole plant carbon allocation. Plant Cell Environ 25:609–616

Arndt SK, Wanek W, Clifford SC, Popp M (2000) Contrasting adaptations to drought stress in fieldgrown *Ziziphus mauritiana* and *Prunus persica* trees, water relations, osmotic adjustment and carbon isotope composition. Aust J Plant Physiol 27:985–996

Ayers RS, Westcot DW (1976) Water quality for agriculture. FAO Irrigation and Drainage Paper 29, pp. 1–96

Bedbabis S, Palese AM, Ben Rouina B, Boukhris M, Rhouma A, Gargouri K (2009) The effect of irrigation by treated wastewater on "Chemlali" olive oil quality. J Food Qual 32(2):141–157

Bellini P (1983) Sovescio. Enciclopedia agraria. REDA, Roma, pp 977–979

Bending GD, Turner MK, Jones JE (2002) Interactions between crop residue and soil organic matter quality and the functional diversity of soil microbial communities. Soil Biol Biochem 34:1073–1082

Benlloch M, Arboleda F, Barranco D, Fernández-Escobar R (1991) Response of young olive trees to sodium and boron excess in irrigation water. HortScience 26: 867–870

Berenguer MJ, Vossen PM, Grattan SR, Connell JH, Polito VS (2006) Tree irrigation levels for optimum chemical and sensory properties of olive oil. HortScience 41:427–432

Berman ME, DeJong TM (1996) Water stress preconditioning to improve drought resistance in young apricot plants. Tree Physiol 16:859–864

Besset J, Génard M, Girard T, Serra V, Bussi C (2001) Effect of water stress applied during the final stage of rapid growth of peach trees (cv. Big-Top). Sci Hortic 91:289–303

Bhantana P, Lazarovitch N (2010) Evapotranspiration, crop coefficient and growth of two young pomegranate (*Punica granatum* L.) varieties under salt stress. Agric Water Manage 97:715–722

Boland AM, Jerie PH, Mitchell PD, Goodwin I (2000) Long-term effects of restricted root volume and regulated deficit irrigation on peach I. Growth and mineral nutrition. J Am Soc Hortic Sci 125:135–142

Borken W, Muhs A, Beese F (2002) Changes in microbial and soil properties following compost treatment of degraded temperate forest soils. Soil Biol Biochem 34: 403–412

Browicz K, Zohary D (1996) The genus *Amygdalus* L. (Rosaceae): species relationships, distribution and evolution under domestication. Genet Resour Crop Evol 43:229–247

Castel JR, Fereres E (1982) Response of young almond trees to two drought periods in the field. J Hortic Sci 57:175–187

Cataldi TR, Margiotta G, Iasi L, Dichio B, Xiloyannis C, Bufo SA (2000) Determination of sugar compounds in olive plant extracts by anion-exchange chromatography with pulsed amperometric detection. Anal Chem 72:3902–3907

Celano G, Palese AM, Xiloyannis C (2002) La gestione sostenibile del suolo nell'albicoccheto. Frutticoltura 3:37–39

Celano G, Palese AM, Xiloyannis C (2003) Gestione del suolo. Olea: Trattato di Olivicoltura. (a cura di Piero Fiorino). Il Sole 24 ORE Edagricole S.r.L., pp. 349–363

Celano G, Palese AM, Parente D, Rizzuti L, Silvestri N (2010) Environmental impact evaluation of olive mill wastewater shedding on cultivated fields. Terr Aquat Environ Toxicol 4(1):39–44

Celano G, Palese AM, Ciucci A, Martorella E, Vignozzi N, Xiloyannis C (2011) Evaluation of soil water content in tilled and cover-cropped olive orchards by the geoelectrical technique. Geoderma 163:163–170

Chalmers DJ, Mitchell PD, van Heek L (1981) Control of peach tree growth and productivity by regulated water supply, tree density and summer pruning. J Am Soc Hortic Sci 106:307–312

Chartzoulakis KS (2005) Salinity and olive: growth, salt tolerance, photosynthesis and yield. Agric Water Manage 78:108–121

Chaves MM, Maroco JP, Pereira JS (2003) Understanding plant responses to drought – from genes to the whole plant. Funct Plant Biol 30:239–264

Crisosto CH, Johnson RS, Luza JG, Crisosto SM (1994) Irrigation regimes affect fruit soluble solids concentration and rate of water loss of 'O'Henry' peaches. HortScience 29:1169–1171

Cuevas J, Cañete ML, Pinillos V, Zapata AJ, Fernández MD, González M, Hueso JJ (2007) Optimal dates for regulated deficit irrigation in 'Algerie' loquat (*Eriobotrya japonica* Lindl.) cultivated in Southeast Spain. Agric Water Manage 89:131–136

d'Andria R, Lavini A, Morelli G, Patumi M, Terenziani S, Calandrelli D, Fragnito F (2004) Effects of water regimes on five pickling and double aptitude olive cultivars (*Olea europaea* L.). J Hortic Sci Biotechnol 79:18–25

d'Andria R, Lavini A, Morelli G, Sebastiani L, Tognetti R (2009) Physiological and productive responses of *Olea europaea* L. cultivars Frantoio and Leccino to a regulated deficit irrigation regime. Plant Biosyst 1:222–231

Dabbou S, Chehab H, Faten B, Dabbou S, Esposto S, Selvaggini R, Taticchi A, Servili M, Montedoro GF, Hammami M (2010) Effect of three irrigation regimes on Arbequina olive oil produced under Tunisian growing conditions. Agric Water Manage 97:763–768

Davies WJ, Bacon MA, Thompson DS, Sobeigh W, Rodriguez LG (2000) Regulation of leaf and fruit growth in plants growing in drying soil: exploitation of the plants' chemical signalling system and hydraulic architecture to increase the efficiency of water use in agriculture. J Exp Bot 51:1617–1626

de Graaff J, Eppink LAAJ (1999) Olive oil production and soil conservation in southern Spain, in relation to EU subsidy policies. Land Use Policy 16:259–267

Dichio B, Romano M, Nuzzo V, Xiloyannis C (2002) Soil water availability and relationship between canopy and roots in young olive trees (cv Coratina). Acta Hortic 586:255–258

Dichio B, Xiloyannis C, Sofo A, Montanaro G (2007) Effects of post-harvest regulated deficit irrigation on carbohydrate and nitrogen partitioning, yield quality and vegetative growth of peach trees. Plant Soil 290: 127–137

Dichio B, Margiotta G, Xiloyannis C, Bufo SA, Sofo A, Cataldi RI (2009) Changes in osmolyte levels and water status in leaves and roots of olive plants (*Olea europaea* L.) subjected to water deficit. Trees Struct Funct 23:247–256

Durán-Zuazo VH, Rodríguez Pleguenzuelo CR, Arroyo Panadero L, Martínez Raya A, Francia Martínez JR, Cárceles Rodríguez B (2009) Soil conservation measures in rainfed olive orchards in south-eastern Spain: impacts of plant strips on soil water dynamics. Pedosphere 19(4):453–464

Dzikiti S, Milford JR, Steppe K, Lemeur R (2006) Water use by young navel orange trees (*Citrus sinensis* (L.) Osbeck) in Zimbabwe. Acta Hortic 707:135–141

Dzikiti S, Steppe K, Lemeur R (2008) Partial rootzone drying of drip irrigated 'Navel' orange trees [*Citrus sinensis* (L.) Osbeck] under semi-arid tropical conditions. Acta Hortic 792:249–256

Egea G, Nortes PA, González-Real MM, Baille A, Domingo R (2010) Agronomic response and water productivity of almond trees under contrasted deficit irrigation regimes. Agric Water Manage 97:171–181

Engin H, Ünal A (2008) The effect of irrigation, gibberellic acid and nitrogen on the occurrence of double fruit in 'Van' sweet cherry. Acta Hortic 795:645–650

Escobar-Gutiérrez AJ, Zipperlin B, Carbonne F, Moing A, Gaudillére JP (1998) Photosynthesis, carbon partitioning and metabolite content during drought stress in peach seedlings. Aust J Plant Physiol 25:197–205

Esparza G, DeJong TM, Weinbaum SA (2001) Effects of irrigation deprivation during the harvest period on nonstructural carbohydrate and nitrogen contents of dormant, mature almond trees. Tree Physiol 21: 1081–1086

Esparza G, DeJong TM, Weinbaum SA, Klein I (2010) Effects of irrigation deprivation during the harvest period on yield determinants in mature almond trees. Tree Physiol 21:1073–1079

FAO (Food and Agriculture Organization) (1985) Water quality for agriculture. FAO Irrigation and Drainage Paper 29, Rome, Italy

FAOSTAT (2010) http://faostat.fao.org

Fereres E, Soriano MA (2007) Deficit irrigation for reducing agricultural water use. Exp Bot 58(2):147–159

Fernández JE, Diaz-Espejo A, Infante JM, Duran P, Palomo MJ, Chamorro V, Giron IF, Villagarica L (2006) Water relations and gas exchange in olive trees under regulated deficit irrigation and partial rootzone drying. Plant Soil 284:273–291

Fernández MD, Hueso JJ, Cuevas J (2010) Water stress integral for successful modification of flowering dates in 'Algerie' loquat. Irrig Sci 28:127–134

Fleskens L, Stroosnijder L, Ouessar M, De Graaff J (2005) Evaluation of the on-site impact of water harvesting in southern Tunisia. J Arid Environ 62:613–630

Foyer CH, Gomez LD, Van Heerdern PDR (2005) Glutathione. In: Smirnoff N (ed) Antioxidants and reactive oxygen species in plants. Blackwell Publishing, Oxford, UK, pp 53–66

García Petillo M, Puppo L, Chamorro A, Hayashi R (2004) Effects of drip irrigation on the amount of water and wetted soil volume on "Washington Navel" orange yield. Acta Hortic 646:101–106

García-Tejero I, Romero-Vicente R, Jiménez-Bocanegra J, Martínez-García G, Duran-Zuazo V, Muriel-Fernández J (2010) Response of citrus trees to deficit irrigation during different phenological periods in relation to yield, fruit quality, and water productivity. Agric Water Manage 97:689–699

Gelly M, Recasens I, Girona J, Mata M, Arbones A, Rufat J, Marsal J (2004) Effects of stage II and postharvest deficit irrigation on peach quality during maturation and after cold storage. J Sci Food Agric 84:561–568

Ghrab M, Sahli A, BenMechli N (1998) Reduction in vegetative growth and fruit quality improvement in the peach variety "Carnival" through moderate watering restrictions. Acta Hortic 465:601–608

Gijón MC, Guerrero J, Couceiro JF, Moriana A (2009) Deficit irrigation without reducing yield or nut splitting in pistachio (*Pistacia vera* cv. Kerman on *Pistacia terebinthus* L.). Agric Water Manage 96:12–22

Gijón MC, Gimenez C, Perez-López D, Guerrero J, Couceiro JF, Moriana A (2010) Rootstock influences the response of pistachio (*Pistacia vera* L. cv. Kerman) to water stress and rehydration. Sci Hortic 125:666–671

Girona J, Mata M, Fereres E, Goldhamer DA, Cohen M (2002) Evapotranspiration and soil water dynamics of peach trees under water deficits. Agric Water Manag 54:107–122

Girona J, Gelly M, Mata M, Arbonès A, Rufat J, Marsal J (2005a) Peach tree response to single and combined deficit irrigation regimes in deep soils. Agric Water Manage 72:97–108

Girona J, Mata M, Marsal J (2005b) Regulated deficit irrigation during the kernel-filling period and optimal irrigation rates in almond. Agric Water Manage 75: 152–167

Godini A, Palasciano M, Camposeo S, Pacifico A (2008) A nine-year study on the performance of twelve cherry rootstocks under non-irrigated conditions in Apulia (southern Italy). Acta Hortic 795:191–198

Goldhamer DA (1999) Regulated deficit irrigation for California canning olives. Acta Hortic 474:369–372

Gomes-Laranjo J, Coutinho JP, Galhano V, Cordeiro V (2006) Responses of five almond cultivars to irrigation: photosynthesis and leaf water potential. Agric Water Manage 83:261–265

Gómez-Rico A, Desamparados Salvador M, La Greca M, Fregapane G (2006) Phenolic and volatile compounds of extra virgin olive oil (*Olea europaea* L. cv. Cornicabra) with regard to fruit ripening and irrigation management. J Agric Food Chem 54:7130–7136

Gonçalbes B, Moutinho-Pereire J, Santos A, Silva AP, Bacelar E, Correia C, Rosa E (2005) Scion-rootstock interaction affects the physiology and fruit quality of sweet cherry. Tree Physiol 26:93–104

Govaerts B, Mezzalama M, Sayre KD, Crossa J, Litcher K, Troch V, Vanherck K, De Corte P, Deckers J (2008) Long-term consequences of tillage, residue management, and crop rotation on selected soil microflora groups in the subtropical highlands. Appl Soil Ecol 38:197–210

Grattan SR, Berenguer MJ, Connell JH, Polito VS, Vossen PM (2006) Olive oil production as influenced by different quantities of applied water. Agr Water Manage 85:133–140

Grossman YL, DeJong TM (1998) Training and pruning system effects on vegetative growth potential, light interception and cropping efficiency in peach trees. J Am Soc Hort Sci 123:1058–1064

Gruhn P, Goletti F, Yudelman M (2000) Integrated nutrient management, soil fertility, and sustainable agriculture: current issues and future challenges. Food, agriculture, and the environment discussion paper 32. International Food Policy Research Institute, Washington, DC

Hepaksoy S, Kukul YS, Engin H, Erogul D, Aksehirli M (2009) Leafwater potential of pomegranate (*Punica granatum* L.) under different irrigation levels. Acta Hortic 818:193–198

Hernández AJ, Lacasta C, Pastor J (2005) Effects of different management practices on soil conservation and soil water in a rainfed olive orchard. Agric Water Manage 77:32–248

Huang X, Huang H, Gao F (2000) The growth potential generated in citrus fruit under water stress and its relevant mechanisms. Sci Hortic 83:227–240

Hueso JJ, Cuevas J (2008) Loquat as a crop model for successful deficit irrigation. Irrig Sci 26:269–276

Iniesta F, Testi L, Orgaz F, Villalobos FJ (2009) The effects of regulated and continuous deficit irrigation on the water use, growth and yield of olive trees. Eur J Agron 30:258–265

Intrigliolo DS, Castel JR (2006) Performance of various water stress indicators for prediction of fruit size response to deficit irrigation in plum. Agric Water Manage 83:173–180

Intrigliolo DS, Nicolas E, Bonet L, Ferrer P, Alarcon JJ, Bartual J (2011) Water relations of field grown Pomegranate trees (*Punica granatum*) under different drip irrigation regimes. Agric Water Manage 98: 691–696

Jagadamma S, Lal R, Hoeft RG, Nafziger ED, Adee EA (2008) Nitrogen fertilization and cropping system impacts on soil properties and their relationship to crop yield in the central Corn Belt, USA. Soil Till Res 98:120–129

Johnson RS, Handley DF, DeJong TM (1992) Long-term response of early maturing peach trees to postharvest water deficits. J Am Soc Hortic Sci 117(881):886

Kaska N (2002) Pistachio nut growing in the Mediterranean basin. Acta Hortic 591:443–455

Kimmins JP (1997) Forest ecology. Pearson Prentice Hall, New York

Klay S, Charef A, Ayed L, Houman B, Rezgui F (2010) Effect of irrigation with treated wastewater on geochemical properties (saltiness, C, N and heavy metals) of isohumic soils (Zaouit Sousse perimeter, Oriental Tunisia). Desalination 253:180–187

Klein I, Esparza G, Weinbaum SA, DeJong TM (2001) Effects of irrigation deprivation during the harvest period on leaf persistence and function in mature almond trees. Tree Physiol 21:1063–1072

Kushwaha CP, Tripathi SK, Singh K (2000) Variations in soil microbial biomass and N availability due to residue and tillage management in a dryland rice agroecosystem. Soil Till Res 56:153–166

Lakso AN (1979) Seasonal changes in stomatal response to leaf water potential in apple. J Am Soc Hortic Sci 104:58–60

Lal R (2004) Soil carbon sequestration to mitigate climate change. Geoderma 123:1–22

Lavee S, Wodner M (1991) Factors affecting the nature of oil accumulation in fruit of olive (*Olea europaea* L.) cultivars. J Hortic Sci 66(5):583–591

Lawand BT, Patil VK, Patil PV (1992) Effect on different water regimes on fruit quality of pomegranate (*Punica granatum* L.). Acta Hortic 321:677–683

Lecouls AC, Bergougnoux V, Rubio-Cabetas MJ, Bosselut N, Voisin R, Poessel JL, Faurobert M, Bonnet A, Salesses G, Dirlewanger E, Esmenjaud D (2004) Marker-assisted selection for the wide-spectrum resistance to root-knot nematodes conferred by the *Ma* gene from Myrobalan plum (*Prunus cerasifera*) in interspecific *Prunus* material. Mol Breed 13: 113–124

Lipecki J, Berbe] S (1997) Soil management in perennial crops: orchards and hop gardens. Soil Till Res 43: 169–184

Lo Bianco R, Rieger M, Sung SS (2000) Effect of drought on sorbitol and sucrose metabolism in sinks and sources of peach. Physiol Plant 108:71–78

Lopez A, Pollice A, Lonigro A, Masi S, Palese AM, Cirelli GL, Toscano A, Passino R (2006) Agricultural wastewater reuse in southern Italy. Desalination 187: 323–334

Luo HJ, Zheng ZB, Luo S, Pan YS, Liu XH (2007) Changes in leaf characters of loquat under repeated drought stress. Acta Hortic 750:417–422

Magliulo V, d'Andria R, Lavini A, Morelli G, Patumi M (2003) Yield and quality of two rain-fed olive cultivars following shifting to irrigation. J Hortic Sci Biotechnol 78:15–23

Marquat C, Vandamme M, Gendraud M, Pétel G (1998) Dormancy in vegetative buds of peach, relation between carbohydrate absorption potentials and carbohydrate concentration in the bud during dormancy and its release. Sci Hortic 79:151–162

Marschner P, Kandeler E, Marschner B (2003) Structure and function of the soil microbial community in a long-term fertilizer experiment. Soil Biol Biochem 35:453–461

Martín-Vertedor AI, Pérez Rodríguez JM, Losada EP, Fereres Castiel E (2011) Interactive responses to water deficits and crop load in olive (*Olea europaea* L., cv. Morisca). II: Water use, fruit and oil yield. Agric Water Manage, 98:950–958

Matos MC, Rebelo E, Lauriano J, Semedo N, Marques N, Campos PS, Matos A, Veira-da-Silva J (2004) CO_2 assimilation and water relations of almond tree (*Prunus amygdalus* Batsch) cultivars grown under field conditions. Photosynthetica 42:473–476

Melgar JC, Mohamed Y, Serrano N, García-Galavís PA, Navarro C, Parra MA, Benlloch M, Fernández-Escobar R (2009) Long term responses of olive trees to salinity. Agric Water Manage 96:1105–1113

Morales F, Abadía A, Abadía J (2006) Photoinhibition and photoprotection under nutrient deficiencies, drought and salinity. In: Demmig-Adams B, Adams WW III, Matoo AK (eds) Photoprotection, photoinhibition, gene regulation, and environment, advances in photosynthesis and respiration. Springer, The Netherlands, pp 65–85

Moriana A, Orgaz F, Pastor M, Fereres E (2003) Yield responses of a mature olive orchard to water deficits. J Am Soc Hortic Sci 128:425–431

Mulas M, D'hallewin G (1997) Fruit quality of four cactus pear (*Opuntia ficus-indica* Mill.) cultivars as influenced by irrigation. Acta Hortic 438:115–122

Mulas M, Mulas G (2004) Utilizzo strategico di piante dei generi *Atriplex* e *Opuntia* nella lotta alla desertificazione. Università degli Studi di Sassari, Nucleo di Ricerca sulla Desertificazione. Short and Medium-Term Priority Environmental Action Programme (SMAP), February 2004. Available on: http://desa.uniss.it/mulas/desertIT.pdf

Naor A, Hupert H, Greenblat Y, Peres M, Kaufman A, Klein I (2001) The response of nectarine fruit size and midday stem water potential to irrigation level in stage III and crop load. J Am Soc Hortic Sci 126:140–143

Neilsen GH, Neilsen D, Kappel F, Toivonen P (2005) Nutrient and water management of high density sweet cherry. Acta Hortic 667:337–344

Nieddu G, Chessa I, Deidda P, Tekle Z (1997) Changes in CAM activity, ABA and PAS in Opuntia ficus-indica as response to drought. Acta Hortic 438:97–104

Nortes PA, Pérez-Pastor A, Egea G, Conejero W, Domingo R (2005) Comparison of changes in stem diameter and water potential values for detecting water stress in young almond trees. Agric Water Manage 77:296–307

Nortes PA, Gonzalez-Real MM, Egea G, Baille A (2009) Seasonal effects of deficit irrigation on leaf photosynthetic traits of fruiting and non-fruiting shoots in almond trees. Tree Physiol 29:375–388

Olesen J, Bindi M (2002) Consequences of climate change for European agricultural productivity, land use and policy. Eur J Agron 16:239–262

Orgaz F, Fereres E (2004) Riego. In: Barranco D, Fernández-Escobar R, Rallo L (eds) El cultivo del olivo. Mundi Prensa, Madrid, pp 323–346

Ortuno MF, García-Orellana Y, Conejero W, Pérez-Sarmiento F, Torrecillas A (2009) Assessment of maximum daily trunk shrinkage signal intensity threshold values for deficit irrigation in lemon trees. Agric Water Manage 96:80–86

Ozpinar S, Cay A (2006) Effects of different tillage systems on the quality and crop productivity of a clay-loam soil in semi-arid north-western Turkey. Soil Till Res 88:95–106

Padulosi S, Van Mele P, Caruso T, Kaska N, Barone E (1998) IPRIG's initiatives for the promotion of better conservation and use of *Pistacia* spp. genetic resources. Acta Hortic 470:138–142

Palese AM, Celano G, Masi S, Xiloyannis C (2006) Treated municipal wastewater for irrigation of olive trees: effects on yield and oil quality. In: Caruso T, Motisi A, Sebastiani L (eds) Proceedings Olivebioteq 2006 – second international seminar – biotechnology and quality of olive tree products around the Mediterranean Basin – Mazara del Vallo, Italy, November 5th–10th, 2006, Vol. II. pp. 123–129

Palese AM, Celano G, Xiloyannis C (2008) Acque per le fertirrigazioni: reflue, urbane e depurate. Olivo Olio 5:48–53

Palese AM, Agnelli AE, Celano G, Platinetti M, Vignozzi N, Xiloyannis C (2009a) Soil management and soil water balance in a mature rainfed olive orchard. Proceedings Olivebioteq 2009 – Sfax (Tunisia), December 15th – 19th, 2009, in press

Palese AM, Pasquale V, Celano G, Figliuolo G, Masi S, Xiloyannis C (2009b) Irrigation of olive groves in Southern Italy with treated municipal wastewater: effects on microbiological quality of soil and fruits. Agric Ecosyst Environ 129:43–51

Palese AM, Nuzzo V, Favati F, Pietrafesa A, Celano G, Xiloyannis C (2010) Effects of water deficit on the vegetative response, yield and oil quality of olive trees (*Olea europaea* L., cv Coratina) grown under intensive cultivation. Sci Hortic 125(3):222–229

Pallardy SG (2008) Physiology of woody plants. Academic, San Diego, CA

Pardini A, Faiello C, Longhi F, Mancuso S, Snowball R (2002) Cover crop species and their management in vineyards and olive groves. Adv Hortic Sci 16(3–4): 225–234

Pastor M, Castro J, Humanes MD, Muñoz J (2000) Gestione del suolo nell'olivicoltura dell'Andalusia. L'Informatore Agrario 3:83–92

Patumi M, d'Andria R, Fontanazza G, Morelli G, Giorio P, Sorrentino G (1999) yield and oil quality of intensively trained trees of three cultivars of olive (*Olea europaea* L.) under different irrigation regimes. J Hortic Sci Biotechnol 74:729–737

Patumi M, d'Andria R, Marsilio V, Fontanazza G, Morelli G, Lanza B (2002) Olive and olive oil quality after intensive monocone olive growing (*Olea europaea* L., cv. Kalamata) in different irrigation regimes. Food Chem 77:27–34

Pérez-Pérez JG, Robles JM, Botía P (2009) Influence of deficit irrigation in phase III of fruit growth on fruit quality in 'lane late' sweet orange. Agric Water Manage 96:969–974

Pimienta-Barrios E, Zanudo J, Yepez E, Pimienta-Barrios E, Nobel PS (2000) Seasonal variation of net CO_2 uptake for cactus pear (*Opuntia ficus-indica*) and pitayo (*Stenocereus queretaroensis*) in a semi-arid environment. J Arid Environ 44:73–83

Ramirez-Fuentes E, Lucho-Constantino C, Escamilla-Silva E, Dendooven L (2002) Characteristics, and carbon and nitrogen dynamics in soil irrigated with wastewater for different lengths of time. BioresourTechnol 85:179–187

Ranjbarfordoei A, Samson R, Van Damne P, Lemeur E (2000) Effects of drought stress induced by polyethylene glicol on pigment content and photosynthetic gas exchange of *Pistacia khinjuk* and *P. mutica*. Photosynthetica 38:443–447

Ranjbarfordoei A, Samson R, Lemeur E, Van Damne P (2002) Effects of osmotic drought stress induced by a combination of NaCl and polyethylene glycol on leaf water status, photosynthetic gas exchange, and water use efficiency of *Pistacia khinjuk* and *P. mutica*. Photosynthetica 40:165–169

Rato AE, Agulheiro AC, Barroso JM, Riquelme F (2006) Soil and rootstock influence on fruit quality of plums (*Prunus domestica* L.). Sci Hortic 118:218–222

Rieger M (1995) Offsetting effects of reduced root hydraulic conductivity and osmotic adjustment following drought. Tree Physiol 16:379–385

Rinaldelli E, Mancuso S (1996) Response of young mycorrhizal and non-mycorrhizal plants of olive tree (Olea europaea L.) to saline conditions. I. Short-term electrophysiological and long-term vegetative salt effect. Adv Hort Sci 10:126–134

Rieger M, Duemmel MJ (1992) Comparison of drought resistance among *Prunus* species from divergent habitats. Tree Physiol 11:369–380

Rieger M, Lo Bianco R, Okie WR (2003) Responses of *Prunus ferganensis*, *Prunus persica* and two interspecific hybrids to moderate drought stress. Tree Physiol 23:51–58

Romero P, Botia P, Garcia F (2004a) Effects of regulated deficit irrigation under subsurface drip irrigation conditions on water relations of mature almond trees. Plant Soil 260:155–168

Romero P, Botia P, Garcia F (2004b) Effects of regulated deficit irrigation under subsurface drip irrigation conditions on vegetative development and yield of mature almond trees. Plant Soil 260:169–181

Romero P, Navarro JM, García F, Botía Ordaz P (2004c) Effects of regulated deficit irrigation during the pre-harvest period on gas exchange, leaf development and crop yield of mature almond trees. Tree Physiol 24:303–312

Rouhi V, Samson R, Lemeur R, Van Damme P (2007) Photosynthetic gas exchange characteristics in three different almond species during drought stress and subsequent recovery. Environ Exp Bot 59:117–129

Rugini E, Fedeli E (1990) Olive as an oilseed crop. In: Bajaj YPS (ed) Legumes and oilseed crops. Springer, Berlin, pp 593–641

Ruiz-Sánchez MC, Domingo R, Torrecillas A, Pérez-Pastor A (2000a) Water stress preconditioning to improve drought resistance in young apricot plants. Plant Sci 156:245–251

Ruiz-Sánchez MC, Torrecillas A, Pérez-Pastor A, Domingo R (2000b) Regulated deficit irrigation in apricot trees. Acta Hortic 537:759–766

Ryan D, Antolovich M, Prenzler P, Robards K, Lavee S (2002) Biotransformations of phenolic compounds in *Olea europaea* L. Sci Hortic 92:147–176

Scebba F, Sebastiani L, Vitagliano C (2001) Activities of antioxidant enzymes during senescence of *Prunus armeniaca* leaves. Biol Plant 44:41–46

Sedaghat R (2008) Sustainable irrigation for pistachio farms in Iran: an economic analysis. Acta Hortic 769:201–206

Sepaskhah AR, Ahmadi SH (2010) A review on partial root-zone drying irrigation. Int J Plant Prod 4(4): 241–258

Snyman HA (2006) A greenhouse study on root dynamics of cactus pears, *Opuntia ficus-indica* and *O. robusta.* J Arid Environ 65:529–542

Sofo A, Dichio B, Xiloyannis C, Masia A (2004a) Effects of different irradiance levels on some antioxidant enzymes and on malondialdehyde content during rewatering in olive tree. Plant Sci 166:293–302

Sofo A, Dichio B, Xiloyannis C, Masia A (2004b) Lipoxygenase activity and proline accumulation in leaves and roots of olive tree in response to drought stress. Physiol Plantarum 121(1):58–65

Sofo A, Tuzio AC, Dichio B, Xiloyannis C (2005) Influence of water deficit and rewatering on the components of the ascorbate-glutathione cycle in four interspecific *Prunus* hybrids. Plant Sci 169:403–412

Sofo A, Dichio B, Montanaro G, Xiloyannis C (2009) Shade effect on photosynthesis and photoinhibition in olive during drought and rewatering. Agric Water Manage 96:1201–1206

Sofo A, Celano G, Ricciuti P, Curci M, Dichio B, Xiloyannis C, Crecchio C (2010a) Changes in composition and activity of soil microbial communities in peach and kiwifruit Mediterranean orchards under an innovative management system. Aust J Soil Res 48:266–273

Sofo A, Palese AM, Casacchia T, Celano G, Ricciuti P, Curci M, Crecchio C, Xiloyannis C (2010b) Genetic, functional, and metabolic responses of soil microbiota in a sustainable olive orchard. Soil Sci 175(2): 81–88

Sorkheh K, Shiran B, Rouhi V, Khodambashi M, Sofo A (2011) Regulation of the ascorbate-glutathione cycle in wild almond during drought stress. Russ J Plant Physiol 58:76–84

Stefanoudaki E, Williams M, Chartzoulakis K, Harwood J (2009) Effect of irrigation on quality attributes of olive oil. J Agric Food Chem 57:7048–7055

Stevenson MT, Shackel KA (1998) Alternate bearing in pistachio as masting phenomenon: construction cost of reproduction versus vegetative growth and storage. J Am Soc Hortic Sci 123:1069–1075

Stikic R, Popovic S, Srdic M, Savic D, Jovanovic, Z, Prokic LJ, Zdravkovic J (2003) Partial root drying (PRD): a new technique for growing plants that saves water and improves the quality of fruit. Bulg J Plant Physiol Special Issue: 164–171.

Stoll M, Loveys B, Dry P (2000) Hormonal changes induced by partial rootzone drying of irrigated grapevine. J Exp Bot 51:1627–1634

Strudley MW, Green TR, Ascough JC II (2008) Tillage effects on soil hydraulic properties in space and time: state of the science. Soil Till Res 99:4–48

Sulochanamma BN, Yellamanda Reddy T, Subbi Reddy G (2005) Effect of basin and drip irrigation on growth, yield and water use efficiency in pomegranate cv. Ganesh. Acta Hortic 696:277–279

Tajabadipour A, Sepaskhah AR, Maftoun M (2006) Plant water relations and seedlings growth of three pistachio cultivars as influences by irrigation frequency and applied potassium. Acta Hortic 726:317–324

Tarchouna Gharbi L, Merdy P, Raynaud M, Pfeifer HR, Lucas Y (2010a) Effects of long-term irrigation with treated wastewater. Part 1: evolution of soil physicochemical properties. Appl Geochem 25:1703–1710

Tarchouna Gharbi L, Merdy P, Lucas Y (2010b) Effects of long-term irrigation with treated wastewater. Part II: role of organic carbon on Cu, Pb and Cr behaviour. Appl Geochem 25:1711–1721

Tataranni G (2009) Water uptake and transport in olive tree under stress conditions. Thesis in International PhD course. Università degli Studi della Basilicata, Italy

Tattini M, Gucci R, Coradeschi MA, Ponzio C, Everard JD (1995) Growth, gas exchange and ion content in Olea europaea plants during salinity stress and subsequent relief. Physiol Plantarum 95:203–210

Therios I (2009) Olives, vol 18, Crop production science in horticulture. CABI, Wallingford, p 409

Tognetti R, d'Andria R, Lavini A, Morelli G (2006) The effect of deficit irrigation on crop yield and vegetative development of *Olea europaea* L. (cvs. Frantoio and Leccino). Eur J Agron 25:356–364

Tognetti R, d'Andria R, Sacchi R, Lavini A, Morelli G, Alvino A (2007) Deficit irrigation affects seasonal changes in leaf physiology and oil quality of Olea europaea (cultivars Frantoio and Leccino). Ann Appl Biol 150:169–186

Torrecillas A, Alarcón JJ, Domingo R, Planes J, Sánchez-Blanco MJ (1996) Strategies for drought resistance in leaves of two almond cultivars. Plant Sci 118: 135–143

Tovar MJ, Motilva MJ, Romero PM (2001) Changes in the phenolic composition of virgin olive oil from young trees (Olea europaea L. cv. Arbequina) grown under linear irrigation strategies. J Agric Food Chem 49:5502–5508

Tovar MJ, Romero PM, Girona J, Motilva MJ (2002) L-Phenylalanine ammonia-lyase activity and concentration of phenolics in developing olive (*Olea europaea* L cv Arbequina) fruit grown under different irrigation regimes. J Sci Food Agric 82:892–898

Tubeileh A, Bruggeman A, Turkelboom F (2009) Effect of water harvesting on growth of young olive trees in degraded Syrian dryland. Environ Dev Sustain 11: 1073–1090

Visioli F, Romani A, Mulinacci N, Zarini S, Conte D, Franco F, Galli C (1999) Antioxidant and other biological activities of olive mill waste waters. J Agric Food Chem 47:3397–3401

Visioli F, Galli C, Mogani P (2006) Is olive good for you? In: Caruso T, Motisi A, Sebastiani L (eds) Olivebioteq 2006 – second international seminar – recent advances in olive industry – special seminars and invited lectures. Mazara del Vallo, Italy, November 5th–10th, 2006, pp 255–264

Wahbi S, Wakrim R, Aganchich B, Tahi H, Serraj R (2005) Effects of partial rootzone drying (PRD) on adult olive tree (*Olea europaea*) in field conditions under arid climate I. Physiological and agronomic responses. Agric Ecosyst Environ 106:289–301

Wang J, Liu J (2009) Change in free polyamine contents and expression profiles of two polyamine biosynthetic genes in citrus embryogenic callus under abiotic stresses. Biotechnol Biotechnolog Equip 23:1289–1293

Weaver LM, Herrmann KM (1997) Dynamic of the shikimate pathway in plants. Trends Plant Sci 2(9): 346–351

Widmer F, Rasche F, Hartmann M, Fliessbach A (2006) Community structures and substrate utilization of bacteria in soils from organic and conventional farming systems of the DOK long-term field experiment. Appl Soil Ecol 33:294–307

Wiesman Z, Itzhak D, Ben Dom N (2004) Optimization of saline water level for sustainable Barnea olive and oil production in desert conditions. Sci Hortic 100: 257–266

Wright GC (2008) Evaluation of ten Navel orange (*Citrus sinensis*) selections under Arizona desert conditions. Acta Hortic 773:77–82

Xiloyannis C, Dichio B, Nuzzo V, Celano G (1999) Defence strategies of olive against water stress. Acta Hortic 474:423–426

Xiloyannis C, Gucci R, Dichio B (2003) Irrigazione. In: Fiorino P (ed) Olea: Trattato di Olivicoltura. Il Sole 24 ORE Edagricole S.r.l., Bologna, Italy, pp 365–389

Xiloyannis C, Massai R, Dichio B (2005) L'acqua e la Tecnica dell'Irrigazione. In: Fideghelli C, Sansavini S (eds) Il Pesco. Edagricole, Bologna, pp 145–171

Xiloyannis C, Dichio B, Montanaro G (2010) Sustainable apricot orchard management to improve soil fertility and water use efficiency. Acta Hortic 862: 419–424

Yadav RK, Goyal B, Shamara RK, Dubey SK, Minhas PS (2002) Post-irrigation impact of domestic sewage effluent on composition of soils, crops and ground water – a case study. Environ Int 28:481–486

Yakubov B, Barazani O, Shachack A, Rowland LJ, Shoseyov O, Golan-Goldhirsh A (2005) Cloning and expression of a dehydrin-like protein from *Pistacia vera* L. Trees 19:224–230

Yangui T, Dhouib A, Rhouma A, Sayadi S (2009) Potential of hydroxytyrosol-rich composition from olive mill wastewater as a natural disinfectant and its effect on seeds vigour response. Food Chem 117:1–8

Yoshita K, Takahara T (2004) Effect of water stress on flower-bud formation and plant hormone content of satsuma mandarin (*Citrus unshiu* Marc.). Sci Hortic 99:301–307

Zeng DQ, Brown PH, Rosecrance RC (1998) The effects of alternate bearing, soil moisture and gypsum on potassium nutrition of pistachio (*Pistacia vera* L.). Acta Hortic 470:412–420

Drought Stress Induced Reactive Oxygen Species and Anti-oxidants in Plants

7

S.M. Impa, S. Nadaradjan, and S.V.K. Jagadish

Abstract

Aerobic metabolism in plants results in the generation of reactive oxygen species (ROS). ROS are produced constantly in plants under physiological steady state condition, and plants have evolved to efficiently scavenge and maintain the levels of ROS at non-damaging levels. However, plants when exposed to either abiotic or biotic stress conditions, the production of ROS exceeds their scavenging capacity, leading to an outburst of highly reactive oxidative species capable of inflicting significant damage to the membranes, DNA, and proteins. On the other hand, these reactive molecules when maintained under non-damaging levels are useful signalling molecules involved in relaying stress signal to activate acclimation and defence mechanism. Drought or water deficit stress is one of the major abiotic stresses which induces the production of different kinds of ROS including both free radicals such as superoxide ($O_2^{\cdot-}$), hydroxyl radicals (·OH), perhydroxy radical ($HO_2^{\cdot}$) and alkoxy radicals (RO·) and non-radical (molecular) forms, that is, singlet oxygen ($^1O_2$), and hydrogen peroxide (H_2O_2). (.OH) is the most reactive chemical species among the known ROS. Generally ROS are produced in cell organelles that are involved in active electron transport like, chloroplast, mitochondria, peroxisomes, apoplast, and their membranes. These organelles also harbour various anti-oxidative enzymes like superoxide dismutase (SOD), catalase (CAT), ascorbate peroxidase (APX), glutathione peroxidase (GPX), etc., having the potential to quench the highly reactive oxidative species to maintain overall plant

S.M. Impa • S.V.K. Jagadish (✉)
Crop and Environmental Sciences Division,
International Rice Research Institute, DAPO Box 7777,
Metro Manila, Philippines
e-mail: k.jagadish@cgiar.org

S. Nadaradjan
Crop Physiology Unit, Department of Plant
Breeding and Genetics, Pandit Jawaharlal Nehru
College of Agriculture and Research Institute, Karaikal,
UT of Puducherry, India

P. Ahmad and M.N.V. Prasad (eds.), *Abiotic Stress Responses in Plants: Metabolism, Productivity and Sustainability*, DOI 10.1007/978-1-4614-0634-1_7,

homeostasis. Additionally the plants are also equipped with the known non-enzymatic scavengers like ascorbate, glutathione, carotenoids, tocopherols, flavonoids, and alkaloids. Crop varieties having an efficient antioxidant quenching mechanism aided by increased production of enzymatic and non-enzymatic scavengers have developed the ability to withstand adverse conditions including drought stress.

Keywords

Drought stress • Reactive oxygen species • Production • Scavenging • Signalling

1 Introduction

Plants are often faced with a wide range of biotic (pathogens and diseases) and abiotic stresses including high temperature, salinity, drought, ozone, flooding, etc., among which drought stress is considered to be a major threat to sustaining food security under current and more so in future climates. Although plants to a certain extent can withstand limited water conditions, a wide genotypic variability in response to drought stress is seen in most of the cultivated crops (for rice – Rang et al. 2011; Kumar et al. 2007; Venuprasad et al. 2008; maize – Atteya 2003; barley – Samarah 2005). Drought stress or water limited condition is a result of insufficient amount of water available for the basic up keep and maintenance of normal physiological processes such as photosynthesis and overall cell, tissue, organ, and plant homeostasis. With the predicted increase in precipitation variability across major cropping regions in the world, the intensity and duration of extreme drought stress are likely to increase, putting additional pressure to meet the demographic demand for food. Drought stress moreover occurs across regions, countries, and continents, for example in South Asia; severe drought years during 1987 and 2002/2003 affected more than 50% of the total cropped area and almost 300 million people in India (Pandey et al. 2007), in south East Asia; drought in 2004 affected 20% of the rice land and more than eight million people in Thailand (Pandey et al. 2007) and among the globally recorded disasters over the past three decades, 20% are accounted by Africa with nearly half of them caused by extreme weather, particularly drought (Cornford 2003). The World Economic Forum (2009) at Davos published a "Water Initiative" report, which estimated global crop production losses up to 30% by 2025 compared to current yields due to water shortage, if unsustainable use of water for agriculture continues (Zhang 2011). In addition, the steeply increasing temperatures could lead to a rapid loss of soil water, bringing forward severe water stress to coincide with critical developmental stages like flowering mainly due to increased evapotranspiration losses.

Crop plants in order to survive have evolved to withstand water limited conditions by (a) escaping, that is, completion of fertilization and life cycle before the on set of stress (b) avoidance through minimized water loss (e.g. stomatal regulation) or through efficient exploration of soil moisture reserves (e.g. deep root system) and (c) tolerance by maintaining normal photosynthetic activity under water deficit conditions or efficient water uptake (e.g. hydraulic conductivity). Breeding for drought tolerant crop varieties having the ability to withstand harsh conditions and simultaneously producing yields (i.e. grains in rice and wheat, lint in cotton, etc.), closer to stress-free conditions are constantly increasing in demand. The demand for these developed products is going to get bigger as we begin to see devastating effects of droughts at a much higher frequency. A wide range of negative impacts of drought on the rice crop during different developmental stages is

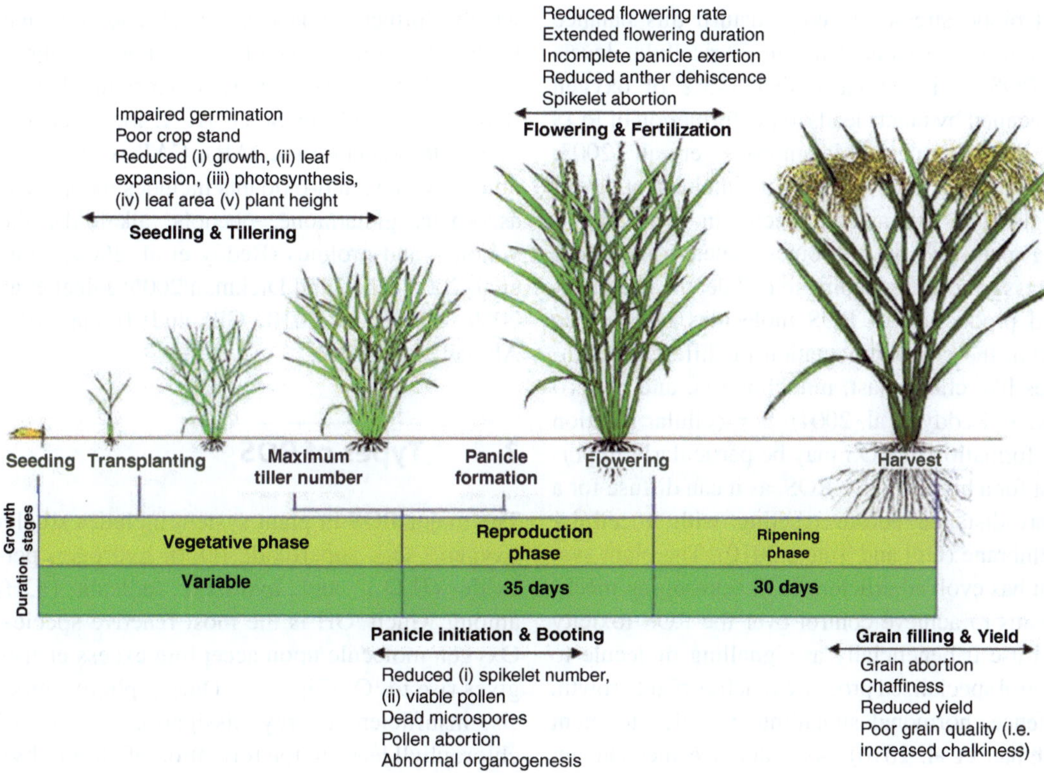

Fig. 7.1 Negative effects of drought stress during different developmental stages. Studies from which the information has been derived are listed below. Seedling and Tillering – Singh et al. (1996), Boonlertnirun et al. (2007), Harris et al. (2002), Tripathy et al. (2000), Manickavelu et al. (2006); Panicle Initiation and Booting – Nguyen et al. (2009), Sheoran and Saini (1996), Liu and Bennett (2010); Flowering and Fertilization – Jongdee et al. (2006), Rang et al. (2011), Liu et al. (2006); Grain filling and yield – Liu et al. (2006), Kumar et al. (2007), Venuprasad et al. (2008)

presented in Fig. 7.1 and similar effects have been documented with other crops (pea – Okcu et al. 2005; sunflower – Kaya et al. 2006; Hussain et al. 2008; maize – Cattivelli et al. 2008; wheat – Wardlaw and Willenbrink 2000; Ahmadi and Baker 2001). One of the immediate responses of crops to drought stress is partial stomatal closure which limits the entry of CO_2 for photosynthesis, further increasing the production of reactive oxygen species (ROS) (Sgherri et al. 1993; Loggini et al. 1999; Boo and Jung 1999). It has been extensively documented that crop varieties having the ability to enhance their potential to reduce the harmful effects of the oxidative damage caused by drought and other abiotic stresses result in significantly higher yields. Hence the main aim of this chapter is to discuss the different aspects of ROS production, scavenging, and signalling which results from drought induced oxidative stress in plants.

2 Abiotic Stresses and ROS in Crop Plants

ROS are free radicals of oxygen that are chemically active. Presence of unpaired electrons in the valence shell of these molecules renders them highly reactive, resulting in damage of cell structure and function. Plant cells continuously produce ROS even under normal condition, since they play an important role in cell signalling, but when produced in excess leads to oxidative stress. Plants maintain a steady state balance between ROS production and anti-oxidant defence mechanism under normal condition, but various abiotic and biotic stresses heavily disturb this balance leading to a sudden burst in intracellular levels of ROS. It is estimated that 1–3% of oxygen consumed by plants leads to the formation of ROS in plant tissues (Matamorous et al. 2003; Bhattachrjee 2005). When maintained at lower levels, ROS acts as components in stress signalling pathway but becomes deleterious under excess conditions leading to cell death. The localized production of ROS molecules takes place within the compartmentation of different organelles like chloroplast, mitochondria, and peroxisomes (Reddy et al. 2004). Sub-cellular location for formation of ROS may be particularly important for a highly active ROS, as it can diffuse for a short distance before reacting with a cellular membrane (Gill and Tuteja 2010). The plant system has evolved efficient ROS scavenging mechanisms to achieve control over the ROS toxicity and use it beneficially as signalling molecule to control specialized processes such as plant growth, defence, hormonal signalling, and development (Ahmad et al. 2010). As a defence mechanism, activity of anti-oxidant enzymes such as ascorbate peroxidase (APX), catalase (CAT), and superoxide dismutase (SOD) is enhanced under abiotic stresses such as drought (Sankar et al. 2007; Jaleel et al. 2008a; Manivannan et al. 2008a), salinity (Jaleel et al. 2008b; Manivannan et al. 2008b), ozone (Puckette et al. 2007), and oxidative stress (Ali et al. 2005; Yin et al. 2010; Ahmad et al. 2010). Unlike other abiotic stresses, elevated temperature reduces the activity of anti-oxidant enzymes as seen in maize (Gong et al. 1997), which further enhances membrane damage, leading to an imbalance between photosynthesis and respiration. The various enzymes involved in scavenging ROS are SOD, APX, monodehydro ascorbate reductase (MDAR), CAT, whereas the non-enzymatic anti-oxidants includes tocopherol, ascorbate, glutathione, phenols, alkaloids, flavonoids, and prolines (Reddy et al. 2004; Gong et al. 2005; Chen and Dickman 2005; Jaleel et al. 2009; Yin et al. 2010; Gill and Tuteja 2010; Ahmad et al. 2010).

3 Types of ROS

The major ROS in plant system includes singlet oxygen ($^1O_2$), superoxide ($O_2^{\cdot -}$), hydrogen peroxide (H_2O_2), and hydroxyl radicals (·OH) among which ·OH is the most reactive species. Oxygen molecule upon accepting excess energy gives rise to $^1O_2$ (Fig. 7.2). During photosynthesis, insufficient energy dissipation by excited chlorophyll leads to the formation of chlorophyll triplet state (^{3}Chl) and this reacts with triplet oxygen ($^3O_2$) to give up highly reactive $^1O_2$ (Gill and Tuteja 2010). Additionally, abiotic stresses including drought, leading to increased stomatal closure results in low internal CO_2 concentration in the chloroplast favouring the production of $^1O_2$. However, $O_2^{\cdot -}$ is the first ROS to be produced by electron transfer to oxygen (Fig. 7.2). This is produced in cell organelles where electron transport occurs in plant system such as chloroplast. $O_2^{\cdot -}$ triggers the production of H_2O_2 and (·OH).

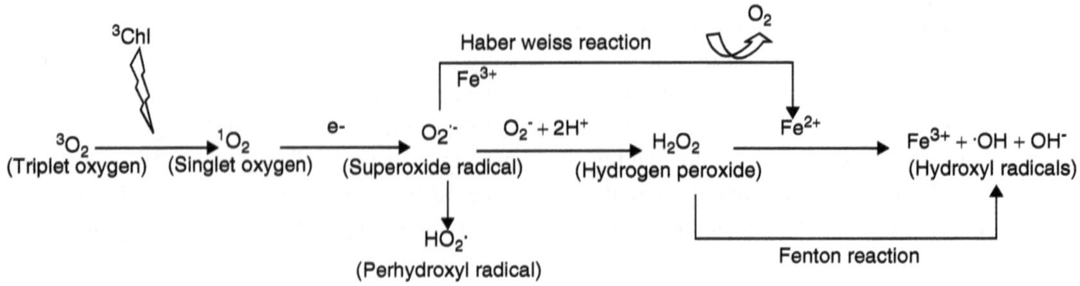

Fig. 7.2 Sequence of ROS synthesis

The univalent reduction of $O_2^{\cdot-}$ in the presence of SOD produces H_2O_2 and a molecule of oxygen (Fig. 7.2). $O_2^{\cdot-}$ can react with ferric ion (Fe^{3+}) reducing it to ferrous ion (Fe^{2+}), through Haber–Weiss reaction (Fig. 7.2). Subsequently, H_2O_2 in the presence of Fe^{2+} gives rise to (·OH), the most reactive ROS identified through Fenton reaction (Fig. 7.2) (Kehrer 2000). H_2O_2 is moderately reactive and inactivates the enzymes by oxidising their thiol groups. By virtue of its long life span and high permeability across membranes H_2O_2 is increasingly seen as a secondary messenger for signals generated by ROS at low concentrations and at high concentrations leads to programmed cell death (Quan et al. 2008). (·OH) is the neutral form of hydroxide ion (OH^-). Hydroxyl radical can potentially react with all biological molecules like DNA, protein, lipids, etc., with no enzymatic mechanism for eliminating this highly reactive radical, in excess levels leads to cell death (Vranova et al. 2002). A list of important biochemical properties of the major ROS known with their nature of damage and their scavengers are given in Table 7.1.

4 Sites of ROS Synthesis

ROS are generally produced in the cell organelles where the electron flow for normal metabolic activity occurs. The common cell compartments in which ROS are produced are chloroplast (Apel and Hirt 2004), mitochondria, and peroxysomes (Mittler et al. 2004). Under light conditions chloroplast and peroxisomes are the dominant sources of ROS production (Foyer and Noctor 2003) while mitochondria appears to replace them as the major source of ROS under dark periods (Moller 2001). Among different cell compartments that produce ROS, thylakoid membranes of the chloroplast are strong source of highly reactive oxidants (Niyogi 1999). During the photosynthetic electron transport from chloroplast the electron moves from water to NADP. Under various abiotic and biotic stress conditions when there is limited regeneration of NADP or less availability of carbon dioxide, the electron in electron transport chain (ETC) are accepted by O_2 instead of NADP, to form superoxide radicals (Flexas and Medrano 2002; Edreva 2005). Further, under stress conditions, particularly drought which leads to inadequate internal CO_2, the photorespiration pathway is activated leading to the production H_2O_2 by the enzyme glycolate oxidase in peroxisomes (Mittler et al. 2004). Mitochondrial electron transport chain (METC) is also another pathway in which ROS are produced. Besides these ROS production can also occur at plasma membrane (Sairam et al. 2005) and apoplast (Bolwell et al. 1999, 2002).

4.1 Chloroplast

Both photo-systems I and II in chloroplast thylakoids are a major source of ROS generation. $^1O_2$ can be produced with input of energy to $^3O_2$ (Krieger-Liszkay 2005). Production of $^1O_2$ in chloroplast takes place primarily, at antenna and reaction centres. Under high light conditions, the light absorbing chlorophyll pigments in antenna gets excited to form singlet exited chlorophyll. These unstable singlet exited chlorophyll further leads to the formation of ^{3}Chl molecules which are more stable (Krieger-Liszkay 2005). The ^{3}Chl can react with $^3O_2$ to produce $^1O_2$, which are highly reactive (Fig. 7.3). Whereas in PSII reaction centre, a block in the forward electron transport, lowers the primary pair charge separation and induces the charge recombination in the primary pair ($P680^+Pheo^-$) there by leading to the production of the triplet state of P_{680} ($^3P_{680}$) further resulting in the production of $^1O_2$ (Durrant et al. 1990). The rate of production of ^{3}Chl is higher in antennae than reaction centres, but in antenna the carotenoids can efficiently quench the $^1O_2$ there by avoiding any further damage. During photosynthesis, chloroplasts release oxygen as by product which is capable of accepting electrons while passing through the photo-systems, resulting in the formation of $O_2^{\cdot-}$. Limited CO_2 entry due to partial stomatal closure under drought stress or exposure to continuous excess light results in direct electron transfer towards

Table 7.1 Biochemical properties of different ROS, their nature of damage and scavengers

Properties	Singlet oxygen ($^1O_2$)	Superoxide (O_2^-)	Hydrogen peroxide (H_2O_2)	Hydroxyl radical ($\cdot OH$)
Reactivity	High	Moderate	Moderate	Extremely high
Half life	1.5 μs	2–4 μs	1 ms	1 to 0.01 μs
Diffusion distance	0.8 μm	8 mm	–	0.5 μm
Nature of damage	Transfers its excitation energy to other biological molecules and damages photosynthetic machinery	Reduces quinones and transition metal complexes (Fe^{2+} and Cu^{2+})	Oxidise the thiol group of enzymes and inactivates them	Reacts with DNA, proteins, lipids, and other constituents of cell, in excess can lead to cell death
Scavenging enzymes	Superoxide dismutase	Superoxide dismutase	Catalases, superoxide dismutase, acsorbate peroxidase, guaiacol peroxidases, glutathione peroxidases	No enzymatic mechanism for elimination
Non-enzymatic anti-oxidant	Tocopherol, β-carotene, plastoquinone	Ascorbic acid	Glutathione, peroxiredoxins, ascorbate, glutathione, flavonoids	Ascorbic acid, glutathione, proline, flavonoids

References for half life and diffusion distance – Pitzschke et al. (2006), de Carvalho (2008), Gill and Tuteja (2010); for nature of damage: Gill and Tuteja (2010) and Gechev et al. (2006); and for scavengers/anti-oxidants – Gechev et al. (2006); Ahmad et al. (2010)

molecular oxygen instead of $NADP^+$, generating a burst of superoxide ions at PSI by Mehler reaction [$PSI^- + O_2 = PSI + O_2^{\cdot -}$] (Asada 2006) (Fig. 7.3). The main electron donors to oxygen in ETC for the production of $O_2^{\cdot -}$ are Fe–S cluster and ferrodoxin in PSI and plastoquinones in PSII (Mehler 1951; Dat et al. 2000; Edreva 2005). The presence of transition metals such as Fe in ferrodoxin and Fe–S clusters or quinoids in plastoquinones assist in easy electron transfer to O_2 (Edreva 2005).

4.2 Peroxisomes

Peroxisomes have oxidative type of metabolism and hence probably one of the major sites of ROS production. Moreover, peroxisomes harbour various ROS producing enzymes including glycolate oxidase, acyl CoA oxidase, uricase, and xanthine oxidase (Rodriguez-Serrano et al. 2009). Under high light or other abiotic stress conditions, under insufficient CO_2 for carboxylation by RuBisCO, would result in oxygenation of RuBP by RuBisCO leading to a process called photorespiration in C_3 plants. Though photorespiration acts as a alternative sink for excess load of light energy, it generates H_2O_2 with the help of glycolate oxidase (Noctor et al. 2002) (Fig. 7.3). The other H_2O_2 production metabolisms in peroxisomes are the fatty acid β oxidation, the enzymatic reaction of flavin oxidases and the disproportionation of $O_2^{\cdot -}$ radicals (Palma et al. 2009). Apart from H_2O_2, $O_2^{\cdot -}$ are also produced in peroxisomes at two sites one inside the matrix in the presence of xanthin oxidase and another at membranes, which is dependent on NADH (del Rio et al. 1989, 1992, 1998, 2002). Xanthine and hypo-xanthine in the presence of enzyme xanthine oxidase gives rise to uric acid and $O_2^{\cdot -}$ (Fridovich 1986; Radi et al. 1992; del Rio et al. 2002) (Fig. 7.3).Whereas, at the peroxisomal membranes a small ETC composing of flavoprotien NADH:ferricyanide reductase and cytochrome b is known to be involved in the production of $O_2^{\cdot -}$ (Fang et al. 1987; Lopez-Huertas et al. 1995; del Rio et al. 2002).

4.3 Mitochondria

Comparatively, mitochondria generates smaller amounts of ROS (Foyer and Noctor 2005). Complex I and complex III of METC are the sites of ROS production (Rhoads et al. 2006; Moller et al. 2007) (Fig. 7.3). In METC, complex I and complex III harbours electrons with sufficient energy to directly reduce oxygen which is very essential for aerobic respiration. Ubisemiquinone intermediates formed at complex I and III donates electrons to generate $O_2^{\cdot -}$ (Raha and Robinson 2000) while under drought, increased mitochondrial respiration could enhance ROS production by transferring an electron from cytochrome to O_2 (Norman et al. 2004). Further, under severe drought, increased demand for mitochondrial ATP compensating for reduced chlorophyll ATP synthesis could also enhance ROS production (Atkin and Macherel 2009). Among the two mitochondrial electron transport pathways, from ubiquinone to oxygen, alternate oxidase activity could result in maintaining normal metabolite levels with reduced ROS production under stress (Farooq et al. 2009).

4.4 Other Cell Organelles

Another source for the production of ROS is the detoxification reactions carried out by cytochrome P450 in cytoplasm and endoplasmic reticulum. In this detoxification reaction, $O_2^{\cdot -}$ is formed (Bolwell and Wojtaszek 1997). ROS is also generated at the plasma membrane or extracellularly in the apoplast in response to abscisic acid (ABA) and drought (Hu et al. 2005, 2006). NADPH oxidase of plasma membrane is also considered as a source of ROS production (Kwak et al. 2003). Moreover, NADPH oxidase has a multimeric flavocytochrome that forms an ETC which has the capacity to reduce oxygen to superoxide. Apart from this, pH dependent cell wall peroxidases, germin-like oxalate oxidase, and amino oxidases in the apoplast are sources of H_2O_2 (Bolwell and Wojtaszek 1997).

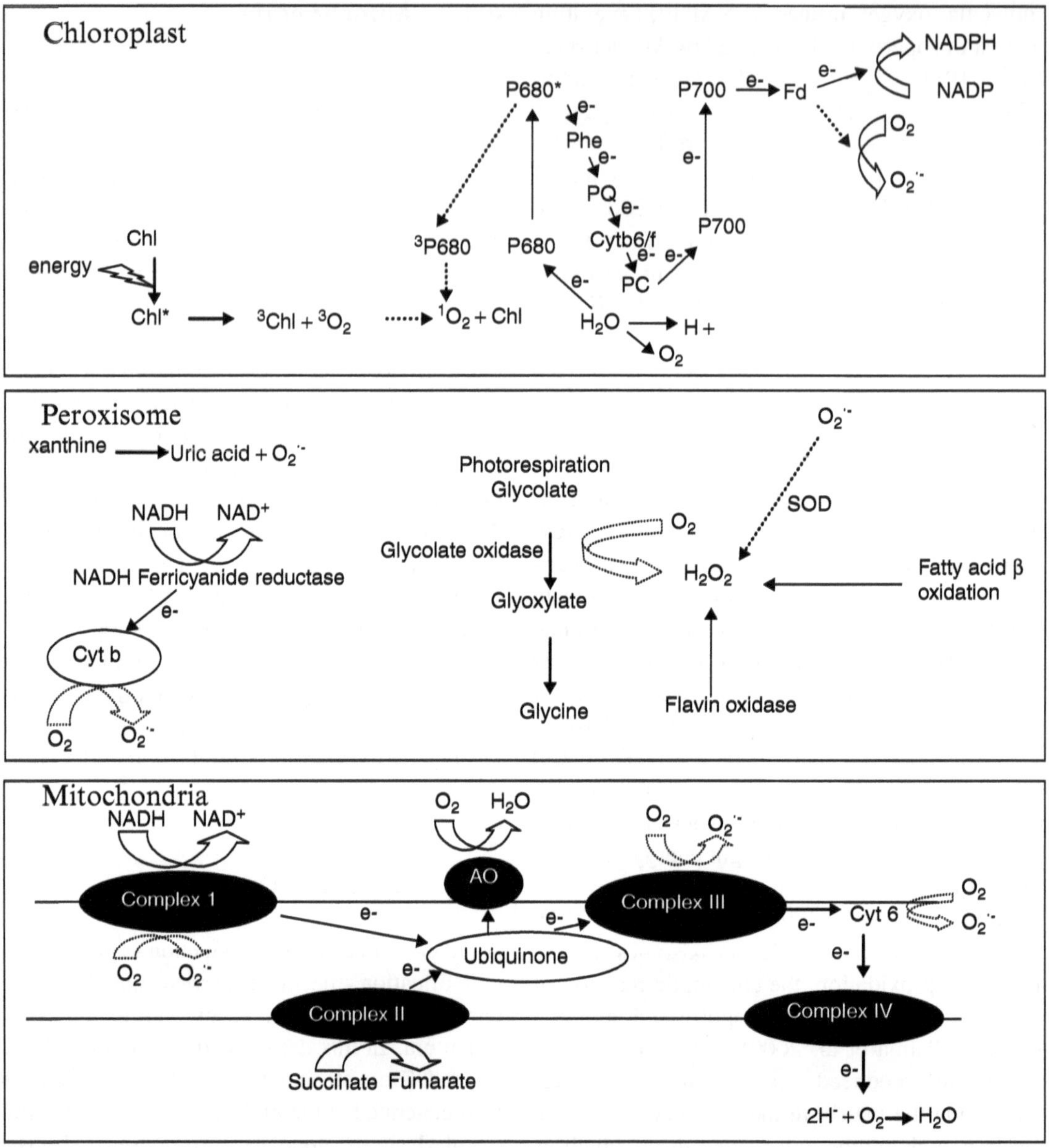

Fig. 7.3 ROS production in chloroplast, peroxisome, and mitochondria. The *dotted arrows* indicate the production of ROS under stress. Phe, pheophytin; PQ, plastoquinone; Cyt, cytochrome; PC, plastocyanin; Fd, ferrodoxin; SOD, superoxide dismutase; AO, alternate oxidase

5 ROS Under Drought

Drought stress invariably leads to decreased photosynthetic rate which is mainly attributed to altered stomatal regulation (Cornic 2000). Shrinkage of cell volume due to water deficit stress makes the cellular contents more viscous which can ultimately result in aggregation and denaturation of proteins (Hoekstra et al. 2001), hindering normal functioning of enzymes involved in photosynthesis. The partial stomatal closure that occur in response to water deficit as a water conserving strategy, also limits the CO_2 entry and in turn its availability for photosynthesis. Further the reduction in carboxylation efficiency of RuBisCO, simultaneously enhances oxygenation and thereby increasing photo-respiratory losses and ROS production (Noctor et al. 2002). The reduction in photosynthesis or CO_2 fixation would result in reduced regeneration of $NADP^+$, the final electron acceptor of ETC in chloroplast. Thus the over reduction of ETC, results in leakage of electron to O_2 and subsequent production of ROS (Smirnoff 1993; Biehler and Fock 1996; Sgherri et al. 1996) (Fig. 7.3). Imbalance between light capture and its utilization (Foyer and Noctor 2000) in the PSII results in changes in photochemistry of chloroplasts in drought stressed leaves, resulting in excess production of highly reactive and dangerous ROS species (Peltzer et al. 2002).

Among different growth and developmental stages, reproductive organ development is extremely sensitive to drought stress across different crop species (Liu et al. 2006; Nguyen et al. 2010; Lalonde et al. 1997). During the male reproductive organ or the anther development in rice, drought stress resulted in the complete suppression of three major ROS scavenging enzymes namely CAT, APX, and dehydroascorbate reductase (DHAR) during the meiosis stage. However, at other stages following the tetrad formation, that is, microspore and vacuolated stage, the scavenging enzyme levels were differentially regulated indicating extreme sensitivity of meiosis stage to drought stress and the need for an efficient ROS scavenging mechanism to overcome stress and set seed (Nguyen et al. 2009). Further, Selote and Chopra (2004) tested the impact of drought stress on the entire panicles of rice and recorded a higher level of relative water content, turgor potential, and lower H_2O_2 in N22, the drought tolerant cultivar compared to the susceptible N118, which was due to significantly higher SOD, GSH (reduced glutathione), APX, ascorbate production. Two drought tolerant entries of rice namely, Xiangzhongxian No. 2 and IR50, recorded a significantly lower level of electrolyte leakage due to a much stable membrane structure and lower H_2O_2 production over the sensitive check. In the above study, all the anti-oxidant enzymes including SOD, CAT, APX, and other non-enzymatic anti-oxidants (ascorbic acid (AA) and reduced glutathione) were enhanced after the initiation of stress and maintained same levels lasting for about 3 days but decreased on a longer timeframe of 5 consecutive days of stress. This study indicated that sensitivity to drought on one hand is related to enhanced anti-oxidant system but to a similar extent to the rate of decline of the system (Guo et al. 2006). Although a general pattern in increase of ROS scavenging enzymes are recorded in different abiotic stresses including drought the actual mechanism for such an indication is not clearly known. It has been speculated that ABA could be involved in inducing anti-oxidant enzyme (Jiang and Zhang 2001; Zhou et al. 2005). Supportive evidence for ABA induced anti-oxidant activity under drought stress in maize has been documented (Jiang and Zhang 2002).

6 Damaging Effects of ROS

The highly reactive $^1O_2$ can react with proteins, pigments, and lipids. In the photosynthetic reaction centre, the $^1O_2$ produced can react with D1 protein there by degrading it (Trebst 2003). This controlled damage of D1 protein helps in destroying the $^1O_2$ at the place of generation, there by avoiding further uncontrolled damage to PSII (Trebst 2003). The damaged D1 protein under normal conditions is, however, re-synthesised resulting in active PSII, called D1 protein turnover (Mattoo et al. 1989). But under high light conditions, the D1 degradation rate exceeds the D1 re-synthesis rate, resulting in complete destruction of PSII there by limiting photosynthesis (Adir et al. 2003; Kruk et al. 2005). Moreover, $^1O_2$ can also damage light harvesting complex II (LHCII) (Lindahl et al. 1995). Among the two photo-systems, PSII is more susceptible to drought with decline in D1 and D2 proteins resulting in deterioration of PSII (Lu and Zhang 1999). As a consequence, significant reduction in CO_2 assimilation and impairing of photosynthetically active ETC induces a rapid burst of ROS species which are responsible for DNA nicking, oxidation of amino acids, and lipid peroxidation (LPO) (Johnson et al. 2003). During the process of LPO small hydrocarbon fragments such as malondialdehyde are formed which are capable of reacting with thiobarbituric acid (TBA) to form coloured products called as thiobarbituric acid reactive substances (TBARS) (Larkindale and Knight 2002). LPO is considered to be the most damaging process in all living organisms (Gill and Tuteja 2010). At the cellular and organelle level, LPO occurs when the ROS levels exceed the thresholds, hence affecting normal cellular functioning but simultaneously aggravating oxidative stress through lipid derived radicals (Montillet et al. 2005). TBARS is a measure of LPO resulting from oxidative stress which interferes with the stability of cell membrane leading to increased electrolyte leakage. An increased TBARS accumulation has been correlated to enhance electrolyte leakage under drought stress indicating injury to plasmalemma in rice (Guo et al. 2006) and maize (Del Longo et al. 1993). The measurement of malondialdehyde is considered as one of the valid indicator to measure the extent of damage occurring to cells under oxidative stress (Hodges et al. 1999; Mahan and Mauget 2005). Interestingly it was shown that the oxidative damage was not uniform across the mesophyll and bundle sheath cells in C_4 plants (Doulis et al. 1997; Foyer 2001) with the oxidative damage restricted to bundle sheath tissue due to lower anti-oxidant protection (Kingston-Smith and Foyer 2000).

7 Detoxification Mechanisms to Quench ROS

To overcome the toxic effects of ROS or to maintain the ROS production under control, the plant cells have evolved highly efficient scavenging mechanisms either through anti-oxidative enzymes or anti-oxidant secondary metabolites (Sharma and Dubey 2005). The main oxidative enzymes present in plant cells are SOD, CAT, APX, glutathione peroxidase (GPX), glutathione reductase (GR), and the important non-enzymatic anti-oxidants are ascorbate, glutathione, thioredoxin, tocopherols, carotenoids, vitamin E, etc.

7.1 Enzymatic Scavenging of ROS

In plant cells various organelles like chloroplast, mitochondria, peroxisome have these anti-oxidant enzymes and their presence is compartment specific. SOD constitutes the first line of defence in response to ROS by converting them to H_2O_2 (Alscher et al. 2002) with APX and CAT majorly responsible for degrading/neutralizing H_2O_2 (Asada 2006; Vanacker et al. 1998) [see (7.1)–(7.4) below].

$$2H_2O \rightarrow 4e^- + O_2 + 4H^+ \ (\text{PSII}) \tag{7.1}$$

$$2O_2 + 2e^- \rightarrow 2O_2^- \ (\text{PSI}) \tag{7.2}$$

$$2O_2^- + 2H+ \rightarrow H_2O_2 + O_2 \ (\text{SOD}) \tag{7.3}$$

$$H_2O_2 + 2\text{AsA} \rightarrow 2H_2O + 2\text{MDA} \ (\text{APX}) \tag{7.4}$$

APX detoxifies H_2O_2 through ascorbate glutathione cycle, for which it needs the reducing agent ascorbate and glutathione regeneration system. In ascorbate glutathione cycle, APX converts H_2O_2 to H_2O by oxidising ascorbate to monodehydroascorbate (MDA). Further MDAR, in the presence of NAD(P)H converts MDA to ascorbate. SOD is the first among the scavengers to react with ROS, wherein strong radicals having half life of 1 s are instantaneously dismutated by SOD into H_2O_2 which is a much more stable product and is detoxified by catalase and peroxidase (Apel and Hirt 2004). A strong correlation between SOD activity and drought induced oxidative stress tolerance has been documented (Pan et al. 2006). Further, the ascorbate–glutathione cycle which includes APX, DHAR, MDAR, and GR allows scavenging of superoxide radicals and H_2O_2 (Fazeli et al. 2007; Farooq et al. 2009). These four enzymes are located in the cytosol, stroma of all the three major ROS producing organelles, that is, chloroplast, mitochondria, and peroxisomes (Jiménez et al. 1998).

7.2 Non-enzymatic Scavenging of ROS

The anti-oxidant metabolites like β-carotene and α-tocopherol acts as the first line of defence against ROS as they can efficiently quench $^1O_2$, provided they are in the close proximity (Krieger-Liszkay 2004). The ^{3}Chl molecules can be quenched by carotenoids in distance less than van der Waals distance (3.6 Å), in the antenna system (Cogdell and Frank 1996; Edge and Truscott 1999), thereby preventing production of singlet oxygen and protecting from oxidative damage. But β-carotene molecules present in PSII, cannot quench the ^{3}Chl as the distance between them is too large (Telfer 2002) and hence they could be involved in quenching the $^1O_2$ produced by ^{3}P680 (Telfer et al. 1994). Among the tocopherols, both α and γ tocopherols act as scavengers of $^1O_2$, especially during photoinhibiton and D1 protein turnover, thereby helping in the maintenance of PSII structure and function (Kruk et al. 2005). Carotenoids have received less importance despite their ROS scavenging capabilities (Deltoro et al. 1998; Wahid et al. 2007; Havaux 1998) probably due to their susceptibility to oxidative destruction (Farooq et al. 2009). Although falvonoids outperform other well-known anti-oxidant scavengers like α-tocopherol, the absence of convincing spatio-temporal correlation with flavonoid oxidation products, has continued the debate on their involvement in reducing oxidative stress (for more information see Hernandez et al. 2009). Another non-enzymatic plant anti-oxidant AA with its ability to react with H_2O_2, $O_2^{\cdot-}$ and is considered to be an important anti-oxidant. Apart from its involvement in regeneration of the ascrobate pool, AA has an additional role of protecting or regeneration of oxidised carotenoids or tocopherols (Imai et al. 1999). Xanthophyll cycle is another alternate pathway for dissipating excess energy with increased xanthophyll pigments, zeaxanthin, and antheraxanthin, at the expense of violoxanthin under drought stress in plants leading to reduced ROS production (Alonso et al. 2001; Foyer 2001). Under conditions of prolonged drought stress wherein plant induced anti-oxidant mechanisms cannot overcome the detrimental effects of ROS the exogenous application of certain compounds are being considered. Chitosan, a cationic polysaccharide, is one such compound which has gained attention due to its ability to scavenge (·OH), $O_2^{\cdot-}$, and H_2O_2 radicals by enhancing the activity of anti-oxidant machinery under drought stress (Yang et al. 2009). The hydroxyl and amino groups present in chitosan allows it to react efficiently with ROS and detoxify it by generating non-toxic macro molecular radicals (Sun et al. 2004).

8 ROS in Signalling

Drought stress signalling is extremely complex with close involvement of ROS, calcium, calcium related proteins, mitogen-activated protein kinase cascades, and cross-talk between different transcriptional factors (Kovtun et al. 2000; Chen et al. 2002). In response to drought stress, different signals are perceived and transmitted by ROS, calcium, plant hormones (induces stress tolerance by activating genome re-programming); and mitogen-activated protein kinases (connecting the perception of external stimuli to cellular responses), etc. (Farooq et al. 2009). Additionally, under drought stress early response effector genes are turned on through both abscisic acid dependent and independent osmotic stress signalling. ROS on one hand are toxic by-products of abiotic stress metabolism in plants while on the other hand are important signal transduction molecules (Miller et al. 2010). Elaborating on the latter part, ROS generated by metabolic imbalances can be utilized as a stress signal to activate acclimation and defence mechanisms to overcome stress associated oxidative stress (Mittler et al. 2004; Miller et al. 2008). Under stress conditions ROS signals from different organelles results in the induction of large transcriptional changes and cellular metabolic re-programming that can either protect the cell or resort to programmed cell death (Umbach et al. 2005; Gadjev et al. 2006; Rhoads et al. 2006). These retrograde signalling has been largely divided into two categories (1) developmental control of organelle biosynthesis and (2) operational control resulting in rapid response to environmental and developmental constraints (Pogson et al. 2008). Interestingly all the chloroplast to nuclei retrograde signalling including

ROS signalling have been shown to converge into a pathway regulated by GUN1 (Genome Uncoupled 1) and ABI4 (ABA insensitive 4) in Arabidopsis seedlings (Koussevitzky et al. 2007). ROS is also involved in the expression of number of genes in plants, indicating their involvement in biological processes (Dalton et al. 1999). H_2O_2 a component of ABA signalling is the only ROS that can cross cell membranes and having the potential to be involved in cell to cell signalling (Pitzschke et al. 2006). Under water deficit conditions the partial stomatal closure induced by ABA takes place via the accumulation of H_2O_2 and Ca^{2+} signalling activity (Pei et al. 2000).

9 Conclusion and Future Perspectives

Drought a major abiotic stress results in rapid production of highly ROS capable of damaging the photosynthetic machinery, membranes, etc., thereby altering the overall plant homeostasis and under severe conditions leading to death of plants. Although a number of enzymatic and non-enzymatic ROS scavengers have been identified to reduce the damaging effects of ROS, with predicted increase in frequency and severity in stress this mechanism alone might not be sufficient. Hence future research is needed to identify signalling pathways wherein ROS could be utilized more efficiently to enhance the internal resistance by perceiving and transmitting stress signals obtained to longer distances to activate acclimation and defence related metabolic mechanisms and genome re-programming, more rapidly and efficiently. Further, detailed understanding and cautious interpretation of the different ROS signalling of cellular stress response is needed to identify if the suggested involvement of a particular compound to be a resultant of general stress response or genuinely stress specific. Moreover, the rapid advances in the "omics" technologies like proteomics and metabolomic analyses will help identify unknown links, cross-talks across different stress signalling pathways that could be exploited to enhance plants tolerance to one of the most damaging abiotic stress.

References

Adir N, Zer H, Shochat S, Ohad I (2003) Photoinhibition – a historical perspective. Photosynth Res 76:343–370

Ahmad P, Jaleel CA, Salem MA, Nabi G, Sharma S (2010) Roles of enzymatic and non-enzymatic antioxidants in plants during abiotic stress. Crit Rev Biotechnol 30: 161–175

Ahmadi A, Baker DA (2001) The effect of water stress on the activities of key regulatory enzymes of the sucrose to starch pathway in wheat. Plant Growth Regul 35:81–91

Ali MB, Hahn EJ, Paek K (2005) Effects of temperature on oxidative stress defense systems, lipid peroxidation and lipoxygenase activity in Phalaenopsis. Plant Physiol Biochem 43:213–223

Alonso R, Elvira S, Castillo FJ, Gimeno BS (2001) Interactive effects of ozone and drought stress on pigments and activities of antioxidative enzymes in *Pinis halpensis*. Plant Cell Environ 24:905–916

Alscher RG, Erturk N, Heath LS (2002) Role of superoxide dismutases (SODs) in controlling oxidative stress in plants. J Exp Bot 53:1331–1341

Apel K, Hirt H (2004) Reactive oxygen species: metabolism, oxidative stress, and signal transduction. Annu Rev Plant Biol 55:373–399

Asada K (2006) Production and scavenging of reactive oxygen species in chloroplasts and their functions. Plant Physiol 141:391–396

Atkin OK, Macherel D (2009) The crucial role of plant mitochondria in orchestrating drought tolerance. Ann Bot 103:581–597

Atteya AM (2003) Alteration of water relations and yield of corn genotypes in response to drought stress. Bulg J Plant Physiol 29:63–76

Bhattachrjee S (2005) Reactive oxygen species and oxidative burst: roles in stress, senescence and signal transduction in plant. Curr Sci 89:1113–1121

Biehler K, Fock H (1996) Evidence for the contribution of the Mehler peroxidase reaction in dissipating excess electrons in drought stressed wheat. Plant Physiol 112:265–272

Bolwell GP, Wojtaszek P (1997) Mechanisms for the generation of reactive oxygen species in plant defense – a broad perspective. Physiol Mol Plant Pathol 51: 347–366

Bolwell GP, Blee KA, Butt VS, Davies DR, Gardner SL, Gerrish C, Minibayeva F, Rowntree EG, Wojtaszek P (1999) Recent advances in understanding the origin of the apoplastic oxidative burst in plant cells. Free Rad Res 31:S137–S145

Bolwell GP, Bindschedler LV, Blee KA, Butt VS, Davies DR, Gardner SL, Gerrish C, Minibayeva F (2002) The apoplastic oxidative burst in response to biotic stress in plants: a three-component system. J Exp Bot 53:1367–1376

Boo YC, Jung J (1999) Water deficit induced oxidative stress and antioxidative defenses in rice plants. J Plant Physiol 155:255–261

Boonlertnirun S, Sarobol ED, Meechoui S, Sooksathan I (2007) Drought recovery and grain yield potential of rice after chitosan application. Kasetsart J Nat Sci 41:1–6

Cattivelli L, Rizza F, Badeck FW, Mazzucotelli E, Mastrangelo AM, Francia E, Mare C, Tondelli A, Stanca AM (2008) Drought tolerance improvement in crop plants: an integrative view from breeding to genomics. Field Crop Res 105:1–14

Chen C, Dickman MB (2005) Proline suppresses apoptosis in the fungal pathogen *Colletotrichum trifolii*. Proc Natl Acad Sci USA 102:3459–3464

Chen W, Provart NJ, Glazebrook J, Katagiri F, Chang HS, Eulgem T, Mauch F, Luan S, Zou G, Whitham SA, Budworth PR, Tao Y, Xie Z, Chen X, Lam S, Kreps JA, Harper JF, Si-Ammour A, Mauch-Mani B, Heinlein M, Kobayashi K, Hohn T, Dangl JL, Wang X, Zhu T (2002) Expression profile matrix of Arabidopsis transcription factor genes suggests their putative functions in response to environmental stresses. Plant Cell 14:559–574

Cogdell R, Frank HA (1996) Carotenoids in photosynthesis. Photochem Photobiol 63:257–264

Cornford SG (2003) The socio-economic impacts of weather events in 2002. WMO Bull 52:269–290

Cornic G (2000) Drought stress inhibits photosynthesis by decreasing stomatal aperture –not by affecting ATP synthesis. Trends Plant Sci 5:187–188

Dalton TP, Shertzer HG, Puga A (1999) Regulation of gene expression by reactive oxygen. Annu Rev Pharmacol Toxicol 39:67–101

Dat J, Vandenabeele S, Vranova E, Van Montagu M, Inzé D, Van Breusegem F (2000) Dual action of the active oxygen species during plant stress. Cell Mol Life Sci 57:779–795

de Carvalho MHC (2008) Drought stress and reactive oxygen species. Plant Signal Behav 3:156–165

Del Longo OT, Gonzalez CA, Pastori GM, Trippi VS (1993) Antioxidant defences under hyperoxygenic and hyperosmotic conditions in leaves of two lines of maize with differential sensitivity to drought. Plant Cell Physiol 34:1023–1028

del Rio LA, Fernandez VM, Ruperez FL, Sandalio LM, Palma JM (1989) NADH induces the generation of super oxide radicals in leaf peroxisomes. Plant Physiol 89:728–731

del Rio LA, Sandalio LM, Palma JM, Bueno P, Corpas FJ (1992) Metabolism of oxygen radicals in peroxisomes and cellular implications. Free Radic Biol Med 13:557–580

del Rio LA, Sandalio LM, Corpas FJ, Lopez-Huertas E, Palma JM, Pastori GM (1998) Activated oxygen-mediated metabolic functions of leaf peroxisomes. Physiol Plant 104:673–680

del Rio LA, Corpas FJ, Sandalio LM, Palma JM, Gomez M, Barroso JB (2002) Reactive oxygen species, antioxidant systems and nitric oxide in peroxisomes. J Exp Bot 53(372):1255–1272

Deltoro VI, Calatayud A, Gimeno C, AbadõÂa A, Barreno E (1998) Changes in chlorophyll a fluorescence, photosynthetic CO_2 assimilation and xanthophylls cycle interconversions during dehydration in desiccation-tolerant and intolerant liverworts. Planta 207:224–228

Doulis AG, Debian N, Kingston-Smith AH, Foyer CH (1997) Differential localization of antioxidants in maize leaves. Plant Physiol 114:1031–1037

Durrant JB, Giorgi LB, Barber J, Klug DR, Porter G (1990) Characterization of triplet states in isolated photosystem II reaction centers: oxygen quenching as a mechanism for photodamage. Biochim Biophys Acta 1017:167–175

Edge R, Truscott TG (1999) Carotenoid radicals and the interaction of carotenoids with active oxygen species. In: Frank HA, Young AJ, Britton G, Cogdell RJ (eds) Advances in photosynthesis: the photochemistry of carotenoids, vol 8. Kluwer, Dordrecht, pp 223–234

Edreva A (2005) Generation and scavenging of reactive oxygen species in chloroplasts: a submolecular approach. Agric Ecosyst Environ 106:119–133

Fang TK, Donaldson RP, Vigil EL (1987) Electron transport in purified glyoxysomal membranes from castor bean endosperm. Planta 172:1–13

Farooq M, Wahid A, Kobayashi N, Fujita D, Basra SMA (2009) Plant drought stress: effects, mechanisms and management. Agron Sustain Dev 29:185–212

Fazeli F, Ghorbanli M, Niknam V (2007) Effect of drought on biomass, protein content, lipid peroxidation and antioxidant enzymes in two sesame cultivars. Biol Plant 51:98–103

Flexas J, Medrano H (2002) Energy dissipation in C3 plants under drought. Funct Plant Biol 29:1209–1215

Foyer CH (2001) Prospects for enhancement the soluble antioxidants ascorbate and glutathione. Biofactors 15:75–78

Foyer CH, Noctor G (2000) Oxygen processing in photosynthesis: regulation and signaling. New Phytol 146: 359–388

Foyer CH, Noctor G (2003) Redox sensing and signaling associated with reactive oxygen in chloroplasts, peroxisomes and mitochondria. Physiol Plant 119: 355–364

Foyer CH, Noctor G (2005) Redox homeostasis and antioxidant signaling: a metabolic interface between stress perception and physiological responses. Plant Cell 17:1866–1875

Fridovich I (1986) Superoxide dismutases. Adv Enzymol Relat Areas Mol Biol 58:61–97

Gadjev I, Vanderauwera S, Gechev TS, Laloi C, Minkov IN, Shulaev V, Apel K, Inze D, Mittler R, Van Breusegem F (2006) Transcriptomic footprints disclose specificity of reactive oxygen species signaling in *Arabidopsis*. Plant Physiol 141:436–445

Gechev TS, van Breusegem F, Stone MJ, Denev L, Laloi C (2006) Reactive oxygen species as signals that modulate plant stress responses and programmed cell death. Bioessays 28:1091–1101

Gill SS, Tuteja N (2010) Reactive oxygen species and antioxidant machinery in abiotic stress tolerance in crop plants. Plant Physiol Biochem 48:909–930

Gong M, Chen SN, Song YQ, Li ZG (1997) Effect of calcium and calmodulin on intrinsic heat tolerance in relation to antioxidant systems in maize seedlings. Austr J Plant Physiol 24:371–379

Gong H, Zhu X, Chen K, Wang S, Zhang C (2005) Silicon alleviates oxidative damage of wheat plants in pots under drought. Plant Sci 169:313–321

Guo Z, Ou W, Lu S, Zhong Q (2006) Differential response of antioxidative system to chilling and drought in four rice cultivars differing in sensitivity. Plant Physiol Biochem 44:828–836

Harris D, Tripathi RS, Joshi A. (2002) On-farm seed priming to improve crop establishment and yield in dry direct-seeded rice pp 231–240. In: Pandey S, Mortimer M, Wade L, Tuong TP, Lopez K, Hardy B, (eds). Proceedings of the International Workshop on Direct Seeding in Asian Rice Systems: Strategic Research Issues and Opportunities, 25–28 January 2000, Bangkok, Thailand. Los Banos (Philippines): International Rice Research Institute pp 383.

Havaux M (1998) Carotenoids as membrane stabilizers in chloroplasts. Trends Plant Sci 3:147–151

Hernandez I, Alegre L, Van Breusegem F, Munne-Bosch S (2009) How relevant are flavonoids as antioxidants in plants? Cell 14:125–132

Hodges DM, DeLong JM, Forney CF, Prange RK (1999) Improving the thiobarbituric acid reactive substances assay for estimating lipid peroxidation in plant tissues containing antocyanin and other interfering compounds. Plants 207:604–611

Hoekstra FA, Golovina EA, Buitink J (2001) Mechanisms of plant desiccation tolerance. Trends Plant Sci 6: 431–438

Hu X, Jiang M, Zhang A, Lu J (2005) Abscisic acid-induced apoplastic H_2O_2 accumulation up-regulates the activities of chloroplastic and cytosolic antioxidant enzymes in maize leaves. Planta 223:57–68

Hu X, Zhang A, Zhang J, Jiang M (2006) Abscisic acid is a key inducer of hydrogen peroxide production in leaves of maize plants exposed to water stress. Plant Cell Physiol 47:1484–1495

Hussain M, Malik MA, Farooq M, Ashraf MY, Cheema MA (2008) Improving drought tolerance by exogenous application of glycinebetaine and salicylic acid in sunflower. J Agron Crop Sci 194:193–199

Imai T, Kingston-Smith AH, Foyer CH (1999) Inhibition of endogenous ascorbate synthesis in potato leaves supplied with exogenous ascorbate. Free Radic Res 31:171–179

Jaleel CA, Sankar B, Murali PV, Gomathinayagam M, Lakshmanan GMA, Panneerselvam R (2008a) Water deficit stress effects on reactive oxygen metabolism in Catharanthus roseus: impacts on ajmalicine accumulation. Colloids Surf B Biointerfaces 62(1):105–111

Jaleel CA, Gopi R, Manivannan P, Gomathinayagam M, Murali PV, Panneerselvam R (2008b) Soil applied propiconazole alleviates the impact of salinity on *Catharanthus roseus* by improving antioxidant status. Pestic Biochem Physiol 90(2):135–139

Jaleel CA, Riadh K, Gopi R, Manivannan P, Ines J, Al-Juburi HJ, Chang-Xing Z, Hong-Bo S, Panneerselvam R (2009) Antioxidant defense responses: physiological plasticity in higher plants under abiotic constraints. Acta Physiol Plant 31: 427–436

Jiang M, Zhang J (2001) Effects of abscisic acid on active oxygen species, antioxidative defense system and oxidative damage in leaves of maize seedlings. Plant Cell Physiol 42:1265–1273

Jiang M, Zhang J (2002) Role of abscisic acid in water stress induced antioxidant defense in leaves of maize seedlings. Free Radic Res 36:1001–1015

Jiménez A, Hernandez JA, Ros Barcelo A, Sandalio LM, del Rio LA, Sevilla F (1998) Mitochondrial and peroxisomal ascorbate peroxidase of pea leaves. Physiol Plant 104:687–692

Johnson SM, Doherty SJ, Croy RRD (2003) Biphasic superoxide generation in potato tubers. A self amplifying response to stress. Plant Physiol 13: 1440–1449

Jongdee B, Pantuwan G, Fukai S, Fischer K (2006) Improving drought tolerance in rainfed lowland rice: an example from Thailand. Agric Water Manage 80: 225–240

Kaya MD, Okçub G, Ataka M, Cikili Y, Kolsar Ö (2006) Seed treatments to overcome salt and drought stress during germination in sunflower (*Helianthus annuus* L.). Eur J Agron 24:291–295

Kehrer JP (2000) The Haber-Weiss reaction and mechanisms of toxicity. Toxicology 149:43–50

Kingston-Smith AH, Foyer CH (2000) Bundle sheath proteins are more sensitive to oxidative damage than those of the mesophyll in maize leaves exposed to paraquat or low temperatures. J Exp Bot 51(342): 123–130

Koussevitzky S, Nott A, Mockler TC, Hong F, Sachetto-Martins G, Surpin M, Lim J, Mittler R, Chory J (2007) Multiple signals from damaged chloroplasts converge on a common pathway to regulate nuclear gene expression. Science 316:715–719

Kovtun Y, Chiu WL, Tena G, Sheen J (2000) Functional analysis of oxidative stress-activated mitogen-activated protein kinase cascade in plants. Proc Natl Acad Sci USA 97:2940–2945

Krieger-Liszkay A (2004) Singlet oxygen production in photosynthesis. J Exp Bot 56:337–346

Krieger-Liszkay A (2005) Singlet oxygen production in photosynthesis. J Exp Bot 56:337–346

Kruk J, Holländer-Czytko H, Oettmeier W, Trebst A (2005) Tocopherol as singlet oxygen scavenger in photosystem II. J Plant Physiol 162:749–757

Kumar R, Venuprasad R, Atlin G (2007) Genetic analysis of rainfed lowland rice drought tolerance under naturally occurring stress in eastern India: heritability and QTL effects. Field Crops Res 103:42–52

Kwak JM, Mori IC, Pei ZM, Leonhardt N, Torres MA, Dangl JL, Bloom RE, Bodde S, Jones JD, Schroeder JI (2003) NADPH oxidase *AtrbohD* and *AtrbohF* genes function in ROSdependent ABA signaling in *Arabidopsis*. EMBO J 22:2623–2633

Lalonde S, Beebe DU, Saini HS (1997) Early signs of disruption of wheat anther development associated with the induction of male sterility by meiotic-stage water deficit. Sex Plant Reprod 10:40–48

Larkindale J, Knight MR (2002) Protection against heat stress-induced oxidative damage in *Arabidopsis* involves calcium, abscisic acid, ethylene, and salicylic acid. Plant Physiol 128:682–695

Lindahl M, Yang DH, Andersson B (1995) Regulatory proteolysis of the major light-harvesting chlorophyll a/b protein of photosystem II by light-induced membrane associated enzymic system. Eur J Biochem 213: 503–509

Liu JX, Bennett J (2010) Reversible and irreversible drought-induced changes in the anther proteome of rice (Oryza sativa L.) genotypes IR64 and Moroberekan. Mol Plant 4(1):59–69

Liu JX, Liao DQ, Oane R, Estenor L, Yang XE, Li ZC, Bennett J (2006) Genetic variation in the sensitivity of anther dehiscence to drought stress in rice. Field Crops Res 97:87–100

Loggini B, Scartazza A, Brugnoli E, Navari-Izzo F (1999) Antioxidative defense system, pigment composition, and photosynthetic efficiency in two wheat cultivars subjected to drought. Plant Physiol 119:1091–1099

Lopez-Huertas E, Sandalio LM, Gomez M, del Rio LA (1995) Integral membrane polypeptides of pea leaf peroxisomes: characterization and response to plant stress. Plant Physiol Biochem 33:295–302

Lu C, Zhang J (1999) Effects of water stress on photosystem II photochemistry and its thermostability in wheat plants. J Exp Bot 50:1199–1206

Mahan JR, Mauget SA (2005) Antioxidant metabolism in cotton seedlings exposed to temperature stress in the field. Crop Sci 45:2337–2345

Manickavelu A, Nadarajan N, Ganesh SK, Gnanamalar RP, Chandra BR (2006) Drought tolerance in rice: morphological and molecular genetic consideration. Plant Growth Regul 50:121–138

Manivannan P, Jaleel CA, Somasundaram R, Panneerselvam R (2008a) Osmoregulation and antioxidant metabolism in drought stressed Helianthus annuus under triadimefon drenching. C R Biol 331(6):418–425

Manivannan P, Jaleel CA, Kishorekumar A, Sankar B, Somasundaram R, Panneerselvam R (2008b) Protection of Vigna unguiculata (L.) Walp. plants from salt stress by paclobutrazol. Colloids Surf B Biointerfaces 61(2):315–318

Matamorous MA, Dalton DA, Ramos J, Clemente MR, Rubio MC, Becana M (2003) Biochemistry and molecular biology of antioxidants in the rhizobia-legume symbiosis. Plant Physiol 133:499–509

Mattoo AK, Marder JB, Edelman M (1989) Dynamics of the photosystem II reaction center. Cell 56:241–246

Mehler AD (1951) Studies on the reactions of illuminated chloroplasts. II. Stimulation on inhibition of the reaction with molecular oxygen. Arch Biochem Biophys 34:339–351

Miller G, Shulaev V, Mittler R (2008) Reactive oxygen signaling and abiotic stress. Physiol Plant 133:481–489

Miller G, Suzuki N, Ciftci-Yilmaz S, Mittler R (2010) Reactive oxygen species homeostasis and signaling during drought and salinity stresses. Plant Cell Environ 33:453–467

Mittler R, Vanderauwera S, Gollery M, Van Breusegem F (2004) Reactive oxygen gene network of plants. Trends Plant Sci 9:490–498

Moller IM (2001) Plant mitochondria and oxidative stress: electron transport, NADPH turnover, and metabolism of reactive oxygen species. Annu Rev Plant Physiol Mol Biol 52:561–591

Moller IM, Jensen PE, Hansson A (2007) Oxidative modifications to cellular components in plants. Annu Rev Plant Biol 58:459–481

Montillet JL, Chamnongpol S, Rustérucci C, Dat J, van de Cotte B, Agnel JP, Battesti C, Inzé D, Van Breusegem F, Triantaphylides C (2005) Fatty acid hydroperoxides and H_2O_2 in the execution of hypersensitive cell death in tobacco leaves. Plant Physiol 138:1516–1526

Nguyen GN, Hailstones DL, Wilkes M, Sutton BG (2009) Drought induced oxidative conditions in rice anthers leading to a programmed cell death and pollen abortion. J Agron Crop Sci 195:157–164

Nguyen GN, Hailstone DL, Wilkes M, Sutton BG (2010) Role of carbohydrate metabolism in drought-induced male sterility in rice anthers. J Agron Crop Sci 196:346–357

Niyogi KK (1999) Photoprotection revisited: genetic and molecular approaches. Annu Rev Plant Physiol Plant Mol Biol 50:333–359

Noctor G, Veljovic-Jovanovic S, Driscoll S, Novitskaya L, Foyer CH (2002) Drought and oxidative load in the leaves of C3 plants: a predominant role for photorespiration? Ann Bot 89:841–850

Norman C, Howell KA, Millar AH, Whelan JM, Day DA (2004) Salicylic acid is an uncoupler and inhibitor of mitochondrial electron transport. Plant Physiol 134:492–501

Okcu G, Kaya MD, Atak M (2005) Effects of salt and drought stresses on germination and seedling growth of pea (*Pisum sativum* L.). Turk J Agric For 29: 237–242

Palma JM, Corpas FJ, del Río LA (2009) Proteome of plant peroxisomes: new perspectives on the role of these organelles in cell biology. Proteomics 9:2301–2312

Pan Y, Wu LJ, Yu ZL (2006) Effect of salt and drought stress on antioxidant enzymes activities and SOD isoenzymes of liquorice (*Glycyrrhiza uralensis* Fisch). Plant Growth Regul 49:157–165

Pandey S, Bhandari H, Ding S, Prapertchob P, Sharan R, Naik D, Taunk SK, Sastri A (2007) Coping with drought in rice farming in Asia: insights from a cross country comparative study. Agric Econ 37:213–224

Pei ZM, Murata Y, Benning G, Thomine S, Klusener B, Allen GJ, Grill E, Schroeder JI (2000) Calcium channels activated by hydrogen peroxide mediate abscisic signaling in guard cells. Nature 406:731–734

Peltzer D, Dreyer E, Polle A (2002) Differential temperature dependencies of antioxidative enzymes in two contrasting species: *Fagus sylvatica* and *Coleus blumei*. Plant Physiol Biochem 40:141–150

Pitzschke A, Forzani C, Hirt H (2006) Reactice oxygen species signalling in plants. Antioxid Redox Signal 8:1757–1764

Pogson BJ, Woo NS, Forster B, Small ID (2008) Plastid signaling to the nucleus and beyond. Trends Plant Sci 13:602–609

Puckette MC, Weng H, Mahalingam R (2007) Physiological and biochemical responses to acute ozone-induced oxidative stress in Medicago truncatula. Plant Physiol Biochem 45:70–79

Quan LJ, Zhang B, Shi WW, Li HY (2008) Hydrogen peroxide in plants: a versatile molecule of the reactive oxygen species network. J Integr Plant Biol 50:2–18

Radi R, Tan S, Prodanov E, Evans RA, Parks DA (1992) Inhibition of xanthine oxidase by uric acid and its influence on superoxide radical production. Biochim Biophys Acta 1122:178–182

Raha S, Robinson BH (2000) Mitochondria, oxygen free radicals, disease and ageing. Trends Biochem Sci 25: 502–508

Rang ZW, Jagadish SVK, Zhou QM, Craufurd PQ, Heuer S (2011) Effect of heat and drought stress on pollen germination and spikelet fertility in rice. Environ Exp Bot 70:58–65

Reddy AR, Chaitanya KV, Vivekanandan M (2004) Drought induced responses of photosynthesis and antioxidant metabolism in higher plants. J Plant Physiol 161:1189–1202

Rhoads DM, Umbach AL, Subbaiah CC, Siedow JN (2006) Mitochondrial reactive oxygen species. Contribution to oxidative stress and inter organellar signaling. Plant Physiol 141:357–366

Rodriguez-Serrano M, Romero-Puertas MC, Sparkes I, Hawes C, del Rio LA, Sandalio LM (2009) Peroxisome dynamics in Arabidopsis plants under oxidative stress induced by cadmium. Free Radic Biol Med 47(11):1632–1639

Sairam RK, Srivastava GC, Agarwal S, Meena RC (2005) Differences in antioxidant activity in response to salinity stress in tolerant and susceptible wheat genotypes. Biol Plant 49(1):85–91

Samarah NH (2005) Effects of drought stress on growth and yield of barley. Agron Sustain Dev 25:145–149

Sankar B, Jaleel CA, Manivannan P, Kishorekumar A, Somasundaram R, Panneerselvam R (2007) Effect of paclobutrazol on water stress amelioration through antioxidants and free radical scavenging enzymes in *Arachis hypogaea* L. Colloids Surf B Biointerfaces 60:229–235

Selote DS, Chopra RK (2004) Drought induced spikelet sterility is associated with an inefficient antioxidant defense in rice panicles. Physiol Plant 121:462–471

Sgherri CLM, Pinzino C, Navari-Izzo F (1993) Chemical changes and O_2^- production in thylakoid membranes under water stress. Physiol Plant 87:211–216

Sgherri CLM, Pinzino C, Navari-Izzo F (1996) Sunflower seedlings subjected to increasing stress by water deficit: changes in O_2 production related to the composition of thylakoid membranes. Physiol Plant 96: 446–452

Sharma P, Dubey RS (2005) Lead toxicity in plants. Braz J Plant Physiol 17:35–52

Sheoran IS, Saini HS (1996) Drought induced male sterility in rice: changes in carbohydrate levels and enzyme activities associated with inhibition of starch accumulation in pollen. Sex Plant Reprod 9:161–169

Singh VP, Singh RK, Singh BB, Seigler RS (1996) Physiology of stress tolerance in rice. In: Proceedings of the international conference on stress physiology of rice, Lucknow, UP, India. International Rice Research Institute, Los Banos, Philippines

Smirnoff N (1993) The role of active oxygen in the response of plants to water deficit and desiccation. New Phytol 125:27–58

Sun T, Xie WM, Xu PX (2004) Superoxide anion scavenging activity of graft chitosan derivatives. Carbohydr Polym 58:379–382

Telfer A (2002) What is β -carotene doing in the photosystem II reaction centre? Philos Trans R Soc Lond B 357:1431–1440

Telfer A, Dhami S, Bishop SM, Philipps D, Barber J (1994) β-carotene quenches singlet oxygen formed in isolated photosystem II reaction center. Biochemistry 33:14469–14474

Trebst A (2003) Function of b-carotene and tocopherol in photosystem II. Z Naturforsch 58c:609–620

Tripathy JN, Zhang J, Robin S, Nguyen TT, Nguyen HT (2000) QTLs for cell-membrane stability mapped in rice (Oryza sativa L.) under drought stress. Theor Appl Genet 100:1197–1202

Umbach AL, Fiorani F, Siedow JN (2005) Characterization of transformed *Arabidopsis* with altered alternative oxidase levels and analysis of effects on reactive oxygen species in tissue. Plant Physiol 139:1806–1820

Vanacker H, Carver TLW, Foyer CH (1998) Pathogen-induced changes in the antioxidant status of the apoplast in barley leaves. Plant Physiol 117:1103–1114

Venuprasad R, Lafitte HR, Atlin GN (2007) Response to direct selection for grain yield under drought stress in rice. Crop Sci 47:285–293

Venuprasad R, Sta Cruz MT, Amante M, Magbanua R, Kumar A, Atlin GN (2008) Response to two cycles of divergent selection for grain yield under drought stress in four rice breeding populations. Field Crops Res 107:232–244

Vranova E, Inze D, Van Brensegem F (2002) Signal transduction during oxidative stress. J Exp Bot 53: 1227–1236

Wahid A, Gelani S, Ashraf M, Foolad MR (2007) Heat tolerance in plants: an overview. Environ Exp Bot 61:199–223

Wardlaw IF, Willenbrink J (2000) Mobilization of fructan reserves and changes in enzyme activities in wheat stems correlate with water stress during kernel filling. New Phytol 148(3):413–422

Yang F, Hu J, Li J, Wu X, Qian Y (2009) Chitosan enhances leaf membrane stability and antioxidant

enzyme activities in apple seedlings under drought stress. Plant Growth Regul 58:131–136
Yin Y, Li S, Liao W, Lu Q, Wen X, Lu C (2010) Photosystem II photochemistry, photoinhibition, and the xanthophylls cycle in heat-stressed rice leaves. J Plant Physiol 167:959–966
Zhang J (2011) China's success in increasing per capita food production. J Exp Bot 62:3707–3711. doi:10.1093/jxb/err132
Zhou B, Guo Z, Xing J, Huang B (2005) Nitric oxide is involved in abscisic acid induced antioxidant activities in *Stylosanthes guianensis*. J Exp Bot 56:3223–3228

Role of Glutathione Reductase in Plant Abiotic Stress

8

Peerzada Yasir Yousuf, Khalid Ul Rehman Hakeem, Ruby Chandna, and Parvaiz Ahmad

Abstract

Abiotic stresses severely affect the growth, development, and ultimately yield of the plant, which results in heavy economic losses and food crisis. Oxidative stress, which is associated with almost all the abiotic stresses, is due to over production of toxic reactive oxygen species (ROS) including superoxide ion, hydrogen peroxide, and hydroxyl radicals. Plants combat the oxidative stress via enzymatic and non-enzymatic machinery. Glutathione reductase (GR) is one of the potential enzymes of the enzymatic antioxidant system, which sustains the reduced status of GSH via Ascorbate–Glutathione pathway and plays a vital role in maintenance of sulfhydryl (–SH) group and acts as a substrate for glutathione-*S*-transferases. GR has been characterised and has been used in the transgenics to provide the plants with tolerance against the oxidative stress.

Keywords

Abiotic stress • GR location • GR structure • ROS scavenging • ROS signalling • Transgenic plants

1 Introduction

Environmental stresses, such as drought, salinity, cold, and heat cause adverse effects on the growth, productivity and trigger a series of morphological, physiological, biochemical, and molecular changes in plants. Abiotic stress is among the primary cause of crop loss worldwide, reducing average yields for most major crop plants by more than 50%. Globally, approximately 25% of the agricultural land is saline (FAO 2009), and areas under drought are expected to increase further (Burke et al. 2006). Often crops are exposed to

P.Y. Yousuf • K.U.R. Hakeem (✉)
Molecular Ecology Lab, Department of Botany, Jamia Hamdard, New Delhi 110062, India
e-mail: Kur.hakeem@gmail.com

R. Chandna
Molecular Ecology Lab, Department of Botany, Jamia Hamdard, New Delhi 110062, India

National Institute for Plant Genomics and Research, New Delhi, India

P. Ahmad
Department of Botany, A.S. College, University of Kashmir, Srinagar 190008, India
e-mail: parvaizbot@yahoo.com

P. Ahmad and M.N.V. Prasad (eds.), *Abiotic Stress Responses in Plants: Metabolism, Productivity and Sustainability*, DOI 10.1007/978-1-4614-0634-1_8, © Springer Science + Business Media, LLC 2012 multiple stresses, and the manner in which a plant senses and responds to different environmental factors appears to be overlapping. The multiple stressful conditions give rise to the production of reactive oxygen species (ROS), indicating that the plant is under oxidative stress (Ahmad et al. 2008, 2010). The consequence of these effects includes molecular damage and adverse cellular effects of the plants, which finally leads to the death of the plant. Gene expression profiles of abiotic stressed plants indicated that although, various genes were differentially regulated in response to different stresses, they possibly induce a similar defence response (Ozturk et al. 2002). When a plant is subjected to abiotic stress, a number of genes are turned on, resulting in increased levels of several metabolites and proteins, some of which may be responsible for conferring a certain degree of protection to these stresses.

The ROS are the partially reduced forms of oxygen. ROS came into the biological existence since the aerobic life evolved from the anaerobic one. There are three main ROS which are produced sequentially in accordance to the degree to which oxygen is reduced:

1. Superoxide (O_2^-) is the primary ROS which is formed when the molecular oxygen undergoes one electron reduction. In this reaction NADPH supplies the electron and NADPH oxidase acts as the reaction catalyst.
2. Hydrogen peroxide (H_2O_2) is the other ROS which is formed when the molecular oxygen undergoes further reduction. The one electron reduction of superoxide first forms peroxide (O_2^{2-}) which is neutralised by two protons to form hydrogen peroxide. The spontaneous dismutation unlikely occurs at physiological pH when superoxide is in anionic form (O_2^-) because there is repulsion of superoxides due to negative charge. On the other hand, when the pH is acidic, the proportion of neutral form (HO_2) rises, and then the spontaneous dismutation starts to largely participate in hydrogen peroxide formation (Koji et al. 2009).
3. Hydroxyl radicals (OH^-) are the third and most toxic ROS which are formed when the hydrogen peroxide undergoes further reduction. The hydroxyl ion formation occurs in the plant by mainly two ways:
   (a) Haber–Weiss reaction: Under normal conditions, this reaction proceeds at a very slower rate and resulting in the low production of OH^- ions.

   $$H_2O_2 + O_2^- \rightarrow OH^- + OH + O_2$$

   (b) Fenton reaction: It is common in biological systems. It occurs in the presence of transition metals like Fe^{2+}, Cu^+, etc.

   $$H_2O_2 + Fe^{2+} \rightarrow Fe^{3+} + OH^- + \cdot OH$$

The ROS are normally produced in plant cells during the reactions involved in the metabolic processes like photosynthesis, respiration, etc., by different enzymes like NADPH oxidases, cell-bound peroxidases, amine oxidase, catalase, and so on, but the antioxidant machinery in the plant cells is strong enough to maintain ROS at a level which does not prove inimical to the plant. For instance, O_2^- is produced in the range of 240 μM S^{-1} and there is a steady-state level of 0.5 μM H_2O_2 in the chloroplasts during the normal conditions. However, during almost all the abiotic stresses like drought, salinity stress, pollutants, herbicides, high light stress, metals, heat shock, chilling UV radiations, etc., there is an over production of the ROS, for example, during abiotic stresses the steady-state level of H_2O_2 increases in the range of 5–15 μM. This imbalance of the poise leads to the unsafe mode making the plant vulnerable to the toxic effects of ROS (Shen et al. 1997).

In recent years, a new role for ROS has been identified: the control and regulation of biological processes, such as growth, cell cycle, programmed cell death, hormone signalling, biotic and abiotic stress responses, and development. These studies extend our understanding of ROS and suggest a dual role for ROS in plant biology as both toxic by products of aerobic metabolism and key regulators of growth, development, and defence pathways (Tateishi et al. 2005). Controlling ROS toxicity while enabling ROS such as H_2O_2 or O_2 to act as signalling molecules appears to require a large gene network composed of at least 152 genes in Arabidopsis. One approach to understanding the ROS-scavenging systems in plant stress tolerance is to manipulate the levels of antioxidant enzyme activities; dehydroascorbate reductase (DHAR), glutathione-*S*-transferase (GST), and glutathione reductase (GR). In this review, we exhibited the predicted changes in enzyme activities, and levels or redox state of glutathione under abiotic stress.

2 Production of ROS and Modulation of ROS Signalling by the Reactive Oxygen Gene Network of Plants

Organelles with a highly oxidizing metabolic activity or with an intense rate of electron flow, such as chloroplasts, mitochondria, and microbodies, are a major source of ROS production in plant cells. Together with an extensive network of oxidases, the plant cell is well armed to fight yet flexible ROS production.

In chloroplasts, the primary sources of ROS production are the Mehler's reaction and the antenna pigments. Production of ROS by drought, salt, and temperature stress is enhanced in plants by conditions limiting CO_2 fixation, as well as by the combination of these conditions with high light stress (Navrot et al. 2007). Limiting CO_2 conditions in C3 plants can also activate the photorespiratory pathway. H_2O_2 being the part of this pathway is generated in peroxisomes by the enzymatic activity of glycolate oxidase. Production of H_2O_2 in microbodies can also occur during lipid catabolism as a side-product of fatty acid oxidation (del Río et al. 2006).

Over-reduction of the electron transport chain in mitochondria is the main source of O_2 production under specific stress conditions. Additional sources of ROS in plant cells include the detoxifying reactions catalysed by cytochromes in both the cytoplasm and the endoplasmic reticulum (Foyer and Noctor 2003). Plasma membrane NADPH-dependent oxidases, plays a key role in ROS signalling in response to different stimuli or developmental signals. It contains a multimeric flavocytochrome that forms an electron transport chain capable of reducing O_2^- to O_2. NADPH oxidases, pH-dependent cell wall peroxidases, germin-like oxalate oxidases, and amine oxidases have been proposed to generate ROS at the apoplast (Noctor et al. 2006).

The intensity, duration, and localization of the different ROS signals are determined by interplay between the ROS-producing and ROS-scavenging pathways of the cell. Different developmental or environmental signals feed into the ROS signalling network and perturb ROS homeostasis in a compartment-specific or even cell-specific manner. Different proteins, enzymes, or receptors response to the rise in ROS levels and modulate different developmental, metabolic, and defence pathways. These whole processes require a regulation and involve amplification and/or feedback inhibition loops. In addition to regulating the intensity and duration of the different ROS signals, the ROS-scavenging pathways are also responsible for maintaining a low steady-state baseline of ROS on which the different signals can be registered. The reactive oxygen gene network therefore modulates the steady-state level of ROS in the different cellular compartments for signalling purposes as well as for protection against oxidative damage (Gao et al. 2008). The use of ROS as versatile signalling molecules originated from their proposed use to sense stress. Most forms of abiotic stress disrupt the metabolic balance of cells, resulting in enhanced production of ROS. Simple organisms, such as bacteria or yeast, sense the enhanced production of ROS using redox-sensitive transcription factors and other molecular sensors, activate different ROS defence pathways, and regulate their metabolic pathways to lower the production rate of ROS (Simova-Stoilova et al. 2010). This 'basic cycle' of ROS metabolism maintains a low steady-state level of ROS in cells.

3 Enzymatic Components of the ROS-Scavenging Pathways of Plants

Major ROS-scavenging enzymes of plants include superoxide dismutase (SOD), ascorbate peroxidase (APX), catalase (CAT), glutathione peroxidase (GPX), and peroxiredoxin (PrxR). Together with the antioxidants ascorbic acid and glutathione, these enzymes provide cells with highly efficient machinery for detoxifying O_2^- and H_2O_2. The balance between SODs and the different $H_2O_2^-$ scavenging enzymes in cells is considered to be crucial in determining the steady-state level of O_2^- and H_2O_2 (Romero-Puertas et al. 2006). This balance, together with the sequestering of metal ions by ferritin and other metal-binding proteins, prevents the formation of the highly toxic HO^- radical via the metal-dependent Haber–Weiss reaction or the Fenton reaction. The cellular pools of the antioxidants ascorbic acid and glutathione are maintained in their reduced state by a set of enzymes capable of using NAD(P)H to regenerate oxidized glutathione or ascorbic acid (e.g. mono-DHAR, DHAR, and GR). DHARs and glutaredoxins are indicated in as capable of reducing dehydroascorbic acid, monodehydroascorbate radicals can be reduced back into ascorbic acid via ferredoxin using electrons diverted from the photosynthetic apparatus in the water–water cycle in chloroplasts. Scavenging of H_2O_2 can also be mediated in plants by 'classical' plant peroxidases (class III) using a range of reductants. These enzymes are encoded by a large gene family of at least 73 genes in Arabidopsis and are found in the cytosol, vacuole, apoplast, or cell wall (Sarvajeet and Tatuja 2010). Transcriptomics analysis of knockout and antisense plants deficient in APX1 or CAT2 indicated that the steady-state level of transcripts that encode certain classical plant peroxidises is elevated in these plants. Work is still being carried out to determine whether specific class III plant peroxidises contribute to the ROS-scavenging capacity of cells and should be included in the ROS gene network of plants. Substrate affinity, reaction rate, and enzyme concentration are important parameters when assessing the relative contribution of the different enzymes shown to ROS detoxification. Membranes are highly susceptible to oxidative stress. In plant cells, they are protected by the activity of specific phospholipid GPXs and by *a*-tocopherol (vitamin E), which is kept in its reduced state by the pool of reduced ascorbic acid. Protection of cells against $^1O_2$ is generally believed to be mediated by carotenoids (Pooja et al. 2008).

Yoshimura et al. (2004) reported that in transgenic tobacco plants expression of GPX protein in the chloroplast was more effective at providing stress tolerance than the expression of the same protein in the cytosol. Studies that aimed at developing transgenic plants expressing several chloroplast-targeted antioxidant enzymes simultaneously showed the effectiveness of manipulating more than one gene in improving plant resistance to stress. Two genes encoding MnSOD and GR were inserted into the plastome of tobacco plants. Transgenic plants over-expressing MnSOD showed an increased resistance to methyl viologen and UV-B stress, while GR over-expressing plants were more tolerant.

The antioxidant system is most commonly modified via nuclear transformation, requiring a transit peptide for translocation into the chloroplast, while plastid engineering provides the advantage of expression at the site of the majority of ROS production in the cell, the chloroplast, without the need for protein import. Chloroplast genetic engineering also offers the advantage of transgene stacking, that is, simultaneous expression of multiple transgenes, creating the opportunity to express different ROS-scavenging enzymes in a single transformation step (Daniell et al. 2005). In this work, three genes were selected for chloroplast transformation experiments: DHAR, GST, and GR. They were chosen based on their roles within the ROS antioxidant pathway and indications from previous nuclear transformation studies that they showed potential for engineering multiple stress tolerance.

The second gene chosen for this investigation was *Escherichia coli gor* encoding GR. From previous studies it seemed clear that GR of bacterial origin could function in plant chloroplasts to increase both the GSH/GSSG ratio and the total glutathione pool (Noctor et al. 1998), and glutathione itself is an attractive target for engineering stress tolerance in plants because of its multiple roles in plant defences against both biotic and abiotic stresses (Foyer et al. 1997). Moreover, this same gene has been inserted in the plastome of tobacco plants and its expression resulted in increases in the enzyme activity and total glutathione levels. GR is also a particularly attractive candidate for combined expression with DHAR, since the two enzymes catalyse consecutive steps in the ascorbate–glutathione cycle in chloroplasts (Mittler et al. 2004). The final target enzyme in this study was a GST. These are evolutionarily conserved detoxification enzymes with the ability to conjugate a broad range of potentially harmful xenobiotics to glutathione, thereby rendering them more susceptible to removal from the cell. Some GSTs have been shown to function as GPXs to detoxify directly the products of oxidative stress. An *E. coli* GST gene was used in this work, which has been shown to exhibit a GSH-dependent peroxidise activity against cumene hydroperoxide and proved to be important for bacterial resistance to oxidative stress generated by hydrogen peroxide. As well as having a potential impact in its own right, this enzyme, combined with GR could significantly influence glutathione homeostasis, further affecting the ROS-scavenging capacity of the chloroplasts. The aim of this work was to produce transplastomic tobacco plants that express these important enzymes of the antioxidant pathway in the plastome both singly and in pairwise combinations and to analyse their influences on enzyme activity, key metabolite (glutathione and ascorbate) contents and plant tolerance to oxidative stresses.

4 Glutathione Reductase (E.C. 1.6.4.2)

GR also known as GSR or GR is a flavo-protein oxidoreductase which catalyses the reduction of glutathione disulphide (GSSG) to the sulphydryl form GSH. This enzyme employs NADPH as a reductant (Fig. 8.1).

4.1 Location

GR is predominantly found in chloroplast. However, a small amount of the enzyme isoforms is also found in mitochondria, cytosol, and peroxisomes (Edwards et al. 1990; Jimmenez et al. 1997). In leaves, bulk of GR activity is found in chloroplast, whereas root plastids exhibit a lower

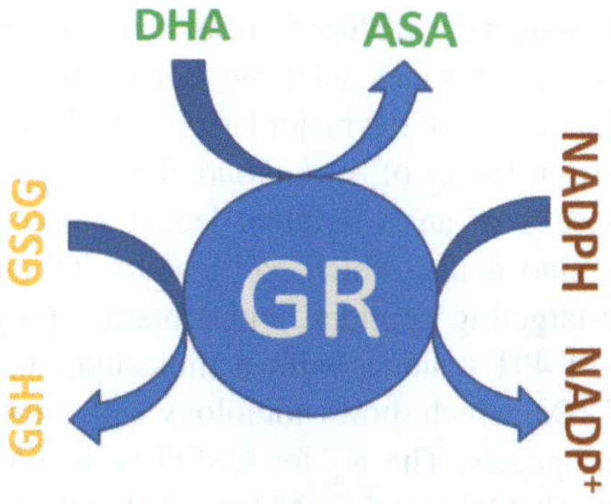

Fig. 8.1 Schematic diagram showing molecular functioning of glutathione reductase (GR)

proportion of enzyme cellular activity (Foyer and Halliwell 1976). In higher plants, three types of GR occur in the cytosol, chloroplast, and mitochondria, respectively (Creissen et al. 1992; Kubo et al. 1993; Tang and Webb 1994; Creissen and Mullineaux 1995; Mullineaux et al. 1996; Kaminaka et al. 1998). Cytosolic isoforms of GRs from rice (RGRC2; Kaminaka et al. 1998) and pea (GOR2; Stevens et al. 2000) have been identified through sub-cellular fractionation; in addition, chloroplastic GRs have been identified from Arabidopsis (AT-2; Kubo et al. 1993) and pea (GOR1; Creissen et al. 1992, 1995).

Changes in the GR isoform population between and within sub-cellular compartments in response to stress were observed in pea (Edwards et al. 1994) and maize (Anderson et al. 1995). Different GR isoforms can be stimulated by different environmental signals and have different functions in the response of plant to stress (Stevens et al. 1997). About 80% of GR activities in leaf tissues are accounted for by chloroplastic isoforms (Edwards et al. 1990). Scavenging of AOS and maintaining a high ratio of reduced to oxidized glutathione by chloroplastic GR (GR1) are necessary in oxygenic photosynthesis (Foyer et al. 1995; Kornyeyev et al. 2003).

4.2 Primary Structure

A cDNA encoding GR was cloned by immunoscreening from *Arabidopsis thaliana*. The amino acid sequence deduced from the nucleotide sequence coincides with the N-terminal amino acid sequence of the major isozyme (GR II) purified from leaves of *A. thaliana*. The polypeptide comprises of an N-terminal leader sequence of 74 amino acids, which has features of chloroplast-targeting peptides, and a mature polypeptide of 491 residues with a molecular mass of 52.7 kDa, which shows homology with GRs from other species. The K_m for GSSG is 44 μM and that for NADPH is 5.0 μM for GR II at 25°C. The pH optimum for GR II was 7.5–8.0. The native molecular mass of GR II was is kDa, indicating that GR II is a homodimer. GR II had an isoelectric point of 4.8. The cDNA hybridizes with a 2.1-kb poly(A)$^+$ RNA from leaves of *A. thaliana*. Genomic Southern analysis indicates that the gene corresponding to the cDNA is likely a single-copy gene (Kubo et al. 1993).

GR is a homodimeric FAD-containing enzyme which belongs to the family of NADPH-dependent oxidoreductases is universal in occurrence and is found in both prokaryotic as well as eukaryotic organisms. This homodimeric protein constitutes of subunits with molecular weight of about 55 kDa. In the absence of thiols GR exhibits a propensity to form tetramers and larger forms. Although these larger forms show catalytic activity, GSH that is its product, maintains the enzyme in its dimeric form under cellular conditions. GR maintains a high GSH/GSSG ratio in cells (Alscher 1989). It forms a salient part of ROS-scavenging system in concert with SOD and the enzymes of the well-known ascorbate–glutathione cycle (Foyer and Halliwell 1976). GR catalyses the reduction of GSSG which consists of two GSH linked by a disulphide bridge to GSH. GSH formed plays a vital role in the ASH–GSH cycle, maintenance of the sulphydryl group and also acts as a substrate for GSTs. Both GR and GSH play key roles in determining the tolerance of a plant under various abiotic stresses. The GSH pool maintained by GR is important for active protein function. In addition, millimolar concentrations of GSH act as an all-important redox buffer, forming a barrier between protein cystine groups and ROS. GSH also functions in limiting the metal-induced oxidative stress as from it are derived the principal heavy-metal complexing peptides of plants – the phytochelatins. One of the main features of GR is that it is thermostable.

Following is the diagram depicting the basic reaction catalysed by GR (Fig. 8.1).

As far as the mechanism of reduction of GSSG to GSH is concerned, GR acts in a ping-pong fashion in which the NADPH binds and transfers a hydride to FAD, then leaves before diglutathione binds. In other words, the two substrates are mutually exclusive. The catalytic cycle of GR comprises of two phases; a reductive half-reaction phase during which FAD, the prosthetic group of GR, is reduced by NADPH, and the oxidative half-reaction phase in which the resulting dithiol reacts with the glutathione disulphide and the final electron acceptor, GSSG, is reduced to two GSH at the GR active site. The H_2O_2 scavenging, in particular, is carried out by catalase, various peroxidases and the ascorbate–glutathione pathway.

The ascorbate–glutathione pathway, also known as Halliwell–Asada pathway, operates in different cellular locations like chloroplast, mitochondria, cytosol, peroxisomes, and apoplast. This cycle involves four main enzymes namely GR, APX, MDHAR, and MDAR. This pathway also includes a network of different metabolites with redox properties for the ROS detoxification which further help in averting the ROS-accrued oxidative damage in plants. The diagrammatic summary of this pathway is indicated in Fig. 8.2.

The studies regarding the GR have shown an increased GR activity in various plant species under different types of abiotic stresses. From the studies using transgenic plants, it has been proved that GR plays a prominent role in conferring resistance to oxidative stress caused by drought, ozone, heavy metals, high light, salinity, cold stress, etc. An increased GR activity has been reported in the roots of *C. arientinum* under salt stress, whereas Eyidogan and Oz (2005) have established elevated GR activity in the leaf tissue of the same plant under the salt stress conditions. There has also been found an enhanced GR activity in *A. thaliana*, *Vigna mungo*, *Triticum aestivum*, *Capsicum annuum*, and *Brassica juncea* following the cadmium treatments. Sharma and Dubey (2005) have found an increased GR

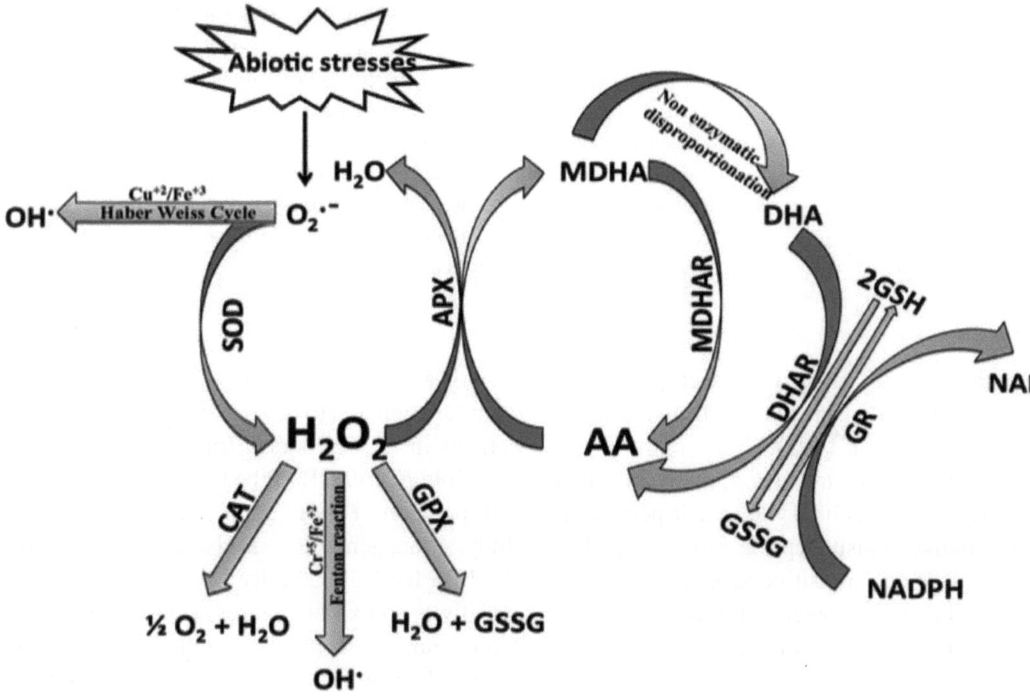

Fig. 8.2 Diagrammatic summary indicating the network of different metabolites with redox properties for the ROS detoxification which further help in averting the ROS-accrued oxidative damage in plants

activity in *Oryza sativa* seedlings during drought conditions.

5 GR: As a Detoxifying Gene in Transgenic Plants

A critical component of the ROI-scavenging system is the maintenance of ascorbate and glutathione pools in a reduced state. Of all the enzymes that are involved in this process, GR has been most extensively studied (Pooja et al. 2008). Initial attempts to enhance the GR expression in transgenic tobacco plants using a GR gene from *E. coli* resulted in a 3.5-fold increase in extractable GR activity and leaves of these plants were found to have reduced visible damage after MV exposure but no increase in ozone tolerance (Aono et al. 1991). Further analyses of transgenic plants that express the *E. coli* GR gene showed that they contained more reduced ascorbate after MV exposure than control plants (Foyer et al. 1991). The introduction of the *E. coli* GR to chloroplasts in transgenic tobacco also consequence in reduced damage after treatment with MV and sulphur dioxide, but not ozone (Aono et al. 1993). Transgenic poplar plants in which the chloroplast-targeted *E. coli* GR was expressed were found to have GR activities up to 500-fold higher than untransformed plants. Leaves of these plants had approximately two times higher levels of both glutathione and ascorbate as compared to control plants or plants that expressed the GR gene without the chloroplast-targeting sequence. Although these plants did not show increased protection from MV-induced inhibition of CO_2 assimilation, but they were found to be more resistant to photoinhibition caused by high light intensity and chilling temperature (Foyer et al. 1995). These authors suggested that the protection from cold-induced photoinhibition in GR1 expressing plants could have been resulted from the increased levels of GSH and ascorbate, and the more rapid cycling of these antioxidants to their reduced forms were found to have an elevated GR activity in cytosolic, chloroplastic, and mitochondrial fractions

Table 8.1 Transgenic plants which show increased GR production have been found to be good abiotic stress tolerant

Source	Target gene	Response in transgenic plants
E. coli	*Triticum aestivum*, cv. Oasis protoplast	Higher GSH content and SH/GSH + GSSG ratio than control, no increase in SOD and GR activities
Arabidopsis thaliana ecotype Columbia	*G. hirsutum* L. cv. Coker	Chilling stress tolerance and photo protection
Pisum sativum	*Nicotiana tabacum*	Reduced O_3 damage
E. coli (*cytosol*)	*N. tabacum*	Reduced MV damage
E. coli (*chloroplast*)	*N. tabacum*	Reduced MV and SO_2 damage
E. coli (*chloroplast*)	Poplar	Reduced photoinhibition

were noticed in the leaves of tobacco plants that expressed a chimeric gene derived from a pea GR cDNA (Creissen et al. 1995).

Interestingly, some transgenic plant lines that expressed pea GR without the transit peptide or with the native transit peptide replaced with a chloroplast-specific transit peptide did show tolerance to MV, but the level of protection did not correlate with the levels of GR expression and none of these plants were more ozone tolerant. Aono et al. (1995) found that decreased expression of GR activity in transgenic tobacco plants increased their susceptibility to MV-induced damage. They also reported that co-expression of cytosolic forms of both GR and SOD in tobacco plants provided more substantial protection from MV treatment than either enzyme alone. These results indicate that the expression of combinations of antioxidant enzymes in transgenic plants may have synergistic effects on stress tolerance.

The expression patterns and enzyme activities of GR were sorted out in plants belonging to poaceae family by Bashir et al. (2007) under iron sufficient and iron deficient conditions. cDNA clones for GR1 and cytosolic GR (GR2) were isolated from *H. vulgare*. In vitro activity was exhibited by the both proteins; however, the specific activity of GR1 was found to be threefold higher than the GR2. By northern blot analysis, the expression patterns of GR1 and GR2 were examined in wheat, barley, rice, and maize. An up-regulation of barley GR1 and GR2, wheat GR2 was found in the conditions of iron deficiency rather than iron sufficient conditions (Melchiorre et al. 2009). Over-expression of a eukaryotic GR from *Brassica campestris* and *E. coli* were inspected in *E. coli* in pET-28a. Better growth and survival rate was found in *E. coli* than the control. However, far better growth was noticed in *E. coli* strain transformed with the inducible *E. coli* GR in the presence of cadmium and paraquat (Yoon et al. 2005). In another study, transgenic *N. tabbacum* plants with 30–70% less GR activity showed an enhanced sensitivity to oxidative stress (Ding et al. 2009). Many transgenic plants which show increased GR production have been found to be good abiotic stress tolerant. A few examples are stated in Table 8.1.

6 Conclusion and Future Perspective

Almost all biotic stresses lead to the overproduction of ROS in plants which are highly reactive and toxic and ultimately results in oxidative stress. Oxidative stress is a condition in which ROS or free radicals are generated extra- or intracellularly, which can exert their toxic effects to the cells. These species may affect cell membrane properties and cause oxidative damage to nucleic acids, lipids, and proteins that may make them non-functional. However, the cells possess well-equipped antioxidant defence mechanisms to detoxify the detrimental effects of ROS. The antioxidant defences could be either non-enzymatic (e.g. glutathione, proline, *a*-tocopherols, carotenoids, and flavonoids) or enzymatic (e.g. SOD, catalase GPX, and GR). It is well known that plant cells and its organelles like chloroplast, mitochondria, and peroxisomes employ antioxidant defence systems to protect themselves against ROS-induced oxidative stress.

GR plays a key role in the response to oxidative stress by maintaining the intracellular glutathione pool mainly in the reduced state and especially functions as an antioxidant that scavenges ROS such as hydrogen peroxide and superoxide. Overexpression of GR results in abiotic stress tolerance in various crop plants due to efficient ROS-scavenging capacity.

References

Ahmad P, Sarwat M, Sharma S (2008) Reactive oxygen species, antioxidants and signaling in plants. J Plant Biol 51:167–173

Ahmad P, Jaleel CA, Salem MA, Nabi G, Sharma S (2010) Roles of enzymatic and non-enzymatic antioxidants in plants during abiotic stress. Crit Rev Biotechnol 30:161–175

Alscher RG (1989) Biosynthesis and antioxidant function of glutathione in plants. Physiol Plant 77:457–464

Anderson MD, Prasad TK, Stewart CR (1995) Changes in isozyme profiles of catalase, peroxidase and glutathione reductase during acclimation to chilling in mesocotyls of maize seedlings. Plant Physiol 109:1247–1257

Aono M, Saji H, Fujiyama K, Sugita M, Kondo N, Tanaka K (1995) Decrease in Activity of Glutathione Reductase Enhances Paraquat Sensitivity in Transgenic Nicotiana tabacum Plant Physiol 107:645–648

Aono M, Kubo A, Saji H, Tanaka K, Kondo N (1993) Enhanced tolerance to photooxidative stress of transgenic Nicotiana tabacum with high chloroplastic glutathione reductase activity. Plant Cell Physiol 34:129–135

Aono M, Kubo A, Saji H, Natori T, Tanaka K, Kondo N (1991) Resistance to active oxygen toxicity of transgenic *Nicotiana tabacum* that expresses the gene for glutathione reductase from *Escherichia coli*. Plant Cell Physiol 32:691–698

Bashir K, Nagasaka S, Itai RN, Kobayashi T, Takahashi M, Nakanishi H, Mori S, Nishizawa NK (2007) Expression and enzyme activity of glutathione reductase is upregulated by Fe-deficiency in graminaceous plants. Plant Mol Biol 65:277–284

Burke EJ, Brown SJ, Christidis N (2006) Modeling the recent evolution of global drought and projections for the twenty-first century with the Hadley centre climate model. J Hydrometeor 7:1113–1125

Creissen GP, Mullineaux PM (1995) Cloning and characterization of glutathione reductase cDNAs and identification of two genes encoding the tobacco enzyme. Planta 197:422–425

Creissen G, Reynolds H, Xue Y, Mullineaux P (1995) Simultaneous targeting of pea glutathione reductase and of a bacterial fusion protein to chloroplasts and mitochondria in transgenic tobacco. The Plant J 8:167–175

Creissen G, Edwards EA, Enard C, Wellburn A, Mullineaux P (1992) Molecular characterization of glutathione reductase cDNAs from pea (*Pisum sativum* L.). Plant J 2:129–131

Daniell H, Kumar S, Duformantel N (2005) Breakthrough in chloroplast genetic engineering of agronomically important crops. Trends Biotechnol 2:238–245

del Río LA, Sandalio LM, Corpas FJ, Palma JM, Barroso JB (2006) Reactive oxygen species and reactive nitrogen species in peroxisomes. Production, scavenging, and role in cell signaling. Plant Physiol 141:330–335

Ding S, Lu Q, Zhang Y, Yang Z, Wen X, Zhang L, Lu C (2009) Enhanced sensitivity to oxidative stress in transgenic tobacco plants with decreased glutathione reductase activity leads to a decrease in ascorbate pool and ascorbate redox state. Plant Mol Biol 69:577–592

Edwards EA, Rawsthorne S, Mullineaux PM (1990) Subcellular distribution of multiple forms of glutathione reductase in leaves of pea (*Pisum sativum* L.). Planta 180:278–284

Edwards EA, Enard C, Creissen GP, Mullineaux PM (1994) Synthesis and properties of glutathione reductase in stressed peas. Planta 192:137–143

Eyidogan F, Oz MT (2005) Effect of salinity on antioxidant responses of chickpea seedlings. Acta Physiol Plant 29:485–493

FAO (Food, Agriculture Organization of the United Nations) (2009) FAO production yearbook. FAO, Rome

Foyer CH, Noctor G (2003) Redox sensing and signaling associated with reactive oxygen in chloroplasts, peroxisomes and mitochondria. Plant Physiol 119: 355–364

Foyer CH, Lopez-Delgado H, Dat JF, Scott IM (1997) Hydrogen peroxide- and glutathione-associated mechanisms of acclimatory stress tolerance and signalling. Physiol Plant 100:241–254

Foyer CH, Souriau N, Perret S, Lelandais M, Kunert KJ, Pruvost C, Jouanin L (1995) Overexpression of glutathione reductase but not glutathione synthetase leads to increases in antioxidant capacity and resistance to photoinhibition in poplar trees. Plant Physiol 109:1047–1057

Foyer CH, Lelandais M, Galap C, Kunert KJ (1991) Effects of elevated cytosolic glutathione reductase activity on the cellular glutathione pool and photosynthesis in leaves under normal and stress conditions. Plant Physiol 97:863–872

Foyer CH, Halliwell B (1976) The presence of glutathione and glutathione reductase in chloroplasts: a proposed role in ascorbic acid metabolism. Planta 133:21–25

Gao C, Xing D, Li L, Zhang L (2008) Implication of reactive oxygen species and mitochondrial dysfunction in the early stages of plant programmed cell death induced by ultraviolet-C overexposure. Planta 227:755–767

Jiménez A, Hernández JA, del Río LA, Sevilla F (1997) Evidence for the presence of the ascorbate-glutathione cycle in mitochondria and peroxisomes of pea (*Pisum sativum* L.) leaves. Plant Physiol 114:275–284

Kaminaka H, Morita S, Nakajima M, Masumura T, Tanaka K (1998) Gene cloning and expression of cytosolic glutathione reductase in rice (*Oriza sativa* L.). Plant Cell Physiol 39:1269–1280

Koji Y, Shiro M, Michio K, Mitsutaka T, Hiroshi M (2009) Antioxidant capacity and damages caused by salinity stress in apical and basal regions of rice leaf. Plant Prod Sci 12:319–326

Kornyeyev D, Logan BA, Payton P, Allen RD, Holaday AS (2003) Elevated chloroplastic glutathione reductase activities decrease chilling-induced photoinhibition by increasing rates of photochemistry, but not thermal energy dissipation, in transgenic cotton. Funct Plant Biol 30:101–110

Kubo A, Sano T, Saji H, Tanaka K, Kondo N, Tanaka K (1993) Primary structure and properties of glutathione reductase from *Arabidopsis thaliana*. Plant Cell Physiol 34:1259–1266

Melchiorre M, Robert G, Trippi V, Racca R, Lascano HR (2009) Superoxide dismutase and glutathione reductase overexpression in wheat protoplast: photooxidative stress tolerance and changes in cellular redox state. Plant Growth Regul 57:57–68

Mittler R, Vanderauwera S, Gollery M, Breusegem FV (2004) Reactive oxygen gene network of plants. Trends Plant Sci 10:490–498

Mullineaux PM, Enard C, Hellens R, Creissen G (1996) Characterization of a glutathione reductase gene and its genetic locus from pea (*Pisum sativum* L.). Planta 200:186–194

Navrot N, Rouhier N, Gelhaye E, Jaquot JP (2007) Reactive oxygen species generation and antioxidant systems in plant mitochondria. Physiol Plant 129:185–195

Noctor G, Paepe RD, Foyer CH (2006) Mitochondrial redox biology and homeostasis in plants. Trends Plant Sci 12:125–134

Noctor G, Arisi ACM, Jouanin L, Foyer CH (1998) Manipulation of Glutathione and Amino Acid Biosynthesis in the Chloroplast. Plant Physiol 118: 471–482

Ozturk ZN, Talame V, Deyholos M, Michalowski CB, Galbraith DW, Gozukirmizi N, Tuberosa R, Bohnert HJ (2002) Monitoring large-scale changes in transcript abundance in drought- and salt stressed barley. Plant Mol Biol 48:551–573

Pooja BM, Vadez V, Sharma KK (2008) Transgenic approaches for abiotic stress tolerance in plants: retrospect and prospects. Plant Cell Rep 27:411–424

Romero-Puertas MC, Corpas FJ, Sandalio LM, Leterrier M, Rodríguez-Serrano M, del Río LA, Palma JM (2006) Glutathione reductase from pea leaves: response to abiotic stress and characterization of the peroxisomal isozyme. New Phytol 170:43–52

Sarvajeet SG, Tatuja N (2010) Reactive oxygen species and antioxidant machinery in abiotic stress tolerance in crop plants. Plant Physiol Biochem 48:909–930

Sharma P, Dubey RS (2005) Modulation of nitrate reductase activity in rice seedlings under aluminium toxicity and water stress: role of osmolytes as enzyme protectant. J Plant Physiol 162:854–864

Shen B, Jensen RG, Bohnert HJ (1997) Increased resistance to oxidative stress in transgenic plants by targeting mannitol biosynthesis to chloroplasts. Plant Physiol 113:1177–1183

Simova-Stoilova L, Vaseva I, Grigorova B, Demirevska K, Feller U (2010) Proteolytic activity and cysteine protease expression in wheat leaves under severe soil drought and recovery. Plant Physiol Biochem 48: 200–206

Stevens RG, Creissen GP, Mullineaux PM (2000) Characterization of pea cytosolic glutathione reductase expressed in transgenic tobacco. Planta 211:537–545

Stevens RG, Creissen GP, Mullineaux PM (1997) Cloning and characterization of a cytosolic glutathione reductase cDNA from pea (*Pisum sativum* L.) and its expression in response to stress. Plant Mol Biol 35:641–654

Tang X, Webb MA (1994) Soybean Root Nodule cDNA Encoding Glutathione Reductase. Plant Physiol 104:1081–1082

Tateishi Y, Nakagawa T, Esaka M (2005) Osmotolerance and growth stimulation of transgenic tobacco cells accumulating free proline by silencing proline dehydrogenase expression with double-stranded RNA interference technique. Physiol Plant 125:224–234

Yoon HS, Lee IA, Lee H, Lee BH, Jo J (2005) Overexpression of a eukaryotic glutathione reductase gene from *Brassica campestris* improved resistance to oxidative stress in *E. Coli*. Biochem Biophys Res Commun 326:618–623

Yoshimura K, Miyaok H, Gaber H, Takeda T, Kanaboshi H, Miyasaka H, Shigeoka S (2004) Enhancement of stress tolerance in transgenic tobacco plants over expressing chlamydomonas glutathione peroxide in chloroplasts of cytosol. Plant J 37:21–33

Flavonoids as Antioxidants in Plants Under Abiotic Stresses

9

Martina Di Ferdinando, Cecilia Brunetti, Alessio Fini, and Massimiliano Tattini

Abstract

Flavonoids make a relevant contribution to the response mechanisms of higher plants to a plethora of abiotic stresses. In addition to the long-reported functions as screeners of damaging short-wave solar radiation, flavonoids have been suggested as playing key functions as antioxidants in stressed plants, by inhibiting the generation and reducing reactive oxygen species (ROS) once formed. The ROS-scavenging properties of flavonoids are restricted to few structures, namely, the dihydroxy B-ring-substituted flavonoid glycosides. This structure–activity relationship conforms to the well-known stress-induced preferential biosynthesis of dihydroxy B-ring-substituted both flavones and flavonols. These flavonoids, especially the derivatives of quercetin, have been shown to greatly affect the movement of auxin at intra- and intercellular levels, and hence to tightly regulate the development of individual organs and the whole plant. The effectiveness of flavonoids to inhibit the activity of the auxin efflux facilitator proteins tightly depends on the chemical features that confer the antioxidant potential. In this review article, we discuss about (1) the effect of different abiotic stresses on the accumulation of individual flavonoids, (2) the potential role served by antioxidant flavonoids in the antioxidant machinery of plants exposed to severe stress conditions, and (3) the function of flavonoids as developmental regulators.

Keywords

Abiotic stresses • Antioxidant enzymes • Auxin movement • Dihydroxy B-ring-substituted flavonoids • Reactive oxygen species • UV-radiation

M. Di Ferdinando • C. Brunetti • A. Fini
Department of Plant, Soil and Environmental Science, University of Florence, Viale delle Idee 30, 50019, Sesto Fiorentino, Firenze, Italy

M. Tattini (✉)
Consiglio Nazionale delle Ricerche, Istituto per la Protezione delle Piante, Via Madonna del Piano 10, I-50019, Sesto Fiorentino, Firenze, Italy
e-mail: m.tattini@ipp.cnr.it

P. Ahmad and M.N.V. Prasad (eds.), *Abiotic Stress Responses in Plants: Metabolism, Productivity and Sustainability*, DOI 10.1007/978-1-4614-0634-1_9, © Springer Science+Business Media, LLC 2012

1 Introduction

Flavonoids, the vast class of secondary metabolites encompassing more than 10,000 structures (Harborne and Williams 2000), have long been reported to display wide range of uses in plant–environment interactions (Winkel-Shirley 2002; D'Auria and Gershenzon 2005; Agati and Tattini 2010). In the recent past, the idea behind functioning of flavonoids primarily as attenuators of short solar wavelengths in plants exposed to UV-B or full solar irradiance has been questioned (Gerhardt et al. 2008; Agati and Tattini 2010; Agati et al. 2011; Akhtar et al. 2010), as flavonoids most responsive to UV-radiation are far from being the most effective UV-B absorbers among the thousands of polyphenol structures (Cockell and Knowland 1999; Harborne and Williams 2000). These suggestions are consistent with the early authoritative views of Swain (1986) and Stafford (1991) of flavonols, the most ancient and widespread of the flavonoids (Rausher 2006; Winkel 2006), having served key antioxidant or "internal regulatory" functions during the evolution of early terrestrial plants.

The biosynthesis of flavonoids is upregulated not only as a consequence of UV-radiation but also in response to a wide range of other abiotic (and biotic) stresses, ranging from nitrogen/phosphorus depletion to cold and salinity/drought stress (Tattini et al. 2004, 2005; Lillo et al. 2008; Olsen et al. 2009; Agati et al. 2011). Since different stresses have in common the generation of reactive oxygen species (Mittler et al. 2004; Mittler 2006), it has been postulated that flavonoids are synthesized to effectively counter the stress-induced oxidative damage. Flavonoids may accomplish their antioxidant functions by both preventing the generation of ROS (through their ability to chelate transition metal ions such as Fe and Cu, Brown et al. 1998; Melidou et al. 2005; Hernández et al. 2009; Agati and Tattini 2010) and scavenging ROS once formed (Ryan et al. 2002; Babu et al. 2003; Tattini et al. 2004; Agati et al. 2007; Jaakola and Hohtola 2010).

How flavonoids may perform their scavenging activity in an in vivo condition is still a matter of conflict, and major criticisms regarding the localization functional relationships (Halliwell 2009; Hernández et al. 2009) have been only partially addressed (Yamasaki et al. 1997; Agati et al. 2007; Agati and Tattini 2010; Akhtar et al. 2010). However, we note that flavonoids do not exclusively occur in the vacuoles of epidermal cells (as long been reported, Saunders and Mc Clure 1976), and hence far enough from the sites of ROS production. Relatively recent experiments have reported a large accumulation of flavonoids in mesophyll cells both in the vacuole (Agati et al. 2002, 2009, 2011; Gould et al. 2002; Tattini et al. 2005, 2006; Kytridis and Manetas 2006) and in the chloroplasts (Agati et al. 2007). This subcellular distribution of flavonoids is, therefore, consistent with a their putative role as ROS-quenchers. Chloroplast flavonoids (chloroplasts have been reported to be both capable of flavonoid biosynthesis and may represent an important site of flavonoid accumulation), (Oettmeier and Heupel 1972; Saunders and Mc Clure 1976; Takahama 1982; Takahama and Oniki 1997; Zaprometov and Nikolaeva 2003) have been shown to effectively quench singlet oxygen generated upon excess blue-light irradiance (Agati et al. 2007). A vacuolar distribution of mesophyll flavonoids may be of key significance in reducing hydrogen peroxide (H_2O_2) that may freely escape from the chloroplast under severe stress conditions, as also hypothesized to occur for flavonoids located in the vacuoles of epidermal cells (Yamasaki et al. 1997; Sakihama et al. 2000). Nevertheless, the matter is far being conclusively elucidated and poses serious concerns from an analytical and technical point of view, mostly concerning the simultaneous visualization of reactive oxygen species and flavonoid distribution at inter and intracellular levels (Hernández et al. 2009; Agati and Tattini 2010).

We highlight that even though polyphenols, particularly flavonoids have long been shown to scavenge various forms of reactive oxygen "in vitro", the flavonoids usually encountered in plants, e.g., in leaf tissues, are the glycosylated structures, as glycosylation both increases the solubility of carbon-based metabolites in an aqueous cellular milieu and preserve the most reactive

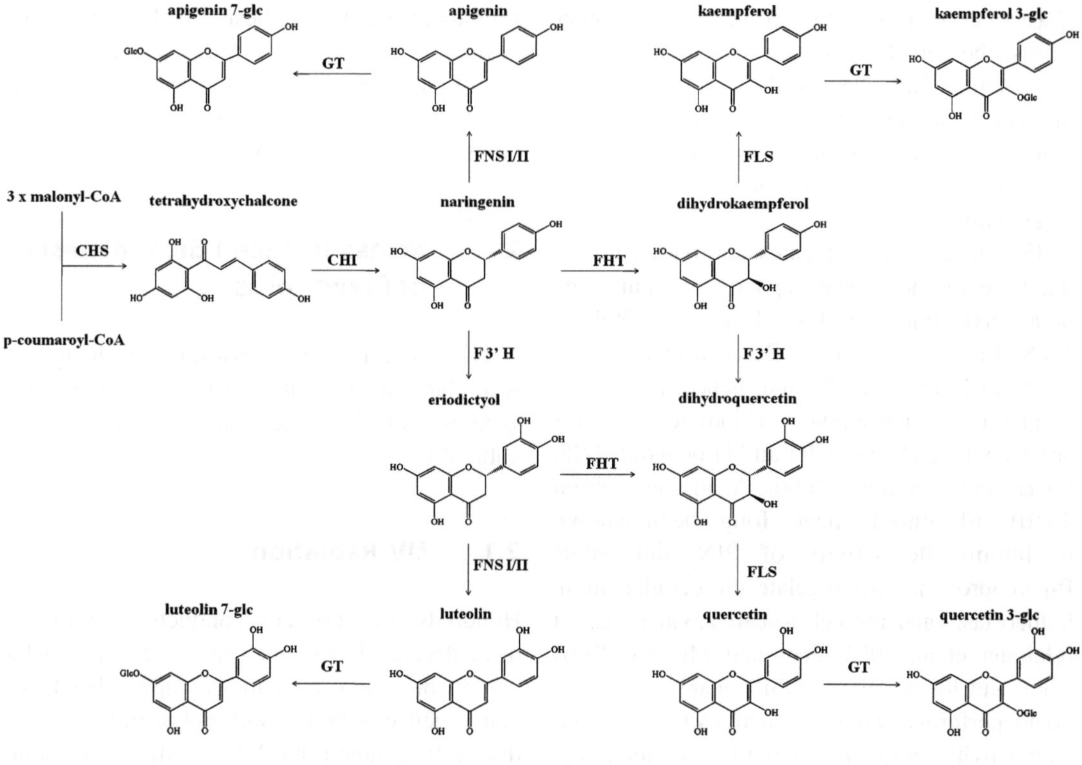

Fig. 9.1 Scheme of the main flavonoid pathway leading to the most common flavanones, flavones, dihydroflavonols, flavonols, and flavonol glycosides. Enzymes – *CHS* chalcone synthase, *CHI* chalcone isomerase, *FNS I/II* flavone synthase I and II, *F3′H* flavonoid 3′-hydroxylase, *FHT* flavanone 3β-hydroxylase, *FLS* flavonol synthase, *GT* glycosyl transferases

functional groups from autooxidation (Pearse et al. 2005). Actually, few glycosylated flavonoids are effective antioxidants, whereas most flavonoid aglycones (i.e., lacking the sugar moiety) are actually capable of quenching ROS (Rice-Evans et al. 1996). Quercetin 3-O- and luteolin 7-O-glycosides that posses a catechol group (ortho-dihydroxy B-ring substitution, Fig. 9.1) in the B-ring of the flavonoid skeleton, but not kaempferol 3-O- or apigenin 7-O-glycosides, display an appreciable antioxidant activity, in the molar concentration-range likely encountered in plant cells (Tattini et al. 2004).

Noticeably, stress-responsive flavonoids have the greatest antioxidant potential, and the ratio of "effective antioxidant" to "poor antioxidant" flavonoids has been conclusively shown to increase steeply in response to a plethora of abiotic stresses (Kolb et al. 2001; Schmitz-Hoerner and Weissenböck 2003; Tattini et al. 2004; Lillo et al. 2008; Kotilainen et al. 2008; Jaakola and Hohtola 2010; Agati et al. 2009, 2011). The issue of "how significant are flavonoids as antioxidants in plants" has been recently explored by Hernández et al. (2009): these authors have suggested of minor significance the contribution of flavonoids to the highly integrated and constitutive antioxidant defense system (which includes antioxidant enzymes, ascorbic acid, and glutathione) operating in plants suffering from various abiotic stresses (as early authoritatively suggested by Halliwell 2009).

Nevertheless, the first line of defense against stress-induced enhancement in ROS concentration – the antioxidant enzymes – have been reported to be ineffective to protect cells from oxidative damage during severe stress conditions (Polle 2001; Apel and Hirt 2004; Hatier and Gould 2008). Hatier and Gould (2008) have recently suggested that the very conditions that lead to the accumulation of flavonoids are those that may inactivate key antioxidant enzymes

(Casano et al. 1997; Streb et al. 1997). In other words, the actual significance of stress-responsive "antioxidant" flavonoids in the context of the well-coordinated antioxidant defenses operating in stressed plants has to be explored on the basis of the stress severity at which plants are faced with.

Flavonoids with the greatest antioxidant potential have the additional capacity to inhibit the polar auxin transport (PAT, Jacobs and Rubery 1988; Brown et al. 2001; Besseau et al. 2007; Peer and Murphy 2007), and, hence, capable of regulating the development of individual organs and the whole plant (Taylor and Grotewold 2005; Lazar and Goodman 2006; Agati and Tattini 2010). Flavonoids have long been shown to inhibit the activity of PIN and MDR P-glycoproteins, that regulate the cellular auxin homeostasis and the cell-to-cell auxin transport (Geissler et al. 2005; Peer and Murphy 2007; Friml and Jones 2010). In this regard, flavonoids do not perform any reducing activity (i.e., antioxidant activity *sensu stricto*), but the chemical features that confer the antioxidant potential are required to effectively interact with the auxin transport proteins (Peer and Murphy 2006; Agati and Tattini 2010). These "internal regulatory" or "physiological" functions of flavonoids, early hypothesized by Stafford (1991) as the most prominent *in planta*, have to be regarded with special attention. In fact, an increasing body of evidence suggests that the health-promoting effect of flavonoids in mammals depends not only upon their ability to scavenge a wide array of reactive oxygen species – free radicals and H_2O_2 – but on their affinity with several proteins (including the mitogen activated protein kinases, MAPK) that supersede key steps of cell growth and differentiation (Williams et al. 2004; Taylor and Grotewold 2005; Peer and Murphy 2006; Lamoral-Theys et al. 2010). Flavonoids might behave, therefore, as "signaling molecules" (Peer and Murphy 2006) or "developmental regulators" (Taylor and Grotewold 2005) in plants, as well as in animals, and their functional roles going well beyond their ability to merely scavenge reactive oxygen.

In this article, we (1) review the pertinent literature on the effect of most common abiotic stresses on the biosynthesis of "UV-absorbing" flavonoids (2) discuss on the potential contribution of flavonoids in the antioxidant machinery of plants under severe stress conditions and (3) their functional roles in the control of plant growth.

2 Stress-Induced Biosynthesis of Flavonoids

A brief summary of stress-induced changes in secondary metabolism, particularly in the biosynthesis of flavonoids has been reported in Table 9.1.

2.1 UV-Radiation

Hundreds of experiments conducted over the last three decades have conclusively shown that flavonoid biosynthesis is mostly upregulated as a consequence of high UV-B and/or full solar irradiance (it is noted that UV-B radiation does not account for more than 5% of the total UV-radiation at mid-latitudes, Ballaré 2003). It is interesting to note that the biosynthesis of other phenolics compounds, both hydroxycinnamic acid derivatives and poly-galloyl derivatives (i.e., hydrolyzable tannins) that are in constitutively greater concentrations than flavonoids under low-light conditions, is very poorly affected by an increase in UV-radiation (Ollson et al. 1999; Burchard et al. 2000; Hofmann et al. 2003; Tattini et al. 2004, 2005, 2006). Since hydroxycinnamic acid derivatives and some hydrolyzable tannins have been reported to display a superior molar extinction coefficient than flavonoids over the 280–320 nm region of the solar spectrum (Harborne and Williams 2000; Tattini et al. 2004), it may be questioned that flavonoids play the primary role as UV-B attenuators in response to UV-B or full solar irradiance (Cockell and Knowland 1999; Agati and Tattini 2010). Similarly, the loss of mycosporin-like aminoacid (MAA) in favor of flavonoid (actually flavonol) metabolism during the colonization of land by plants was likely for fulfilling several uses, as MAA are effective UV-B absorbers (Cockell and Knowland 1999).

Table 9.1 Changes in phenylpropanoid and flavonoid metabolism in response to a variety of abiotic stresses

Species	Treatment	Flavonoids or gene expression affected by treatment	References
Arabidopsis thaliana (hybrid lines differing in freezing tolerance)	Freezing	Upregulation of *CHS*, *CHI*, *DFR*, *FLS1*, and *F3'H* during cold acclimation in the cold tolerant lines	Hannah et al. (2006)
A. thaliana (hybrid lines differing in freezing tolerance)	Freezing	Flavonoid accumulated during cold acclimation and leaf flavonoid content strongly correlated to freezing tolerance	Korn et al. (2008)
A. thaliana	Various treatments	Quercetin biosynthesis enhanced by N deprivation when coupled with another stressor, e.g., cold or high light	Lillo et al. (2008)
A. thaliana	Cold + darkness	Upregulation of flavonoid 3-O-transferase, PAL, CHS and other genes of the phenylpropanoid pathway induced by cold + light treatment	Soitamo et al. (2008)
A. thaliana ecotype Colombia	Nitrogen deprivation + cold	Greater quercetin than kaempferol accumulation	Olsen et al. (2009)
Arnica montana	Cold + UV-exclusion	Ortho-dihydroxy flavonoid biosynthesis was induced by low temperature, but not UV	Albert et al. (2009)
Betula pendula	O_3-fumigation + CO_2 enrichment	O_3 increased the concentration of quercetin derivates at ambient CO_2	Peltonen et al. (2005)
Brassica napus	UV-B enhancement and exclusion	Greater biosynthesis of quercetin glycosides than acylated kaempferol glycoside	Gerhardt et al. (2008)
Brassica oleracea (different cultivars)	Cold	Cold induces higher quercetin to kaempferol ratio in all cultivars, and increases the tissue antioxidant activity	Schmidt et al. (2010) and Zietz et al. (2010)
Fagus sylvatica	O_3-fumigation	Upregulation of all shikimate pathway genes and increased concentration of kaempferol 3-O-glycosides	Betz et al. (2009)
Ginkgo biloba	O_3-fumigation	Ozone steeply increased the quercetin to kaempferol ratio	He et al. (2009)
Lemna gibba	Cu^{2+} + different light environment	Copper and simulated solar radiation induced the biosynthesis of the very same flavonoids, as a consequence of ROS accumulation	Babu et al. (2003)
L. gibba	DMCU/DBMIB + cold + different light environments	Flavonoids accumulate in response to a photosynthetic electron transport chain redox signal, which can operate independently from the ROS signal	Akhtar et al. (2010)
Ligustrum vulgare	High light + drought	Upregulation of quercetin 3-O-rutinoside, luteolin 7-O-glucoside and echinoside in response to high light	Tattini et al. (2004)
L. vulgare	High light + UV-exclusion	Greater accumulation of quercetin and luteolin in response to high light, even in the absence of UV-radiation	Agati et al. (2009)
L. vulgare	High light + NaCl + UV-exclusion	Biosynthesis of quercetin 3-O-rutinoside, luteolin 7-O-glucoside similarly induced by UV-radiation and root zone salinity. Accumulation of di-hydroxylated flavonoids upregulated by excess light even in absence of UV-radiation	Agati et al. (2011)

(continued)

Table 9.1 (continued)

Species	Treatment	Flavonoids or gene expression affected by treatment	References
L. vulgare and *Phyllirea latifolia* (shade tolerant and sun-demanding respectively)	High light	High light induced accumulation of di-hydroxylated flavonoids in the mesophyll to a greater extent in the shade-tolerant than in the sun-demanding species	Tattini et al. (2005)
Mahonia repens	High light	High light induced greater accumulation of anthocyanins and chlorogenic acid	Grace et al. (1998)
Marchantia polymorpha	UV-B enhancement and/or exclusion	Enhancement of UV-B increased luteolin to apigenin ratio	Markham et al. (1998)
Myrtus communis (salt-sensitive) and *Pistacia lentiscus* (salt-tolerant)	NaCl and high light	Greater carbon allocation to myricetin and quercetin glycosides in the salt-sensitive than in the salt-tolerant species	Tattini et al. (2006)
Olea europaea	NaCl + high light	Flavonoid accumulation was increased by high light and salinity, but the effect of high light was greater	Remorini et al. (2009)
Oryza sativa (two genotypes differing in salt tolerance)	NaCl	Upregulation of genes involved in the flavonoid pathway in the salt-sensitive genotype	Walia et al. (2005)
O. sativa (flavonoid deficient mutants vs. parent line)	UV-B enhancement	Accumulation of saponarin and lutonarin reduced DNA damage in the parent line if compared to the flavonoid-knock out mutants	Schmitz-Hoerner and Wissenbock (2003)
Panax ginseng	Cu^{2+}	Upregulation of PAL and increased flavonoid content	Ali et al. (2006)
Petunia (wild type vs. mutants with enhanced anthocyanin production vs. mutants with antisense FLS)	UV-B enhancement	UV-B increased quercetin to kaempferol ratio, and the effect was greater in the mutants with antisense FLS	Ryan et al. (1998)
Petunia (wild type vs. F3H deficient mutants)	UV-B enhancement	Increased quercetin to kaempferol ratio in wild type	Ryan et al. (2002)
Phaseolus vulgaris	O_3-fumigation	Increased *PAL*, *CHS*, and *CHI*	Paolacci et al. (2001)
P. vulgaris	O_3-fumigation	Increased accumulation of kaempferol 3-O-glucuronide	Kanoun et al. (2003)
Pinguicola vulgaris	UV-B enhancement	Supplemental UV-B increased anthocyanis, and this resulted in enhanced high light and cold tolerance	Mendez et al. (1999)
Pinus sylvestris	Cd^{2+}	Root exposition to Cadmium caused an accumulation of soluble phenolics in the cytosol of root cells	Schutzendubel et al. (2001)
Salix (*S. myrsinifolia* and hybrids)	UV-B enhancement + drought	UV-B increased quercetin derivates in all plants, while the effect of drought was clone-specific	Turtola et al. (2005)

Schefflera arboricola	H_2O_2	Evidences for the dramatic superior antioxidant activity of quercetin than kaempferol derivatives	Yamasaki et al. (1997)
Solanum lycopersicon	Nitrogen deprivation + cold + different light intensities	Maximum increases in quercetin biosynthesis were found in the low nitrogen + cold + high light treatment. Cold, nitrogen deprivation and light had a synergetic effect on *PAL*, *CHS*, *F3H*, *FLS* upregulation	Løvdal et al. (2010)
Trifolium pratense	Mild O_3-exposure	Mild ozone stress increased total leaf phenolics	Saviranta et al. (2010)
Vitis vinifera	Different light environments	High light induced the biosynthesis of antioxidant flavonoids in the presence or in the absence of UV	Kolb et al. (2001)
V. vinifera	UV-exclusion + ABA or water control	Quercetin and kaempferol were decreased by UV-exclusion and increased by ABA	Berli et al. (2010)
Various species: *Cistus creticus*, *Photinia x fraseri* (which accumulate anthocyanins in the mesophyll) and *Rosa* spp. and *Ricinus vulgaris* (which accumulate in the epidermis)	Methyl viologen	Evidences for the antioxidant activity of anthocyanins	Kytridis and Manetas (2006)
Various species	Cold + UV-exclusion	Flavonoids accumulation was upregulated by cold in the absence of UV-radiation	Bilger et al. (2007)

The old statement that flavonoids are UV-screening pigments that allow the penetration of photosynthetic active radiation in photosynthetic cells is of limited significance (Caldwell et al. 1983), as all phenolics display the same physical–chemical properties (Harborne and Williams 2000). Instead, the UV-B-induced biosynthesis of anthocyanins (actually flavonoids sensu stricto) is hard to be explained on the basis of their UV-screening features (Mendez et al. 1999; Manetas 2006; Kytridis and Manetas 2006), as most anthocyanins (with the exception of few acylated forms) do not appreciably absorb in the 280–390 (UV-waveband) region of the solar spectrum (Harborne and Williams 2000).

Furthermore, UV-induced biosynthesis of flavonoids is actually restricted to very few structures, i.e., the dihydroxy B-ring-substituted flavonoid glycosides, such as quercetin 3-O or luteolin 7-O-glycosides. UV-induced increase in the ratio of dihydroxy to monohydroxy B-ring substituted flavonoid glycosides, e.g., the luteolin to apigenin or quercetin to kaempferol ratios, has long been reported in different plant species exposed to various proportions of UV-radiation (Markham et al. 1998; Ryan et al. 1998, 2002; Schmitz-Hoerner and Weissenböck 2003; Kotilainen et al. 2008). Gerhardt et al. (2008) have shown that the ratio of quercetin glycosides to acylated kaempferol glycosides, which are capable to effectively absorb in both the UV-B and UV-A region of the solar spectrum, increased greatly as a consequence of UV-B irradiance. This differential accumulation of flavonoid glycosides has been explained by different authors in terms of the large variations in the free radical scavenger properties of mono with respect to dihydroxy B-ring substituted flavonoids (Rice-Evans et al. 1996; Tattini et al. 2004; Gerhardt et al. 2008; Akhtar et al. 2010).

It has been reported that glycosylation, firstly in 3-postion in the case of flavonols or in 7-position in the case of flavones (Fig. 9.1), is necessary to make these compounds soluble in the aqueous cellular milieu, in addition to preserving the most reactive groups from autooxidation (Pearse et al. 2005). The catechol group in the B-ring of the flavonoid skeleton is mostly responsible to conferring "antioxidant" capacity to glycosylated flavonoids in the concentration range that may be actually encountered in cell compartments (Rice-Evans et al. 1996; Yamasaki et al. 1997; Tattini et al. 2004; Agati et al. 2009; Agati and Tattini 2010). Luteolin and quercetin glycosides may effectively chelate Fe and Cu-ions, thus preventing the generation of ROS (Brown et al. 1998; Melidou et al. 2005), in addition to reducing a wide range of ROS, from singlet oxygen (Agati et al. 2007) to superoxide anion (Tattini et al. 2004), and the stable hydrogen peroxide (Takahama and Oniki 1997; Yamasaki et al. 1997). Tattini et al. (2004) reported that upon high solar irradiance the ratio of the antioxidant luteolin 7-O-glucoside to the poor antioxidant luteolin 4′-O-glucoside was steeply enhanced, suggesting a fine tuning exerted by high light in the flavonoid metabolism to specifically synthesize antioxidant metabolites, as also reported for the light-induced increase in chlorogenic to other mono-hydroxy hydroxycinnamates (Grace et al. 1998). Recently, Jaakola and Hohtola (2010) have reported the latitude-induced enhancement in the quercetin to kaempferol ratio in different species was for protecting plants from cold-induced oxidative damage rather than for merely absorbing UV-B-radiation. Flavonoids with the greatest antioxidant potential have been reported to accumulate in response to high solar irradiance, either in the presence or in the absence of UV-radiation (Kolb et al. 2001; Agati et al. 2009, 2011). Berli et al. (2010) have shown that in grape leaves exposed to UV-B radiation some of the key components of the antioxidant machinery, e.g., CAT, APX, and carotenoids were unaltered, while the biosynthesis of quercetin derivatives mostly upregulated. It is noted that hydroxycinnamic acid derivatives, including caffeic acid, were unaffected by UV-B irradiance. These findings once again confirm that UV-B irradiance did not enhance the biosynthesis of effective UV-B attenuators (i.e., hydroxycinnamates) but that of ROS-scavenging compounds, i.e., the flavonoids. On the whole, these findings strongly support the hypothesis of a key antioxidant function served by flavonoids in photoprotection (Agati and Tattini 2010).

It may be hypothesized that the flavonoid biosynthesis is upregulated under high-sunlight irradiance, irrespective of the proportion of various solar wavelengths reaching the leaf surface, as a consequence of changes in ROS and/or REDOX homeostasis. This hypothesis is further corroborated by the observation that R2R3MYB transcription factors that regulate the flavonol biosynthesis are themselves REDOX-controlled (Taylor and Grotewold 2005; Dubos et al. 2010).

2.2 Nitrogen-Depletion and Cold

Low nitrogen availability and cold have been reported to have a strong impact on flavonoid, particularly flavonol metabolism. Bilger et al. (2007) have shown that UV-absorbing compounds, particularly flavonoids, accumulated to a great extent as a consequence of a decrease in temperature – from 19 to 11°C – in the epidermal cells of various species. Since this increase was detected in plants exposed to light irradiance in the absence of UV-radiation, the authors suggested for flavonoids an antioxidant function in response to cold-induced photooxidative damage. Similar conclusions were also drawn by Korn et al. (2008), who observed a positive correlation between the concentration of flavonoids and the cold-tolerance of *Arabidopsis thaliana* accessions, even though these authors did not discharge a protective function of flavonoids in membrane stability (Erlejman et al. 2004). These findings are consistent with those coming from the experiments of Hannah et al. (2006), as the expression of *CHS*, *CHI*, *DFR*, *FLS1*, and *F3′H*, i.e., the whole set of genes involved in the biosynthesis of di-hydroxylated B-ring flavonols (Fig. 9.1) was induced to a substantially greater degree in cold-tolerant than in cold-sensitive accessions, and positively correlated with the biosynthesis of quercetin derivatives and anthocyanins.

Several experiments conducted by the Cathrine Lillo's group have shed new light on the impact of nitrogen depletion and low T on the biosynthesis of flavonoids, particularly flavonols. Kaempferol glycosides were less responsive to nitrogen depletion than corresponding quercetin derivatives, and the interactive effects of nitrogen and low T were significantly greater on the accumulation of quercetin as compared with the kaempferol accumulation (Olsen et al. 2009). It is interesting to note that in *Arabidopsis*, the mere depletion of nitrogen in plants growing under low light irradiance was unable to substantially affect the biosynthesis of quercetin, which was at trace level in leaves growing under normal greenhouse conditions (Lillo et al. 2008). Lillo et al. (2008) have suggested that a second factor, such as low temperature or high light, seems to be actually required to trigger the biosynthesis of the antioxidant quercetin derivatives. This suggestion have to be considered, however, with some cautions, as most experiments have been conducted under growth conditions typical of the understorey or mild-to-deep shaded environments, since light irradiance did not exceed 200 μmol quanta $m^{-2} s^{-1}$. However, results are in partial agreement with the recent suggestion reported in Jaakola and Hohtola (2010), for the effect of latitude on the flavonol biosynthesis. In detail, Albert et al. (2009) and Schmidt et al. (2010) observed that the increase in quercetin to kaempferol ratio was strongly correlated with a decrease in external temperature in *Arnica montana* and *Brassica oleracea*, respectively, under natural conditions. It has been suggested that the enhancement in the quercetin to kaempferol ratio depended more on the decrease of T than on the increase of UV-B radiation because of an increase in latitude (Jaakola and Hohtola 2010).

On the whole, cold and nitrogen deprivation affect the biosynthesis of flavonoids and lead to a preferential biosynthesis of the antioxidant quercetin derivatives with respect to that of the corresponding monohydroxy B-ring-counterparts, i.e., kaempferol glycosides. These effects are very similar to those exerted by high light irradiance, irrespective of the portions of various solar wavebands reaching the leaf surface (Lillo et al. 2008; Jaakola and Hohtola 2010). Even more, the synergic effect of cold and high light in inducing the expression of both the flavonoid and the flavonol-branch biosynthesis is similar to that found in response to nitrogen and cold treatments on the biosynthesis of quercetin derivatives (Soitamo et al. 2008; Løvdal et al. 2010).

The issue, however, deserves further investigation, as an in-depth analysis of the stress-induced effects on the regulation of flavonoid biosynthesis has not been supported with a detailed analysis on the sites of flavonoid accumulation, a prerequisite to conclusively address their "antioxidant" functions in vivo.

2.3 Drought and Salinity Stress

There are few, but interesting experiments that have detailed the impact of "osmotic stress", as drought and salinity, on the flavonoid metabolism in higher plants. Walia et al. (2005) have reported an upregulation of the flavonoid biosynthetic genes, as a consequence of NaCl stress. The increase in the expression of *F3'H*, which leads to the biosynthesis of ortho-dihydroxylated B-ring "antioxidant flavonoids", was superior in the salt-sensitive genotype than in the salt-tolerant rice genotype. The enhancement in flavonoid biosynthesis was paralleled with the increase in glutathione-S-transferase, which is involved in the transport of flavonoids to the vacuole (Zhao and Dixon 2009). The increase in carbon allocated to myricetin and quercetin glycosides, two well-known antioxidant flavonols, was significantly greater in the salt-sensitive *Myrtus communis* than in the salt-tolerant *Pistacia lentiscus* (Tattini et al. 2006), and supposed to be involved in the peroxidase-catalyzed reduction of H_2O_2, based on their large accumulation in the palisade parenchyma cells. Recently, Agati et al. (2011) have shown that root-zone salinity stress had a very similar effect on flavonoid metabolism of that exerted by UV-radiation. In both cases the biosynthesis of quercetin 3-O-glycosides was mostly upregulated, while the biosynthesis of apigenin 7-O-glycosides (mono-hydroxy B-ring flavones) remaining unaffected by mild NaCl stress. It was additionally observed that salinity stress and UV-radiation greatly enhanced the biosynthesis of luteolin 7-O-glycosides. These findings support the idea that different osmotic stressors lead to a similar plant response: the upregulation of the biosynthesis of those flavonoids, as the dihydroxy B-ring-substituted flavones, like luteolin 7-O-glycosides, which are effective antioxidants (Tattini et al. 2004). A fine tuning exerted by root-zone salinity on the flavonoid metabolic pathway was observed, as the ratio of luteolin 7-O-glucoside to the monohydroxy B-ring-substituted luteolin 4′-O-glucoside steeply increased passing from control to NaCl-treated plants. We note that luteolin 4′-O-glucoside lacking the catechol group in the B-ring of the flavonoid skeleton does not display effective free radical scavenger activity in the molar concentration range that may reasonably occur in leaf cells. Vasquez-Robinet et al. (2008) have found that the maximum flavonoid gene expression (*CHS* and *GST*) was observed during the period of more intense drought stress, which suggests for flavonoids a key protective role against water stress. The decrease of maximum carbon assimilation (A_{max}) induced by osmotic stresses exacerbates the deleterious effects of high solar irradiance, as it has been found in some Mediterranean species (Guidi et al. 2008, 2011; Melgar et al. 2009; Agati et al. 2011). In grape berries, Castellarin et al. (2007a, b) found a steep induction of the whole set of genes involved in the biosynthesis and transport of flavonoids because of water stress. These data are consistent with those of Tattini et al. (2004) in *Ligustrum vulgare* leaves, as both water stress and sunlight irradiance led to a steep enhancement in the biosynthesis of flavonoids with an orthodihydroxy B-ring substitution. Although drought stress and sunlight irradiance have been shown to synergistically enhance the expression of flavonoid biosynthetic genes, drought-stressed plants growing at 35 or 100% sunlight irradiance in the field had a significantly smaller concentration of both quercetin 3-O- and luteolin 7-O-glycosides than the well-watered counterparts. It was shown that drought stress depressed steeply the amount of "newly assimilated" or "fresh" carbon available to flavonoid biosynthesis. The matter needs to be explored in depth as, transcript or expression abundance may not directly translate in protein abundance, i.e., enzyme activity (Sweetlove and Fernie 2005), and the tissue flavonoid accumulation may additionally largely depend upon the extent to which different stresses may affect carbon gain and tissue specific concentration in nonstructural carbohydrates. Since it is the molar concentration of flavonoids in different plant tissues coupled with their intracellular distribution that supersede the potential ROS-scavenging functions in vivo, the issue merits further investigation (Turtola et al. 2005; Melgar et al. 2009; Remorini et al. 2009).

2.4 Ozone

Ozone, a secondary pollutant formed in the troposphere by the interaction of hydrocarbons, nitrogen oxides and sunlight, is a powerful oxidizing agent capable of reacting with any biomacromolecule, although it is neither a free radical species nor a reactive oxygen species, such as H_2O_2 (Mustafa 1990). It is not clear whether oxidative burst occurs and whether visible lesions are caused by ozone through programmed cell death (Sandermann 1996), but the upregulation of aromatic secondary metabolism, including the shikimate and the flavonoid pathways, has been commonly reported during O_3-stress (Janzik et al. 2005; Betz et al. 2009; Iriti and Faoro 2009).

In *Arabidopsis*, ozone induced mRNA levels of PAL within 3 h from exposure, whereas increases in the mRNA levels of antioxidant enzymes (e.g., a neutral peroxidase and a cytosolic CuZn-superoxide dismutase) were found after 12 h (Sharma and Davis 1994). Other works report an induction of PAL and glutathione S-transferase (GST, which conjugates most flavonoids and allowed transport to different cellular compartments, Agati and Tattini 2010; Zhao and Dixon 2009) transcription and activity within 2–3 h from treatment (Eckey-Kaltenbach et al. 1994; Sgarbi et al. 2003; Guidi et al. 2009), followed by a twofold increase of flavone glycoside concentration (Eckey-Kaltenbach et al. 1994). An increase of chalcone synthase (CHS) and chalcone isomerase (CHI), the enzymes involved in the first committed steps of flavonoid biosynthesis (Fig. 9.1), has been reported after ozone fumigation (Kangasjarvi et al. 1994; Paolacci et al. 2001) in a variety of plant species. Kanoun et al. (2003) detected a linear relationship between O_3-levels and the accumulation of kaempferol 3-O-glucuronide, and Betz et al. (2009) found an increase of kaempferol 3-O-glycoside in ozone-treated beeches. Higher concentration of quercetin derivates were found in O_3-treated birch growing at ambient CO_2 concentration (Peltonen et al. 2005). Interestingly, ozone both decreased total phenolics and did not display any significant effects on kaempferol biosynthesis, while steeply enhancing the biosynthesis of quercetin derivatives in leaves of *Ginkgo biloba* (He et al. 2009). Saviranta et al. (2010) have recently reported a specific induction of flavonoid biosynthesis in response to mild ozone stress, and supposed to be involved in countering O_3-induced oxidative damage. As previously reported, ozone enters the leaf through stomata, which are located, in most dicotyledonous species, on the lower side of the leaf lamina, where the density of glandular trichomes producing epicuticular flavonoids is also high (Valkama et al. 2003). Moreover, flavone aglycones have antioxidant activity (Rice-Evans et al. 1997) and can react directly with O_3 deposited on leaf surface.

2.5 Heavy Metals

Other potential abiotic stresses that may expose plants to oxidative damage have been investigated with respect to flavonoid metabolism. The functional roles of flavonoids in the response mechanisms to heavy-metal stress have received some attention. Kováĉik et al. (2009) have reported an increase in both flavonoids and caffeic acid in *Matricharia camomilla* leaves as a consequence of high Ni^{2+}-supply, whereas coumaric acid derivatives and phenolic acids did not vary. These findings led authors to hypothesize an antioxidant function of phenolics under heavy-metal stress. Ali et al. (2006) have shown an increase in flavonoid content and free radical scavenging activity in root suspension culture because of excess Cu^{2+}. Excess-Cu^{2+} ions have been reported to stimulate the biosynthesis of flavonoids, mostly luteolin-glycosides, in the absence of UV-irradiance, closely resembling the effect to exposing plants to simulated solar irradiance in leaves of *Lemna gibba* (Babu et al. 2003). These authors showed a positive correlation between the stress-induced increase in ROS concentration and the accumulation of flavonoids with the greatest antioxidant potential. It may be, therefore, postulated that the flavonol metabolism responds to alterations in ROS homeostasis and flavonoids contributed to its restoration, to maintain ROS concentration at a sublethal level. Similar conclusions were drawn by Schützendübel et al. (2001) for the role of root polyphenols in plants suffering from a severe Cd^{2+}-stress. These authors found that the accumulation of polyphenols was inversely related with the activities of key antioxidant enzymes, that declined steeply as Cd^{2+}-stress progressed, and hypothesized for polyphenols a ROS-scavenging functions to compensate for the decrease in the activity of primary antioxidant defenses. More recently, Potters et al. (2007) have suggested that flavonoids may exert their beneficial effects on Cd^{2+}-stress by affecting the movement of auxin and hence exert a tight control on the root architecture.

3 Antioxidant Flavonoids and the Antioxidant Machinery of Plants

3.1 Stress-Induced Alterations in the Antioxidant Enzymes System

There is a large consensus that a well-coordinated system of constitutive antioxidant defenses is activated in plants upon a plethora of abiotic stresses (for a recent review, see Gill and Tuteja 2010). Superoxide dismutase (SOD), the well-known first-line of defense against ROS generation (aimed at removing the highly reactive superoxide anion) (O_2^-), and both ascorbate peroxidase (APX) and catalase (CAT), the enzymes that are devoted at detoxifying the "relatively stable" H_2O_2, have long been reported to play a key role in protecting plants from stress-induced oxidative injuries (Schwanz and Polle 2001a, b; Polle 2001). The extent to which the activity of antioxidant enzymes increases upon stress imposition has been widely reported to correlate positively with tolerance to various abiotic stresses (Hernández et al. 1999, 2000, 2003; Tattini et al. 2005; Sekmen et al. 2007), although it has been also reported that the constitutive activity of antioxidant enzymes is correlated with stress tolerance (Pasqualini et al. 2001; Schwanz and Polle 2001a, b; Guidi et al. 2010). However, under prolonged stress, the activities of key components of the antioxidant machinery of stress-sensitive species have been reported to decline (Hatier and Gould 2008). For example, high doses of UV-B radiation have been reported to decrease the activity of both SOD and APX in *Ulva fasciata* (Shiu and Lee 2005). Several papers of the Andrea Polle's group have conclusively reported of a depression in the activity of antioxidant enzymes in plants exposed to the concomitant action of two or more stresses (Peltzer and Polle 2001; Polle 2001; Peltzer et al. 2002). These findings lead to the hypothesis that the stress severity, which depends on both the intensity and duration of the stress imposed, in addition to the species-specific ability to counter the stress-induced impairments of the photosynthetic machinery, may detrimentally affect the first line of defense against the generation of ROS (Schwanz and Polle 2001a, b; Schützendübel et al. 2001; Wang et al. 2007, 2008).

On the whole, these findings do not support the general view that the whole set of constitutive antioxidant defenses is activated as a consequence of stressful conditions of different origin, and poses the question of the extent to which key components of the antioxidant machinery may actually integrate to counter stress-induced ROS generation. The action of antioxidant enzymes have long been suggested to need of being be complemented by the action of other antioxidant defenses on a long-term basis, when the severity of stress increases (Apel and Hirt 2004).

3.2 Antioxidant Enzymes and Antioxidant Flavonoids: Is There a Relation?

Hatier and Gould (2008) has recently suggested that flavonoids may serve an important antioxidant function when the activity of other components of the antioxidant machinery is steeply depressed under severe conditions of excess light. Their hypothesis is based upon the observation that the very conditions that lead to enzyme inactivation are those responsible for the maximum biosynthesis of anthocyanins.

Indeed, the biosynthesis of antioxidant flavonoids is mostly upregulated under excess light stress, both in the absence and in the presence of UV-irradiance. Kolb et al. (2001) have reported a great increase in the biosynthesis of quercetin derivatives with respect to kaempferol derivatives in response to visible sunlight irradiance. Agati et al. (2009) in *L. vulgare*, a sun-sensitive species (Tattini et al. 2005), detected a sevenfold increase in the concentration of quercetin 3-O-rutinoside passing from 20 to 100% sunlight irradiance in the absence of UV-radiation. Even more, the ratio of phenylpropanoids with catechol group in the B-ring of the flavonoid skeleton (quercetin 3-O- plus luteolin 7-O-glycosides) or in the benzene ring (echinacoside, a caffeic acid derivative) to monohydroxy substituted counterparts (*p*-coumaric, apigenin 7-O-glycosides) increased as much as 360% because of an increase in PAR irradiance. Babu et al. (2003) have reported of a preferential biosynthesis of luteolin as compared with apigenin derivatives in *L. gibba* as a consequence of excess Cu^{2+} in the absence of UV-radiation. In addition to the known ROS-related signaling pathway leading to the induction of flavonoids, a recent work showed the existence of a second retrograde signaling pathway that operates during various stress situation and influences flavonoid biosynthesis (Akhtar et al. 2010). This second signaling pathway has been related to photosynthetic electron transport chain (PETC) redox state. The PETC redox signal can operate independently of ROS and override the effects of ROS on flavonoid biosynthesis (Akhtar et al. 2010).

Few experiments have been conducted to specifically address the issue of flavonoid biosynthesis as a consequence of stress-induced alteration in antioxidant enzyme activity. Xu et al. (2008) have reported a greater accumulation of antioxidant enzyme proteins in a soybean line with reduced flavonoid content. Aguilera et al. (2002) and Shiu and Lee (2005) have shown that in *U. fasciata* exposed to high UV-B doses, a condition that leads to the maximal flavonoid accumulation, the activity of antioxidant enzymes, particularly CAT and APX, declined greatly. Under severe excessive light stress, the expression of genes involved in the biosynthesis and conjugation of flavonoids were mostly upregulated, whereas the activity of SOD was unaltered (Soitamo et al. 2008). In two *Oleaceae* species differing in their ability to withstand excessive sunlight radiation (estimated in terms of chlorophyll loss and the leaf lipid peroxidation), which exhibited a constitutively different antioxidant enzyme activity, the activity of phenylalanine ammonia-lyase and the biosynthesis of antioxidant quercetin glycosides was steeply greater in the sun-sensitive than in the sun-tolerant species. Agati et al. (2011) have recently reported that the accumulation of epidermal flavonoids was completed after 2 weeks of treatment, when the activities of key antioxidant enzymes declined, as a consequence of UV-irradiance (Guidi et al. 2011).

On the whole, these findings may lead to the hypothesis that the biosynthesis of flavonoids is mostly upregulated under severe stress conditions, when the activities of antioxidant enzymes decline, and, hence, flavonoids may complement the action of other ROS-scavenging systems. Flavonoids have, therefore, to be regarded as a "secondary" antioxidant system, activated upon a severe ROS/REDOX unbalance because of the depletion of primary antioxidant defense systems. This hypothesis is supported by the observation that the greatest antioxidant flavonoid biosynthesis is correlated with the greatest oxidative damage, which conforms to the depletion of antioxidant defenses primarily, also in terms of time, aimed at detoxifying ROS. Under high solar radiation, a greater increase of di-hydroxylated flavonoids was in plants of *L. vulgare*, a shade-tolerant species, when compared to *Phillyrea latifolia*, a sun-requiring species (Tattini et al. 2005). The greater shift toward the flavonoid biosynthetic pathway detected in *L. vulgare* than in *P. latifolia* was related to the greater need of the former species to counter oxidative damage. Similarly, when *M. communis* and *P. lentiscus* were exposed to high solar radiation and root-zone salinity, the allocation of carbon to flavonoid metabolism increased more in the former than in the latter species, and appeared to be related to leaf oxidative damage (Tattini et al. 2006).

It may not be a mere coincidence that flavonols, the most ancient and widespread of flavonoids, are effective antioxidants. Excess light stress is experienced by plants not only on a seasonal, but also on a daily basis, under natural conditions (Li et al. 2009), and the flavonol metabolism has been highly conserved from the colonization of land by plants, even though the evolution of flavonoid metabolism have produced more than 10,000 structures.

The issue of how flavonoids may perform their reducing activity if confined in cellular compartments, the vacuole, far from the chloroplast, the main source of ROS still generates conflict (Hernández et al. 2009). The old view that flavonoids accumulate almost exclusively in the vacuole of epidermal cells has been recently confuted by a series of experiments in which flavonoids have been shown to largely accumulate in the vacuole of mesophyll cells, and, hence, in the proximity of ROS generation centers. Gould et al. (2002) have shown that vacuolar anthocyanins in the mesophyll may quench H_2O_2 generated upon mechanical injury. Antioxidant quercetin and luteolin glycosides have been reported to accumulate in the vacuole of mesophyll cells exposed to excess PAR irradiance, and it has been speculated that they may help reducing hydrogen peroxide. It is noted that ascorbic acid is a very poor substrate for vacuolar peroxidases that may act to reduce H_2O_2 using flavonoids as preferential substrates (Yamasaki et al. 1997). Ascorbic acid has long been reported to serve as a secondary reducing agent to recycle the flavonoid radicals to their original forms (Sakihama et al. 2000). In a recent experiment Zechmann et al. (2011) have interestingly noted that the pool of vacuolar ascorbate increased dramatically as a consequence of excess light stress, and it may be speculated to be involved in the peroxidase-catalyzed reduction of H_2O_2 using flavonoids as substrates.

Recently, Agati et al. (2007) have reported of a chloroplastic distribution of antioxidant flavonoids using three-dimensional deconvolution microscopy. These findings conform to early views of chloroplast localization of flavonoids (Oettmeier and Heupel 1972; Saunders and Mc Clure 1976; Takahama 1982), and of chloroplast being capable of flavonoid biosynthesis (Zaprometov and Nikolaeva 2003). Chloroplast quercetin and luteolin 7-O-glycosides were effective in quenching singlet oxygen generated upon excess visible light in vivo. Feucht et al. (2004) and Polster et al. (2006) have reported of the occurrence of flavonoids in the nucleus of emerging leaflets of various species, which conforms to a nuclear distribution of both chalcone synthase (CHS) and chalcone isomerase (CHI) (Saslowsky et al. 2005). Hernández et al. (2009) have suggested nuclear flavonoids being capable to chelate transition metal ions and, hence, to inhibit the generation of H_2O_2. This function is to be considered as an antioxidant function, even though there is not a reducing activity here, as it prevents the oxidative damage (Halliwell 2009).

Nevertheless, the issue of the subcellular distribution of flavonoids is far from being conclusively addressed. Flavonoids do not display fluorescence when dissolved in the aqueous cellular milieu and, hence, they need to become pseudofluorescent upon the addition of "fluorescent" probes. The largely used Naturstoff reagent (NR), 2-amino ethyl diphenyl borinic acid, has long been reported to have difficulty to enter cellular compartments, because of its acidic nature (Shehan et al. 1998), and, it is highly specific for dihydroxy B-ring-substituted flavonoid glycosides (Agati et al. 2009). By contrast, the alkalinization of flavonoid solutions with NH_3 under UV-excitation is not specific for flavonoid fluorescence (Kolb et al. 2001; Agati et al. 2002). We recall here that the fluorescence signature of NR-stained tissue under blue light excitation at 488 nm, as commonly used in confocal laser scanning microscopy (CLSM), has to be considered with some cautions.

4 "Antioxidant" Flavonoids as Developmental Regulators

The term "developmental regulators" for flavonoids has been proposed recently by Taylor and Grotewold (2005), and resemble closely that of "internal regulators" early coined by Stafford (1991). Stafford speculated that in early terrestrial plants the concentration of flavonoids had to be relatively low, and, hence, their UV-B screening functions of relatively scarce significance, as flavonoid concentrations capable to effectively attenuating UV-radiation need to be in the mM range (Edwards et al. 2008; Agati and Tattini 2010). By contrast, flavonoid concentration in the high nM to low μM range may be effective in regulating the auxin movement (Besseau et al. 2007) and quenching reactive oxygen species (Tattini et al. 2004). Recent findings of the localization of the auxin efflux facilitator protein, PIN5, to the endoplasmic reticulum (Friml and Jones 2010), the site of flavonoid biosynthesis, are of particular interest. The localization of the moss PIN proteins to the endoplasmic reticulum leads to the hypothesis that the ancestral functions of these "short" PIN proteins was likely to mediate the cellular auxin homeostasis, which further reinforces the Stafford's hypothesis.

Flavonoids are, in fact, well-known endogenous regulators of auxin movement (Jacobs and Rubery 1988; Brown et al. 2001; Peer and Murphy 2007), and, interestingly, antioxidant flavonoids display the greatest ability to regulate the transport of auxin in vivo (Taylor and Grotewold 2005; Besseau et al. 2007). Quercetin aglycone (which lacks the glycosyl moiety in the 3-postion of the flavonoid skeleton, and hence, display the greatest antioxidant potential, Rice-Evans et al. 1996) is more effective than both kaempferol aglycone and quercetin 3-O-rutinoside to inhibiting the basipetal transport of auxin (PAT, Jacobs and Rubery 1988; Besseau et al. 2007). Flavonoids may profoundly alter the tissue- and cell-specific auxin concentrations by tightly affecting the IAA-oxidation (Mathesius 2001), not just by modulating its intra- and intercellular movements. Monohydroxy B-ring flavonoids have long been shown to behave as cofactors and dihydroxy B-ring flavonoids as inhibitors of the peroxidase-mediated oxidation of auxin (Galston 1969). Jansen (2002) has provided strong evidence that the increase in quercetin to kaempferol ratio may confer UV-tolerance because of the strikingly different affinities of the two flavonols on class III peroxidases (Yamasaki et al. 1997).

Plants suffering from different abiotic stresses display a marked redistribution of growth (Baena González 2010), the so-called stress-induced morphogenic responses (SIMR) (Potters et al. 2007). This "unspecific" response of plants suffering from different stressful conditions is a part of their acclimation strategy to "flight" away from unfavorable environments (Potters et al. 2007). It has become clear that ROS-production and altered phytohormone transport and/or metabolism, that are traits common to different stresses, are involved in SIMR (for review articles, Pritzschke and Hirt 2006; Peer and Murphy 2006; Beveridge et al. 2007), thus making flavonoids as ideal candidates to greatly impact on stress-induced redistribution of growth (Thibaud-Nissen et al. 2003; Lazar and Goodman 2006). The involvement of phenolics in the tolerance mechanisms to Cd^{2+}-stress (Schützendübel et al. 2001; Potters et al. 2007) depends on their ability to both scavenging ROS (as a consequence of Cd^{2+}-induced depression in the activities of antioxidant enzymes, Schützendübel et al. 2001) and inhibiting the basipetal transport of auxin and hence the redirection of root growth (Potters et al. 2007).

It may not be a coincidence, therefore, that stress-responsive flavonoids are effective antioxidants. The chemical features that confer reducing ability against a wide array of free radicals and H_2O_2 allow flavonoids to also display the greatest affinity for proteins involved in key processes of growth and development (Peer and Murphy 2006). Flavonoids have long been reported to behave as transcript regulators in eukaryotic cells mostly through the inhibition of phosphorylation signaling cascades or specific kinases (Peer and Murphy 2006; Lamoral-Theys et al. 2010). This effect is believed to be responsible for the flavonoid-induced inhibition in the activity of PIN/MDR-PGP auxin transport proteins (Muday and DeLong 2001; Taylor and Grotewold 2005), and the catechol group in the B-ring of the flavonoid skeleton is the key feature responsible for the high affinity of flavonoids for different protein kinases (Williams et al. 2004; Lamoral-Theys et al. 2010). As a consequence, stress-responsive dihydroxy B-ring-substituted flavonoids may have a dual role on SIMR, behaving as regulators of both ROS (H_2O_2) concentration and ROS-induced MAPK signaling cascades.

5 Conclusion and Future Perspective

Flavonoids have long been shown to be involved in the response of plants to a plethora of stressful agents of both abiotic and biotic origin. This general finding is consistent with flavonoids being capable of displaying a wide range of functional roles in stressed plants. However, few flavonoid structures are capable of multiple functions, ranging from UV-screening, ROS scavenging, and inhibition of the activity of auxin efflux facilitator proteins. These flavonoids, which are the most responsive to various abiotic stresses, display the greatest potential to behave as antioxidants. The flavonol metabolism, which was at work in early terrestrial plants, has remained intact for millions of years despite the evolution of flavonoid metabolism and has produced more than 12,000 structures.

To serve such a variety of functional roles, antioxidant flavonols have to be distributed in different tissues and cellular compartments. Flavonols have been detected in the nucleus and suggested to protect DNA from damage, in the vacuole of mesophyll cells as well as in the chloroplasts, and suggested to scavenge highly reactive free radicals and the relatively stable H_2O_2. Flavonols have also been detected at the plasma membrane and hence optimally located to interfere with PIN/PGP-glycoproteins that mediate the cell-to-cell movement of auxin. Even more, flavonoids, which are synthesized in the endoplasmic reticulum, may exert a tight control on the cellular auxin homeostasis, through their interaction with ER-located "short" PIN proteins.

Nevertheless, there are still relevant issues that need to be deeply explored to address the actual relevance of stress-responsive flavonoids in an *in planta* situation. In our opinion, the relative contribution of flavonoids into the well-coordinated antioxidant machinery "activated" as a consequence of different abiotic stress is to be primarily assessed. In this regard, time-course experiments aimed not only at determining the transcript abundance or the gene expression of different antioxidant components but also at quantifying their activities or concentrations are actually required. At the same time, the stress-induced alterations on the inter- and intracellular distribution of key components of the antioxidant machinery, with special emphasis to flavonoids and ascorbic acid, should be routinely estimated as the severity of stress increases.

References

Agati G, Tattini M (2010) Multiple functional roles of flavonoids in photoprotecion. New Phytol 186:786–793

Agati G, Galardi C, Gravano E, Romani A, Tattini M (2002) Flavonoid distribution in tissues of *Phyllyrea latifolia* as estimated by microspectrofluorometry and multispectral fluorescence microimaging. Photochem Photobiol 76:350–360

Agati G, Matteini P, Goti A, Tattini M (2007) Chloroplast-located flavonoids can scavenge singlet oxygen. New Phytol 174:77–89

Agati G, Stefano G, Biricolti S, Tattini M (2009) Mesophyll distribution of antioxidant flavonoids in *Ligustrum vulgare* leaves under contrasting sunlight irradiance. Ann Bot 104:853–861

Agati G, Biricolti S, Guidi L, Ferrini F, Fini A, Tattini M (2011) The biosynthesis of flavonoids is enhanced similarly by UV radiation and root zone salinity in *L. vulgare* leaves. J Plant Physiol 168:204–212

Aguilera J, Dummermuth A, Karsten U, Schriek R, Wiencke C (2002) Enzymatic defenses against photo-oxidative stress induced by ultraviolet radiation in Arctic marine macroalgae. Polar Biol 25:432–441

Akhtar TA, Lees HA, Lampi MA, Enstone D, Brain RA, Greenberg BM (2010) Photosynthetic redox imbalance influences flavonoid biosynthesis in *Lemna gibba*. Plant Cell Environ 33:1205–1219

Albert A, Sareedenchai V, Heller W, Seidlitz HK, Zidorn C (2009) Temperature is the key to altitudinal variation of phenolics in *Arnica montana* L. c.v. ARBO. Oecologia 160:1–8

Ali RM, Singh N, Shohael AM, Hahn EJ, Paek K-Y (2006) Phenolics metabolism and lignin synthesis in root suspension cultures of *Panax ginseng* in response to copper stress. Plant Sci 17:147–154

Apel K, Hirt H (2004) Reactive oxygen species: metabolism, oxidative stress, and signal transduction. Annu Rev Plant Biol 55:373–399

Babu TS, Akhtar TA, Lampi MA, Tripuranthakam S, Dixon DG, Greenberg BM (2003) Similar stress responses are elicided by copper and ultraviolet radiation in the aquatic plant *Lemna gibba*: implication of reactive oxygen species as common signals. Plant Cell Physiol 44:1320–1329

Baena González E (2010) Energy signaling in the regulation of gene expression during stress. Mol Plant 3:300–313

Ballaré CL (2003) Stress under the sun: spotlight on ultraviolet-B responses. Plant Physiol 132:1725–1727

Berli FJ, Moreno D, Piccoli P, Hespanhol-Viana L, Silva MF, Bressan-Smith R, Cavagnaro JB, Bottini R (2010) Abscisic acid is involved in the response of grape (*Vitis vinifera* L.) cv. Malbec leaf tissues to ultraviolet-B radiation by enhancing ultraviolet-absorbing compounds, antioxidant enzymes and membrane sterols. Plant Cell Environ 33:1–10

Besseau S, Hoffmann L, Geoffroy P, Lapierre C, Pollet B, Legrand M (2007) Flavonoid accumulation in *Arabidopsis* repressed in lignin synthesis affects auxin transport and plant growth. Plant Cell 19:148–162

Betz GA, Gerstner E, Stich S, Winkler B, Welzl G, Kremmer E, Langebartels C, Heller W, Sandermann H, Ernst D (2009) Ozone affects shikimate pathway genes and secondary metabolites in saplings of European beech (*Fagus sylvatica* L.) grown under greenhouse conditions. Trees 23:539–555

Beveridge CA, Mathesius U, Rose RJ, Gresshoff PM (2007) Common regulatory themes in meristem development and whole-plant homeostasis. Curr Opin Plant Biol 10:44–51

Bilger W, Rolland M, Nybakken L (2007) UV screening in higher plants induced by low temperature in the absence of UV-B radiation. Photochem Photobiol Sci 6:190–195

Brown JE, Khodr H, Hider RC, Rice-Evans CA (1998) Structural dependence of flavonoid interactions with Cu^{2+} ions: implication for their antioxidant properties. Biochem J 330:1173–1178

Brown DE, Rashotte AM, Murphy AS, Normanly J, Tague BW, Peer WA, Taiz L, Muday GK (2001) Flavonoids act as negative regulators of auxin transport in vivo in *Arabidopsis*. Plant Physiol 126:524–535

Burchard P, Bilger W, Weissenböck G (2000) Contribution of hydroxycinnamates and flavonoids to epidermal shielding of UV-A and UV-B radiation in developing rye primary leaves as assessed by ultraviolet-induced chlorophyll fluorescence measurements. Plant Cell Environ 23:1373–1380

Caldwell MM, Robberecht R, Flint SD (1983) Internal filters: prospects for UV-acclimation in higher plants. Physiol Plant 58:445–450

Casano LM, Gómez LD, Lascano HR, Gonzáles CA, Trippi VS (1997) Inactivation and degradation of CuZn–SOD by active oxygen species in wheat chloroplasts exposed to photooxidative stress. Plant Cell Physiol 38:433–440

Castellarin S, Matthews MA, Gaspero GD, Gambetta GA (2007a) Water deficits accelerate ripening and induce changes in gene expression regulating flavonoid biosynthesis in grape berries. Planta 227:101–112

Castellarin S, Pfeiffer A, Sivilotti P, Degan M, Peterlunger E, Di Gaspero G (2007b) Transcriptional regulation of anthocyanin biosynthesis in ripening fruits of grapevine under seasonal water deficit. Plant Cell Environ 30:1381–1399

Cockell CS, Knowland J (1999) Ultraviolet radiation screening compounds. Biol Rev 74:311–345

D'Auria JC, Gershenzon J (2005) The secondary metabolism of *Arabidopsis thaliana*: growing like a weed. Curr Opin Plant Biol 8:308–316

Dubos C, Stracke R, Grotewold E, Weisshaar B, Martin C, Lepiniec L (2010) MYB transcription factors in *Arabidopsis*. Trends Plant Sci 15:573–581

Eckey-Kaltenbach HE, Ernst D, Heller W, Sandermann H (1994) Biochemical plant response to ozone (IV. Cross-induction of defensive pathways in parsley (*Petroselinum crispum* L.) plants). Plant Physiol 104:67–74

Edwards WR, Hall JA, Rowlan AR, Schneider-Barfield T, Sun TJ, Patil MA, Pierce ML, Fuclher RG, Bell AA, Essenberg M (2008) Light filtering by epidermal flavonoids during the resistant response of cotton to *Xanthomonas* protects leaf tissue from light-dependent phytoalexin toxicity. Phytochemistry 69:2320–2328

Erlejman AG, Verstraiten SV, Fraga CG, Oteiza PI (2004) The interaction of flavonoids with membranes: potential determinant of flavonoid antioxidant effects. Free Radic Res 38:1311–1320

Feucht W, Treutter D, Polstre J (2004) Flavanol binding of nuclei from tree species. Plant Cell Rep 22:430–436

Friml J, Jones AR (2010) Endoplasmic reticulum: the rising compartment in auxin biology. Plant Physiol 154: 458–462

Galston AW (1969) Flavonoids and photomorphogenesis in peas. In: Harborne JB, Swain T (eds) Perspectives in phytochemistry. Academic Press, New York, pp 193–204

Geissler M, Blakeslee JJ, Bouchard R, Lee OR, Vincenzetti V, Bandyopadhyay A, Peer WA, Bailly A, Richards EL, Edjendal KF, Smith AP, Baroux C, Grossniklaus U, Muller A, Hrycyna CA, Dudler R, Murphy AS, Martinoia E (2005) Cellular efflux of auxin by catalyzed by the *Arabidopsis* MDR/RGP transporter AtPGP1. Plant J 44:179–194

Gerhardt KE, Lampi MA, Greenberg BM (2008) The effects of far-red on plant growth and flavonoid accumulation in *Brassica napus* in the presence of ultraviolet B radiation. Photochem Photobiol 84:1445–1454

Gill SS, Tuteja N (2010) Reactive oxygen species and antioxidant machinery in abiotic stress tolerance in crop plants. Plant Physiol Biochem 48:909–930

Gould KS, McKelvie J, Markham KR (2002) Do anthocyanins function as antioxidant in leaves? Imaging of H_2O_2 in red and green leaves after mechanical injury. Plant Cell Environ 25:1261–1269

Grace SC, Logan BA, Adams WW (1998) Seasonal differences in foliar content of chlorogenic acid, a phenylpropanoid antioxidant in *Mahonia repens*. Plant Cell Environ 21:513–521

Guidi L, Degl'Innocenti E, Remorini D, Massai R, Tattini M (2008) Interaction of water stress and solar irradiance on the physiology and biochemistry of *Ligustrum vulgare*. Tree Physiol 28:873–883

Guidi L, Degl'Innocenti E, Martinelli F, Piras M (2009) Ozone effects on carbon metabolism in sensitive and insensitive *Phaseolus* cultivars. Environ Exp Bot 66: 117–125

Guidi L, Degl'Innocenti E, Giordano C, Biricolti S, Tattini M (2010) Ozone tolerance in *Phaseolus vulgaris* depends on more than one mechanism. Environ Pollut 158:3164–3171

Guidi L, Degl'Innocenti E, Remorini D, Biricolti S, Fini A, Ferrini F, Nicese FP, Tattini M (2011) The impact of UV-radiation on the physiology and biochemistry of *Ligustrum vulgare* exposed to different visible-light irradiance. Environ Exp Bot 70:88–95

Halliwell B (2009) The wanderings of a free radical. Free Radic Biol Med 46:531–542

Hannah MA, Weise D, Freund S, Fiehn O, Heyer AG, Hincha DK (2006) Natural genetic variation of freezing tolerance in *Arabidopsis*. Plant Physiol 142:98–112

Harborne JB, Williams CA (2000) Advances in flavonoid research since 1992. Phytochemistry 55:481–504

Hatier JHB, Gould KS (2008) Foliar anthocyanins as modulators of stress signals. J Theor Biol 253:625–627

He X, Huang W, Chen W, Dong T, Liu C, Chen Z, Xu S, Ruan Y (2009) Changes of main secondary metabolites in leaves of *Ginkgo biloba* in response to ozone fumigation. J Environ Sci 21:199–203

Hernández JA, Campillo A, Jimenez A, Alarcon JJ, Sevilla F (1999) Response of antioxidant systems and leaf water relations to NaCl stress in pea plants. New Phytol 141: 241–251

Hernández JA, Jimenez A, Mullineaxu P, Sevilla F (2000) Tolerance of pea (*Pisum sativum* L.) to long-term salt stress is associated with induction of antioxidant defences. Plant Cell Environ 23:853–862

Hernández JA, Aguilar AB, Portillo B, Lopez-Gomez E, Beneyto JM, Garcia Legaz MF (2003) The effect of calcium on the antioxidant enzymes from salt-treated loquat and anger plants. Funct Plant Biol 30: 1127–1137

Hernández I, Alegre L, van Breusegem F, Munné-Bosch S (2009) How relevant are flavonoids as antioxidant in plants? Trends Plant Sci 14:125–132

Hofmann RW, Campbell BD, Bloor SJ, Swinny EF, Markham KR, Ryan KG, Fountain DW (2003) Responses to UV-B radiation in *Trifolium repens* L. – physiological links to plant productivity and water availability. Plant Cell Environ 26:603–612

Iriti M, Faoro F (2009) Chemical diversity and defence metabolism: how plants cope with pathogens and ozone pollution. Int J Mol Sci 10:3371–3399

Jaakola L, Hohtola A (2010) Effect of latitude on flavonoid biosynthesis in plants. Plant Cell Environ 33:1239–1247

Jacobs M, Rubery PH (1988) Naturally occurring auxin transport regulators. Science 241:346–349

Jansen M (2002) Ultraviolet-B radiation effects on plants: induction of morphogenic responses. Physiol Plant 116:423–439

Janzik I, Preiskowski S, Kneifel H (2005) Ozone has dramatic effects on the regulation of the prechorismate pathway in tobacco. Planta 223:20–27

Kangasjarvi J, Talvinen J, Utriainen M, Karjalainen R (1994) Plant defence system induced by ozone. Plant Cell Environ 17:783–794

Kanoun M, Goulas P, Biolley JP (2003) Describing and modelling ozone-dependent variation in flavonoid content of bean (*Phaseolus vulgaris* cv. Bergamo) leaves: a particular dose–response relationship analysis. Funct Plant Biol 30:561–570

Kolb CA, Käser MA, Kopecky J, Zotz G, Riederer M, Pfündel EE (2001) Effects of natural intensities of visible and ultraviolet radiation on epidermal ultraviolet screening and photosynthesis in grape leaves. Plant Physiol 127:863–875

Korn M, Peterek S, Petermock H, Heyer AG, Hincha DK (2008) Heterosis in the freezing tolerance, and sugar and flavonoid contents of crosses between *Arabidopsis thaliana* accessions of widely varying freezing tolerance. Plant Cell Environ 31:313–327

Kotilainen T, Tegelberg R, Julkunen-Tiitto R, Lindfors A, Aphalo PJ (2008) Metabolic specific effects of solar UV-A and UV-B on alder and birch leaf phenolics. Glob Chang Biol 14:1294–1304

Kováĉik J, Klejdus B, Baĉkor M (2009) Phenolic metabolism of *Matricaria chamomilla* plants exposed to nickel. J Plant Physiol 166:1460–1464

Kytridis VP, Manetas Y (2006) Mesophyll versus epidermal anthocyanins as potential in vivo antioxidants: evidence linking the putative antioxidant role to the proximity of oxy-radical source. J Exp Bot 57:2203–2210

Lamoral-Theys D, Pottier L, Dufrasne F, Nève J, Dubois J, Kornienko A, Kiss R, Ingrassia L (2010) Natural polyphenols that display anticancer properties through inhibition of kinase activity. Curr Med Chem 17: 812–825

Lazar G, Goodman HM (2006) MAX1, a regulator of flavonoid pathway, controls vegetative bud outgrowth in *Arabidopsis*. Proc Natl Acad Sci U S A 103:472–476

Li Z, Wakao S, Fusher BB, Niyogi KK (2009) Sensing and responding to excess light stress. Annu Rev Plant Biol 50:239–260

Lillo C, Lea US, Ruoff P (2008) Nutrient depletion as a key factor for manipulating gene expression and product formation in different branches of the flavonoid pathway. Plant Cell Environ 31:587–601

Løvdal T, Olsen KM, Slimestad R, Verheul M, Lillo C (2010) Synergetic effects of nitrogen depletion, temperature, and light on the content of phenolic compounds and gene expression in leaves of tomato. Phytochemistry 71:605–613

Manetas Y (2006) Why some leaves are anthocyanic and why most anthocyanic leaves are red? Flora 201:163–177

Markham KR, Ryan KG, Bloor SJ, Mitchell KA (1998) An increase in luteolin: apigenin ratio in *Marchantia polymorpha* on UV-B enhancement. Phytochemistry 48:791–794

Mathesius U (2001) Flavonoids induced in cells undergoing nodule organogenesis in white clover are regulators of auxin breakdown by peroxidase. J Exp Bot 52:419–426

Melgar JC, Guidi L, Remorini D, Agati G, Degl'Innocenti E, Castelli S, Baratto MC, Faraloni C, Tattini M (2009) Antioxidant defences and oxidative damage in salt-treated olive plants under contrasting sunlight irradiance. Tree Physiol 29:1187–1198

Melidou M, Riganakos K, Galaris D (2005) Protection against nuclear DNA damage offered by flavonoids in cells exposed to hydrogen peroxide: the role of iron chelation. Free Radic Biol Med 39:1591–1600

Mendez M, Gwynn Jones D, Manetas Y (1999) Enhanced UV-B radiation under field conditions increases anthocyanin and reduces the risk of photoinhibition but does not affect growth in the carnivorous plant *Pinguicula vulgaris*. New Phytol 144:275–282

Mittler R (2006) Abiotic stress, the field environment and stress combination. Trends Plant Sci 11:15–19

Mittler R, Vanderauwera S, Gollery M, Van Breusegem F (2004) The reactive oxygen gene network in plants. Trends Plant Sci 9:490–498

Muday GK, DeLong A (2001) Polar auxin transport: controlling where and how much. Trends Plant Sci 6:535–542

Mustafa MG (1990) Biochemical basis of ozone toxicity. Free Radic Biol Med 9:245–265

Oettmeier W, Heupel A (1972) Identification of flavonoids and cinnamic acid derivatives from spinach chloroplast preparations. Zeitschrift für Naturforschung B 27:177–183

Ollson LC, Veit M, Bornman JF (1999) Epidermal transmittance and phenolic composition of atrazine-tolerant and atrazine-sensitive cultivars of *Brassica napus* grown under enhanced UV-B radiation. Physiol Plant 107:259–266

Olsen KM, Slimestad R, Lea US, Brede C, Løvdal T, Ruoff P, Verheul M, Lillo C (2009) Temperature and nitrogen effects on regulators and products of the flavonoid pathway: experimental and kinetic model studies. Plant Cell Environ 32:286–299

Paolacci AR, D'ovidio R, Marabottini R, Nali Lorenzini G, Abanavoli MR, Badiani M (2001) Ozone induces a differential accumulation of phenylalanine ammonia-lyase, chalcone synthase and chalcone isomerase RNA transcripts in sensitive and resistant bean cultivars. Aust J Plant Physiol 28:425–428

Pasqualini S, Batini P, Ederli L, Porceddu A, Piccioni C, De Marchis F, Antonielli F (2001) Effects of short-term ozone fumigation on tobacco plants: response of the scavenging system and expression of the glutathione reductase. Plant Cell Environ 24:245–252

Pearse IS, Heath KD, Cheeseman JM (2005) Biochemical and ecological characterization of two peroxidise isoenzymes from the mangrove, *Rhizophora mangle*. Plant Cell Environ 28:612–622

Peer WA, Murphy AS (2006) Flavonoids as signal molecules. In: Grotewold E (ed) The science of flavonoids. Springer, New York, pp 239–267

Peer WA, Murphy AS (2007) Flavonoids and auxin transport: modulators or regulators. Trends Plant Sci 12:556–563

Peltonen PA, Vapaavouri E, Julkunen-Tiitto R (2005) Accumulation of phenolic compounds in birch leaves is changed by elevated carbon dioxide and ozone. Glob Chang Biol 11:1305–1324

Peltzer D, Polle A (2001) Diurnal fluctuations of antioxidant systems in leaves of field-grown beech trees (*Fagus sylvatica*): response to light and temperature. Physiol Plant 111:158–164

Peltzer D, Dreyer E, Polle A (2002) Differential temperature dependencies of antioxidative enzymes in two contrasting species: *Fagus sylvatica* and *Coleus blumei*. Plant Physiol Biochem 40:141–150

Polle A (2001) Dissecting the superoxide dismutase–ascorbate–glutathione–pathway in chloroplasts by metabolic modeling. Computer simulations as a step towards flux analysis. Plant Physiol 126:445–462

Polster J, Dithmar H, Burgemeister R, Friedemann G, Feucht W (2006) Flavonoids in plant nuclei: detection by laser microdissection and pressure catapulting (LMPC), in vivo staining, and UV–visible spectroscopic titration. Plant Physiol 128:163–174

Potters G, Pasternak TP, Guisez Y, Palme KJ, Jansen MA (2007) Stress-induced morphogenic responses: growing out of trouble? Trends Plant Sci 12:98–105

Pritzschke A, Hirt H (2006) Mitogen-activated protein kinases and reactive oxygen species signaling in plants. Plant Physiol 141:351–356

Rausher MD (2006) The evolution of flavonoids and their genes. In: Grotewold E (ed) The science of flavonoids. Springer, New York, pp 175–211

Remorini D, Melgar JC, Guidi L, Degl'Innocenti E, Castelli S, Traversi ML, Massai R, Tattini M (2009) Interaction effects of root-zone salinity and solar irradiance on the physiology and biochemistry of *Olea europaea*. Environ Exp Bot 65:210–219

Rice-Evans CA, Miller NJ, Papanga G (1996) Structure–antioxidant activity relationships of flavonoids and phenolic acids. Free Radic Biol Med 20:933–956

Rice-Evans CA, Miller NJ, Paganga G (1997) Antioxidant properties of phenolic compounds. Trends Plant Sci 2:152–159

Ryan KG, Markham KR, Bloor SJ, Bradley JM, Mitchell KA, Jordan BR (1998) UV-B radiation induces increase in quercetin: kaempferol ratio in wild-type and transgenic lines of *Petunia*. Photochem Photobiol 68:323–330

Ryan KG, Swinny EE, Markham KR, Winefield C (2002) Flavonoid gene expression and UV photoprotection in transgenic and mutant *Petunia* leaves. Phytochemistry 59:23–32

Sakihama Y, Mano J, Sano S, Asada K, Yamasaki H (2000) Reduction of phenoxyl radicals mediated by monodehydroascorbate reductase. Biochem Biophys Res Commun 279:949–954

Sandermann H (1996) Ozone and plant health. Annu Rev Phytopathol 34:347–366

Saslowsky DE, Warek U, Winkel BSJ (2005) Nuclear localization of flavonoid enzymes in *Arabidopsis*. J Biol Chem 25:23735–23740

Saunders JA, Mc Clure JN (1976) The distribution of flavonoids in chloroplasts of twenty five species of vascular plants. Phytochemistry 15:809–810

Saviranta NMM, Julkunen-Tiitto R, Oksanen E, Karjalainen RO (2010) Leaf phenolic compounds in red clover (*Trifolium pratense* L.) induced by exposure to moderately elevated ozone. Environ Pollut 158:440–446

Schmidt S, Zietz M, Schreiner M, Rohn S, Kroh LW, Krumbein A (2010) Genotypic and climatic influences on the concentration and composition of flavonoids in kale (*Brassica oleracea var. sabellica*). Food Chem 119:1293–1299

Schmitz-Hoerner R, Weissenböck G (2003) Contribution of phenolic compounds to the UV-B screening capacity of developing barley primary leaves in relation to DNA damage and repair under elevated UV-B levels. Phytochemistry 64:243–255

Schützendübel A, Schwanz P, Teichmann T, Gross K, Langenfeld R, Heyser D, Godbold L, Polle A (2001) Cadmium induced changes in antioxidative system, hydrogen peroxide content, and differentiation in Scots pine roots. Plant Physiol 127:878–898

Schwanz P, Polle A (2001a) Differential stress response of oxidative system to drought in pedunculate oak (*Quercus robur*) and maritime pine (*Pinea pineaster*) grown under high CO_2 concentration. J Exp Bot 52:133–143

Schwanz P, Polle A (2001b) Growth under elevated CO_2 ameliorates defenses against photo-oxidative stress in poplar (*Populus alba x tremula*). Environ Exp Bot 45:43–53

Sekmen AH, Turkan I, Takio S (2007) Differential responses of antioxidative enzymes and lipid peroxidation to salt stress in salt-tolerant *Plantago maritima* and salt-sensitive *Plantago media*. Physiol Plant 131: 399–411

Sgarbi E, Baroni Fornasiero R, Lins AP, Medeghini Bonatti P (2003) Phenol metabolism is differentially affected by ozone in two cell lines from grape (*Vitis vinifera* L.) leaf. Plant Sci 165:951–957

Sharma YK, Davis KR (1994) Ozone-induced expression of stress related genes in *Arabidopsis thaliana*. Plant Physiol 105:1089–1096

Shehan JJ, Cheong H, Rechnitz GA (1998) The colorless flavonoids of *Arabidopsis thaliana* (*Brassicaceae*). I. A model system to study the orthodihydroxy structure. Am J Bot 85:467–475

Shiu CT, Lee TM (2005) Ultraviolet-B-induced oxidative stress and responses of the ascorbate–glutathione cycle in a marine macroalga *Ulva fasciata*. J Exp Bot 56:2851–2865

Soitamo A, Piippo M, Allahverdiyeva Y, Battchikova N, Aro EM (2008) Light has a specific role in modulating *Arabidopsis* gene expression at low temperature. BMC Plant Biol 8:13

Stafford HA (1991) Flavonoid evolution: an enzymic approach. Plant Physiol 96:680–685

Streb PF, Feierebend J, Bigney R (1997) Resistance to photoinhibition of photosystem II and catalase and antioxidative protection in high mountain plants. Plant Cell Environ 20:1030–1040

Swain T (1986) Plant flavonoids in biology and medicine. In: Cody V, Middleton E Jr, Harborne JB (eds) Progress in clinical and biological research 213. Liss, New York, NY, pp 1–14

Sweetlove LJ, Fernie AR (2005) Regulation of metabolic networks: understanding metabolic complexity in the systems biology era. New Phytol 168:9–24

Takahama U (1982) Suppression of carotenoids photobleaching by kaempferol in isolated chloroplasts. Plant Cell Physiol 23:859–864

Takahama U, Oniki T (1997) A peroxidase/phenolics/ascorbate system can scavenge hydrogen peroxide in plant cells. Physiol Plant 101:845–852

Tattini M, Galardi C, Pinelli P, Massai R, Remorini D, Agati G (2004) Differential accumulation of flavonoids and hydroxycinnamates in leaves of *Ligustrum vulgare* under excess light and drought stress. New Phytol 163:547–561

Tattini M, Guidi L, Morassi Bonzi L, Pinelli P, Remorini D, Degl'Innocenti E, Giordano C, Massai R, Agati G (2005) On the role of flavonoids in the integrated mechanisms of response of *Ligustrum vulgare* and *Phyllirea latifolia* to high solar radiation. New Phytol 167:457–470

Tattini M, Remorini D, Pinelli P, Agati G, Saracini E, Traversi ML, Massai R (2006) Morpho-anatomical, physiological and biochemical adjustments in response to root zone salinity stress and high solar radiation in two Mediterranean evergreen shrubs, *Myrtus communis* and *Pistacia lentiscus*. New Phytol 170:779–794

Taylor LP, Grotewold E (2005) Flavonoids as developmental regulators. Curr Opin Plant Biol 8:317–323

Thibaud-Nissen F, Shealy RT, Khanna A, Vodkin LO (2003) Clustering of microarray data reveals transcript patterns associated with somatic embryogenesis in soybean. Plant Physiol 132:118–136

Turtola S, Rousi M, Pusenius J, Yamaji K, Heiska S, Tirkkonens V, Meier B, Julkunen Tiitto R (2005) Clone-specific responses in leaf phenolics of willows exposed to enhanced UVB radiation and drought stress. Glob Chang Biol 11:1655–1663

Valkama E, Salminen JP, Koricheva J, Pihlaia H (2003) Comparative analysis of leaf trichome structure and composition of epicuticular flavonoids in Finnish birch species. Ann Bot 91:643–655

Vasquez-Robinet C, Mane SP, Ulanov AV, Watkinson JI, Stromberg VK, de Koeyer DD, Schafleitner R, Willmot DB, Bonierbale M, Bohnert HJ, Grene R (2008) Physiological and molecular adaptations to drought in Andean potato genotypes. J Exp Bot 59: 2109–2123

Walia H, Wilson C, Condamine P, Liu X, Ismail AM, Zeng L, Wamaker SI, Mandal J, Xu J, Cui X, Close TM (2005) Comparative transcriptional profiling of two contrasting rice genotypes under salinity stress during the vegetative growth stage. Plant Physiol 139:822–835

Wang R, Chen S, Deng L, Fritz E, Hüttermann A, Polle A (2007) Leaf photosynthesis, fluorescence response to salinity and the relevance to chloroplast salt compartmentation and anti-oxidative stress in two poplars. Trees 21:581–591

Wang R, Chen S, Zhou X, Shen X, Deng L, Zhu H, Shao J, Shi Y, Dai S, Fritz E, Hüttermann A, Polle A (2008) Ionic homeostasis and reactive oxygen species control in leaves and xylem sap of two poplars subjected to NaCl stress. Tree Physiol 28:947–957

Williams RJ, Spencer JPE, Rice-Evans CA (2004) Flavonoids: antioxidants or signalling molecules? Free Radic Biol Med 36:838–849

Winkel BSJ (2006) The biosynthesis of flavonoids. In: Grotewold E (ed) The science of flavonoids. Springer, New York, pp 71–95

Winkel-Shirley B (2002) Biosynthesis of flavonoids and effect of stress. Curr Opin Plant Biol 5:218–223

Xu C, Sullivan JH, Garrett WM, Caperna TJ, Natarajan S (2008) Impact of solar Ultraviolet-B on the proteome in soybean lines differing in flavonoid contents. Phytochemistry 69:38–48

Yamasaki H, Sakihama Y, Ikeara N (1997) Flavonoid-peroxidase reaction as a detoxification mechanism of plant cells against H_2O_2. Plant Physiol 115: 1405–1412

Zaprometov MN, Nikolaeva TN (2003) Chloroplast isolated from kidney bean leaves are capable of phenolic compound biosynthesis. Russ J Plant Physiol 50:623–626

Zechmann B, Stumpe M, Mauch F (2011) Immunocytochemical determination of the subcellular distribution of ascorbate in plants. Planta 233:1–12

Zhao J, Dixon RA (2009) The 'ins' and 'outs' of flavonoid transport. Trends Plant Sci 14:72–80

Zietz M, Weckmuller A, Schmidt S, Rohn S, Schreiner M, Krumbein A, Kroh LW (2010) Genotypic and climatic influence on the antioxidant activity of flavonoids in kale (*Brassica oleracea* var. *sabellica*). J Agric Food Chem 58:2123–2130

Proteomic Markers for Oxidative Stress: New Tools for Reactive Oxygen Species and Photosynthesis Research

10

Ruby Chandna, Khalid Ul Rehman Hakeem, and Parvaiz Ahmad

Abstract

Abiotic stresses are the primary causes of crop loss worldwide. Photosynthesis is an important phenomenon that is particularly affected towards reactive oxygen species, generated during any stress condition in an organism. Oxidative stresses in plants leads to debilitation and death or to response and tolerance. The sub-cellular energy organelles (chloroplast, mitochondria and peroxisomes), responsible for major metabolic processes including photosynthesis, photorespiration, oxidative phosphorylation, β-oxidation and the tricarboxylic acid cycle, are much affecting centers in a plant cell by oxidative stresses. Plant adaptation to environmental stresses is dependent upon the activation of cascades of molecular networks involved in stress perception, signal transduction, and the expression of specific stress related genes and metabolites. Progress in genomics, proteomics and metabolomics results in more understanding of global cellular responses to oxidative stress on transcript, protein, and metabolite levels. Elucidating the function of proteins expressed by genes in stress tolerant and susceptible plants would advance our understanding of plant adaptation and tolerance to environmental stresses, but also may provide important information for designing new strategies for crop improvement.

R. Chandna
Molecular Ecology Lab, Department of Botany, Jamia Hamdard, New Delhi 110062, India

National Institute of Plant Genomics and Research, Aruna Asif Ali Marg, New Delhi 110070, India

K.U.R. Hakeem (✉)
Molecular Ecology Lab, Department of Botany, Jamia Hamdard, New Delhi 110062, India
e-mail: Kur.hakeem@gmail.com

P. Ahmad
Department of Botany, A.S. College, University of Kashmir, Srinagar 190008, India
e-mail: parvaizbot@yahoo.com

P. Ahmad and M.N.V. Prasad (eds.), *Abiotic Stress Responses in Plants: Metabolism, Productivity and Sustainability*, DOI 10.1007/978-1-4614-0634-1_10, © Springer Science+Business Media, LLC 2012

Keywords

Abiotic stress • Proteomics • Plant responses • ROS • Gene expression • Transgenics

1 Introduction

Plants are exposed to different abiotic stresses, such as high temperature, cold, drought, water deficit, salinity, heavy metals and mechanical wounding under field conditions. It is estimated that the yield of crop plants can potentially reduce by more than 50% under such stress conditions. Plants encountered by various stress conditions acts differently to reduce or minimise the adverse effects caused by these unfavourable conditions. Thus, molecular, biochemical and physiological processes activated by a specific stress condition might differ from those activated by a slightly different composition of environmental parameters. Therefore, a plant builds specific response in the given environmental conditions that it encounters. Transcriptome profiling studies of plants subjected to different abiotic stress conditions illustrated that various stress condition prompts a somewhat unique response, and little overlap in transcript expression could be found between the responses of plants to abiotic stress (Friso et al. 2004). Generally with the advent of any stress condition, reactive oxygen species/intermediates (ROS/ROI) are produced. This is a family of many oxidative radicals which normally damage the cells and the whole plant. Reactive oxygen intermediates (ROIs) results from the excitation of O_2 to form singlet oxygen or from the transfer of one, two or three electrons to O_2 to form, respectively, a superoxide radical (O_2^-), hydrogen peroxide (H_2O_2) or a hydroxyl radical (HO^-) (Shulaev and Oliver 2006). ROIs are capable of unrestricted oxidation of various cellular components irrespective of atmospheric conditions and can lead to the oxidative destruction of the cell (Melgar et al. 2009). There are many potential sources of ROIs in plants (Table 10.1). NADPH oxidases, amine oxidases and cell-wall-bound peroxidases are some of the new sources of ROIs have been identified in plants recently. They are involved and in the production of ROIs during processes such as programmed cell death (PCD) and pathogen defense (Mittler 2002). In addition to the enhanced production of ROIs during stress that can pose a threat to cells, ROIs also act as signals for the activation of stress-response and defence pathways (Mittler et 2006). Thus, ROIs can be viewed as cellular indicators of stress and as secondary messengers involved in the stress-response signal transduction pathway. ROI-induced cell death can result from oxidative processes such as protein oxidation, membrane lipid peroxidation, enzyme inhibition and nucleic acid damage. ROIs participate in signaling events by at least two different mechanisms, to regulate their intracellular ROI concentrations by scavenging of ROIs: fine modulation of low levels of ROIs for signaling purposes, and second that will enable the detoxification of excess ROIs, especially during stress (Melgar et al. 2009). Therefore, ROI scavenging mechanisms can change drastically depending upon the physiological condition of the plant and the integration of different environmental, developmental and biochemical stimuli.

In plants, the reaction centres of photosystem I (PSI) and photosystem II (PSII) in chloroplast thylakoids which are the major generation site of reactive oxygen species (ROS). PSII is the catalytic centre that generates the oxygen in the atmosphere from light-driven oxidation of water (Barber 2008). Therefore, the photo production of ROS is largely affected by physiological and environmental factors and a target of many abiotic stress conditions such as heavy metals, photoinhibition, drought, ozone, high and low temperature. The photosynthesis rate is enhanced under the conditions where photon intensity (P) is in excess of that required for the CO_2 assimilation. More recently some workers have shown that pathogen, such as virus and fungi, disturb the PSII photochemistry and induce photoprotective

Table 10.1 Producing, scavenging and avoiding reactive oxygen intermediates in plants

Mechanism	Localization	Primary ROI	References
Production			
Photosynthesis ET and PSI or II	Chlorophyll	O_2^-	Asada (1999)
Respiration ET	Mitochondria	O_2^-	Dat (2000)
Glycolate oxidase	Peroxisomes	H_2O_2	Asada and Takahashi (1987)
Excited chlorophyll	Chlorophyll	O_2^-	Corpas (2001)
NADPH oxidase	Plasma membrane	O_2^-	Grant and Loake (2000)
Fatty acid β-oxidation	Peroxisomes	H_2O_2	Corpas (2001)
Peroxidases, Mn^{2+} and NADH	Cell wall	H_2O_2, O_2^-	Grant and Loake (2000)
Scavenging			
Superoxide dismutase	Chlorophyll, peroxisomes, cytochrome, apoplast	O_2^-	Bowler (1992)
Ascorbate peroxidase	Chlorophyll, peroxisomes, cytochrome, apoplast	H_2O_2	Asada (1999)
Catalase	Peroxisomes	H_2O_2	Willekens (1997)
Glutathione peroxidise	Cytosol	H_2O_2, ROOH	Dixon (1998)
Peroxidases	Cell wall, cytosol, vacuoles	H_2O_2	Asada and Takahashi (1987)
Ascorbic acid	Chlorophyll, peroxisomes	H_2O_2, O_2	Dixon (1998)
α-Tocopherol	Membranes	ROOH, O_2^1	Asada and Takahashi (1987)
Carotenoids	Chlorophyll	O_2^1	Asada and Takahashi (1987)
Avoidance			
Anatomical adaptations	Epidermis	O_2^-, H_2O_2	Mullineaux and Karpinski (2002)
C4 or CAM metabolism	Chlorophyll	O_2^-, H_2O_2	Mittler et al. (2001)
Chl movement	Chlorophyll	O_2^-, H_2O_2	Mullineaux and Karpinski (2002)
Suppression of photosynthesis	Chlorophyll	O_2^-, H_2O_2	Mullineaux and Karpinski (2002)
PS and antenna modulations	Chlorophyll	O_2^-, O_2^1	Mittler et al. (2001)
Alternative oxidases	Chlorophyll	O_2	Mullineaux and Karpinski (2002)

Mittler (2002); Trends in Plant Science

mechanisms in order to preserve the integrity of this complex in the host plant (Asada 2006). Thus, PSII is a target within the photosynthetic apparatus for both biotic and abiotic stress conditions. Under both stress situations a common response pattern occurs consisting of the enhancement of non photochemical-quenching related to energy dissipation before being trapped by the PSII reaction centre (Hideg et al. 2002).

Photosynthetic chain supplies metabolic energy and redox power for CO_2, nitrate and sulphate assimilation by plants, thus a healthy photosynthetic activity is critical for the plant growth and nutrition (Chaves et al. 2009). Under normal growth conditions active oxygen species (AOS) are formed at low rate in photosynthetic cells as by-products of aerobic metabolism, but many stresses can produce a dramatic increase in the rate of AOS production, being the chloroplasts the major AOS source in the leaf cells. An efficient removal of AOS from chloroplasts is critical since H_2O_2 concentrations as low as 10 mM can inhibit photosynthesis by 50% (Tuberosa and Salvi 2006; Sanchez et al. 2007).

Chloroplasts in higher plants contain photosynthetic thylakoid membranes with four multi-subunit protein complexes (PSI, PSII, ATP synthase, and cytochrome b6f complexes), each with multiple cofactors (Chaves et al. 2009). These four complexes are composed of at least 70 different proteins that perform the biogenesis, maintenance, and regulated breakdown of the photosynthetic complexes (Wollman et al. 1999). Abiotic and biotic stresses bring about numerous changes in the protein expression. The thylakoid membrane system also must adjust to changes in stress conditions. This requires short-term responses, such as state transitions and a build up of quenching components, whereas long-term responses bring change of PSI/PSII ratios (Aro and Andersson 2001). These regulators are typically expressed at much lower levels than components of the photosynthetic apparatus, making their biochemical identification quite challenging (Zhou et al. 2007).

Thus, the thylakoid membrane system must be protected against abiotic stresses, of which oxidative stress is one of the most prominent (Friso et al. 2004). Oxidative damage results from incomplete detoxification of ROS. To prevent and respond to oxidative stress, an antioxidative defense system is expressed in the chloroplast, consisting of proteins and scavenging molecules (Froehlich et al. 2003). Several of these proteins like thylakoid ascorbate peroxidase (Yabuta et al. 2002), three M-type thioredoxins (Issakidis-Bourguet et al. 2001), peroxiredoxins (Konig et al. 2002), and superoxide dismutases (SODs) are associated with the thylakoid membrane. These proteins are with expression levels at two to three orders of magnitude lower than the photosynthetic apparatus (Peltier et al. 2002). If protection of the thylakoid membrane system is not complete, damage to components such as proteins, lipids, and cofactors will occur, leading to irreversible changes. Cold stress or drought, combined with high light conditions, result in enhanced production of ROS by the photosynthetic apparatus because these conditions limit the availability of CO_2 for the dark (Sharkey 2005).

So, there is a need to understand the basis of stress tolerance, the diversity of the stress response and its utility for the survival of plants. As the plants sense the change in environmental conditions there is an initial perception, a signal is relayed via several signal transduction cascades (Melgar et al. 2009). Various strategies have been employed to isolate the genes that are involved in the stress response. Information about stress-responsive genes has been obtained largely using conventional approaches. However, the challenge still remains to integrate the function of these genes logically to generate a global understanding of the stress response process (Valliyodan and Nguyen 2006).

Plant cells defend against stresses by altering their expression of genes and, consequently, their proteome. Gene expression varies according to the type and severity of the stress and the developmental stage of the plant. Analysis of the proteome is a powerful tool for linking gene expression to cell metabolism. It also provides the possibility of studying individual organelles within the cell (Vinocur and Altman 2005). Proteomics, with its growing collection of technologies for extraction and identification of proteins and for studies of their interactions, permits the elucidation of the mechanisms that are involved in the responses of cells to abiotic stresses. In this review, the use of such proteome and genome-wide strategies in understanding the basis of abiotic stress tolerance is discussed.

2 Tailoring a Response to a Particular Stress Situation

Global expression profiling reveals the information of the initial repertoire of genes involved in the abiotic stress response that can be used to establish the roles of genes in abiotic stress and to identify their mechanisms of action (Lin et al. 2004). This step acts as a sieve to identify a smaller subset of genes which can be employed to draw up a final list of candidate genes using a multipronged strategy dependent on several check points, including the association with abiotic stress-related QTLs, mutant phenotypes and alleles from wild relatives (Valliyodan and Nguyen 2006). Several components of regulatory systems are activated on the perception of stress. This leads to the production of effecter molecules which are directly involved in mitigating stress. Regulatory systems are represented by transcription factors and signal transduction components such as kinases and phosphatases (Yamaguchi-Shinozaki and Shinozaki 2006).

Studies done by Liu et al. (1998) and Nakashima et al. (2000) showed that the ABA-independent regulatory cascade is mainly represented by the DREB1 (cold) and DREB2 (salt, dehydration) families of proteins. Another major abiotic stress signal transduction pathway is the ABA-dependent pathway. This pathway is mainly associated with the bZIP class of transcription factors called ABRE-binding factors, ABFs (Choi et al. 2000; Vinocur and Altman 2005). Vij and Tyagi (2007) studied the expression profiling of mutants and transgenics overexpressing transcription factors can aid in an understanding of their mechanism of action. They showed that the analysis of the CBF regulon in *Arabidopsis* revealed 306 cold-responsive genes; only 70% were part of the CBF regulon which included 15 transcription factors. ZAT12 has been identified as a new cold-responsive regulon which functions as a negative regulator of CBF2 (Vogel et al. 2005). In addition, studies with another cold-responsive regulon, i.e. ICE1, have shown that it acts as a master switch controlling many CBF-dependent and CBF-independent regulons (Lee et al. 2005).

Microarray analysis performed for transgenic *Arabidopsis* plants constitutively expressing DREB2A, a key transcription factor involved in salt and drought stress regulation showed 21 genes were upregulated by DREB2A over-expression (Sakuma et al. 2006). Such microarray analyses to decipher abiotic stress regulation have also been performed for signalling components, such as kinases and phosphatases (Umezawa et al. 2004; Osakabe et al. 2005). The significance of this approach is highlighted by the availability of several reports on contrasting stress-tolerant relatives of *Arabidopsis*, tomato, barley and maize (Bohnert et al. 2006). Similar studies performed by Vinocur and Altman (2005) to compare the gene expression patterns in desiccation-tolerant (ice plant) and non-tolerant plants have shown that the difference is in the expression pattern, irrespective of presence or absence of particular genes.

The complete genome sequence of rice and *Arabidopsis* and emerging sequence information for several other plant genomes, have given rise to the use of tools which can aid in the determination of the function of many genes simultaneously (The *Arabidopsis* Genome Initiative 2000; International Rice Genome Sequencing Project 2005). Functional genomics employs multiple parallel approaches, with the use of mutants and transgenics coupled with transcript profiling, to study gene function in a high throughput mode (Kamal et al. 2010). This change in the approach to the study of the abiotic stress response has undoubtedly reduced the time of completion of an otherwise arduous task in plants with well-established enomics platforms, such as *Arabidopsis* (Denby and Gehring 2005; Yamaguchi-Shinozaki and Shinozaki 2006).

Besides molecular approaches, proteomics, based on the recent developments of 2DE, Mass spectroscopy and bioinformatics, offers a complementary insight into protein expression and regulation within abiotic and biotic stresses in plants (Kamal et al. 2010). Proteins are physically and chemically much more diverse than nucleic acids, which hinders the quantitative analysis of complex samples of proteins. In addition, due to different RNA splicing and post-translational modifications, it is expected that for a given organism the number of protein species exceeds several folds the number of genes. Another level of complexity arises when considering the potential number of protein–protein interactions in an organism modified by developmental events and physiological constraints (Shulaev and Oliver 2006). After reviewing past and recent studies, the potential of proteomics for an integrated understanding of the processes involved will provide a basis for future proteome comparisons of biotically and abiotically challenged plants

3 Proteomics: The Analysis of Genomic Complements of Proteins

Proteomics, the systematic analysis of (differentially) expressed proteins, is a tool for the identification of proteins involved in cellular processes – has burst onto the scientific scene with stunning rapidity over the past few years, perhaps befitting a discipline that can enjoy the virtually instantaneous conversion of a genome sequence to a set of predicted proteins (Toorchi et al. 2009). Every fragment of DNA behaves biochemically much like any other, proteins possess unique properties, and such individuality creates an enormous hurdle for methodologies that seek to assign an activity to sets of proteins that may number in the thousands (Nouri et al. 2011).

Proteomics is a powerful tool for investigating the molecular mechanism/s of plant response/s towards different stress conditions, and it provides a path for increasing the efficiency of indirect selection for inherited traits (Fig. 10.1). Besides the enzymes, transport and regulatory proteins are also involved in combating any adverse condition, which makes the proteome an essential topic for studying metabolic pathways. Proteomics technology is based on high-throughput techniques for the

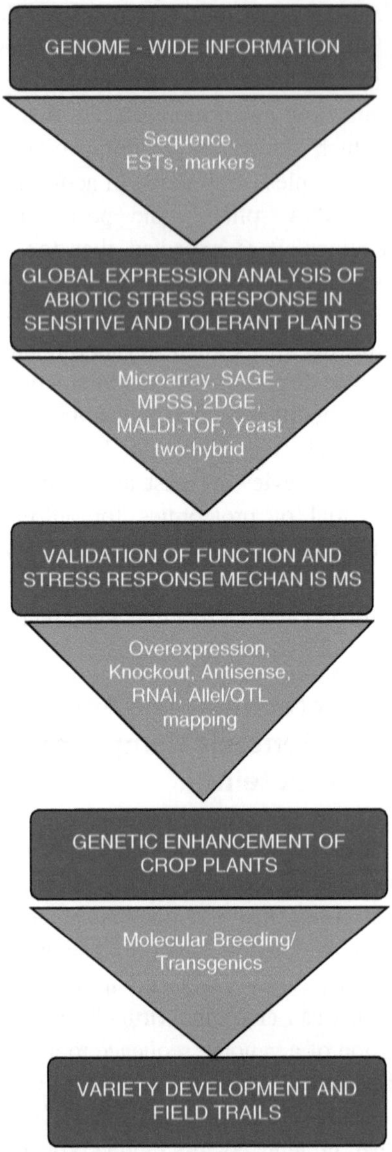

Fig. 10.1 A typical functional genomics approach to improve crop performance under abiotic stress conditions. *2GE* two-dimensional gel electrophoresis, *EST* expression sequence tag, *MALDI-TOF* matrix assisted laser desorption/ionization-time of flight, *MPSS* massively parallel signature sequencing, *QTL* quantitative trait locus, *SAGE* serial analysis of gene expression (Vij and Tyagi 2007)

separation and identification of proteins, allowing an integral study of many proteins at the same time. For the separation of protein mixtures the most powerful technique available is two-dimensional polyacrylamide gel electrophoresis (2D-PAGE): after separation, proteins can be subsequently identified by mass spectrometry (mass spectrometry) (Vij and Tyagi 2007). The increasing amount of genome sequence data has to be followed by deciphering the function of the genes and proteins. Studying differential expression by proteomics is a complementary tool for functional analysis.

Several proteomic studies have been performed on the responses of plants to various abiotic stresses, knowledge about the stresses leading to improving plant tolerance is inadequate (Table 10.1). In studies on stress, it should be noted that under natural conditions adverse environmental factors are almost never present as individual entities; on the contrary, they tend to occur together, a factor that should be considered when designing comprehensive studies (Sobhanian et al. 2010). The involvement of several signalling molecules and the activation of signalling pathways in the cells under stress conditions have been partly studied, but an understanding of the signal transduction pathways that mediate responses to stresses remains a challenge (Ahsan et al. 2010). The application of high-throughput proteomics approaches is expected to accelerate progress in our understanding of these signalling elements. Elucidation of the signalling networks through proteomics should pave the way for more rational engineering of stress tolerant soybean plants (Nouri and Komatsu 2010).

Proteomics experimental plan consists of four steps: sample preparation, protein separation, identification and function analysis (Fig. 10.2). The important features of proteome analysis are high-throughput techniques for protein separation and identification. 2D-PAGE is the method of choice for separating proteins from a complex mixture in a fast and reproducible way. This technique is able to separate thousands of proteins in single experiment. Subsequently, the separated proteins can be analysed by MS for protein identification. In order to reduce the manual handling steps and to increase the high-throughput of proteomics, combinations of capillary electrophoresis, High Performance Liquid Chromatography (HPLC) and MS are being developed (Rohde et al. 1998; Manabe 1999).

Two-dimensional (2D) electrophoresis (O'Farrell 1975), separates polypeptides according to their pI

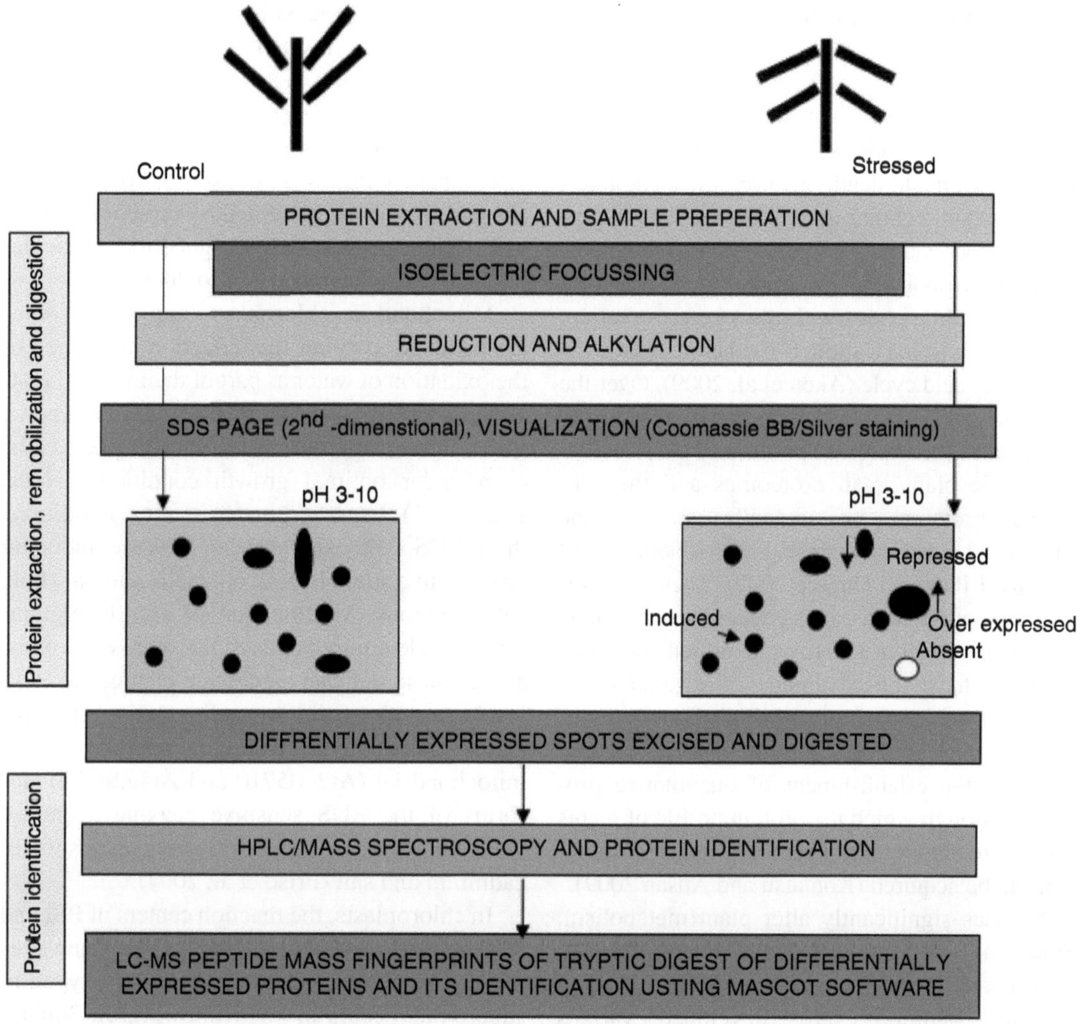

Fig. 10.2 Schematic overview of proteomics

in the first dimension (isoelectro-focalization) and to their molecular weight in the second dimension (SDS-PAGE), has been the most powerful technique for the separation of complex mixtures of proteins. Methods of identification MALDI-TOF (matrix-assisted laser desorption-ionization-time of flight) instruments allow the high-precision measurement of the masses of peptides resulting from the digestion of a protein by an endoprotease. The masses of several peptides can be compared with those predicted from sequences in databases. This technique of fingerprinting is particularly applicable for proteins from organisms with genomes or cDNAs entirely or largely sequenced. The peptides isolated after proteolytic cleavage are fragmented by collision in the MS instrument, and the masses of the fragments are measured. They can be compared to the masses expected from database-stored sequences, and sequence stretches can be deduced from the mass differences between fragments ranked in order of size. Although ESI-MS/MS is not yet a high-throughput technique, it is more appropriate than MALDI-TOF for the identification of proteins from organisms in which few cDNAs have been sequenced: micro-sequences allow cross-species identification, while sequence polymorphism limits the efficiency of MALDI-TOF fingerprinting (Nouri and Komatsu 2010).

3.1 Sub-cellular Proteome Analysis of Plants

An adverse abiotic environmental condition is cause of severe oxidative stress in plants, leading to debilitation and death or to response and tolerance. The sub-cellular energy organelles (chloroplast, mitochondria and peroxisomes) in plants those are responsible for major metabolic processes including photosynthesis, photorespiration, oxidative phosphorylation, β-oxidation and the tricarboxylic acid cycle (Aken et al. 2009). Over the past few years a number of studies have begun to characterize the expressed proteins of plant including whole plant/organ proteomes and the sub-cellular proteomes of organelles such as the chloroplast mitochondrion, peroxisome and nucleus (Lilley and Dupree 2007). These sub-cellular proteomic studies are a greater dynamic range of proteins are often able to be identified since they generally have the advantage of assessing less complex protein subsets than whole tissue extracts. This range of proteomics studies has provided a basis for the establishment of quantitative proteomic through which the protein profile of plants exposed to abiotic stress, pathogen attack or mutation can be acquired (Komatsu and Ahsan 2009).

Stresses significantly alter plant metabolism, growth and development and ultimately lead to plant death. Plants in order to survive in these adverse conditions have developed various responses. The signal transduction pathways that elicit these responses, or the way in which plants perceive these environmental stresses, are not well understood. Proteomics is already proving to be a valuable tool in the functional characterization of plants to interpret the stress response of plants (Hongsthong et al. 2009). A number of proteomics studies have assessed the affect of environmental stress on organelle proteomes. Each organelle of a plant cell generally has its own specific functions in addition to communicating with other parts of the cell. A study of an enriched fraction of a desired organelle offers many advantages in proteome research. The analysis of proteome of organelles and sub-cellular fractions is one of the most informative approaches to the functional analysis of living cells. The study of cell organelles is usually subject to two major constraints: the availability of an appropriate purification technique, and the verification of the purity of extract (Nouri et al. 2011). The evaluation of the purity or the degree of contamination of organelle extracts is necessary step; otherwise any novel proteins identified by proteomics analyses cannot be definitely assigned to a particular organelle (Komatsu and Ahsan 2009).

Mitochondria, chloroplasts and peroxisomes are involved in either the reduction of oxygen or the oxidation of water as part of their normal metabolic activity. Because of this, these are known to be significant sources of ROS in plant cells, even under optimal growth conditions (Nouri et al. 2011). However, under extreme conditions these ROS synthesis rates can increase, inducing an oxidative stress in both organelle and on whole cell functions. Modification of ascorbate/glutathione cycle components, SODs, peroxiredoxins, and catalase indicate significant changes in ROS levels in each of the organelles during environmental stress. Changes in abundances of the mitochondrial (At2g05710 and At4g26970) isoforms of the ROS sensitive enzyme aconitase were observed to occur following exposure to cadmium and salt (Friso et al. 2004).

In chloroplasts, the reaction centers of PSI and PSII in the thylakoid membrane are the main sites of ROS generation. Photoreduction of oxygen to superoxide occurs in PSI (following the Mehler reaction), in PSII, oxygen of the ground (triplet) state of oxygen ($^3O_2$) is excited to the excited singlet state of oxygen ($^1O_2$) by the P680 reaction center chlorophyll (Chl) (Bae et al. 2003). The photo production of ROS is affected by physiological and environmental factors. The rate is enhanced under the conditions where photon intensity is in excess of that required for the CO_2 assimilation. During condition of photon excess, a variety of systems suppress ROS production in chloroplasts including photorespiration, the cyclic electron flow through PSI or PSII, and the down-regulation of PSII quantum yield by the xanthophyll cycle and the proton gradient across the thylakoid membrane (Kamal et al. 2010). Thus the chloroplast has the potential to be a variable source of ROS during environmental stress, and much of the damage to chloroplast proteins during stress may be linked to its changed rate of ROS synthesis (Taylor et al. 2009).

Superoxide produced in mitochondria by peripheral single electron transfers from reduced components in the respiratory electron transport chain (ETC) to oxygen, the ubiquinone pool and components in complex I and III have been implicated in mitochondrial superoxide production (Aken et al. 2009). It is observed that 3–5% of oxygen consumption by mitochondria is due to single electron superoxide formation, while the majority of oxygen consumption is four electron reduction of oxygen to water (Xu et al. 2006). The rate of superoxide production by mitochondria depends on the redox poise of ETC components, and on the concentration of oxygen. During hypoxic conditions, ROS production by mitochondria is low, and is elevated when respiration inhibitors block the ETC and cause over-reduction of earlier components, and can be altered by environmental factors of chemicals that alter the rate of these peripheral electron transfer reactions. Notably, nitric oxide is a potent inhibitor of the mitochondrial ETC and its generation during plant stress may be critical in the elevation of ROS production from mitochondria in plants, just as it is in animals (Taylor et al. 2009).

Peroxisome is a house of variety of oxidases that reduce oxygen to hydrogen peroxide and concomitantly oxidize a variety of compounds; these enzymes include urate oxidase, glycolate oxidase, xanthine oxidase and acyl-CoA oxidases. Peroxisome ROS formation is a stoichiometric and linear consequence of the rate of the metabolic pathways containing hydrogen peroxide-producing enzymes. Acyl-CoA oxidases operate in β-oxidation of fatty acids and hormones which will vary depend on the stage of plant development, while glycolate oxidase operates in the photorespiration pathway which depends on light, the photosynthetic rate, and the CO_2 concentration in the tissue (Agrawal et al. 2010). Studies done by Hoa et al. (2004), on soybean organelles performed for mitochondrial fractions from roots and nodules, peribacteroid membranes (Panter et al. 2000), and etiolated cotyledon peroxisomes (Arai et al. 2008). In relation to abiotic stress responsive proteins in plant organelles and subcellular fractions, the effects of osmotic stress on the plasma membrane, of flooding on the cell wall and plasma membrane, and of ozone on chloroplasts have also been focussed.

In proteome analysis of sub-cellular organelles, the purification of protein extracts and the verification of their purity determine the validity of the results. Plasma membrane extracts were purified by using a two-phase partitioning method and their purity was verified by measurement of the activity of P-type ATPase (Komatsu et al. 2009; Nouri and Komatsu 2010). A cell wall fraction was obtained by using calcium chloride, and its purity was confirmed by assaying the activity of glucose-6-phosphate dehydrogenase (Komatsu et al. 2010). Chloroplasts from soybean leaves were purified by using a Percoll gradient, and their purity was assayed by immunoblot analysis using specific antibodies (Ahsan et al. 2010). Although techniques for organelle proteomics continue to be improved, achieving to a pure fraction free of contaminants from other parts of the cell remains a challenging problem. This aspect is particularly important under stress conditions where several proteins, such those related to quality control, defense, or metabolism, can migrate through secretary pathways in the cell to cope with the imposed stress. In such cases, protein localization may be capable of verifying the existence of a given protein in a specific fraction (Lilley and Dupree 2007). Completion of annotation of the important crop plants genome and the corresponding database information should result in further improvements and the analysis of the proteome of subcellular organelles.

3.2 Proteomics to Identify Targets Beyond the Gene

Analysis of the proteome provides a direct link of genome sequence with biological activity (Pandey and Mann 2000; Agrawal and Rakwal 2006). It helps us to understand analysis of the proteome including the knowledge of entire protein repertoire as well as the expression levels, post-translational modifications and interactions, to understand the cellular processes at the protein level (Peck 2005). The combination of (MS) with two-dimensional gel electrophoresis (2DGE) in the 1990s proved to be useful for proteome analysis. Matrix-assisted laser desorption/ionization-time of flight (MALDI-TOF) and electrospray ionization (ESI) are the two most commonly used MS techniques (Mann and Pandey 2001). Bae et al. (2003) used 2DGE and MALDI-TOF-MS to study the *Arabidopsis* nuclear proteome and changes in the nuclear proteome in response to cold stress. One hundred and eighty-four protein spots were identified, with the expression of almost 30% of these proteins altered in response to cold stress. These included several proteins previously reported to be involved in stress, including heat-shock proteins, transcription factors (AtMYB2 and OBF4), DNA-binding proteins (DRT102 and Dr1), catalytic enzymes (phosphoglycerate kinase, serine acetyltransferase and glyceraldehyde-3-phosphate dehydrogenase), syntaxin, calmodulin and germin like proteins. Salekdeh et al. (2002) studied the proteome analysis of rice and the changes in response to drought stress. More than 1,000 protein spots were identified on 2DGE, with the expression of 42 proteins altered by stress. The identities of 16 of these drought-responsive proteins were established by MS.

Similarly, studies performed by Cui et al. (2005) on rice cold stress proteome, proteins from unstressed seedlings were compared with those from seedlings exposed to temperatures of 15, 10 and 5°C. Of a total of 1,700 protein spots separated by 2DGE, 60 proteins were up-regulated with a decrease in temperature. The identities of 41 of these proteins were established by MALDITOF-MS or ESI/MS/MS, and these mainly included chaperones, proteases, detoxifying enzymes, and enzymes linked to cell wall biosynthesis, energy pathways and signal transduction. The results obtained emphasize the importance of maintaining protein quality control via chaperones and proteases, together with an increase in cell wall components, during the cold stress response. The availability of a rice proteome database (http://gene64.dna.affrc.go.jp/RPD) which catalogues information from 23 reference maps of 2DGE analysis of proteins from diverse biological samples. The database contains, in total, 13,129 identified proteins and the amino acid sequences of 5,092 proteins (Komatsu 2005).

Salt stress-responsive proteins were also studied in the rice root proteome. Around 54 proteins, whose expression changed in response to salt stress, were obtained. Twelve among them were identified by MS, 50% of which were novel salt-responsive proteins, such as UGPase, Cox6b-1, GS root isozyme, α-NAC, putative splicing factor-like protein and putative ABP (Yan et al. 2005). In the study of tobacco leaf apoplast proteome in response to salt stress, 20 proteins were identified whose expression has changed in response to stress. These included several well-known stress associated proteins, together with chitinases, germin-like protein and lipid transfer proteins (Dani et al. 2005).

In addition to this, a large number of other interactions were established for proteins, involved in different stress conditions and, on many occasions, this interaction network also overlapped with proteins involved in development, showing the complexity of the stress response using proteome analysis studies (Vij and Tyagi 2007; Kamal et al. 2010). A large number of bioinformatics tools are available for plant proteome analysis. These include the Proteins of *Arabidopsis thaliana* Database (PAT) (http://www.pat.sdsc.edu/), MIPS *Arabidopsis thaliana* Database (MAtDB) (http://mips.gsf.de/proj/thal/db) and Rice Proteome Database (RPD) (http://gene64.dna.affrc.go.jp/RPD/main_en.html).

4 Towards Making Plants Abiotic Stress Tolerant

It must be kept in mind that the basic task of the identification of key gene(s), whose manipulation will ultimately affect crop performance in response to abiotic stress, is highly complex and difficult to decipher because of the polygenic nature of the abiotic response. In addition, the plant's response to each stress is unique, and thus the response to multiple stresses will also be different. Indeed, global expression profiling of a plant's response to abiotic stress conditions has shown that, although overlap may occur for different abiotic stresses, such as cold, salt, dehydration, heat, high light and mechanical stress, a set of genes unique to each stress response is also seen. Even in the case of ROS, which are known to play a central role in both biotic and abiotic stress conditions, it has been shown that different genes of the ROS network respond differently to different stress treatments (Mittler 2006; Noctor 2006; Melgar et al. 2009). However, most of the studies carried out to investigate the performance of plants under abiotic stress conditions have not focused on this aspect, making it an important area of concern, especially as it is known that plants are exposed to multiple environmental stresses in the field. Further, the response to abiotic stress is also developmentally regulated (Vinocur and Altman 2005; Nouri et al. 2011). It has been observed, in plant species such as rice, wheat, tomato, barley and corn, salt tolerance increases with an increase in plant age. Also it has been studied that QTLs (quantitative trait loci) associated with salt tolerance in the germination stage in barley, tomato and *Arabidopsis* are different from QTLs associated with the early stage of growth. In transgenic studies on crop plants such as rice, the majority have not evaluated the effect of stress on grain yield. This aspect becomes especially important in plants such as rice, in which stresses such as salinity do not affect the vegetative growth as much as the grain yields (Yamaguchi and Blumwald 2005). It is apparent that an understanding of the abiotic stress responsive network will require a considerable amount of time and resources, but a systematic and concerted effort will ensure that only the most suitable genes are identified for crop improvement.

The task can be shortened by integrating the information already available and by avoiding the repetition of effort or branching away from the main focus. The work performed by already existing list of candidate genes and their alleles identified through this approach with a handful of positive results, but there is still a long way to go.

5 Stress-Associated Genes and Proteins Expression

With the availability of the *Arabidopsis* and rice genome sequences, together with improvements in the methods used to detect DNA polymorphisms, has made map-based cloning a viable option for global functional genomics (Friso et al. 2004). There are 37,344 single nucleotide polymorphisms (SNPs), 18,579 insertions/ deletions (InDels) and 747 large InDels available at TAIR website (The *Arabidopsis* Information Resource website). A large number of markers are also available for rice, including 3,267 markers released in 2000, 332 PCR-based genetic markers released in 2002 (rgp.dna.affrc.go.jp) and over 18,000 simple sequence repeats (SSRs) (International Rice Genome Sequencing Project 2005). A large number of genes in the abiotic stress response pathway have been identified using the map-based cloning approach, including *SOS1*, *SOS2*, *SOS3*, *SOS4*, *SOS5*, *HOS1*, *SpI7*, *STT3*, *FRO1*, *LOS5/ABA3* and *AtCesA8/IRX1* (Liu and Zhu 1998; Liu et al. 2000; Shi et al. 2000, 2002, 2003; Lee et al. 2001, 2002; Xiong et al. 2001; Yamanouchi et al. 2002; Koiwa et al. 2003; Chen et al. 2005).

Targeting induced local lesions in genomes (TILLING) is another high-throughput technology helpful in identifying mutations in a selected gene or a variant allele (Henikoff and Comai 2003). TILLING projects are underway for plant species such as *Arabidopsis*, lotus, maize, wheat and *Brassica* (Gilchrist and Haughn 2005). In rice, mutants generated in the IR64 background using EMS and diepoxybutane have been used to generate DNA pools for TILLING, and were screened for mutations in selected genes (Wu et al. 2005). With the advantage of amenability to high throughput, together with the capability of generating an allelic series at a given locus, TILLING will also find application in the identification of useful alleles for abiotic stress tolerance.

QTLs are specific genetic loci in the genome associated with a particular trait. Stress tolerance is a complex trait, and dissection of its QTLs would be of immense value in understanding the stress response and would be useful for plant breeders (Gorantla et al. 2005). Several QTLs involved in the stress response have been reported recently (Xu et al. 2005; Tuberosa and Salvi 2006). Most of the plant QTLs cloned to date have been obtained using a map-based cloning strategy. Lin et al. (2004) mapped eight QTLs responsible for variation in K^+ or Na^+ content from an F2 population derived from a cross between a salt-tolerant *indica* variety (Nona Bokra) and a susceptible *japonica* variety (Koshihikari) Of these, *SKC1*, a major QTL for shoot K+ content, was mapped to chromosome 1 (Ren et al. 2005). A map-based approach was also demonstrated to be useful in the identification of an important submergence tolerance QTL present on chromosome 9 of rice. Three genes belonging to the ethylene response-factor (ERF) family were identified on this locus, designated as *Sub1*, the variation of one of which, *Sub1A*, results in tolerance/susceptibility to submergence. *Sub1* from the submergence-tolerant variety was introgressed in flooding susceptible local rice varieties. The new varieties showed submergence tolerance without compromising on yield or other agronomic traits, demonstrating the efficacy of this locus (Xu et al. 2006). The expression profile of genes in a QTL interval associated with the abiotic stress response is also being used to identify target genes. Gorantla et al. (2005) used information from ESTs sequenced from drought-stressed cDNA libraries to generate a transcript map of rice.

Silencing of a target gene and targeted gene inactivation have also been performed by homologous recombination using a transgenic approach with respect to understanding abiotic stress response in plants (An et al. 2005; Iida and Terada 2005). Gene inactivation using anti-sense, co-suppression or RNAi strategies is largely based on a single gene approach and, at present, it is difficult to use these methods at the genome-wide level (Parinov and Sundaresan 2000). A major part of our present understanding of the plant stress responsive network comes from the functional analysis of *Arabidopsis* genes in transgenic systems. A number of transgenic studies have been performed in tobacco in this respect, and interestingly, the majority of such studies have aimed to decipher the function of genes encoding downstream components (effectors), such as those coding for antiporters, heat-shock proteins, SODs and LEA proteins, rather than upstream components (regulators), such as those coding for transcription factors and kinases.

The functions of genes representing QTLs for abiotic stress have also been confirmed by employing transgenics (Xu et al. 2006). Most of the stress-responsive genes have *cis*-acting conserved elements in their promoter region involved in regulating the stress response. A transgenic approach has also been used to dissect the role of stress-responsive promoters (Yamaguchi-Shinozaki and Shinozaki 2005). The major *cis*-acting elements present in several abiotic stress-responsive promoters include the ABA responsive element (ABRE) from the promoter of ABA-responsive genes (Ingram and Bartels 1996; Grover et al. 2001), dehydration responsive element (DRE) from the promoter of cold- and drought-inducible genes (Yamaguchi-Shinozaki and Shinozaki 1994; Thomashow 1999), anaerobic response element (ARE) from the promoter of genes responsive to low oxygen conditions, such as maize *Adh1*, *LDH1* and *PDC1* (Dolferus et al. 1994; Dennis et al. 2000), and heat-shock element (HSE) from the promoter of heat stress-inducible genes (Schoffl et al. 1998).

With so many positive implications of the transgenic approach, still it is not useful for large-scale functional analysis. Therefore, a possible way of employing transgenics for functional analysis would be the use of plant artificial chromosomes (PLACs) containing large-sized genomic fragments. This would significantly reduce the number of transgenic plants required for functional analysis (Somerville and Somerville 1999).

6 Conclusion and Future Prospects

Development of crop plants, tolerant to environmental stresses, is considered a promising approach, which may persuade the growing food demands of the developing and under-developed countries. Development of crop plants with stress tolerance, however, requires, knowledge of the physiological mechanisms and genetic controls of the contributing traits at different plant developmental stages. While, we are at a beginning to gain a skeletal understanding of the steady state changes occurring in plant proteomes during a number of environmental stresses. With proteome analysis we have been able to define the selectivity and functional category of these changes but we are still away from having real insight into the functional significance of many of the changes observed on plant cell stress susceptibility or tolerance. It is also important to mention that the effect of stresses on proteomes depends on the level of tolerance of the plants and thus, sensing and signalling processes can be a primary response of significance. More work will be required to demonstrate if temporal patterns in proteome stress responses exist. Further, even if temporal separation of events exists, it is still not clear how to mechanistically represent the cascade of chemicals events involving ROS, metals and lipid peroxidation with the responses in protein abundance and stability in a unified framework that allows sophisticated engineering of stress tolerance to commence. Also, there is still much work to be done on spatial resolution of stress responses and the impact on the proteome. Future proteomic studies offer the promise of new data to provide better temporal and spatial resolution of the stress response and recognition of damaged or unfolded proteins and the selectivity of metal induced oxidative modifications of proteins.

References

Agrawal GK, Rakwal R (2006) Rice proteomics, a cornerstone for cereal food crop proteomes. Mass Spectrom Rev 25:1–53

Agrawal GK, Bourguignon J, Rolland J, Ephritikhine G, Ferro M, Jaquinod M, Alexiou KG, Chardot T, Chakraborty N, Jolivet P, Doonan JH, Rakwal R (2010) Plant organelle proteomics: collaborating or optimal cell function. Mass Spectro Rev 30:772–853

Ahsan N, Donnart T, Nouri MZ, Komatsu S (2010) Tissue-specific defense and proteomic approach. J Proteome Res 9:4189–4204

Aken OV, Zhang B, Carrie C, Uggalla V, Paynter E, Giraud E, Whelan J (2009) Defining the mitochondrial stress response in Arabidopsis thaliana. Mol Plant 2:1310–1324

An G, Jeong DH, Jung KH, Lee S (2005) Reverse genetic approaches for functional genomics of rice. Plant Mol Biol 59:111–123

Arai Y, Hayashi M, Nishimura M (2008) Proteomic analysis of highly purified peroxisomes from etiolated soybean cotyledons. Plant Cell Physiol 49:526–539

Aro EM, Andersson B (2001) Regulation of photosynthesis. Kluwer Academic, Dordrecht, The Netherlands

Asada K (1999) The water–water cycle in chloroplasts: scavenging of active oxygen and dissipation of excess photons. Annu Rev Plant Physiol Plant Mol Biol 50: 601–639

Asada K (2006) Production and scavenging of reactive oxygen species in chloroplasts and their functions. Plant Physiol 141:391–396

Asada K, Takahashi M (1987) Production and scavenging of active oxygen in photosynthesis. In: Kyle DJ et al (eds) Photoinhibition. Elsevier, Amsterdam, pp 227–287

Bae MS, Cho EJ, Choi EY, Park OK (2003) Analysis of the Arabidopsis nuclear proteome and its response to cold stress. Plant J 36:652–663

Barber J (2008) Photosynthetic generation of oxygen. Philos Trans R Soc Lond B Biol Sci 363:2665–2674

Bohnert HJ, Gong Q, Li P, Ma S (2006) Unraveling abiotic stress tolerance mechanisms – getting genomics going. Curr Opin Plant Biol 9:180–188

Bowler C (1992) Superoxide dismutases and stress tolerance. Annu Rev Plant Physiol Plant Mol Biol 43: 83–116

Chaves MM, Flexas J, Pinheiro C (2009) Photosynthesis under drought and salt stress: regulation mechanisms from whole plant to cell. Ann Bot 103:551–560

Chen M, Hillier RG, Howe CJ, Larkum AW (2005) Unique origin and lateral transfer of prokaryotics chlorophyll-b and chlorophyll-d light-harvesting systems. Mol Biol Evol 22:21–28

Choi H, Hong J, Ha J, Kang J, Kim SY (2000) ABFs, a family of ABA-responsive element binding factors. J Biol Chem 275:1723–1730

Corpas FJ (2001) Peroxisomes as a source of reactive oxygen species and nitric oxide signal molecules in plant cells. Trends Plant Sci 6:145–150

Cui S, Huang F, Wang J, Ma X, Cheng Y, Liu J (2005) A proteomic analysis of cold stress responses in rice seedlings. Proteomics 5:3162–3172

Dani V, Simon WJ, Duranti M, Croy RR (2005) Changes in the tobacco leaf apoplast proteome in response to salt stress. Proteomics 5:737–745

Dat J (2000) Dual action of the active oxygen species during plant stress responses. Cell Mol Life Sci 57:779–795

Denby K, Gehring C (2005) Engineering drought and salinity tolerance in plants: lessons from genome-wide expression profiling in Arabidopsis. Trends Biotechnol 23:547–552

Dennis ES, Dolferus R, Ellis M, Rahman M, Wu Y, Hoeren FU, Grover A, Ismond KP, Good AG, Peacock WJ (2000) Molecular strategies for improving waterlogging tolerance in plants. J Exp Bot 51:89–97

Dixon DP (1998) Glutathione-mediated detoxification systems in plants. Curr Opin Plant Biol 1:258–266

Dolferus R, Jacobs M, Peacock WJ, Dennis ES (1994) Differential interactions of promoter elements in stress responses of the Arabidopsis Adh gene. Plant Physiol 105:1075–1087

Friso G, Giacomelli L, Ytterberg AJ, Peltier JB, Rudella A, Sun Q et al (2004) In-depth analysis of the thylakoid membrane proteome of Arabidopsis thaliana chloroplasts: new proteins, new functions, and a plastid proteome database. Plant Cell 16:478–499

Froehlich JE, Wilkerson CG, Ray WK, McAndrew RS, Osteryoung KW, Gage DA, Phinney BS (2003) Proteomic study of the Arabidopsis thaliana chloroplastic envelope membrane utilizing alternatives to traditional two-dimensional electrophoresis. J Proteome Res 2:413–425

Gilchrist EJ, Haughn GW (2005) TILLING without a plough: a new method with applications for reverse genetics. Curr Opin Plant Biol 8:211–215

Gorantla M, Babu PR, Lachagari VBR, Feltus FA, Paterson A, Reddy AR (2005) Functional genomics of drought stress response in rice: transcript mapping of annotated unigenes of an indica rice (Oryza sativa L. cv. Nagina 22). Curr Sci 89:496–514

Grant JJ, Loake GJ (2000) Role of reactive oxygen intermediates and cognate redox signalling in disease resistance. Plant Physiol 124:21–29

Grover A, Kapoor A, Satya Lakshmi O, Agarwal S, Sahi C, Katiyar-Agarwal S, Agarwal M, Dubey H (2001) Understanding molecular alphabets of the plant abiotic stress response. Curr Sci 80:206–216

Henikoff S, Comai L (2003) Single-nucleotide mutations for plant functional genomics. Annu Rev Plant Biol 54:375–401

Hideg E, Barta C, Kalai T, Vass M, Hideg K, Asada K (2002) Detection of singlet oxygen and superoxide with fluorescence sensors in leaves under stress by photoinhibition or UV radiation. Plant Cell Physiol 43:1154–1164

Hoa LT, Nomura M , Kajiwara H, Day DA, Tajima S (2004) Proteomic analysis on symbiotic differentiation of mitochondria in soybean nodules. Plant Cell Physiol 45:300–308

Hongsthong A, Sirijuntarut M, Yutthanasirikul R, Senachak J, Kurdrid P, Cheevadhanarak S, Tanticharoen M (2009) Subcellular proteomic characterization of the high-temperature stress response of the cyanobacterium Spirulina platensis. Proteome Sci 7:33–42

Iida S, Terada R (2005) Modification of endogenous natural genes by gene targeting in rice and other higher plants. Plant Mol Biol 59:205–219

Ingram J, Bartels D (1996) The molecular basis of dehydration tolerance in plants. Annu Rev Plant Biol 47: 377–403

International Rice Genome Sequencing Project (2005) The map-based sequence of the rice genome. Nature 436:793–800

Issakidis-Bourguet E, Mouaheb N, Meyer Y, Miginiac-Maslow M (2001) Heterologous complementation of yeast reveals a new putative function for chloroplast m-type thioredoxin. Plant J 25:127–135

Kamal AHM, Kim KH, Shin KH, Shin DH, Seo HS, Park DS, Woo SH, Lee MS, Chung KM (2010) Functional proteome analysis of wheat: systematic classification of abiotic stress responsive proteins. World Congress of Soil Science, Soil Solutions for a Changing World

Koiwa H, Li F, McCully MG, Mendoza I, Koizumi N, Manabe Y, Nakagawa Y, Zhu J, Rus A, Pardo JM, Bressan RA, Hasegawa PM (2003) The STT3a subunit isoform of the Arabidopsis oligosaccharyltransferase controls adaptive responses to salt/osmotic stress. Plant Cell 15:2273–2284

Komatsu S (2005) Rice proteome database: a step toward functional analysis of the rice genome. Plant Mol Biol 59:179–190

Komatsu S, Ahsan N (2009) Soybean proteomics and its application to functional analysis. J Proteomics 72: 325–336

Komatsu S, Wada T, Abalea Y, Nouri MZ, Nanjo Y, Nakayama N, Shimamura S, Yamamoto R, Nakamura T, Furukawa K (2009) Analysis of plasma membrane proteome in soybean and application to flooding stress response. J Proteome Res 8:4487–4499

Komatsu S, Kobayashi Y, Nishizawa K, Nanjo Y, Furukawa K (2010) Comparative proteomics analysis of differentially expressed proteins in soybean cell wall during flooding stress. Amino Acids 39:1435–1449. doi:10.1007/s00726-010-0608-1

Konig J, Baier M, Horling F, Kahmann U, Harris G, Schurmann P, Dietz KJ (2002) The plant-specific function of 2-Cys peroxiredoxin-mediated detoxification of peroxides in the redox-hierarchy of photosynthetic electron flux. Proc Natl Acad Sci USA 99:5738–5743

Lee H, Xiong L, Gong Z, Ishitani M, Stevenson B, Zhu JK (2001) The Arabidopsis HOS1 gene negatively regulates cold signal transduction and encodes a RING finger protein that displays cold-regulated nucleo-cytoplasmic partitioning. Genes Dev 15:912–924

Lee BH, Lee H, Xiong L, Zhu JK (2002) A mitochondrial complex I defect impairs cold-regulated nuclear gene expression. Plant Cell 14:1235–1251

Lee BH, Henderson DA, Zhu JK (2005) The Arabidopsis cold-responsive transcriptome and its regulation by ICE1. Plant Cell 17:3155–3175

Lilley KS, Dupree P (2007) Plant organelle proteomics. Curr Opin Plant Biol 10:594–599

Lin HX, Zhu MZ, Yano M, Gao JP, Liang ZW, Su WA, Hu XH, Ren ZH, Chao DY (2004) QTLs for Na+ and K+ uptake of the shoots and roots controlling rice salt tolerance. Theor Appl Genet 108:253–260

Liu J, Zhu JK (1998) A calcium sensor homolog required for plant salt tolerance. Science 280:1943–1945

Liu Q, Kasuga M, Sakuma Y et al (1998) Two transcription factors, DREB1 and DREB2, with an EREBP/AP2 DNA binding domain separate two cellular signal transduction pathways in drought- and low-temperature-responsive gene expression, respectively, in Arabidopsis. Plant Cell 10:1391–1406

Liu J, Ishitani M, Halfter U, Kim CS, Zhu JK (2000) The Arabidopsis thaliana SOS2 gene encodes a protein kinase that is required for salt tolerance. Proc Natl Acad Sci USA 97:3730–3734

Manabe T (1999) Capillary electrophoresis of proteins for proteomic studies. Electrophoresis 20:3116–3121

Mann M, Pandey A (2001) Use of mass spectrometry-derived data to annotate nucleotide and protein sequence databases. Trends Biochem Sci 26:54–61

Melgar JC, Guidi L, Remorini D, Agati G, Degl'innocenti G, Castelli S, Baratto MC, Faraloni C, Tattini M (2009) Antioxidant defences and oxidative damage in salt-treated olive plants under contrasting sunlight irradiance. Tree Physiol 29:1187–1198

Mittler R, Merquiol E, Hallak-Herr E, Kaplan A, Cohen M (2001) Living under a "dormant" canopy: a molecular acclimation mechanism of the desert plant Retama raetam. Plant J 25:407–416

Mittler R (2002) Oxidative stress, antioxidants, and stress tolerance. Ternds Plant Sci 7:405–410

Mittler R (2006) Abiotic Stress, the Field Environment and Stress Combination. Trends Plant Sci 11:15–19

Mullineaux P, Karpinski S (2002) Signal transduction in response to excess light: getting out of the chloroplast. Curr Opin Plant Biol 5:43–48

Nakashima K, Shinwari ZK, Sakuma Y, Seki M, Miura S, Shinozaki K, Yamaguchi-Shinozaki K (2000) Organization and expression of two Arabidopsis DREB2 genes encoding DREbinding proteins involved in dehydration- and high-salinity responsive gene expression. Plant Mol Biol 42:657–665

Noctor G (2006) Metabolic signalling in defence and stress: the central roles of soluble redox couples. Plant Cell Environ 29:409–425

Nouri MZ, Komatsu S (2010) Comparative analysis of soybean plasma membrane proteins under osmotic stress using gel-based and LC MS/MS-based proteomics approaches. Proteomics 10:1930–1945

Nouri MZ, Toorchi M, Komatsu S (2011) Proteomics approach for identifying abiotic stress responsive proteins in soybean. Soybean Mol Asp Breed 9:187–214

O'Farrell PH (1975) High resolution two-dimensional electrophoresis of protein. J Biol Chem 250: 4007–4021

Osakabe Y, Maruyama K, Seki M, Satou M, Shinozaki KA, Yamaguchi-Shinozaki K (2005) Leucine-rich repeat receptor-like kinase1 is a key membrane-bound regulator of abscisic acid early signaling in Arabidopsis. Plant Cell 17:1105–1119

Pandey A, Mann M (2000) Proteomics to study genes and genomes. Nature 405:837–846

Panter S, Thomson R, de Bruxelles G, Laver D, Trevaskis B, Udvardi M (2000) Identification with proteomics of novel proteins associated with thermo-adaptive mechanisms of soybean seedlings under heat stress revealed by the peribacteroid membrane of soybean root nodules. Mol Plant Microbe Interact 13:325–333

Parinov S, Sundaresan V (2000) Functional genomics in Arabidopsis: large-scale insertional mutagenesis complements the genome sequencing project. Curr Opin Biotechnol 11:157–161

Peck SC (2005) Update on proteomics in Arabidopsis. Where do we go from here? Plant Physiol 138: 591–599

Peltier JB, Emanuelsson O, Kalume DE, Ytterberg J, Friso G, Rudella A et al (2002) Central functions of the luminal and peripheral thylakoid proteome of Arabidopsis determined by experimentation and genome-wide prediction. Plant Cell 14:211–236

Ren ZH, Gao JP, Li LG, Cai XL, Huang W, Chao DY, Zhu MZ, Wang ZY, Luan S, Lin HX (2005) A rice quantitative trait locus for salt tolerance encodes a sodium transporter. Nat Genet 37:1141–1146

Rohde E, Tomlinson AJ, Johnson DH, Naylor S (1998) Comparison of protein mixtures in aqueous humor by membrane pre-concentration, capillary electrophoresis mass spectrometry. Electrophoresis 19:2361–2370

Sakuma Y, Maruyama K, Osakabe Y, Qin F, Seki M, Shinozaki K, Yamaguchi-Shinozaki K (2006) Functional analysis of an Arabidopsis transcription factor, DREB2A, involved in drought responsive gene expression. Plant Cell 18:1292–1309

Salekdeh GH, Siopongco J, Wade LJ et al (2002) A proteomic approach to analyzing drought- and salt-responsiveness in rice. Field Crops Res 76:199–219

Sanchez DH, Siahpoosh MR, Roessner U, Udvardi M, Kopka J (2007) Plant metabolomics reveals conserved and divergent metabolic responses to salinity. Physiol Planta 132:209–219

Schoffl F, Prandl R, Reindl A (1998) Regulation of the heat shock response. Plant Physiol 17:1135–1141

Sharkey TD (2005) Effects of moderate heat stress on photosynthesis: importance of thylakoid reactions, rubisco deactivation, reactive oxygen species, and thermotolerance provided by isoprene. Plant Cell Environ 28:269–277

Shi H, Ishitani M, Kim C, Zhu JK (2000) The Arabidopsis thaliana salt tolerance gene SOS1 encodes a putative Na+/H+ antiporter. Proc Natl Acad Sci USA 97: 6896–6901

Shi H, Xiong L, Stevenson B, Lu T, Zhu JK (2002) The Arabidopsis salt overly sensitive 4 mutants uncover a critical role for vitamin B6 in plant salt tolerance. Plant Cell 14:575–588

Shi H, Kim Y, Guo Y, Stevenson B, Zhu JK (2003) The Arabidopsis SOS5 locus encodes a putative cell surface adhesion protein and is required for normal cell expansion. Plant Cell 15:19–32

Shulaev V, Oliver DJ (2006) Metabolic and proteomic markers for oxidative stress. New tools for reactive oxygen species research. Plant Physiol 141:367–372

Sobhanian H, Motamed N, Rastgar Jazii F, Nakamura T, Komatsu S (2010) Salt stress induced differential proteome and metabolome response in the shoots of aeluropus lagopoides (poaceae), a halophyte C4 plant. J Proteome Res 9:2882–2897

Somerville C, Somerville S (1999) Plant functional genomics. Science 285:380–383

Taylor NL, Tan NF, Jacoby RP, Millar AH (2009) Abiotic environmental stress induced changes in the Arabidopsis thaliana chloroplast, mitochondria and peroxisome proteomes. J Proteomics 72:367–378

The Arabidopsis Genome Initiative (2000) Analysis of the genome sequence of the flowering plant Arabidopsis thaliana. Nature 408:796–815

Thomashow MF (1999) Plant cold acclimation, freezing tolerance genes and regulatory mechanisms. Annu Rev Plant Biol 50:571–599

Toorchi M, Yukawa K, Nouri MZ, Komatsu S (2009) Proteomics approach for identifying osmotic-stress-related proteins in soybean roots. Peptides 30: 2108–2117

Tuberosa R, Salvi S (2006) Genomics-based approaches to improve drought tolerance of crops. Trends Plant Sci 11:405–412

Umezawa T, Yoshida R, Maruyama K, Yamaguchi-Shinozaki K, Shinozaki K (2004) SRK2C, a SNF1-related protein kinase 2, improves drought tolerance by controlling stress-responsive gene expression in Arabidopsis thaliana. Proc Natl Acad Sci USA 101:17306–17311

Valliyodan B, Nguyen HT (2006) Understanding regulatory networks and engineering for enhanced drought tolerance in plants. Curr Opin Plant Biol 9:189–195

Vij S, Tyagi AK (2007) Emerging trends in the functional genomics of the abiotic stress response in crop plants. Plant Biotechnol J 5:361–380

Vinocur B, Altman A (2005) Recent advances in engineering plant tolerance to abiotic stress: achievements and limitation. Curr Opin Biotechnol 16:123–132

Vogel JT, Zarka DG, Van Buskirk HA, Fowler SG, Thomashow MF (2005) Roles of the CBF2 and ZAT12 transcription factors in configuring the low temperature transcriptome of Arabidopsis. Plant J 41: 195–211

Willekens H (1997) Catalase is a sink for H2O2 and is indispensable for stress defence in C-3 plants. EMBO J 16:4806–4816

Wollman FA, Minai L, Nechushtai R (1999) The biogenesis and assembly of photosynthetic proteins in thylakoid membranes. Biochim Biophys Acta 1411:21–85

Wu JL, Wu C, Lei C, Baraoidan M, Bordeos A, Madamba MR, Ramos-Pamplona M, Mauleon R, Portugal A, Ulat VJ, Bruskiewich R, Wang G, Leach J, Khush G, Leung H (2005) Chemical- and irradiation-induced mutants of indica rice IR64 for forward and reverse genetics. Plant Mol Biol 59:85–97

Xiong L, Ishitani M, Lee H, Zhu JK (2001) The Arabidopsis LOS5/ABA3 locus encodes a molybdenum cofactor sulfurase and modulates cold stress- and osmotic stress-responsive gene expression. Plant Cell 13:2063–2083

Xu Y, McCouch SR, Zhang Q (2005) How can we use genomics to improve cereals with rice as a reference genome? Plant Mol Biol 59:7–26

Xu K, Xu X, Fukao T, Canlas P, Maghirang-Rodriguez R, Heuer S, Ismail AM, Bailey-Serres J, Ronald PC, Mackill DJ (2006) Sub1A is an ethylene-response-factor-like gene that confers submergence tolerance to rice. Nature 442:705–708

Yabuta Y, Motoki T, Yoshimura K, Takeda T, Ishikawa T, Shigeoka S (2002) Thylakoid membrane-bound ascorbate peroxidise is a limiting factor of antioxidative systems under photooxidative stress. Plant J 32:915–925

Yamaguchi T, Blumwald E (2005) Developing salt-tolerant crop plants: challenges and opportunities. Trends Plant Sci 10:615–620

Yamaguchi-Shinozaki K, Shinozaki K (1994) A novel cis-acting element in an Arabidopsis gene is involved in responsiveness to drought, low-temperature, or high-salt stress. Plant Cell 6:251–264

Yamaguchi-Shinozaki K, Shinozaki K (2005) Organization of cis-acting regulatory elements in osmotic- and cold-stress-responsive promoters. Trends Plant Sci 10:88–94

Yamaguchi-Shinozaki K, Shinozaki K (2006) Transcriptional regulatory networks in cellular responses and tolerance to dehydration and cold stresses. Annu Rev Plant Biol 57:781–803

Yamanouchi U, Yano M, Lin H, Ashikari M, Yamada K (2002) A rice spotted leaf gene, Spl7, encodes a heat stress transcription factor protein. Proc Natl Acad Sci USA 99:7530–7535

Yan S, Tang Z, Su W, Sun W (2005) Proteomic analysis of salt stress-responsive proteins in rice root. Proteomics 5:235–244

Zhou J, Wang X, Jiao Y, Qin Y, Liu X, He K et al (2007) Global genome expression analysis of rice in response to drought and high-salinity stresses in shoot, flag leaf, and panicle. Plant Mol Biol 63:591–608

Environmental Stress and Role of Arbuscular Mycorrhizal Symbiosis

11

Anna Fusconi and Graziella Berta

Abstract

Mycorrhizae are widespread symbioses between plant roots and soil fungi, involved in the absorption of nutrients from the soil. Arbuscular mycorrhizae (AMs) are made up of a wide range of land plants, including at least 80% angiosperms, and fungi belonging to the glomeromycota. They generally have a positive effect on plant growth and nutrition, and improve the absorption of relatively immobile nutrients, such as phosphate, particularly in low nutrient soils or under drought. Moreover, AMs have been shown to promote plant fitness under a variety of stress conditions. In this chapter, we provide an overview of the effects of AM colonization on host plants subjected to drought stress and soil pollution, two of the most common types of stress that limit plant growth, and of the possible mechanisms involved in the beneficial effects of AM fungi.

Keywords

Arbuscular mycorrhizae • Drought stress • Water relations • Heavy metals • Arsenic • Hydrocarbons

1 Introduction

Plants in the field are frequently exposed to different stress conditions. High or low levels of water, extreme temperature, high alkalinity or acidity, low nutrient availability and anthropogenic stressors, such as metals and toxic organic pollutants, reduce plant growth and productivity, and stress the plant. Plants may respond to stress with modifications that allow them to avoid the stress (Ruiz-Lozano et al. 2006), or may implement stress tolerance through an array of morphological, physiological and biochemical responses that limit damage or facilitate the repair of damaged systems (Potters et al. 2007). Moreover, most plants possess an additional mechanism that helps them to tolerate stress, which consists of an association with soil fungi and rhizospheric micro-organisms that can help them to survive

A. Fusconi (✉)
Dipartimento di Biologia Vegetale, Università degli Studi di Torino, Torino 10125, Italy
e-mail: anna.fusconi@unito.it

G. Berta
Dipartimento di Scienze dell'Ambiente e della Vita, Università del Piemonte Orientale "Amedeo Avogadro", Alessandria 15121, Italy

P. Ahmad and M.N.V. Prasad (eds.), *Abiotic Stress Responses in Plants: Metabolism, Productivity and Sustainability*, DOI 10.1007/978-1-4614-0634-1_11, © Springer Science+Business Media, LLC 2012

and ameliorate their productivity. Among these associations, mycorrhizal symbiosis (the term "mycorrhiza" comes from the Greek words for "fungus" and "root") are widespread, so much that, in the natural environments, mycorrhizal is the norm rather than the exception (Hodge et al. 2009). Mycorrhizal associations are involved in the absorption of nutrients from the soil and, according to Smith and Read (2008), mycorrhizas, and not roots, are the normal nutrient-absorbing organs of most plant species. Furthermore, as previously mentioned, association with mycorrhizal fungi increases plant productivity and encourages resistance to stress.

Mycorrhizal fungi are specialized members of a vast population of micro-organisms that colonize the rhizosphere. They penetrate the tissues of the root with different patterns and, with few exceptions, are completely dependent on the plant for organic carbon. The location of the fungal symbiont in the root, and the hyphal connection with the soil, ensure the absorption of soil-derived nutrients, through the fine exploration of the rhizosphere and nearby soil, and the translocation of nutrients, which would otherwise not be available to the plant. The bidirectional transfer of nutrients is the basis of mutualism in most mycorrhizal symbiosis (Smith and Read 2008).

2 The Arbuscular Mycorrhizae

Of all the mycorrhizal associations, AMs are the most widespread and ancient. They are made up by a wide range of land plants, including at least 80% angiosperms, and fungi belonging to glomeromycota (Smith and Read 2008; Fitter et al. 2011). After contact between the symbionts is established (Fig. 11.1a), the fungus forms an appressorium on the root surface and enters the root. Colonization of the root system, by the fungal partner, is restricted to the parenchymatous cells of the cortex, where it grows intercellularly and produces long intercellular hyphae. Lateral branches of the intercellular hyphae penetrate the cortical cells and repeatedly branch dichotomously giving rise to the "arbuscules", which give the name to this type of mycorrhiza (Fig. 11.1b, c). The AM symbiosis is a highly compatible association, and in phosphate-limiting conditions, intraradical development of the fungus can occur over more than 80% of the root length (Harrison 2005).

Arbuscules form inside the cortical cell, but remain separated from the plant cell cytoplasm by the invaginated host plasma membrane, and are surrounded by a metabolically active plant cytoplasm. A new compartment, known as the interfacial compartment, arises and consists of the invaginated host membrane, cell wall-like material, fungal wall and plasma membrane (Fig. 11.1c) (Bonfante and Anca 2009). The development of the AM interaction is, therefore, accompanied by significant alterations of the cellular morphology of both symbionts. Reciprocal exchanges of metabolites are thought to occur at this interface (Harrison 2005; Hodge et al. 2009). Besides these cellular alterations, AM root systems can modify their architecture by increasing the number of lateral roots of different orders, and can change the specific root length, according to the plant species (Scannerini et al. 2001; Hodge et al. 2009).

As in other mycorrhizas, in addition to the intraradical growth phase, the AM fungus maintains a living extraradical mycelium (Fig. 11.1a) that can extend several centimetres from the root. The fungal hyphae within the root are connected to the extraradical mycelium and form a single continuum (Harrison 2005). These extraradical hyphae absorb nutrients, mainly phosphate (Pi), that are transferred to the host root at the symbiotic interface. Phosphorus is one of the most important elements for plants; however, it is also one of the least available of all essential nutrients in the soil (Vance et al. 2003). A slow diffusion of Pi in the soil solution and rapid absorption results in the development of depletion zones around the roots (Smith and Read 2008). The hyphae of AM fungi extend into soil far beyond the depletion zone. Due to their small diameter, in the range 2–20 μm, they may grow into soil pores that roots cannot penetrate (Smith and Read 2008). Furthermore, hyphae appear to have the ability to regulate their diameter, depending on the soil pore size (Smith et al. 2010b).

Following AM symbiosis, the P status of the plant tissues is generally higher than that of non-AM plants grown on the same medium (Fusconi

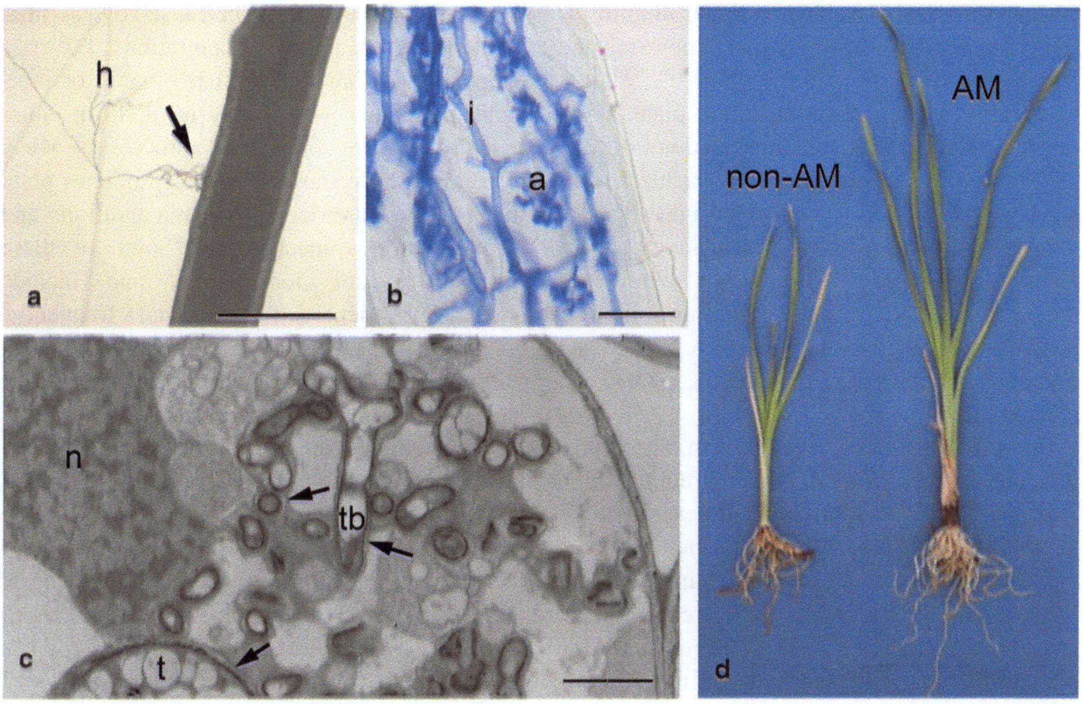

Fig. 11.1 (**a**) Extraradical hyphae (h) and initiation of colonization (*arrow*) of a transformed carrot root with *Gigaspora rosea*. *Bar*=500 μm. (**b**) Colonization of the root cortex by the AM fungus *Glomus mosseae*. Intercellular hyphae (i), arbuscules (a). *Bar*=50 μm. (**c**) Transmission electron micrograph of a *Glomus* sp. arbuscule inside a cortex cell of *Allium porrum*. The main trunk of the arbuscule (t) and the thinner branches (tb) are surrounded by the perifungal interfaces (*arrows*). Nucleus of the cortex cell (n). *Bar*=1.5 μm. (**d**) *Typha latifolia* plants mycorrhized or not with a mix of *Glomus* sp. Note the increased growth of AM plants following colonization

et al. 2005) and a number of plants increase in growth following colonization (Fig. 11.1d), showing a higher shoot-to-root ratio than the non-colonized controls (Scannerini et al. 2001).

However, the AM pathway through extraradical hyphae also plays a major role in the Pi uptake in non-responsive AM plants (Smith and Read 2008). Recently, it has been demonstrated that the development of an AM association changes the pathway of plant phosphate uptake. In functional AM symbiosis, in fact, the direct plant Pi influx via root epidermis and root hairs is reduced, while genes involved in the AM P uptake pathway, through the plant–fungus interfaces, are up-regulated (Grace et al. 2009).

Although usually considered important primarily because of Pi uptake, there is now evidence that AM fungi may also increase the efficiency of the uptake of other nutrients. The hyphae of AM fungi are involved in the uptake and transfer of inorganic N, although it is not clear whether this always occurs in amounts that are significant for whole plant nutrition (Smith and Read 2008; Fitter et al. 2011). It has been calculated that up to 42 and 30% of plant N is absorbed via the AM pathway in tomato and carrot, respectively (Smith and Read 2008). AM fungi are especially important with respect to the uptake of relatively immobile nutrients that form depletion zones around roots (Cavagnaro 2008): the uptake of Cu has been confirmed for a number of plant–fungus combinations, and Zn uptake, via the AM pathway, has inequivocally been demonstrated. However, the uptake of other micronutrients, such as Mn and K, via external hyphae is not so well established (Smith and Read 2008).

The positive effects of AM fungi on plant nutrition are of particular significance in low nutrient status soils, and where the distribution of the soil nutrients is heterogeneous (Cavagnaro 2008). Moreover, the positive effects of AM fungi on host plant nutrient uptake are noticeable in drought environments, since nutrient mobility is limited under drought conditions (Boomsma and Vjn 2008).

Improvements in the nutrition of plants colonized by AM fungi can be attributed not only to the uptake of nutrients via the mycorrhizal pathway, but also to indirect effects brought about by morphological and physiological changes in roots due to colonization. AM fungi may also influence nutrient availability via their effects on soil physicochemical properties, nutrient cycling and microbial communities (Cavagnaro 2008 and references therein).

In AM symbiosis, fungal growth, spore production and mycelial transport require high amounts of energy in the form of hexoses and a significant proportion of the photosynthesis products are therefore delivered to the fungus, which leads to an increase in the sink strength of the root (Feddermann et al. 2010). The cost for the plant to sustain the associated mycelium, in terms of carbon loss, varies according to the plant and fungal species, plant age, and AM developmental stage. Moreover, the respiratory cost of AM symbiosis can vary according to the environmental conditions. It has been calculated that AM can consume between 2 and 20% of the daily photosynthate production of the host (Boomsma and Vjn 2008). Therefore, AM colonization is generally advantageous in poor-nutrient soils, while colonization by AM in fertile agricultural soils can actually reduce crop productivity, because the carbon costs associated with AM colonization may exceed the carbon-production benefits that are derived from the symbiotic relationship (Boomsma and Vjn 2008).

In the last few years, several publication have dealt with the contribution of AM symbiosis to the tolerance of plants growing under one or more different types of environmental stress (Hildebrandt et al. 2007; Evelin et al. 2009; Gamalero et al. 2009; Garg and Chandel 2010; Koltai and Kapulnik 2010; Miransari 2010; Smith et al. 2010b). Water deficit, salinity and elevated temperatures that cause dehydration in plant tissues are the most common environmental stress factors experienced by plants. Another important form of environmental stress is represented by contamination, which can be caused by essential elements in excessive concentrations, by toxic elements and ions or by other organic or inorganic pollutants, which may result from either human activities or natural processes. Many researchers have shown that AM promotes plant fitness under a variety of stress conditions and through various mechanisms. Some of these mechanisms are non-specific, and include enhanced nutrient acquisition and, frequently, enhanced plant growth, regulation of the plant hormone balance, and improving rhizospheric and soil conditions. In addition, AM improves plant growth tolerance with more specific mechanisms in relation to the kind of stress.

In this chapter, we provide an overview of the effects of AM colonization on host plants subjected to drought stress and soil pollution, and of the possible mechanisms involved in the beneficial effects of AM fungi.

3 Arbuscular Mycorrhizae and Drought Tolerance

Water deficiency, or drought stress (DS), is a major abiotic factor that limits agricultural crop production. Plants experience DS either when the water supply to the roots becomes difficult or when the transpiration rate becomes very high (Reddy et al. 2004). Under arid and semi-arid conditions, drought states are one of the most common stresses that affect plant growth and yield (Miransari 2010).

In response to drought stress, isohydric plants close their stomata and reduce transpiration to avoid dehydration (Maurel et al. 2010), and, as an early response to water deficit, may reduce their growth, to save and redistribute resources that can become limited under extreme stress (Skirycz and Inzé 2010). On the other hand, anisohydric plants have the ability to maintain high transpiration and to grow under water-limiting conditions (Maurel et al. 2010). Furthermore, plants can accumulate solutes in order to maintain a low water potential and synthesize protective proteins, such as dehydrins and antioxidants (Skirycz and Inzé 2010).

Mycorrhizal colonization has been shown to increase the drought tolerance, and this has been shown in many plant species, including maize, wheat, barley, soybean, onion, lettuce and tomato (Augé 2001; Khalvati et al. 2010).

Since AM colonization generally modifies root features, e.g. the specific root length and root architecture, and can increase the uptake of diffusion-limited nutrients in relation to non-AM plants, AM frequently increases the size of the host plant under DS (Wu and Xia 2006; Wu et al. 2008; Khalvati et al. 2010; Ruiz-Sánchez et al. 2010). The plant size and the features of the root system can affect the water relations and the drought tolerance of a plant (Augé 2001) and hence may represent, by themselves, a mechanism that increases plant tolerance to stress. AM symbiosis generally increases the photosynthetic rate of the host plant (Augé 2001, Wu and Xia 2006, Ruiz-Sánchez et al. 2010) and may modify any of the mechanisms that the plants use to overcome DS. Therefore, the potential mechanisms proposed to explain the induction of DS tolerance by AM symbiosis include an improved regulation of the plant water relations, a greater osmotic adjustment, an enhanced antioxidant defence and the production of protective molecules (Aroca et al. 2007; Ruiz-Sánchez et al. 2010). Finally, the external soil mycelium may stabilize the soil aggregates, which increase moisture retention and improve the absorption of water (Marulanda et al. 2003; Rillig and Mummey 2006).

3.1 Water Relations

There is no doubt that AM colonization affects the water relations of plants under both well watered and DS conditions. The effect of AM fungi on the water balance and the possible causal mechanisms involved have been thoroughly revised by Augé (2001) who has quoted more than 200 articles on the subject and, more recently, by Smith and Read (2008) and Smith et al. (2010b).

The water content of a plant mainly depends on the amount of water lost through transpiration, and the root water uptake and conductivity. These may be influenced by AM, with consequent effects on tissue hydration and leaf physiology.

(1) The leaf transpiration and stomatal conductance (g_s) of AM plants are often different from those of non-mycorrhizal plants. When they are different, they are frequently higher in AM plants, even when the size of AM and non-AM plants are similar (Augé 2001).

The increase in g_s in AM plants can be directly ascribed to the increased P leaf concentration. A good correlation between g_s and the P concentration has in fact been found in *Trifolium pratense*, regardless of whether the differences in P nutrition were induced by fertilization or AM colonization (Fitter 1988).

However, there are many exceptions. For example, tomato plants under well-watered conditions have been reported to possess a similar transpiration rate, regardless of the AM treatments, and the decline in transpiration caused by DS occurs to similar extents in AM and non-AM plants (Aroca et al. 2008a). The effect of AM on g_s may differ according to the changing water availability. A detailed study on rice plants has recently been published by Ruiz-Sánchez et al. (2010). Under well-watered conditions, the g_s of non-AM rice plants was almost two times higher than that of AM plants. After a period of DS, the g_s was reduced more in the non-AM plants, and similar values were found for all the treatments. However, after recovery from drought, the stomatal conductance of the AM plants was considerably enhanced, and at this stage the non-AM plants always exhibited lower stomata conductance than their AM counterparts (Ruiz-Sánchez et al. 2010).

It is well known that ABA rises in response to water deficit and that it protects plants by promoting stomatal closure to minimize transpirational water loss (see, for example, Aroca et al. 2008a). The effects of AM colonization on the endogenous levels of ABA have been analyzed in depth in AM and non-AM *Lactuca sativa* plants. These plants showed a similar transpiration rate under well-watered conditions. However, under DS conditions, the AM plants showed a lower transpiration rate than the non-AM plants that has been related with a significant increase of the leaf ABA (Aroca et al. 2008b). These data point to an important role of AM in the regulation of ABA levels. AM fungi, indeed, allow for a more adequate plant hydric balance during drought and recovery, compared to non-AM plants (Aroca et al. 2008b).

(2) Roots sense changes in the soil water content, and consequently activate a combination of hydraulic and chemical signals that are directed to the leaves, where a number of tolerance mechanisms are triggered (Maurel et al. 2010; Skirycz and Inzé 2010). Drought usually induces a marked drop in the hydraulic conductivity of the root (Lp_r) which plays a central role in maintaining the plant water status (Maurel et al. 2010) but the effect of AM on Lp_r is unclear, and increases and decreases have been reported as a result of colonization (Augé 2001). Recently, an Lp_r increase has been found in lettuce plants, in AM and non-AM plants of comparable size (Aroca et al. 2008b).

The short-term regulation of root hydraulics is related to changes in aquaporin activity and/or content. Aquaporins are water channel proteins that facilitate and regulate the passive movement of water molecules down a water potential gradient. They are abundantly expressed in roots, where they mediate most soil water uptake (Ruiz-Lozano et al. 2006). The contribution of aquaporin genes to the enhanced tolerance to drought in AM plants has been investigated in the recent years (Porcel et al. 2005; Ruiz-Lozano et al. 2006; Aroca et al. 2007, 2008a). It was shown that the impairment of a plasma membrane intrinsic protein (*PIP*) gene in an antisense tobacco mutant reduced the symbiotic efficiency of two AM fungi under DS conditions (Porcel et al. 2005), thus pointing to the role of aquaporin regulation in the protective effect of AM colonization. In *P. vulgaris*, it has been shown that the regulation of the hydraulic properties of roots by AM symbiosis is closely correlated to the regulation of the PIP2 protein amount and phosphorylation state, and the expression of different analyzed *PIP* genes responds differently to DS in AM and non-AM plants (Aroca et al. 2007).

However, the data concerning Lp_r regulation in AM and non-AM plants under DS are conflictual, they highlight the complexity of the influence of AM colonization, and may be partly explained by the different water use strategies of the plants (Hernández et al. 2010). Another important source of variability is that of the different strategies of AM fungal species and strains to protect the host plant against DS. Marulanda et al. (2003) have shown that *Glomus intraradices* is one of the most efficient fungi in improving plant water uptake in lettuce plants while *Glomus mosseae* has shown a reduced ability. These differences have been related to the different regulation of plant *PIP* aquaporin genes by the fungi. Up-regulation of the *PIP* gene expression induced by *G. intraradices* enhances the water uptake of root and the root water movement. On the other hand, down-regulation induced by *G. mosseae* decreases membrane water permeability and leads to the conservation of cellular water (Ruiz-Lozano et al. 2006).

In the long term, the adjustment of Lp_r through aquaporins is accompanied, in some species, by root architecture changes, and by the differentiation of suberized exodermis or endodermis with low apoplastic permeability, which modifies the hydraulic conductivity (Maurel et al. 2010). Mycorrhization can modify apoplastic permeability, and a lower deposition of suberin has been observed in the endoderm and esoderm of AM leek roots (Fusconi, unpublished data). AM symbiosis can also modify root hydraulics by altering root morphology, increasing the size of the root stele and influencing the root turnover (Fusconi et al. 1994; Hodge et al. 2009).

3.2 Osmolyte Accumulation

The adjustment of the osmotic potential by AM is probably one of the most important reasons for the improved ability of the host plant to grow under water stress. Plants lower the cell osmotic potential by accumulating higher amounts of organic products, e.g. proline, glycine betaine and carbohydrates, such as sucrose, mannitol and non-organic ions, and this allows a higher water retention during drought (Medina et al. 2010; Miransari 2010). Of all these metabolites, proline is probably the most widespread in plants, and it has been shown that proline accumulates under water shortage (Ruiz-Lozano et al. 2006).

The investigations carried out on proline in AM symbiosis, however, are somewhat contradictory (Ruiz-Lozano et al. 2006 and references therein). While some studies have shown an increase in proline accumulation in the leaves/shoots of plants subjected to DS (Medina et al. 2010), others have shown a lower proline accumulation (Aroca et al. 2008a; Ruiz-Sánchez et al. 2010; Manoharan et al. 2010). AM and non-AM tomato plants, for example, show almost the same leaf proline content under well-watered conditions and accumulate more proline under DS, with a greater increase in non-AM than in AM plants. According to Aroca et al. (2008a), since mycorrhization protects host plants from dehydration stress, AM plants may need to accumulate less proline than their non-AM counterparts.

However, the analysis of the proline content in the leaves and roots of soybean has shown that AM roots accumulate more proline than non-AM roots while the opposite occurs in the shoots. The enhanced osmotic adjustment in AM roots could contribute to maintaining a favourable water potential gradient for water absorption (Porcel and Ruiz-Lozano 2004).

Carbohydrates or non-organic ions possibly play a role in the osmoregulation of AM plants. In *Citrus* seedlings, for example, the enhanced host plant drought tolerance induced by AM colonization has not been correlated to proline, which decreased in AM seedlings, but to an increased cell concentration of the total non-structural carbohydrates, K, Ca and Mg (Wu and Xia 2006).

3.3 Antioxidant Molecules/Enzymes

In higher plants, reactive oxygen species (ROS) are continuously produced in chloroplasts, mitochondria and peroxisomes. The production and removal of ROS are strictly controlled under well-watered conditions. When higher plants are subjected to water stress, the equilibrium between the production and scavenging of ROS is broken, and this results in oxidative damage. As a consequence, antioxidant systems are induced. These consist of enzymatic and non-enzymatic antioxidants which are designed to minimize the concentrations of ROS (Wu et al. 2006).

Under well-watered conditions, AM formation has shown contrasting results on oxidative damage and the production of antioxidant molecules in soybean, trifoliate orange and rice. Instead, in the same plants under DS, colonization reduces the concentrations of the superoxide anion radical, $O_2^{\bullet-}$, and H_2O_2 and/or oxidative damage to lipids compared to non-AM (Porcel and Ruiz-Lozano 2004; Wu et al. 2006; Wu and Zou 2009; Ruiz-Sánchez et al. 2010). However, this reduction is not always accompanied by a rise in the antioxidant activity, and in some cases the protective activity of AM is not related to any antioxidant activity induced by AM fungi (Table 11.1). It has also been shown that different AM fungal species can influence the production of antioxidant compounds in different ways (Ruiz-Sánchez et al. 2010). In agreement with these findings, Marulanda et al. (2007) have shown that, under DS conditions, lavender plants accumulate less H_2O_2, glutathione (GSH) and ascorbate when inoculated with strains of *G. intraradices* and *G. mosseae* isolated from dry sites, compared to the plants inoculated with non-adapted strains. They concluded that the low cell accumulation of these compounds in AM plants is an indication of a high drought tolerance induced by colonization. It would seem that the induction of protective antioxidant molecules is less important if AM colonization actually protects plants through other mechanisms.

3.4 Dehydrines and Other Protective Molecules

Late embryogenesis abundant proteins (LEA) proteins accumulate in plant seeds during their maturation phase and also accumulate in vegetative plant tissues during periods of water deficit. It has been proposed that, during cell dehydration, LEA

Table 11.1 Effects of AM colonization on the superoxide dismustase (SOD), catalase (CAT), ascorbate peroxidase (APX), glutathione reductase (GR) and guaiacol peroxidase (G-POD) activities, and in the ascorbate (ASC) and glutathione (GSH) contents in roots and shoots of plants subjected to drought-stressed conditions

			SOD	CAT	APX	GR	G-POD	ASC	GSH	
Glycine max	*G. intraradices*	Root	<	=	<	<				Porcel and Ruiz-Lozano (2004)
		Shoot	>	=	<	=				
Poncirus trifoliata	*G. versiforme*	Root	<	=	>	=	=	>	>	Wu et al. (2006)
		Leaf	=	>	>	>	>	>	>	
Citrus sinensis/ *P.trifoliata*	*G. versiforme*	Root	>	>			>	>	>	Wu and Zou (2009)
		Leaf	>	>			>	>	>	
Cucumis melo	*G. mosseae*	Root	>	>			=			Huang et al. (2011)
		Shoot	>	>			>			
	G. versiforme	Root	>	>			=			
		Shoot	>	>			=			
	G. intraradices	Root	>	=			=			
		Shoot	>	=			=			

=, unaltered; >, increased; <, reduced activity/content in AM plants compared to non-AM plants

proteins play different functions: maintenance of the structure of other proteins, vesicles or endomembranes, sequestration of ions, such as calcium, binding or replacement of water and functioning as molecular chaperones (Ruiz-Lozano et al. 2006). LEA proteins include dehydrins, which represent the most conspicuous soluble proteins induced by a dehydration stress (Ruiz-Lozano et al. 2006 and references therein). However, the scarce available literature suggests that the accumulation of LEA proteins seems unlikely to be a general protective mechanism by which AM symbiosis protects the host plant. In tomato plants, colonization with *G. intraradices* slightly increases the expression of *Sldhn,* a gene encoding for a dehydrin, under DS, but without any or with little differences between AM and non-AM plants (Aroca et al. 2008a), and in soybean and lettuce plants colonized by either *G. mosseae* or *G. intraradices*, the levels of *lea* transcripts were considerably lower than those of the corresponding non-AM plants (Ruiz-Lozano et al. 2006).

Other proteins with a protective effect against DS include the luminal-binding protein, an important component of endoplasmic reticulum which shows chaperone-like activity (BiP). BiP expression in plants has been shown to respond to water stress, and it has been demonstrated that the constitutive overexpression of BiP in tobacco confers tolerance to water stress. Recent analyses have shown the up-regulation of the *G. intraradices* *BiP* gene under DS conditions in soybean, lettuce, maize and tobacco. It has been proposed that the GiBiP protein can facilitate the proper folding and maturation of the water stress-induced secretory proteins involved in the osmotic response mechanism (Porcel et al., 2007 and references therein).

3.5 Interactions Among AM Fungi, Roots and Soil

In addition to influencing plants directly by colonizing the plant tissues, AM symbiosis may also affect drought responses by changing the soil in which the plants are growing (Augé et al. 2007). Studies on sorghum and squash have shown that the soil density of AM fungi has a stronger influence on stomata behaviour and the extent of soil drying than root colonization (Augé et al. 2007). Therefore, the extension of the AM fungal mycelium in the soil is considered an important determinant on the contribution of the AM symbiosis to plant performance under DS.

Soil dryness has been shown to increase or decrease the amount of extraradical AM fungal hyphae (Neumann et al. 2009). Moreover, a great variability in the degree of root colonization, which may increase (Aroca et al. 2008b; Khalvati et al. 2010; Ruiz-Sánchez et al. 2010) or decrease (Wu et al. 2006, 2008; Kohler et al. 2008; Wu and

Zou 2009; Manoharan et al. 2010) under DS has also been reported.

The reasons for these observations are not well understood. Fungal hyphae have access to fine soil pores which remain filled with soil solutions even under low soil moisture regimes. This might render the soil AM fungi less sensitive to decreasing soil water potentials than plant roots (Neumann et al. 2009). However, the decreased plant photosynthetic capacity under DS might reduce the supply of carbohydrates to the fungal symbiont, limiting its growth (Neumann et al. 2009). Moreover, drought modifies root system features which, in turn, may influence the degree of colonization and the frequency of the different AM fungal structures. In grapevine, for example, drought causes a reduction in the fine roots, which are easily colonized by AM fungi during the early deficit treatment (Schreiner et al. 2007).

However, different fungal species show different behaviour under DS, both when they colonize the root or grow in the soil. Some experiments on citrus, for example, have shown that DS reduced *G. versiforme* and *G. mosseae* root infection to a great extent but did not significantly affect the colonization by *G. diaphanum* while, among the three fungi, it only markedly decreased the soil hyphal density of *G. mosseae* (Wu et al. 2008).

Evidence exists that the mycorrhizal fungal mycelium influences soil aggregation through different biological, biochemical and physical mechanisms (Rillig and Mummey 2006; Wu et al. 2008). Since soil aggregates regulate the water flow and moisture in soil, it has been proposed that extraradical AM fungi may improve the water relations of plants (Wu et al. 2008) and, indirectly, mineral nutrition.

Fungal hyphae, especially those of AM fungi, grow in the soil matrix and create a skeletal structure that holds primary soil particles together via physical entanglement (Wu et al. 2008). Moreover, glomalin, a glycoprotein produced by AM fungi, accumulates in soil contributing to the stabilization of soil water stable macro-aggregates (Bedini et al. 2009). Glomalin has formerly been hypothesized to act as a "glue" and play a main role in increasing soil structural stability (Rillig and Mummey 2006). However, more recently, glomalin has been shown to be bounded above all in the fungal hyphae. The putative gene for glomalin has been sequenced and has shown homology to a class of stress-induced proteins with a known cellular function, thus pointing to a role of glomalin in the living mycelium (Rillig and Mummey 2006). Therefore, the beneficial effects of AM fungi through the production of glomalin are probably long-term effects, like other indirect effects on soil aggregation. These latter include the influence of AM fungi on plant community composition, on growth and root morphogenesis of the host plants which, in turn, can alter the amount of carbon that may eventually enter the soil, for example through rhizodeposition and root decomposition (Rillig and Mummey 2006).

4 Arbuscular Mycorrhizae and Soil Pollution

Heavy metal (HM) pollution is one of the main problems that negatively affects both human and environmental health. Some HMs are important as micronutrients (Fe, Mo and Mn). Some toxic HMs are trace elements (e.g., Zn, Ni and Cu): they are required in numerous enzyme catalyzed or redox reactions and in electron transfer as well as playing structural function in nucleic acid metabolism (Zenk 1996). Although some of these metals are essential for life, when they are present in excess, they induce macroscopic effects concerning plant growth and leaf morphology (Rout and Das 2003; Todeschini et al. 2011), as well as root development, via large alterations of the mitotic activity (Liu et al. 2009) and via genotoxic damage. Their toxicity is mainly related to oxidative and (or) genotoxic mechanisms (Gamalero et al. 2009). They can be taken up in different ways: Cu uptake and translocation by plants is strictly regulated, resulting in very low leaf concentrations (Todeschini et al. 2007), whereas Zn is primarily accumulated in leaves, with concentrations in the order of hundreds of ppm (Rosselli et al. 2003); by contrast, two ferns, *Polypodium cambricum* and *Pteris vittata* differently respond to Zn pollution: the latter accumulates relatively high metal concentrations in the fronds, causing a progressive deterioration of anatomical structures and functions, while the former shows a saturation mechanism in the root/rhyzome, which is evident already at non-toxic Zn concentrations, so ensuring the preservation of frond anatomy and function even under the exposure to sub-lethal Zn doses (Roccotiello et al. 2010).

Finally, several HMs have no known nutritional function, but are toxic for plants and microorganisms (Hg, Ag, Cd, Pb and U) (Fusconi et al. 2007). Non-essential metals are usually toxic at lower concentrations than essential ones (Clemens 2006). In addition to these, there are also a number of toxic metalloids, including arsenic.

4.1 Heavy Metals and Essential Metals in Excessive Concentration

AM fungi are a direct link between soil and roots, and they can therefore be very important for HM availability and toxicity to plants. The AM symbiotic status changes the chemical composition of root exudates (Barea et al. 2002), thus quantitatively and qualitatively affecting the microbial populations in the rhizosphere and/or in the rhizoplane (Offre et al. 2008). All these factors, alone or in combinations, can influence metal mobility or availability; nevertheless, the role of AM fungi in the uptake and in the transfer of HMs to the plant is not yet completely understood and the literature is conflicting. Some reports indicate that AM fungi enhance plant accumulation and the tolerance of HMs while reduced HM concentrations were instead found in a number of mycorrhizal associations (Citterio et al. 2005; Lingua et al. 2008). However, the bulk of evidence seems to suggest a species specific effect of AM associations on root metal uptake.

Spores and pre-symbiotic hyphae are generally sensitive to HMs in the absence of the host plants (Göhre and Paszkowski 2006). Anyway, AM symbiosis has been observed in plants growing in soils containing HMs (Vallino et al. 2006; Bothe et al. 2010; Massa et al. 2010). Specifically adapted HM tolerant plants grow in HM-contaminated soils, and some have been reported to harbour AM fungi though with a low degree of colonization (Bothe et al. 2010). Zinc violets are clear examples of the exploitation of AM fungi by plants for heavy metal tolerance as their roots are intensely colonized (Tonin et al. 2001).

It is well known that AM fungi can alter metal concentrations and induce increased tolerance in plants, and different mechanisms have been proposed to explain this: the binding of metals to fungal cell walls and subsequently being accumulated in the vacuoles; sequestration by siderophores, deposited in the root apoplasm or in the soil, and possibly taken up by plant ferrisiderophore receptors, or complexing of metals to metallothioneins or phytochelatins synthesized by the fungus or the plant, as well as organic acids, amino acids, and metal-specific chaperons (shown for plants, but assumed for AM fungi) (Miransari, 2010) or metal chelation by fungal compounds, such as glomalin (Hildebrandt et al. 2007; Bedini et al. 2009).

Different plant species (and even different clones of the same species) respond in different ways to metal stress and to AM colonization. AM symbiosis can be either beneficial or ineffective, under metal pollution conditions, in relation to the host plant, and have proved to alleviate HM stress in the more sensitive species, probably by improving P nutrition (Todeschini et al. 2007; Lingua et al. 2008; Castiglione et al. 2009). An example is that of the results obtained by Lingua et al. (2008) on two poplar clones (Villafranca and Jean Pourtet), which responded differently to Zn and Cu addition and to AM symbiosis.

Under HM stress, unfavourable oxidative effects adversely influence plant growth. However, AMs are able to enhance the production of antioxidant enzymes, including glutathione S-transferase, superoxide dismutase, cytochrome P450 and thioredoxin, which can alleviate the stress of HMs (Hildebrandt et al. 2007). The enhanced tolerance of AM plants is related to the simultaneous regulation of AM stress genes and plant tolerance genes (Hildebrandt et al. 2007; Gamalero et al. 2009). The improvement in plant tolerance to HMs may be related to changes in gene expression as well as protein synthesis induced by the symbiosis itself. As an example, the Zn transporter MtZIP2 from *Medicago truncatula* is up-regulated by the presence of Zn and down-regulated by AM colonization, leading to a lower content of Zn within the host plant tissues (Burleigh et al. 2003). HM stress increases the transcript levels of some *LeNramp2* (encoding a broad-range HM transporters) and *Lemt1, Lemt3* and *Lemt4* genes (encoding metallothioneins). On the other hand, AM fungal colonization results in the down-regulation of other HM transporter genes, presumably because the content of HM is lower in AM plants than in non-mycorrhizal ones. However, the down-regulation of plant mRNA (Ouziad et al. 2005; Burleigh et al. 2003) may be related to the "dilutive effect" of HM that occurs when plant growth improves as a result of AM colonization (Burleigh et al. 2003).

Among the HMs, cadmium (Cd) is of great environmental concern. Even in trace concentrations, this metal can cause serious health hazards to most living organisms of both the eukariotic and prokariotic kingdoms (Fusconi et al. 2007). Cd interacts with various functional groups of proteins, mainly with SH groups, which results in the alteration of the reactive centre of many enzymes, the reduction of the photosynthetic rate and chlorophyll content, alterations of membrane permeability, oxidative damage, increases in the cell polyamine pool (Sharma and Dietz 2006; Lingua et al. 2008). In addition, Cd influences protein–protein and protein–DNA interactions (Freedman et al. 1988). Different mechanisms have been proposed to explain mycorrhiza alleviation of Cd stress (Rivera-Becerril et al. 2002; Aloui et al. 2009). In pea, the buffering effect of AM symbiosis vis-à-vis Cd pollution has been linked to the modulation of root protein profiles. A protein band of about 30 kDa, a short-chain alcohol dehydrogenase (ADH), a UTP-1-phosphate uridylyltransferase (UDP-glucose pyrophosphorylase, UDPGP or UGPase), and a protein with a high homology to subunit B from a vacuolar H^+–ATP synthase (V-ATPase), were all induced in pea plants by Cd treatment but down-regulated by inoculation with *G. mosseae* (Repetto et al. 2003). An increase in the highest ploidy nucleus populations which is possibly related to an increased transcription of genes, leading to the synthesis of the proteins involved in response/detoxification mechanisms to Cd toxicity, has also been observed (Repetto et al. 2007).

A more recent study has again shown that AM colonization in plants exposed to Cd stress can modulate the pattern of protein expression (Aloui et al. 2009). This study has shown that Cd-induced root proteome changes in *M. truncatula* plants are buffered by AM symbiosis. There is evidence of down accumulation of Cd stress-plant responsive proteins and the concomitant accumulation of mycorrhiza-related proteins putatively involved in reducing Cd toxicity in AM plants. More precisely, seven specifically up-accumulated proteins in AM roots, and whose expression was not modified upon Cd exposure, were detected and identified. All these proteins, which corresponded to a cyclophilin (s1), a guanine nucleotide-binding protein (s2), a ubiquitin carboxyl-terminal hydrolase (s3), a thiazole biosynthetic enzyme (s4), an annexin (s8), a GST-like protein (s13) and an SAM synthase (s14), had functions putatively relevant in alleviating Cd toxicity (Aloui et al. 2009)

Increased accumulation has also been found at the proteome level of antioxidant enzymes and non-enzymatic antioxidants, and this accumulation is probably involved in the protection against oxidative damage, thus reducing Cd toxicity (Aloui et al. 2009).

Among the essential metals, Zn and Cu have been particularly investigated. In particular, the negative effects of Cu on plant development (Wang et al. 2002) and modifications in the protein profile have been described (Bona et al. 2007). The Zn toxicity mechanisms take place through a number of biochemical processes, as described for Cd.

Defence mechanisms, based on antioxidant enzymes and on small antioxidant molecules, including proline, may protect the plant cell from ROS (Sharma and Dietz 2006). Although metal-induced proline accumulation in plant tissues has been observed (Andrade et al. 2009; Fariduddin et al. 2009), reports on the effects of mycorrhizal symbiosis in proline or soluble amino acid contents are scarce. Recently, however, high proline accumulation has been shown in response to Cu in

AM jack bean leaves. One of the proposed roles of proline is to reduce the level of free radicals generated as a result of toxicity in a similar manner to other molecules like glutathione, ascorbic acid or tocopherol (Andrade et al. 2010).

Other defence- or stress-related compounds are polyamines (PAs), which are present in all living organisms and which are essential in higher plants for growth and development (Bagni et al. 1993). The up-regulation of the PA metabolism has been reported in response to several environmental stress conditions (Urano et al. 2003), including HMs, in a number of plant species (Pirintsos et al. 2004; Scoccianti et al. 2006). Modifications in the content of a precursor of PAs in plants (arginine) have also been observed. The arginine contents were the most striking difference in the amino acid composition of mycorrhizal and non-mycorrhizal jack bean plants, grown in the presence of high Cu concentrations: the latter consistently exhibited arginine concentrations of between 35 and 50% of the total amino acid pool (Andrade et al. 2010). Lingua et al. (2008) observed levels of both free and conjugated PAs in poplar colonized by *G. mosseae* and grown in a Zn-polluted soil, similar to those that occur in plants grown without Zn pollution, suggesting that, in the presence of this AM fungus, given that the amount of zinc accumulated was very high, the toxicity of the metal was reduced. The same effect was not observed in *G. intraradices* inoculated plants, in which the growth inhibition due to zinc was not alleviated and the PA profile was altered in comparison with that of the controls.

4.2 Arsenic

Metalloids can also be extremely toxic for plants. An example is arsenic (As), which is heavily toxic for all living organisms and the environment, where it is released by natural and human activities. Arsenic is mainly present in the soil as arsenate (AsV) and arsenite (AsIII). AsV is an analogue of phosphate (Pi), and it competes with the latter for plant uptake by Pi transporters (Smith et al. 2010a). AsIII is more mobile and its uptake is believed to occur passively through membrane aquaporins (Ma et al. 2008). Once inside the plant, AsV can interfere with the phosphate metabolism, substituting it in the ATP, while AsIII, due to its high affinity for thiols, can inactivate several enzymes. Plants contrast As toxicity by reducing AsV to AsIII, and the latter is then eliminated from free cell circulation by complexation with thiolic peptides, such as GSH, and phytochelatins. The AsIII-thiol complexes can be segregated into the vacuole, by means of some glutathione-conjugated transporters (Smith et al. 2010a).

Plants from As contaminated soils are generally mycorrhizal (Cairney and Meharg 1999; Leung et al. 2007), indicating that fungal symbionts can evolve arsenic tolerance. As arsenate As(V) is an analogue of Pi, it might be expected that AMs would enhance the uptake of both. Because of the As(V)/Pi analogy, it could also be expected that the role of AMs in As tolerance would be different than that in HM tolerance (Smith et al. 2010a).

As mentioned in Sect. 2, it is well recognized that AM plants have two pathways through which Pi is absorbed from the soil solution (1) the direct pathway, in which Pi is taken up by roots as in non-AM plants, and can result in depletion of Pi in the soil solution close to the root system. The low concentration in the rhizosphere reduces subsequent Pi influx and may increase competition from As(V); (2) the AM pathway that involves uptake by the external mycelium and translocation to the plant through the fungal hyphae and transfer across the arbuscules (Grace et al. 2009; Smith and Read 2008). This pathway overcomes severe diffusion limitation of Pi uptake, as the external hyphae scavenge Pi at long distances from the roots (Smith et al. 2010a). AM and direct Pi uptake pathways are integrated and controlled, and, although data are not always consistent, it appears that the AM plants, compared to non-AM, take up relatively more Pi than As(V), and this results in a higher P/As ratio in AM plants (Smith et al. 2010a).

Some data have been reported on the influence of *G. mosseae* on As acquisition in *Medicago sativa* (Chen et al. 2007), on the effect of AM fungi of the *Glomus* spp. on biomass production and As accumulation in *Pityrogramma calomelanos, Tagetes erecta* and *Melastoma malabathricum* (Jankong and Visoottiviseth 2008) and on the As hyperaccumulation in the fern *Pteris vittata* (Liu et al. 2005; Trotta et al. 2006). Besides, Gonzalez-Chavez et al. (2002) reported that AM enhanced As resistance (through As exclusion) in *Holcus lanatus*.

In a study on As hyperaccumulation, in the absence and presence of *G. mosseae* and *Gigaspora margarita* in *P. vittata*, the expression in the fronds of the enzymes involved in photosynthesis and carbon fixation (i.e. RuBisCO, RuBisCO activase and ATP synthase) and sugar metabolism and bioenergetics (e.g. glyceraldehyde-phosphate dehydrogenase and triosephosphate isomerase) were especially affected. AM symbiosis also modulated the enzymes involved in the biosynthesis of S compounds (considering the role of some sulphurous compounds as non-toxic osmolytes or protective antioxidant agents) and some antioxidant enzymes, such as thioredoxin peroxidase and glutathione peroxidase (Berta et al. 2008; Bona et al. 2010). Two proteins in the roots of *P. vittata* showed a specific expression pattern in response to As and mycorrhization: glutamine synthetase, the key enzyme controlling the use of nitrogen inside the cells, and S-adenosylmethionine (SAM) synthase, which catalyzes SAM formation from methionine and ATP. These two proteins increased when non-AM plants were treated with As, while the same arsenic up-regulation did not occur in the presence of *G. mosseae* colonization; mycorrhization alleviated the metalloid effect and a decrease in expression was detected when the AM plants treated with As were compared with the non-AM As plants (Bona et al. 2010). Since *P. vittata* is an arsenic hyperaccumulator, proteomic analysis did not detect any enzymes involved in ROS scavenging, and the only response to oxidative stress was the up-regulation of aldehyde dehydrogenase (ALDH). ALDHs have been considered as general detoxifying enzymes that eliminate toxic biogenic and xenobiotic aldehydes (Gao and Han 2009). The up-regulation of ALDH was detected in cadmium-exposed poplar plants (Kieffer et al. 2008) and in aluminium-stressed tomato roots (Zhou et al. 2009).

4.3 PAHs and BTEX

Most of the information available in literature concerning organic pollutants examines the effect of AMs on polycyclic aromatic hydrocarbons (PAHs) in polluted soils (Leyval and Binet 1998; Binet et al. 2001; Joner et al. 2001; Liu and Dalpé 2009). The phytotoxic effect of these contaminants is often due to their hydrophobicity, which compromises the uptake of water by plant roots. The uptake of nutrients, which are usually dissolved in the aqueous phase of the soil, is also considerably decreased (Volante et al. 2005). A study by Joner et al. (2001), on clover and ryegrass colonized by *G. mosseae*, has shown the beneficial effect of the mycorrhization on plant growth in a soil artificially polluted with anthracene, crysene and benzanthracene. The decrease in PAHs in the soil was mainly attributed to the enhanced nutrient uptake by AMF (Leyval et al. 2002), which leads to improved plant growth, which, in turn, may stimulate soil microbial activity (Rabie 2005; Liu and Dalpé 2009). The contribution of the mycorrhizosphere to PAH biodegradation in the presence of ryegrass inoculated with *G. mosseae* has been studied by a number of authors (e.g. Corgie et al. 2006; Korade and Fulekar 2008; Liu and Dalpé 2009).

While there is increasing interest in AM fungi as bioremediators for PAH polluted soil, little is known about their possible use for the reclamation of sites contaminated by aromatic hydrocarbons as benzene, toluene, ethylbenzene, *meta- para-* e *ortho*-xylene (BTEXs), which have mutagenic and carcinogenic properties as well as relatively high hydrosolubility. The effect of three AM fungal species on the persistence of BTEXs in artificially contaminated substrates was evaluated using leek as the host plant. A specifically designed mesocosm system, in which the internal air and substrate samples were analyzed for the BTEX content by means of gas chromatography, was used. Important reductions were observed in the BTEX concentration in the substrates in the presence of AM plants. The residual BTEX content ranged between almost total disappearance (<2%) and 40% of the original concentration (Volante et al. 2005).

5 Conclusions and Future Perspective

AM colonization has been shown to increase plant tolerance to drought and pollution, and often reduces the typical responses of plants to stress, such as the production of antioxidant molecules or enzymes. Soil modifications induced by AM fungi have also been observed to play a role in protecting plants. However, despite the large number of studies on this topic, the underlying mechanisms of AM protection are not jet fully understood, and contradictory results have sometimes been reported. One reason for this, apart from the heterogeneity of the experimental procedures followed by different authors, may be connected to the different morphological and functional strategies of plants to decrease stress exposure or tolerate stress, and to the different efficiency of the associated AM fungi.

Therefore, the molecular, physiological and morphological mechanisms that regulate the responses of the AM plant to stress should be implemented by data on AM fungal biodiversity, and on functional diversity between AM fungal species and strains. In this respect, autochthonous AM fungal populations have been shown to be more effective in alleviating plant stress, and therefore autochthonous strains would seem to be preferable as inoculum to increase plant performances. The knowledge of the different strategies of plants to cope with stressful situations, the degree of plant dependence on mycorrhization, and the degree of specificity of individual plant–fungus association are equally important. Finally, plants and AM fungi in the field interact with pathogenic and beneficial micro-organisms. Among the latter, plant growth-promoting rhizobacteria (PGPR) may contribute to plant performances under stress, in some cases enhancing the effect of AM fungi (Gamalero et al. 2009, Marulanda et al. 2009). All these factors should be considered to obtain more detailed knowledge on the basic protection mechanisms and in order to apply AM fungi for phytoremediation purposes and for environmental and agriculture management programmes.

Acknowledgements This work has been partially supported by the Italian Ministero dell'Istruzione, dell'-Università e della Ricerca (MIUR)

References

Aloui A, Recorbet G, Gollotte A, Robert F, Valot F, Gianinazzi-Pearson V, Aschi-Smiti S, Dumas-Gaudot E (2009) On the mechanisms of cadmium stress alleviation in *Medicago truncatula* by arbuscular mycorrhizal symbiosis: a root proteomic study. Proteomics 9:420–433

Andrade SAL, Gratao PL, Schiavinato MA, Silveira APD, Azevedo RA, Mazzafera P (2009) Zn uptake, physiological response and stress attenuation in mycorrhizal jack bean growing in soil with increasing Zn concentrations. Chemosphere 75:1363–1370

Andrade SAL, Silveira AP, Mazzafera P (2010) Arbuscular mycorrhiza alters metal uptake and the physiological response of *Coffea arabica* seedlings to increasing Zn and Cu concentrations in soil. Sci Total Environ 408: 5381–5391

Aroca R, Porcel R, Ruiz-Lozano JM (2007) How does arbuscular mycorrhizal symbiosis regulate root hydraulic properties and plasma membrane aquaporins in *Phaseolus vulgaris* under drought, cold or salinity stresses? New Phytol 173:808–816

Aroca R, del Mar Alguacil M, Vernieri P, Ruiz-Lozano JM (2008a) Plant response to drought stress and exogenous ABA application are modulated differently by mycorrhization in tomato and an ABA-deficient mutant (*Sitiens*). Microb Ecol 56:704–719

Aroca R, Vernieri P, Ruiz-Lozano JM (2008b) Mycorrhizal and non-mycorrhizal *Lactuca sativa* plants exhibit contrasting responses to exogenous ABA during drought stress and recovery. J Exp Bot 59:2029–2041

Augé RM (2001) Water relations, drought and vesicular-arbuscular mycorrhizal symbiosis. Mycorrhiza 11: 3–42

Augé RM, Toler HD, Moore JL, Cho K, Saxton AM (2007) Comparing contributions of soil versus root colonization to variations in stomatal behavior and soil drying in mycorrhizal *Sorghum bicolor* and *Cucurbita pepo*. J Plant Physiol 164:1289–1299

Bagni N, Altamura MM, Biondi S, Mengoli M, Torrigiani P (1993) Polyamines and morphogenesis in normal and transgenic plant cultures. In: Roubelakis-Angelakis KA, van Tran Thanh K (eds) Morphogenesis in plants. Plenum, New York, pp 89–111

Barea J, Azcón R, Azcón-Aguilar C (2002) Mycorrhizosphere interactions to improve plant fitness and soil quality. Antonie Van Leeuwenhoek 81: 343–351

Bedini S, Ramasamy CK, Giovanetti M (2009) Glomalin-related soil protein as an indicator of the impact of different agronomical practices on soil quality. Aspect Appl Biol 98:75–80

Berta G, Bona E, Cattaneo C, Marsano F, D'Agostino G, Lingua G, Fusconi A, Cesaro P, Bonelli G, Cavaletto M (2008) AM symbiosis improved As tolerance in the As-hyperaccumulating brake fern *Pteris vittata*: cytological and molecular study. Physiol Plantarum 133:3

Binet P, Portal JM, Leyval C (2001) Application of GC-MS to the study of anthracene disappearance in the rhizosphere of ryegrass. Org Geochem 32:217–222

Bona E, Marsano F, Cavaletto M, Berta G (2007) Proteomic characterization of copper stress response in *Cannabis sativa* roots. Proteomics 7:1121–1130

Bona E, Cattaneo C, Cesaro P, Marsano F, Lingua G, Cavaletto M, Berta G (2010) Proteomic analysis of *Pteris vittata* fronds: two arbuscular mycorrhizal fungi differentially modulate protein expression under arsenic contamination. Proteomics 10:3811–3834

Bonfante P, Anca IA (2009) Plants, mycorrhizal fungi, and bacteria: a network of interactions. Annu Rev Microbiol 63:363–83

Boomsma CR, Vjn TJ (2008) Maize drought tolerance: potential improvements through arbuscular mycorrhizal symbiosis? Field Crops Res 108:14–31

Bothe H, Turnau K, Regvar M (2010) The potential role of arbuscular mycorrhizal fungi in protecting endangered plants and habitats. Mycorrhiza 20:445–457

Burleigh SH, Kristensen BK, Bechmann IE (2003) A plasma membrane zinc transporter from *Medicago truncatula* is up-regulated in roots by Zn fertilization, yet down-regulated by arbuscular mycorrhizal colonization. Plant Mol Biol 52:1077–1088

Cairney JWG, Meharg AA (1999) Influences of anthropogenic pollution on mycorrhizal fungal communities. Environ Pollut 106:169–182

Castiglione S, Todeschini V, Franchin C, Torrigiani P, Gastaldi D, Cicatelli A, Rinaudo C, Berta G, Biondi S, Lingua G (2009) Clonal differences in survival capacity, copper and zinc accumulation, and correlation with leaf polyamine levels in poplar: a large-scale field trial on heavily polluted soil. Environ Pollut 157:2108–2117

Cavagnaro TR (2008) The role of arbuscular mycorrhizas in improving plant zinc nutrition under low soil zinc concentrations: a review. Plant Soil 304:315–325

Chen BD, Xiao XY, Zhu YG, Smith FA, Xie ZM, Smith SE (2007) The arbuscular mycorrhizal fungus *Glomus mosseae* gives contradictory effects on phosphorus and arsenic acquisition by *Medicago sativa* Linn. Sci Total Environ 379:226–234

Citterio S, Prato N, Fumagalli P, Aina R, Massa N, Santagostino A, Sgorbati S, Berta G (2005) The arbuscular mycorrhizal fungus *Glomus mosseae* induces growth and metal accumulation changes in *Cannabis sativa* L. Chemosphere 59:21–29

Clemens S (2006) Toxic metal accumulation, responses to exposure and mechanisms of tolerance in plants. Biochimie 88:1707–1719

Corgie SC, Fons F, Beguiristain T, Leyval C (2006) Biodegradation of phenanthrene, spatial distribution of bacterial populations and dioxygenase expression in the mycorrhizosphere of *Lolium perenne* inoculated with *Glomus mosseae*. Mycorrhiza 16:207–212

Evelin H, Kapoor R, Giri B (2009) Arbuscular mycorrhizal fungi in alleviation of salt stress: a review. Ann Bot 104:1263–1280

Fariduddin Q, Yusuf M, Hayat S, Ahmad A (2009) Effect of 28-homobrassinolide on antioxidant capacity and photosynthesis in *Brassica juncea* plants exposed to different levels of copper. Environ Exp Bot 66: 418–424

Feddermann N, Finlay R, Boller T, Elfstrand M (2010) Functional diversity in arbuscular mycorrhiza – the role of gene expression, phosphorous nutrition and symbiotic efficiency. Fungal Ecol 3:1–8

Fitter AH (1988) Water relations in red clover, *Trifolium pratense* L., as affected by VA mycorrhizal infection and phosphorus supply before and during drought. J Exp Bot 39:595–604

Fitter AH, Helgason T, Hodge A (2011) Nutritional exchanges in the arbuscular mycorrhizal symbiosis: implications for sustainable architecture. Fungal Biol Rev 25:68–72

Freedman LP, Luisi BF, Korszun ZR, Basavappa R, Sigler PB, Yamamoto KR (1988) The function and structure of the metal coordination sites within the glucocorticoid receptor DNA binding domain. Nature 334:543–546

Fusconi A, Berta G, Tagliasacchi AM, Scannerini S, Trotta A, Gnavi E, De Padova S (1994) Root apical meristems of arbuscular mycorrhizae of *Allium porrum*. Environ Exp Bot 34:181–193

Fusconi A, Lingua G, Trotta A, Berta G (2005) Effects of AM colonization and phosphorus application on nuclear ploidy in *Allium porrum* plants. Mycorrhiza 15:313–321

Fusconi A, Gallo C, Camusso W (2007) Effects of cadmium on root apical meristems of *Pisum sativum* L.: cell viability, cell proliferation and microtubule pattern as suitable markers for assessment of stress pollution. Mutat Res-Gen Tox En 632:9–19

Gamalero E, Lingua G, Berta G, Glick BR (2009) Beneficial role of plant growth promoting bacteria and arbuscular mycorrhizal fungi on plant responses to heavy metal stress. Can J Microbiol 55:501–514

Gao C, Han B (2009) Evolutionary and expression study of the aldehyde dehydrogenase (ALDH) gene superfamily in rice (*Oryza sativa*). Gene 431:86–94

Garg N, Chandel S (2010) Arbuscular mycorrhizal networks: process and functions. A review. Agron Sustain Dev 30:581–599

Göhre V, Paszkowski U (2006) Contribution of the arbuscular mycorrhizal symbiosis to heavy metal phytoremediation. Planta 223:1115–1122

Gonzalez-Chavez C, Harris PJ, Dodd J, Meharg AA (2002) Arbuscular mycorrhizal fungi confer enhanced arsenate resistance on *Holcus lanatus*. New Phytol 155:163–171

Grace EJ, Cotsaftis O, Tester M, Smith FA, Smith SE (2009) Arbuscular mycorrhizal inhibition of growth in barley cannot be attributed to extent of colonization, fungal phosphorus uptake or effects on expression of plant phosphate transporter genes. New Phytol 181: 938–949

Harrison MJ (2005) Signaling in the arbuscular mycorrhizal symbiosis. Annu Rev Microbiol 59:19–42

Hernández EI, Vilagrosa A, Pausas JG, Bellot J (2010) Morphological traits and water use strategies in seedlings of Mediterranean coexisting species. Plant Ecol 207:233–244

Hildebrandt U, Regvar M, Bothe H (2007) Arbuscular mycorrhiza and heavy metal tolerance. Phytochemistry 68:139–146

Hodge A, Berta G, Doussan C, Merchan F, Crespi M (2009) Plant root growth, architecture and function. Plant Soil 321:153–187

Huang Z, Zou Z, He C, He Z, Zhang Z, Li J (2011) Physiological and photosynthetic responses of melon (*Cucumis melo* L.) seedlings to three *Glomus* species under water deficit. Plant Soil 339:391–399

Jankong P, Visoottiviseth P (2008) Effects of arbuscular mycorrhizal inoculation on plants growing on arsenic contaminated soil. Chemosphere 72:1092–1097

Joner EJ, Johansen A, Loibner AP, De la Cruz MA, Szolar OHJ, Portal JM, Leyval C (2001) Rhizosphere effects on microbial community structure and dissipation and toxicity of polycyclic aromatic hydrocarbons (PAHs) in spiked soil. Environ Sci Technol 35:2773–2777

Khalvati M, Bartha B, Dupigny A, Schroder P (2010) Arbuscular mycorrhizal association is beneficial for growth and detoxification of xenobiotics of barley under drought stress. J Soils Sediments 10:54–64

Kieffer P, Dommes J, Hoffmann L, Hausman JF, Renaut J (2008) Quantitative changes in protein expression of cadmium-exposed poplar plants. Proteomics 8: 2514–2530

Kohler J, Hernández JA, Caravaca F, Roldán A (2008) Plant-growth-promoting rhizobacteria and arbuscular mycorrhizal fungi modify alleviation biochemical mechanisms in water-stressed plants. Funct Plant Biol 35:141–151

Koltai H, Kapulnik Y (2010) Arbuscular mycorrhizal symbiosis under stress conditions: benefits and costs. Symbioses Stress 17:339–356

Korade DL, Fulekar MH (2008) Remediation of anthracene in mycorrhizospheric soil using ryegrass. Afr J Environ Sci Technol 2:249–256

Leung HM, Ye ZH, Wong MH (2007) Survival strategies of plants associated with arbuscular mycorrhizal fungi on toxic mine tailings. Chemosphere 66:905–915

Leyval C, Binet P (1998) Effect of polyaromatic hydrocarbons in soil on arbuscular mycorrhizal plants. J Environ Qual 27:402–407

Leyval C, Joner EJ, del Val C, Haselwandter K (2002) Potential of arbuscular mycorrhizal fungi for bioremediation. In: Gianinazzi S, Schuepp H, Barea JM, Haselwandter K (eds) Mycorrhizal technology in agriculture. Birkhauser-Verlag, Basel, Switzerland

Lingua G, Franchin C, Todeschini V, Castiglione S, Biondi S, Burlando B, Parravicini V, Torrigiani P, Berta G (2008) Arbuscular mycorrhizal fungi differentially affect the response to high zinc concentrations of two registered poplar clones. Environ Pollut 153:137–147

Liu Y, Zhu YG, Chen BD, Christie P, Li XL (2005) Influence of the arbuscular mycorrhizal fungus *Glomus mosseae* on uptake of arsenate by the As hyperaccumulator fern *Pteris vittata* L. Mycorrhiza 15:187–192

Liu A, Dalpé Y (2009) Reduction in soil polycyclic aromatic hydrocarbons by arbuscular mycorrhizal leek plants. Int J Phytoremed 11:39–52

Liu D, Jiang W, Meng Q, Zou J, Gu J, Zeng M (2009) Cytogenetical and ultrastructural effects of copper on root meristem cells of *Allium sativum* L. Biocell 33: 25–32

Ma JF, Yamaji N, Mitani N, Xu XY, Su YH, McGrath SP, Zhao FJ (2008) Transporters of arsenite in rice and their role in arsenic accumulation in rice grain. Proc Natl Acad Sci USA 105:9931–9935

Manoharan PT, Shanmugaiah V, Balasubramanian N, Gomathinayagam S, Sharma MP, Muthuchelian K (2010) Influence of AM fungi on the growth and physiological status of *Erythrina variegata* Linn. grown under different water stress conditions. Eur J Soil Biol 46:151–156

Marulanda A, Azcón R, Ruiz-Lozano JM (2003) Contribution of six arbuscular mycorrhizal fungal isolates to water uptake by *Lactuca sativa* L. plants under drought stress. Physiol. Plantarum 119:526–533

Marulanda A, Porcel R, Barea JM, Azcón R (2007) Drought tolerance and antioxidant activities in lavender plants colonized by native drought-tolerant or drought-sensitive *Glomus* species. Microbial Ecol 54: 543–552

Marulanda A, Barea JM, Azcón R (2009) Stimulation of plant growth and drought tolerance by native microorganisms (AM fungi and bacteria) from dry environments: mechanisms related to bacterial effectiveness. J Plant Growth Regul 28:115–124

Massa N, Andreucci F, Poli M, Aceto M, Barbato R, Berta G (2010) Screening for heavy metal accumulators amongst autochtonous plants in a polluted site in Italy. Ecotox Environ Safe 73:1988–1997

Maurel C, Simonneau T, Sutka M (2010) The significance of roots as hydraulic rheostats. J Exp Bot 61: 3191–3198

Medina A, Roldán A, Azcón R (2010) The effectiveness of arbuscular-mycorrhizal fungi and *Aspergillus niger* or *Phanerochaete chrysosporium* treated organic amendments from olive residues upon plant growth in a semi-arid degraded soil. J Environ Manage 91:2547–2553

Miransari M (2010) Contribution of arbuscular mycorrhizal symbiosis to plant growth under different types of soil stress. Plant Biol 12:563–569

Neumann E, Schmid B, Römheld V, George E (2009) Extraradical development and contribution to plant performance of an arbuscular mycorrhizal symbiosis exposed to complete or partial root zone drying. Mycorrhiza 20:13–23

Offre P, Pivato B, Mazurier S, Siblot S, Berta G, Lemanceau P, Mougel C (2008) Compared diversity of bacteria associated with mycorrhized and non-mycorrhized roots of *Medicago truncatula*. FEMS Microbiol Ecol 65:180–192

Ouziad F, Hildebrandt U, Schmelzer E, Bothe H (2005) Differential gene expressions in arbuscular mycorrhizal-colonized tomato grown under heavy metal stress. J Plant Physiol 162:634–649

Pirintsos SA, Kotzabasis K, Loppi S (2004) Polyamine production in lichens under metal pollution stress. J Atmos Chem 49:303–315

Porcel R, Ruiz-Lozano JM (2004) Arbuscular mycorrhizal influence on leaf water potential, solute accumulation, and oxidative stress in soybean plants subjected to drought stress. J Exp Bot 55:1743–1750

Porcel R, Gomez M, Kaldenhoff R, Ruiz-Lozano JM (2005) Impairment of NtAQP1 gene expression in tobacco plants does not affect root colonisation pattern by arbuscular mycorrhizal fungi but decreases their symbiotic efficiency under drought. Mycorrhiza 15: 417–423

Porcel R, Aroca R, Cano C, Bago A, Ruiz-Lozano JM (2007) A gene from the arbuscular mycorrhizal fungus *Glomus intraradices* encoding a binding protein is up-regulated by drought stress in some mycorrhizal plants. Environ Exp Bot 60:251–256

Potters G, Pasternak TP, Guisez Y, Palme KJ, Jansen MAK (2007) Stress-induced morphogenic responses: growing out of trouble? Trends Plant Sci 12:98–105

Rabie GH (2005) Role of arbuscular mycorrhizal fungi in phytoremediation of soil rhizosphere spiked with poly aromatic hydrocarbons. Mycobiology 33:41–50

Reddy AR, Chaitanya KV, Vivekanandan M (2004) Drought-induced responses of photosynthesis and antioxidant metabolism in higher plants. J Plant Physiol 161:1189–1202

Repetto O, Bestel-Corre' G, Dumas-Gaudot E, Berta G, Gianinazzi-Pearson V, Gianinazzi S (2003) Targeted proteomics to identify cadmium-induced protein modifications in *Glomus mosseae*-inoculated pea roots. New Phytol 157:555–568

Repetto O, Massa N, Gianinazzi S, Gianinazzi-Pearson V, Dumas-Gaudot E, Berta G (2007) Cadmium effects on populations of root nuclei in two pea genotypes inoculated or not with the arbuscular mycorrhizal fungus *Glomus mosseae*. Mycorrhiza 17:111–120

Rillig MC, Mummey DL (2006) Mycorrhizas and soil structure. New Phytol 171:41–53

Rivera-Becerril F, Calantzis C, Turnau K, Caussanel J-P, Belimov AA, Gianinazzi S, Strasser RJ, Gianinazzi-Pearson V (2002) Cadmium accumulation and buffering of cadmium-induced stress by arbuscular mycorrhiza in three *Pisum sativum* L. genotypes. J Exp Bot 53:1177–1185

Roccotiello E, Manfredi A, Drava G, Minganti V, Mariotti M, Berta G, Cornara L (2010) Zinc tolerance and accumulation in the ferns *Polypodium cambricum* L. and *Pteris vittata* L. Ecotox Environ Safe 73: 1264–1271

Rosselli W, Keller C, Boschi K (2003) Phytoextraction capacity of trees growing on a metal contaminated soil. Plant Soil 256:265–272

Rout GR, Das P (2003) Effect of metal toxicity on plant growth and metabolism: I. Zinc. Agronomie 23:3–11

Ruiz-Lozano JM, Porcel R, Aroca R (2006) Does the enhanced tolerance of arbuscular mycorrhizal plants to water deficit involve modulation of drought-induced plant genes? New Phytol 171:688–693

Ruiz-Sánchez M, Aroca R, Muñoz Y, Polón R, Ruiz-Lozano JM (2010) The arbuscular mycorrhizal symbiosis enhances the photosynthetic efficiency and the antioxidative response of rice plants subjected to drought stress. J Plant Physiol 167:862–869

Scannerini S, Fusconi A, Mucciarelli M (2001) The effect of endophytic fungi on host plant morphogenesis. In: Seckback J (ed) Symbiosis: organisms and model systems. Kluwer, Dordrecht, The Netherlands, pp 427–447

Schreiner RP, Tarara JM, Smithyman RP (2007) Deficit irrigation promotes arbuscular colonization of fine roots by mycorrhizal fungi in grapevines (*Vitis vinifera* L.) in an arid climate. Mycorrhiza 17:551–562

Scoccianti V, Crinelli R, Tirillini B, Mancinelli V, Speranza A (2006) Uptake and toxicity of Cr(III) in celery seedlings. Chemosphere 64:1695–1703

Sharma SS, Dietz KJ (2006) The significance of amino acids and amino acid-derived molecules in plant responses and adaptation to heavy metal stress. J Exp Bot 57:711–726

Skirycz A, Inzé D (2010) More from less: plant growth under limited water. Curr Opin Biotechnol 21:197–203

Smith SE, Read DJ (2008) Mycorrhizal symbiosis, 3rd edn. Academic, London, UK

Smith SE, Christophersen HM, Pope S, Smith FA (2010a) Arsenic uptake and toxicity in plants: integrating mycorrhizal influences. Plant Soil 327:1–21

Smith SE, Facelli E, Pope S, Smith FA (2010b) Plant performance in stressful environments: interpreting new and established knowledge of the roles of arbuscular mycorrhizas. Plant Soil 326:3–20

Todeschini V, Franchin C, Castiglione S, Burlando B, Biondi S, Torrigiani P, Berta G, Lingua G (2007) Responses of two registered poplar clones to copper, after inoculation, or not, with arbuscular mycorrhizal fungi. Caryologia 60:146–155

Todeschini V, Lingua G, D'Agostino G, Carniato F, Roccotiello E, Berta G (2011) Effects of high zinc concentration on poplar leaves: a morphological and biochemical study. Environ Exp Bot 71:50–56

Tonin C, Vandenkoornhuyse P, Joner EJ, Straczek J, Leyval C (2001) Assessment of arbuscular mycorrhizal fungi diversity in the rhizosphere of *Viola calaminaria* and effect of these fungi on heavy metal uptake by clover. Mycorrhiza 10:161–168

Trotta A, Falaschi P, Cornara L, Minganti V, Fusconi A, Drava G, Berta G (2006) Arbuscular mycorrhizae increase the arsenic translocation factor in the As hyperaccumulating fern *Pteris vittata* L. Chemosphere 65:74–81

Urano K, Yoshiba Y, Nanjo T, Igarashi Y, Seki M, Sekiguchi F, Yamaguchi-Shinozaki K, Shinozaki K (2003) Characterization of *Arabidopsis* genes involved in biosynthesis of polyamines in abiotic stress responses and developmental stages. Plant Cell Environ 26:1917–1926

Vallino M, Massa N, Lumini E, Bianciotto V, Berta G, Bonfante P (2006) Assessment of arbuscular mycorrhizal fungal diversity in roots of *Solidago gigantea* growing in a polluted soil in northern Italy. Environ Microbiol 8:971–983

Vance CP, Uhde-Stone C, Allan DL (2003) Phosphorus acquisition and use: critical adaptations by plants for securing a nonrenewable resource. New Phytol 157:423–447

Volante A, Lingua G, Cesaro P, Cresta A, Puppo M, Ariati L, Berta G (2005) Influence of three species of arbuscular mycorrhizal fungi on the persistence of aromatic hydrocarbons in contaminated substrates. Mycorrhiza 16:43–50

Wang JR, Zhao FJ, Meharg AA, Raab A, Feldmann J, McGrath SP (2002) Mechanisms of arsenic hyperaccumulation in *Pteris vittata*. Uptake kinetics, interactions with phosphate, and arsenic speciation. Plant Physiol 130:1552–1561

Wu QS, Xia RX (2006) Arbuscular mycorrhizal fungi influence growth, osmotic adjustment and photosynthesis of citrus under well-watered and water stress conditions. J Plant Physiol 163:417–425

Wu QS, Xia RX, Zou YN (2006) Reactive oxygen metabolism in mycorrhizal and non-mycorrhizal citrus (*Poncirus trifoliata*) seedlings subjected to water stress. J Plant Physiol 163:1101–1110

Wu QS, Xia RX, Zou YN (2008) Improved soil structure and citrus growth after inoculation with three arbuscular mycorrhizal fungi under drought stress. Eur J Soil Biol 44:122–128

Wu QS, Zou YN (2009) Mycorrhiza has a direct effect on reactive oxygen metabolism of drought-stressed citrus. Soil Environ 55:436–442

Zenk MH (1996) Heavy metal detoxification in higher plants - a review. Gene 179:21–30

Zhou SP, Sauvé R, Thannhauser TW (2009) Proteome changes induced by aluminium stress in tomato roots. J Exp Bot 60:1849–1857

Effects of Exogenous Application of 5-Aminolevulinic Acid in Crop Plants

12

Ahmet Korkmaz

Abstract

Studies on compounds capable of enhancing tolerance of crop species to abiotic stresses are of great importance. Although first discovered as possessing herbicidal properties, latest studies have demonstrated that 5-aminolevulinic acid (ALA) has both growth promoting and stress enhancing capabilities. ALA is known as the precursor of all porphyrins compounds such as vitamin B12, chlorophyll, heme, and phytochrome, and it is naturally found in plants, animals, algae, and photosynthetic bacteria. As a precursor of chlorophyll, exogenous application of ALA at low concentrations increases chlorophyll content of plants, promoting photosynthetic capacity and yield of crops. When applied in proper concentrations, ALA is also reported to provide significant tolerance against abiotic stresses such as chilling, salinity, and drought. The ability of ALA to boost abiotic stress tolerance is due to elevated activities of enzymatic or nonenzymatic antioxidant system providing significant protection to the membranes against harmful reactive oxygen species within tissues. This review provides a comprehensive coverage of the effects exerted by ALA in relation to plant productivity and abiotic stress tolerance.

Keywords

5-Aminolevulinic acid • Abiotic stress • Plant productivity • Antioxidant system

A. Korkmaz (✉)
Department of Horticulture, Faculty of Agriculture,
Kahramanmaras Sutcu Imam University,
Kahramanmaras 46100, Turkey
e-mail: akorkmaz@ksu.edu.tr

P. Ahmad and M.N.V. Prasad (eds.), *Abiotic Stress Responses in Plants: Metabolism, Productivity and Sustainability*, DOI 10.1007/978-1-4614-0634-1_12, © Springer Science+Business Media, LLC 2012

1 Introduction

Stress is defined as an external factor that exerts a detrimental effect on overall growth of a plant. Plants being sessile are subjected to various types of environmental stresses caused by both abiotic and biotic factors (Srivastava 2002). Abiotic stresses are caused by complex environmental conditions such as high and low temperatures, freezing, drought, salinity, heavy metals, strong light, UV, or hypoxia (Hirayama and Shinozaki 2010). These conditions will significantly decrease crop productivity in the near future due to global climate change, according to reports from the Intergovernmental Panel of Climate Change (http://www.ipcc.ch). For example, during the European heat wave of 2003, crop production was approximately decreased by 30% (Ciais et al. 2005); therefore, development of methods to induce stress tolerance in plants is vital and still receives considerable attention.

The approaches employed to develop stress-tolerant plants include breeding, genetic engineering, and the use of plant growth regulators (PGRs) (Vettakkorumakankav et al. 1999). Improving stress tolerance in sensitive species through breeding is difficult to accomplish for three reasons. First, stress tolerance traits are quantitative in nature and controlled by more than one gene. Second, breeding with distant or wild relatives which are tolerant to a specific stress suffers from the risk of introducing undesirable traits. And lastly, although improving stress tolerance by breeding is a path with solid results at the end, it could be a very long process requiring large amount of resources. Genetic engineering on the other hand is attractive since it avoids the problem of important traits loses as only one or two tolerance genes are introduced into the target crop after the transformation (Srivastava 2002). However, the problem with genetic engineering is the lack and the difficulty of identifying tolerance gene(s) to a specific stress factor in crop species. The roles of plant hormones or in broader sense, PGRs are being explored and identified through ongoing research by plant scientists. Many molecules naturally produced by plants, for example, jasmonic acid, abscisic acid, salicylic acid, and polyamines have been suggested as signal transducers and messengers (Klessig and Malamy 1994). Increasing evidence indicates that plants with high levels of these compounds have increased tolerance to various abiotic stress conditions, and an increase in endogenous concentrations of these molecules before exposure to stressful conditions might be an essential step to activate a protection mechanism against the stress. The application of PGRs or various chemicals to plants prior to onset of stress conditions has had considerable success in enhancing abiotic stress tolerance in recent years. For example, exogenous application of abscisic acid (ABA) can substitute for low temperature to initiate acclimation in a number of species including watermelon (Korkmaz 2002), chickpea (Kumar et al. 2008), and maize (Janowiak et al. 2002) and is known to enhance tolerance to salinity (Etehadnia et al. 2008), high temperature (Ding et al. 2010), and drought stresses (Lu et al. 2009). Similarly, the application of chemicals in the triazole family including paclobutrazol, uniconizole, and triadimefon protects plants from many stresses including chilling (Baninasab 2009), drought (Gilley and Fletcher 1997), and salinity (Jaleel et al. 2008). Even though much of the research on these compounds has been limited to the laboratory conditions, it is particularly important to identify their potential mode of action for manipulation by molecular breeding or genetic engineering in order to develop stress tolerant or resistant lines in the future. One such compound that has attracted great attention lately is the 5-aminolevulinic acid (ALA) and this review encompasses an overview of the current work reported on the roles of ALA in plant productivity under optimum conditions as well as in improving abiotic stress tolerance of crop species.

2 5-Aminolevulinic Acid

2.1 Properties and Biosynthesis

ALA is a keto-amino acid with a molecular weight of 131 kDa (Fig. 12.1). ALA is known as a precursor of all porphyrins compounds such as vitamin

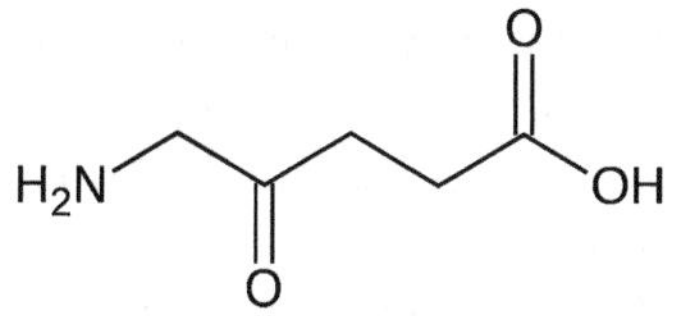

Fig. 12.1 Structure of 5-aminolevulinic acid (ALA)

B12, chlorophyll, heme, and phytochrome, and it is naturally found in plants, animals, algae, and photosynthetic bacteria (Wang et al. 2005; Tabuchi et al. 2009). It is generally assumed that in plants, algae, bacteria (except for the α-proteobacteria group), and archaea, ALA is synthesized via light-dependent C5 pathway (also known as Beale pathway) in plastids from glutamic acid via glutamyl-tRNA and glutamate-1-semialdehyde (Beale 1990; Reinbothe and Reinbothe 1996). The enzymes involved in this pathway are glutamyl-tRNA synthetase, glutamyl-tRNA reductase, and glutamate-1-semialdehyde aminotransferase. In nonphotosynthetic eukaryotes such as animals, insects, fungi, and protozoa, as well as the α-proteobacteria group of bacteria, ALA is formed by a reaction known as Shemin pathway (C4 pathway) via the condensation of succinyl COA and glycine by the enzyme ALA synthase (Beale and Weinstein 1989; Bisbis et al. 1997). It has also been proposed that exposure of plants to low and high temperatures is associated with blockage of ALA biosynthesis suggesting that ALA biosynthesis is a temperature-sensitive process (Hodgins and van Huystee 1986; Tewari and Tripathy 1998). Low temperature exposure of leaf tissues impaired ALA biosynthesis in maize seedlings resulting in reduced porphyrin synthesis and chlorosis (Hodgins and van Huystee 1986). Moreover, illumination of cucumber seedlings under chilling and heat stress conditions resulted in inhibition of chlorophyll biosynthesis by 90 and 60%, respectively, which demonstrated that inhibition of chlorophyll biosynthesis is higher under chilling stress than heat stress conditions (Tewari and Tripathy 1998). Same authors also found that reduced chlorophyll biosynthesis was partly resulted from the impairment of ALA biosynthesis since ALA biosynthesis was inhibited to a similar extent both under chilling (78%) and heat (70%) stress conditions.

It is generally accepted that angiosperms can produce chlorophyll only under the influence of light. On the other hand, evidence has been accumulating that several species of angiosperms may also be capable of forming chlorophyll in the dark and that ALA could have a stimulatory role in chlorophyll synthesis in dark (Yang et al. 2003). Treating the leaves of etiolated angiosperm plants with ALA caused the accumulation of protochlorophyllide (Pchlide) and it was concluded that all enzymes required for Pchlide synthesis were already active and present in nonlimiting amounts and that only the activity and amount of enzymes involved in ALA synthesis limited the synthesis rate (Papenbrock and Grimm 2001). Pchlide accumulates in the dark because angiosperms reduce Pchlide to chlorophyllide (Chlide) by a photo-enzyme, light-dependent protochlorophyllide oxidoreductase and this reaction represents a key regulatory step in the strictly light-dependent biosynthesis of chlorophyll in angiosperms (Pavlovic et al. 2009). On the contrary, some photosynthetic organisms such as cyanobacteria, algae, mosses, ferns, and gymnosperms could operate a second chlorophyll biosynthesis pathway and they are capable of synthesizing chlorophyll and bacteriochlorophlls in the dark by light-independent protochlorophyllide oxidoreductase (Armstrong 1998).

ALA formation is the rate-limiting step in chlorophyll biosynthesis in plants (Beale, 1990; Pavlovic et al., 2009) and ALA concentration is firmly controlled to less than 50 nmol/g FW in plant tissues (Stobart and Ameen-Bukhari 1984). At high concentrations (5–40 mmol/L), ALA undergoes enolization and further metal-catalyzed anaerobic oxidation at physiological pH to form reactive oxygen species (ROS) such as superoxide radical (O_2^-), hydrogen peroxide (H_2O_2), and hydroxyl radical ($HO^\cdot$) (Kumar et al. 1999; Balestrasse et al. 2010). Thus, higher concentrations of ALA in tissues enhance ROS production leading to oxidative stress, acting as a herbicide. When green plants are exposed to light after ALA treatment in darkness, excess chlorophyll intermediates such as Pchlide and protoporphyrin IX are photosensitized and consequent photodynamic reactions destroy the plant by oxidizing the unsaturated fatty acids in the cell membranes (Rebeiz et al. 1984; Chakraborty and Tripathy 1992). ALA, therefore, can be used as a safe substitute for highly toxic herbicides such as paraquat (Sasaki et al. 2002). Similar mechanism was also observed when ALA was used as a biodegradable insecticide to combat cabbage looper (*Trichopusia ni*) (Rebeiz et al. 1988). On the other hand, low ALA concentrations (0.06–0.6 mM) appear to promote rather than damage growth of several crops and vegetables (Watanabe et al. 2000; Hotta et al. 1997a). In addition, ALA is known to induce tolerance to various abiotic stress conditions in a variety of crop species (Korkmaz et al. 2010; Naeem et al. 2010). Furthermore, ALA has low mammalian toxicity (Kennedy et al. 1990) and it is biodegradable in soil (Hotta et al. 1997b). All of these properties suggest that ALA may have a great application potential in agriculture as a new nontoxic endogenous substance (Wang et al. 2003).

2.2 Effects of ALA on Chlorophyll Content and Crop Productivity

The trend in present day agricultural crop production is to increase the economic yield through more efficient use and partitioning of photosynthates as well as by improving net photosynthesis in crop plants under ever-changing conditions (Bindu Roy and Vivekanandan 1998a; Wardlaw 1980). Yield of crop plants can be improved through a number of ways such as by promoting branching, increasing leaf area index, enhancing tolerance to various biotic and abiotic stress conditions, and by manipulating partitioning of photosynthetic assimilates. PGRs alter plant growth and development by triggering numerous physiological responses, and ALA is listed among these PGRs that are used to manipulate plant growth and yield (Watanabe et al. 2006; Hotta et al. 1997c). ALA applied through different methods such as seed soaking, root drenching, or foliar application is known to enhance plant growth and productivity.

ALA has a variety of physiological effects on chlorophyll biosynthesis, photosynthesis, and plant growth. The efficacy of ALA application on overall plant growth and yield depends on ALA concentration used, application method as well as application time and crop species (Kantha et al. 2010). Soaking the seeds of several legume species in ALA solutions at concentrations ranging from 10 to 100 $g\,m^{-3}$ resulted in enhanced chlorophyll accumulation and photosynthetic rate which, in turn, caused an increase in yield (Bindu Roy and Vivekanandan 1998a). The authors concluded that the action of ALA cannot be simply explained by the fact that ALA is the precursor of chlorophyll and that the increment in leaf photosynthetic rate and plant leaf area significantly contributed to plant photosynthetic carbon assimilation and total biomass production. Similar confirmatory findings were also reported by others that ALA stimulates photosynthesis and decreases respiration in the dark promoting the yield of several crops such as kidney beans, garlic and spinach (Tanaka et al. 1992), and pakcohoi (Memon et al. 2009). Although positive effects of ALA treatment in low concentrations on growth, chlorophyll content, and photosynthetic rate of angiosperms has been well documented by several studies, the effects of exogenously applied ALA on chlorophyll formation in seedlings of gymnosperms is somewhat ambiguous. Treating pine (*Pinus nigra* Arn.) seedlings with ALA at 10^{-5} M concentration did not have a notable effect on chlorophyll accumulation in the light whereas chlorophyll formation was significantly enhanced in dark-grown seedlings (Drazic and Mihailovic 1998). It was presumed that in the light, ALA was synthesized endogenously in optimal amounts which may explain the lack of response to exogenous ALA application. On the other hand, ALA application promoted chlorophyll accumulation in darkness probably due to compensation of insufficient production of endogenous ALA. On the contrary, application of 1 mM ALA to the Norway spruce (*Picea abies*) seedlings in light resulted in increased chlorophyll accumulation while higher concentrations of ALA had negative effect on growth and chlorophyll accumulation (Pavlovic et al. 2009).

Treating 2-year-old grapevines with ALA through foliar application (30–300 mg L^{-1}) or root drench (0.1–10 mg L^{-1}) significantly increased photosynthetic rate by up to 22%, fruit sugar content by 2.7%, and cluster weight by 53% (Watanabe et al. 2006). Enhanced photosynthesis was reported to cause higher sugar content in berries, and increased berry weight in the cluster may have given the cluster a stronger sink strength which could be an indirect cause for the enhanced photosynthesis besides increased stomatal aperture. It was also reported that ALA applied at 0.06–6 μM through root soaking improved the growth of rice seedlings in light while ALA at 0.06 μM elicited chlorophyll accumulation in addition to increased photosynthetic rate in pothos lime (*Epiprenunum aureus*) plants (Hotta et al. 1997c). Additionally, in another study where wide range of ALA concentrations were tested in order to identify the optimum concentration for different application methods, it was found that up to 50% increase in plant mass of rice seedlings was observed when the optimum concentrations were applied at the rate of 0.1–1, 30–100, and 10–100 ppm for root soaking, foliar spray, and soil treatment, respectively (Hotta et al. 1997b). However, when concentrations above the optimum were used, deleterious effects were reported.

The most comprehensive results on the effects of exogenous ALA application on photosynthetic activity and crop's yield were published by Hotta et al. (1997a). They found that ALA positively affected the growth and yield of several crops and vegetables at concentrations lower than those eliciting herbicidal effect (i.e., less than 1.8 mM). Foliar application of ALA at the range of 0.06–1.8 mM significantly increased dry weight of radish roots but injured radish seedlings at 6 mM, while concentrations of 0.18 and 0.6 mM increased CO_2 assimilation in light and decreased the release of CO_2 in darkness. ALA application at low concentrations (0.18 and 0.6 mM) increased the growth and yield by 10–60% over non-ALA-treated plants on kidney bean, barley, potato, and garlic. In addition, ALA applied at the range of 10–50 ppm increased the yield of wheat by about 15% over the control after foliar application, and it was found that the effects of ALA depended on the timing of application and its concentration (Bingshan et al. 1998).

Exogenous application of ALA to fruits and vegetables is also reported to influence positively the quality of crop species. Spraying date palm fruits with 100 ppm ALA at different stages of fruit development increased fruit weight, fruit flesh percentage, fruit volume, and total and reducing sugar content although the positive response was dependent upon the application time or fruit developmental stage (Al-Khateeb et al. 2006). Application of 300 ppm ALA to apple fruits 20 days before harvest increased the total soluble solid content and decreased titratable acidity with no negative effect on fruit firmness and shelf life (Wang et al. 2004a). It was also noted that no significant residue was found on the fruits which suggested that ALA could be used to improve apple quality. ALA application at the rate of 300 mg L^{-1} to 'Fuji' apple fruits 43 days before harvest significantly increased anthocyanin accumulation rate doubling the final anthocyanin content present in the fruit in comparison to untreated control fruits (Wang et al. 2006). Moreover, exogenous application of ALA to tomato fruits decreased respiration rate, the malondialdehyde (MDA) content, relative membrane permeability and titratable acidity and increased total soluble solid content, all of which resulted in improved fruit quality and prolonged shelf life (Wang et al. 2009). Additionally, exogenous application of ALA caused significant enhancement in glucose content and starch degrading enzyme, amylase activity in radish (*Raphanus sativus*) taproot (Hara et al. 2011). ALA-based fertilizer "Pentakeep" applied at the rate of 0.3% significantly increased fruit dry matter, sugar, and citric acid contents of the hydroponically grown strawberries (Iwai et al. 2005), while an increase in N content by ALA treatment has also been reported in spinach (Yoshida et al. 1995).

Plants themselves can synthesize phytohormones, but they can also utilize exogenous sources such as exogenously applied phytohormones by humans or microbially produced phytohormones, and this may be one of the mechanisms of plant growth promotion by microorganisms. There have been many reports on the microbial production of phytohormones. Photosynthetic bacteria (PB) which are widely distributed in nature especially in submerged conditions such as paddy fields, riverbeds, seashores, and sewage disposable plants (Kobayashi and Kobayashi 2000) are also able to synthesize tetraphyrroles. Some PB species such as *Rhodopseudomonas palustris* and *Rhodobacter sphaerides* can produce relatively large amounts of physiologically active substances such as vitamin B12, ubiquinone, and ALA (Sasaki et al. 2002), and they can be considered to be one of natural fertilizers (Kantha et al. 2010). For example, Koh and Song (2007) reported that two PB strains of *Rhodopseudomonas* sp. produced as much as 8.75 mg L^{-1} ALA within 48 h of inoculation which caused efficient growth enhancement of tomato seedlings under axenic conditions. The germination percentage of PB-inoculated tomato seeds, total length, and dry mass of germinated tomato seedlings increased by 30.2, 71.1, and 270.8%, respectively, compared to those of the uninoculated control. It was also reported that when soil and straw products were inoculated with different strains of *Rhodopseudomonas palustris* for 4 weeks with microaerobic-dark conditions, the ALA content increased with time to achieve levels of 2.96 mM depending on the PB strain, and it was concluded that PB could be practically applied to organic saline paddy fields and increase growth and yields of rice (Kantha et al. 2010). Moreover, application of PB also enhanced growth, fruit formation, yield and fruit quality in tomato plants grown in greenhouse (Lee et al. 2008), increased mushroom (*Agariscus bisporus*) yield (Han 1999), and controlled the root rot on rice seedlings (Kobayashi and Kobayashi 2000).

Iwai et al. (2003) found that response to exogenously applied ALA was amplified when plants were supplied with higher rates of N, which may be partially attributed to the role of N in chlorophyll synthesis and plant growth. Hydroponically grown paprika type pepper plants treated with ALA yielded up to 9% more than control plants which may have been due to the fact that ALA-treated plants utilized 16% more NO_3 from the nutrient solution. Similar results were also reported in papaya (*Carica papaya* L.) where simultaneous application of N and ALA increased vegetative growth and reduced the time from papaya seedling emergence to the transplanting stage (Morales-Payan and Stall 2005). Increased Ca^{2+} content in spinach plants were also reported when ALA-based fertilizer "Pentakeep V" was applied simultaneously with N fertilizers (Smoleñ and Sady 2010).

Plant tissue culture is an important technique in plant propagation and since ALA possesses PGR properties, it is reported to play an important role in plant tissue culture. ALA treatment of explants of *Laminaria japonica* sporophyte was found to be useful to produce and propagate callus-like cells stably (Tabuchi et al. 2009). ALA treatment at the rate of 50–500 mg L^{-1} was more effective in inducing callus formation than control (0 mg L^{-1}) and cell division rate was the highest when explants were cultured with 500 mg L^{-1} ALA. Same concentration of ALA also promoted the growth of photoautotrophically growing cells of *Spirulina platensis* causing intracellular accumulations of phycocyanin and chlorophyll followed by enhancement of the photosynthetic activities of photosystems I and II (Sasaki et al. 1995). In vitro studies with *Vigna unguiculata* L. confirmed the hormonal role of ALA by striking proliferation of callus and pari-passu induction of rooting and shooting with a profound effect of the former than the latter, and ALA was therefore reported to exhibit both auxin and cytokinin properties in the induction of callusing and rooting and shooting, respectively (Bindu Roy and Vivekanandan 1998b). Also, ALA-based fertilizer "Pentakeep" applied at the rate of 0.04–0.08% shortened the required program to acclimatize the tissue culture-derived date palm seedlings by about 4–5 months compared to untreated plants by enhancing the growth of the seedlings via increasing nutrient uptake, chlorophyll concentration, and photosynthetic assimilation (Awad 2008).

One of the physiological roles of ALA in plant growth was recently reported by Maruyama-Nakashita et al. (2010). They demonstrated that exogenously applied ALA at the rate of 0.3–1 mmol L^{-1} increased the transcript levels of sulfur transport and assimilatory genes causing significant enhancements of sulfate uptake under both sulfur-sufficient and sulfur-deficient conditions in *Arabidopsis thaliana*. In addition, ALA application also increased the accumulation of cysteine and glutathione, particularly in the shoots all of which suggest a new role for ALA in regulating the sulfur assimilatory pathway.

The chemical stability of ALA in aqueous solutions was reported to be a function of its concentration, pH, and storage temperature with the higher the concentration, pH, and storage temperature were, the faster the rate of ALA in aqueous solution degraded. Thus, when ALA solutions are prepared, the concentration and the pH of the solution should be as low as possible according to different application purposes, and the solution should be stored at low temperatures (<–20°C) (Bunke et al. 2000; Gadmar et al. 2002). It was also suggested that the final solution of pH 5.5–7.4 would have to be prepared a maximum of 1 h before use.

2.3 ALA and Plants Under Stress

2.3.1 Effects of Exogenous ALA on Plants Under Chilling Stress

Low temperature is one of the major factors limiting the productivity and geographical distribution of many species, including several important agricultural crops. Reductions in temperatures can substantially slow the velocity of many metabolic pathways, which leads to the natural deterioration and loss of crop quality. There are two types of injuries a plant faces under exposure to low temperatures. The first type of injury is called freezing injury which occurs when the external temperature drops below the freezing point of water. When a plant freezes, this causes ice formation within the tissues and ruptures cell walls causing loss of cellular integrity and ultimate death of the tissue. Freezing-tolerant plants have several strategies to reduce the probability of this phenomenon occurring, even when air temperature drops below zero, including maintaining high intracellular solute concentrations which reduces probability of freezing inside cells and encouraging ice nucleation outside the cells (Allen and Ort 2001). Many plants, primarily young plants or seedlings that are native to tropics or warm climates, are very sensitive to low temperatures, showing abrupt reductions in the rates of physiological processes and exhibiting signs of injury following exposure to temperatures less than 15°C and they are called chilling-sensitive plants. Chilling injury can be defined as injury resulting from temperature that is cool enough to cause damage but not cold enough to freeze or to kill the plant (Levitt 1980). Chilling injury depends not only on the species and tissue type, but also on the severity and duration of exposure to low temperature (Lynch 1990). The temperature below which chilling injury can occur varies with species and regions of origin, ranging from 0 to 4°C for temperate fruits, 8°C for subtropical fruits, and about 12°C for tropical fruits such as banana (Lyons 1973). Major crop species including maize (*Zea mays*), cotton (*Gossypium hirsatum*), and rice (*Oryza sativa*) are sensitive to chilling temperatures. Warm season vegetables such as those belonging to Cucurbitaceae and Solanaceae also suffer heavily from chilling stress and their growth and development can be adversely affected by temperatures below 15°C resulting in yield loss and crop failure.

Sudden exposure to low temperatures especially to temperatures around or below 0°C may result in extensive and irreversible damages on plant tissues since it causes the membranes to lose their semipermeability and thus their active ion transporting ability (Janda et al. 2007). During chilling stress, the phospholipids in the membranes start to decompose, phase transition takes place, the distribution of the membrane proteins changes and the first visible symptom of low temperature injury, wilting, occurs. Freshly imbibed seeds of chill-sensitive species tend to be also very sensitive, as does the pollen development stage. Imbibitional chilling injury occurs in sensitive seeds such as soybean or cotton during the early stages of imbibition. If soil temperatures are very low at planting, water entering into the seed disrupts membrane integrity, increases electrolyte leakage, and blocks germination. However, if chilling stress follows a brief period of imbibition at warm temperatures, then no damage occurs. The initial reorganization of the membranes from the dry to the hydrated state, therefore, is the critical cellular process (Crowe et al. 1989). Imbibitional chilling injury may also take place in the pollen of sensitive species and the lipid phase properties of membranes in pollen are sensitive to both hydration and temperature. Normally, lipids in the membranes are in a fluid or liquid-crystalline phase, but at either low moisture or low temperature, they form the more rigid gel phase. If rehydration occurs when the membrane lipids are locked in the gel phase due to low temperature exposure, they cannot reorganize and they become leaky and dysfunctional. On the other hand, if rehydration occurs when the membrane lipids are in the liquid-crystalline phase, membranes can reorganize successfully and become tolerant to a subsequent exposure to low temperatures.

One of the earliest works to report on the protective effects of ALA against abiotic stress factors dealt with low temperature stress. The pretreatment of rice seedlings at three-leaf stage by root soaking with ALA solution at 0.1–1 ppm concentration reduced the ratio of leaf rolling and tissue electrolyte leakage after cold treatment (Hotta et al. 1998). Seedlings pretreated with 1 ppm had 85% survival rate and 111.8 mg shoot dry weight per aerial part of seedling while the untreated plants had 65% survival rate and 65 mg dry weight 30 days after the cold treatment at 5°C for 5 days. It was also found that the protection obtained from ALA application was similar to that caused by brassinolide (BR) application, but differed from the ABA pretreatment in terms of leaf rolling or visual appearance of cold damage, since ABA protected younger leaves while ALA and BR were more effective on the protection of older leaves.

To investigate the effect of exogenous application of ALA on chilling tolerance of melon plants (*Cucumis melo*), grown under low light conditions which mimics the typical growing conditions in the greenhouses in the northern hemisphere during the winter, melon seedlings at four-leaf stage were treated with 10 or 100 mg L^{-1} ALA after which they were exposed to chilling stress at 8°C for up to 6 h (Wang et al. 2004b). Although there were no significant differences between the control and ALA-treated plants after chilling treatment at 8°C for 2 h, the control plants were completely dehydrated and dead after chilling at 8°C for 6 h, whereas plants pretreated with 10 mg L^{-1} ALA only exhibited some injury symptoms in a few leaves. Moreover, after the plants had recovered from the stress for 20 h, photosynthesis of ALA-treated leaves had almost recovered to the comparable levels of the control plants before chilling, whereas the photosynthesis of non-ALA-treated plants was only 37–47% of the control plants, suggesting that chilling stress lasted for 4 h led to an irreversible damage on the photosynthetic apparatus. ALA treatment also caused significant increase in soluble sugar levels of melon leaves under chilling stress, which might be helpful for elevating the chilling tolerance of melon seedlings as an important osmotic solute. It was concluded that even though the protection obtained from ALA application was similar to that caused by ABA, it differed significantly from ABA application in such a way that ALA did not inhibit but rather improved plant photosynthesis and growth as well as chilling tolerance without any adverse effect.

In a latter work conducted to identify the optimum ALA application method and concentration, the chilling tolerance of pepper (*Capsicum annuum* L.) seedlings was significantly increased by exogenous application of ALA (Korkmaz et al. 2010). Before exposing to chilling stress at 3°C for 2 days, pepper seedlings were treated with ALA in a range of 1–50 ppm through three different methods (seed treatment, foliar spray, and soil drench). ALA application was very effective in reducing visual injury symptoms of pepper seedlings after the plants had recovered from the stress for 3 days and among the application methods, foliar spray resulted in the least visual damage symptoms followed by the seed treatment (Fig. 12.2). ALA application increased leaf chlorophyll, sucrose, and proline contents and improved relative water content, stomatal conductance, and SOD enzyme activity while reducing membrane permeability. Even though all ALA application methods increased chilling tolerance

Fig. 12.2 Protection of ALA-pretreatment of pepper seedlings against chilling stress at 3°C for 2 days. Plants pretreated with ALA via seed soaking or foliar spray 3 days before the onset of stress and they were allowed to recover from stress for 3 days

of pepper seedlings, seed treatment and foliar spray provided better protection compared to the soil drench while plants treated with 25 ppm ALA had the highest chilling tolerance compared to rest of the ALA concentrations. Similar results were reported in other studies in which additional evidence for mechanisms underlying the protective role of ALA in low concentrations against cold stress was provided. For example, higher antioxidant enzyme (e.g., SOD, CAT, and etc.) activities with increased ascorbic acid, proline, and soluble sugar content were also documented in cucumber (*Cucumis sativus* L.) seedlings pretreated with 0.5 mg L^{-1} ALA before chilling stress at 5°C for 4 days compared to control plants (Yin et al. 2007). Treating soybean (*Glycine max* L.) seedlings with ALA in low concentrations (5–10 μM) prior to a cold stress at 4°C for 2 days resulted in elevated levels of tolerance to cold stress (Balestrasse et al. 2010). ALA pretreatment increased chlorophyll content, relative water content, and catalase and heme oxygenase-1 enzyme activities, and prevented membrane damage by reducing the thiobarbituric acid reactive species. The highest cold tolerance was obtained with 5 μM ALA pretreatment, while higher ALA concentrations (15–40 μM) resulted in a dose-dependent increase of membrane peroxidation.

Seed germination in chilling-sensitive species is slowed or reduced at temperatures below 20°C resulting in poor stand establishment and is usually prevented totally at temperatures lower than 15°C (Korkmaz et al. 2004; Korkmaz 2005). The problem is exacerbated as the length of time to emergence increases because the probability of soil crust formation becomes greater.

Delayed emergence also increases the chances of germinating seeds and seedlings to be infected by damping-off causing pathogens such as *Fusarium* and *Pythium* (Hendrix and Campbell 1973). Therefore, obtaining ideal plant stands requires fast and uniform emergence to avoid these problems. Presoaking the seeds of chilling-sensitive species with ALA could also be an effective way of improving germination or emergence performance under chilling conditions. When the pepper (*Capsicum annuum* L.) seeds were immersed in ALA solutions with varying concentrations for 24 h after which they subjected to emergence tests under chilling (15°C) and optimum (25°C) conditions, emergence was significantly enhanced by ALA treatment (Fig. 12.3a). ALA pretreatment of seeds also enhanced seedling shoot fresh weight (Fig. 12.3b) and chlorophyll *a* content (Fig. 12.3c). Seedlings raised from seeds treated with 50 ppm ALA had significantly lower levels of H_2O_2 (Fig. 12.4a) and MDA contents (Fig. 12.4b) and elevated SOD enzyme activity (Fig. 12.4c) in the leaves. Improvement of pepper seedling emergence performance under chilling stress conditions may have resulted from reduced lipid peroxidation and elevated SOD enzyme activity, all of which is an indication of membrane protection. The efficacy of seed treatment with ALA was also reported to last when seeds were stored after the treatment. For example, priming seeds in 25 or 50 ppm ALA incorporated into the KNO_3 solution improved low temperature performance of red pepper seeds even after the pretreated seeds were stored for 1 month at 4°C or 25°C (Korkmaz and Korkmaz 2009).

2.3.2 Effects of Exogenous ALA on Plants Under Salinity Stress

Soil salinity presents a major limitation to agricultural production since it restrains crop yield and restricts use of land previously uncultivated. Salinity affects almost 20% of all irrigated land and 2.1% of dry-land agriculture worldwide (FAO 2005). Each year there is a deterioration of 2 million ha (about 1%) of world agricultural lands to salinity, leading to reduced or no crop productivity (Szabolcs 1994). In addition to natural causes such as salty raining waters near and around the coasts and weathering of native rocks, low precipitation, high surface evaporation and poor growing practices have also aggravated growing concentration of salts in the rhizosphere (Mahajan and Tuteja 2005). Secondary salinization, in particular, worsens the situation where once productive agricultural lands are becoming unsuitable to cultivation due to poor quality irrigation water (Ashraf and Foolad 2007).

Salts in the soil water may be detrimental to plant growth for two reasons. First, the presence of salt in the soil solution decreases the ability of the plant to take up water and nutrients causing significant reductions in plant growth rate. This is referred to as the osmotic or water-deficit effect of salinity. The salts in the soil solution reduce shoot growth more than root growth, and decrease stomatal conductance and thereby photosynthesis (Munns 1993). The rate at which new leaves are produced depends heavily on the water potential of the soil solution around the roots. Second, if excessive amounts of salt are taken up by the plant via the transpiration stream, there will be injury to cells or tissues in the transpiring leaves and this may cause further reductions in growth. This is called the salt-specific or ion-excess effect of salinity (Greenway and Munns 1980). The definition of salt tolerance is usually the percent biomass production in saline soil relative to plants in nonsaline soil, after growth for an extended period of time. For slow-growing, perennial, or uncultivated species it is often difficult to determine the reduction in biomass production, so percent survival is generally used (Munns 2009).

It has been shown that ALA at low concentrations (10–100 mg L^{-1}) has the potential to improve salinity tolerance in cotton seedlings through foliar application (Watanabe et al. 2000). Cotton seedlings treated with ALA survived in the soil containing 1.5% NaCl while untreated seedlings died. The ALA treatment significantly counteracted the negative effects of salinity and ALA pretreated seedlings weighed as much as those that were not exposed to salinity stress. The analysis of mineral composition of seedlings revealed that Na^+ concentrations in the roots of plants treated with ALA were significantly lower compared to control plants, and it was presumed that

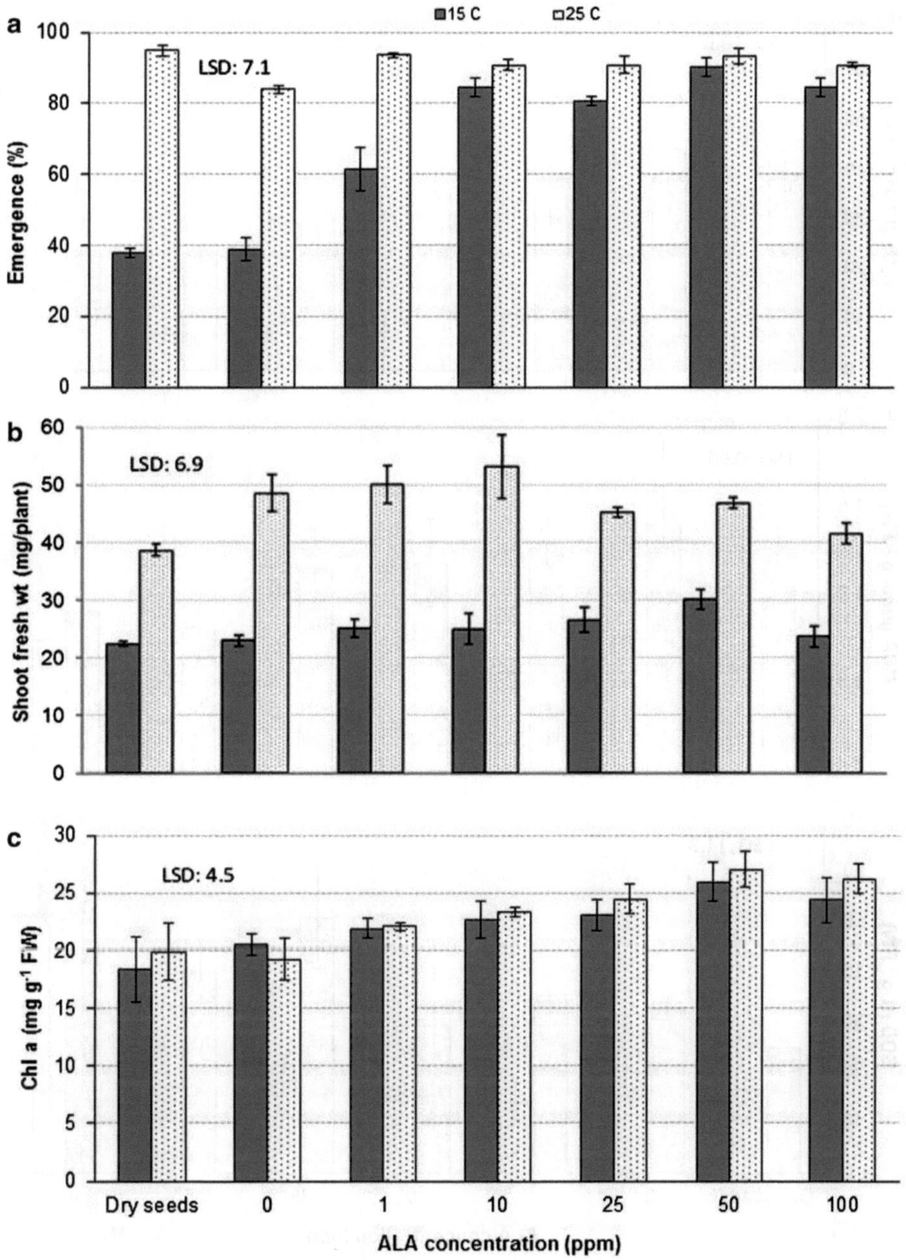

Fig. 12.3 Effect of presowing seed treatment with ALA on pepper seedling emergence percentage (**a**), seedling shoot fresh weight (**b**), and Chl *a* content (**c**). *Vertical bars* represent mean ± SE ($n=8$). Pepper seeds were treated with various concentrations of ALA for 1 day before sowing after which they were subjected to emergence test at 15°C (chilling stress) or 25°C (optimum conditions). Dry seeds are the untreated seeds

the presence of ALA may cause a reduction of Na^+ uptake and may suppress water deficiency caused by osmotic stress resulting from high Na^+ concentration around the roots in growth media. Consistent with the results of Watanabe et al. (2000) in cotton seedlings, foliar application of ALA at the rate of 30 mg L^{-1} to the oilseed rape (*Brassica napus* L.) seedlings grown under high salt stress (up to 200 mM NaCl) significantly reduced the accumulation of Na^+ and K^+, leading

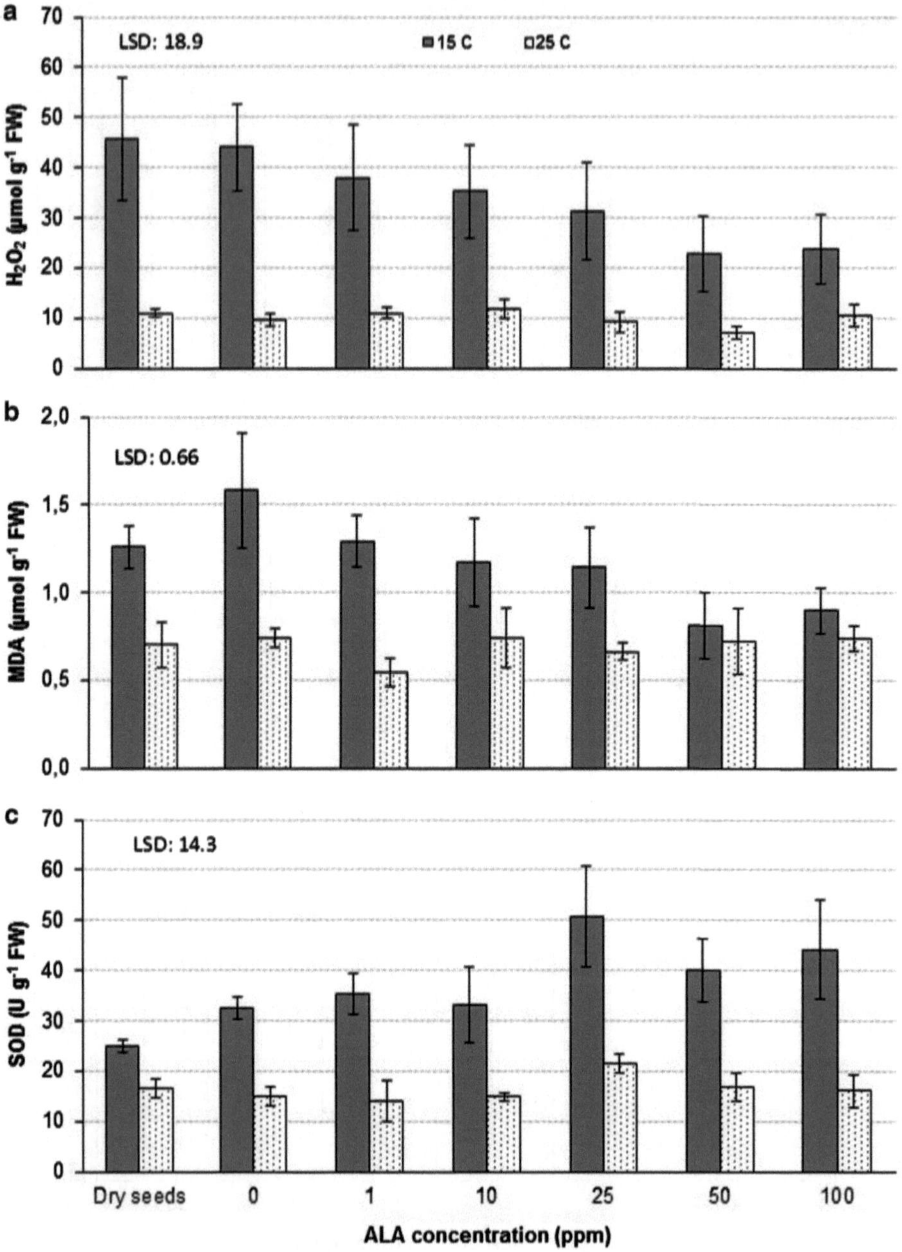

Fig. 12.4 Effect of presowing seed treatment with ALA on H_2O_2 content (**a**), MDA concentration (**b**), and SOD enzyme activity (**c**) of pepper seedlings. *Vertical bars* represent mean ± SE ($n=8$). Pepper seeds were treated with various concentrations of ALA for 1 day before sowing after which they were subjected to emergence test at 15°C (chilling stress) or 25°C (optimum conditions). Dry seeds are the untreated seeds

to a reduction of Na^+/K^+ ratio both in roots and leaves compared to control plants (Naeem et al. 2010). It was also reported that higher Na^+ accumulation in the leaves compared to roots suggests that ALA does not influence Na^+ transport from the roots to the shoots but it might rather suppress the uptake of Na^+ from the growth media to the roots. Reduced Na^+ accumulation in the leaves and reduced K^+ uptake by the roots which led to a concomitant reduction in K^+/Na^+ ratio were reported when date palm (*Phoenix dactylifera* L.) seedlings exposed to 30 mS cm^{-1} salinity stress were treated with ALA-based fertilizer "Pentakeep-v" (Youssef and Awad 2008).

Addition of ALA at low concentrations (0.3–3 mg L^{-1}) to the culture media promoted development and growth of potato (*Solanum tuberosum* L.) microtubers in vitro by increasing average number, diameter, and fresh weight of microtubers under 0.5% NaCl stress conditions (Zhang et al. 2006), but further increase in ALA concentration caused significant reduction in microtuber yield under salt stress. The microtubers treated with low concentrations of ALA exhibited 73% more peroxidase and 28% more polyphenoloxidase activity compared to the untreated control plants which implied that ALA functions as a protectant against oxidative damages of membranes. ALA concentrations of 30 mg L^{-1} or higher induced oxidative damage probably via formation and accumulation of photooxidative porphyrins compounds; therefore, it is important to determine optimal application rates and timing of ALA for growth promotion of crops. The authors concluded that it was important to clarify optimal rates for the promotion of crop growth; optimal rates might be slightly higher for crops exposed to stressful conditions and the safety margin of application rates of ALA would rather wide since ALA is rapidly metabolized in both plants and their surrounding environment (Zhang et al. 2006).

To investigate the role of exogenously applied ALA in spinach (*Spinacia oleracea*) plants grown under salinity stress, seedlings exposed to two levels of salt stress (50 and 100 mM NaCl) were treated with ALA with varying concentrations (0, 0.18, 0.60, and 1.80 mM) (Nishihara et al. 2003). Plants treated with 0.6 and 1.80 mM ALA showed marked increases in photosynthetic rate under both salinity stress levels, while photosynthesis continued to decline in control (0 mM ALA) plants. With regard to antioxidant enzyme activities in the leaves, catalase, ascorbate peroxidase, and glutathione reductase activities were enhanced significantly 3 days after ALA treatment at the rate of 0.60 and 1.80 mM. Other results also indicate that foliar application of 30 mg L^{-1} ALA to oilseed rape plants exposed to salinity stress improved the growth of shoots and roots and increased leaf chlorophyll content and net photosynthetic rate (Naeem et al. 2010, 2011). ALA-treated plants grown under 100 mM NaCl also maintained similar levels of leaf water potential as that of plants grown under optimal conditions. ALA treatment also triggered accumulation of osmolytes such as soluble sugars, free amino acids, and proline in the leaves of salt-stressed plants as well as enzymatic (APX, CAT, and SOD) and nonenzymatic (glutathione and ascorbate) antioxidants activity while decreasing membrane permeability, MDA content and ROS production. On the contrary, no enhancement in chlorophyll content of rice (*Oryza sativa* L.) seedlings under salt stress treated with ALA was observed, suggesting that the growth recovery was not due to the increased chlorophyll content but rather due to reduced lipid peroxidation caused by increased antioxidant enzyme activity (Wongkantrakorn et al. 2009). Augmented activities of antioxidant enzymes such as SOD, APX, CAT, and GR after ALA treatment were also reported in other species grown under various conditions such as *Gingko biloba* seedlings grown under optimal conditions (Xu et al. 2009), oilseed rape plants under herbicide toxicity stress (Zhang et al. 2008), and watermelon seedlings exposed to low light conditions (Sun et al. 2009).

The ability of barley plants to synthesize ALA was reported to increase in response to increasing salt concentrations and maximum amount of ALA accumulated in plants grown at 100 mM NaCl was two fold higher than in control plants grown under optimum (0 mM NaCl) conditions (Averina et al. 2010). When salt concentration was further increased to 200 mM, the rate of ALA accumulation was decreased and the reduced ability to synthesize ALA was accompanied by an increase in proline content. Thus, the impairment in ALA-synthesizing ability was reported to redirect metabolic conversion of glutamic acid from chlorophyll synthesis to the proline synthesis pathway, which would stimulate proline biosynthesis and enhance salt tolerance.

The efficacy of ALA-based fertilizer "Pentakeep-v" applied at the rate of 0.08% in inducing tolerance to salinity stress was also tested in date palm (*Phoenix dactylifera* L.) seedlings (Youssef and Awad 2008). Application of Pentakeep-v significantly improved chlorophyll *a* content of plants, leading to improved total chlorophyll content and chlorophyll *a*/*b* ratios. Pentakeep-v also enhanced the biochemical efficiency of carbon fixation of Rubisco enzyme and the rate of electron transport required for RuBP regeneration over untreated plants exposed to salinity stress at 15 mS cm^{-1}. In addition, Pentakeep-v reduced the percentage contribution of stomatal factor (gas phase limitation) to the apparent reduction in photosynthetic gas exchange to values similar to those of control plants and lowered CO_2 compensation points by reducing respiratory CO_2 loss with increasing salinity to the 30 mS cm^{-1}. It was concluded that ALA-based fertilizer improved salt tolerance of date palm seedlings by increasing photosynthetic assimilation via boosting light-harvesting capabilities of the treated plants by enhancing chlorophyll *a* content and by reducing stomatal limitation to photosynthetic gas exchange (Youssef and Awad 2008).

The soaking of pakchoi (*Brassica campestris* ssp. *chinensis* var. *communis* Tsenet Lee) seeds in ALA solution before sowing reduced the damaging effects of salinity during seed germination (Wang et al. 2005). Treating the seeds with ALA, at concentrations ranging from 0.01 to 10 mg L^{-1}, promoted seed germination when seeds were stressed by 150 mM NaCl. However, levulinic acid, an inhibitor of ALA dehydrase, significantly prevented seed germination and seedling growth, suggesting that ALA was necessary for seed germination, and that the effect of ALA is dependent upon its conversion into porphyrin. ALA pretreatment of seeds also caused significant increases in respiration rates during seed germination under salt stress compared to untreated seeds which maybe further supported by the fact that salt-tolerant pakchoi cultivars contained higher endogenous ALA and heme under salt stress conditions compared to salt-sensitive cultivars. Additionally, promotive effects of exogenous application of ALA were also reported in watermelon (*Citrullus lanatus*) seed germination under salt stress (Liu et al. 2006). Treatment of watermelon seeds with 15–30 mg L^{-1} ALA improved seed germination and seedling growth under 125 mM NaCl stress, and the promotion of ALA treatment on germination under salt stress might be associated with the enhanced activities of antioxidant enzymes, especially POD and decreased activity of lipoxygenase and lowered levels of MDA in hypocotyls and radicles.

2.3.3 Effects of Exogenous ALA on Plants Under Water Stress

Plants are exposed to a variety of environmental stresses including extreme temperatures, unfavorable chemical and physical soil conditions, and various pests and diseases. However, in the long term, water deficit affects negatively the growth and yield of crop plants more than all the other stresses combined, because it is ubiquitous. More than one-third of the earth's surface is classified as arid or semiarid because it is subjected to permanent drought. Equally important is the fact that most of the humid temperate regions, where most of the world's agricultural production takes place, are frequently subjected to periods of severe drought. Water deficit or water stress refers to situations where plant water potential and turgor are reduced enough to interfere with normal growth of plants (Kramer 1983). The exact cell water potential at which this takes place is dependent upon the crop species, the stage of development, and the process under consideration. For example, cell division and enlargement usually stops at a water potential of only −0.2 to −0.4 MPa, whereas stomatal closure does not begin until the water potential falls below −0.8 to −1.0 MPa. The first and the most obvious symptom of water deficiency is wilting due to reduced turgor, causing significant retardation of growth processes especially lengthwise growth. Photosynthesis is also reduced for a number of reasons including reduced CO_2 uptake due to stomatal closure, damaged cytoplasmic ultrastructure and impaired enzyme activity, reduced canopy absorption of photosynthetically active radiation, and decreased radiation-use efficiency (Janda et al. 2007; Farooq et al. 2009). Another outcome of water stress is the disrupted nutrient uptake due to impaired root growth and reduction in water migration which results in a significant decrease in the quantity of ions transported by water.

Even though considerable amount of data has been accumulated with regard to the roles of ALA on enhancing the tolerance to chilling and salinity stresses during the last two decades, very little effort has been put in to investigate the role of ALA in plants under water stress. Among the very few, first study to demonstrate the effect of ALA on drought tolerance showed that spraying the barley (*Hordeum vulgare* L.) plants from tillering to milk-ripe stage with ALA enhanced tolerance to water stress (Al-Khateeb 2006). ALA treatment promoted the yield of barley plants under optimal conditions (weekly irrigation) causing as much as 45% increase in the grain yield and 27% in straw yield. Irrigation frequency of 21 days significantly reduced grain and straw yields per hectare while ALA application at the rate of 100 ppm increased the grain and straw yields by 35 and 41%, respectively, compared to untreated plants. The increase in yield under severe water stress (3 weeks irrigation interval) in response to ALA treatment was associated with the fact that ALA treatment enhanced photosynthesis, stomatal conductance, and intercellular CO_2 concentration by 41, 29 and 50%, respectively. Confirming results were also reported in wheat (*Triticum aestivum* L.) that spraying the plants with ALA solutions enhanced the tolerance to water stress (Al-Thabet 2006). Wheat plants that were irrigated every 2 weeks and foliar sprayed with ALA at the rate of 50–100 ppm ha^{-1} out yielded and surpassed in water use efficiency compared with untreated plants grown under normal conditions (irrigated every week). Moreover, plants treated with 100 ppm ha^{-1} ALA yielded more than untreated plants when irrigated every 3 weeks although yield was considerably lower than that of plants irrigated every week.

To assess the physiological and biochemical changes within the plant caused by ALA application on plants, water-stressed oilseed rape seedlings were fed with ALA containing solutions (Liu et al. 2011). ALA treatment at low concentrations (1 mg L^{-1}) significantly improved plant biomass and chlorophyll content, but reduced MDA content and ROS production. On the contrary, ALA applied at moderately high concentrations (10 mg L^{-1}) hampered plant growth while higher concentrations (100 mg L^{-1}) killed the plants. Application of ALA at low concentrations also enhanced reduced/oxidized glutathione and ascorbic acid ratios while boosting the activity of antioxidant enzymes such as APX, CAT, GR, and POD by inducing the expression of the specific antioxidant genes.

3 Conclusions and Future Perspectives

It may be deduced from the above discussion that ALA could be a very promising plant growth regulating compound for increasing plant productivity and enhancing abiotic stress tolerance of crop plants since under certain conditions it has been found to mitigate the deleterious effects of various abiotic stress factors in numerous crop species (Fig. 12.5). The fact that it is the precursor of chlorophyll and all other porphyrins molecules, exogenous application of ALA enhances chlorophyll content of crop plants which translates into increased photosynthetic capacity thus yield. It is also documented that ALA application improves the quality of food crops by promoting the accumulation of antioxidant molecules and sugars and balancing the acidity within the fruits. Different plant species may vary in their responses to exogenous application of ALA and for any given species there is no consensus as to the optimal concentration or application method for ALA to maximize crop productivity under stressful environments. However, even though exogenous application of the lower concentrations of ALA proved to be beneficial in enhancing the plant growth and various other physiological and biochemical characteristics of plants exposed to abiotic stress factors, it should be noted that at higher concentrations, ALA itself may cause a high level of stress in plants acting as a herbicide. The effectiveness of exogenous applications of ALA on enhancing tolerance to various abiotic stress conditions may be resulted from boosted activities of enzymatic or nonenzymatic antioxidant system

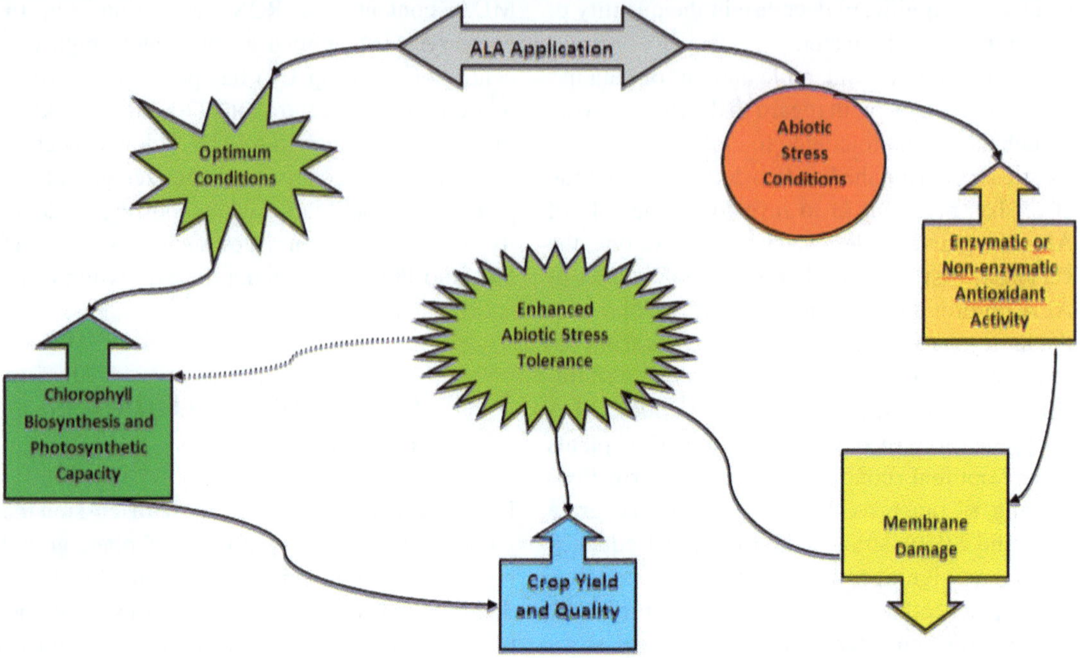

Fig. 12.5 Schematic model of the action of ALA on plant productivity and the induction of abiotic stress tolerance. See the text for details

providing significant protection to the membranes against harmful ROS within tissues.

The recently introduced PGR, however, still requires a lot of work to be carried out to elucidate the key regulatory points of biosynthesis, mechanism of action, and other specific and collaborative regulatory roles played by ALA since there are still large gaps in current knowledge at both theoretical and practical level. By looking at the results documented above, although it has become clear that ALA enhances the activities of the enzymatic and nonenzymatic antioxidants that scavenge the ROS and stabilizing the cell membranes, the mechanism(s) underlying this improvement remains quite unknown. For example, we have no information whether exogenous application of ALA could compensate for the imbalance in other plant growth substances normally caused by abiotic stress factors, interact and/or being regulated by the cross-talk in harmony with other established phytohormones and PGRs such as ABA or upregulate specific defense mechanisms against the stress factors. Such determinations will require extensive molecular and physiological examination of ALA treated and untreated plants subjected to stress conditions. Also, most of the currently known information comes from laboratory studies, therefore, extensive field studies are required to determine the optimal concentration and application method to exhibit the beneficial effects of ALA application in plant productivity and decreasing the sensitivity of crops species to abiotic stress factors. Additionally, considerable amount of data has accumulated on the effects of ALA applications on plants exposed to chilling and salinity stresses and partially to water stress but there is no information about how ALA application will affect the plants exposed to other abiotic stress factors such as high temperatures, ozone, or toxic metals. Thus, more work is necessary to determine the effectiveness of ALA applications on plants exposed to such stress conditions. The future applications of this PGR holds a great promise as a management tool for enhancing the productivity and protecting our agricultural crops against the aforesaid constrains ultimately aiding to increase potential crop yield in near future.

Acknowledgments This work was supported by a grant (project no: 107O611) from The Scientific and Technical Research Council of Turkey (TUBITAK) and the author is grateful for the financial support. He also thank Dr. Ferit Kocaçınar for his critical review of the manuscript and Dr Yasemin Korkmaz for her technical help with the figures.

References

Al-Khateeb SA (2006) Promotive effect of 5-aminolevulinic acid on growth, yield and gas exchange capacity of barley (*Hordeum vulgare* L.) grown under different irrigation regimes. J King Saud Univ 18:103–111

Al-Khateeb AA, Al-Khateeb SA, Okawara R, Al-Abdoulhady IA (2006) Promotive effects of 5-aminolevulinic acid (5-ALA) on fruit yield and quality of date palm cv. Khalas J Biol Sci 6:1118–1121

Allen DJ, Ort DR (2001) Impacts of chilling temperatures on photosynthesis in warm-climate plants. Trends Plant Sci 6:36–42

Al-Thabet SS (2006) Promotive effect of 5-aminolevulinic acid on growth and yield of wheat grown under dry conditions. J Agron 5:45–49

Armstrong GA (1998) Greening in the dark: light-independent chlorophyll biosynthesis from anoxygenic photosynthetic bacteria to gymnosperms. J Photochem Photobiol B: Biol 43:87–100

Ashraf M, Foolad MA (2007) Improving plant abiotic-stress resistance by exogenous application of osmoprotectants glycine betaine and proline. Environ Exp Bot 59:206–216

Averina NG, Gritskevich ER, Vershilovskaya IV, Usatov AV, Yaronskaya EB (2010) Mechanisms of salt stress tolerance development in barley plants under the influence of 5-aminolevulinic acid. Russ J Plant Physiol 57:792–798

Awad MA (2008) Promotive effects of a 5-aminolevulinic acid-based fertilizer on growth of tissue culture-derived date palm plants (*Phoenix dactylifera* L.) during acclimatization. Sci Hortic 118:48–52

Balestrasse KB, Tomaro ML, Batlle A, Noriega GO (2010) The role of 5-aminolevulinic acid in the response to cold stress in soybean plants. Phytochemistry 71:2038–2045

Baninasab B (2009) Amelioration of chilling stress by paclobutrazol in watermelon seedlings. Sci Hortic 121:144–148

Beale SI, Weinstein JD (1989) Tetrapyrrole metabolism in photosynthetic organisms. In: Dailey HA (ed) Biosynthesis of hemes and cholorophylls. Mc-Graw Hill, New York, pp 287–391

Beale SI (1990) Biosynthesis of tetrapyrrole pigment precursor, *d*-aminolevulinic acid, from glutamate. Plant Physiol 93:1273–1279

Bindu Roy C, Vivekanandan M (1998a) Role of aminolevulinic acid in improving biomass production in *Vigna catjung, V. mungo,* and *V. radiata.* Biol Plant 41:211–215

Bindu Roy C, Vivekanandan M (1998b) Hormonal activities of 5-aminolevulinic acid in callus induction and micropropagation. Plant Growth Regul 26:15–18

Bingshan L, Hotta Y, Yinglan Q, Jinsong Z, Tanaka T, Takeuchi Y, Konnai M (1998) Effects of 5-aminolevulinic acid on the growth and ripening of wheat. J Pestic Sci 23:300–303

Bisbis B, Billard JP, Huault C, Kevers C, Dily FL, Gaspar T (1997) Biosynthesis of 5-aminolevulinic acid via the Shemin pathway in a green sugar beet callus. Biol Plant 40:493–497

Bunke A, Zerbe O, Schmid H, Burmeister G, Merkle HP, Gander B (2000) Degradation mechanism and stability of 5-aminolevulinic acid. J Pharm Sci 89:1335–1341

Chakraborty N, Tripathy BC (1992) Involvement of singlet oxygen in 5-aminolevulinic acid-induced photodynamic damage of cucumber (*Cucumis sativus* L.) chloroplasts. Plant Physiol 98:7–11

Ciais Ph, Reichstein M, Viovy N et al (2005) Europe-wide reduction in primary productivity caused by the heat and drought in 2003. Nature 437:529–533

Crowe JH, Hoekstra FA, Crowe LM (1989) Membrane phase transitions are responsible for imbibitional damage in dry pollen. Proc Natl Acad Sci USA 86: 520–523

Ding W, Song L, Wang X, Bi Y (2010) Effect of abscisic acid on heat stress tolerance in the calli from two ecotypes of *Phragmites communis*. Biol Plant 54:607–613

Drazic G, Mihailovic N (1998) Chlorophyll accumulation in black pine seedlings treated with 5-aminolevulinic acid. Biol Plant 41:277–280

Etehadnia M, Waterer DR, Tanino KK (2008) The method of ABA application affects salt stress responses in resistant and sensitive potato lines. J Plant Growth Regul 27:331–341

F.A.O. 2005. Global network on integrated soil management for sustainable use of salt-affected soils. Rome, Italy: FAO Land and Plant Nutr. Manag. Service. http://www.fao.org/ag/agl/agll/spush

Farooq M, Wahid A, Kobayashi N, Fujita D, Basra SMA (2009) Plant drought stress: effects, mechanisms and management. Agron Sustain Dev 29:185–212

Gadmar QB, Moan J, Scheie E, Ma LW, Peng Q (2002) The stability of 5-aminolevulinic acid in solution. J Photochem Photobiol B: Biol 67:187–193

Gilley A, Fletcher RA (1997) Relative efficacy of paclobutrazol, propiconazole and tetraconazole as stress protectants in wheat seedlings. Plant Growth Regul 21:169–175

Greenway H, Munns R (1980) Mechanisms of salt tolerance in nonhalophytes. Annu Rev Plant Physiol 31: 149–190

Hara M, Takahashi I, Yamori M, Tanaka T, Funada S, Watanabe K (2011) Effects of 5-aminolevulinic acid on growth and amylase activity in the radish taproot. Plant Growth Regul 64:287–291

Hendrix FF Jr, Campbell WA (1973) Pythiums as plant pathogens. Annu Rev Phytopathol 11:77–98

Hirayama T, Shinozaki K (2010) Research on plant abiotic stress responses in the post-genome era: past, present and future. Plant J 61:1041–1052

Hodgins RR, van Huystee RB (1986) Porphyrin metabolism in chill stressed maize (*Zea mays* L.). J Plant Physiol 125:325–336

Hotta Y, Tanaka T, Takaoka H, Takeuchi Y, Konnai M (1997a) Promotive effects of 5-aminolevulinic acid on the yield of several crops. Plant Growth Regul 22: 109–114

Hotta Y, Watanabe K, Tanaka T, Takeuchi Y, Konnai M (1997b) Effects of 5-aminolevulinic acid on growth of plant seedlings. J Pestic Sci 22:102–107

Hotta Y, Tanaka T, Takaoka H, Takeuchi Y, Konnai M (1997c) New physiological effects of 5-aminolevulinic acid in plants: the increase of photosynthesis, chlorophyll content, and plant growth. Biosci Biotech Biochem 61:2025–2028

Hotta Y, Tanaka T, Bingshan L, Takeuchi Y, Konnai M (1998) Improvement of cold resistance in rice seedlings by 5-aminolevulinic acid. J Pestic Sci 23:29–33

Han J (1999) The influence of photosynthetic bacteria treatments on the crop yield, dry matter content, and protein content of the mushroom *Agaricus bisporus*. Sci Hortic 82:171–178

Iwai K, Takeuchi Y, Kuramochi H (2003) Fertiliser containing 5-aminolevulinic acid affects growth of creeping bentgrass. In: Proceedings of Spring 2003 Meet. Japanese Soc. Turfgrass Sci. p. 1

Iwai K, Saito A, van Leeuwen J, Tanaka T, Takeuchi Y (2005) A new functional fertilizer containing 5-aminolevulinic acid promoted hydroponically-grown vegetables in the Netherlands. Acta Hortic 697:351–355

Jaleel CA, Lakshmanan GMA, Gomathinayagam M, Panneerselvam R (2008) Triadimefon induced salt stress tolerance in *Withania somnifera* and its relationship to antioxidant defense system. South Afr J Bot 74:126–132

Janda T, Hovarth E, Szalai G, Paldi E (2007) Role of salicylic acid in the induction of abiotic stress tolerance. In: Hayat S, Ahmad A (eds) Salicylic acid – a plant hormone. Springer, Dordrecht, The Netherlands, pp 91–150

Janowiak F, Maas B, Dorffling K (2002) Importance of abscisic acid for chilling tolerance of maize seedlings. J Plant Physiol 159:635–643

Kantha T, Chaiyavat C, Kantachote D, Sukrong S, Muangprom A (2010) Selection of photosynthetic bacteria producing 5-aminolevulinic acid from soil of organic saline paddy fields from the Northeast region of Thailand. Afr J Microbiol Res 4:1848–1855

Kennedy JC, Pottier RC, Pross DC (1990) Photodynamic therapy with endogenous protoporphyrin IX: basic principle and present clinical experience. J Photochem Photobiol 6:143–148

Klessig DF, Malamy J (1994) The salicylic acid signal in plants. Plant Mol Biol 26:1439–1458

Kobayashi M, Kobayashi M (2000) Waste remediation and treatment using anoxygenic phototrophic bacteria. In: Blankenship RE, Madigan MT, Bauer CE (eds) Anoxygenic photosynthetic bacteria. Kluwer, New York, pp 1269–1282

Koh RH, Song HG (2007) Effects of application of *Rhodopseudomonas sp.* on seed germination and growth of tomato under axenic conditions. J Microbiol Biotechnol 17:1805–1810

Korkmaz A (2002) Amelioration of chilling injuries in watermelon seedlings by abscisic acid. Turk J Agric For 26:17–20

Korkmaz A, Tiryaki I, Nas MN, Ozbay N (2004) Inclusion of plant growth regulators into priming solution improves low temperature germination and emergence of watermelon seeds. Can J Plant Sci 84:1161–1165

Korkmaz A (2005) Inclusion of acetyl salicylic acid and methyl jasmonate into the priming solution improves low temperature germination and emergence of sweet pepper. Hort Sci 40:197–200

Korkmaz A, Korkmaz Y (2009) Promotion by 5-aminolevulenic acid of pepper seed germination and seedling emergence under low temperature stress. Sci Hortic 119:98–102

Korkmaz A, Korkmaz Y, Demirkiran AR (2010) Enhancing chilling stress tolerance of pepper seedlings by exogenous application of 5-aminolevulinic acid. Environ Exp Bot 67:495–501

Kramer PJ (1983) Water relations of plants. Academic, New York

Kumar AM, Chaturvedi S, Söll D (1999) Selective inhibition of HEMA gene expression by photooxidation in *Arabidopsis thaliana*. Phytochemistry 51:847–851

Kumar S, Kaur G, Nayyar H (2008) Exogenous application of abscisic acid improves cold tolerance in chickpea (*Cicer arietinum* L.). J Agron Crop Sci 194: 449–456

Lee KH, Koh RH, Song HG (2008) Enhancement of growth and yield of tomato by *Rhodopseudomonas sp.* under greenhouse conditions. J Microbiol 46: 641–646

Levitt J (1980) Responses of plants to environmental stresses, 2nd edn, vol 1. Academic, New York

Liu H, Lang KL, Wang LJ (2006) Promotion of 5-aminolevunlinic acid on seed germination of watermelon (*Citrullus lanatus*) under salt stress. J Fruit Sci 6:854–859

Liu D, Pei ZF, Naeem MS, Ming DF, Liu HB, Khan F, Zhou WJ (2011) 5-aminolevulinic acid activates antioxidative defence system and seedling growth in *Brassica napus* L. under water deficit stress. J Agron Crop Sci 197:284–295

Lynch DV (1990) Chilling injury in plants: the relevance of membrane lipids. In: Katterman F (ed) Environmental injury to plants. Academic, New York, pp 17–34

Lyons JM (1973) Chilling injury in plants. Annu Rev Plant Physiol 24:445–466

Lu S, Su W, Li H, Guo Z (2009) Abscisic acid improves drought tolerance of triploid bermuda grass and involves H_2O_2- and NO- induced antioxidant enzyme activities. Plant Physiol Biochem 47:132–138

Mahajan S, Tuteja N (2005) Cold, salinity and drought stresses: an overview. Arch Biochem Biophy 444: 139–158

Maruyama-Nakashita A, Hirai MY, Funada S, Fueki S (2010) Exogenous application of 5-aminolevulinic acid increases the transcript levels of sulfur transport and assimilatory genes, sulfate uptake, and cysteine and glutathione contents in *Arabidopsis thaliana*. Soil Sci Plant Nutr 56:281–288

Memon SA, Hou X, Wang L, Li Y (2009) Promotive effect of 5-aminolevulinic acid on chlorophyll, antioxidative enzymes and photosynthesis of pakchoi (*Brassica campestris* ssp. *chinensis* var. *communis* Tsen et Lee). Acta Physiol Plant 31:51–57

Morales-Payan JP, Stall WM (2005) Papaya transplant growth as affected by 5- aminolevulinic acid and nitrogen fertilization. Proc FL State Hort Soc 118: 263–265

Munns R (1993) Physiological processes limiting plant growth in saline soil: some dogmas and hypotheses. Plant Cell Environ 16:15–24

Munns R (2009) Strategies for crop improvement in saline soils. In: Ashraf M, Ozturk M, Athar HR (eds) Task for vegetation science-44: salinity and water stress improving crop efficiency. Springer, Dordrecth, The Netherlands, pp 99–110

Naeem MS, Jin ZL, Wan GL, Liu D, Liu HB, Yoneyama K, Zhou WJ (2010) 5-Aminolevulinic acid improves photosynthetic gas exchange capacity and ion uptake under salinity stress in oilseed rape (*Brassica napus* L.). Plant Soil 332:405–415

Naeem MS, Rasheed M, Liu D, Jin ZL, Ming DF, Yoneyama K, Takeuchi Y, Zhou WJ (2011) 5-Aminolevulinic acid ameliorates salinity-induced metabolic, water-related and biochemical changes in *Brassica napus* L. Acta Physiol Plant 33:517–528

Nishihara E, Kondo K, Parvez MM, Takahashi K, Watanabe K, Tanaka K (2003) Role of 5-aminolevulinic acid (ALA) on active oxygen-scavenging system in NaCl-treated spinach (*Spinacia oleracea*). J Plant Physiol 160:1085–1091

Papenbrock J, Grimm B (2001) Regulatory network of tetrapyrrole biosynthesis-studies of intracellular signalling involved in metabolic and developmental control of plastids. Planta 213:667–81

Pavlovic A, Demko V, Durchan M, Hudák J (2009) Feeding with aminolevulinic acid increased chlorophyll content in Norway spruce (*Picea abies*) in the dark. Photosynthetica 47:631–634

Rebeiz CA, Montazer-Zouhoor A, Jopen HJ, Wu SM (1984) Photodynamic herbicides. I Concepts and phenomenology. Enzyme Microb Technol 6:390–401

Rebeiz CA, Juvik JA, Rebeiz CC (1988) Porphyric insecticides. I. Concept and phenomenology. Pestic Biochem Physiol 30:11–27

Reinbothe S, Reinbothe C (1996) Regulation of chlorophyll biosynthesis in angiosperms. Plant Physiol 111:1–7

Sasaki K, Marquez FJ, Nishio N, Nagai S (1995) Promotive effect of 5-aminolevulinic acid and the growth and photosynthesis of *Spirulina platensis*. J Ferment Bioeng 79:453–457

Sasaki K, Watanabe M, Tanaka T, Tanaka T (2002) Biosynthesis, biotechnological production and applications of 5-aminolevulinic acid. Appl Microbiol Biotechnol 58:23–29

Smoleñ S, Sady W (2010) Effect of plant biostimulation with Pentakeep v fertilizer and nitrogen fertilization on the content of macro-and micronutrients in spinach. J Elementol 15:343–354

Srivastava LM (2002) Plant growth and development. Hormones and the environment. Academic Press, Oxford

Stobart AK, Ameen-Bukhari J (1984) Regulation of δ-aminolevulinic acid synthesis and protochlorophyllide regeneration in the leaves of dark-grown barley (*Hordeum vulgare*) seedlings. Biochem J 222:419–426

Sun YP, Zhang ZP, Wang LJ (2009) Promotion of 5-aminolevulinic acid treatment on leaf photosynthesis is related with increase of antioxidant enzyme activity in watermelon seedlings grown under shade condition. Photosynthetica 47:347–354

Szabolcs I (1994) Soils and salinisation. In: Pessarakali M (ed) Handbook of plant and crop stress. Marcel Dekker, New York, pp 3–11

Tabuchi K, Mizuta H, Yasui H (2009) Promotion of callus propagation by 5-aminolevulinic acid in a *Laminaria japonica* sporophyte. Aquacult Res 41:1–10

Tanaka T, Takahashi K, Hotta Y, Takeuchi Y, Konnai M (1992) Promotive effects of 2-aminolevulinic acid on yield of several crops. 19th Meeting of PGR Soc. Amer. San Fransisco. pp 3–9

Tewari AK, Tripathy BC (1998) Temperature-stress-induced impairment of chlorophyll biosynthetic reactions in cucumber and wheat. Plant Physiol 117: 851–858

Vettakkorumakankav NN, Falk D, Saxena P, Fletcher RA (1999) A crucial role for gibberellins in stress protection of plants. Plant Cell Physiol 40:542–548

Wang JJ, Jiang WB, Zhang Z, Yao QH, Matsui H, Ohara H (2003) 5-aminolevinilic acid and its potential application in agriculture. Plant Physiol Commun 39:185–192

Wang LJ, Wang ZH, Li ZQ, Li H, Liu W, Chen ZQ, Yan P, Sun D (2004a) Effects of 5-aminolevulinic acid on enhancing apple fruit coloration. J Fruit Sci 21:512–515

Wang LJ, Jiang WB, Huang BJ (2004b) Promotion of 5-aminolevulinic acid on photosynthesis of melon (*Cucumis melo*) seedlings under low light and chilling stress conditions. Physiol Plant 121:258–264

Wang LJ, Jiang WB, Liu H, Liu WQ, Kang L, Hou XL (2005) Promotion by 5-aminolevulinic acid of germination of pakchoi (*Brassica campestris* ssp. *chinensis* var. *communis* Tsen et Lee) seeds under salt stress. J Integr Plant Biol 47:1084–1091

Wang Z, Tang G, Li Z, Wang L (2006) Promotion of 5-aminolevulinic acid and genistein on anthocyanin accumulation in apples. Acta Hort Sin 33:1055–1058

Wang T, Rao J, Song Y, Zhang H, Zou Z (2009) Effects of foliar application 5-aminolevulinic acid on post harvest physiological of tomato during cold storage period. Acta Agric Boreali-Occidentalis Sin 1:243–247

Wardlaw IF (1980) Translocation and source-sink relationship. In: Carlson PS (ed) The biology of crop productivity. Acedemic Press, New York, pp 297–339

Watanabe K, Tanaka T, Hotta Y, Kuramochi H, Takeuchi Y (2000) Improving salt tolerance of cotton seedlings with 5-aminolevulinic acid. Plant Growth Regul 32: 97–101

Watanabe K, Nishihara E, Watanabe S, Tanaka T, Takahashi K, Takeuchi Y (2006) Enhancement of growth and fruit maturity in 2-year-old grapevines cv. Delaware by 5-aminolevulinic acid. Plant Growth Regul 49:35–42

Wongkantrakorn N, Sunohara Y, Matsumoto H (2009) Mechanism of growth amelioration of NaCl-stressed rice (*Oryza sativa* L.) by δ-aminolevulinic acid. J Pestic Sci 34:89–95

Xu F, Chang J, Cheng SY, Zhu J, Li LL, Wang Y, Cheng H (2009) Promotive effect of 5-aminolevulinic acid on the antioxidant system in *Ginkgo biloba* leaves. Afr J Biotech 8:3769–3776

Yang CM, Tsai HM, Yang JH (2003) Glucose and delta-aminolevulinic acid stimulate the dark chlorophyll synthesis of rice seedlings. Acta Botanica Sinica 45: 422–426

Yin L, Yu XC, Wamg YH, Xu ZH, Li K, Han DJ (2007) Effect of 5-aminolevulinic acid on chilling tolerance in cucumber seedlings. Acta Agric Boreali-Occidentalis Sin 4:166–169

Yoshida R, Tanaka T, Hotta Y (1995) Physiological effects of 5-aminolevulinic acid in vegetable crops. Abstract of the 15th Int Conf Plant Growth Substances, Minneapolis, USA, 32:417

Youssef T, Awad MA (2008) Mechanisms of enhancing photosynthetic gas exchange in date palm seedlings (*Phoenix dactylifera* L.) under salinity stress by a 5-aminolevulinic acid-based fertilizer. J Plant Growth Regul 27:1–9

Zhang ZJ, Li HZ, Zhou WJ, Takeuchi Y, Yoneyama K (2006) Effect of 5-aminolevulinic acid on development and salt tolerance of potato (*Solanum tuberosum* L.) microtubers in vitro. Plant Growth Regul 49:27–34

Zhang WF, Zhang F, Raziuddin R, Gong HJ, Yang ZM, Lu L, Ye QF, Zhou WJ (2008) Effects of 5-aminolevulinic acid on oilseed rape seedling growth under herbicide toxicity stress. J Plant Growth Regul 27:159–169

Abiotic Stress and Role of Salicylic Acid in Plants

13

Miyuki Hara, Jun Furukawa, Aiko Sato, Tsuyoshi Mizoguchi, and Kenji Miura

Abstract

Salicylic acid (SA), is an important phytohormones that plays a role in response to biotic stresses and pathogenesis. Apart from this role, recent studies have demonstrated that SA also participates in the signaling of abiotic stress responses, such as drought, high and low temperature, salinity, ozone, UV radiation, and heavy metals. In addition, abiotic stresses also induce endogenous SA accumulation. The appropriate application of SA could provide protection against several types of environmental stresses. SA may cause oxidative stress, partially through accumulation of hydrogen peroxide. A low concentration of hydrogen peroxide also improves the antioxidative capacity of plants and stimulates the synthesis of protective compounds, leading to enhanced tolerance to abiotic stresses. The effect of SA application depends on numerous factors such as the species and developmental stage of the plant, the mode of application, and the concentration of applied and endogenous SA levels. This chapter reviews the effects of SA on different abiotic stresses, and possible mechanisms for abiotic stress responses controlled by SA.

Keywords

Abiotic stress • Biotic stress • Flowering • Reactive oxygen species • Salicylic acid • Stomatal closure

M. Hara • J. Furukawa • A. Sato • T. Mizoguchi • K. Miura (✉)
Graduate School of Life and Environmental Sciences, University of Tsukuba, Tsukuba 305-8572, Japan
e-mail: kmiura@gene.tsukuba.ac.jp

P. Ahmad and M.N.V. Prasad (eds.), *Abiotic Stress Responses in Plants: Metabolism, Productivity and Sustainability*, DOI 10.1007/978-1-4614-0634-1_13, © Springer Science+Business Media, LLC 2012

1 Introduction

Salicylic acid (SA) is one of plant hormone, whose function in biotic stress response has been well studied (Vlot et al. 2009). SA mediates plant defense against biotrophic pathogens, such as *Pseudomonas syringae* by induction of pathogenesis-related proteins. The name of SA is derived from the word *Salix*, the scientific name of the willow tree. In the fourth century bc, Hippocrates encouraged woman to chew willow leaves to relieve the pain of childbirth (Raskin 1992). The plants containing SA were used for therapeutic purpose throughout the ancient world. In 1828, salicin, the glucoside of salicylic alcohol, was purified from willow bark. Then, salicin is splitted into a sugar and an aromatic compound, salicylice. One of widely used drugs, aspirin is a synthetic derivative of SA.

SA is generally present in plants in quantities of few μg/g freshweight or less (Raskin et al. 1990), either in a free state or in the form of glycosylated, methylated, glucose-ester, or amino acid conjugates (Lee et al. 1995). SA is synthesized via two distinct pathways (Fig. 13.1). Chorismate can be converted to isochorismate by ICS (isochorismate synthase), and IPL (isochorismate pyruvate lyase) catalyzes the conversion of isochorismate to SA (Fig. 13.1; Wildermuth et al. 2001; Strawn et al. 2007). This is the major pathway for the synthesis of SA in *Arabidopsis, Nicotiana benthamiana,* and tomato (Wildermuth et al. 2001; Uppalapati et al. 2007; Catinot et al. 2008). Because an *Arabidopsis ics1 ics2* double mutant contains residual SA (Garcion et al. 2008), the ICS pathway is not the only source of SA. Another pathway starts from phenylalanine, which is deaminated by PAL (phenylalanine ammonia lyase) to yield cinnamic acid. Cinnamic acid is converted to SA via *O*-coumaric acid or benzoic acid. BA2H (benzoic acid 2-hydroxylase) catalyzes the conversion from benzoic acid to SA (Fig. 13.1; Garcion and Métraux 2006).

SA is a well-known plant hormone for plant immunity and SAR (systemic acquired resistance) (Vernooij et al. 1994). After infection with biotrophic pathogens, plants increase accumulation of endogenous SA in the necrotic lesion and surrounding tissues (Enyedi et al. 1992).

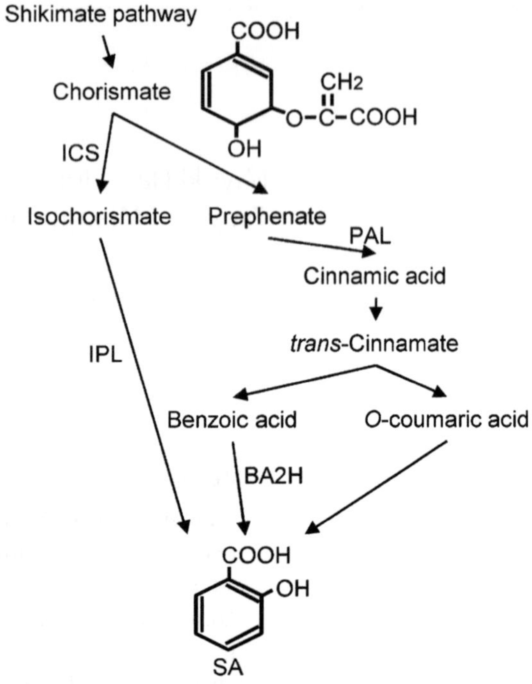

Fig. 13.1 Simplified pathways for SA biosynthesis. *BA2H* benzoic acid-2-hydroxylase, *ICS* isochorismate synthase, *IPL* isochorismate pyruvate lyase, *PAL* phenylalanine ammonia lyase, *SA* salicylic acid

Accumulation of SA induces HR (hypersensitive response)-like cell death and the expression of *PR* (pathogenesis-related) genes (Vlot et al. 2009). Development of SAR requires SA because transgenic tobacco plants expressing *nahG,* which encodes salicylate hydroxylase to catalyze the conversion of SA to catechol, were unable to develop SAR (Gaffney et al. 1993). But, the first plant physiological processes in which SA was shown to be involved are growth regulation (DeKock et al. 1974) and flowering induction (Cleland 1974; Cleland and Ajami 1974). Evidence suggesting that SA is involved in abiotic stresses has also been increasing.

2 Effect of Salicylic Acid on Various Abiotic Stress Responses

Besides plant defense, SA is involved in the response to abiotic stresses. However, the actual role of SA in abiotic stresses remains unresolved.

Generally, deficiency of SA or a very high level of SA increases plant susceptibility to abiotic stresses. The optimal concentration (0.1–0.5 mM for most plants) enhances abiotic stress tolerance. Because SA is also involved in redox regulation, a high concentration of SA may decrease stress tolerance through disturbance of redox status. The relationship between SA and each abiotic stress is discussed below.

2.1 Drought and Stomatal Closure

Plants have developed two strategies to resist drought: drought avoidance and dehydration tolerance (Blum 2005). Drought avoidance refers to plant's abilities to maintain high water status when water is scarce. For instance, plants grow long roots to reach deep soil moisture or reduce water loss by closing stomata on the leaf surface. When moisture is limited, the stomata close to slow transpiration and conserve water. However, at that time, CO_2 supply is decreased, consequently, reducing photosynthesis. Stomata play a major role in plant adaptation to stresses. Drought tolerance refers to plant's ability to withstand loss of water content and regrow when moist conditions return. Resurrection plants can withstand about 90% water loss, whereas most other plants can withstand moderate dehydration (about 30% water loss).

ABA (abscisic acid) is well-known plant hormone that plays an important role in plant responses to abiotic stresses, including stomatal closure and resistance to drought. Water deficit promotes ABA biosynthesis, accumulation, and redistribution in the plant body. ABA promotes stomatal closure via the production of ROS (reactive oxygen species) generated by NADPH oxidase (Acharya and Assmann 2009). SA also has a role in stomatal closure via the production of ROS, which is mediated by a peroxidase-catalyzed reaction (Dong et al. 2001; Mori et al. 2001) and regulates inward-rectifying potassium channel inactivation (Khokon et al. 2011). Stomatal closure is triggered by pathogen-associated molecular patterns to prevent penetration of pathogenes through these pores (Gudesblat et al. 2009). Because bacterial pathogen-induced stomatal closure is compromised in *nahG* plants and the SA biosynthesis mutant *eds16-2*, SA is required for stomatal defense (Melotto et al. 2006, 2008).

SA is also involved in the regulation of drought responses. However, the effect of SA on drought tolerance is still unclear. Because SA potentiates ROS generation in photosynthetic tissues (Borsani et al. 2001), high concentrations of applied SA (0.5 mM) decrease the drought tolerance of maize plants (Németh et al. 2002). However, several reports demonstrate that SA increases drought tolerance. Application of acetyl SA at a range of 0.1–1 mM through seed soaking or foliar spray provides protection against drought stress in muskmelon seedlings (Korkmaz et al. 2007). Similarly, soaking wheat seeds in 100 ppm acetyl SA enhanced plant resistant to drought stress (Hamada and Al-Hakimi 2001). Treatment with ascorbic acid or thiamine had a similar protective effect, which is attributed to the protection of the photosynthetic apparatus from oxidation and the retardation of dark respiration (Hamada and Al-Hakimi 2001). When tomato and bean seeds were imbibed in 0.1–0.5 mM SA or acetyl SA, plants were more tolerant to heat, chilling, and drought stresses (Senaratna et al. 2000). During drought stress, endogenous SA increased up to five-fold in *Phillyrea angustifolia* (Munne-Bosch and Penuelas 2003), suggesting that the role of endogenous SA is possibly in the induction of a protective mechanism during water stress. In barley, SA content in the root was increased by water deficit, whereas that in the leaves was not altered (Bandurska and Stroinski 2005). When barley was treated with SA before drought stress, the damaging effect of water deficit on the cell membranes in the leaves was reduced (Bandurska and Stroinski 2005). SA treatment increases the ABA content and proline levels in leaves of barley, suggesting that SA-induced ABA and proline may contribute to the development of the antistress reaction. The endogenous SA accumulation mutant *adr1* or *myb96-1d* also exhibited both SA-dependent disease resistance and drought tolerance (Grant et al. 2003; Chini et al. 2004; Seo et al. 2009; Seo and Park 2010). Transgenic *Arabidopsis* plants expressing the pepper pathogen-induced gene *CAPIP2* exhibited enhanced disease resistance and drought tolerance (Lee et al. 2006).

ABA and SA antagonistically affect SAR development; exogenous application of ABA suppresses the induction of SAR, while activation of SAR suppresses ABA signaling (Asselbergh et al. 2008; Yasuda et al. 2008). However, ABA and SA play positive roles in stomatal closure. It is suggested that there are condition-specific positive/negative interactions between ABA and SA. Some components that are involved in SA-ABA cross-talk have been identified. *AHG2*, encoding poly(A)-specific ribonuclease, regulates ABA sensitivity, and the mutant exhibited higher levels of SA-inducible gene expression in *Arabidopsis* (Nishimura et al. 2005). The *siz1* mutant, which is impaired in SUMO (small ubiquitin-related modifier) E3 ligase, confers ABA hypersensitivity (Miura et al. 2009) and enhanced SA accumulation and expression of SA-regulated genes (Lee et al. 2007).

2.2 Cold Stress

Temperature is one of the major determinants of agricultural yield and crop productivity, and cold limits the spread of natural plant association (Guy 1990). Cold triggers cell death by cytoplasmic dehydration and ice formation in the cell wall. Plants of tropical or subtropical origin suffer severe damage at temperatures between 0 and 15°C. However, plants from temperate regions become cold tolerant and are still able to grow near the freezing point. Plants develop adaptation to cold stress. Plant cells synthesize and accumulate cryoprotectant solutes and cryoprotective proteins that stabilize cellular membranes and enhance antioxidative mechanisms (Mahajan and Tuteja 2005). Low temperature triggers rigidification of membranes by increasing desaturated phospholipids in membranes (Anchorodoguy et al. 1987). Furthermore, cells also accumulate osmolytes, such as sugars, polyalcohols, aminoacids, polyamines, quaternary ammonium compounds, and antifreezing proteins because sucrose and proline-rich proteins trap water by creating hydrogen bonds to prevent dehydration of the cytoplasm, and accumulation of these compounds causes the freezing point within the cell to drop. In addition to sucrose and proline, some pathogen-related proteins, which may interact with ice to prevent damage from ice crystal dynamical growth, are also induced to function in cold tolerance (Thomashow 1999).

Several reports demonstrate that exogenous application of SA enhances the cold tolerance of various species such as maize, cucumber, rice, potato, pepper, and wheat (Kang and Saltveit 2002; Tasgin et al. 2003; Fung et al. 2004; Korkmaz 2005; Mora-Herrera et al. 2005). Pretreatment of the leaves of chilling-sensitive banana seedlings with 0.5 mM SA solution by spraying the foliage or irrigating the roots for 1 day increased chilling tolerance (Kang et al. 2003). Presoaking seeds before sowing improved cold tolerance in tomato and bean plants (Senaratna et al. 2000). SA treatment enhances the accumulation of H_2O_2. Microarray analyses demonstrate that a transient increase in intracellular H_2O_2 is a primary trigger for the transcriptional regulatory network, which is critical for prolonged survival under chilling stress in rice (Yun et al. 2010). It is plausible that H_2O_2 metabolism is involved in enhanced chilling tolerance.

Higher concentrations and continual application of SA cause severe damage to growth and decrease cold tolerance. The addition of exogenous SA to hydroponic solution inhibits root growth (Pál et al. 2002). Several SA overaccumulating *Arabidopsis* mutants, such as *agd2-1, acd6, cpr5,* and *siz1*, exhibit dwarfism (Rate et al. 1999; Rate and Greenberg 2001; Kirik et al. 2001; Miura et al. 2005). This dwarfism is associated with decreases in cell elongation and proliferation (Kirik et al. 2001; Miura et al. 2010). Plants grown from seeds imbibed in a high concentration of SA (1 mM) did not display enhanced tolerance to chilling stress, whereas 0.1–0.5 mM SA promoted chilling tolerance in bean and tomato (Senaratna et al. 2000). When SA is hydroponically applied, both winter and spring wheat are severely damaged by freezing stress (Horváth et al. 2007a, b), even though freezing tolerance of winter wheat was promoted when SA was sprayed (i.e., was transiently applied) onto the leaves (Tasgin et al. 2003). Endogenous accumulation of SA causes continual application of SA to plants. During chilling conditions, free SA and glucosyl SA accumulate in *Arabidopsis* shoots (Scott et al. 2004) as well as in wheat (Janda et al. 2007) and grape berry (Wan et al. 2009). The growth of *nahG* and *eds5* (enhanced disease susceptibility) mutants is greater than that of wild-type plants at 5°C, even though both plants grow at similar rates at 23°C. *nahG* plants displayed relative growth rates about one-third greater than wild-type plants; thus, by 2 months, *nahG* plants were typically 2.7-fold larger than wild-type plants, and *eds5* behaved similar to *nahG* (Scott et al. 2004). However, the SA-accumulation mutant *cpr1* (constitutive expresser of pathogenesis-related gene) exhibits a very high accumulation of SA and a strong reduction in growth under low temperature conditions (Scott et al. 2004). The other SA accumulating *Arabidopsis* mutants, *siz1* and *acd6* (accelerated cell death), exhibit freezing sensitivity, whereas *nahG siz1* and *nahG acd6* plants have similar survival rates to those of wild-type plants (Miura and Ohta 2010). *DEAR1* (DREB and EAR motif protein) overexpression causes SA accumulation and freezing sensitivity in *Arabidopsis* (Tsutsui et al. 2009). Overexpressing OsWRKY13 enhances disease resistance but suppresses salt and cold tolerance in rice (Qiu et al. 2008). These data suggest that SA controls growth inhibition at low temperatures. Because *cpr1* exhibited severe oxidative damage (Scott et al. 2004), the endogenous constitutive accumulation of SA may enhance ROS production, leading to cold sensitivity.

ICE1 is a transcription factor that binds to the promoter of *CBF3/DREB1A* to control cold signaling (Lissarre et al. 2010). ICE1 may play a role in the integration of cold signaling and SA because several cold-responsive genes are downregulated (Chinnusamy et al. 2003) and SA-inducible genes are upregulated in the *ice1* mutant (Miura and Ohta 2010). ICE1 also associates with the promoter of *BAP1* (BON1-associated protein), which encodes a membrane-associated C2-domain protein that negatively regulates defense responses (Hua et al. 2001; Yang et al. 2006, 2007). The *ice1* mutant exhibits enhanced resistance to bacterial pathogenesis, as does the *bap1* mutant (Zhu et al. 2010). CAMTA3/AtSR1, a member of the calmodulin binding transcription activator family, recognizes the *CBF2/DREB1C* promoter to positively regulate *CBF2/DREB1C* expression for cold tolerance (Doherty et al. 2009) but also interacts with the promoter of *EDS1* to repress its expression and disease resistance (Du et al. 2009). These results suggest that cold signaling and SA signaling are interrelated and that the effect of SA might be tissue-specific and dependent on the organism, concentration, and duration.

2.3 Salinity and Osmotic Stress

Salinity stress in plants is multifactorial, including osmotic stress and cellular sodium toxicity, such as inhibition of vital enzymes and metabolic processes (Ward et al. 2003). Simply, in response to salinity stress, the reduction in growth occurs in two phases: a rapid response to the increase in external osmotic pressure and a slower response due to the accumulation of Na^+ in leaves (Munns and Tester 2008). Salt stress destroys the ionic homeostasis in the water potential and ionic distribution. Photosynthetic processes are severely affected by salinity; thus, salt stress directly reduces carbon fixation and biomass production in plants (Munns, 2007). More than 20% of irrigated lands are affected by high salt content. Because NaCl is the most soluble and widespread salt, plants have evolved mechanisms to regulate its accumulation and to select other nutrients commonly present in low concentrations, such as K^+ and NO_3^-. In most plants, Na^+ and Cl^- are effectively excluded by roots (Munns 2007). Because salinity is a common feature of arid and semiarid lands, plants have evolved mechanisms to tolerate the low soil water potential caused by salinity and drought.

Generally, SA provides tolerance to multiple stresses; soaking wheat seeds in SA solution enhances protection against salinity stress and drought stress (Hamada and Al-Hakimi 2001). Salinity induces increases in endogenous SA levels and the activity of the SA biosynthesis enzyme, benzoic acid 2-hydroxylase, in rice seedlings (Sawada et al. 2006). SA alleviates salt-induced decreases in photosynthesis and decreases the content of leaf Na^+, Cl^-, and H_2O_2 in mungbean, even though high concentration of SA (1.0 mM) is a growth inhibitor (Nazar et al. 2011). Exogenous application of SA improves salt tolerance in sunflower plants, increasing profitable yield production and oil content (Noreen and Ashraf 2010), and it also improves grain yield under salt stress in wheat (Arfan et al. 2007). In these species, antioxidant compound levels were increased because salinity stress can generate ROS in plants to a great extent (Munns and Tester 2008). SA application to barley induced a pre-adaptive response to salt stress, promoted protective reactions to the photosynthetic pigments, and maintained membrane integrity, leading to improvement of plant growth (El Tayeb 2005). The application of 0.05 mM SA improved plant growth after salt stress and caused the accumulation of ABA and proline (Shakirova et al. 2003). SA-treated maize exhibited increased growth under salinity stress resistance and decreased lipid peroxidation and membrane permeability, which were increased by salt stress (Gunes et al. 2007). The application of SA to tomato via root drenching protected against NaCl stress and increased photosynthetic rates under salt stress (Stevens et al. 2006; Poór et al. 2011).

However, other articles demonstrate that application of SA may promote the formation of ROS in the photosynthetic tissues and increase oxidative damage during salt and osmotic stresses. NaCl treatment enhances necrotic lesions on the shoots of wild-type plants, but these lesions were not observed in SA deficient *nahG* plants (Borsani et al. 2001). In *nahG* plants, the glutathione/oxidized glutathione ratio and the ascorbate/dehydroascorbate ratio were higher during salt stress than that of wild type, probably leading to a better adaptation of *nahG* plants to moderate salt stress (Cao et al. 2009).

2.4 Thermotolerance

High temperature stress is a serious threat to plants because the stress causes membrane integrity loss, production of ROS, aggregation and inactivation of proteins, and metabolic and cellular disequilibria, ultimately leading to cell death (Los and Murata 2000; Iba 2002). Photochemical reactions in thylakoid lamellae in the chloroplast stroma are thought to be the primary sites of injury during heat stress (Wise et al. 2004); thus, one critical aspect of heat tolerance in plants is the continual maintenance of photosynthesis. If water is sufficient, plants increase transpirational cooling, but if the temperature continues to rise, membrane compositional changes, the activation of the oxidative defensive system through ethylene and SA, and the production of heat shock proteins (HSPs) are stimulated for cellular protection (Clarke et al. 2004; Larkindale et al. 2005). HSPs are molecular chaperones that reduce protein denaturation, target denatured proteins for proteasome degradation, and facilitate protein folding necessary for proper maturation (Kotak et al. 2007).

SA may improve heat tolerance in a concentration-dependent manner. Low concentrations (0.01–0.1 mM) of SA treatment to mustard plants via spraying increased heat tolerance (Dat et al. 1998). Tobacco treated with 0.01 mM SA exhibited enhanced thermotolerance, whereas a treatment with 0.1 mM SA had no protective effect (Dat et al. 2000). In grape leaves, SA pretreatment alleviated the decrease of photosynthesis rate under heat stress, in part through maintaining a high Rubisco activation state and rapid recovery of PSII function (Wang et al. 2010). Application of AIT (2-amino-2-indanophosphonic acid, a highly specific inhibitor of PAL) or ABT (1-aminobenzotriazole, an inhibitor of BA2H) reduced the endogenous SA content and heat tolerance in pea plants (Pan et al. 2006). Heat-inducible SA enhances the activity and amount of plasma membrane H^+-ATPase in pea leaves during heat acclimation, suggesting that the plasma membrane H^+-ATPase is important to maintain the integrity of the plasma membrane during heat stress (Liu et al. 2009). SA is also increased by heat stress in the creeping bentgrass *Agrostis stolonifera*, grapevine, and pea (Larkindale and Huang 2005; Pan et al. 2006; Wang and Li 2006).

Plants may tolerate elevated temperatures without heat acclimation or any chemical treatment, which is called basal thermotolerance.

Acquired thermotolerance means that plants subjected to mild heat stress transiently acquire tolerance to previously lethal high temperatures (Clarke et al. 2004). The *Arabidopsis* mutants with defective signaling pathways and *nahG* were investigated for the importance of different processes in the development of basal and acquired thermotolerance (Larkindale et al. 2005). ABA, SA, and ROS play roles in the development of acquired thermotolerance (Larkindale et al. 2005), although another study demonstrated that SA is essential for basal but not for acquired thermotolerance (Clarke et al. 2004), possibly due to the complexity of the heat stress response, in which ethylene, ABA, UV, NADP oxidases, and HSPs as well as SA are involved (Kotak et al. 2007). One study suggested that SA acts downstream of ABA and upstream of a phosphatidyl-inositol-4,5-bisphosphate-specific phospholipase C (Liu et al. 2006). However, like ABA, SA does not appear to be required for HSP synthesis during heat stress (Larkindale et al. 2005; Liu et al. 2006).

2.5 Ultraviolet Radiation and Ozone Stress

Stratospheric ozone protects life from detrimental ultraviolet (UV) radiation, but depletion of the ozone layer may increase the amount of UV radiation reaching the surface of the earth. UV is traditionally divided into UV-A (320–400 nm), UV-B (280–320 nm), and UV-C (200–280 nm), which have increasing levels of energy and harmful effects. UV radiation increases the production of ROS. Exposure of leaves to UV increases SA accumulation (Yalpani et al. 1994). Increased SA levels are accompanied by the accumulation of an SA conjugate and by an increase in the activity of BA2H (Yalpani et al. 1994). In pepper leaves, SA treatment moderated an increase in the activities of some antioxidant enzymes, including peroxidase, ascorbate peroxidase, catalase, and glutathione reductase (Mahdavian et al. 2008). SA application increases α-tocopherol concentration and SOD and catalase activity and alleviates UV-B damage (Ervin et al. 2004).

Ozone is formed through photochemical reactions between nitrogen oxides and UV light (Oltmans et al. 1998; Mauzerall and Busconi 2001). Ozone is toxic to plants and animals because it is a powerful oxidizing agent and is able to react directly with lipids and proteins, leading to the production of other ROS such as the hydroxyl radical, singlet oxygen, and H_2O_2 (Kanofsky and Sima 1991; Evans et al. 2005). Ozone and ozone-induced ROS in plants directly harm cells, resulting in leaf necrosis or chlororsis (Castagna and Ranieri 2009). Ozone-derived ROS also work as signal molecules to stimulate the programmed cell death pathway (Overmyer et al. 2005).

Like UV radiation, ozone also induces SA accumulation (Yalpani et al. 1994). In *Arabidopsis*, SA synthesis genes and SA-regulated genes are induced by ozone (Tosti et al. 2006). *nahG* plants are more sensitive to the damaging effect of ozone because of the lack of a satisfactory antioxidant response (Sharma et al. 1996; Rao and Davis 2001). The SA overaccumulating *Arabidopsis* ecotype Cvi-0 is also sensitive to ozone because a large amount of SA induces oxidative processes during ozone stress, leading to cell death (Rao and Davis 2001). These results suggest that optimal concentrations of SA are required to induce antioxidant defense responses and maintain an optimal cellular redox state.

During ozone exposure, SA, ethylene, and JA do not act independently, but in mutually antagonistic or coordinated ways (Tamaoki 2008). The ozone sensitive ecotype Cvi-0, which accumulates a large amount of SA, exhibits low JA sensitivity, suggesting that SA accumulation in response to ozone is negatively regulated by JA signaling (Rao et al. 2000). Ethylene is also negatively regulated by JA signaling. The *oji1* (ozone-sensitive and jasmonate-semi-insensitive) *Arabidopsis* mutant exhibits a high level of ethylene production, and ozone sensitivity, but a reduction of JA sensitivity (Kanna et al. 2003). The interaction between ethylene and salicylic acid has also been studied in ozone-exposed plants. Ethylene and SA together promote continuous ROS production and cell death in response to ozone (Overmyer et al. 2003). An increase in ozone-inducible ethylene production was observed in plants pretreated with SA, and ethylene production decreases in *nahG* during ozone exposure (Rao et al. 2002). Transgenic tobacco plants with inhibited ethylene biosynthesis reduced their levels of SA in response to ozone (Ogawa et al. 2005). The *Arabidopsis rcd1, eto1,* and *eto3* mutants and tobacco cultivar Bel-W3, which accumulate high levels of both ethylene and SA, exhibit ozone sensitivity (Overmyer et al. 2000, 2005; Pasqualini et al. 2002).

2.6 Heavy Metals Stress

In this decade, mechanisms of heavy metal toxicity and tolerance have been investigated energetically by many research groups. In heavy metal sensitive plants, cell damage is caused by a number of mechanisms directly or indirectly (1) reduction of photosynthetic efficiency by inhibiting PSII activity, (2) inhibition of enzymatic activity by the exchange of essential metal ions in the active center of enzymes (e.g., Mg^{2+}, Zn^{2+}, Cu^{2+}, and Co^{2+}) by heavy metal ions, and (3) generation of ROS induced by heavy metal ions, leading to lipid peroxidation in cell membranes. Recent studies focused on the metal hyperaccumulators, such as plants that can accumulate Mn, Co, Ni, Cu, Zn, As, Se, Cd, and Pb, have increased the information about heavy metal tolerance drastically. Combining the mechanisms of heavy metal toxicity and tolerance, plant adaptation to heavy metal stress has been gradually revealed.

As an early phase of research focused on SA alleviation of heavy metal toxicity, SA application reduced the effects of Hg^{2+} and Pb^{2+} on the seed germination and growth of rice seedlings (Mishra and Choudhuri, 1997). SA increased shoot and root biomass under heavy metal stress conditions. In barley, treatment of SA was found to prevent the lipid peroxidation induced by 25 μM Cd^{2+}, resulting in an increase in shoot and root biomass (Metwally et al. 2003). Furthermore, exogenous SA treatment alleviated Mn^{2+}, Ni^{2+}, Cd^{2+}, and Hg^{2+} toxicity through antioxidant systems in many plant species (Shi and Zhu 2008; Wang et al. 2009; Popova et al. 2009; Zhou et al. 2009). For example, SA pretreatment elevated enzymatic and nonenzymatic antioxidants and the concentrations of glutathione and nonprotein thiols, molecules that bind Cd with their SH–groups for Cd-detoxification, in roots and shoots in Cd-treated rice (Guo et al. 2009). These increases in antioxidant systems lead to the alleviation of oxidative damage as indicated by the lowered H_2O_2 and malondialdehyde, which is an end product of lipid peroxidation.

Considering whether these mechanisms are conserved plant defense systems for heavy metals, research using metal hyperaccumulators and genetically manipulated plants is important. In Ni-hyperaccumulating *Thlaspi* species, glutathione-involved Ni tolerance is activated by the constitutively high SA levels (Freeman et al. 2005). Both exogenous feeding and genetic manipulations that increase SA in *Arabidopsis thaliana* plants mimicked the glutathione-related Ni resistance (Freeman et al. 2005). These data suggest that SA might be one of the key factors related to the differences between hyperaccumulators and nonaccumulators. Additionally, *nahG* exhibits a significant decrease in superoxide dismutase activity, which might result in the prevention of H_2O_2 increase after Cd treatment, suggesting that endogenous SA may function in *Arabidopsis* as a signaling molecule necessary to generate, sustain, or amplify Cd-induced oxidative stress (Zawoznik et al. 2007). To reinforce the involvement of endogenous SA in plant heavy metal tolerance, the content of SA should be clarified under heavy metal stress conditions. Cd treatment induces the accumulation of free and conjugated SA in pea (Popova et al. 2009) and maize plants (Pál et al. 2005; Krantev et al. 2008). Similar to the involvement of SA in oxidative stress, the relationship between the alleviation of heavy metal stress by SA and endogenous SA and related molecules should be investigated in more detail.

3 Relationship Between SA and ROS for Abiotic Stress Responses

As described above, SA is involved in several abiotic stress responses in plants. Most abiotic stresses increase the *in planta* concentration of SA. The effect of exogenous SA application is not

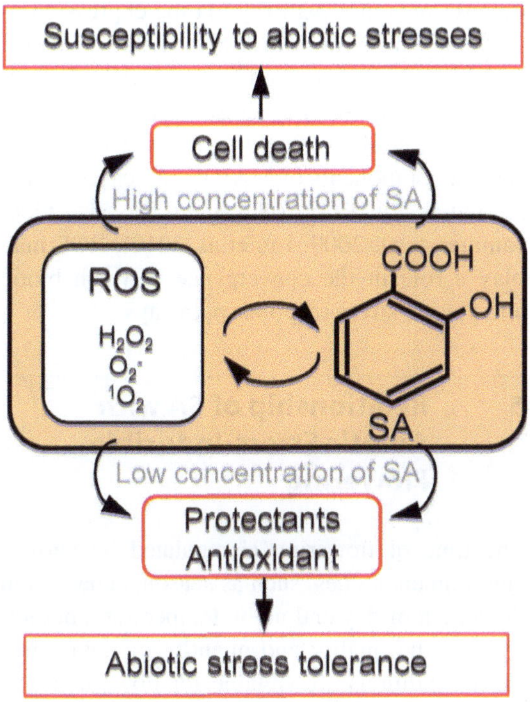

Fig. 13.2 Schematic model of a self-amplifying feedback loop between SA and H_2O_2 for response to abiotic stresses

obvious, but it may depend on the concentration of applied SA, mode of application, and state (developmental stage, oxidative balance, and acclimation) of the plants. Generally, low concentrations of SA application alleviate susceptibility to abiotic stresses and high concentrations of SA cause high levels of oxidative stress, leading to decreased abiotic stress tolerance (Fig. 13.2). Similarly, H_2O_2, which is stable, relatively long-lived, and highly permeable across membranes, acts as a signal molecule involved in acclamatory signaling at low concentrations, triggering tolerance to various abiotic and biotic stresses, but at high concentrations, it leads to programmed cell death (Quan et al. 2008). The conversion of benzoic acid into SA is catalyzed by the H_2O_2-mediated activation of BA2H (Dempsey and Klessig 1995). SA pretreatment also results in the accumulation of H_2O_2 (Agarwal et al. 2005; Harfouche et al. 2008). Thus, it is proposed that SA and H_2O_2 form a self-amplifying feedback loop in response to abiotic and biotic stresses (Fig. 13.2); H_2O_2 induces SA accumulation, and SA enhances H_2O_2 level (van Camp et al. 1998).

Redox homeostasis in plant cells is maintained by the appropriate balance between ROS generation and scavenging mechanisms (Apel and Hirt 2004). Biotic and abiotic stress conditions produce an increase in ROS levels, leading to an alteration in the cellular redox homeostasis. A number of studies demonstrate that the efficiency of antioxidative systems is correlated with tolerance to abiotic stresses (Athar et al. 2008; Munns and Tester 2008). Application of SA at low concentrations enhances accumulation of H_2O_2, which induces antioxidant defense systems, including enzymatic antioxidants such as SOD (superoxide dismutase), CAT (catalase), APX (ascorbate peroxidase), and GPX (glutathione peroxidase) and nonenzymatic antioxidants such as glutathione, ascorbic acid, carotenoids, and tocopherols (Ahmad et al. 2010; Gill and Tuteja 2010). Several reports demonstrate that SA also stimulates the activity of SOD, GPX, glutathione reductase, and peroxidase (Janda et al. 1999; Milla et al. 2003; Azevedo et al. 2004; Noreen et al. 2009). The activation of ROS scavenging systems may contribute to the equilibration of ROS homeostasis to enhance abiotic stress tolerance.

High levels of SA treatment cause not only production of ROS (Kawano 2003), but also reductions in APX and CAT activity, leading to an over-accumulation of ROS (Durner and Klessig 1996; Janda et al. 2003). SA is capable of binding directly to catalase enzyme, inhibiting its activity in tobacco, *Arabidopsis*, tomato, maize, and cucumber (Chen et al. 1993; Sánchez-Casas and Klessig 1994; Conrath et al. 1995; Horváth et al. 2002). Over-accumulation of ROS causes oxidative damage (Ahmad et al. 2010; Gill and Tuteja 2010) and triggers both apoptosis-like and autophagic cell death (Love et al. 2008). Mitochondrial AOX (alternative oxidase) can significantly reduce electron build-up, which leads to the reduction of redox stress and ROS accumulation (Millenaar et al. 1998). Downregulation of AOX stimulates programmed cell death (Maxewll et al. 2002). Both H_2O_2 and SA were found to disrupt normal mitochondrial function, resulting in decreased rates of electron transport and lowering of cellular ATP levels (Norman et al. 2004). These findings suggest that the mitochondrion may play an important role in conveying intracellular stress signals to the nucleus, leading to alterations in gene expression.

4 Crosstalk Between Abiotic and Biotic Stress Responses

Phytohormones such as SA, jasmonic acid, ethylene, and ABA regulate the protective responses of plants against biotic and abiotic stresses via synergistic and antagonistic actions (Fujita et al. 2006). The generation of ROS has been proposed as a key process that is shared among biotic and abiotic stress responses (Quan et al. 2008; Ahmad et al. 2010; Gill and Tuteja 2010). In many case, ABA acts as a negative regulator of disease resistance (Mauch-Mani and Mauch 2005). High ABA concentrations inhibit the SA-dependent defense response. In nature, simultaneous exposure of plants to drought and pathogen attack is actually rare, as successful pathogen infection requires relatively humid conditions, and water deficiency more severely threatens plant survival than does pathogen infection. The phenotypes of several lesion-mimicking mutants, such as *ssi4* (*SA-insensitivity*), *cpn1, slh1,* and overexpression of the *R* gene, were suppressed by environmental cues, including high humidity and high temperature (Jambunathan et al. 2001; Xiao et al. 2003; Zhou et al. 2004; Noutoshi et al. 2005), suggesting that *R*-gene-mediated disease resistance responses and abiotic stress responses cross-talk.

Protein phosphorylation and dephosphorylation significantly influence both the regulation of physiological morphology and the gene expression that is associated with basic cellular activities. *Arabidopsis* MEKK1 (MAP kinase kinase kinase) is induced by cold, salt, drought, and wounding (Mizoguchi et al. 1996). Cold and salt stress signaling is regulated by MEKK1-MKK2 (MAP kinase kinase)–MPK4/MPK6 (MAP kinase) (Teige et al. 2004). The *mekk1* and *mpk4* mutants accumulate SA, increasing resistance to *Pseudomonas syringae* (Petersen et al. 2000; Suarez-Rodriguez et al. 2007). MEKK1-MKK4/MKK5-MPK3/MPK6 cascades have been reported to regulate the pathogen defense response pathway via the expression of *WRKY22* and *WRKY29* (Asai et al. 2002). MPK4 and MPK6 are activated by cold, dehydration, touch, hyper-osmotic stresses, and cadmium (Ichimura et al. 2000; Liu et al. 2010). Because several components of MAP kinase cascades are regulated by ROS (Nakagami et al. 2006; Jammes et al. 2009; Liu et al. 2010), ROS may play a role in the convergence between biotic and abiotic stress response pathways.

5 Relationship of SA with Abiotic Stress-Inducible Flowering

The time of flowering is regulated by various environmental cues, such as seasonal changes in the length of day and night, temperature, humidity, and the quality and quantity of light when plant conditions are suitable for flowering. The photoperiod, autonomous, vernalization, and gibberelic acid pathways are known to play key roles in the control of flowering time (Mouradov et al. 2002; Kolář and Seňková 2008).

SA has been proposed to induce flowering (Cleland 1974). For example, such induction resulted from an experiment that was focused on florigen transition through phloem. After aphid feeding in vegetative and flowering plants of *Xanthium strumarium* L., the different honeydew produced by aphids was applied to *Lemna gibba G3* for flowering (Cleland 1974). This extracted compound was identified as SA, suggesting that SA might have a phloem-mobilized activity capable of inducing flowering in *L. gibba* (Cleland and Tanaka 1979). Furthermore, SA also induced flowering in *Lamna paucicostata 6746* (Wada 1974). When exposed to abiotic stresses, flowering is induced (Shinozaki and Takimoto 1983; Kolář and Seňková 2008; Wada et al. 2010a, b), which is referred to as stress-induced flowering (Hatayama and Takeno 2003). Nitrogen limitation accelerates flowering in *Perillacrispa*, even under continuous light conditions (Shinozaki and Takimoto 1982). In *Pharbitis nil*, nutrient starvation, low temperature, and highlight intensity also induce flowering

(Wada et al. 2010a; Shinozaki and Takimoto 1982; Shinozaki 1985; Hirai et al. 1993; Tanada et al. 1997). Flowering induction by such stress conditions is often accompanied by an increase in PAL activity (Fig. 13.1; Hatayama and Takeno 2003; Wada and Takeno 2010). The application of aminooxyacetic acid, an inhibitor of PAL, suppresses the accumulation of phenylpropanoids and the induction of flowering in *P. nil* (Hirai et al. 1995). The exogenous application of SA induces flowering in *P. nil* under low temperature conditions, suggesting that SA may function as an inducer of stress-promoted flowering (Shinozaki and Takimoto 1983). However, SA application alone does not induce flowering under nonstress conditions in *P. nil* (Wada et al. 2010a). UV-C light induces SA biosynthesis and accelerated flowering in *Arabidopsis* (Martinez et al. 2004). Investigation of flowering time after irradiation of UV-C with various mutants impaired in SA biosynthesis, photoperiodic flowering response, or autonomous flowering suggested that SA is involved in the negative regulation of the photoperiodic and autonomous pathways and *FLC* expression (Martinez et al. 2004). However, application of SA was not sufficient to accelerate stress-induced flowering under nonstress conditions (Martinez et al. 2004). *PCC1* (pathogen and circadian controlled 1; Sauerbrunn and Schlaich 2004), which is induced by SA, RNAi lines exhibited late flowering and were defective in UV-C light acceleration of flowering (Segarra et al. 2009).

These studies support the idea that SA induces flowering in many plant species. Stress-induced flowering has been reported to be associated with changes not only in SA biosynthesis but also in other factors (Cleland and Tanaka 1979; Martinez et al. 2004; Kolář and Seňková 2008; Wada et al. 2010a), because SA application alone does not induce flowering under nonstress conditions in some plant species (Cleland and Tanaka 1979; Martinez et al. 2004; Wada et al. 2010a), and this flowering regulation seems to involve the photoperiodic and autonomous pathways (Martinez et al. 2004). Further studies will be necessary to demonstrate the molecular mechanisms underlying the SA-dependent regulation of flowering time.

6 Conclusions and Future Perspective

SA is a very promising compound for regulation of abiotic stresses because application of the appropriate concentration of SA enhances tolerance to abiotic stresses as described above. However, in certain cases, contradictory results are obtained in investigating the effect of exogenous SA application and that of endogenous SA levels. The effects of SA are dependent not only on the concentration of applied SA but also on the method of application. Still, several questions remain unanswered. It is unclear if exogenous SA application directly or indirectly increases endogenous SA levels and whether the effect of SA is connected with ROS production. Furthermore, no SA receptor has been identified. SA signaling and abiotic stress signaling could be connected and may contribute to the complex signal transduction network. The clarification of these questions could lead to a better understanding of the precise role of SA in abiotic stress responses.

Acknowledgements This work was supported, in part, by Special Coordination Funds for Promoting Science and Technology, and by a Grant-in-Aid for Young Scientists (B, No. 20780046; B, No. 21770032) from the Ministry of Education, Culture, Sports, Science and Technology.

References

Acharya BR, Assmann SM (2009) Hormone interactions in stomatal function. Plant Mol Biol 69:451–462

Agarwal S, Sairam RK, Srivastava GC, Tyagi A, Meen RC (2005) Role of ABA, salicylic acid, calcium and hydrogen peroxide on antioxidant enzymes induction in wheat seedlings. Plant Sci 169:559–570

Ahmad P, Jaleel CA, Salem MA, Nabi G, Sharma S (2010) Roles of enzymatic and non-enzymatic antioxidants in plants during abiotic stress. Crit Riv Biotechnol 30:161–175

Anchorodoguy TJ, Rudolph AS, Carpenter JF, Crowe JH (1987) Modes of interaction of cryoprotectants with membrane phospholipids during freezing. Cryobiology 24:324–331

Apel K, Hirt H (2004) Reactive oxygen species: metabolism, oxidative stress, and signal transduction. Annu Rev Plant Biol 55:373–399

Arfan M, Athar HR, Ashraf M (2007) Does exogenous application of salicylic acid through the rooting medium modulate growth and photosynthetic capacity in differently adapted spring wheat cultivated under salt stress? J Plant Physiol 6:685–694

Asai T, Tena G, Plotnikova J, Willmann MR, Chiu WL, Gomez-Gomez L, Boller T, Ausubel FM, Sheen J (2002) MAP kinase signaling cascade in *Arabidopsis* innate immunity. Nature 415:977–983

Asselbergh B, Vleesschauwer D, Höfte M (2008) Global switches and fine-tuning – ABA modulates plant pathogen defense. Mol Plant Microb Int 21:709–719

Athar HR, Khan A, Ashraf A (2008) Exogenously applied ascorbic acid alleviates salt-induced oxidative stress in wheat. Environ Exp Bot 63:224–231

Azevedo H, Lino-Neto T, Tavares RM (2004) Salicylic acid up-regulates the expression of chloroplastic Cu, Zn-superoxide dismutase in needles of maritime pine (*Pinus pinaster*Ait.). Ann For Sci 61:847–850

Bandurska H, Stroinski A (2005) The effect of salicylic acid on barley response to water deficit. Acta Physiol Plant 27:379–386

Blum A (2005) Drought resistance, water-use deficiency, and yield potential – are they compatible, dissonant, or mutually exclusive? Aust J Agric Res 56:1159–1168

Borsani O, Valpuesta V, Botella MA (2001) Evidence for a role of salicylic acid in the oxidative damage generated by NaCl and osmotic stress in *Arabidopsis* seedlings. Plant Physiol 126:1024–1030

Cao Y, Zhang Z-W, Xue L-W, Du J-B, Shang J, Xu F, Yuan S, Lin HH (2009) Lack of salicylic acid in *Arabidopsis* protects plants against moderate salt stress. Z Naturforsch C 64:231–238

Castagna A, Ranieri A (2009) Detoxification and repair process of ozone injury: From O_3 uptake to gene expression adjustment. Environ Pollut 157:1461–1469

Catinot J, Buchala A, Abou-Mansour E, Métraux JP (2008) Salicylic acid production in response to biotic and abiotic stress depends on isochorismate in *Nicotiana benthamiana*. FEBS Lett 582:473–478

Chen Z, Ricigliano JR, Klessig DF (1993) Purification and characterization of a soluble salicylic acid binding protein from tobacco. Proc Natl Acad Sci USA 90: 9533–9537

Chini A, Grant JJ, Seki M, Shinozaki K, Loake GJ (2004) Drought tolerance established by enhanced expression of the *CC-NBS-LRR* gene, *ADR1*, requires salicylic acid, EDS1 and ABI1. Plant J 38:810–822

Chinnusamy V, Ohta M, Kanrar S, Lee BH, Hong X, Agarwal M, Zhu JK (2003) ICE1: A regulator of cold-induced transcriptome and freezing tolerance in *Arabidopsis*. Genes Dev 17:1043–1054

Clarke SM, Mur LA, Wood JE, Scott IM (2004) Salicylic acid dependent signaling promotes basal thermotolerance but is not essential for acquired thermotolerance in *Arabidopsis thaliana*. Plant J 38:432–447

Cleland CF (1974) Isolation of flower-inducing and flower-inhibitory factors from aphid honeydew. Plant Physiol 54:899–903

Cleland CF, Ajami A (1974) Identification of the flower-inducing factor isolated from aphid honeydew as being salicylic acid. Plant Physiol 54:904–906

Cleland CF, Tanaka O (1979) Effect of daylenght on the ability of salicylic acid to induce flowering in the Long day plant *Lemna gibba* G3 and the short-day plant *Lemna paucicosta 6746*. Plant Physiol 64:421–424

Conrath U, Chen Z, Ricigliano JR, Klessig DF (1995) Two inducers of plant defense responses, 2,6-dichloroisonicotininc acid and salicylic acid, inhibit catalase activity in tobacco. Proc Natl Acad Sci USA 92:7143–7147

Dat JF, Lopez-Delgado H, Foyer CH, Scott IM (1998) Parallel changes in H_2O_2 and catalase during thermotolerance induced by salicylic acid or heat acclimation in mustard seedlings. Plant Physiol 116:1351–1357

Dat JF, Lopez-Delgado H, Foyer CH, Scott IM (2000) Effects of salicylic acid on oxidative stress and thermotolerance in tobacco. J Plant Physiol 156:659–665

DeKock PC, Grabowsky FB, Innes AM (1974) The effect of salicylic acid on the growth of *Lemma gibba*. Ann Bot 38:903–908

Dempsey DA, Klessig DF (1995) Signals in plant disease resistance. Bull Inst Pasteur 93:167–186

Doherty CJ, van Buskirk HA, Myers SJ, Thomasho MF (2009) Roles for *Arabidopsis* CAMTA transcription factors in cold-regulated gene expression and freezing tolerance. Plant Cell 21:972–984

Dong F-C, Wang PT, Song CP (2001) The role of hydrogen peroxide in salicylic acid-induced stomatal closure in *Vicia faba* guard cells. Acta Phytophysiol Sinica 27:296–302

Du L, Ali GS, Simons KA, Hou J, Yang T, Reddy ASN, Poovaiah BW (2009) Ca^{2+}/calmodulin regulates salicylic-acid-mediated plant immunity. Nature 457:1154–1158

Durner J, Klessig DF (1996) Salicylic acid is a modulator of tobacco and mammalian catalases. J Biol Chem 271:28492–28501

El Tayeb MA (2005) Response of barley grains to the interactive effect of salinity and salicylic acid. Plant Growth Regul 45:215–224

Enyedi AJ, Yalpani N, Silverman P, Raskin I (1992) Localization, conjugation, and function of salicylic acid in tobacco during the hypersensitive reaction to tobacco mosaic virus. Proc Natl Acad Sci USA 89: 2480–2484

Ervin EH, Zhang X-Z, Fike JH (2004) Ultraviolet-B radiation damage on Kentucky Bluegrass II: Hormone supplement effects. Hort Sci 39:1471–1474

Evans NH, McAinsh MR, Hetherington AM, Knight MR (2005) ROS perception in *Arabidopsis thaliana:* the ozone-induced calcium response. Plant J 41:615–626

Freeman JL, Garcia D, Kim D, Hopf A, Salt DE (2005) Constitutively elevated salicylic acid signals glutathione-mediated nickel tolerance in Thlaspi nickel hyperaccumulators. Plant Physiol 137:1082–1091

Fujita M, Fujita Y, Noutoshi Y, Takahashi F, Narusaka Y, Yamaguchi-Shinozaki K, Shinozaki K (2006) Crosstalk between abiotic and biotic stress responses: a current view from the points of convergence in the stress signaling networks. Curr Opin Plant Biol 9: 436–442

Fung RWM, Wang C-Y, Smith DL, Gross KC, Tian MS (2004) MeSA and MeJA increase steady-state transcript levels of alternative oxidase and resistance against chilling injury in sweet peppers (*Capsicum annum* L.). Plant Sci 166:711–719

Gaffney T, Friedrich L, Vernooij B, Negrotto D, Nye G, Uknes S, Ward E, Kessmann H, Ryals J (1993) Requirement of salicylic acid for the induction of systemic acquired resistance. Science 261:754–756

Garcion C, Métraux JP (2006) Salicylic acid. In: Plant hormone signaling. 24: Oxford: Blackwell Publishing Ltd, pp 229–255

Garcion C, Lohmann A, Lamodière E, Catinot J, Buchala A, Doermann P, Métraux JP (2008) Characterization and biological function of the *ISOCHORIMATE SYNTHASE 2* gene of *Arabidopsis*. Plant Physiol 147:1279–1287

Gill SS, Tuteja N (2010) Reactive oxygen species and antioxidant machinery in abiotic stress tolerance in crop plants. Plant Physiol Biochem 48:909–930

Grant JJ, Chini A, Basu D, Loake GJ (2003) Targeted activation tagging of the *Arabidopsis NBS LRR* gene, ADR1, conveys resistance against virulent pathogens. Mol Plant Microbe Interact 16:669–680

Gudesblat GE, Torres PS, Vojnov AV (2009) Stomata and pathogens. Plant Signal Behav 4:1114–1116

Gunes A, Inal A, Alpaslan M, Eraslan F, Bagci EG, Cicek N (2007) Salicylic acid induced changes on some physiological parameters symptomatic for oxidative stress and mineral nutrition in maize (*Zea mays* L.) grown under salinity. J Plant Physiol 164:728–736

Guo B, Liang Y, Zhue Y (2009) Does salicylic acid regulate antioxidant defense system, cell death, cadmium uptake and partitioning to acquire cadmium tolerance in rice? J Plant Physiol 166:20–31

Guy CL (1990) Cold acclimation and freezing stress tolerance: role of protein metabolism. Annu Rev Plant Physiol Plant Mol Biol 41:187–223

Hamada AM, Al-Hakimi AMA (2001) Salicylic acid versus salinity-drought-induced stress on wheat seedlings. Rostl Vyr 47:444–450

Harfouche AL, Rugini E, Mencarelli F, Botondi R, Mulco R (2008) Salicylic acid induces H_2O_2 production and endochitinase gene expression but not ethylene biosynthesis in *Castanea sativa in vitro* model system. J Plant Physiol 165:734–744

Hatayama T, Takeno K (2003) The metabolic pathway of salicylic acid rather than of chlorogenic acid is involved in the stress-induced flowering of *Pharbitis nil*. J Plant Physiol 160:461–467

Hirai N, Kojima Y, Koshimizu K, Shinozaki M, Takimoto A (1993) Accumulation of phenylpropanoids in cotyledons of morning glory (*Pharbitis nil*) seedlings during the induction of flowering by poor nutrition. Plant Cell Physiol 34:1039–1044

Hirai N, Kuwano Y, Kojima Y, Koshimizu K, Shinozaki M, Takimoto A (1995) Increase in the activity of phenylalanine ammonia-lyase during the non photoperiodic induction of flowering in seedlings of Morning glory (*Pharbitis nil*). Plant Cell Physiol 36:291–297

Horváth E, Janda T, Szalai G, Páldi E (2002) *In vitro* salicylic acid inhibition of catalase activity in maize: differences between the isozymes and a possible role in the induction of chilling tolerance. Plant Sci 163: 1129–1135

Horváth E, Pál M, Szalai G, Páldi E, Janda T (2007b) Exogenous 4-hydroxybenzoic acid and salicylic acid modulate the effect of short-term drought and freezing stress on wheat plants. Biol Plant 51:480–487

Hua J, Grisafi P, Cheng SH, Fink GR (2001) Plant growth homeostasis is controlled by the *Arabidopsis BON1* and *BAP1* genes. Genes Dev 15:2263–2272

Iba K (2002) Acclimative response to temperature stress in higher plants: approaches of gene engineering for temperature tolerance. Annu Rev Plant Biol 53: 225–245

Ichimura K, Mizoguchi T, Yoshida R, Yuasa T, Shinozaki K (2000) Various abiotic stresses rapidly activate *Arabidopsis* MAP kinases ATMPK4 and ATMPK6. Plant J 24:655–665

Jambunathan N, Siani JM, McNellis TW (2001) A humidity-sensitive Arabidopsis copine mutant exhibits precocious cell death and increased disease resistance. Plant Cell 13:2225–2240

Jammes F, Song C, Shin D, Munemasa S, Takeda K, Gu D, Cho D, Lee S, Giordo R, Sritubtim S, Leonhardt N, Ellis BE, Murata Y, Kwak JM (2009) MAP kinases *MPK9* and *MPK12* are preferentially expressed in guard cells and positively regulate ROS-mediated ABA signaling. Proc Natl Acad Sci USA 106:20520–20525

Janda T, Szalai G, Tari I, Páldi E (1999) Hydroponic treatment with salicylic acid decreases the effect of chilling injury in maize (*Zea mays* L.) plants. Planta 208: 175–180

Janda T, Szalai G, Rios-Gonzalez K, Veisz O, Páldi E (2003) Comparative study of frost tolerance and antioxidant activity in cereals. Plant Sci 164:301–306

Janda T, Szalai G, Leskó K, Yordanova R, Apostol S, Popova LP (2007) Factors contributing to enhanced freezing tolerance in wheat during frost hardening in the light. Phytochemistry 68:1674–1682

Kang HM, Saltveit ME (2002) Chilling tolerance of maize, cucumber and rice seedling leaves and roots are differentially affected by salicylic acid. Physiol Plant 115:571–576

Kang GZ, Wang CH, Sun GC, Wang ZX (2003) Salicylic acid changes activities of H_2O_2-metabolizing enzymes and increases the chilling tolerance of banana seedlings. Environ Exp Bot 50:9–15

Kanna M, Tamaoki M, Kubo A, Nakajima N, Rakwal R, Agrawal GK, Tamogami S, Ioki M, Ogawa D, Saji H, Aono M (2003) Isolation of an ozone-sensitive and jasmonate-semi-insensitive *Arabidopsis* mutant (*oji1*). Plant Cell Physiol 44:1301–1310

Kanofsky JR, Sima P (1991) Singlet oxygen production from the reactions of ozone with biological molecules. J Biol Chem 266:9039–9042

Kawano T (2003) Roles of the reactive oxygen species-generating peroxidase reactions in plant defense and growth induction. Plant Cell Rep 21:829–837

Khokon A, Okuma E, Hossain M, Munemasa S, Uraji M, Nakamura Y, Mori IC, Murata Y (2011) Involvement of extracellular oxidative burst in salicylic acid-induced stomatal closure in *Arabidopsis*. Plant Cell Environ 34:434–443

Kirik V, Bouyer D, Schöbinger U, Bechtold N, Herzogm M, Bonneville JM, Hülskamp M (2001) *CPR5* is involved in cell proliferation and cell death control and encodes a novel transmembrane protein. Curr Biol 11:1891–1895

Kolář J, Seňková J (2008) Reduction of mineral nutrient availability accelerates flowering of *Arabidopsis thaliana*. J Plant Physiol 165:1601–1609

Korkmaz A (2005) Inclusion of acetyl salicylic acid and methyl jasmonate into the priming solution improves low temperature germination and emergence of sweet pepper. Hort Sci 40:197–200

Korkmaz A, Uzunlu M, Demirkiran AR (2007) Treatment with acetyl salicylic acid protects muskmelon seedlings against drought stress. Acta Physiol Plant 29:503–508

Kotak S, Larkindale J, Lee U, von Koskull-Döring P, Vierling E, Scharf KD (2007) Complexity of the heat stress response in plants. Curr Opin Plant Biol 10: 310–316

Krantev A, Yordanova R, Janda T, Szalai G, Popova L (2008) Treatment with salicylic acid decreases the effect of cadmium on photosynthesis in maize plants. J Plant Physiol 165:920–931

Larkindale J, Huang BR (2005) Effects of abscisic acid, salicylic acid, ethylene and hydrogen peroxide in thermotolerance and recovery for creeping bentgrass. Plant Growth Regul 47:17–28

Larkindale J, Hall JD, Knight MR, Verling E (2005) Heat stress phenotypes of Arabidopsis mutants implicate multiple signaling pathways in the acquisition of thermotolerance. Plant Physiol 138:882–897

Lee H, León J, Raskin I (1995) Biosynthesis and metabolism of salicylic acid. Proc Natl Acad Sci USA 92: 4076–4079

Lee SC, Kim SH, An SH, Yi SY, Hwang BK (2006) Identification and functional expression of the pepper pathogen-induced gene, *CAPIP2,* involved in disease resistance and drought and salt stress tolerance. Plant Mol Biol 62:151–164

Lee J, Nam J, Park HC, Na G, Miura K, Jin JB, Yoo CY, Baek D, Kim DH, Jeong JC, Kim D, Lee SY, Salt DE, Mengiste T, Gong Q, Ma S, Bohnert HJ, Kwak SS, Bressan RA, Hasegawa PM, Yun DJ (2007) Salicylic acid-mediated innate immunity in *Arabidopsis* is regulated by SIZ1 SUMO E3 ligase. Plant J 49:79–90

Lissarre M, Ohta M, Sato A, Miura K (2010) Cold-responsive gene regulation during cold acclimation in plants. Plant Signal Behav 5:948–952

Liu H, Liu Y, Pan Q, Yang H, Zhan J, Huang W (2006) Novel interrelationship between salicylic acid, abscisic acid, and PIP_2-specific phospholipase C in heat acclimation-induced thermotolerance in pea leaves. J Exp Bot 57:3337–3347

Liu Y, Liu H, Pan Q, Yang H, Zhan J, Huang W (2009) The plasma membrane H^+-ATPase is related to the development of salicylic acid-induced thermotolerance in pea leaves. Planta 229:1087–1098

Liu XH, Kim KE, Kim KC, Nguyen XC, Han HJ, Jung MS, Kim HS, Kim SH, Park HC, Yun DJ, Chung WS (2010) Cadmium activates *Arabidopsis* MPK3 and MPK6 via accumulation of reactive oxygen species. Phytochemistry 71:614–618

Los DA, Murata N (2000) Regulation of enzymatic activity and gene expression by membrane fluidity. Sci. STKE 2000

Love AJ, Milner JJ, Sadanandom A (2008) Timing is everything: regulatory overlap in plant cell death. Trends Plant Sci 13:589–595

Mahajan S, Tuteja N (2005) Cold, salinity and drought stresses: an overview. Arch Biochem Biophys 444:139–158

Mahdavian K, Ghorbanli M, Kalantari KM (2008) Role of salicylic acid in regulating ultraviolet radiation-induced oxidative stress in pepper leaves. Russ J Plant Physiol 55:560–563

Martinez C, Pons E, Prats G, Leon J (2004) Salicylic acid regulates flowering time and links defence responses and reproductive development. Plant J 37:209–217

Mauch-Mani B, Mauch F (2005) The role of abscisic acid in plant-pathogen interactions. Curr Opin Plant Biol 8:409–414

Mauzerall DL, Busconi L (2001) Protecting agricultural crops from the effects of tropospheric ozone exposure: reconciling science and standard setting in the United states. Europe and Asia. Annu Rev Energy Environ 26:237–268

Maxewll DP, Nickels R, McIntosh L (2002) Evidence of mitochondrial involvement in the transduction of signals required for the induction of genes associated with pathogen attack and senescence. Plant J 29:269–279

Melotto M, Underwood W, Koczan J, Nomura K, He SY (2006) Plant stomata function in innate immunity against bacterial invasion. Cell 126:969–980

Melotto M, Underwood W, He SY (2008) Role of stomata in plant innate immunity and foliar bacterial diseases. Annu Rev Phytopathol 46:101–122

Metwally A, Finkemeier I, Georgi M, Dietz KJ (2003) Salicylic acid alleviates the cadmium toxicity in barley seedlings. Plant Physiol 132:272–281

Milla MAR, Mauer A, Huete AR, Gustafson JP (2003) Glutathione peroxidase genes in *Arabidopsis* are ubiquitous and regulated by abiotic stresses through diverse signaling pathways. Plant J 36:602–615

Millenaar FF, Benschop JJ, Wagner AM, Lambers H (1998) The role of the alternative oxidase in stabilizing the in vivo reduction state of the ubiquinone pool and the activation state of the alternative oxidase. Plant Physiol 118:599–607

Mishra A, Choudhuri MA (1997) Ameliorating effects of salicylic acid on lead and mercury-induced inhibition of germination and early seedling growth of two rice cultivars. Seed Sci Technol 25:263–270

Miura K, Ohta M (2010) SIZ1, a small ubiquitin-related modifier ligase, controls cold signaling through regulation of salicylic acid accumulation. J Plant Physiol 167:555–560

Miura K, Rus A, Sharkhuu A, Yokoi S, Karthikeyan AS, Raghothama KG, Baek D, Koo Y D, Jin J B, Bressan

RA, Yun D J, Hasegawa PM (2005) The *Arabidopsis* SUMO E3 ligase SIZ1 controls phosphate deficiency responses. Proc Natl Acad Sci USA 102:7760–7765

Miura K, Lee J, Jin JB, Yoo CY, Miura T, Hasegawa PM (2009) Sumoylation of ABI5 by the *Arabidopsis* SUMO E3 ligase SIZ1 negatively regulates abscisic acid signaling. Proc Natl Acad Sci USA 106: 5418–5423

Miura K, Lee J, Miura T, Hasegawa PM (2010) *SIZ1* controls cell growth and plant development in *Arabidopsis* through salicylic acid. Plant Cell Physiol 51:103–113

Mizoguchi T, Irie K, Hirayama T, Hayashida N, Yamaguchi-Shinozaki K, Matsumoto K, Shinozaki K (1996) A gene encoding a mitogen-activated protein kinase kinasekinase is induced simultaneously with genes for a mitogen-activated protein kinase and an S6 ribosomal protein kinase by touch, cold and water stress in *Arabidopsis thaliana*. Proc Natl Acad Sci USA 93:765–769

Mora-Herrera ME, Lopez-Delgado H, Castillo-Morales A, Foyer CH (2005) Salicylic acid and H_2O_2 function by independent pathways in the induction of freezing tolerance in potato. Physiol Plant 125:430–440

Mori IC, Pinontoan R, Kawano T, Muto S (2001) Involvement of superoxide generation in salicylic acid-induced stomatal closure in *Vicia faba*. Plant Cell Physiol 42:1383–1388

Mouradov A, Cremer F, Coupland G (2002) Control of flowering time: interacting pathways as a basis for diversity. Plant Cell 14:111–130

Munne-Bosch S, Penuelas J (2003) Photo- and antioxidative protection, and a role for salicylic acid during drought and recovery in field-grown *Phillyrea angustifolia* plants. Planta 217:758–766

Munns R (2007) Prophylactively parking sodium in the plant. New Phytol 176:501–504

Munns R, Tester M (2008) Mechanisms of salinity tolerance. Annu Rev Plant Biol 59:651–681

Nakagami H, Soukupová H, Schikora A, Zárský V, Hirt H (2006) A mitogen-activated protein kinase kinasekinase mediates reactive oxygen species homeostasis in *Arabidopsis*. J Biol Chem 281:38697–38704

Nazar R, Iqbal N, Syeed S, Khan NA (2011) Salicylic acid alleviates decreases in photosynthesis under salt stress by enhancing nitrogen and sulfur assimilation and antioxidant metabolism differentially in two mungbean cultivars. J Plant Physiol 168:807–815

Németh M, Janda T, Horváth E, Páldi E, Szalai G (2002) Exogenous salicylic acid increases polyamine content but may decrease drought tolerance in maize. Plant Sci 162:569–574

Nishimura N, Kitahata N, Seki M, Narusaka Y, Narusaka M, Kuromori T, Asami T, Shinozaki K, Hirayama T (2005) Analysis of ABA hypersensitive germination2 revealed the pivotal functions of PARN in stress response in *Arabidopsis*. Plant J 44:972–984

Noreen S, Ashraf M (2010) Modulation of salt (NaCl)-induced effects on oil composition and fatty acid profile of sunflower (*Helianthus annuus* L.) by exogenous application of salicylic acid. J Sci Food Agric 90: 2608–2616

Noreen S, Ashraf M, Hussain M, Jamil A (2009) Exogenous application of salicylic acid enhances antioxidative capacity in salt stressed sunflower (*Helianthus annuus* L.) plants. Pak J Bot 41:473–479

Norman C, Howell KA, Millar AH, Whelan JM, Day DA (2004) Salicylic acid is an uncoupler and inhibitor of mitochondrial electron transport. Plant Physiol 134: 492–501

Noutoshi Y, Ito T, Seki M, Nakashita H, Yoshida S, Marco Y, Shirasu K, Shinozaki K (2005) A single amino acid insertion in the WRKY domain of the *Arabidopsis* TIR-NBS-LRR-WRKY-type disease resistance protein SLH1 (senseitive to low humidity 1) causes activation of defense responses and hypersensitive cell death. Plant J 43:873–888

Ogawa D, Nakajima N, Sano T, Tamaoki M, Aono M, Kubo A, Kanna M, Ioki M, Kamada H, Saji H (2005) Salicylic acid accumulation under O_3 exposure is regulated by ethylene in tobacco plants. Plant Cell Physiol 46:1062–1072

Oltmans DJ, Lefohn AS, Scheel HE, Harris JM, Levy H II, Galbally IE, Brunke EG, Meyer CP, Lathrop JA, Johnson BJ, Shadwick DS, Cuevas E, Schmidlin FJ, Tarasick DW, Claude H, Kerr JB, Uchino O, Mohnen V (1998) Trends of ozone in the troposphere. Geophys Res Lett 25:139–142

Overmyer K, Tuominen H, Kettunen R, Betz C, Langebartels C, Sandermann H Jr, Kangasjärvi J (2000) The ozone-sensitive *Arabidopsis rcd1* mutant reveals opposite roles for ethylene and jasmonate signaling pathways in regulating superoxide-dependent cell death. Plant Cell 12:1849–1862

Overmyer K, Brosch M, Kangasjärvi J (2003) Reactive oxygen species and hormonal control of cell death. Trends Plant Sci 8:335–342

Overmyer K, Brosché M, Pellinen R, Kuittinen T, Tuominen H, Ahlfors R, Keinänen M, Saarma M, Scheel D, Kangasjärvi J (2005) Ozone-induced programmed cell death in the *Arabidopsis* radical-induced cell death1 mutant. Plant Physiol 137:1092–1104

Pál M, Szalai G, Horváth E, Janda T, Páldi E (2002) Effect of salicylic acid during heavy metal stress. Acta Biol Szegediensis 46:119–120

Pál M, Horváth E, Janda T, Páldi E, Szalai G (2005) Cadmium stimulates the accumulation of salicylic acid and its putative precursors in maize (*Zea mays*) plants. Physiol Plant 125:356–364

Pan Q, Zhan J, Liu H, Zhang J, Chen J, Wen P, Huang W (2006) Salicylic acid synthesized by benzoic acid 2-hydroxylase participate in the development of thermotolerance in pea plants. Plant Sci 171:226–233

Pasqualini S, Torre GD, Ferranti F, Ederli L, Piccioni C, Reale L, Antonielli M (2002) Salicylic acid modulates ozone-induced hypersensitive cell death in tobacco plants. Physiol Plant 115:204–212

Petersen M, Brodersen P, Naested H, Andreasson E, Lindhart U, Johansen B, Nielsen HB, Lacy M, Austin MJ, Parker JE, Sharma SB, Klessig DF, Martienssen R, Mattsson O, Jensen AB, Mundy J (2000) *Arabidopsis* MAP kinase 4 negatively regulates systemic acquired resistance. Cell 103:1111–1120

Poór P, Gémes K, Horváth F, Szepesi A, Simon ML, Tari I (2011) Salicylic acid treatment via the rooting medium interferes with stomatal response, CO_2 fixation rate and carbohydrate metabolism in tomato, and decreases harmful effects of subsequent salt stress. Plant Biol 13:105–114

Popova LP, Maslenkova LT, Yordanova RY, Ivanova AP, Krantev AP, Szalai G, Janda T (2009) Exogenous treatment with salicylic acid attenuates cadmium toxicity in pea seedlings. Plant Physiol Biochem 47:224–231

Qiu D, Xiao J, Xie W, Liu H, Li X, Xiong L, Wang S (2008) Rice gene network inferred from expression profiling of plants overexpressing OsWRKY13, a positive regulator of disease resistance. Mol Plant 3:538–551

Quan LJ, Zhang B, Shi WW, Li HY (2008) Hydrogen peroxide in plants: a versatile molecule of the reactive oxygen species network. J Integr Plant Biol 50: 2–18

Rao MV, Davis KR (2001) The physiology of ozone-induced cell death. Planta 213:682–690

Rao MV, Lee H, Creelman RA, Mullet JE, Davis KR (2000) Jasmonic acid signaling modulates ozone-induced hypersensitive cell death. Plant Cell 12:1633–1646

Rao MV, Lee H, Davis KR (2002) Ozone-induced ethylene production is dependent on salicylic acid, and both salicylic acid and ethylene act in concert to regulate ozone-induced cell death. Plant J 32:447–456

Raskin I (1992) Role of salicylic acid in plants. Annu Rev Plant Physiol 43:439–463

Raskin I, Skubatz H, Tang W, Meeuse BJD (1990) Salicylic acid levels in thermogenic and non-thermogenic plants. Ann Bot 66:369–373

Rate DN, Greenberg JT (2001) The *Arabidopsis aberrant growth and death* 2 mutant show resistance to *Pseudomonas syringae* and reveals a role for NPR1 in suppressing hypersensitive cell death. Plant J 27: 203–211

Rate DN, Cuenca JV, Bowman GR, Guttman DS, Greenberg JT (1999) The gain-of-function *Arabidopsis acd6* mutant reveals novel regulation and function of the salicylic acid signaling pathway in controlling cell death, defense, and cell growth. Plant Cell 11: 1695–1708

Sánchez-Casas P, Klessig DF (1994) A salicylic acid-binding activity and a salicylic acid-inhibitable catalase activity are present in a variety of plant species. Plant Physiol 106:1675–1679

Sauerbrunn N, Schlaich NL (2004) PCC1: a merging point for pathogen defence and circadian signalling in *Arabidopsis*. Planta 218:552–561

Sawada H, Shim IS, Usui K (2006) Induction of benzoic acid 2-hydroxylase and salicylic acid biosynthesis-Modulation by salt stress in rice seedlings. Plant Sci 171:263–270

Scott IM, Clarke SM, Wood JE, Mur LAJ (2004) Salicylate accumulation inhibits growth at chilling temperature in Arabidopsis. Plant Physiol 135:1040–1049

Segarra S, Mir R, Martinez C, Leon J (2009) Genomic wide analysis of the transcriptomes of salicylic acid-deficient versus wild-type plants uncover pathogen and circadian controlled 1 (PCC1) as a regulator of flowering time in *Arabidopsis*. Pland Cell Environ 33:11–22

Senaratna T, Touchell D, Bunn E, Dixon K (2000) Acetyl salicylic acid (Aspirin) and salicylic acid induce multiple stress tolerance in bean and tomato plants. Plant Growth Regul 30:157–161

Seo PJ, Park CM (2010) MYB96-mediated abscisic acid signals induce pathogen resistance response by promoting salicylic acid biosynthesis in *Arabidopsis*. New Phytol 186:471–483

Seo PJ, Xiang F, Qiao M, Park JY, Lee YN, Kim SG, Lee YH, Park WJ, Park CM (2009) The MYB96 transcription factor mediates abscisic acid signaling during drought stress response in Arabidopsis. Plant Physiol 151:275–289

Shakirova FM, Sakhabutdinova AR, Bezrukova MV, Fatkhutdinova RA, Fatkhutdinova DR (2003) Changes in the hormonal status of wheat seedlings induced by salicylic acid and salinity. Plant Sci 164:317–322

Sharma YK, León J, Raskin I, Davis KR (1996) Ozone-induced responses in *Arabidopsis thaliana*: The role of salicylic acid in the accumulation of defense-related transcripts and induced resistance. Proc Natl Acad Sci USA 93:5099–5104

Shi Q, Zhu Z (2008) Effects of exogenous salicylic acid on manganese toxicity, element contents and antioxidative system in cucumber. Environ Exp Bot 63:317–326

Shinozaki M (1985) Organ correlation in long day flowering of *Pharbitis nil*. Biol Plant 27:382–385

Shinozaki M, Takimoto A (1982) The role of cotyledons in flower initiation of *Phabitis nil* at low temperatures. Plant Cell Physiol 23:403–408

Shinozaki M, Takimoto A (1983) Effects of some growth regulators and benzoic acid derivatives on flower intiation and root elongation of *Pharbitis nil* strain Kidachi. Plant Cell Physiol 24:433–439

Stevens J, Senaratna T, Sivasithamparam K (2006) Salicylic acid induces salinity tolerance in tomato (*Lycopersicon esculentum* cv. Roma): associated changes in gas exchange, water relations and membrane stabilisation. Plant Growth Regul 49:77–83

Strawn MA, Marr SK, Inoue K, Inaba N, Zubieta C, Wildermuth MC (2007) *Arabidopsis* isochorismate synthase functional in pathogen-induced salicylate biosynthesis exhibits properties consistent with a role in diverse stress responses. J Biol Chem 282:5919–5933

Suarez-Rodriguez MC, Adams-Phillips L, Liu Y, Wang H, Su SH, Jester PJ, Zhang S, Bent AF, Krysan PJ (2007) MEKK1 is required for flg22-induced MPK4 activation in Arabidopsis plants. Plant Physiol 143:661–669

Tamaoki M (2008) The role of phytohormone signaling in ozone-induced cell death in plants. Plant Signal Behav 3:166–174

Tanada O, Nakayama Y, Emori K, Takeba G, Sato K, Sugino M (1997) Flower-inducing activity of lysine in *Lemna paucicostata* 6746. Plant Cell Physiol 38:124–128

Tasgin E, Atici O, Nalbantoglu B (2003) Effects of salicylic acid and cold on freezing tolerance in winter wheat leaves. Plant Growth Regul 41:231–236

Teige M, Scheikl E, Eulgem T, Dóczi R, Ichimura K, Shinozaki K, Dangl JL, Hirt H (2004) The MKK2 pathway mediates cold and salt stress signaling in *Arabidopsis*. Mol Cell 15:141–152

Thomashow MF (1999) Plant cold acclimation: freezing tolerance genes and regulatory mechanisms. Annu Rev Plant Physiol Plant Mol Biol 50:571–599

Tosti N, Pasqualini S, Borgogni A, Ederli L, Falistocco E, Crispi S, Paolocci F (2006) Gene expression profiles of O_3-treated *Arabidopsis* plants. Plant Cell Environ 29:1686–1702

Tsutsui T, Kato W, Asada Y, Sako K, Sato T, Sonoda Y, Kidokoro S, Yamaguchi-Shinozaki K, Tamaoki M, Arakawa K, Ichikawa T, Nakazawa M, Seki M, Shinozaki K, Matsui M, Ikeda A, Yamaguchi J (2009) DEAR1, a transcriptional repressor of DREB protein that mediates plant defense and freezing stress responses in *Arabidopsis*. J Plant Res 122:633–643

Uppalapati SR, Ishiga Y, Wangdi T, Kunkel BN, Anand A, Mysore KS, Bender CL (2007) The phytotoxincoronatine contributes to pathogen fitness and is required for suppression of salicylic acid accumulation in tomato inoculated with *Pseudomonas syringae* pv. *tomato* DC3000. Mol Plant Microbe Interact 20:955–965

van Camp W, van Montagu M, Inze D (1998) H_2O_2 and NO: redox signals in disease resistance. Trends Plant Sci 3:330–334

Vernooij B, Friedrich L, Morse A, Reist R, Kolditz-Jawhar R, Ward E, Uknes S, Kessmann H, Ryals J (1994) Salicylic acid is not the translocated signal responsible for inducing systemic acquired resistance but is required in signal transduction. Plant Cell 6:959–965

Vlot AC, Dempsey DA, Klessig DF (2009) Salicylic acid, a multifaceted hormone to combat disease. Annu Rev Phytopathol 47:177–206

Wada K (1974) Floral initiation under continuous light in *Phabitis nil*, a typical short day plant. Plant Cell Physiol 15:381–384

Wada KC, Takeno K (2010) Stress-induced flowering. Plant Signal Behav 5:1–4

Wada KC, Kondo H, Takeno K (2010a) Obligatory short-day plant *Perilla frutescens* var. crispa can flower in response to low-intensity light stress under long-day conditions. Physiol Plant 138:339–345

Wada KC, Yamada M, Shiraya T, Takeno K (2010b) Salicylic acid and the flowering gene FLOWERING LOCUS T homolog are involved in poor-nutrition stress- induced flowering of *Pharbitis nil*. J Plant Physiol 167:447–452

Wan SB, Tian L, Tian RR, Pan QH, Zhan JC, Wen PF, Chen JY, Zhang P, Wang W, Huang WD (2009) Involvement of phospholipase D in the low temperature acclimation-induced thermotolerance in grape berry. Plant Physiol Biochem 47:504–510

Wang LJ, Li SH (2006) Thermotolerance and related antioxidant enzyme activities induced by heat acclimation and salicylic acid in grape (*Vitis vinifera* L.) leaves. Plant Growth Regul 48:137–144

Wang H, Feng T, Peng X, Yan M, Tang X (2009) Up-regulation of chloroplastic antioxidant capacity is involved in alleviation of nickel toxicity of *Zea mays L.* by exogenous salicylic acid. Ecotoxicol Environ Saf 72:1354–1362

Wang LJ, Fan L, Loescher W, Duan W, Liu GJ, Cheng JS, Luo HB, Li SH (2010) Salicylic acid alleviates decreases in photosynthesis under heat stress and accelerates recovery in grapevine leaves. BMC Plant Biol 10:34

Ward JM, Hirschi KD, Sze H (2003) Plants pass the salt. Trends Plant Sci 8:200–201

Wildermuth MC, Dewdney J, Wu G, Ausubel FM (2001) Isochorismate synthase is required to synthesize salicylic acid for plant defense. Nature 414:562–565

Wise RR, Olson AJ, Schrader SM, Sharkey TD (2004) Electron transport is the functional limitation of photosynthesis in field-grown Pima cotton plants at high temperature. Plant Cell Environ 25:717–724

Xiao S, Brown S, Patrick E, Brearley C, Turner JG (2003) Enhanced transcription of the *Arabidopsis* disease resistance genes *RPW8.1* and *RPW8.2* via a salicylic acid-dependent amplification circuit is required for hypersensitive cell death. Plant Cell 15:33–45

Yalpani N, Enyedi AJ, León J, Raskin I (1994) Ultraviolet light and ozone stimulate accumulation of salicylic acid, pathogenesis-related proteins and virus resistance in tobacco. Planta 193:372–376

Yang H, Li Y, Hua J (2006) The C2 domain protein BAP1 negatively regulates defense responses in *Arabidopsis*. Plant J 48:238–248

Yang H, Yang S, Li Y, Hua J (2007) The *Arabidopsis* BAP1 and BAP2 genes are general inhibitors of programmed cell death. Plant Physiol 145:135–146

Yasuda M, Ishikawa A, Jikumaru Y, Seki M, Umezawa T, Asami T, Maruyama-Nakashita A, Kudo T, Shinozaki K, Yoshida S, Nakashita H (2008) Antagonistic interaction between systemic acquired resistance and the abscisic acid-mediated abiotic stress response in *Arabidopsis*. Plant Cell 20:1678–1692

Yun KY, Park MR, Mohanty B, Herath V, Xu F, Mauleon R, Wijaya E, Bajic VB, Bruskiewich R, de los Reyes BG (2010) Transcriptional regulatory network triggered by oxidative signals configures the early response mechanisms of japonica rice to chilling stress. BMC Plant Biol 10:16

Zawoznik MS, Groppa MD, Tomaro ML, Benavides MP (2007) Endogenous salicylic acid potentiates cadmium-induced oxidative stress in *Arabidopsis thaliana*. Plant Sci 173:190–197

Zhou F, Menke FL, Yoshioka K, Moder W, Shirano Y, Klessig DF (2004) High humidity suppresses *ssi4*-mediated cell death and disease resistance upstream of MAP kinase activation, H_2O_2 production and defense gene expression. Plant J 39:920–932

Zhou ZS, Guo K, Elbaz AA, Yang ZM (2009) Salicylic acid alleviates mercury toxicity by preventing oxidative stress in roots of *Medicago sativa*. Environ Exp Bot 65:27–34

Zhu Y, Yang H, Mang H-G, Hua J (2010) Induction of *BAP1* by a moderate decrease in temperature is mediated by *ICE1* in *Arabidopsis*. Plant Physiol 155:580–588

Trehalose and Abiotic Stress Tolerance

14

Miguel López-Gómez and Carmen Lluch

Abstract

Trehalose is a nonreducing disaccharide present in diverse organisms ranging from bacteria and fungi to invertebrates, in which it serves as an energy source as well as an osmolyte and/or protein/membrane protectant. Until recently, trehalose was not thought to be of any real significance in plants, although genetic studies have confirmed the existence of surprising abundance of genes for trehalose metabolism in plants, which have led to propose trehalose pathway as a central metabolic regulator. Multiple studies have linked trehalose to abiotic stress tolerance in plants and different research groups have attempted to create stress tolerant plants by introducing trehalose biosynthetic genes in important crops such as rice, tomato, and potato. Particular cases of the trehalose metabolism are plant symbiotic interactions such as the *rhizobia*–legume symbiosis, where trehalose has been described as a major carbohydrate in root nodules of some species. The discovery of trehalose metabolism in the recent years has pointed out the importance of trehalose biosynthesis in stress responses in plants.

Keywords

Abiotic stress • Mycorrhizal symbiosis • Osmoprotector • *Rhizobium*–legume symbiosis • Trehalose metabolism • Trehalose biosynthesis

M. López-Gómez (✉) • C. Lluch
Departamento de Fisiología Vegetal,
Facultad de Ciencias, Universidad de Granada,
Campus de Fuentenueva s/n, Granada 18071, Spain
e-mail: mlgomez@ugr.es

P. Ahmad and M.N.V. Prasad (eds.), *Abiotic Stress Responses in Plants: Metabolism, Productivity and Sustainability*, DOI 10.1007/978-1-4614-0634-1_14,

1 Introduction: Role of Nonreducing Disaccharides in Plants

Nonreducing disaccharides provide a soluble energy source in the form of a stable molecule that can also function as a protectant compound under stress conditions in all organisms except vertebrates. Trehalose and sucrose are the two sugars that perform this role.

The components of reducing sugars, glucose in the case of trehalose and glucose and fructose in the case of sucrose, are linked at their reducing ends. In naturally occurring trehalose, the two glucose units are linked in a α,α-1,1 configuration. Isomers include neotrehalose with an α,β link, and isotrehalose, which has a β,β link. In sucrose, fructose and glucose are linked in an α,β-1,2 configuration. Both configurations produce stable energy molecules (Fig. 14.1).

Plants are unique in that they can synthesize both the nonreducing disaccharides, but sucrose performs the main role of translocated sugar in plants. Trehalose is found in millimolar amounts in only a few plants, namely, resurrection species, where it is thought to protect against desiccation.

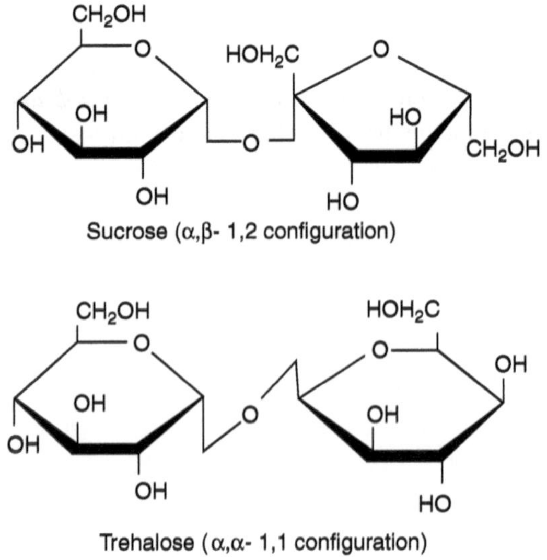

Fig. 14.1 Chemical structures of sucrose and trehalose

In the vast majority of plants, trehalose is only present in trace amounts.

The different chemistries of trehalose and sucrose dictate their biology, the functions they perform, and the mechanisms that determine the concentration of these compounds in vivo. Several arguments can be put forward to explain the prevalence of sucrose as translocated sugar in plants:

1. Sucrose is more soluble than trehalose, particularly at low temperatures, and hence may be more suited as a transport sugar in plant phloem at concentrations as high as 1 M.
2. Sucrose can be cleaved by invertase into glucose and fructose, and by sucrose synthase into uridine diphosphoglucose (UDPG) and fructose, preserving energy as UDPG.
3. Cell wall polysaccharides are synthesized from UDPG; thus, the ability to liberate UDPG directly from sucrose to synthesize cell wall polysaccharides may be the main reason that sucrose dominates in plants.

The importance of trehalose in stress conditions compared with other sugars can be explained by several unique physical properties, which include high hydrophilicity, chemical stability, and the absence of internal hydrogen bond formation that account for the principal ability of trehalose for protein stabilization. It has been proposed that in the absence of water, trehalose preserves membrane or protein structures by forming an amorphous glass structure and interacting through hydrogen bonds with polar phospholipids head groups or with amino acids (Crowe et al. 1984). Thereby trehalose is helping the protein to keep in shape and concentrate the remaining water next to the protein (Schiraldi et al. 2002). Trehalose is among the most chemically unreactive sugars and its strong stability is result of the very low energy (1 kcal mol^{-1}) of the glycoside oxygen bond joining the two hexose rings. In comparison, sucrose has an energy bond of 27 kcal mol^{-1} (Paiva and Panek 1996). Therefore, trehalose does not dissociate into two reducing monosaccharidic constituents unless exposed to extreme hydrolytic conditions or to the action of trehalose.

2 Occurrence of Trehalose in Different Organisms

Trehalose was first identified as a constituent of the ergot fungus of rye in 1832. The name trehalose was introduced in 1858 when it was found in the cocoons or "trehala" of the desert beetles of the Middle East, *Laurinus nidificans* and *L. maculates*.

The occurrence of trehalose has been described in all kingdoms where it is assumed to play a similar role in vivo as its demonstrated properties in vitro. With the exception of plants, trehalose has a central role as an energy source and stress response molecule in microorganisms and invertebrates and as a starting point for chitin synthesis in fungi. Apart from the starting point for chitin synthesis, trehalose is synthesized at the onset of reduced growth periods *in fungi* to protect the cell's integrity against stress damage and then is rapidly mobilized during recovery and during the early germination of spores, where the trehalose content can constitute a 10% on a dry-weight basis (Arguelles 2000). When these spores germinate, the trehalose rapidly disappears, suggesting that this sugar is stored as a source of carbon and/or energy.

In *bacteria*, trehalose is widely distributed among different genus that include *Streptomyces* (Martin et al. 1986), *Mycobacterium* (Elbein and Mitchell 1973), and *Corinebacterium* (Shimakata and Minatagawa 2000) where this disaccharide has a structural role as component of the cell wall. It has also been found in many other bacteria including *Escherichia coli* (Kaasen et al. 1994) and *Rhizobium* sp. (Maruta et al. 1996) where trehalose can constitute the sole carbon source, be used as a compatible osmolyte, or form part of the cell wall structure.

In the *animal* kingdom, trehalose was initially described in insects, where it constitutes a source of glucose for energy on the basis that the levels of trehalose decrease rapidly during certain energy-consuming activities such as flight (Becker et al. 1996). In the case of the nematode *Aphelenchus avenae*, trehalose is involved in desiccation protection with a strong correlation between trehalose synthesis and drought constituting trehalose as much as 20% of its dry weight when dehydrated (Madin and Crowe 1975).

3 Trehalose Metabolism

So far, five different biosynthetic pathways for trehalose have been described (Fig. 14.2). The most extensively distributed was discovered about 50 years ago by Cabib and Lenoir (1958), and since then, it has been described in diverse organisms including eubacteria, archaea, fungi, insects, and plants:

1. This pathway is constituted by two enzymatic steps catalyzed by *trehalose-6-phosphate synthase* (TPS) and *trehalose-phosphatase* (TPP). TPS transfer glucose from UDP-glucose to glucose 6-phosphate generating trehalose 6-phosphate (T6P) and UDP, while TPP dephosphorylates T6P to trehalose and inorganic phosphate (De Smet et al. 2000; Elbein et al. 2003).
2. Reported in thermophilic archaea of the genus *Sulfolobus* the second pathway consists in the conversion of maltodextrines (maltooligosaccharides, glycogen and starch) to trehalose. This pathway is catalyzed by *maltooligosyl trehalose synthase* (TreY), coded by the *treY* gene, and *maltooligosyl trehalose trehalohydrolase* (TreZ), coded by the *treZ* gene that catalyzes the hydrolytic release of trehalose from maltooligosyltrehalose, which contains a trehalose moiety at the end of the polymer (Elbein et al. 2003; Streeter and Bhagwat 1999).
3. First reported in *Pimelobacter* sp., orthologs of the third pathway has been found in other eubacteria. In this pathway maltose is isomerized to trehalose by the enzyme *trehalose synthase* (TS) (Higashiyama 2002; Elbein et al. 2003).
4. The fourth pathway consists in the reversible hydrolysis of trehalose in the presence of inorganic phosphate by *trehalose phosphorylase* (TreP), present in some fungi. In this reaction the transfer of a glucose molecule to a phosphate generates glucose 1-phosphate and releases the other glucose residue. It is

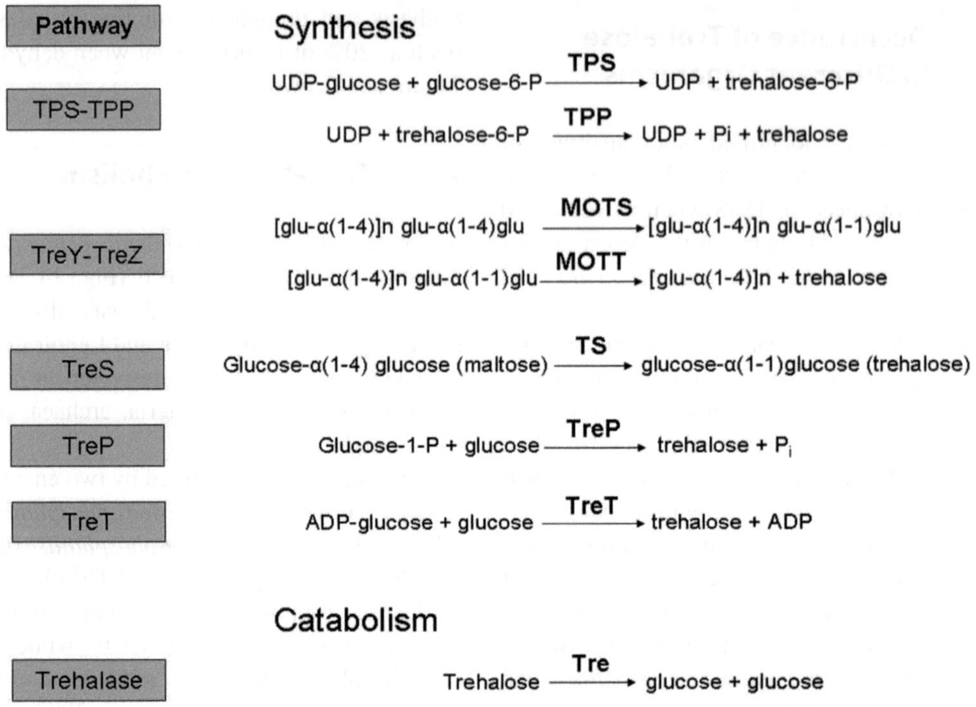

Fig. 14.2 Known pathways of trehalose synthesis and catabolism in eukaryotes and prokaryotes. *UDP* uridine diphosphate, *Glucose-6-P* glucose 6-phosphate, *TPS* trehalose phosphate synthase, *Trehalose-6-P* trehalose 6-phosphate, *TPP* trehalose phosphate phosphatase, *MOTS* maltooligosyl trehalose synthase, *MOTT* maltooligosyl trehalose tetrahydrolase, *TS* trehalose synthase, *Glucose-1-P* glucose 1-phosphate, *TreP* trehalose phosphorylase, *TreT* trehalose glycosyltransferring synthase, *Tre* trehalase

unclear whether TreP enzyme participates in the synthesis or degradation of trehalose, since the biosynthetic reaction has only been shown in vitro (Wannet et al. 1998; Schiraldi et al. 2002).

5. The last biosynthetic pathway for trehalose consists in the reversible formation of trehalose from ADP-glucose and glucose and was discovered in the hyperthermophilic archaeon *Thermococcus litoralis* (Qu et al. 2004; Ryu et al. 2005). This reaction is catalyzed by the *trehalose glycosyltransferring synthase* (TreT), which can also use UDP-glucose and GDP-glucose, although it is less efficient with these substrates.

Trehalose *catabolism* in two glucose molecules is catalyzed by *trehalase* enzyme activity (Tre) that has been found in a variety of organisms including prokaryotic and eukaryotic (Elbein 1974). Trehalase is ubiquitous in higher plants, and it is likely that trehalase is the sole route of trehalose breakdown in plants, as trehalose accumulates in the presence of specific trehalase inhibitor validamycin A (Müller et al. 2001).

4 Trehalose in Plants

Until recently, trehalose was not thought to be of any real significance in plants; instead, it was regarded as a rare obscure sugar found in some marginal resurrection species such as *Selaginella* species (Anselmino and Gilg 1913) and *Myrothamnus flabellifolia* (Bianchi et al. 1993) where its accumulation to high levels enables protection for desiccation. The lack of trehalose detection in the majority of plants led to the assumption that trehalose function had died out and had been replaced by sucrose. Later, trehalose was detected in the model plant *Arabidopsis thaliana* using validamycin A (Gooddijn et al. 1997),

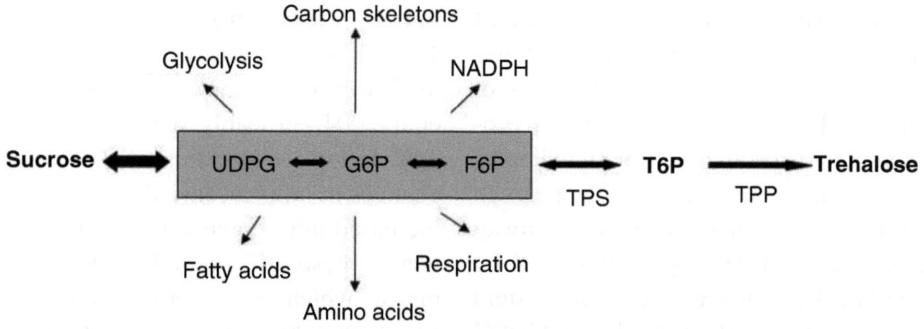

Fig. 14.3 Summary of the role of trehalose metabolism in plants. Trehalose-6-phosphate (T6P) act as a signal metabolite because of its close proximity to the pool size of hexose phosphates and uridine diphosphoglucose (UDPG), which are the starting point of different basic processes in the plant cell metabolism. *G6P* glucose 6-phosphate, *F6P* fructose 6-phosphate, *TPS* trehalose phosphate synthase, *TPP* trehalose phosphate phosphatase

an inhibitor of the trehalose-degrading enzyme, trehalase. More recently, trehalose has been detected in crops, such as rice (*Oryza sativa*) (Garg et al. 2002) and tobacco (*Nicotiana tabacum*) (Karim et al. 2007). The publication of the full genomic sequence database for *A. thaliana* confirmed the existence of a surprising abundance of genes for the trehalose synthesis (Leyman et al. 2001) and trehalose and trehalose 6-phosphate (T6P) were subsequently detected in this specie (Schluepmann et al. 2003; Vogel et al. 2001).

Eastmond et al. (2002) were the first to demonstrate the indispensability of a plant trehalose pathway gene, *AtTPS1*, which codify for the enzyme trehalose phosphate synthase (TPS) involved in the synthesis of T6P from glucose 6-phosphate (G6P) and UDP-glucose (UDPG) in *A. thaliana*. Numerous effects of altering the trehalose pathway on metabolism and development (Ramon and Rolland 2007), possibly all due to modification of T6P content, have been reported. These include embryo (Eastmond et al. 2002) and leaf (Pellny et al. 2004) development, cell division, and cell wall synthesis (Gomez et al. 2006), inflorescence architecture (Satoh-Nagasawa et al. 2006), seedling biomass (Schluepmann et al. 2003), adult plant biomass and photosynthesis (Pellny et al. 2004), sucrose utilization (Schluepmann et al. 2003), starch metabolism (Kolbe et al. 2005), and tolerance to abiotic stresses, particularly drought (Almeida et al. 2007; Garg et al. 2002; Karim et al. 2007; Miranda et al. 2007; Pilon-Smits et al. 1998).

4.1 Importance of Trehalose 6-Phosphate

In plants the major and possibly only role of the trehalose pathway, except in specialized resurrection plants, is as a central metabolic regulator. This regulatory function is performed, at least in part, by T6P (Kolbe et al. 2005; Lunn et al. 2006; Schluepmann et al. 2003).

The initial discovery that the trehalose pathway and T6P in particular, has a powerful function in metabolic regulation is, on reflection, perhaps not surprising. T6P and trehalose are made from UDPG and G6P, so T6P and trehalose synthesis is drawn from a metabolite pool at the center of metabolism, but because trehalose is not a major end product in plants, it is removed from major metabolic flux (Fig. 14.3). This means that the synthesis of T6P and trehalose can act as an effective indicator of G6P and UDPG pool size without compromising any other function of the trehalose pathway. T6P is a low-abundance molecule and responds rapidly to the sucrose supply (Lunn et al. 2006). This rapid and large response is a consequence of a yet not fully elucidated control features (transcriptional and post-translational control, including phosphorylation) that regulate T6P synthesis and breakdown. However, T6P synthesis mediated through constitutive *TPS1* expression likely reflects the availability of hexose phosphates, UDPG, and sucrose, which feeds into this pool. T6P, therefore, has all the characteristics of a signaling molecule.

The low abundance and dynamic response of T6P could potentiate specific and rapid communication of metabolic status that reflects pool sizes of G6P, UDPG, and sucrose and hence provide a different and specific kind of signaling to that of other sugars. *TPS1* expression appears to be constitutive, but other trehalose pathway enzymes are regulated developmentally and by stress, providing the basis for a regulatory system linking the hexose phosphate pool and UDPG with development and the environment.

The wide range of phenotypes observed in transgenic plants with a modified trehalose pathway clearly suggests that T6P is involved in coordinating UDPG and G6P with growth and development in different tissues.

4.1.1 Embryos

T6P is an essential factor in embryo maturation based on studies performed with a transposon insertion mutant of *Arabidopsis AtTPS1* gene (Schluepmann et al. 2004). Modifications of T6P levels cause dramatic effects on carbohydrate metabolism and partitioning as well as on morphogenesis and development in *Arabidopsis* (Schluepmann et al. 2003). Interestingly, *AtTPS*1 is an essential gene, and knocking the gene out results in an embryo lethal phenotype (Eastmond et al. 2002). In vegetative stage, *AtTPS1* is essential for normal growth in particular for the development in the flowers, buds, and ripening fruits (Van Dijken et al. 2004).

4.1.2 Photosynthesis

In transgenic plants expressing *E. coli* TPS and TPP genes, researchers observe large changes in vegetative development, which correlate with T6P content (Paul 2007; Paul et al. 2001; Pellny et al. 2004). Quite remarkably, photosynthetic capacity per unit leaf area is enhanced in transgenic plants expressing TPS. This enhancement is due to a specific increase in Rubisco activity, because of increased amounts of Rubisco protein, chlorophyll, and the light-harvesting apparatus. Nevertheless, targeting the trehalose pathway provides another means to alter plant photosynthesis for improved yield.

4.1.3 Flowering

TPS1 is also necessary for the normal transition to flowering (Gomez et al. 2006; Van Dijken et al. 2004), probably again through provision of T6P. Ectopic expression of *TPS1* also leads to changes in inflorescence development, including increased inflorescence branching. Recently, the genetic basis of *ramosa3*, a classical mutant of maize, which causes large changes in inflorescence branching, has been determined to be a TPP that metabolizes T6P (Satoh-Nagasawa et al. 2006). RAMOSA3 is part of a distinct clade of TPPs found in monocots and is expressed in discrete domains subtending axillary inflorescence meristems. Overaccumulation in T6P in these meristems may cause the large change in inflorescence phenotype. Meristems are characterized by the need to coordinate the supply of intermediates, UDPG and G6P, with cell growth and development, and hence the possibility of a crucial role for T6P as in embryos and leaves.

4.1.4 Starch

Starch accumulation is one of the most striking examples of metabolic regulation by the trehalose pathway since it has been shown a strong accumulation of starch in response to trehalose feeding and transcriptional regulation of ADP-glucose pyrophosphorylase (AGPase), the key enzyme of starch synthesis (Wingler et al. 2000). T6P activates AGPase through a thioredoxin-dependent redox activation mechanism (Kolbe et al. 2005). This activation mechanism operates under conditions of high sucrose, which induce high T6P levels (Lunn 2007; Lunn et al. 2006). This finding again supports the concept that T6P reflects conditions of high assimilate supply and in this case communicates these conditions to the chloroplast to activate starch synthesis.

4.2 Trehalose Metabolism in Plants

Although five different trehalose synthesis pathways exist in bacteria, fungi, yeast, and algae (Paul et al. 2008; Avonce et al. 2006), trehalose biosynthesis in higher plants only occurs in the trehalose phosphate synthase (TPS) trehalose phosphate phosphatase (TPP) pathway (also known as OtsA–OtsB pathway) (Fig. 14.2).

The description of trehalose biosynthesis genes in plants began in the late 1990s with the characterization of the *A. thaliana* TPS (*AtTPS1*) and TPP encoding genes (*AtTPPA* and *AtTPPB*) (Blázquez et al. 1998; Vogel et al. 1998). These three genes encode functional TPS and TPP proteins because they complement tps1 and tps2 yeast mutants, which are deficient in TPS and TPP activities, respectively. A large number of TPS genes have been found in *A. thaliana* (Leyman et al. 2001). Using in silico analysis, ten homologs of *AtTPS1* have been identified; these can be divided into two classes: class I genes (*AtTPS1–AtTPS4*) encode proteins that have a TPS domain closely related to *Saccharomyces cerevisiae ScTPS1*; class II genes (*AtTPS5–AtTPS11*) encode proteins with a TPP domain, exhibiting a strong homology to *AtTPPA* and *AtTPPB*, but only 30% homology with class I proteins (AtTPS1–AtTPS4). The function of these class II genes remains largely unknown.

Ten homologs of TPP have been found in *A. thaliana* (*AtTPPA–AtTPPJ*) (Avonce et al. 2006; Leyman et al. 2001). Interestingly, TPS and TPP form multigenic families in other plants. For example, rice (*Oryza Sativa*) contains at least nine *OsTPS* and nine *OsTPP* genes (Avonce et al. 2006). According to most authors, this redundancy is an indicator that either trehalose or T6P has an important metabolic role. By contrast, trehalase is encoded by a unique gene in *A. thaliana*, rice, and soybean (*Glycine max*) (Frison et al. 2007). The complete trehalose biosynthesis gene expression pattern has been extensively reviewed (Paul et al. 2008).

The presence of trehalase in higher plants has been puzzling because its substrate appeared to be absent although it has been postulated that trehalases could play a role in defense mechanisms (Fernandez et al. 2010). In addition, it has been suggested that trehalase could be involved in the degradation of trehalose derived from plant-associated microorganisms, such as *rhizobia* in root nodules or fungi involved in mycorrhizal symbiosis (Müller et al. 1995a, b).

4.3 Trehalose Under Abiotic Stress in Plants

The role of trehalose in abiotic stress tolerance was first demonstrated in resurrection plants, such as *Myrothamnus flabellifolius*, *Selaginella tamariscina*, or *Selaginella lepidophylla*. These desiccation tolerant plants can withstand almost complete dehydration and upon rehydration regain complete viability (Liu et al. 2008). Interestingly, in these three species, trehalose is the main soluble sugar, with levels reaching 3 mg g^{-1} FW in *M. flabellifolius* and 12 mg g^{-1} FW in *S. lepidophylla*. During dehydration, trehalose concentration increases only slightly and it acts as a protector for both proteins and membranes (Liu et al. 2008). Trehalose accumulation under abiotic stresses is not restricted to resurrection plants.

In *A. thaliana*, the trehalose level doubled within 4 h of heat stress (40°C) and increased eightfold 4 days after cold exposure (4°C) (Kaplan et al. 2004). A recent study (Suzuki et al. 2008) revealed the involvement of TPS (*AtTPS5*) in *A. thaliana* thermotolerance.

In rice, trehalose accumulation has been shown in roots 3 days following salt stress (García et al. 1997), and two different TPPs were transiently induced in response to multiple abiotic stresses as well as exogenous ABA applications, which suggest a regulation of trehalose biosynthesis (Pramanik and Imai 2005; Shima et al. 2007).

In model legumes *Medicago truncatula* and *Lotus japonicus*, nodular trehalose content increased by 50 and 100%, respectively, under salt stress conditions (López et al. 2008a). Previously in alfalfa (*Medicago sativa* L.), it has been also described an increase of trehalose accumulation in roots and bacteroids upon salt stress (Fougère et al. 1991).

Furthermore, a microarray analysis has revealed that a wide range of abiotic stresses, such as cold, salt and UV, cause a response in the genes involved in trehalose metabolism in *A. thaliana* (Iordachescu and Imai 2008). This finding indicates that trehalose and/or T6P are involved in the response to abiotic environmental fluctuations.

4.3.1 Trehalose: Compatible Solute in Plants?

It has been questioned the role of trehalose as compatible solute in plants under abiotic stress since compatible solutes are nontoxic molecules able to accumulate at high concentrations in the cytoplasm, participating in turgor maintenance and/or the protection of macromolecular structures against the destabilizing effect of anhydrobiotic conditions (Gibon et al. 1997). However, if trehalose concentration in resurrection plants reaches levels consistent with the compatible solutes definition, the concentration of trehalose in other plants is low, which suggests that trehalose is not a compatible solute (Avonce et al. 2004; Schluepmann et al. 2003; Grennan 2007). Furthermore, trehalose genetically engineered plants exhibit altered morphology, possibly caused by toxicity of high trehalose concentrations, indicating that trehalose is a noncompatible solute (Schluepmann et al. 2003; Cortina and Culianez-Maciá 2005). In many organisms, trehalose has been reported as a better stabilizer than other sugars for protecting membranes and biomolecules (Elbein et al. 2003; Purvis et al. 2005; Crowe 2007).

4.4 Engineering Trehalose Biosynthesis Pathway in Crops

The introduction of trehalose biosynthetic genes has been used as an strategy for different research groups to create stress tolerant plants in tobacco (*Nicotiana tabaccum*) (Romero et al. 1997; Pilon-Smits et al. 1998; Han et al. 2005), rice (Garg et al. 2002), tomato (*Solanum lycopersicum*) (Cortina and Culianez-Maciá 2005), potato (Stiller et al. 2008), and *Arabidopsis* (Karim et al. 2007; Miranda et al. 2007). First attempts with yeast *TPS1* or *E. coli OtsA* genes induced trehalose accumulation, although at a low level. However, T6P accumulation produced abnormal phenotypes (Romero et al. 1997; Pilon-Smits et al. 1998; Cortina and Culianez-Maciá 2005). Nevertheless, this problem were overcome directing the gene product into chloroplast by using a TPS–TPP fusion gene together with a stress inducible promoter (Karim et al. 2007; Miranda et al. 2007), or using trehalose biosynthetic genes (trehalose phosphorylase) that avoid the T6P formation (Han et al. 2005). These improved methods resulted in stress tolerance without the phenotypic alterations.

Drought tolerance was one of the first traits obtained by constitutive overexpression of the yeast *ScTPS1* gene in tobacco (Romero et al. 1997) and the *AtTPS1* gene in *A. thaliana* (Avonce et al. 2004). Improved drought tolerance was also achieved with stress inducible and chloroplast-targeted expression of the plastid *TPS1* gene in tobacco (Karim et al. 2007) and by expression of bifunctional fusion genes, *OtsA–OtsB* and *ScTPS–ScTPP*, in rice and tobacco, respectively (Garg et al. 2002, Karim et al. 2007).

Transgenic tomatoes overexpressing the *ScTPS1* gene are more resistant to salt, drought, and oxidative stresses (Cortina and Culianez-Maciá 2005). Improved freezing and heat stress tolerance have been obtained in *A. thaliana* by constitutive or stress-inducible expression of a bifunctional yeast *ScTPS1–ScTPS2* gene, leading to a significant accumulation of trehalose (Miranda et al. 2007).

Rice overexpressing the *E. coli* trehalose synthesis genes (*OtsA* and *OtsB*) becomes tolerant to salt and low-temperature stresses. These plants are characterized by trehalose accumulation (increased three- to tenfold, when compared with the nontransgenic controls), stronger photosynthetic activity, and global accumulation of carbohydrates (Garg et al. 2002).

5 The Particular Case of the Plant Symbiosis Interactions

5.1 Trehalose in the *Rhizobia*–Legume Symbiosis

It is the domain of some prokaryotes to reduce atmospheric nitrogen to ammonia that afterward can be assimilated into biological material. A large group of nitrogen fixing soil bacteria is able to establish symbiotic associations with plants to obtain the energy necessary for nitrogen fixation from compounds that they receive from their host plants. The symbiosis between legumes and rhizobia occurs within specialized organs called nodules situated mainly on the root. The architecture of these nodules provides the specific physiological and anatomical requisites for the activity of nitrogenase (the key enzyme of this symbiosis) and the nutrient exchange between the symbiotic partners.

In 1980 was first described the appearance of trehalose as a major carbohydrate in soybean root nodules at the onset of nitrogen fixation (Streeter 1980). Trace amounts of trehalose were also detected in other plant organs, but the bulk of trehalose was specific to the symbiotic organs (Streeter 1980). At this time trehalose was thought to be uncommon in plants, so it was speculated that the trehalose measured in other parts of the plant was nodule-derived. In later studies, trehalose appeared to be a common carbohydrate in almost all nodules tested. Phillips et al. (1984) detected trehalose accumulation in nodules of white clover (*Trifolium repens*), *Pueraria thunbergiana*, *Albizia julibrissin*, and even in the nodules of the nonlegume *Alnus glutinosa* and *Elaeagnus angustifolia*. Streeter (1985) discovered trehalose accumulation in the field grown nodules of peanut (*Arachis hypogenus*), alfalfa (*Medicago sativa*), common bird's-foot-trefoil (*Lotus corniculatus*), and again in white clover (*Trifolium repens*).

When trehalose was discovered as a carbohydrate in root nodules, it was predicted that it was synthesized by bacteroids since the presence of trehalose in uninfected higher plants was not known at that time. This prediction was supported by the findings that (a) trehalose was synthesized in bacteroids isolated from soybean nodules (Streeter 1985), (b) trehalose was not depleted in senescing soybean nodules but accumulated while the concentration of other compounds declined (Müller et al. 2001), and (c) trehalose concentration in bacteroids varied greatly depending on the rhizobial strain (Streeter 1985).

The fact that nodules of soybean infected by *Bradyrhizobium japonicum* and *B. elkanii* contains three independent pathways for trehalose biosynthesis such as TS, MOTS and TPS suggests the importance of trehalose in this microorganism in symbiosis and free-living (Streeter and Gómez 2006).

5.1.1 Impact of Trehalose on Nodule Metabolism

The impact of trehalose on nodule metabolism has been further examined by the addition of the trehalase inhibitor validamycin A, or the external supply of trehalose. The addition of validamycin A to *Medicago truncatula* (López et al. 2009) and *Lotus japonicus* (López et al. 2006) caused an increase in the amount of trehalose that improved the response to salinity in both legumes by increasing the biomass production under stress conditions, although nitrogen fixation was not affected in both cases. The addition of trehalose to the nutrient solution of soybean roots growing in sterile conditions caused a strong impact on sucrose metabolism by an increase of the sucrose synthase and to a lesser extent alkaline invertase (Müller et al. 1997). In addition, soybean nodules with naturally occurring high levels of trehalose, had significantly lower levels of sucrose than nodules with low levels of trehalose, and higher levels of the catalytic activities sucrose synthase and alkaline invertase (Müller et al. 1997). With all this information, it has been hypothesized that trehalose turns out to have the same effect as sucrose suggesting that trehalose might be secreted by plant-associated microorganisms as an instrument to influence assimilates allocation by inducing a sink in the surrounding cells or tissues.

5.1.2 Role of Trehalose in Nodules Under Abiotic Stress Conditions

Trehalose accumulation in nodules can further be altered by diverse abiotic factors such as salt stress, drought or nitrate. Therefore it has been tested whether trehalose is involved in stress protection in the *rhizobia*–legume symbiosis. The overexpression of TPS in the symbiotic bacteria *Rhizobium etli* has been used as strategy to increase drought tolerance in legumes. The host plant (*Phaseolus vulgaris*) inoculated with this transformed bacteria displayed higher resistance to drought stress. Besides, plants inoculated with the mutant strain showed higher nodule number and hence increased nitrogenase activity and higher biomass compared with plants inoculated with the wild-type *R. etli* (Suárez et al. 2008).

In soybean root nodules, an increase of the trehalose pools size upon water stress was reported, but this accumulation was dependent on the rhizobial strain that was used for infection (Müller et al. 1996). In addition, an increase of sucrose and pinitol pools was measured in these experiments.

In common bean (*Phaseolus vulgaris*), the increase in nodule trehalose content during drought stress differed among rhizobial strains, exhibiting higher leaf relative water contents and more drought resistance those cultivars with higher nodule trehalose levels (Farías-Rodriguez et al. 1998).

In the model legumes *Lotus japonicus* and *Medicago truncatula* exposed to salt stress, an increase of trehalose concentration of about 40 and 100%, respectively, has been detected (López et al. 2008a). Similarly, in *Medicago sativa* plants infected with *Rhizobium meliloti*, maltose and trehalose concentrations were significantly enhanced upon 0.15 M sodium chloride stress, especially in roots and bacteroids (Fougère et al. 1991). All these data support a role for trehalose as osmoprotectant under stress conditions in the *Rhizobium*–legume symbiosis.

5.2 Trehalose Impact in the Mycorrhizal Symbiosis

Around 90–95% of land plants maintain some type of colonization of their roots by soil borne filamentous fungi known as arbuscular mycorrhiza (AM), due to the tree-like structures that fungi form into the plant cell. This symbiosis increases plant biomass and photosynthesis since the extraradical mycelium, spreading into the soil, is responsible of the mineral nutrient and water uptake that benefits its plant host. In return, the fungi direct the flow of a significant fraction of the host plant photoassimilate. Glycogen and trehalose have been described as the dominant storage carbohydrates in AM fungal hyphae and spores (Pfeffer et al. 1999).

In the mycorrhizal roots of two maize cultivars exposed to drought, the trehalose content increased fivefold (Schellenbaum et al. 1998). AM fungi seem to accumulate trehalose upon stress in the same way as other microorganisms (Müller et al. 1995b), allowing them to survive freezing overwintering conditions (Addy et al. 1994) and periods of drought (Jasper et al. 1993). Hence, accumulation of trehalose could be an important determinant for sustained viability under stress and for successful colonization of plants after frost or drought at the beginning of the growing season.

6 Conclusions and Perspectives

The recently discovered trehalose metabolism in plants has a high significance in different aspects of the plant physiology and development, since trehalose metabolism, and especially the intermediate T6P, has been shown to play a central role in the plant cell carbon metabolism as well as in the sugar distribution between different tissues. In that sense, most of the recent studies have focused on T6P and its function on plant development. Although transgenic plants with microbial trehalose biosynthesis often lead to developmental aberrations, diverse studies have shown that trehalose accumulation is involved in protecting plants from different forms of abiotic stresses such as desiccation or salinity. These results are promising for the generation of crops resistant to abiotic stresses, although to better understand the role of trehalose in plant protection further studies are still required including transcriptomics, proteomics, and metabolomics approaches for a complete view of trehalose role in plants.

References

Addy HD, Schaffer GF, Miller MH, Peterson RL (1994) Survival of the extraradical mycelium of a VAM fungus in frozen soil over winter. Mycorrhiza 5:1–5

Almeida AM, Silva AB, Aráujo SS, Cardoso LA, Santos DM (2007) Responses to water withdrawal of tobacco plants genetically engineered with the *AtTPS1* gene: a

special reference to photosynthetic parameters. Euphytica 154:113–126
Anselmino O, Gilg E (1913) Ueber das Vorkommen von Trehalose in *Selaginella lepidophylla*. Ber Deut Pharm Ges 23:326–330
Arguelles JC (2000) Physiological roles of trehalose in bacteria and yeasts: a comparative analysis. Arch Microbiol 174:217–224
Avonce N, Leyman B, Mascorro-Gallardo JO, Van Dijck P, Thevelein JM, Iturriaga G (2004) The Arabidopsis trehalose-6-P synthase AtTPS1 gene is a regulator of glucose, abscisic acid, and stress signaling. Plant Physiol 136:3649–3659
Avonce N, Mendoza-Vargas A, Morett E, Iturriaga G (2006) Insights on the evolution of trehalose biosynthesis. BMC Evol Biol 6:109–113
Becker A, Scholeder P, Wegener G (1996) The regulation metabolism in insects. Experientia 52:433–439
Bianchi G, Gamba A, Limiroli R, Pozzi N, Elster R, Salamini F, Bartels D (1993) The unusual sugar composition in leaves of the resurrection plant *Myrothamnus flabellifolia*. Physiol Plant 87:223–226
Blázquez MA, Santos E, Flores CL, Martínez-Zapater JM, Salinas J, Gancedo C (1998) Isolation and molecular characterization of the Arabidopsis TPS1 gene, encoding trehalose-6-phosphate synthase. Plant J 13: 685–689
Cabib E, Lenoir LF (1958) The biosynthesis of trehalose phosphate. J Biol Chem 231:259–275
Cortina C, Culianez-Maciá FA (2005) Tomato abiotic stress enhanced tolerance by trehalose biosynthesis. Plant Sci 169:75–82
Crowe JH (2007) Trehalose as a "chemical chaperone": fact and fantasy. Adv Exp Med Biol 594:143–158
Crowe JH, Crowe LM, Chapman D (1984) Preservation of membranes in anhydrobiotic organisms: the role of trehalose. Science 223:701–703
De Smet KA, Weston A, Brown IN, Young DB, Robertson BD (2000) Three pathways for trehalose biosynthesis in mycobacteria. Microbiology 146:199–208
Eastmond PJ, van Dijken AJ, Spielman M, Kerr A, Tissier AF, Dickinson HG, Jones JD, Smeekens SC, Graham IA (2002) Trehalose-6-phosphate synthase 1, which catalyses the first step in trehalose synthesis, is essential for Arabidopsis embryo maturation. Plant J 29: 223–235
Elbein AD (1974) The metabolism of α,α-trehalose. Adv Carbohydr Chem Biochem 30:227–256
Elbein AD, Mitchell M (1973) Levels of glycogen and trehalose in *Mycobacterium smegmatis* and the purification and properties of the glycogen synthetase. J Bacteriol 113(2):863–873
Elbein AD, Pan YT, Pastuszak I, Carroll D (2003) New insights on trehalose: a multifuntional molecule. Glycobiology 13:17–27
Farías-Rodriguez R, Mellor RB, Arias C, Peña-Cabriales JJ (1998) The accumulation of trehalose in nodules of several cultivars of common bean (*Phaseolus vulgaris*) and its correlation with resistance to draught stress. Physiol Plant 102:353–359
Fernandez O, Béthencourt L, Quero A, Sangwan RS, Clément C (2010) Trehalose and plant stress responses: friend or foe? Trends Plant Sci 15:409–417
Fougère F, Le Rudulier D, Streeter JG (1991) Effects of salt stress on amino acid, organic acid, and carbohydrate composition of roots, bacteroids, and cytosol of alfalfa (*Medicago sativa* L.). Plant Physiol 96:1228–1236
Frison M, Parrou JL, Guillaumot D, Masquelier D, François J, Chaumont F, Batoko H (2007) The *Arabidopsis thaliana* trehalase is a plasma membrane-bound enzyme with extracellular activity. FEBS Lett 581:4010–4016
García B, de Almeida EJ, Iyer S, Gerats T, Van Montagu M, Caplan AB (1997) Effects of osmoprotectants upon NaCl stress in rice. Plant Physiol 115:159–169
Garg AK, Kim JK, Owens TG, Ranwala AP, Choi YD, Kochian LV, Wu RJ (2002) Trehalose accumulation in rice plants confers high tolerance levels to different abiotic stresses. Proc Natl Acad Sci USA 99:15898–15903
Gibon Y, Bessieres MA, Larher F (1997) Is glycine betaine a non-compatible solute in higher plants that do not accumulate it? Plant Cell Environ 20:329–340
Gomez LD, Baud S, Gilday A, Li Y, Graham I (2006) Delayed embryo development in the *Arabidopsis* trehalose 6-phosphate synthase 1 mutant is associated with altered cell wall structure, decreased cell division and starch accumulation. Plant J 46:69–84
Gooddijn OJM, Verwoerd TC, Voogd E, Krutwagen PWHH, de Graaf PTHM, Poels J, Van Dun K, Ponstein AS, Damm B, Pen J (1997) Inhibition of trehalase activity enhances trehalose accumulation in transgenic plants. Plant Physiol 113:181–190
Grennan AK (2007) The role of trehalose biosynthesis in plants. Plant Physiol 144:3–5
Han SE, Park SR, Kwon HB, Yi BY, Lee GB, Byun MO (2005) Genetic engineering of drought-resistant tobacco plants by introducing the trehalose phosphorylase (TP) gene from *Pleurotus sajor-caju*. Plant Cell Tissue Organ Cult 82:151–158
Higashiyama T (2002) Novel functions and applications of trehalose. Pure Appl Chem 74:1263–1269
Iordachescu M, Imai R (2008) Trehalose biosynthesis in response to abiotic stresses. J Integr Plant Biol 50: 1223–1229
Jasper DA, Abbott LK, Robson AD (1993) The survival of infective hyphae of vesicular arbuscular mycorrhizal fungi in soil: an interaction with sporulation. New Phytol 124:73–479
Kaasen I, McDougall J, Strom AR (1994) Analysis of the otsBA operon for osmoregulatory trehalose synthesis in *Escherichia coli*. Gene 145:9–15
Kaplan F, Kopka J, Haskell DW, Zhao W, Schiller K, Gatzke N, Sung DY, Guy CL (2004) Exploring the temperature–stress metabolome of Arabidopsis. Plant Physiol 136:4159–4168
Karim S, Aronsson H, Ericson H, Pirhonen M, Leyman B, Welin B, Mäntylä E, Tapio Palva E, Van Dijck P, Holmström KO (2007) Improved drought tolerance without undesired side effects in transgenic plants producing trehalose. Plant Mol Biol 64:371–386

Kolbe A, Tiessen A, Schluepmann H, Paul M, Ulrich S, Geigenberger P (2005) Trehalose 6-phosphate regulates starch synthesis via post-translational redox activation of ADP-glucose pyrophosphorylase. Proc Natl Acad Sci USA 102:11118–11123

Leyman B, Van Dijck P, Thevelein JM (2001) An unexpected plethora of trehalose biosynthesis genes in *Arabidopsis thaliana*. Trends Plant Sci 6:510–513

Liu MS, Chien CT, Lin TP (2008) Constitutive components and induced gene expression are involved in the desiccation tolerance of *Selaginella tamariscina*. Plant Cell Physiol 49:653–663

López M, Herrera-Cervera JA, Lluch C, Tejera NA (2006) Trehalose metabolism in root nodules of the model legume *Lotus japonicus* in response to salt stress. Physiol Plant 128(4):701–709

López M, Herrera-Cervera JA, Iribarne C, Tejera NA, Lluch C (2008a) Growth and nitrogen fixation in *Lotus japonicus* and *Medicago truncatula* under NaCl stress: Nodule carbon metabolism. J Plant Physiol 165:641–650

López M, Tejera NA, Iribarne C, Lluch C, Herrera-Cervera JA (2008b) Trehalose and trehalase in root nodules of *Medicago truncatula* and *Phaseolus vulgaris* in response to salt stress. Physiol Plant 134:575–582

López M, Tejera NA, Lluch C (2009) Validamycin A improves the response to salt stress of *Medicago truncatula* by inducing trehalose accumulation in the root nodules. J Plant Physiol 166:1218–1222

Lunn JE (2007) Gene families and evolution of trehalose metabolism in plants. Funct Plant Biol 34:550–563

Lunn JE, Feil R, Hendriks JHM, Gibon Y, Morcuende R (2006) Sugar induced increases in trehalose 6-phosphate are correlated with redox activation of ADP-glucose pyrophosphorylase and higher rates of starch synthesis in *Arabidopsis thaliana*. Biochem J 397:139–148

Madin KAC, Crowe JH (1975) Anhydrobiosis in nematodes: carbohydrate and lipid metabolism during dehydration. J Exp Zool 193:335–342

Martin MC, Diaz LA, Manzanal MB, Hardisson C (1986) Role of trehalose in the spores of *Streptomyces*. FEMS Microbiol Lett 35:49–54

Maruta K, Hattori K, Nakada T, Kubota M, Sugimoto T, Kurimoto M (1996) Cloning and sequencing of trehalose biosynthesis genes from *Rhizobium* sp. Biosci Biotechnol Biochem 60:717–720

Miranda JA, Avonce N, Suarez R, Thevelein JM, Van Dijck P, Iturriaga G (2007) A bifunctional TPS–TPP enzyme from yeast confers tolerance to multiple and extreme abiotic-stress conditions in transgenic *Arabidopsis*. Planta 226:1411–1421

Müller J, Boller T, Wiemken A (1995a) Effects of validamycin A, a potent trehalase inhibitor, and phytohormones on trehalose metabolism in roots and root nodules of soybean and cowpea. Planta 197:362–368

Müller J, Boller T, Wiemken A (1995b) Trehalose and trehalase in plants: recent developments. Plant Sci 112: 1–9

Müller J, Boller T, Wiemken A (1996) Pools of non-structural carbohydrates in soybean root nodules during water stress. Physiol Plant 98:723–730

Müller J, Boller T, Wiemken A (1997) Trehalose affects sucrose synthase and invertase activities in soybean (*Glycine max* (L.) Merr.) roots. J Plant Physiol 153: 255–257

Müller J, Aeschbacher RA, Wingler A, Boller T, Wiemken A (2001) Trehalose and trehalase in Arabidopsis. Plant Physiol 125:1086–1093

Paiva CLA, Panek AD (1996) Biotechnological applications of the disaccharide trehalose. Biotechnol Annu Rev 2:293–314

Paul M (2007) Trehalose 6-phosphate. Curr Opin Plant Biol 10:303–309

Paul MJ, Pellny TK, Goddijn O (2001) Enhancing photosynthesis with sugar signals. Trends Plant Sci 6:197–200

Paul MJ, Primavesi LF, Jhurreea D, Zhang Y (2008) Trehalose metabolism and signaling. Annu Rev Plant Biol 59:417–441

Pellny TK, Ghannoum O, Conroy JP, Schluepmann H, Smeekens S (2004) Genetic modification of photosynthesis with *E. coli* genes for trehalose synthesis. Plant Biotechnol J 2:71–82

Pfeffer PE, Douds DD, Bécard G, Shachar-Hill Y (1999) Carbon uptake and the metabolism and transport of lipids in an arbuscular mycorrhiza. Plant Physiol 120:587–592

Phillips DV, Wilson DO, Dougherty DE (1984) Soluble carbohydrates in legumes and nodulated nonlegumes. J Agric Food Chem 32:1289–1291

Pilon-Smits EAH, Terry N, Seors T, Kim H, Zayed A, Seongbin H, Van Dun K, Voogd E, Verwoerd TC, Krutwagen RW, Goddijn OJM (1998) Trehalose-producing transgenic tobacco plants show improved growth and performance under drought stress. J Plant Physiol 152:525–532

Pramanik MHR, Imai R (2005) Functional identification of a trehalose-6-phosphatase gene that is involved in transient induction of trehalose biosynthesis during chilling stress in rice. Plant Mol Biol 58:751–762

Purvis JE, Yomano LP, Ingram LPO (2005) Enhanced trehalose production improves growth of *Escherichia coli* under osmotic stress. Appl Environ Microbiol 71:3761–3769

Qu Q, Lee SJ, Boss W (2004) TreT, a novel trehalose glycosyltransferring synthase of the hyperthermophilic archeon *Thermococcus litoralis*. J Biol Chem 279: 47890–47897

Ramon M, Rolland F (2007) Plant development: introducing trehalose metabolism. Trends Plant Sci 12:185–188

Romero C, Bellés JM, Vayá JL, Serrano R, Culiáñez-Macià A (1997) Expresion of the yeast trehalose-6-phosphate synthase gene in transgenic tobacco plants: pleiotropic phenotypes include drought tolerance. Planta 201:293–297

Ryu SI, Park CS, Cha J, Woo EJ, Lee SB (2005) A novel trehalose-synthesizing glycosyltransferase from *Pyrococcus horikoshii*: molecular cloning and characterization. Biochem Biophys Res Commun 329:429–436

Satoh-Nagasawa N, Nagasawa N, Malcomber S, Sakai H, Jackson D (2006) A trehalose metabolic enzyme controls inflorescence architecture in maize. Nature 441: 227–230

Schellenbaum L, Müller J, Boller T, Wiemken A, Schüepp H (1998) Effects of drought on non-mycorrhizal and mycorrhizal maize: changes in the pools of non-structural carbohydrates, in the activities of invertase and trehalase, and in the pools of amino acids and imino acids. New Phytol 138:59–66

Schiraldi C, Di Lernia I, De Rosa M (2002) Trehalose production: exploiting novel approaches. Trends Biotechnol 20:420–425

Schluepmann H, Pellny T, Van Dijken A, Smeekens S, Paul MJ (2003) Trehalose 6-phosphate is indispensable for carbohydrate utilisation and growth in *Arabidopsis thaliana*. Proc Natl Acad Sci USA 100: 6849–6854

Schluepmann H, Van Dijken A, Aghdasi M, Wobbes B, Paul M, Smeekens S (2004) Trehalose mediated growth inhibition of Arabidopsis seedlings is due to trehalose 6-phosphate accumulation. Plant Physiol 135:879–890

Shima S, Matsui H, Tahara S, Imai R (2007) Biochemical characterization of rice trehalose-6-phosphate phosphatases supports distinctive functions of these plant enzymes. FEBS J 274:1192–1201

Shimakata T, Minatagawa Y (2000) Essential role of trehalose in the synthesis and subsequent metabolism of corynomycolic acid in *Corynebacterium matruchotii*. Arch Biochem Biophys 380:331–338

Stiller I, Dulai S, Kondrák M, Tarnai R, Szabó L, Toldi O, Bánfalvi Z (2008) Effects of drought on water content and photosynthetic parameters in potato plants expressing the trehalose-6-phosphate synthase gene of *Saccharomyces cerevisiae*. Planta 227(2):299–308

Streeter JG (1980) Carbohydrates in soybean nodules. Distribution of compounds in seedling during the onset of nitrogen fixation. Plant Physiol 66:471–476

Streeter JG (1985) Accumulation of α,α-trehalose by *Rhizobium* bacteria and bacteroids. J Bacteriol 164:78–84

Streeter JG, Bhagwat A (1999) Biosynthesis of trehalose from maltooligosaccharides in Rhizobia. Can J Microbiol 45:716–721

Streeter JG, Gómez ML (2006) Three enzymes for trehalose synthesis in *Bradyrhizobium* cultured bacteria and bacteroids from soybean nodules. Appl Environ Microbiol 72:4250–4255

Suárez R, Wong A, Ramírez M, Barraza A, Orozco C, Cevallos MA, Lara M, Hernández G, Iturriaga G (2008) Improvement of drought tolerance and grain yield in common bean by overexpressing trehalose-6-phosphate synthase in rhizobia. Mol Plant Microbe Interact 21:958–966

Suzuki N, Bajad S, Shuman J, Shulaev V, Mittler R (2008) The transcriptional co-activator mbf1c is a key regulator of thermotolerance in *Arabidopsis thaliana*. J Biol Chem 283:9269–9275

Van Dijken AJH, Schluepmann H, Smeekens SCM (2004) Arabidopsis trehalose 6-phosphate synthase 1 is essential for normal vegetative growth and transition to flowering. Plant Physiol 135:969–977

Vogel G, Aeschbacher RA, Müller J, Boller T, Wiemken A (1998) Trehalose-6-phosphate phosphatases from *Arabidopsis thaliana*: identification by functional complementation of the yeast tps2 mutant. Plant J 13:673–683

Vogel G, Fiehn O, Jean-Richard L, Boller T, Wiemken A (2001) Trehalose metabolism of Arabidopsis: occurrence of trehalose and molecular cloning and characterisation of trehalose 6-phosphate synthase homologues. J Exp Bot 52:1817–1826

Wannet WJB, Op den Camp HJM, Wisselink HW, Van der Drift C, Van Griensven LJLD, Vogels GD (1998) Purification and characterization of trehalose phosphorylase from the commercial mushroom *Agaricus bisporus*. Biochim Biophys Acta 1425: 177–188

Wingler A, Fritzius T, Wiemken A, Boller T, Aeschbacher A (2000) Trehalose induces the ADP-glucose pyrophosphorylase gene and starch synthesis in Arabidopsis. Plant Physiol 124:105–114

Uptake of Mineral Elements During Abiotic Stress

15

Fatih Duman

Abstract

Several environmental factors influence the mineral uptake of plants, including pH, redox potential, and the presence of xenobiotics as well as the temperature and salinity. Changes in environmental conditions affect the biological and physiological response of plants. Most important targets of plants are maintained by the ion homeostasis and mineral uptake in stress conditions. Plants may use the different procedures to regulate homeostasis. In general, Na salt stress decreases the levels of cationic nutrients such as K, Ca, and Mg due to competition for ion transport sites. Drought and salinity stress are physiologically related and the tolerance mechanisms overlap. Metals can interfere with mineral nutrition and change the concentration and composition of plant nutrients. Besides, metals may also alter the conformation of proteins, including transporters, or regulator proteins. Herbicides may disrupt the function and integrity of the cell membrane, and significant ion losses can occur. However, molecular mechanisms and genetic basis of interactions between abiotic stress and mineral uptake is lacking. Thus, future studies will focus on these aspects. In this chapter, the effects of some common stressors, such as salinity, drought, metals, herbicides, and on nutrient uptake are elucidated.

Keywords

Abiotic stress • Salinity • Drought • Heavy metal • Herbicide • Mineral uptake • Metal

F. Duman (✉)
Department of Biology, Faculty of Science,
Erciyes University, Kayseri 38039, Turkey
e-mail: fduman@erciyes.edu.tr

P. Ahmad and M.N.V. Prasad (eds.), *Abiotic Stress Responses in Plants: Metabolism, Productivity and Sustainability*, DOI 10.1007/978-1-4614-0634-1_15,

1 Introduction

Sixteen chemical elements are known to be important for plant growth and survival. Mineral nutrients are inorganic elements in food, which the body cannot synthesize. These chemicals may be divided into two groups, namely, minerals and non-minerals. Non-mineral nutrients include hydrogen (H), oxygen (O), and carbon (C). These nutrients are found in the air and water. The other 13 mineral nutrients are found in the soil and dissolved in water, so they are absorbed through the roots. Mineral nutrients are classified as macronutrients and micronutrients. There are two groups of macronutrients, namely, primary and secondary nutrients. The primary nutrients are nitrogen (N), phosphorus (P), and potassium (K), which are needed in large amounts for plant growth and survival. The secondary nutrients are calcium (Ca), magnesium (Mg), and sulphur (S). Since the amount of these minerals in the soil is usually sufficient for growth, fertilization is not always needed. Micronutrients are elements, which are essential for plant growth but are required in much smaller amounts than primary nutrients. The micronutrients are boron (B), copper (Cu), iron (Fe), manganese (Mn), molybdenum (Mo), zinc (Zn), and chloride (Cl). Although micronutrients only comprise 5% of the biomass of a plant, they are essential for growth. For example, Zn plays a key role in enzymatic catalysis of reactions, which require an electrophile. In addition, Fe, Mn, Cu, and Mo are responsible for redox transformations (Merchant 2006). Although the exact functions of B are not fully understood, it is known to be necessary for numerous important processes, such as protein synthesis, sugar transport, and respiration (Hänsch and Mendel 2009). The uptake, translocation, and exclusion of these mineral nutrients are related to many environmental factors, such as salinity, acidity, water shortage, and the presence of xenobiotics. In this chapter, we discuss the effects of some common stressors, such as salinity, drought, metals, herbicides, and on nutrient uptake.

2 Abiotic Stressors Effecting Mineral Uptake

2.1 Salt Stress

Salinity is the major environmental factor, which limits plant growth and productivity. Salinity changes the biological properties of plants, such as facilitating their retention and acquisition of water and maintenance of ion homeostasis (Parida and Das 2005). Grattan and Grieve (1999) showed that soils have extremely high ratios of Na^+/Ca^{2+}, Na^+/K^+, Ca^{2+}/Mg^{2+}, and Cl^-/NO_3^- under saline conditions, which leads to ion toxicities (e.g., Na^+ and Cl^-) and ionic imbalance. In Parida and Das' review (2005) of salt tolerance and the effects of salinity in plants, they discussed three main issues, which are related to ion homeostasis during salt stress (1) selective accumulation or exclusion of ions, (2) control of ion uptake by roots and transport into leaves, and (3) compartmentalization of ions at both the cellular and whole-plant levels. However, plants do not have the ability to tolerate high concentrations of salt in the cytoplasm. As a result, they sequester excess salts in vacuoles or compartmentalize the ions in different tissues (Zhu 2003).

Khan (2001) showed that plants, which are exposed to salt stress exhibit increased levels of Na^+ and Cl^- and decreased levels of Ca^{2+}, K^+, and Mg^{2+}. Similarly, Lee and Liu (1999) demonstrated that Na^+ and Cl^- concentrations in *Ulva fasciata* increased when exposed to increasing salinity but decreased the total and water-soluble concentrations of Ca^{2+}. However, Ferreira et al. (2001) reported that salinity did not affect the Mg^{2+} content in the stems and roots of *Psidium guajava* L., but decreased it in the leaves. Although K is not incorporated into chemical compounds in plants, it is an important regulator of development. K regulates not only the electrical and water balance but also enzymatic processes, such as protein and carbohydrate synthesis. High levels of Na are cytotoxic. During salt stress, K^+ uptake is blocked due to accumulation of Na^+. Under normal conditions, the cytosolic concentration of K^+ is higher than that of Na^+. Maintaining this balance is essential for plant life (Zhu 2003). Na^+ competitively inhibits K^+ uptake (Grattan and Grieve 1998). During salt stress, K^+ uptake is disrupted in root cells. Intracellular K^+ and Na^+ homeostasis is important for many biological activities, such as enzyme activity, maintenance of the membrane potential, and regulation of cell volume. Unlike animal cells, plant cells do not have Na^+-ATPases or Na^+/K^+-ATPases (Hasegawa et al. 2000). Therefore, excess Na^+ must be extruded or compartmentalized in vacuoles. During salt stress, plants also need a mechanism to regulate turgidity. Many studies have shown the importance of K in regulating turgidity (Guardia and Benlloch, 1980; Mengel and Arneke, 1982; Hsiao and Läuchli 1986).

Low levels of Ca in growth media cause defects, such as deterioration of the cell membrane, loss of cellular components, and eventually cell and tissue death. Kaya et al. (2002) showed that Ca deficiency induces high concentrations of NaCl in strawberry. Calcium ions ameliorate the effect of salt stress by competing with sodium ions for membrane-binding sites. Tuna et al. (2007) studied the effects of calcium sulphate ($CaSO_4$) on nutrient uptake in tomato plants, which were grown in pots under salt stress and concluded that $CaSO_4$ increased concentration of Ca^{2+}, N, and K^+ and reduced the concentration of Na^+ in the leaves. Patel et al. (2011) also tested the effects of supplemental Ca on the nutrient levels in *Caesalpinia crista* L. (Fabaceae) in salinized soil in a greenhouse. They demonstrated that salt stress reduces N, P, K, and Ca content in tissues; however, the addition of Ca restored the levels of these nutrients. In general, as external Ca concentrations increase, Na uptake and concentrations decrease while Ca uptake and concentrations increase because Ca^{2+} interferes with non-selective cation channels and restricts Na^+ uptake. In addition, as the salt concentration in the root zone increases, the requirement for Ca^{2+} increases. However, the uptake of Ca^{2+} from the soil may be reduced as a result of ion interactions, precipitation, and increased ionic strength. These factors reduce the activity of Ca^{2+} in solution, which reduces the availability of Ca^{2+} (Grattan and Grieve 1998).

In plants, nitrogen comprises 80% of the total absorbed nutrients (Marschner 1995). In general, salinity reduces N accumulation and saline environments are associated with nitrogen deficiency (Siddiqui et al. 2010). For example, tomato (Cerda and Martinez 1988) and barley (Shen et al. 1994) plants have reduced nitrogen content when they are grown under salt stress. This is not surprising since an increase in Cl^- uptake and accumulation is often accompanied by a decrease in NO_3^- concentration in shoots. Tuna et al. (2007) suggested that this decrease may be due to inhibition of nitrogen uptake by NO_3^-/Cl^- interaction at ion transport sites. Many nitrogen-containing compounds, such as amino acids (e.g., proline and glycine betaine), protect against salt stress. For example, Siddiqui et al. (2010) showed that supplemental N limited the effects of salt stress on *Brassica* genotypes in salinized environments. In addition, the chemical form of N, such as NO_3 or NH_4, affects the interaction between salt and N content. Martinez and Cerdá (1989) demonstrated that Cl^- uptake decreased in cucumber (*Cucumis sativus*) when only NO_3^- was added but Cl^- accumulation increased when half the NO_3^- was replaced with NH_4^+.

We do not have enough data about the interaction between salinity and P to understand the underlying mechanism of changes in P uptake (Patel et al. 2009). The availability of P depends on the salinity and ionic strength of soil. In addition, P concentrations in the soil solution are tightly controlled by sorption processes and by the low solubility of Ca–P minerals (Violante et al. 2002). Champagnol (1979) showed that the addition of P to saline soils increased crop growth and yield. However, Grattan and Grieve (1999) reviewed studies about the relationship between salinity and P concentration in tissues and concluded that salinity increased the P concentration in tissues in sand or solution cultures but not soil.

In general, Na salt stress decreases the levels of cationic nutrients (K, Ca, and Mg). Ruiz et al. (1997) studied the effects of salt stress on mineral uptake in citrus plants and showed that NaCl salinity reduced Mg^{2+} concentrations in the leaves. There is a strong competition between Mg^{2+} and Ca^{2+} for plasma membrane binding sites in roots; however, Ca has a higher affinity for these sites than Mg^{2+} (Marschner 1995). As a result, plants grown in the presence of high levels of Ca have lower Mg content. Loupassaki et al. (2002) reported similar effects of salt stress on olive cultivars. However, Barhoumi et al. (2007) demonstrated that salinity did not affect Mg levels in *Aeluropus littoralis*, a perennial halophyte that retains Mg.

Several factors influence the metal uptake capacity of plants, including pH, redox potential, and concentration of surrounding metals as well as the temperature and salinity of the surrounding water (Leblebici et al. 2011). In addition, Na^{+} ions release Cd from the sediment into water, thereby increasing the concentration of soluble Cd (Greger et al. 1995). Greger et al. (1995) also showed that when *Potamogeton pectinatus* grows in highly saline water, low levels of free Cd ions in the water correlate with a low Cd uptake capacity. At high salinity, increased concentrations of NaCl reduce the uptake of metals in *Spirodela polyrrhiza* (Leblebici et al. 2011). Many other studies also have shown that high salinity decreases metal uptake (Munda and Hudnik 1988; Wang and Dei 1999), which may be due to the formation of complexes of metal and chloride ions (Mamboya et al. 2009). Similarly, Greger et al. (1995) designed an experimental study to determine the influence of salinity on Cd uptake in submerged macrophytes by using *Potamogeton pectinatus*, and showed that increasing salinity decreases Cd uptake in *P. pectinatus* from water, most likely due to the formation of Cd complexes with chloride and sulphate. In contrast, in the presence of sediment, increasing salinity increases Cd uptake because Na^{+} and Mg^{2+} displace Cd^{2+} in sediment colloids. However, Fritioff et al. (2005) did not find any effect of salinity on Pb accumulation. This may be because Pb does not form a complex with chloride and has a high binding affinity for organic matter. Manousaki et al. (2008) also reported a similar effect of increased soil salinity on increased cadmium uptake in *Tamarix smyrnensis*, which is a halophytic plant, and speculated that this effect may be related to a higher mobility of metals in the sediment or higher water uptake, which would increase the metal flux into the plant. In another study on *T. smyrnensis*, Kadukova and Kalogerakis (2007) observed that high salt concentrations in soil decreased the accumulation of Pb in the roots but increased it in the leaves. Collectively, these studies show that salt–metal interactions are very complex. As a result, further research is needed to elucidate these interactions in more detail.

2.2 Drought Stress

Drought (continuous water deficit) is one of the most important factors that affect the growth, development, and survival of plants. Drought stress usually occurs because of insufficient water availability in the soil; however, it can also occur due to excessive loss of water by transpiration or evaporation (Jaleel et al. 2009). Drought stress causes significant physiological changes, such as stomatal closure, decreased photosynthetic activity, and changes in cell wall elasticity. Drought and salinity stress are physiologically related and the tolerance mechanisms overlap. Drought disturbs the nutritional status of plants by altering ion concentrations in tissues. Since soil moisture is reduced during drought stress, the rate of diffusion of nutrients from the soil matrix to the absorbing root surface is also reduced (Hu et al. 2007). In addition, nutrient transport from the roots to the shoots decreases due to reduced transpiration rates.

Ca^{2+} is involved in plant drought resistance. When Ca is applied to leaves, it enhances their ability to conserve water (Shao et al. 2008). In addition, Ma et al. (2009) showed that Ca^{2+} alters the degree of hydration of the plasma membrane and improves the cohesion of cell walls, which increases the viscosity of the protoplasm and the resistance of cells to dehydration. Berkowitz et al. (2000) demonstrated that Ca^{2+} inhibits the influx of K^{+} to guard cells during water stress by affecting inward rectifying K^{+} channels.

In addition to Ca, K is responsible for osmoregulation in plant cells during drought stress (Roberts 1998). Mohsenzadeh et al. (2006) showed that K levels in *Aeluropus lagopoides* increased significantly under mild and moderate drought stress. Zhang et al. (2006) demonstrated that K transport increases during the early stages of adaption to stress in plants. K channels are important to regulate the water status of plants (Roberts 1998). However, there are some contradictory reports that drought stress decreases K levels. For example, Thiec and Manninen (2003) suggested that K levels decrease during drought stress due to changes in nutrient compartmentalization.

Martínez et al. (2004) proposed that Na^+ may play a positive role in response to water stress, because Na^+ absorption increases in plants that are subjected to drought stress on a non-saline substrate. Similar observations have been reported (leaf Na^+ concentration was significantly increased by simultaneous exposure to drought) in halophyte species, such as *Sesuvium portulacastrum*, *Atriplex halimus*, and *Ipomoea pes-caprae* (Slama et al. 2008; Martínez et al. 2005; Sucre and Suárez 2011). These studies suggest that Na^+ may have a direct or indirect positive influence on the accumulation of other elements, which are involved in osmotic regulation. In addition, Na plays a specific role in parasitic plants, such as *Cuscuta attenuate*. Kelly and Horning (1999) showed that parasitic plants accumulate higher levels of Na than their hosts do, to draw water and nutrients from their hosts.

After drought, there is usually a nitrogen deficiency in the ground. There is a strong relationship between water availability and N absorption. In general, the availability, uptake, and utilization of N by plants increases as the soil moisture content increases (Albrizio et al. 2010; Kibe et al. 2006). However, Payne et al. (1995) reported that total nitrogen and NO_3 levels increase during drought stress in pearl millet (*Pennisetum glaucum*). Song et al. (2010) demonstrated that N and P have strong interactive effects on plant growth and that the water supply was the primary determinant of this interaction. Water deficit reduces the net CO_2 assimilation rate, which may be due to inhibition of ribulose bisphosphate synthesis and ATP synthase activity (Tezara et al. 1999). Decreased ATP synthesis in chloroplasts may also be caused by a low availability of free inorganic phosphate in the cytoplasm (Guida Dos Santos et al. 2006). Thus, water deficit decreases the levels of free phosphate in the cytoplasm (Pieters et al. 2001).

Increased levels of Mg in leaves can help maintain water content during drought stress. Mahouachi (2009) investigated the changes in the concentration of mineral nutrients in *Musa acuminate* plants, which were grown in increasingly dry soil. These drought-stressed plants showed a significant accumulation of Mg (higher 28% compared to control). Since Mg^{2+} is an essential element in chloroplasts, a deficiency can decrease photosynthetic activity. Mahouachi (2009) proposed two possible explanations for this accumulation (1) the increased concentration of mineral nutrients may be associated with the translocation of ions from old to young leaves or (2) the increased ion concentration in the root zone may be due to a concentration effect, which was produced by the reduction in soil moisture. However, in this case, xylem flux should still be active enough to move ions to functioning leaves.

The effects of metals on water content in soil solutions are controversial. Tipping et al. (2003) argued that the interaction between metals, such as Cu and Zn, and drought is the most important factor in soil acidification. Since the strength of metal binding to natural organic matter is inversely related to pH, acidification may release metals into the soil. On the other hand, Zn availability may promote water conservation by decreasing the transpiration rate. Disante et al. (2011) showed that, at least in the short term, Zn may promote water conservation. Similarly, Gadallah (2000) demonstrated that Zn stimulates the accumulation of osmotically active solutes, such as soluble sugars, to facilitate water uptake. In addition, Disante et al. (2011) suggested that Zn affects stomatal dynamics by inhibiting water channels and reducing K^+ uptake.

2.3 Heavy Metal Stress

2.3.1 Cadmium

Cadmium (Cd) is a phytotoxic element because it can interfere with mineral nutrition and change the concentration and composition of plant nutrients. Cd^{2+} may interfere with nutrient uptake by altering plasma membrane permeability and may also alter the conformation of proteins, including enzymes, transporters, or regulator proteins, due to its strong affinity for sulfhydryl and carboxylic groups (Assche and Clijsters 1990; Gonçalves et al. 2009). Many studies have focused on the effect of Cd on nutrient uptake and translocation in plants (Zhang et al. 2002; Dong et al. 2006; Gonçalves et al. 2009). However, there is no consensus about the effects of Cd on mineral nutrient uptake because there are contradictory reports about species or cultivar differences, and interactions between metals and plant tissues (Liu et al. 2003). For example, Cui et al. (2008) showed that copper (Cu), iron (Fe), and zinc (Zn) interfere with Cd in rice. The addition of Cu significantly decreased Cd uptake by the shoots and roots of rice. However, Zn uptake decreased significantly as the amount of Cd and Cu increased. In another study, Liu et al. (2003) studied the interaction of Cd with five mineral nutrients (Fe, Zn, Cu, Mn, and Mg) and found significant differences between rice cultivars. In addition, Gonçalves et al. (2009) studied the interaction between Cd^{2+} and mineral nutrients in potato (*Solanum tuberosum*) both in vitro and in hydroponic culture. Although Cd^{2+} did not affect the content of mineral nutrients in hydroponically grown plantlets, it decreased the content in in vitro plantlets. Nada et al. (2007) conducted an experimental study with hydroponically grown almond (*Prunus dulcis*) with 25–150 μM Cd^{2+} for 14 days and showed that all concentrations of Cd^{2+} reduced the concentration of Ca^{2+}, Mg^{2+}, and K^{+} in the leaves, but only the highest concentration of Cd^{2+} decreased the concentration of K^{+} and Mg^{2+} in the roots.

The concentration of Cd^{2+} also affects its interaction with minerals. For example, when *Brassica chinensis* was exposed to 0.1 μg mL^{-1} Cd^{2+}, the concentration of Zn increased; however, at higher concentrations of Cd^{2+}, the concentration of Zn decreased (Wong et al. 1984). At low levels, Cd may hyperpolarize the plasma membranes on the surface of roots, thereby increasing the transmembrane potential (Gonçalves et al. 2009). This hypothesis is consistent with some studies with metal transporters, which belong to the natural resistance-associated macrophage protein (NRAMP) and zinc-regulated transporter/iron-regulated transporter-related protein (ZIP) families (Guerinot 2000). These transporters may increase the concentration of mineral nutrients. Dong et al. (2006) showed that the main toxic effects of Cd^{2+} result from its interaction with essential elements, especially those with the same valence, such as Cu^{2+}, Fe^{2+}, Mn^{2+}, and Zn^{2+}. In contrast, Liu et al. (2003) demonstrated close relationships between Cd^{2+} and Fe^{2+}, Cd^{2+} and Zn^{2+}, and Cd^{2+} and Cu^{2+} in both roots and leaves of rice.

Yu and Zhou (2009) revealed an antagonistic relationship between Cd and P in a study on *Mirabilis palaja*. As a result, they advised that the addition of P to Cd-contaminated soil may be an effective way to immobilize Cd. Similarly, Jiang et al. (2007) suggested that external P can decrease the bioavailability of Cd and Zn. Yang et al. (1999) determined the effect of P on the accumulation of Cd and Zn in suspension-cultured wheat cells. They observed that the addition of P into the culture medium reduces Cd and Zn bioaccumulation. Earlier studies showed that Cd exposure decreases nitrate reductase activity in *Nicotiana tabacum* (Dguimi et al. 2009) and *Oryza sativa* (Huang and Xiong 2009), which increases the concentration of free amino acids and decreases the concentration of soluble proteins and nitrates in plant tissues. Although few studies have focused on the interactions between sulphur availability and Cd exposure, it is known that Cd plays an important role in the synthesis of thiol-based complexing substances of phytochelatins by upregulating glutathione biosynthesis (Astolfi et al. 2004). Fan et al. (2010) showed that excessive S significantly decreases the accumulation of Cd in brown rice by decreasing Cd availability and increasing glutathione in rice leaves. In contrast, Nocito et al. (2002) demonstrated that S can increase Cd availability and concentration in plants. Eventually, relationship between S and Cd is controversial. Previous studies showed that the presence of Cd in the culture medium decreases the concentration of K and Ca in different plant organs (Yang and Lee 2002; Ghnaya et al. 2005). This decrease may be due to the competition of Cd^{2+} with Ca^{2+} and other cations for entry into plant cells.

2.3.2 Arsenic

Inorganic arsenic (As) compounds are used in industrial and agricultural applications as well as aquatic weed control. Previous studies have shown that As may compete directly with nutrients or alter metabolic processes in plants. Since inorganic As(V) and phosphate are chemically similar, As(V) can act as a phosphate analogue and be transported into the cell (Meharg and Macnair 1990). Intracellular As(V) can interfere with essential cellular processes, such as oxidative phosphorylation and ATP synthesis (Tripathi et al. 2007). Many studies have demonstrated competition between As and P (Mkandawire et al. 2004; Mkandawire and Dudel 2005; Karadjova et al. 2008). Meharg and Hartley-Whitaker (2002) explained that this competition arises because As and P use the same transport system.

Tu and Ma (2005) tested the effects of As on essential macronutrients (P, K, Ca, and Mg) and micronutrients (Fe, Mn, Cu, Zn, B, and Mo) in *Pteris vittata*, which is an As hyperaccumulator. At low levels of As, the levels of P and K increased along with As. These authors reported a negative correlation between As accumulation and the Ca level in *P. vittata*, which may reflect the ability of Ca to prevent As toxicity. In addition, they showed that Mg has similar patterns of accumulation and distribution as Ca in the fronds. Although As reduced the concentration of Fe and Zn in the fronds, As did not significantly affect the concentration of Cu and Mn. Vance et al. (2003) reported that P can combine with other cations, especially aluminium (Al) and Fe, in complexes under acidic conditions. As a result, the As-induced reduction in micronutrients (excluding Mn) in the fronds was probably due to As phytotoxicity. Another study in tomato (*Lycopersicum esculentum*) showed that inorganic As(V) significantly decreased the concentration of both macronutrients (K, Ca, and Mg) and micronutrients (B, Cu, Mn, and Zn) (Carbonell-Barrachina et al. 1998a). Carbonell-Barrachina et al. (1998b) determined the effects of different chemical forms and concentrations of As on its uptake and nutrition in an aquatic plant (*Spartina alterniflora*). The application of organic As species significantly increased the concentration of Ca in leaves. These increased Ca levels conferred protection from metal and metalloid toxicity. However, the application of inorganic As species significantly increased the concentration of Cu in the roots and shoots of *S. alterniflora*. Due to the high phytotoxicity of monomethylarsonic acid (MMAA), the concentration of several essential macronutrients (e.g., P, K, Ca, and Mg) and micronutrients (e.g., B, Cu, and Fe) decreased significantly. The increase in Ca levels could have been due to the protective action of Ca against As toxicity.

2.3.3 Mercury

Previous studies have mostly focused on mercury (Hg) accumulation and the effect of Hg on the antioxidant activities of plant enzymes. As a result, few studies have investigated the effects of Hg on uptake of nutrient elements (Rodríguez et al. 2009). Hg has a high affinity for sulfhydryl groups, which can disrupt the function of essential proteins and, consequently, alter plant development. Hg also can restrict water channels in higher plants by altering membrane permeability (Patra et al. 2004). In plants, Hg can replace some nutritional elements, such as Mg, Zn, and Mn. This may be the main effect of Hg on nutrient uptake (Patra et al. 2004). In previous studies, Hg alters the uptake and translocation of mineral nutrients (Gupta and Chandra 1998); however, Hg uptake decreases as nutrient levels increase (Göthberg et al. 2004). Gupta and Chandra (1998) showed that increasing concentrations of Hg in the culture medium of *Vallisneria spiralis* significantly decreases the concentration of N, P, and K. In addition, the translocation of P, S, and K to the leaves increased as the concentration of Hg increased.

Rodríguez et al. (2009) investigated the uptake of Hg and levels of some micronutrients and macronutrients in hydroponically grown *Chilopsis linearis*. Hg only affected Fe, Mn, and Zn micronutrients and K, P, and S macronutrients. The concentration of Zn in the roots was reduced by 62% and 49% by 50 and 100 μM Hg, respectively. In addition, 50 μM Hg increased the concentration of Fe in the roots whereas 100 μM Hg decreased the concentration of Fe in the roots. Moreno-Jiménez et al. (2006) reported that Hg increased the concentration of Fe in the roots of *Marrubium vulgare* by more than 40% but reduced the translocation of Fe.

2.3.4 Chromium

Chromium (Cr) is a non-essential and toxic element for plants. It does not have any specific uptake mechanisms. However, Cr uptake is dependent on its chemical form. For example, Cr(VI) is taken up actively, whereas Cr(III) is taken up passively through the carriers for essential anions, such as sulphate, and stored in the cell wall (Zayed and Terry 2003). Shanker et al. (2005) reviewed the known relationships between Cr and plants and concluded that a high concentration of Cr is associated with a low concentration of mineral nutrients, such as Ca, K, Mg, P, B, and Cu. Cr can interfere with the uptake of other ionically similar elements, such as Fe and S. Moral et al. (1996) studied the effect of Cr on the concentration of mineral nutrients in tomato and showed that Cr negatively affects Fe absorption. In the case of Fe, the reduction in its uptake may be due to competition with chemically similar ions. In addition, Shanker et al. (2003) suggested that nutrient uptake may decrease due to inhibition of the activity of plasma membrane H^+ ATPase. Despite the general consensus in the literature, there is still some disagreement about whether Cr increases Fe content (Barceló et al. 1993).

Cr has a significant effect on the N content of plants. Kumar and Joshi (2008) concluded that Cr (VI) adversely affects N content by interfering with key enzymes in nitrogen metabolism. Similar findings have been reported for *Miscanthus sinensis* (Arduini et al. 2006) and *Nelumbo nucifera* (Vajpayee et al. 1999). Dube et al. (2003) studied the interaction between Cr and P in citrullus. They showed that increasing concentrations of Cr were associated with increasing concentrations of P in citrullus leaves. This accumulation of P might be due to the direct interference of Cr with the metabolism of P in plants.

2.3.5 Nickel

Although nickel (Ni) is essential for plants at low concentrations (Gajewska and Sklodowska 2007; Baccouh et al. 2001), it is phytotoxic at high concentrations (Duman and Ozturk 2010). Excess Ni also affects nutrient absorption by roots (Rahman et al. 2005). Ahmad et al. (2011) assessed the effect of Ni on the accumulation of macronutrients (K, Ca, and Mg) and micronutrients (Zn, Mn, Fe, and Cu) in different parts of sunflower (*Helianthus annuus*). They showed that Ni stress substantially decreases all macronutrients and micronutrients in sunflower leaves and achenes. Specifically, high concentrations of Ni decreased the concentrations of Ca, Mn, and Fe in achenes. In addition, increasing concentrations of Ni decreased the concentration of N, K, Zn, Mn, and Cu in achenes. However, Ni did not affect the concentration of P or Mg. Similarly, Ali et al. (2009) reported that Ni reduced the N, P, and K content in *Brassica napus*. Kähkönen and Kairesalo (1998) also demonstrated that Ni inhibits nutrient metabolism in *Elodea canadensis*. Moreover, Gajewska and Sklodowska (2007) suggested that Ni competitively displaces Ca ions from the Ca binding site in the oxygen-evolving complex. Ni has similar chemical characteristics as other mineral nutrients, such as Ca, Mg, Mn, Fe, Cu, and Zn. In addition, Ni is absorbed and transported by the same transport system as that for some other micronutrients, such as Cu and Zn (Ahmad et al. 2011). As a result, high levels of Ni may inhibit the absorption of these nutrients.

2.3.6 Lead

Lead (Pb) competes with divalent cations for transport into roots. This might be due to direct competition between Pb and other essential nutrients for the same binding site. Therefore, the concentration of micronutrients, such as Mn and Cu, may decrease in the presence of Pb. Sinha et al. (2006) studied the effects of Pb on the uptake and translocation of essential nutrients in cabbage (*Brassica oleracea*). They demonstrated that as the concentration of Pb increased, the concentration of Zn increased whereas those of P, S, Fe, Mn, and Cu decreased in various parts of the cabbage plant. Geebelen et al. (2002) and Diaz-Aguilar et al. (2001) observed a similar relationship between Pb and P and suggested that Pb forms insoluble complexes with P. Many other studies also have shown that excess Pb decreases the concentration of Fe in plants (Kannan and Keppel 1976; Paivoke 2002).

2.3.7 Copper

Previous studies have shown that the addition of Cu to plant growth media may affect the uptake of other mineral nutrients (Ke et al. 2007; Puig et al. 2007). For example, Bouazizi et al. (2010) investigated the accumulation and toxicity of Cu and determined the relationship between Cu accumulation and plant nutrients, such as Fe, K, Ca, and Zn in *Phaseolus vulgaris*. They concluded that the Fe, Zn, and K content decreased as a result of Cu accumulation, which reflected a change in nutrient homeostasis. Ke et al. (2007) compared the effects of Cu and other mineral nutrients on mineral uptake in a population of *Rumex japonicas* that grew near a copper mine with another population that grew in an uncontaminated area. The population that grew in copper-contaminated soil evolved a tolerance of not only high levels of Cu but also a lack of nutrients. As Yang and Romheld (2002) suggested in a study on *Elsholtzia splendens* that Cu tolerance may be related to the ability to maintain high levels of other mineral nutrients while under Cu stress. Indeed, the nutrient composition of plants that grew in the contaminated area exhibited less variation than those growing in the uncontaminated area.

Several studies have reported that mineral nutrients can affect the uptake and accumulation of Cu in plants (Nenova and Stoyanov 1999; Xiong et al. 2002). For example, Fe deficiency in the growth medium of pea (*Pisum sativum*) increased the concentration of Cu (Cohen et al. 1998). Similarly, Xiong et al. (2002) showed that Fe deficiency in the culture medium of plants stimulates Cu accumulation, while excess P reduces Cu accumulation. These results suggest that Cu has an antagonistic relationship with various mineral nutrients. Likewise, P antagonizes the absorption and metabolism of several trace elements. However, some studies do not support this hypothesis. For example, Cambrollé et al. (2011) reported a positive correlation between the accumulation of Cu and P in *Glaucium flavum*, which suggested that P plays an important role in controlling Cu accumulation and transport. In addition, Cu significantly inhibits nitrogen metabolism. Xiong et al. (2006) studied Cu-induced disruption of nitrogen metabolism in Chinese cabbage (*Brassica pekinensis* Rupr.) and demonstrated that Cu exposure increases Cu concentration and decreases nitrate reductase (NR) activity in the roots and shoots. In addition, Mazen (2004) showed that Cu increases the concentration of free amino acids. Alaoui-Sossé et al. (2004) investigated the effect of Cu on ion concentrations and growth in cucumber (*Cucumis sativus*) and showed that Cu inhibits leaf expansion and reduces the net assimilation rate, which may be due to decreased levels of K and Mg, respectively.

2.3.8 Zinc

Zinc (Zn) is essential for plant growth; however, it is phytotoxic at elevated levels. Extracellular Zn stress can disrupt the nutrient balance in plant cells. Jiang et al. (2007) showed that external application of P effectively protects plants from Zn toxicity by forming P–Zn complexes. Thus, the Zn-induced decrease in P content might enhance Zn toxicity in plants. In general, earlier studies showed that Zn exposure decreases nutrient content (Wang et al. 2009; Bonnet et al. 2000) and suggested that excess Zn might competitively inhibit the uptake of these elements. Since excess Zn kills root cells (Chang et al. 2005), injured roots might have a reduced capability to assimilate nutrients.

2.4 Herbicide Stress

In modern agriculture, herbicides are widely used to control weeds. Although herbicides are generally used in small amounts, they are potent, and, consequently, they have potentially significant risks in aquatic ecosystems. In addition, there are many different kinds of herbicide that are used in different areas. Furthermore, intensive herbicide use has caused significant soil and water pollution. In addition to their desired effects on target organisms, herbicides also have undesirable effects on non-target organisms (Duman et al. 2010). For example, Pandey et al. (2005) showed that hydroquinone, which is a phytotoxin, disrupts the cellular membrane integrity of *Chara zeylanica*, which is a non-target organism.

The severity of the toxic effects of pesticides can be mitigated by nutrients in the local environment. For example, Battah et al. (2001) showed that high concentrations of phosphate alleviated the toxic effects of thiobencarb (*S*-4-chlorobenzyl diethyl (thiocarbamate)) on the growth and photosynthetic activity of *Anabaena variabilis*. Conversely, Das and Debnath (2006) showed that herbicides stimulate the growth and activity of aerobic non-symbiotic N_2-fixing bacteria and increase the amount of N and P in the rhizosphere. Qian et al. (2009) also reported that exogenous nitric oxide protects *Chlorella vulgaris* from the toxicity of herbicides by reducing the damaging effects of oxidants and increasing the transcription of related genes. In addition, the duration of herbicide treatment is another important factor affecting their toxicity. For instance, Pandey et al. (2005) demonstrated that the N, P, and K content of plants decreases as the duration of herbicide treatment increases.

Previous studies showed that the application of high concentrations of herbicides to plants may harm their cell membrane integrity (Duman et al. 2010; Wendt-Rasch et al. 2003). Specifically, reactive oxygen species may disrupt the function and integrity of the cell membrane and cause irreparable damages to cellular functions (Nemat-Alla and Hassan 2006). As a result, significant ion losses can occur. Sinha (2002) studied the effect of hexachlorocyclohexane (HCH) alone and in combination with Fe on the cellular integrity of *Hydrilla verticillata*. Increasing concentrations of HCH increased K^+ leakage; however, this leakage was lessened in the presence of Fe. In addition, the accumulation of HCH decreased in the presence of high concentrations of Fe, while the accumulation of Fe increased in the presence of high concentrations of HCH.

The interactions between herbicides and metals in soil are very complex. For example, the amount of dissolved organic carbon compounds in growth media affects these interactions. The presence or absence of herbicides in soil can affect the uptake of mineral nutrients. For example, Chen et al. (2004) showed that 2,4-dichlorophenol (2,4-DCP) increases the concentration of water-soluble Cu and Zn. Teisseire et al. (1999) investigated a synergistic effect between Diuron (3-(3,4-dichlorophenyl)-1,1-dimethylurea) and Cu and revealed that the combination may prevent the toxic effects of Diuron. Some herbicides, such as glyphosate can create complex with divalent ions, such as Cu^{2+} (Sheals et al. 2003), Zn^{2+}, Mn^{2+}, Ca^{2+}, and Mg^{2+}. Therefore, herbicides may decrease the availability of nutrients. In addition, Rengel and Wheal (1997) demonstrated that chlorsulfuron decreases the uptake of micronutrients (Zn, Cu, and Mn) in wheat genotypes. Similarly, Azmat et al. (2006) showed that atrazine decreases the levels of Na and K in *Vigna radita*.

3 Conclusion and Future Perspective

Plant cultivation and consumption are important for human survival. However, water and soil resources around the world are increasingly contaminated by xenobiotic compounds. For example, in many parts of the world, soil is becoming increasingly arid as a result of many factors, such as wild irrigation and misuse of fertilizers. Consequently, hunger is a significant global problem. Every year, 15 million children die of hunger. It may be possible to solve, or at least alleviate, this problem by using soil and water resources more effectively. Unless plants are able to obtain sufficient nutrients and water from their growth media, they cannot survive. In general, environmental stresses negatively affect plant growth. Consequently, understanding the interaction between plant nutrients and stressors is critical. Currently, there is ample knowledge about the interactions between stressors and mineral nutrients; however, information about the underlying molecular mechanisms and genetic basis of these interactions is lacking. Thus, future studies should focus on these aspects.

References

Ahmad MSA, Ashraf M, Hussain M (2011) Phytotoxic effects of nickel on yield and concentration of macro- and micro-nutrients in sunflower (*Helianthus annuus* L.) achenes. J Hazard Mater 185:1295–1303

Alaoui-Sossé B, Genet P, Vinit-Dunand F, Toussaint ML, Epron D, Badot PM (2004) Effect of copper on growth in cucumber plants (*Cucumis sativus*) and its relationships with carbohydrate accumulation and changes in ion contents. Plant Sci 166:1213–1218

Albrizio R, Todorovic M, Matic T, Stellacci AM (2010) Comparing the interactive effects of water and nitrogen on durum wheat and barley grown in a Mediterranean environment. Field Crops Res 115:179–190

Ali MA, Ashraf M, Athar HR (2009) Influence of nickel stress on growth and some important physiological/biochemical attributes in some diverse canola (*Brassica napus* L.) cultivars. J Hazard Mater 172 (2–3): 964–969

Arduini I, Masoni A, Ercoli L (2006) Effects of high chromium applications on miscanthus during the period of maximum growth. Environ Exp Bot 58:234–243

Assche FV, Clijsters H (1990) Effects of metals on enzyme activity in plants. Plant Cell Environ 13:195–206

Astolfi S, Zuchi S, Passera C (2004) Role of sulphur availability on cadmium-induced changes of nitrogen and sulphur metabolism in maize (*Zea mays* L.) leaves. J Plant Physiol 161:795–802

Azmat R, Perveen R, Naqvi II (2006) Effect of chromium combined with atrazine on potassium, sodium, manganese, iron and phosphate in roots and shoots in bean *Vigna radita* (L.) Wilczek. J Saudi Chem Soc 11:111–120

Baccouh S, Chaoui A, El Ferjani E (2001) Nickel toxicity induces oxidative damage in *Zea mays* roots. J Plant Nutr 24:1085–1097

Barceló J, Poschenrieder C, Vázquez MD, Gunsé B (1993) Beneficial and toxic effects of chromium in plants: solution culture, pot and field studies in environmental contamination. In: Vernet JP (ed) Studies in environmental science. Elsevier, Amsterdam, pp 147–171

Barhoumi Z, Djebali W, Smaoui A, Chaïbi W, Abdelly C (2007) Contribution of NaCl excretion to salt resistance of *Aeluropus littoralis* (Willd) Parl. J Plant Physiol 164:842–850

Battah MG, Shabana EF, Kobbia IA, Eladel HM (2001) Differential effects of thiobencarb toxicity on growth and photosynthesis of *Anabaena variabilis* with changes in phosphate level. Ecotox Environ Safe 49:235–239

Berkowitz G, Zhang X, Mercier R, Leng Q, Lawton M (2000) Co-expression of calcium-dependent protein kinase with the inward rectified guard cell K+channel KAT1 alters current parameters in *Xenopus laevis* oocytes. Plant Cell Physiol 41:785–790

Bonnet M, Camares O, Veisseire P (2000) Effects of zinc and influence of *Acremonium lolii* on growth parameters, chlorophyll a fluorescence and antioxidant enzyme activities of ryegrass (*Lolium perenne* L. cn Apollo). J Exp Bot 51:945–953

Bouazizi H, Joudaouili H, Geitmann A, El Ferjani E (2010) Copper toxicity in expanding leaves of *Phaseolus vulgaris* L.: antioxidant enzyme response and nutrient element uptake. Ecotox Environ Safe 73:1304–1308

Cambrollé J, Mateos-Naranjo E, Redondo-Gómez S, Luque T, Growth FME (2011) Reproductive and photosynthetic responses to copper in the yellow-horned poppy. Glaucium flavum Crantz. Environ Exp Bot 71:57–64

Carbonell-Barrachina AA, Burló F, López E, Mataix J (1998a) Tomato plant nutrition as affected by arsenic concentration. J Plant Nutr 21:235–244

Carbonell-Barrachina AA, Aarabi MA, DeLaune RD, Gambrell RP, Patrick WH Jr (1998b) The influence of arsenic chemical form and concentration on *Spartina patens* and *Spartina alterniflora* growth and tissue arsenic concentration. Plant Soil 198:33–43

Cerda A, Martinez V (1988) Nitrogen fertilization under saline conditions in tomato and cucumber plants. J Hortic Sci 63:451–458

Champagnol F (1979) Relationships between phosphate nutrition of plants and salt toxicity. Phosphorus Agri 76:35–43

Chang HB, Lin CW, Huang HJ (2005) Zinc-induced cell death in rice (*Oryza sativa* L.) roots. Plant Growth Regul 46:261–266

Chen YX, Lin Q, He YF, Tian GM (2004) Behavior of Cu and Zn under combined pollution of 2,4-dichlorophenol in the planted soil Department of Environmental Engineering, Zheji. Plant Soil 261:127–134

Cohen CK, Fox TC, Garvin DF, Kochian LV (1998) The role of iron-deficiency stress responses in stimulating heavy-metal transport in plants. Plant Physiol 116: 1063–1072

Cui Y, Zhang X, Zhu Y (2008) Does copper reduce cadmium uptake by different rice genotypes? J Environ Sci 20:332–338

Das AC, Debnath A (2006) Effect of systemic herbicides on N_2-fixing and phosphate solubilizing microorganisms in relation to availability of nitrogen and phosphorus in paddy soils of West Bengal. Chemosphere 65:1082–1086

Dguimi HM, Debouba M, Ghorbel MH, Gouia H (2009) Tissue-specific cadmium accumulation and its effects on nitrogen metabolism in tobacco (*Nicotiana tabaccum*, Bureley v. Fb9). CR Biol 332:58–68

Diaz-Aguilar MV, Larque-Saavedra G, Alcantar-Gonzalez R, Carrillo-Gonzalez R, Vazquez-Alarcon A (2001) Alteration of some physiological processes in wheat by lead additions. Revista Int de Contam Ambiental 17:79–90

Disante KB, Fuentes D, Cortina J (2011) Response to drought of Zn-stressed *Quercus suber* L. seedlings. Environ Exp Bot 70:96–103

Dong J, Wu F, Zhang G (2006) Influence of cadmium on antioxidant capacity and four microelement concentrations in tomato seedlings (*Lycopersicon esculentum*). Chemosphere 64:1659–1666

Dube BK, Tewari K, Chatterjee J, Chatterjee C (2003) Excess chromium alters uptake and translocation of certain nutrients in citrullus. Chemosphere 53: 1147–1153

Duman F, Ozturk F (2010) Nickel accumulation and its effect on biomass, protein content and antioxidative enzymes in roots and leaves of watercress (*Nasturtium officinale* R. Br.). J Environ Sci 22(4):1–7

Duman F, Urey E, Temizgul R, Bozok F (2010) Biological responses of a non-target aquatic plant (*Nasturtium officinale* R.Br.) to the herbicide tribenuron-methyl. Weed Biol Manag 10:81–90

Fan JL, Hu ZY, Ziadi N, Xia X, Wu CYH (2010) Excessive sulfur supply reduces cadmium accumulation in brown rice (*Oryza sativa* L.). Environ Pollut 158:409–415

Ferreira RG, Tavora FJAF, Hernandez FFF (2001) Dry matter partitioning and mineral composition of roots, stems and leaves of guava grown under salt stress conditions. Pesqui Agropecu Bras 36:79–88

Fritioff A, Kautsky L, Greger M (2005) Influence of temperature and salinity on heavy metal uptake by submersed plants. Environ Pollut 133:265–274

Gadallah MAA (2000) Effects of indole-3-acetic acid and zinc on the growth, osmotic potential and soluble carbon and nitrogen components of soybean plants growing under water deficit. J Arid Environ 44:451–467

Gajewska E, Sklodowska M (2007) Effect of on ROS content and antioxidative enzyme activities in wheat leaves. Biometals 20:27–36

Geebelen W, Vangronsveld J, Adriano DC, Van Poucke LC, Clijsters H (2002) Effects of Pb-EDTA and EDTA on oxidative stress reactions and mineral uptake in *Phaseolus vulgaris*. Physiol Plant 115:377–384

Ghnaya T, Nouairi I, Slama I, Messedi D, Grignon C, Abdelly C, Ghorbel MH (2005) Cadmium effects on growth and mineral nutrition of two halophytes: *Sesuvium portulacastrum* and *Mesembryanthemum crystallinum*. J Plant Physiol 162:1133–1140

Gonçalves JF, Antes FG, Maldaner J, Pereira LB, Tabaldi LA, Rauber R, Rossato LV, Bisognin DA, Dressler VL, Flores ME, Nicoloso FT (2009) Cadmium and mineral nutrient accumulation in potato plantlets grown under cadmium stress in two different experimental culture conditions. Plant Physiol Biochem 47:814–821

Göthberg A, Greger M, Holm K, Bergtsson BE (2004) Influence of nutrient levels on uptake and effects of mercury, cadmium and lead in water spinach. J Environ Qual 33:1247–1255

Grattan SR, Grieve CM (1998) Salinity mineral nutrient relations in horticultural crops. Sci Hortic 78:127–157

Grattan SR, Grieve CM (1999) Mineral nutrient acquisition and response by plants grown in saline environments. In: Pessarakli M (ed) Handbook of plant and crop stress. Marcel Dekker, New York, pp 203–226

Greger M, Kautsky L, Sandberg TA (1995) Tentative model of Cd uptake in *Potamogeton pectinatus* in relation to salinity. Environ Exp Bot 35:215–225

Guardia MD, Benlloch M (1980) Effects of potassium and giberellic acid on stem growth of whole sunflower plants. Physiol Plant 49:443–448

Guerinot ML (2000) The ZIP family of metal transporters. Biochim Biophys Acta 1465:190–198

Guida dos Santos M, Ribeiro RV, Ferraz de Oliveira R, Machado EC, Pimente C (2006) The role of inorganic phosphate on photosynthesis recovery of common bean after a mild water deficit. Plant Sci 170:659–664

Gupta M, Chandra P (1998) Bioaccumulation and toxicity of mercury in rooted-submerged macrophyte *Vallisneria spiralis*. Environ Pollut 103:327–332

Hänsch R, Mendel RR (2009) Physiological functions of mineral micronutrients (Cu, Zn, Mn, Fe, Ni, Mo, B, Cl). Curr Opin Plant Biol 12:259–266

Hasegawa PM, Bressan RA, Zhu JK, Bohnert HJ (2000) Plant cellular and molecular responses to high salinity. Annu Rev Plant Physiol Plant Mol Biol 51:463–499

Hsiao TC, Läuchli A (1986) Role of potassium in plant-water relations. In: Tinker B, Läuchli A (eds) Advances in plant nutrition. Preager Scientific, New York, pp 281–312

Hu Y, Burucs Z, Tucher SV, Schmidhalter U (2007) Short-term effects of drought and salinity on mineral nutrient distribution along growing leaves of maize seedlings. Environ Exp Bot 60:268–275

Huang H, Xiong Z (2009) Toxic effects of cadmium, acetochlor and bensulfuron-methyl on nitrogen metabolism and plant growth in rice seedlings. Pestic Biochem Phys 94:64–67

Jaleel CA, Manivannan P, Wahid A, Farooq M, Al-Juburi HJ, Somasundaram R, Panneerselvam R (2009) Drought stress in plants: a review on morphological characteristics and pigments composition. Int J Agr Biol 11:100–105

Jiang HM, Yang JC, Zhang JF (2007) Effects of external phosphorus on the cell ultrastructure and the chlorophyll content of maize under cadmium and zinc stress. Environ Pollut 147:750–756

Kadukova J, Kalogerakis N (2007) Lead accumulation from non-saline and saline environment by *Tamarix smyrnensis* Bunge. Eur J Soil Biol 43:216–223

Kähkönen MA, Kairesalo T (1998) The effects of nickel on the nutrient fluxes and on the growth of *Elodea Canadensis*. Chemosphere 37:1521–1530

Kannan S, Keppel H (1976) Absorption and transport of Pb^{2+} in young pea seedlings. Z Naturforsch 31:393–396

Karadjova IB, Slaveykova VI, Tsalev DL (2008) The biouptake and toxicity of arsenic species on the green microalga *Chlorella salina* in seawater. Aquat Toxicol 87:264–271

Kaya C, Ak BE, Higgs D, Murillo-Amador B (2002) Influence of foliar applied calcium nitrate on strawberry plants grown under salt stress conditions. Aust J Exp Agric 42:631–636

Ke W, Xiong Z, Chen S, Chen J (2007) Effect of copper and mineral nutrition on growth, copper accumulation and mineral element uptake in two *Rumex japonicus* populations from a copper mine and an uncontaminated filed sites. Environ Exp Bot 59:59–67

Kelly CK, Horning K (1999) Acquisition order and resource value in *Cuscuta attenuate*. Proc Natl Acad Sci USA 96:13219–13222

Khan MA (2001) Experimental assessment of salinity tolerance of *Ceriops tagal* seedlings and saplings from the Indus delta. Pakistan Aquat Bot 70:259–268

Kibe AM, Singh S, Kalra N (2006) Water-nitrogen relationships for wheat growth and productivity in late sown conditions. Agric Water Manage 84:221–228

Kumar S, Joshi UN (2008) Nitrogen metabolism as affected by hexavalent chromium in sorghum (*Sorghum bicolor* L.). Environ Exp Bot 64:135–144

Leblebici Z, Aksoy A, Duman F (2011) Influence of salinity on the growth and heavy metal accumulation capacity of *Spirodela polyrrhiza* (Lemnaceae). Turk J Biol 35: 215–220

Lee TM, Liu CH (1999) Correlation of decreased calcium contents with proline accumulation in the marine green macroalga *Ulva fasciata* exposed to elevated NaCl contents in seawater. J Exp Bot 50:1855–1862

Liu J, Li K, Xu J, Liang J, Lu X, Yang J, Zhu Q (2003) Interaction of Cd and five mineral nutrients for uptake and accumulation in different rice cultivars and genotypes. Field Crops Res 83:271–281

Loupassaki MH, Chartzoulakis KS, Digalaki NB, Androulakis II (2002) Effects of salt stress on concentration of nitrogen, phosphorus, potassium, calcium, magnesium, and sodium in leaves, shoots, and roots of six olive cultivars. J Plant Nutr 25:2457–2482

Ma Y, Song W, Liu Z, Zhang H, Guo H, Shao H, Ni F (2009) The dynamic changing of Ca^{2+} cellular localization in maize leaflets under drought stress. CR Biol 332:351–362

Mahouachi J (2009) Changes in nutrient concentrations and leaf gas exchange parameters in banana plantlets under gradual soil moisture depletion. Sci Hort 120:460–466

Mamboya F, Lyimo TJ, Landberg T, Björk M (2009) Influence of combined changes in salinity and copper modulation on growth and copper uptake in the tropical green macroalga *Ulva reticulate*. Estuar Coast Shelfs 84(3):326–330

Manousaki E, Kadukova J, Papadantonakis N (2008) Phytoextraction and phytoexcretion of Cd by the leaves of *Tamarix smyrnensis* growing on contaminated non-saline and saline soils. Environ Res 106:326–332

Marschner H (1995) Mineral nutrition of higher plants. Academic, London

Martínez JP, Kinet JM, Bajji M, Lutts S (2005) NaCl alleviates polyethylene glycol-induced water stress in the halophyte species*Atriplex halimus*L. J Exp Bot 56:2421–2131

Martínez JP, Lutts S, Schanck A, Bajji M, Kinet JM (2004) Is osmotic adjustment required for water-stress resistance in the Mediterranean shrub *Atriplex halimus* L.? J Plant Physiol 161:1041–1051

Martinez V, Cerdá A (1989) Influence of N source on rate of Cl, N, Na, and K uptake by cucumber seedlings grown in saline conditions. J Plant Nutr 12:971–983

Mazen AMA (2004) Accumulation of four metals in tissues of *Corchorus olitorius* and possible mechanisms of their tolerance. Biol Plant 48:267–272

Meharg AA, Hartley-Whitaker J (2002) Arsenic uptake and metabolism in arsenic resistant and nonresistant plant species. New Phytol 154:29–43

Meharg AA, Macnair MR (1990) An altered phosphate uptake system arsenate-tolerant *Holcus lanatus* L. New Phytol 116:29–35

Mengel K, Arneke WW (1982) Effects of potassium on the water potential, the osmotic potential and cell elongation in leaves of *Phaseolus vulgaris*. Physiol Plant 54:402–408

Merchant S (2006) Trace metal utilization in chloroplasts. In: Wise RR, Hoober JK (eds) The structure and function of plastids. Springer, Dordrecht, pp 199–218

Mkandawire M, Dudel EG (2005) Accumulation of arsenic in *Lemna gibba* L. (duckweed) in tailing waters of two abandoned uranium mining sites in Saxony, Germany. Sci Total Environ 336:81–89

Mkandawire M, Lyubun YV, Kosterin PV, Dudel EG (2004) Toxicity of arsenic species to *Lemna gibba* L. and the influence of phosphate on arsenic bioavailability. Environ Toxicol 19:26–34

Mohsenzadeh S, Malboobi MA, Razavi K, Farrahi-Aschtiani S (2006) Physiological and molecular responses of *Aeluropus lagopoides* (Poaceae) to water deficit. Environ Exp Bot 56(3):314–322

Moral R, Gomez I, Pedreno JN, Mataix J (1996) Absorption of Cr and effects on micronutrient content in tomato plant (*Lycopersicum esculentum* M.). Agrochimica 40:132–138

Moreno-Jiménez E, Gamarra R, Carpena-Ruiz RO, Millán R, Peñalosa JM, Esteban E (2006) Mercury bioaccumulation and phytotoxicity in two wild plant species of Almadén area. Chemosphere 63:1969–1973

Munda IM, Hudnik V (1988) The effects of Zn, Mn, and Co accumulation on growth and chemical composition of *Fucus vesiculosus* L. under different temperature and salinity conditions. Mar Ecol 9:213–225

Nada E, Ferjani BA, Ali R, Bechir BR, Imed M, Makki B (2007) Cadmium-induced growth inhibition and alteration of biochemical parameters in almond seedlings grown in solution culture. Acta Physiol Plant 29: 57–62

Nemat Alla MM, Hassan NM (2006) Changes of antioxidants levels in two maize lines following atrazine treatments. Plant Physiol Biochem 44:202–210

Nenova V, Stoyanov I (1999) Physiological and biochemical changes in young maize plants under iron deffi.ciency. 3. Concentration and distribution of some nutrient elements. J Plant Nutr 22:565–578

Nocito FF, Pirovano L, Cocucci M, Sacchi GA (2002) Cadmium-induced sulfate uptake in maize roots. Plant Physiol 129:1872–1879

Paivoke AEA (2002) Soil lead alters phytase activity and mineral nutrient balance of *Pisum sativum*. Environ Exp Bot 48:61–73

Pandey DK, Mishra N, Singh P (2005) Relative phytotoxicity of hydroquinone on rice (*Oryza sativa* L.) and associated aquatic weed green musk chara (*Chara zeylanica* Willd.). Pestic Biochem Physiol 83:82–96

Parida AK, Das AB (2005) Salt tolerance and salinity effects on plants: a review. Ecotoxicol Environ Safe 60:324–49

Patel AD, Bhensdadia H, Pandey AN (2009) Effect of salinisation of soil on growth, water status and general nutrient accumulation in seedlings of *Delonix regia* (Fabaceae). Acta Ecol Sin 29:109–115

Patel NT, Vaghela PM, Patel AD, Pandey AM (2011) Implications of calcium nutrition on the response of *Caesalpinia crista* (Fabaceae) to soil salinity. Acta Ecol Sin 31:24–30

Patra M, Bhowmik N, Bandopadhyay B, Sharma A (2004) Comparison of mercury, lead and arsenic with respect to genotoxic effects on plants systems and the development of genetic tolerance. Environ Exp Bot 52:199–223

Payne WA, Hossner LR, Onken AB, Wendt CW (1995) Nitrogen and phosphorus uptake in pearlmillet and its relation to nutrient and transpiration efficiency. Agro J 87:425–431

Pieters AJ, Paul MJ, Lawlor DW (2001) Low sink demands limits photosynthesis under Pi deficiency. J Exp Bot 52:1083–1091

Puig S, Andrés-Colás N, García-Molina A, Peñarrubia L (2007) Copper and iron homeostasis in *Arabidopsis*: responses to metal deficiencies, interactions and biotechnological applications. Plant Cell Environ Mar 30: 271–90

Qian H, Chen W, Li J, Wang J, Zhou Z, Liu W, Fu Z (2009) The effect of exogenous nitric oxide on alleviating herbicide damage in Chlorella vulgaris. Aquat Toxicol 92:250–257

Rahman H, Sabreen S, Alam S, Kawai S (2005) Effects of nickel on growth and composition of metal micronutrients in barley plants grown in nutrient solution. J Plant Nutr 28:393–404

Rengel Z, Wheal MS (1997) Herbicide chlorsulfuron decreases growth of fine roots and micronutrient uptake in wheat genotypes. J Exp Bot 48:927–934

Roberts SK (1998) Regulation of K^+ channels in maize roots by water stress and abscisic acid. Plant Physiol 116:145–153

Rodríguez E, José R, Peralta-Videa JR, Israr M, Shivendra V, Pelayo H, Sánchez-Salcido B, Jorge L (2009) Effect of mercury and gold on growth, nutrient uptake, and anatomical changes in *Chilopsis linearis*. Environ Exp Bot 65:253–262

Ruiz D, Martínez V, Cerdá A (1997) Citrus response to salinity: growth and nutrient uptake. Tree Physiol 17: 141–150

Shanker AK, Cervantes C, Loza-Tavera H, Avudainayagam S (2005) Chromium toxicity in plants. Environ Int 31:739–753

Shao HB, Chu LY, Jaleel CA, Zhao CX (2008) Water-deficit stress-induced anatomical changes in higher plants. CR Biol 331:215–225

Sheals J, Granström M, Sjöberg S, Persson P (2003) Coadsorption of Cu(II) and glyphosate at the water-goethite (α-FeOOH) interface: molecular structures from FTIR and EXAFS measurements. J Colloid Interf Sci 262:38–47

Shen Z, Shen Q, Liang Y, Liu Y (1994) Effect of nitrogen on the growth and photosynthetic activity of salt-stressed barley. J Plant Nutr 17:787–789

Siddiqui MH, Mohammad F, Khan MN, Al-Whaibi MH, Bahkali AHA (2010) Nitrogen in relation to photosynthetic capacity and accumulation of osmoprotectant and nutrients in brassica genotypes grown under salt stress. Agr Sci China 9:671–680

Sinha P, Dube BK, Srivastava P, Chatterjee C (2006) Alteration in uptake and translocation of essential nutrients in cabbage by excess lead. Chemosphere 65:651–656

Sinha S (2002) Oxidative stress induced by iron in *Hydrilla verticillta* (l.f.) royle: modulation in uptake and toxicity due to iron. Chemosphere 46:281–288

Slama I, Ghnaya T, Savouré A, Abdelly C (2008) Combined effects of long-term salinity and soil drying on growth, water relations, nutrient status and proline accumulation of*Sesuvium portulacastrum*. CR Biol 331:442–451

Song CJ, Ma KM, Qu LY, Liu Y, Xu XL, Fu BJ, Zhong JF (2010) Interactive effects of water, nitrogen and phosphorus on the growth, biomass partitioning and water-use efficiency of *Bauhinia faberi* seedlings. J Arid Environ 74:1003–1012

Sucre B, Suárez N (2011) Effect of salinity and PEG-induced water stress on water status, gas exchange, solute accumulation, and leaf growth in *Ipomoea pes-caprae* Environ. Exp Bot 70:192–203

Teisseire H, Couderchet M, Verne G (1999) Phytotoxicity of diuron alone and in combination with copper or folpet on duckweed (*Lemna minor*). Environ Pollut 106:39–45

Tezara W, Mitchell VJ, Driscol SD, Lawlor DW (1999) Water stress zinhibits plant photosynthesis by decreasing coupling factors and ATP. Nature 401:914–917

Thiec DL, Manninen S (2003) Ozone and water deficit reduced growth of Aleppo pine seedlings. Plant Physiol Biochem 41:55–63

Tipping E, Smith EJ, Lawlor AJ, Hughes S, Stevens PA (2003) Predicting the release of metals from ombrotrophic peat due to drought-induced acidification. Environ Pollut 123:239–253

Tripathi RD, Srivastava S, Mishra S, Singh N, Tuli R, Gupta DK, Maathuis FJM (2007) Arsenic hazards: strategies for tolerance and remediation by plants. Trends Biotechnol 25:158–165

Tu C, Ma LQ (2005) Effects of arsenic on concentration and distribution of nutrients in the fronds of the arsenic hyperaccumulator *Pteris vittata* L. Environ Pollut 135:333–340

Tuna AL, Kaya C, Ashraf M, Altunlu H, Yokas I, Yagmur B (2007) The effects of calcium sulphate on growth, membrane stability and nutrient uptake of tomato plants grown under salt stress. Environ Exp Bot 59:173–178

Vajpayee P, Sharma SC, Tripathi RD, Rai UN, Yunus M (1999) Bioaccumulation of chromium and toxicity to photosynthetic pigments, nitrate reductase activity and protein content of *Nelumbo nucifera* gaertin. Chemosphere 39:2159–2169

Vance CP, Uhde-Stone C, Allan DL (2003) Phosphorus acquisition and use: critical adaptations by plants for securing a nonrenewable resources. New Phytol 157:423–447

Violante A, Pigna M, Ricciardella M, Gianfreda L (2002) Adsorption of phosphate on variable charge minerals and soils as affected by organic and inorganic ligands. Dev Soil Sci 28:279–295

Wang C, Zhang SH, Wang PF, Hou J, Zhang WJ (2009) The effect of excess Zn on mineral nutrition and antioxidative response in rapeseed seedlings. Chemosphere 75:1468–1476

Wang WX, Dei RCH (1999) Kinetic measurements of metal accumulation in two marine macroalgae. Mar Biol 135:11–23

Wendt-Rasch L, Pirzadeh P, Woin P (2003) Effects of metsulfuron methyl and cypermethrin exposure on freshwater model ecosystems. Aquat Toxicol 63:243–256

Wong MK, Chuah GK, Koh LL, Ang KP, Hew CS (1984) The uptake of cadmium by *Brassica chinensis* and its effect on plant zinc and iron distribution. Environ Exp Bot 119:33–44

Xiong ZT, Li YH, Xu B (2002) Nutrition influence on copper accumulation by *Brassica pekinensis* Rupr. Ecotox Environ Safe 53:200–205

Xiong ZT, Liu C, Geng B (2006) Phytotoxic effects of copper on nitrogen metabolism and plant growth in *Brassica pekinensis* Rupr. Ecotoxicol Environ Saf 64:273–280

Yang XE, Romheld V (2002) Growth and nutrient composition of *Elsholtzia splendens* Nakai under copper toxicity. Plant Nutr 25:1359–1375

Yang YY, Lee Y (2002) Pb and Cd uptake in rice roots. Physiol Plant 116:368–372

Yang Z, Zheng S, Zhao X, Hu A (1999) Effect of phosphorus on the accumulation of cadmium and zinc in suspension-cultured wheat cells. Acta Sci Circum 19:693–695

Yu Z, Zhou O (2009) Growth responses and cadmium accumulation of *Mirabilis jalapa* L. under interaction between cadmium and phosphorus. J Hazard Mater 167:38–43

Zayed AM, Terry N (2003) Chromium in the environment: factors affecting biological remediation. Plant Soil 249:139–156

Zhang G, Fukami M, Sekimoto H (2002) Influence of cadmium on mineral concentrations and yield components in wheat genotypes differing in Cd tolerance at seedling stage. Field Crops Res 7:93–98

Zhang Y, Wang Z, Zhang L, Cao Y, Huang D, Tang K (2006) Molecular cloning and stress-dependent regulation of potassium channel gene in Chinese cabbage (*Brassica rapa* ssp. pekinensis). J Plant Physiol 163:968–978

Zhu JK (2003) Regulation of ion homeostasis under salt stress. Curr Opin Plant Biol 6:441–445

Effect of Micronutrient Deficiencies on Plants Stress Responses

16

R. Hajiboland

Abstract

Low productivity because of limited mineral nutrient supply is common under various environmental conditions. Micronutritional disorders are common nutritional imbalance in plants and affect greatly plants performance and their response to surrounding environment. Micronutrients deficiencies exert secondary, often unpredicted influences on the growth of plants by changes in growth pattern, chemical composition, and antioxidant defense capacity of plants and particularly decrease the resistance of plants to biotic and abiotic environmental stresses. In this chapter, we discuss mechanisms involve in the intensification of damages due to environmental stress factors caused by micronutritional imbalances. Because of the significant effect of beneficial elements on plants growth and productivity under marginal environmental conditions, we present detailed information on how these elements alleviate plants stress injuries.

Keywords

Antioxidants • Beneficial elements • Environmental stresses • Micronutrients • Plant responses • Reactive oxygen species

1 Introduction

The effects of mineral nutrients on plant growth and productivity are usually explained in terms of the functions of these elements in plant metabolism. However, mineral nutrition may also exert secondary, often unpredicted influences on the growth of plants by changes in growth pattern, plant morphology and anatomy, and particularly chemical composition. Mineral nutrients may also increase or decrease the resistance or the tolerance of plants to biotic stress factors, e.g., pathogens and pests, and abiotic environmental stress factors, e.g., high light, UV radiation, drought, salinity, flooding, and chilling. Our knowledge on the effect of nutritional status on the response of plants to abiotic environmental stress factors is still inadequate.

Effect of a given nutrient on plants stress response could be expressed under deficiency

R. Hajiboland (✉)
Plant Science Department, University of Tabriz,
Tabriz 51666-14779, Iran
e-mail: ehsan@tabrizu.ac.ir

P. Ahmad and M.N.V. Prasad (eds.), *Abiotic Stress Responses in Plants: Metabolism, Productivity and Sustainability*, DOI 10.1007/978-1-4614-0634-1_16, © Springer Science+Business Media, LLC 2012

conditions of this nutrient. Macronutrients are mainly important structural components of plants and the effect of macronutrients deficiency on the increase of plants susceptibility to stress factors is easily explainable. For micronutrients, by contrast, change in the activity of enzymes and/or production of some metabolites involving in the plants response to their surrounding environment or even modulations in the signal transduction pathways brought about by deficiency conditions may influence directly or indirectly the susceptibility of plants to environmental stresses.

In this chapter, we present and discuss on the evidences showing effect of micronutrients deficiency on plants response to environmental stresses. First, we briefly describe the role of micronutrients in plants metabolism and then focus on the main mechanisms involving in the changes in plants response to stress under nutrients starvation. In this regard, we especially emphasize on the antioxidant defense system as influenced by micronutrients deficiency. Not only micronutritional status of plants affects plants responses to environmental stress factors but also environmental factors particularly those related to soil affect strongly availability, uptake, transport, and utilization of micronutrients. We deal with this subject briefly in this chapter. Finally, detailed information on the response of micronutrients-starved plants to environmental stresses is presented for every micronutrient according to the evidences provided by plant physiologists during the last decades. Because of the significant effect of some mineral elements that have been defined so far as beneficial elements, and because of emerging evidences for their functional roles in plants growth and productivity under marginal environmental conditions, we include these elements in this chapter.

2 Micronutrients in Plants

Supply of chemical compounds required for growth and metabolism is defined as nutrition and chemical compounds are needed by an organism termed nutrients. For an element to be described as essential three criteria must be met. These are as follows: (1) A given plant must be unable to complete its life cycle in the absence of the mineral element; (2) The function of the element must not be replaced by another mineral element; (3) The element must be directly involved in plant metabolism, e.g., as a component of an essential plant constituent such as an enzyme, or it must be required for a distinct metabolic step such as enzyme reaction (Marschner 1995). Based on these criteria, the following chemical elements are known to be essential for higher plants: nitrogen (N), sulfur (S), phosphorus (P), potassium (K), calcium (Ca), magnesium (Mg), zinc (Zn), iron (Fe), manganese (Mn), copper (Cu), molybdenum (Mo), boron (B), chlorine (Cl), and nickel (Ni). Plant nutrients have been divided into macronutrients and micronutrients. Macronutrients are needed in relatively higher amounts than micronutrients. Using this classification based on the element concentration in plant material, the following elements are defined as macronutrients: N, S, P, K, Ca, and Mg. The micronutrients are: Zn, Fe, Mn, Cu, Mo, B, Cl, and Ni. The main physiological functions of micronutrients are presented in Table 16.1.

3 Effect of Micronutrients Deficiency on Oxidative Stress in Plants

Increasing evidences have indicated that much of the injury to plants due to various environmental stresses is associated with oxidative damage through direct or indirect formation of reactive oxygen species (ROS) (Apel and Hirt 2004). Production and scavenging of ROS is tightly linked with the presence of micronutrients such as Zn, Fe, Mn, and Cu in plant tissues. These micronutrients are components of antioxidant defense enzymes and modulation in the activity of these enzymes in nutrient-deficient plants is well documented. In addition, deficiency of micronutrients may affect other physiological attributes of plants, such as electron transport, water relations, and gas exchange that could directly or indirectly influence ROS metabolism in plants.

Table 16.1 Physiological functions of micronutrients in higher plants

Nutrient	Physiological function
Zinc	Catalytic, cocatalytic, and structural role in more than 300 enzymes. Constituent of carbonic anhydrase, carboxypeptidase, alcohol dehydrogenase, CuZn superoxide dismutase, Zn-finger motif class of transcription factors, alkaline phosphatase (microorganisms), phospholipase, RNA polymerase, Zn-PPiase (tonoplast), fructose 1,6 bisphosphatase, aldolase. Role in the integrity of ribosomes. Role in protein, RNA, DNA, and carbohydrates metabolism. Role in membrane integrity and metabolism of reactive oxygen species.
Iron	Role in biological redox systems. Constituent of prosthetic group of heme proteins: cytochromes, leghemoglobin, catalase, peroxidases, ascorbate peroxidase, nitrogenase (microorganisms). Constituent of nonheme groups: ferredoxin, aconitase, Fe-superoxide dismutase, xanthine oxidase, ACC oxidase (ethylene forming enzyme), lipooxygenase. Role in chlorophyll synthesis: synthesis of δ-aminolevulinic acid, protoporphyrinogen, and protochlorophyllide. Role in chloroplast development, photosynthesis, and lignin biosynthesis.
Manganese	Role in biological redox systems. Constituent of enzymes: water splitting (oxygen evolving) complex of PSII and Mn-superoxide dismutase. Role in TCA cycle in oxidative and nonoxidative carboxylations. Role in the activity of NADPH-dependent malic enzyme, malate dehydrogenase, and isocitrate dehydrogenase. Role in the activity of PEP carboxykinase, phenylalanine ammonia lyase, peroxidases, IAA oxidase, phytoene synthase, allantoate amidohydrolases, arginase, RNA polymerase. Role in development of thylakoid membranes, lipids and carotenoids synthesis.
Copper	Role in biological redox systems. Constituent of oxidases: cytochrome oxidase, diamine oxidase, phenol oxidase, DOPA oxidase, tyrosinase, phenolase, polyphenol oxidase, laccase, plastocyanin, CuZn superoxide dismutase. Role in pollen formation and viability, pollination, desaturation of lipids, role in biosynthesis of lignin, quinones, carotenoids
Nickel	Constituent of urease, hydrogenases (microorganisms). Role in nitrogen metabolism.
Molybdenum	Role in biological redox systems. Constituent of nitrogeanse (microorganism), nitrate reductase, xanthine dehydrogenase (oxidase), sulfite oxidase. Role in metabolism of nitrogeneous compounds (purine, ureides, protein)
Boron	Constituent of cell wall in cross-links of rhamnogalacturonan II. Role in cell wall synthesis and cell extension, integrity of membranes, sugar transport, pollen germination and pollen tube elongation. Effective on the metabolism of phenolics, IAA, carbohydrates, proteins, RNA.
Chlorine	Role in water splitting (oxygen evolving) complex of PSII. Role in osmosis and water relations. Regulation of tonoplast proton pumps. Role in plant organs and stomata movements.

3.1 Factors Involve in Production of ROS

The photosynthetic electron transport system is the major source of active oxygen in plant tissues, having potential to generate singlet oxygen ($^1O_2$) and superoxide ($O_2^{\bullet -}$). The major oxygen-consuming processes associated with photosynthesis are the oxygenase reaction of ribulose-1, 5-bisphosphate carboxylase (Rubisco), which is the initiating reaction of the photorespiratory pathway, and direct reduction of molecular oxygen by the photosystem I (PSI) electron transport chain:

$$O_2 + PSI(\text{reduced}) \rightarrow O_2^{\bullet -} + PSI(\text{oxidized})$$

In addition, certain photosystem II (PSII) components are also capable of converting molecular O_2 to high-energy singlet oxygen. A cyanide-insensitive respiratory pathway in chloroplasts that competes for electrons with photosynthetic electron transport may also reduce oxygen (Asada 1999). Induced production of $O_2^{\bullet -}$ is also catalyzed by NAD (P) H-oxidizing enzyme systems localized in different cell compartments, such as cell walls, plasma membranes, cytosol and microsomes, peroxisomes, and mitochondria (Murphy and Auh 1996). High levels of Fe accumulation in plant cells are also responsible for the initiation of severe oxidative stress because they produce ROS by various cellular reactions (Becanne et al. 1998).

The reduced ferrous (Fe II) compounds can be oxidized causing production of H_2O_2 or $O_2^{\bullet-}$:

$$2Fe(II) + O_2 + 2H^+ \rightarrow 2Fe(III) + H_2O_2$$

$$Fe(II) + O_2 \rightarrow Fe(III) + O_2^{\bullet-}$$

The powerful oxidant hydroxyl radical ($^\bullet OH$) is produced by the oxidation of Fe (II) by H_2O_2 to $^\bullet OH$, which is known as the Fenton reaction:

$$H_2O_2 + Fe(II) \rightarrow {}^\bullet OH + OH^- + Fe(III)$$

Increased Fe accumulation has been reported for plants subjected to various stress conditions such as Zn deficiency, root anoxia, drought and light (Becanne et al. 1998). Under these stress conditions, accumulation of Fe is associated with enhanced lipid peroxidation and chlorophyll (Chl) damage. In the case of drought stress, chloroplast membranes enhance their production of $O_2^{\bullet-}$ in response to Fe accumulation, and Fe-catalyzed formation of O_2 radicals has been considered as a major factor contributing to drought damage in plant cells (Price and Hendry 1991). Plants grown under flooded conditions accumulate high levels of Fe (Sahrawat et al. 1996) and oxidation of Fe (II) leads to production of $O_2^{\bullet-}$, which is suggested as a cause of flooding damage and post-anoxic injury to plants (Neue et al. 1998).

3.2 Antioxidant Defense System

Oxidative stress is essentially a regulated process; the equilibrium between the oxidative and antioxidative capacities determines the fate of the plant. Plants possess very efficient scavenging systems for ROS that protect them from destructive oxidative reactions. Under nonstressful conditions the antioxidant defense system provides adequate protection against active oxygen and free radicals. Under stress conditions, however, an imbalance between production and scavenging of ROS causes oxidative damage (Apel and Hirt 2004).

3.2.1 Superoxide Dismutase

By catalyzing detoxification of $O_2^{\bullet-}$ to O_2 and H_2O_2 and blocking $O_2^{\bullet-}$ driven cell damage, SODs are a major component of the antioxidative defense system of plant cells (Bowler et al. 1994).

$$2O_2^{\bullet-} + 2H^+ \rightarrow H_2O_2 + O_2$$

In many cases, the protective effect of SOD is enhanced by the presence of H_2O_2 scavenging enzymes (see below), which prevent the production of hydroxyl radicals that may otherwise be formed in the presence of $O_2^{\bullet-}$ as reductant and an appropriate catalyst such as metal ions, quinones, and Ferredoxin via the Haber–Weiss reaction:

$$O_2^{\bullet-} + H_2O_2 + H^+ \rightarrow {}^\bullet OH + H_2O + O_2^{\bullet-}$$

According to their metal cofactor, SODs are classified into three types containing either Mn (Mn-SOD), Fe (Fe-SOD), or Cu and Zn (CuZn-SOD). In general, Mn-SOD is located in mitochondria, Fe-SOD in chloroplasts, and CuZn-SOD in chloroplasts and the cytosol (Bowler et al. 1994). Accordingly, all three isozymes of SOD have a metal cofactor, and under deficiency conditions decline in the activity of related isozymes is expected to occur. Activity of isoforms of SOD is useful for determining the micronutrient status of plants, for example Zn (Cakmak et al. 1997; Hajiboland 2000), Fe (Iturbe-Ormaetxe et al. 1995), and Mn (Yu et al. 1998) nutritional status.

3.2.2 H_2O_2 Scavenging Enzymes

Enzymes involved in the metabolism of H_2O_2, namely, catalase, peroxidases, and ascorbate peroxidase are heme proteins. Catalase (CAT) catalyzes the conversion of hydrogen peroxide into water and O_2:

$$2H_2O_2 \rightarrow 2H_2O + O_2$$

Peroxidases (POD) catalyze the conversion of hydrogen peroxide to water. For example, ascorbate peroxidase (APX) is involved in the

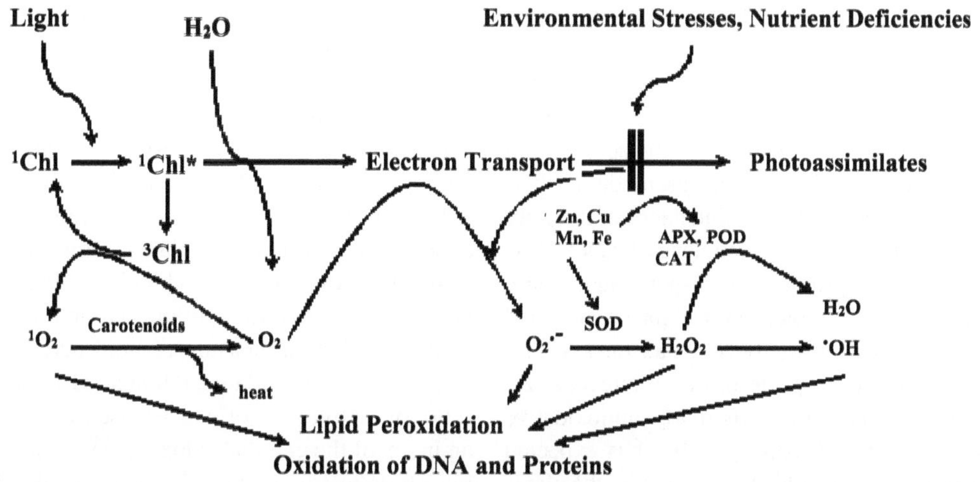

Fig. 16.1 Production and scavenging of reactive oxygen species (ROS) in plants. Light energy absorbed by chlorophyll (Chl) is converted to electrochemical potential, which oxidizes H_2O to O_2 and generates the electrons for CO_2 reduction and production of photoassimilates. Singlet oxygen ($^1O_2$) is formed in chloroplasts when photoexcited Chl in the triplet state (^{3}Chl) reacts with dioxygen (O_2). Carotenoids react with $^1O_2$ and dissipate excess light energy as heat. Production of $^1O_2$ and $O_2^{\bullet-}$ is inevitable under light conditions. Impairment in utilization of electrons and absorbed light energy for CO_2 fixation in nutrient-deficient plants or under environmental stresses increases possibility of ROS generation. $O_2^{\bullet-}$ is converted to hydrogen peroxide (H_2O_2) by superoxide dismutase (SOD) which contains Zn, Cu, Mn or Fe as metal component. In the case of reduction in the SOD activity under deficiency conditions, H_2O_2 produces hydroxyl radical (•OH), a highly reactive and potent ROS. Hydrogen peroxide is scavenged by Fe-containing ascorbate peroxidase (APX), peroxidases (POD) and catalase (CAT). Under Fe deficiency, scavenging reaction is impaired and H_2O_2 is accumulated. Reaction oxygen species reacts with membranes, DNA, and proteins and causes cell damage sand death

detoxification of H_2O_2 through conversion of ascorbate to dehydroascorbate:

$$H_2O_2 + AH_2 \rightarrow A + 2H_2O$$

Catalase is involved in the protection of chloroplasts from free radicals produced during the water-splitting reaction of photosynthesis. Activity of APX has mainly been reported from chloroplast and cytosol. In the chloroplasts, SOD and APX enzymes exist in both soluble and thylakoid-bound forms (Arora et al. 2002).

3.2.3 Antioxidant Metabolites

Antioxidant metabolites such as glutathione (GSH), α-tocopherol, carotenoids, and proline are also involved in scavenging ROS. A relationship between lipid peroxidation and proline accumulation has been reported in plants subjected to diverse kinds of stress (Molinari et al. 2007; Ashraf and Foolad 2007). Proline acts as a free radical scavenger to protect plants away from damage by oxidative stress (Wang et al. 2009) and is an important compatible osmosolute serves as a protectant for enzymes and cellular structures under osmotic and drought stress (Hasegawa et al. 2000; Banu et al. 2009). There are evidences to demonstrate that accumulation of proline under Zn (Hajiboland and Amirazad 2010b) and B (Hajiboland, unpublished data) deficiency conditions may play a role in the tolerance of deficient plants to the other environmental stresses such as drought.

Factors involved in the production of ROS, their reactions, and scavenging are presented schematically in Fig. 16.1.

4 Effect of Zinc Deficiency on Plants Stress Responses

There are approximately 300 enzymes in which Zn is an integral component of the enzyme structure. In these enzymes Zn has catalytic (e.g., carbonic anhydrase, carboxypeptidase, alkaline phosphatase, and phospholipase) or structural (e.g., alcohol dehydrogenase, CuZn-SOD, and RNA polymerase) functions. Besides Zn-containing enzymes, Zn is either required for the activity of a large number of enzymes, or at least modulate their activity. Zn is required for maintenance of integrity of biomembranes and Zn deficiency leads to loss of membrane integrity and corresponding increase in leakage of low molecular weight organic and inorganic solutes from roots (Marschner 1995). Regarding the wide spectrum of the effects of Zn on plants metabolism, particularly antioxidative defense of plants (Cakmak 2000), it is expected that Zn deficiency would affect markedly the plants responses to every kind of environmental stress factors.

4.1 Effect of Zn Deficiency on the Induction of Oxidative Stress

Zinc is the most important micronutrient that is involved both directly and indirectly in the metabolism of ROS as well as protection of structural components of cell against ROS. In leaves of different wheat and rice cultivars, deficient supply of Zn decreased total SOD activity and, more distinctly, CuZn-SOD activity, whereas Mn-SOD activity was not affected by Zn deficiency (Cakmak et al. 1997; Hajiboland 2000). Interestingly, change in the activity of CuZn-SOD could be used for interpreting different susceptibility of various wheat (Yu et al. 1999b) and rice (Hajiboland 2000) genotypes to Zn deficiency (Table 16.2). Zinc may indirectly be required for high activity of the enzymes involved in H_2O_2 detoxification. Reports on the effect of Zn deficiency on the activity of CAT, APX and glutathione reductase (GR) are contradictory. Some reports demonstrated increase (Tewari et al. 2008; Hajiboland and Beiramzadeh 2008) and others reduction of their activity under Zn starvation (Yu et al. 1998). Reduction of CAT activity with Zn deficiency was assumed to be related to inhibition of CAT by $O_2^{\bullet-}$. Because Zn deficiency strongly reduces protein synthesis, biosynthesis of H_2O_2 scavenging enzymes can be impaired as a result of Zn deficiency. Accordingly, in bean plants Zn deficiency reduced the activities of GR and APX in young leaves, although these enzymes do not need Zn for their activity (Cakmak 2000). Consequently, because of the reduced activity of enzymes scavenging $O_2^{\bullet-}$ and H_2O_2, enhanced production of $^{\bullet}OH$ (via the Haber–Weiss reaction) can be expected in Zn-deficient plants. However, concentrations of $^{\bullet}OH$ in Zn-deficient plants have not been reported. In addition, malondialdehyde content (MDA) a measure of lipid peroxidation as indication of oxidative damage, H_2O_2 and the ratio of dehydroascorbate to ascorbic acid as an index of the cellular redox state was increased in Zn-deficient plants (Tewari et al. 2008; Hajiboland and Beiramzadeh 2008). In plant cells, Zn interferes with NADPH-dependent $O_2^{\bullet-}$ generation. Activity of $O_2^{\bullet-}$-generating NADPH oxidase was higher under low Zn supply particularly in Zn-inefficient genotypes (Hajiboland 2000) (Table 16.2) and resupply of Zn to Zn-deficient plants rapidly decreased the generation of $O_2^{\bullet-}$ in plasma membrane vesicles (Pinton et al. 1994).

Increased Fe accumulation that is associated with enhanced lipid peroxidation and Chl damage, has been reported for plants subjected to Zn deficiency. Accumulation of Fe in Zn-deficient plants is particularly pronounced under high light intensity (Cakmak 2000). Such high levels of accumulation of Fe in plant tissues can be responsible for excessive production of toxic $O_2^{\bullet-}$ species and extensive cellular damage. In most cases, Fe and Zn compete for the binding sites on proteins and phospholipids. The existence of adequate amounts of Zn in the vicinity of Fe-binding sites blocks binding of Fe and thus inhibits Fe-catalyzed site-specific generation of $^{\bullet}OH$. Hydroxyl radicals produced at Fe binding sites on membranes can rapidly initiate peroxidation of phospholipids, causing site-specific damage (Bray and Bettger 1990).

4.2 Zinc Deficiency-Induced Changes in the Photochemistry of Leaves

For nutrients without direct involvement in the electron transport or Chl synthesis such as Zn,

Table 16.2 Dry matter production and activity of CuZnSOD, NADPH oxidase and generation rate of $O_2^{\bullet -}$ in Zn-efficient (IR34 and IR36) and Zn-inefficient (IR26 and IR54) rice genotypes grown in hydroponic medium with adequate (130 pM free Zn^{2+} activity) and low (9 pM free Zn^{2+} activity) Zn supply

	Dry weight (mg plant^{-1})		Activity of CuZnSOD (Unit mg^{-1} protein)		Activity of NADPH oxidase (nmol NADPH min^{-1} mg^{-1} protein)		$O_2^{\bullet -}$ Generation (nmol $O_2^{\bullet -}$ min^{-1} mg^{-1} protein)	
	Adequate Zn	Low Zn	Adequate Zn	Low Zn	Adequate Zn	Low Zn	Adequate Zn	Low Zn
	Leaves							
IR 34	166 ± 15^a	118 ± 10^b	351 ± 99^a	336 ± 88^a	0.85 ± 0.1^a	0.98 ± 0.1^a	0.42 ± 0.02^a	0.39 ± 0.02^a
IR 36	160 ± 12^a	106 ± 9^b	510 ± 13^a	483 ± 87^a	17.1 ± 4.1^b	24.6 ± 3.9^a	6.9 ± 3.4^a	11.0 ± 2.5^a
IR 26	149 ± 18^a	109 ± 15^b	378 ± 50^a	287 ± 54^b	4.8 ± 0.6^b	14.9 ± 0.2^a	4.0 ± 2.8^b	12.4 ± 0.5^a
IR 54	135 ± 11^a	52 ± 8^b	342 ± 72^a	233 ± 59^b	6.3 ± 2.6^b	17.2 ± 3.5^a	6.6 ± 0.2^b	8.9 ± 0.2^a
	Roots							
IR 34	36 ± 9^a	26 ± 6^a	248 ± 48^a	216 ± 67^a	18.8 ± 1.2^a	18.0 ± 1.9^a	5.6 ± 0.5^a	4.8 ± 0.8^a
IR 36	50 ± 9^a	30 ± 9^b	945 ± 61^a	660 ± 13^b	23.7 ± 1.6^a	16.1 ± 1.2^b	7.8 ± 0.3^a	7.3 ± 0.8^a
IR 26	32 ± 7^a	14 ± 6^b	414 ± 203^a	120 ± 21^b	13.8 ± 3.0^b	23.5 ± 1.3^a	5.2 ± 0.8^b	10.4 ± 1.3^a
IR 54	37 ± 5^a	9 ± 3^b	820 ± 135^a	556 ± 98^b	57.4 ± 4.0^b	83.7 ± 7.3^a	9.9 ± 1.2^b	21.0 ± 4.5^a

Data of two Zn supply levels within each genotype and parameter followed by the same letter are not significantly different ($P < 0.05$). Definition of Zn-efficient genotypes refers to a classification based on the ability of plants to grow under low Zn supply with less Zn deficiency symptoms and less reduction in dry matter production compared with Zn-inefficient genotypes (Hajiboland 2000)

a close linkage between nutritional status of leaves and spectral characteristics seems unlikely (Adams et al. 2000). However, there are reports on significant reduction of maximum quantum efficiency of PSII (Wang and Jin 2005) and severe damage to the ultrastructure of chloroplasts in plants subjected to inadequate Zn supply (Chen et al. 2007). Photooxidative damage in Zn-deficient leaves can be expected as a result of impaired photosynthetic CO_2 fixation and reduced activity of SOD. Reduction in photosynthesis induced by Zn deficiency is associated with a decrease in intercellular CO_2 concentration and stomatal conductance (Hajiboland and Amirazad 2010a). Sharma et al. (1995) reported a significant role of Zn in the regulation of the stomatal aperture. This role of Zn was ascribed to maintenance of a high K concentration in guard cells. A decrease in carbonic anhydrase activity due to Zn deficiency is well known (Hajiboland 2000), and may be a factor contributing to reduced photosynthesis (Cakmak and Engels 1999). Inhibition of photosynthesis in Zn-deficient plants can also be a consequence of a Zn-deficiency-induced reduction in phloem sap sink demand. Additionally, in Zn-deficient plants there is an enhanced accumulation of carbohydrates, possibly resulting from either impaired phloem export of carbohydrates or decreased sink demand (Marschner et al. 1996). Impairment in utilization of electrons and absorbed light energy for CO_2 fixation in Zn-deficient plants may accentuate photogeneration of ROS and photooxidative damage to chloroplasts (Fig. 16.1).

4.3 Zinc Deficiency-Enhanced Susceptibility to Excess Light

Enhancements in chlorosis and necrosis due to increased light intensity are very typical in Zn-deficient source leaves, reflected in a massive accumulation of sucrose and starch (Marschner and Cakmak 1989) causing a high potential for photooxidative damage of chloroplast constituents. In accordance with this suggestion, enhancements in light intensity markedly stimulated appearance of leaf chlorosis under Zn deficiency, but not at adequate Zn supply. Also, partial shading of Zn-deficient leaves prevented or strongly delayed appearance of chlorosis in the shaded areas (Cakmak 2000). Increased severity of leaf chlorosis under high light intensity in Zn-deficient conditions is not caused by lower Zn concentration in leaves but is a consequence of photooxidative damage to chloroplast pigments catalyzed by ROS. Photooxidative damage of the chloroplast constituents under Zn deficiency can also be aggravated by reduced activity of enzymes scavenging $O_2^{\bullet-}$ and H_2O_2 in chloroplasts (Hajiboland and Amirazad 2010b).

4.4 Zinc Deficiency-Induced Susceptibility to Drought Stress

It was reported that Zn deficiency is prone to occur in arid and semi-arid regions where soils, particularly top soil, are usually deficient in water (Cakmak et al. 1996). Under drought conditions, Zn mobility in the soil is extremely low, therefore, Zn uptake is usually reduced by low water availability in the substrate. In addition, strongly inhibited root growth in Zn-deficient plants reduces markedly the soil volume exploited by roots and impairs nutrients uptake particularly those are dependent more on spatial availability such as Zn (Marschner 1995). Within plants, there are also some interactions between Zn nutritional status and water relations. It was shown that the ability of plants to cope with water stress during early vegetative stage could be enhanced with adequate Zn supply (Grewal and Williams 2000). Sensitivity to Zn deficiency stress became more pronounced when plants were drought-stressed (Bagci et al. 2007). Impairment of growth in Zn-deficient plants was markedly higher when they were subjected to drought stress and in turn, the effect of drought stress on the inhibition of dry matter production was greater in Zn-deficient compared with Zn-sufficient plants (Hajiboland and Amirazad 2010b).

Because of the effect of Zn deficiency on increasing stomatal limitation (Sharma et al. 1995), plants under low Zn supply are more conservative in relation to water economy than sufficient plants when grown under drought stress as indicated by lower water loss, greater water and osmotic potential (Hajiboland and Amirazad 2010b). However, higher growth impairment under combinative effects of Zn deficiency and drought stress is due to damage to photosynthesis apparatus, greater ROS production and remarkable reduction of whole plant photosynthesis following stomatal limitation. In addition, under low Zn and drought stress only a small part of Zn taken up by plants is transported into leaves due to significantly lower stomatal opening and transpiration (Hajiboland and Amirazad 2010b).

4.5 Reduction of Plants Resistance to Flooding

It was reported that symptoms of Zn deficiency normally appear shortly after flooding (Van Breemen and Castro 1980). Flooding conditions may not only influence Zn availability in soil but also alter plants performance under Zn starvation. In soils, flooded conditions lead to a high concentration of bicarbonate and organic matters which in turn reduce the Zn concentration in soil solution. On the contrary, under flooding conditions concentration of Zn in soil solution decreases through formation of insoluble Zn sulfide (Marschner 1995). Alcohol dehydrogenase (ADH), an enzyme which contains two Zn atoms per molecule, involves in the reaction of acetaldehyde to ethanol. In Zn-deficient plants particularly under anaerobic conditions, ADH activity decreased, which might lead to accumulation of acetaldehyde up to toxic levels (Marschner 1995). Moore and Patrick (1988) reported a decreased root ADH activity in flooded rice plants that was correlated with Zn concentration in the roots and resulted in less ATP production, thereby reducing vital metabolic activities of the roots. Accordingly, a correlation would be expected between susceptibility of plants to Zn deficiency and flooding. In a work with four rice genotypes, however, it was found that ADH activity rather increased in some genotypes under low Zn supply irrespective to their Zn-efficiency trait (Table 16.3). It could be suggested that, Zn availability at molecular level was not necessarily a limiting factor for ADH holoenzyme in these genotypes. The main cause for negative effect of flooding may be disruption in the function of antioxidant defense system required particularly for Zn-deficient plants under waterlogged conditions. Accordingly, a close correlation has been observed between flooding tolerance and accumulation of ROS in rice genotypes

Table 16.3 Plants dry matter production and activity of alcohol dehydrogenases (ADH) in roots of Zn-efficient (IR34 and IR36) and Zn-inefficient (IR26 and IR54) rice genotypes grown in hydroponic medium with adequate (130 pM free Zn^{2+} activity) and low (9 pM free Zn^{2+} activity) Zn supply with (control) or without (hypoxia) aeration

		Shoot dry weight (mg plant^{-1})		Root dry weight (mg plant^{-1})		Activity of ADH (nmol NADH min^{-1} mg^{-1} protein)	
Genotypes	Treatments	Adequate Zn	Low Zn	Adequate Zn	Low Zn	Adequate Zn	Low Zn
IR 34	Control	153±11^{a}	89±3^{b}	89±5^{a}	42±6^{c}	119±36^{c}	434±47^{a}
	Hypoxia	147±6^{a}	85±9^{b}	80±4^{b}	33±5	241±36^{b}	502±40^{a}
IR36	Control	144±10^{a}	80±13^{b}	84±11^{a}	30±5^{b}	71±29^{b}	101±11^{b}
	Hypoxia	131±11^{a}	72±3^{b}	82±3^{a}	27±3^{b}	92±29^{b}	242±46^{a}
IR26	Control	118±15^{a}	71±6^{b}	82±12^{a}	23±5^{b}	296±68^{a}	117±39^{b}
	Hypoxia	121±11^{a}	58±9^{b}	80±7^{a}	15±3^{b}	313±93^{a}	150±22^{b}
IR54	Control	136±7^{a}	58±5^{b}	89±5^{a}	13±2^{b}	181±30^{b}	243±60^{a,b}
	Hypoxia	126±14^{a}	45±2^{c}	87±7^{a}	9±1^{c}	171±53^{b}	302±19^{a}

Data of each parameter within each genotype followed by the same letter are not significantly different ($P<0.05$) (Hajiboland 2000)

under hypoxic conditions. Content of $O_2^{\bullet-}$ and MDA decreased in the lowland rice genotype due to hypoxia while increased significantly in the upland genotype in both low- and adequately-Zn supplied plants (Hajiboland and Beiramzadeh 2008).

5 Effect of Iron Deficiency on Plants Stress Responses

Iron deficiency is the most intensively studied plant responses to micronutrients deficiencies in the past 30 decades. Due to a known low Fe availability in soils of different types, Fe uptake and homeostasis are tightly regulated in plants ensure a sufficient supply of Fe from the soil. At the root level, Fe deficiency induces various responses aimed at increasing the availability of the metal in the rhizosphere. Strategy I plants (dicotyledonous and nongraminaceous plants) are able to respond to a lack of Fe in the soil by increasing the capacity of root tissues to reduce apoplastic Fe, the acidification of the rhizosphere, accumulation and release of organic acids (mainly citrate) to increase Fe solubility in soil and mobility within plants and the Fe uptake activities in rhizodermal root cells. In Strategy II plants, members of the mugineic acid family of phytosiderophores are secreted into the rhizosphere by the roots, where mugineic acids chelate and help solubilize Fe^{3+}. The Fe^{3+}-mugineic acid complex is then taken up by root cells through the action of Yellow Stripe 1 (YS1) proteins (Curie and Briat 2003). In contrast to strategies at the root level, our knowledge is much more limited on the mechanisms to cope with Fe deficiency acting in the aboveground part of plants.

5.1 Iron Deficiency-Induced Oxidative Stress

Iron deficiency is expected to reduce activity of CAT and PODs, the ubiquitous haem-containing enzymes. Activity of CAT reduces under conditions of Fe deficiency and therefore is an indicator of Fe nutritional status of plants (Marschner 1995). The high levels of H_2O_2 in Fe-deficient plants (Marschner 1995) suggest that Fe starvation may cause reduction of capacity for peroxide detoxification and/or active production of peroxide and consequently rise of oxidative cell status. An increase in SOD activity following Fe deficiency and reduction of capacity for detoxification of overproduced H_2O_2 may induce a secondary oxidative stress. Activity of POD isoforms involving in the polymerization of phenols to lignin is depressed in the roots of Fe-deficient plants that leads to the accumulation and/or release of phenolics and probably development of rhizodermal cells (Marschner 1995). In the leaves, however, Fe deficiency affects the activity of different PODs to different extents (Ranieri et al. 2001).

Activity of APX decreased in the apoplastic fluid as well as at an intracellular level likely because of the high request for Fe from the APX molecule. This enzyme in addition to the haem group contains also a nonheme Fe atom. In contrast to APX, soluble and ionically cell wall bound PODs did not undergo any change in Fe-deprived plants in comparison to control ones (Ranieri et al. 2001). These results seem to suggest that under Fe-starvation plants retain the functionality of the PODs mainly involved in the maintenance of cell wall structure, cell homeostasis and turgor. It is well established that APX plays a key role in the removal of H_2O_2 in the chloroplast and cytosol of higher plants under stress conditions such as drought (Mittler and Zilinskas 1994). Drought stress, as indicated by a decrease in leaf water potential and stomatal closure, resulted in an increase in APX and CuZn-SOD gene expression and in an increase in CAT activity. In addition, plants recovering from drought showed a dramatic increase in APX and CuZn-SOD steady-state transcript levels (Mittler and Zilinskas 1994). These results suggest that optimized Fe nutrition could have a significant role in the protection of plants against oxidative stress during the progression of drought and recovery from drought. The regulation of cytosolic (Mittler and Zilinskas 1994) and chloroplastic (Tanaka et al. 1990) APX has been shown at the level of steady-state transcript accumulation and by regulation of protein synthesis. More investigations are required to understand the functional significance of CAT and POD isoforms under combinative effects of Fe-deficiency and environmental stress factors such as drought and low temperatures.

5.2 Iron Deficiency-Induced Damage to Photosynthetic Apparatus

Iron is important in the synthesis of Chl in higher plants, and leaves suffering from Fe deficiency show damaged chloroplast structure and decreased Chl content. A low Chl leaf not only has a reduced photosynthetic capacity but also absorbs more light per Chl. The light absorbed and not used in photosynthesis could lead potentially to photoinhibitory and photooxidative processes (Abadía et al. 1999). This could be especially important under high light intensities found in field conditions (Jiang et al. 2001). At high photosynthetic photon flux density (PPFD), the accumulation of excitation energy in the PSII antenna favors the production of triplet Chl that can interact with O_2, generating reactive singlet oxygen ($^1O_2$). In addition, overreduction of the photosynthetic electron carrier chain would also favor the direct reduction of O_2 by PSI, and the subsequent generation of the ROS superoxide, hydrogen peroxide H_2O_2 and the hydroxyl radical. Plants suffering a PPFD excess may show sustained photoinhibitory damage, which requires de novo protein biosynthesis to be overcome (Aro et al. 1993).

5.3 Molecular and Biochemical Adaptations of Fe-Starved Plants to Environmental Stresses

Under Fe starvation, plants develop mechanisms for improvement their resistance to the second environmental stress factor such as excess light and drought.

5.3.1 An Efficient Thermal Dissipation in Fe-Deficient Leaves

To avoid photodamage, the excess radiant energy must be dissipated properly in Fe-starved plants particularly under higher light intensities and drought stress. The excess of light over than that can be used in photosynthesis could be especially important in field conditions, where PPFD is as high as 2,200 μmol m^{-2} s^{-1} (Abadía et al. 1999). Nonphotochemical quenching is one of the mechanisms that prevent or alleviate damage to the photosynthetic apparatus. In this mechanism, excess radiation energy is dissipated as heat in the light harvesting antenna of PSII (Müller et al. 2001). The excessive light energy due to reduced photosynthetic carbon metabolism imposed by lower stomatal opening and/or lower demand because of impaired growth, could be dissipated as heat through nonphotochemical quenching. The xanthophyll cycle is an important dissipation mechanism and it may play an important role in

Fe-deficient leaves. Compared with control, the ratio of xanthophyll to Chl in Fe-deficient leaves is 1.2 times higher. Higher degree of deepoxidation and higher carotenoid content was clearly a protective response to excess irradiance caused by lower photosynthesis in Fe-deficient leaves. In addition, energy dissipation depending upon D1 protein turnover is probably very important in Fe-deficient leaves and there is a close relationship between xanthophyll cycle and D1 protein turnover (Jiang et al. 2001). The constitutive level of thermal dissipation in the leaves is approximately 20% but increased up to 70–74% of the absorbed light under moderate water stress (Morales et al. 2000). Accordingly, Fe-deficient leaves could remain without apparent damage in the field for months, which indicates that they have very efficient protective mechanisms (Fig. 16.2).

5.3.2 An Efficient Protection of Plants Against Reactive Oxygen Species

The H_2O_2 content undergoes a significant rise following Fe deficiency treatment. Although it can be reduced to extremely reactive $^{\bullet}OH$ radicals through the Fenton reaction, the content of catalytic Fe is extremely low in Fe-deficient plants and the Fenton reaction is unlike to occur. Experimental evidences indicated the absence of oxidative damages and suggested that Fe-deficient plants are sufficiently protected against oxidative stress (Ranieri et al. 2001). In addition, some of the enzymes involved in the detoxification of ROS found to be over expressed in the roots under Fe-limiting conditions (Zaharieva and Abadía 2003). An improved protection of membranes in Fe-deficient plants is another strategy for survival of plants under prolonged Fe deficiency. Thylakoids are very sensitive targets for photodestruction by ROS, because of their unique lipid composition containing highly unsaturated (C18:3) fatty acids. Iron deficiency alters such lipid composition by decreasing the concentration of unsaturated (C18:3) and increasing those of saturated fatty acids (C18:0 and C16:0) (Abadía et al. 1988). This makes thylakoids from Fe-deficient plants less susceptible to be degraded by ROS. In fact, oxidatively damaged lipids (and proteins) do not accumulate in Fe-deficient leaves (Iturbe-Ormaetxe et al. 1995).

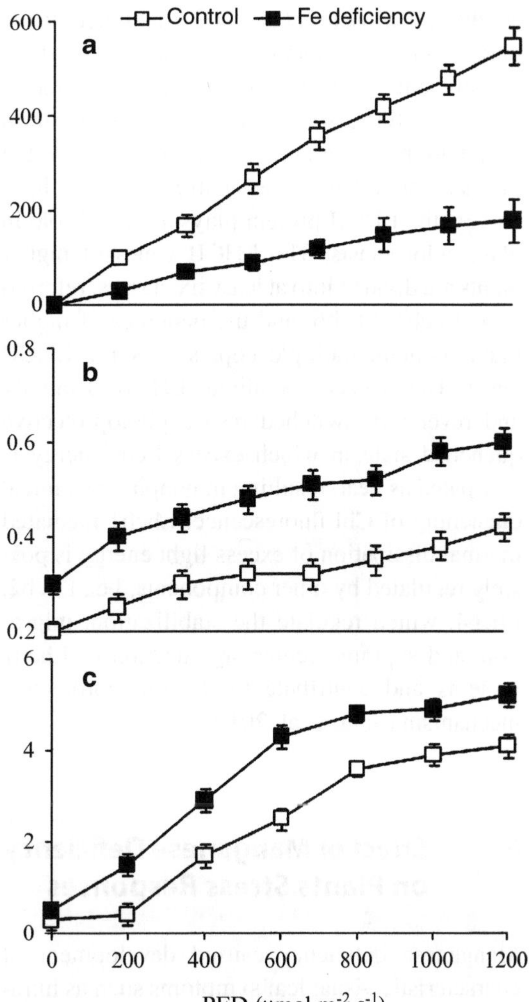

Fig. 16.2 Photochemical reaction rate (μmol m^{-2} s^{-1}) (**a**), fraction of photon absorbed in PSII antenna that is removed via thermal energy dissipation (**b**) and nonphotochemical quenching (**c**) in control and iron-deficient maize leaves exposed to different photon flux density (PFD) (Jiang et al. 2001)

5.3.3 Remodeling of Light-Harvesting Complex

Recent molecular and biological analyses revealed that several photosynthetic organisms remodel their photosynthetic apparatus under prolonged Fe deficiency. These structural changes minimize photooxidative stress to the thylakoid membrane of algae grown under Fe-deficient conditions (Moseley et al. 2002). In higher plants, photosynthetic apparatus is not damaged in Fe-deficient leaves after prolonged high irradiation suggesting also a mechanism for a long-term acclimation to Fe-deficient conditions. In the mechanism of thermal dissipation of excess light energy, the LHCII protein plays a critical role in plant chloroplasts. The LHCII genes of higher plants are divided into at least six classes, referred to as Lhcb1–Lhcb6, and the genomes of higher plants contain multiple copies of some LHCII genes. Under excess sunlight, LHCII is rapidly and reversibly switched into a photoprotective quenched state in which excess light energy is dissipated as heat, resulting in nonphotochemical quenching of Chl fluorescence. Lhcb1-mediated thermal dissipation of excess light energy is possibly regulated by other components, i.e., Lhcb2, Lhcb4, which regulate the stabilization, migration, and supramolecular organization of Lhcb1 proteins and contribute to the photoprotective mechanism (Saito et al. 2010).

6 Effect of Manganese Deficiency on Plants Stress Responses

Manganese deficiency caused development of characteristic visual leaf symptoms such as intravenous chlorosis and subsequently the development of necrotic spots, which are supposed to be related to disorganization of the thylakoid system and loss of PSII reaction centers (Papadakis et al. 2007). Perturbations in the photosynthetic apparatus clearly involved a markedly reduced efficiency of PSII under Mn deficiency due to loss of the PSII core protein (PsbA) (Husted et al. 2009). Even mild Mn deficiency without any distinct leaf symptoms may induce damages to PSII and consequently limit harvest yields, adaptability, and survivability of plants under field conditions. Regarding the fact that Mn is a key component of PSII, the relationship between differential Mn efficiency of plant genotypes and the resistance of the photosynthetic apparatus to perturbations induced by Mn deficiency have been reported (Husted et al. 2009).

6.1 Increased Production of ROS

The improper function under Mn deficiency not only of the water splitting reaction but also of the photosynthetic electron transport chain increases the probability of oxidative stress for leaf chloroplasts. Under such stress, molecular O_2 operates as an alternative acceptor for nonutilized electrons and photon energy, resulting thus in the generation of ROS (Papadakis et al. 2007). In addition, Mn is a component of Mn-SOD, and reduction of ROS scavenging due to reduced Mn-SOD activity is expected. However, reports on the change of Mn-SOD activity in Mn-deficient plants are contradictory, reduction (Yu et al. 1999a), no change or even increase (Shenker et al. 2004) of Mn-SOD activity have been reported. Nevertheless, oxidative stress caused by Mn deficiency has been reported in various species (Yu et al. 1998). Although Mn is not a structural component of Chl molecule and is not directly involve in the biochemical pathways of biosynthesis of Chl, leaves with low Mn content are chlorotic because of Chl losses (Henriques 2003). The ability of ROS to cause photooxidative damages to membranes, organic molecules including Chl under Mn deficiency explains this phenomenon.

6.2 Increased Susceptibility to Higher Light Intensity

The values of electron transport rate and effective quantum yield of PSII are considerably lower in Mn-deficient than in the Mn-sufficient plants (Papadakis et al. 2007). Under low or intermediate irradiances high percentage of excess photon energy of PSII produced in Mn-deficient leaves is dissipated as heat via the xanthophyll cycle (Papadakis et al. 2007). Accordingly, reduction of effective quantum yield of PSII and increase of nonphotochemical quenching are associated with the xanthophyll pigment cycle that provides photoprotection of photosystem by the dissipation of excess absorbed photon energy (Müller et al. 2001). By contrast, under higher light intensities, the reduced values of effective quantum yield of PSII observed in Mn-deficient plants, could be

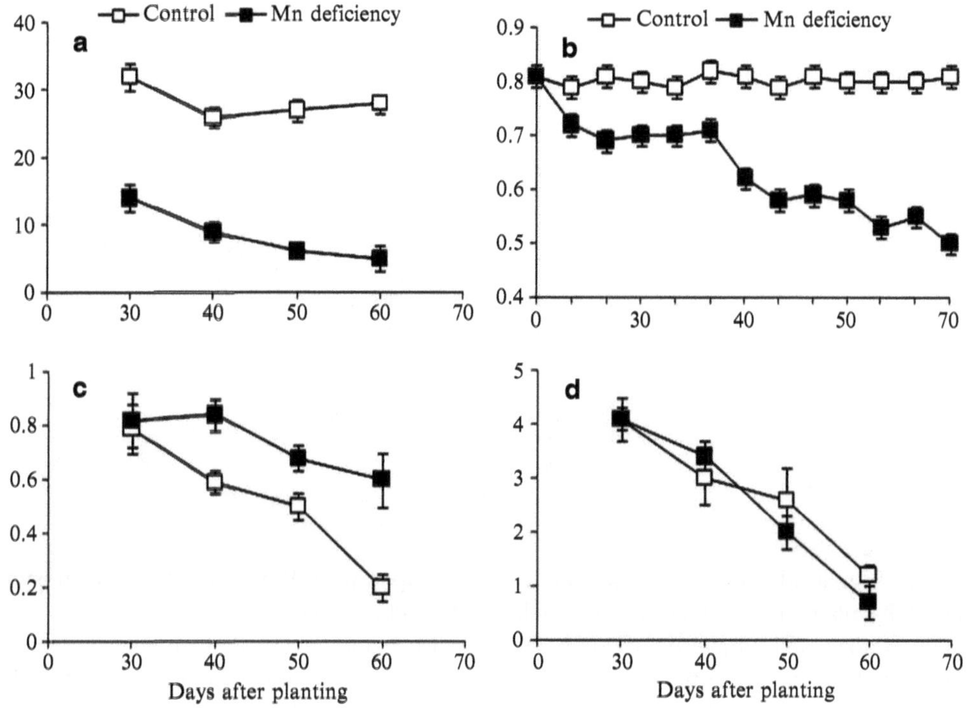

Fig. 16.3 Concentration of Mn (μg g^{-1} DW) (**a**), quantum yield efficiency of PSII (F_v/F_m) (**b**) transpiration (mmol m^{-2} s^{-1}) (**c**) and CO_2 exchange with the atmosphere (μmol m^{-2} s^{-1}) (**d**) during growth under Mn deficiency in barley (*Hordeum vulgare* L.) plants (Hebbern et al. 2009)

ascribed to the increase of the percentage of close-reduced PSII reaction centers (decreased values of photochemical quenching) due to photoinhibition (Papadakis et al. 2007). Xanthophyll cycle is not fully stimulated in Mn-starved leaves and large amount of excess photons is not successfully dissipated (Jiang et al. 2002). It was suggested that the deficiency in transthylakoid pH gradient is responsible for the decrease of the xanthophyll cycle-dependent nonradiative dissipation in Mn-starved leaves. Therefore, under high photon flux density much more active oxygen is produced in Mn-starved leaves and PSII reaction centers are damaged by active oxygen in Mn-starved leaves, which resulted in serious photoinhibition (Jiang et al. 2002).

6.3 Increased Susceptibility to Drought and High Temperatures

Very limited information is currently available on how Mn deficiency affects plant water relations. Although some authors reported lower transpiration rates in some plant species with visible symptoms of Mn deficiency (Singh et al. 2001), other reports demonstrated that Mn deficiency leads to a marked increase in transpiration (Hebbern et al. 2009) (Fig. 16.3). Elevated transpiration and lower water use efficiency in Mn-starved plants can be related to more open stomata during the daytime and imperfect nocturnal closure of stomata. Ion leakage from guard cells observed in Mn-deficient leaves will inevitably lead to an imperfect stomatal closure during both night and daytime, resulting in poor water use efficiency (Hebbern et al. 2009). These evidences implied that drought will put additional stress on Mn-deficient plants that are already suffering from disturbances in key metabolic processes. Thus, drought and the poor water use efficiency will attenuate Mn deficiency because Mn-deficient plants exhibit a reduction in root–shoot ratio, which further restricts plants from exploring the soil for available water resources and for plant available Mn.

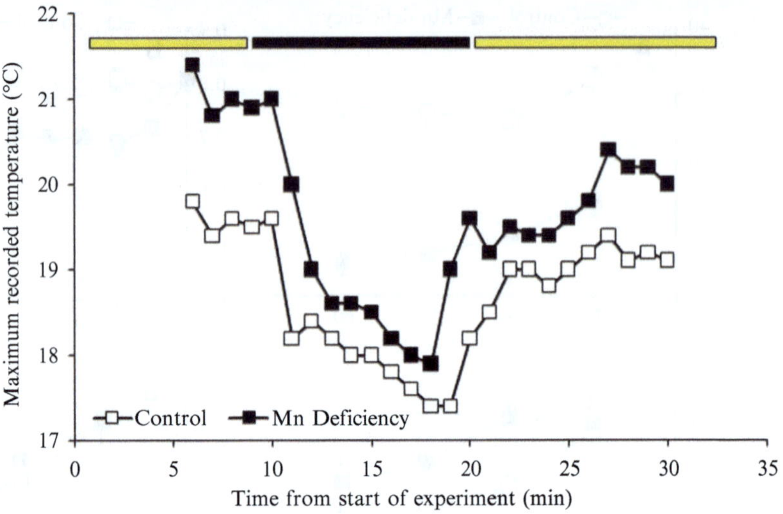

Fig. 16.4 Leaf temperature changes in plants with adequate or low Mn supply exposed to light/dark/light transition 50 days after planting (Hebbern et al. 2009)

In turn, water availability in the soil affects Mn uptake by plants (Marschner 1995). Under flooded conditions, soils show low redox potentials, thus Mn deficiency is virtually unknown in plants adapted to waterlogged conditions such as lowland rice. However, when common lowland-rice varieties are cultivated under nonflooded conditions, higher soil redox potential leads to strong reduction of plant Mn availability (Snyder et al. 1990). It was shown that Mn uptake of lowland rice was affected by reduced soil water content under nonflooded conditions. By contrast, soil moisture had little effects on P, Fe, Zn, and Cu nutrition (Tao et al. 2007).

6.3.1 Reduction of Leaf Cuticle: Causes and Results

The hydrophobic cuticle of leaves serves many fundamental purposes, including the prevention of excessive water loss and lowering of light absorptance to reduce risks of heat stress and overexcitation of the photosystems (Chaerle et al. 2007). Manganese deficiency results in strong decrease in the epicuticular wax content, which in combination with imperfect stomatal closure are likely to be the main reasons for the markedly higher transpiration observed from Mn-deficient relative to control plants of similar physiological age (Hebbern et al. 2009). The relationship between tissue concentrations of Mn and the wax content of leaves is presently not understood, but there are several indications that Mn might play an important role in the fatty acid biosynthesis (Marschner 1995).

Manganese-starved plants transpire up to fourfold more water on a leaf area basis than do control plants (Fig. 16.3), expecting that higher water loss would lead to a significant cooling of leaves and lower temperatures. Using infrared thermography and measurement of leaf optical properties, however, it has been shown that maximum temperatures of leaves from Mn-starved plants are up to 2°C warmer than those of control plants (Hebbern et al. 2009) (Fig. 16.4). It was suggested that decreased epicuticular wax content will reduce the reflectance of light and result in higher light absorption, increased heat load and higher leaf temperatures (Holmes and Keiller 2002). In Mn-starved plants changes in the epicuticular wax might influence canopy temperatures via modifications of leaf optical properties. Even marginal and insignificant changes in leaf reflectance may be sufficient to induce substantial changes in leaf temperature (Hebbern et al. 2009).

6.4 Increased Susceptibility to Low Temperatures

Manganese-deficient plants are more susceptible to damage by freezing temperatures (Marschner 1995). Exposure of palms to temperatures below that to which they are acclimated causes injury include extensive necrosis of the foliage. These symptoms are associated with Mn deficiency (Larcher and Winter 1981).

6.5 Increased Susceptibility to Pathogens

Mn-deficient leaves have a slack appearance probably caused by poor lignification due to reduced phenylalanine ammonia-lyase (PAL) activity (Marschner 1995). Of the various defense mechanisms available to plants the phenolics and lignin have the most important roles in plants resistance to pathogens. Lower soluble phenolics and lignin contents in Mn-deficient plants is a reflection of the requirement for Mn in various steps of phenolics and lignin biosynthesis and is an important factor responsible for the lower resistance of Mn-deficient plants particularly to root infecting pathogens (Marschner 1995). The close relationship between Mn nutritional status of plants and their disease infection leads often to the misdiagnosis of disease infection as Mn deficiency or toxicity (Humphries et al. 2007). In the case of root pathogens, reduction in root growth and absorption surface causes in turn reduction of plant Mn uptake, thus intensifying the Mn deficiency symptoms (Marschner 1995). In addition to impaired lignification and lower phenolics content, higher activity of amino peptidase, which supplies essential amino acids for fungal growth, and pectin methyl esterase, which is a fungal enzyme for degrading host cell walls under Mn-deficient conditions (Sadasivan 1965), and inhibition of photosynthesis, leading to a decrease in root exudates and thus becoming more susceptible to invasion by root pathogens (Humphries et al. 2007), are mechanisms for increased susceptibility of Mn-deficient plants to pathogens.

7 Effect of Copper Deficiency on Plants Stress Responses

Copper is required as a cofactor for several processes including photosynthesis, respiration, ethylene perception, oxidative stress reduction, cell expansion, and cell-wall lignification (Marschner 1995). In plant chloroplasts, two major Cu-containing proteins are found, plastocyanin (PC) and CuZn-SOD. Copper deficiency was found to reduce PSI electron transport due to decreased formation of plastocyanin which is the major target site of Cu deficiency in photosynthesis. Copper-deficient plants show disintegration of the thylakoid membranes as well as decreased pigments (Chl and carotenoids) content, reduced plastoquinone synthesis, and lower unsaturated C18 fatty acid contents (Ayala et al. 1992). Copper deficiency also depresses CO_2 fixation, electron transport, and thylakoid prenyl lipid synthesis relative to control plants (Bussler 1981). In contrast to other micronutrients, studies on the effect of Cu deficiency on plants responses to environmental stresses are rare.

7.1 Increased Oxidative Damage

Copper ion as the constituent of CuZn-SOD plays an important role in scavenging $O_2^{\bullet-}$ radicals. The Cu atom represents the catalytic metal component and Zn the structural (Marschner 1995). The CuZn-SOD localized near the PSI complex, considerably accelerates $O_2^{\bullet-}$ decomposition which results in the formation of H_2O_2 and O_2. Weakened function of CuZn-SOD can lead to the formation of highly reactive $^{\bullet}OH$ radical. Reduction in the activity of CuZn-SOD in Cu-deficient plants, similar with that in Zn-deficient ones is well documented (Marschner 1995). Since Cu is a constituent of CuZn-SOD, it is expected that Cu deficiency would accentuate the effect of other stress factors that produce ROS. However, reports on the intensifying effect of Cu deficiency on oxidative damage caused by other stresses are rare. Interestingly, plants downregulate CuZn-SOD and upregulate Fe-SOD under Cu limitation, thus can save Cu for use in PC, which is essential for photosynthesis. This plasticity in response to Cu nutrition allows plants to always have an SOD enzyme ready to scavenge superoxide radicals in the stroma and at the same time maintain PC activity for photosynthetic electron transport (Cohu and Pilon 2007).

7.2 Increased Susceptibility to Environmental Stresses via Changes in Membrane Composition

Maintaining membrane structure and function is common aspect of plants response to both abiotic and biotic stresses. The ability to adjust membrane lipid fluidity by changing levels of unsaturated fatty acids is a feature of acclimating plants to stress, e.g., cold stress, provided mainly by the regulated activity of fatty acid desaturases (Upchurch 2008). In stress tolerant plants, the degree of membrane lipid unsaturation increases in response to low temperature, salt, and pathogen stresses, but decreases in response to elevated temperature and heavy metal stresses (Upchurch 2008). There are new evidences regarding Cu effects at the plasma membrane level and on its lipid composition in particular (Quartacci et al. 2001). The analysis of total plasma membrane fatty acids showed a lower unsaturation level in the roots grown without Cu supply which is the indication that a degradative process occurred. Under Cu deficiency NADPH-dependent superoxide producing oxidase but not membrane-bound lipooxygenase activities increased (Quartacci et al. 2001). Copper deficiency affects the photosynthetic apparatus indirectly via changes in the saturation degree of the particular acyllipids of thylakoid membrane as well as ultrastructural alterations in chloroplasts (Maksymiec 1997). More investigations are needed on the effect of Cu deficiency at the membrane level and on its effect in the tolerance of plants to other stresses such as cold and high temperatures that trigger membrane injury.

7.3 Increased Susceptibility to Environmental Stresses via Reduction of Lignifications

Lignin provides the mechanical support and structural strength for plants plays a crucial role in xylem vessels for conducting water and nutrients from roots into shoot and is constituent of mechanical defense against herbivory (Boudet 1998). Lignin biosynthesis is the result of the activity of several enzymes and responds differently to biotic and abiotic factors (Rogers and Campbel 2004). There are evidences on the changes in lignin content and composition due to various stress factors, suggesting complex genetic and physiological control. Different types of abiotic stresses, such as nutrient deficiency, drought, UV-B radiation, and chilling, as well as infection by fungi, bacteria, and viruses, cause changes in the lignin contents of plants (Moura et al. 2010). One of the most typical anatomical changes induced by Cu deficiency is impaired lignifications of cell walls. The inhibition of lignifications in Cu-deficient tissues is related to a direct role of at least two Cu-enzymes in lignin biosynthesis, polyphenol oxidase catalyzing the oxidation of phenolics as precursors of lignin, and diamine oxidase provides the H_2O_2 required for oxidation by peroxidations (Marschner 1995). The chemical composition of cell wall changes also in Cu-deficient plants. The ratio of cell wall material to the total dry matter decreases, simultaneously the proportion of α-cellulose increases and the lignin content is only about half that of leaves adequately supplied with Cu. A decrease in lignifications occurs even with mild Cu deficiency and lignifications respond rapidly to Cu deficiency. In plants suffering from severe Cu deficiency the xylem vessels are also insufficiently lignified. Reduction of lignification gives rise to the characteristic increase in the lodging susceptibility of cereals and increases infection of plants with pathogens (Marschner 1995). Study of expression pattern of genes involve in biosynthesis of lignin revealed that low temperatures (Hausman et al. 2000), drought (Bok-Rye et al. 2007), high light intensity (Kimura et al. 2003), and UV radiation (Hilal et al. 2004) cause increased lignifications in plants. It is expected that, therefore, Cu-deficient plants with lower lignification would be more susceptible to these stress factors. More investigation is needed to assess the response of Cu-deficient plants to these stress factors.

8 Effect of Boron Deficiency on Plants Stress Responses

Boron is directly or indirectly involved in many physiological and biochemical processes during plant growth, such as cell elongation and division, cell wall biosynthesis, membrane function, leaf photosynthesis, and N metabolism (Marschner 1995). Function of B is mainly related to cell wall stability caused by borate cross-linking of apiose residues in the pectic polysaccharide rhamnogalacturonan II (O'Neill et al. 2004). Boron is required particularly at the initial phases of organogenesis for eliciting mechanisms leading to cell differentiation. It has been proposed that membrane glycoproteins stabilized by borate are involved in cell-to-cell signaling during plant organogenesis and symbiosis with microorganisms (Redondo-Nieto et al. 2008). The finding of B being an essential part of a signal molecule in bacteria highlights the possibility that, besides being an indispensable factor for RGII cross-linking, B may play further roles in plant metabolism (Goldbach and Wimmer 2007).

8.1 Boron Deficiency and Plant Responses to Drought

Evidences have been demonstrated that low soil water availability often accentuates the B deficiency symptoms. Boron deficiency is a major cause of wheat (*Triticum aestivum*) sterility, and it is evident from the several studies that wheat sterility is often accentuated by environmental stresses, particularly drought and extreme temperatures (Perkasem and Jamjod 2004). In Norway spruce, the height growth of young plants at low B is reduced when they are subjected to drought stress, whereas at adequate B it is not affected by drought (Möttönen et al. 2005). Repeated drought stress results in visible suddenly occurring symptoms of B deficiency in plants in the field studies (Möttönen et al. 2005). These findings suggest that B nutritional status and drought stress may have interactions both at the level of soil and within plant.

8.1.1 Boron-Deficiency Induced Impairment in Root Growth

Boron deficiency may affect plant water balance by reducing water uptake through a reduced number of root tips and root length (Möttönen et al. 2001). Roots play a major role in the regulation of water and nutrient uptake, and in the maintenance of plant water balance. Therefore, an extensive and efficient root system is important in drought resistance of plants. The importance of B, especially for root growth, has been well documented (Marschner 1995). In addition, B deficiency reduces root hydraulic conductivity (Apostol and Zwiazek 2004). Boron-deficiency-induced disturbance in the formation of functional xylem vessels (Marschner 1995) is possibly one mechanism. A decline in root hydraulic conductance may also be caused by a reduction in new root growth in B-deficient plants. In addition, B is required for function and activity of H^+ATPase (Roldan et al. 1992) and expression of H^+ATPase gene (Camacho-Cristóbal and González-Fontes 2007). Therefore, it is plausible that B deficiency causes reduction in the uptake of K^+ and other solutes into cells required for water uptake, cell turgidity, and expansion. In tree species with ectomycorrhizal associations such as Norway spruce, reduction of mycorrhizal colonization in B-deficient plants (Möttönen et al. 2005) could be involved in reducing capacity of whole plants for water extraction from dry soil.

8.1.2 Drought-Induced Reduction of Soil B Availability, Uptake and Transport

Effect of drought on reduction of B uptake could be observed not only under low but also under adequate B supply (Hajiboland and Farhanghi 2011). Root B uptake is mostly a passive process of boric acid (H_3BO_3) permeation across the membrane and is mainly determined by the rate of water uptake by root cells (Hu and Brown 1997) and the flow through water channels (Dordas et al. 2000). Reduced soil solution in connection with reduced mass flow and diffusion rate are involved in the reduction of B availability in a drying substrate. In addition, impairment

Table 16.4 Dry weight (mg plant^{-1}) and boron content (μg plant^{-1}) of shoot and root, leaf chlorophyll and carotenoids content (mg g^{-1} FW), photochemical quenching (*qP*), nonphotochemical quenching (*qN*), net photosynthetic rate (*A*, μmol m^{-2} s^{-1}), transpiration rate (*E*, mmol m^{-2} s^{-1}), stomatal conductance to water vapor (*gs*, mol m^{-2} s^{-1}), and water potential (MPa) in the leaves of turnip (*Brassica rapa* L.) plants grown for 3 months with adequate (+B) and low (−B) boron supply under control or drought conditions

	Control		Drought	
Parameters	+B	−B	+B	−B
Shoot DW	5051 ± 416^{a}	3684 ± 180^{b}	1975 ± 205^{c}	957 ± 175^{d}
Root DW	284 ± 35^{a}	$149\pm28^{a,b}$	127 ± 33^{c}	39 ± 12^{d}
Shoot B	1313 ± 219^{a}	214 ± 73^{b}	421 ± 174^{b}	41 ± 10^{c}
Root B	58 ± 13^{a}	15 ± 2^{b}	22 ± 1^{b}	2 ± 1^{c}
Chl a+b	3.02 ± 0.11^{a}	$2.34\pm0.21^{b,c}$	$2.65\pm0.97^{a,b}$	1.97 ± 0.34^{c}
Carotenoids	934 ± 29^{a}	669 ± 32^{c}	715 ± 39^{b}	589 ± 33^{d}
q_P	0.99 ± 0.01^{a}	0.95 ± 0.01^{b}	$0.93\pm0.02^{a,b}$	0.92 ± 0.02^{b}
q_N	0.19 ± 0.06^{a}	0.17 ± 0.06^{a}	0.04 ± 0.01^{b}	0.07 ± 0.01^{b}
A	10.2 ± 0.3^{a}	5.9 ± 0.2^{c}	8.5 ± 0.4^{b}	2.8 ± 0.5^{d}
E	4.6 ± 0.1^{a}	3.1 ± 0.6^{b}	4.4 ± 0.4^{a}	2.1 ± 1.1^{b}
gs	0.97 ± 0.1^{a}	0.35 ± 0.1^{b}	0.76 ± 0.2^{a}	0.19 ± 0.1^{b}
Water potential	-0.06 ± 0.01^{a}	-0.18 ± 0.01^{b}	-0.15 ± 0.02^{b}	-0.27 ± 0.02^{c}

Data of each row followed by the same letter are not significantly different ($P<0.05$) (Hajiboland and Farhanghi 2011)

of shoot–root translocation of B under drought conditions caused reduction of shoot B content under drought stress (Hajiboland and Farhanghi 2011). The transport of B from root to shoot is mainly driven by transpiration in the leaves, and the distribution of B in the shoot follows the gradient of transpiration rate in leaves (Brown and Shelp 1997). There has been an active debate concerning a relationship between the ion and water transport in plants (Dalton et al. 2000). Water loss through transpiration in B-deficient plants has been studied in various plant species. In mustard (*Brassica campestris* L.), B-deficient plants had lower water potentials, decreased stomatal pore opening, and reduced transpiration relative to B-sufficient plants (Sharma and Ramchandra 1990). Reduction of stomatal conductance has been reported in other plant species under B deficiency conditions (Hajiboland and Farhanghi 2011). Interestingly, though a more conservative behavior of B-deficient plants regarding water economy following elevated stomatal limitations, water potential was lower in B-deficient plants irrespective to the watering regime (Hajiboland and Farhanghi 2011) (Table 16.4). It implies that disturbance in water economy was due to lower water uptake and not greater water loss.

Lowered stomatal conductance by B deficiency and drought results in reduction of net CO_2 assimilation rate (Han et al. 2008). Mechanism of a role for B in stomatal opening has not been investigated. Boron is required for integrity of membranes (Cakmak and Römheld 1997), and function as well as activity of H^+ATPase (Roldan et al. 1992). Therefore, it is plausible that B deficiency causes reduction of K^+ uptake into guard cells and following loss of membrane integrity, stimulates passive leakage of K^+ from guard cells. Increased K^+ leakage from leaf tissues was observed in B-deficient turnip plants (Hajiboland and Farhanghi 2010). Not only root–shoot transport but also retranslocation of B is influenced by drought conditions. Lower phloem transport of B under drought conditions was reflected in significantly lower ratio of B content of young to old leaf (Hajiboland et al. unpublished results). A dramatic effect of drought on reduction of B retranslocation evidenced that water-stressed plants had significantly lower B remobilization from old to young growing leaves. It was reported that B-deficient Norway spruce plants were not able to retranslocate B from other plant parts to the meristematic tissues and developing buds (Möttönen et al. 2005). The visible damage to the apical growth of the B-deficient plants observed after drought period may have been caused by inadequate B supply to the developing buds of mature Norway spruce trees growing in soil poor in B (Möttönen et al. 2005).

8.1.3 Boron-Deficiency Induced Photoinhibition Under Drought Conditions

In addition of stomatal limitation, nonstomatal limitation of photosynthesis has been reported under combination of B-deficiency and drought stress that could be explained by inhibited leaf photochemistry as well as metabolic impairment. Photochemical quenching decreased significantly in plants subjected to combination of B deficiency and drought stresses (Hajiboland and Farhanghi 2011). Reduction of photochemical quenching could be related to photoinhibition rather than to a direct damage to PSII (Baker and Bowyer 1994). In addition, leaves under B deficiency and drought showed lower capacity for heat dissipation as indicated by significantly lower nonphotochemical quenching (Table 16.4). The activation of the carbon reduction cycle may be especially sensitive to drought (Reddy et al. 2004). Reduction of the activities of ribulose-1, 5-bisphosphate carboxylase/oxygenase, NADP-glyceraldehyde-3-phosphate dehydrogenase, and stromal fructose-1,6-bisphosphatase in B-deficient citrus leaves has been reported (Han et al. 2008).

8.1.4 Boron-Deficiency Induced Oxidative Stress Under Drought Conditions

A significant rise of H_2O_2 titer and increased lipid peroxidation were detected under combinative effect of B deficiency and drought (Hajibland et al. unpublished data) implied occurrence of oxidative damage. However, proline accumulation was observed in B-starved plants when subjected to drought stress (Hajiboland et al. unpublished results). Boron deficiency-induced accumulation of proline may be regarded as a strategy for plants to counteract with the oxidative stress provoked after subjection of plants to the second stress factor such as drought.

8.2 Boron Deficiency and Plant Responses to Low Temperatures

Low temperature is a common threat faced by crop species, particularly those of tropical or subtropical origin (Huang et al. 2005), fruit trees, eucalyptus, ash, and Norway spruce (Räisänen et al. 2007). Cultivation areas of many crop species of tropical or subtropical origin have been expanded into temperate regions for growth in the warm season. These plants encounter chilling stress in early spring and/or in late autumn. In addition, for perennial species, freezing damage in wintering buds and dieback of apical shoots is a common problem in forestry in boreal climates (Räisänen et al. 2009).

Boron deficiency with characteristic symptoms, dieback of leaders and a bushy appearance of crown, is often reported in forest trees. Dieback of leaders has been observed after winter and therefore, winter hardiness is assumed to be impaired in B deficiency (Räisänen et al. 2006a). The susceptibility to summer frost of fruit trees, ash and spruce seedlings (Räisänen et al. 2007), and many cereals (Huang et al. 2005) decreased also when leaves were treated with borax solution some days before frost occurred. In chilling-sensitive crop species, but not in chilling-tolerant ones, exposure to low temperature induced severe B deficiency symptoms (Huang et al. 2005). In contrast to the interaction between drought stress and B deficiency, extensive studies have been performed on the chilling stress under low B supply. These studies suggested that there are interactions between chilling temperature and B nutrition at the organ level as well as the whole plant level.

8.2.1 Boron Deficiency-Induced Impairment of Cold Acclimation

In plants supplied with adequate B and in which the buds are structurally normal, freezing tolerance increases properly during the cold acclimation in the autumn. However, if the buds are deformed because of B deficiency, they are not able to deep supercool and are unable to cold harden well accordingly (Räisänen et al. 2006b). The deformation is usually visible (under a stereomicroscope) as poor development of the primordial shoot, collenchymatic plate, and bud cavity in buds cut in half, even though all buds looked normal outward. Deep supercooling is the mechanism of survival of winter temperatures in most perennials in boreal forests. It means cooling of the buds to temperatures down to even −40°C without ice crystal formation in the primordial shoot. Deep supercooling of buds is dependent on the structures within the bud, as the collenchymatic plate in the bud axis functions as a barrier for the spread of ice (Räisänen et al. 2006b). Possible mechanisms that can affect the properties of this barrier include changes in the pectic compounds of the collenchymatic plate (Fleischer et al. 1999) or membrane–cell wall interactions (Bassil et al. 2004). Such changes might occur due to B deficiency even in buds that are not visibly deformed. Boron is a regulative element in pectin orientation in cell walls (Hu and Brown 1994). In stems and buds with normally developed structure, pectin-rich ice barriers with small micropores prevent lethal ice invasion into the sensitive tissues in deep supercooling stems and buds (Wisnievski 1995). B-deficient cells have been found to be nonelastic with increased pore-sizes in cell walls (Fleischer et al. 1999). If the structure of micropores in cell walls is irregular, the deep-supercooling ability may decline. A decrease in photoperiod initiates the cold acclimation of trees, which further proceeds at low temperatures with several physiological and anatomical changes in cells. Chilling temperatures are also needed for hardening of roots (Räisänen et al. 2009).

8.2.2 Chilling-Induced Reduction of B Uptake, Root–Shoot Translocation, and Partitioning

In chilling-sensitive cassava plants, exposure to root temperature of 18°C for 28 days at a relatively high B supply induced severe B deficiency symptoms due to the inhibition of B absorption its transport into the shoot. Plants grown at 22, 28, and 33°C, by contrast, did not show any B deficiency symptoms at the same level of B supply (Huang et al. 2005). However, in chilling-tolerant species, such as wheat and oilseed rape exposure of roots to 10°C did not decrease B uptake (Huang et al. 2005). In addition, reduction of B uptake under low temperatures was the result of impairment in root growth. In contrast to susceptible species, in the chilling-tolerant species, root growth was tolerant to low root temperature that led to an increased root: shoot ratio at 10°C in the root zone temperature (Huang et al. 2005). In turn, increased mortality of roots during soil freezing and thawing cycles due to low-B conditions and increase in the risk of freezing injury due to structural damages has been reported for Norway spruce roots (Räisänen et al. 2009). In a long-term field study in a Norway spruce stand with low B status, the proportion of dead fine roots decreased due to B fertilization (Möttönen et al. 2003).

8.2.3 Boron-Deficiency Induced Disruption of Water Relations Under Chilling Stress

Variation among species and genotypes in chilling tolerance is closely related to the effects of chilling on water uptake by roots and transpiration by leaves (Bloom et al. 2004).

Transpiration

In chilling-sensitive species such as cucumber, tomato, and sunflower, dysfunction of stomatal control expressed as delayed closure or closure failure causes an excessive transpiration (Allen and Ort 2001). Root low temperatures reduce B partitioning into new leaves and increase the sensitivity of growth of young leaves to low B supply (Ye et al. 2000).

Reduction of root hydraulic conductance in chilling-sensitive species like cassava, sunflower and tomato, reduction of root hydraulic conductance and water absorption under low temperatures are likely the cause of observed reduction in the B absorption and enhanced B deficiency. By contrast, chilling tolerant species such as oilseed rape are able to maintain a high root hydraulic conductance at similar root zone temperature (10–15°C) (Huang et al. 2005).

Reduction of Water and B Uptake via Water Channels in B-Deficient Plants

Water channels (aquaporins) in the plasma membrane of root cells play an important role in plant–water relationships particularly under changing environmental conditions (Tyerman et al. 2002). Evidences suggest that closing of water channels in chilling-sensitive species occurs in response to root chills. Water channels in the plasma membrane may be also reversibly closed by ·OH radicals (Henzler et al. 2004) that are accumulated under B deficiency conditions (Cakmak and Römheld 1997). In addition, water channels also play an important role in the B uptake across the plasma membrane particularly when external B concentrations are low (Dordas et al. 2000). Severe B deficiency causes reduction in the amount of the plasma membrane water channel proteins (ZmPIP1 aquaporins) in the roots of maize and transgenic tobacco (Goldbach et al. 2002).

Boron Xylem Loading

When external B levels are low, loading of B into the xylem requires the function of B transporters (Takano et al. 2002). The activity of B-transporters is sensitive to low temperature and chilling stress may inhibit active B transport across the pericycle into the stele of the root cylinder (Huang et al. 2005).

Plasma Membrane Function and Permeability

Instability of cell membranes has been recently suggested to explain decreased chilling tolerance of crop species in B deficiency (Huang et al. 2005). Since the plasma membrane is a primary site of freezing injury, B deficiency may increase the susceptibility of cells to frost damage (Räisänen et al. 2007). Changes in the proportions of sterols and longer chain fatty acids in the plasma membrane of root cells alters significantly B uptake in *Arabidopsis thaliana* mutants (Dordas and Brown 2000). Reduction in membrane fluidity and permeability of root cells following chilling stress may also contribute to the inhibition of B uptake in chilling-sensitive species such as *Coffea arabica* (Queiroz et al. 1998).

Function of H^+ATPase

Low temperature (6°C) during acclimation period increases transcription and translation of plasma membrane H^+ATPase gene (Ahn et al. 1999). Boron-deficiency-induced impairment of H^+ATPase activity (Roldan et al. 1992), which needs to be more activated under cold conditions, could be one of the reasons for the exacerbating effect of B deficiency under low temperatures.

8.2.4 Boron-Deficiency Induced Photoinhibition and Oxidative Stress Under Chilling Temperatures

Boron deficiency intensifies chilling-induced photooxidative damage and reduces plants antioxidant capacity during recovery from photoinhibition. In plants grown in low B soil, there is a relationship between chilling temperature and leaf tissue damage (bleached patches) (Ye 2005). One of the possible mechanisms is the role of B in the integrity and functions of thyalkoid membranes. In a chilling-sensitive cucumber cultivar, low temperatures enhanced effect of B-deficiency on membrane leakage (K^+) and chloroplast disruption and plasmolysis of mesophyll cells (Wang et al. 1999). Another reason for B-deficiency induced photoinhibition under cold conditions is the substrate feedback inhibition and starch accumulation in chloroplasts (Huang et al. 2005).

In addition, suboptimal B nutrition reduces the activity of antioxidant enzymes and the level of antioxidants such as reduced form of ascorbic acid, SH-compounds and GR (Cakmak and Römheld 1997). Moreover, B-deficiency-induced accumulation of soluble phenolics due to enhanced PAL activity and subsequent oxidation by polyphenol oxidase located in the thylakoid membrane and in cell walls (Cara et al. 2002) causes permanent damage to the cells and photosynthetic membranes. In addition, chilling temperature increases production of H_2O_2, which with superoxide forms highly reactive ·OH radicals in leaf cells (Saruyama and Tanida 1995). Accordingly, B deficiency increases the sensitivity of leaf cells to chilling through enhanced generation of ROS and weakened antioxidation capacity.

9 Effect of Molybdenum Deficiency on Plants Stress Responses

Molybdenum as component of specific plant enzymes participates in reduction and oxidative reactions. Molybdenum is a component of some bacterial nitrogenases and therefore is especially important for plants that live in symbiosis with N-fixing bacteria (Marschner 1995). Molybdenum deficiency influences plant metabolism at various levels. These responses are related mainly to the requirement of Mo for different types of molybdoenzymes in plants. Molybdenum is an integral part of an organic pterin complex called the Mo cofactor (MoCo). Plant molybdoenzymes include nitrate reductase (NR), xanthine dehydrogenase/oxidase (XDH), sulfite oxidase (SO) and those involved in abscisic acid (ABA) and indole-3 acetic acid (IAA) synthesis (aldehyde oxidase; AO). Nitrate reductase and SO contain a dioxo-Mo cofactor, while xanthine dehydrogenase/oxidase and AO have a monoxo-Mo cofactor which requires MoCo insertion and then subsequent sulfuration (Mendel and Haensch 2002). Since Mo is involved in various enzymatic processes and hormone biosynthesis, Mo deficiency can alter vegetative and reproductive growth and susceptibility to biotic (Graham and Stangoulis 2005) and abiotic stress factors. Indeed, Mo plays an important role in plants adaptation to environmental stresses through its effect on the activity of aldehyde oxidase.

9.1 Relationship Between Molybdenum Nutrition and Responsiveness of Plants to Stresses

During last decades there have been numerous reports of the increased susceptibility of Mo-deficient plants to environmental stresses such as salinity, drought and low temperature. These findings could be explained in recent years after elucidation of the role of the gene (LOS5) in the converting desulfo/dioxyo form of MoCo to the sulfide form of MoCo, a cofactor of aldehyde oxidase that catalyzes the last step of abscisic acid (ABA) biosynthesis. The los5 mutant (low expression of osmotically responsive genes) is highly sensitive to drought and salinity as well as cold stress due to the lack of ability for ABA biosynthesis (Xiong et al. 2001). Drought, salinity, and to some extent cold stress cause an increased biosynthesis and accumulation of ABA mainly by the induction of genes coding for ABA biosynthetic enzymes (Rock 2000). The first gene encodes zeaxanthin epoxidase (ZEP), which converts zeaxanthin to epoxycarotenoid and is defective in the Arabidopsis mutants aba1 and los6. The cleavage enzyme 9-*cis*-epoxycarotenoid dioxygenase (NCED) catalyzes the conversion of epoxycarotenoids to xanthoxin. Abscisic acid aldehyde oxidase (AAO) converts ABA aldehyde to ABA and is defective in the Arabidopsis aao3 mutant (Schwartz et al. 2003). Aldehyde oxidase is a MoCo enzyme and requires sulfuration for activation. This step is catalyzed by a MoCo sulfurase (MCSU), which is encoded by the ABA3/LOS5 locus in Arabidopsis (Fig. 16.5). Mutations in the aldehyde oxidase apoprotein and MoCo biosynthetic enzymes would lead to ABA deficiency in plants (Xiong et al. 2002a).

9.2 Increased Susceptibility of Plants to Salt and Drought Stress

Osmotic stress resulting from either high salinity or water deficit induces the expression of numerous stress-responsive genes in plants (Hasegawa et al. 2000). The role of ABA in the signal transduction of ionic and nonionic stresses has been extensively studied (Xiong et al. 2002b). Low ABA levels result in wilty plants with excessive transpiration and without stomatal control, altered seed dormancy, and impaired defense responses (Mendel and Haensch 2002). The ABA-deficient mutants flacca and aba3/los5, which are disrupted in the MoCo sulfuration step, have wilty phenotypes and increased transpirational water loss (Bittner et al. 2001; Sagi et al. 2002). In turn, one of the distinct responses of Mo-deficient plants is

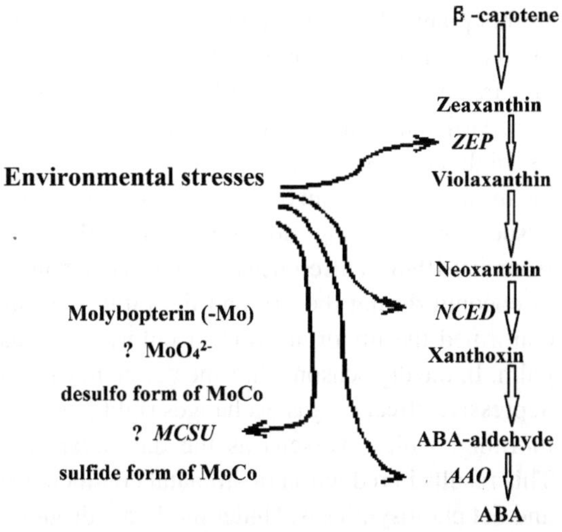

Fig. 16.5 The role of Mo in the ABA signaling pathway under environmental stresses. ABA is synthesized from β-carotene via the oxidative cleavage of neoxanthin and a two-step conversion of xanthoxin to ABA via ABA-aldehyde. Environmental stress such as drought, salt and, to a lesser extent, cold stimulates the biosynthesis and accumulation of ABA by activating genes coding for ABA biosynthetic enzymes including *ZEP* (*LOS6/ABA1* in *Arabidopsis*, codes for zeaxanthin epoxidase), *NCED* (*NCED3* in *Arabidopsis*, codes for 9-*cis*-epoxycarotenoid dioxygenase), *AAO* (*AAO3* in *Arabidopsis*, codes for ABA aldehyde oxidase), and *MCSU* (*LOS5/ABA3* in *Arabidopsis*, codes for molybdate cofactor sulfurase)

flaccid and cupped leaves similar with flacca and aba3/los5 mutants (Robinson and Burne 2000).

9.3 Increased Frost Damage

It has been shown that addition of Mo to acid soils will protect plants against damage caused by low temperature or water logging (Marschner 1995). Under acidic soil conditions, the molybdate anion is adsorbed strongly to the surface of Fe and Al oxides by a ligand exchange mechanism and the Mo concentration in the soil solution can be reduced greatly (Hamlin 2007). Molybdenum deficiency is a micronutritional disorder that has been reported frequently in plants grown on acidic soils (Marschner 1995). los5 mutant plants are not only susceptible to salinity and drought but also sensitive to freeze-induced damages. The impairment of low temperature gene regulation is specific to the los5/aba3 mutation and the LOS5/ABA3 gene is expressed in different plant parts and is a key regulator of ABA biosynthesis in response to stresses (Xiong et al. 2001).

10 Effect of Chlorine Deficiency on Plants Stress Responses

Chlorine is classified as a micronutrient, but it is often taken up by plants at levels comparable to a macronutrient. Supplies of chlorine in nature are often plentiful, and obvious symptoms of deficiency are seldom observed. However, some plant species such as members of Palmaceae and kiwifruit (*Actinidia deliciosa*) have a much higher chlorine requirement, thus, chlorine deficiency can readily be induced in these species. Chlorine appears to be required for optimal enzyme activity of asparagine synthethase, amylase and ATPase. In photosynthesis, chlorine is an essential cofactor for the activation of the oxygen-evolving enzyme associated with PSII. Chlorine binds to the polypeptides associated with the water-splitting complex of PSII and stabilize the oxidized state of Mn by acting as a bridging ligand (Marschner 1995).

10.1 Functions of Chloride in Plants Water and Ion Balance

Most of the chlorine in plants is not incorporated into organic molecules or dry matter, but remains in solution as the monovalent ion chloride (Cl^-). Chlorine concentrations required for biochemical functions are relatively low in comparison to concentrations required for osmoregulation. Chlorine concentration in plants exceed the critical deficiency level (~6 mM) by two orders of magnitude, therefore, is important in osmotic adjustment and plant water relations including role in xylem volume flow and root pressure, phloem loading and unloading (Marschner 1995).The accumulation of chloride in plant cells increases tissue hydration and turgor pressure. This osmotic function of Cl^- works closely with K^+ to facilitate cell elongation and growth. Relative differences in the uptake of cations and anions by plants require the maintenance of electroneutrality in plant cells as well as in the external soil solution. As an anion, chloride serves to balance charges from cations (Heckman and Strick 1996). However, it seems likely that specified function of chlorine in osmoregulation is mainly restricted to distinct organs (e.g., extension zones, the stigma of grasses, pulvini of *Mimosa pudica* during seismonastic leaf movement) or cells (e.g., guard cells) (Marschner 1995). This osmoregulatory function in specific tissues requires also concentrations of chloride that are not typical of a micronutrient (Flowers 1988). In rapidly expanding tissues such as elongating cells of roots and shoots and expanding leaves, chloride accumulates in the tonoplast, to function as an osmotically active solute. Chloride is essential for stomatal functioning in some plant species such as onion (*Allium cepa* L.), which lacks functional chloroplasts for malate synthesis (Schnabl and Raschke 1980). Members of the Palmaceae such as coconut (*Cocus nucifera* L.) and oil palm (*Elaeis guineensis* Jacq.) also need Cl^- for stomatal functioning (Marschner 1995).

10.2 Drought Tolerance

The most commonly described symptom of chlorine deficiency is wilting of leaves, especially at the margins. As the deficiency becomes more severe, the leaves may exhibit curling, shriveling, and necrosis. Roots of chlorine-deficient plants become stubby with club tips. In chlorine-deficient wheat, the symptoms are expressed as chlorotic or necrotic lesions on leaf tissue (Engel et al. 2001). In coconut palm, the symptoms are exhibited as wilting and premature senescence of leaves, frond fracture, and stem cracking and bleeding (Marschner 1995). Coconut palm is of great economic importance in the tropics and subtropics and drought is one of the main environmental factors that limit coconut productivity. In this plant, chloride is an importance factor in the mechanisms governing stomatal opening and closure and is also important for stomatal regulation, particularly during the dry season. Moreover, its high concentration in coconut leaf tissues means that it acts as an osmoticum in maintaining tissue turgor during drought (Braconnier and Bonneau 1998). Differences in the gas exchanges in coconut during the dry and the rainy seasons confirmed the important role of chloride in this palm. In the dry season, chlorine deficiency has a depressive effect on gas exchanges right from the morning, which worsens as the day wears on. This results in reduction of stomatal conductance and net photosynthesis. Under moderate drought, coconut palms not suffering from a chlorine deficiency respond to higher evaporative demand by increasing their stomatal conductance and transpiration, and by maintaining a reasonable level of net photosynthesis. Under the same conditions, deficient palms react by reducing their stomatal conductance and net photosynthesis, hence expressing a state of stress. The chloride therefore enables coconut palms to withstand the dry season, by maintaining a relatively high level of leaf gas exchanges (Braconnier and Bonneau 1998).

10.3 Resistance Against Pathogens

Addition of chlorine has been reported to reduce the severity of at least 15 different foliar and root diseases on 11 different crops (Heckman 2007). Several possible mechanisms may explain the effects of chlorine nutrition on disease suppression and host resistance. In acid soils, chloride inhibits nitrification (Rosenberg et al. 1986). Keeping N in the ammonium form can lower rhizosphere pH and influence microbial populations and nutrient availability in the rhizosphere (Heckman and Strick 1996). Competition between chloride and nitrate for uptake also tends to reduce nitrate concentrations in plant tissues. When plants take up more ammonium and less nitrate, it usually causes rhizosphere acidification, which in turn, may enhance Mn availability (Thompson et al. 1995).

Chlorine can also enhance Mn availability by promoting Mn-reducing microorganisms in soil. Factors which increase Mn availability have been associated with improved host resistance to diseases in grain crops (Huber 1989). Higher concentrations of chlorine in plant tissues can also enhance water retention and turgor when roots have been attacked by pathogens. The amount of organic acids, such as malate, in plant tissues and exuded from roots, decreases with chlorine supply. This action deprives pathogens of an organic substrate (Goos et al. 1987).

11 Effect of Nickel Deficiency on Plants Stress Responses

Nickel is the latest element to be classified as essential for plant growth, however, its biological role is poorly understood. This is mainly because plants require low levels of Ni while it is relatively abundant in soil (Marschner 1995). The knowledge of Ni uptake by plants is indeed very limited, and apart from the observation that Ni is quite mobile as compared to other heavy metals, little is known about the uptake mechanism and translocation under Ni-limiting conditions (Brown 2007). There are several enzyme systems (NiFe-hydrogenase, carbon monoxide dehydrogenase, acetyl-CoA decarbonylase synthase, methyl-coenzyme M reductase, superoxide dismutase, Ni-dependent glyoxylase, acireductone dioxygenase, and methyleneurease) in bacteria and lower plants (Mulrooney and Hausinger 2003) that are activated by Ni. Activation of urease with two Ni ions at the active site (Ciurli 2001), however, is the only known biological function of Ni in higher plants (Gerendás et al. 1999).

The metabolic effects of Ni deficiency have been reported in cereals (Brown et al. 1990), legumes (Gerendás and Sattelmacher 1997) and perennial species (Bai et al. 2006). These include reduced urease activity, induced metabolic N deficiency, disruption of N metabolism because of alterations in the ureide catabolism and metabolism of amino acids and ornithine cycle intermediates. Function of citric acid cycle disrupts also in Ni-deficient plants. Under Ni-deficiency conditions, leaves contained low levels of citrate compared to Ni-sufficient leaves and accumulated lactic and oxalic acids (Bai et al. 2006).

According to these results, Ni deficiency results in distinct biochemical symptoms even before development of morphological symptoms and disruption of vegetative growth. The wide spectrum of metabolic disruption in Ni-deficient plants is an evidence for the existence of unidentified physiological roles for Ni in plants. This finding in combination with the diverse known functions of Ni in bacteria suggests that Ni may indeed play a role in many, yet undiscovered processes in higher plants (Brown 2007).

Improvement of our knowledge of the biochemical role of Ni in plants may bring new insights into how Ni nutrition affects plants stress responses. Genetics and molecular biology approaches may be useful in identification of the roles of Ni in the biochemical processes particularly under stressful conditions similar with the studies on Mo and its effect on the plants stress response via ABA metabolism.

12 Effect of Beneficial Elements on Plants Stress Responses

Mineral elements that either stimulate growth but are not essential or that are essential only for certain plant species or under specific conditions, are defined as beneficial elements. This definition applies to sodium (Na), silicon (Si), selenium (Se), cobalt (Co), and aluminum (Al) (Marschner 1995). The main physiological functions of beneficial elements are presented in Table 16.5.

For elements defined as beneficial, instead of application of the word "deficiency", it seems to be more practical to focus on the plant responses in the presence of these elements. In this section, we summarize evidences showed plants response to supplementation with beneficial elements when grown under various stressful environmental conditions.

Table 16.5 Physiological functions of beneficial elements in higher plants

Element	Physiological function
Sodium	Essential nutrient for some species from Chenopodiaceae, Amaranthaceae, and Cyperaceae. Substitution for K in some metabolic and nonmetabolic roles. Osmosis regulation in halophytes. Role in photosynthesis of some C_4 species: in carbon shuttle between bundle sheet and mesophyll cells, in uptake of pyruvate into plastids of mesophyll cells, in maintenance of chloroplasts ultrastructure of mesophyll cells. Growth stimulation in natrophilic species: increased cell elongation and expansion, production of greater photosynthetic area, acceleration of stomatal response to drought, improvement of water relations, succulence, and water use efficiency.
Silicon	Essential nutrient for species from Equisetaceae and wetland Poaceae. Prevention of toxicity of P, Mn and Fe. Roles in: stability of plants, cell wall rigidity and elasticity, increasing leaf erectness, decreasing effects of mutual shading, reduction of susceptibility to lodging, increasing the volume and rigidity of aerenchyma and root oxidizing power of wetland plants, reduction of cuticular transpiration. Role in increasing plants resistance against fungi and pest attacks.
Selenium	Activates antioxidant defense system. Involves in salicylic acid and jasmonic acid pathway of plants stress responses. Stimulates growth under unstressed conditions. Protects plants against UV stress, low and high temperatures, heavy metal toxicity and delays senescence. In accumulator species protects plants from fungal infection and herbivory.
Cobalt	As constituent of cyanocobalamin (vitamin B_{12}) is effective on the activity of methionine synthase, ribonucleotide reductase and methylmalonyl-CoA mutase in soil microorganisms, nitrogen fixing bacteria and rumen microflora in ruminants. Role in leghemoglobin synthesis in nodulated legumes. Inhibition of ACC oxidase activity.
Aluminum	In accumulator species stimulates growth, particularly of root. Activates H^+ATPase activity. Alleviates H^+ toxicity, increases plants adaptation to P and B deficiency. Activates antioxidant defense system and improves membrane integrity. Increases plant resistance to pathogens.

13 Effects of Sodium Supplementation on Plants Stress Responses

Sodium has been studied more for its negative effect at excess levels (salt stress) than as a beneficial or essential element. Sodium is essential only for some C_4 species, but is undoubtedly beneficial to the growth of euhalophytes. It may stimulate the growth of some species with an evolutionary history in saline environments, and even of apparently totally glycophytic species under certain conditions.

Although Na has not been shown to be an "essential nutrient" for most plants, there is a high degree of Na utilization in many plants and some utilization in most if not all plants. The criteria described by Arnon and Stout (1939) that must be met for an element to be considered as an "essential nutrient" for plants are based exclusively on ecological considerations for survival and reproduction; high yield or biomass production may or may not be an important aspect, and may not even be associated with nutrient essentiality. For example, some mineral elements such as Na, Se and Si may promote increased biomass production, but may not be required for the species to survive. To overcome some of the limitations and difficulties associated with a strict definition of "essentiality," the term "functional or metabolism nutrient," has been suggested (Nicholas 1961) which is defined as "any mineral element that functions in plant metabolism irrespective of whether or not its action is specific." Recently (Subbarao et al. 2003), this term has been defined as "an element that is essential for maximal biomass production or can reduce the critical level of an essential element by partially replacing it in an essential metabolic process." This section deals with this issue and presents evidence to support the notion that Na should be considered as a "functional nutrient," based on the above definition.

Because of the chemical similarity between K and Na, it is generally assumed that K and Na compete for common absorption sites in the root. Sodium, even in 20-fold excess, fails to compete significantly with K under mechanism I, while Mechanism II does not discriminate K from Na and thus Na can competitively inhibit the absorption of K. Recently, K_{in} channels (inward rectifying K channels) have been reported in different root cells, including cortical, root hair, stelar, and xylem parenchyma cells, which can sense K concentrations (Blumwald et al. 2000). Although it is widely believed that mechanism I does not have much affinity to transport Na in the presence of adequate K, for some crops such as beet this mechanism may be transporting Na independent of the external concentration. Several Atriplex species take up Na in preference to K. In these species, Na competes with K during uptake, but K does not compete with Na. Thus, specific mechanisms of Na transport at low concentrations and in the presence of K is open to further investigation (Subbarao et al. 2003).

13.1 Effects in C_4 Species

In some C_4 species such as *Atriplex vesicaria* (Chenopodiaceae), *Amaranthus tricolor* (Amaranthaceae), and *Panicum miliaceum* (Poaceae), Na is required for the function of CO_2 concentration mechanism, plays a critical role in the regeneration of phosphoenolpyruvate (PEP) in mesophyll chloroplasts, has a role in Chl synthesis, pyruvate uptake into chloroplasts via Na^+/pyruvate cotransport system and in nitrate assimilation (Marschner 1995). Sodium deficiency impairs conversion of pyruvate to PEP in the mesophyll chloroplasts, leads to a reduction in PSII activity and ultrastructural changes in mesophyll but not bundle sheath chloroplasts, and reduction of nitrate-reductase activity (Marschner 1995). In sorghum species (*Sorghum* L.), there is a specific effect of higher concentrations of Na on the kinase that regulates the activity of PEP carboxylase, the primary carbon-fixing enzyme in C_4 and crassulacean acid metabolism (CAM) plants (Monreal et al. 2003).

13.2 Drought Tolerance

In natrophilic species such as sugar beet when the availability of water in the substrate is high, Na decreases the total dry mass per unit water consumption, i.e., water use efficiency. If, however, the availability of water in the substrate is low, water use efficiency remained unchanged in plants supplied with Na but increases sharply in plants receiving a K supply only. Improvement of water balance of plants when the water supply is limited is obviously occurs via stomatal regulation. With a sudden decrease in the availability of water in the substrate (drought stress), the stomata of plants supplied with Na close more rapidly than plants supplied with K only and, after stress release, exhibit a substantial delay in opening. As a consequence, in plants supplied with Na the relative leaf water content is maintained at a higher level even at low substrate water availability (drought periods, saline soils) (Marschner 1995).

13.3 Salt Tolerance

An improved osmotic adjustment is a major factor in growth stimulation of halophytes by high Na supply. Growth responses of halophytes to Na under saline conditions reflect the need for an osmoticum during osmotic adjustment to salinity stress. Many halophytes osmotically compensate for high external osmotic potential by accumulating Na salts, often NaCl from the environment. Growth stimulation by Na is particularly apparent in the Chenopodiaceae and among nonchenopods, some cultivars of tomato adapted to saline soils has been reported to respond positively to additional Na (Hajiboland et al. 2010). In the presence of Na, cell expansion in natrophilic species is maintained and water balance is even improved. In these species, not only can Na replace K in its contribution to the solute potential in the vacuoles and in the generation of turgor and cell expansion, it may surpass K in this respect since it accumulates preferentially in the vacuoles. The superiority of Na can be demonstrated by the expansion of sugar beet leaf segments in vitro as well as in intact sugar beet plants, where leaf area, thickness and succulence are distinctly greater when a high proportion of K is replaced by Na (Marschner 1995). In sugar beet, mild salinity (5.5 dS m^{-1}) caused significant improvement in the yield and sugar content of storage roots (Hajiboland et al. 2009). Sugar beet cultivars differ in the response to low (50 mM) salinity (Hajiboland and Joudmand 2009). In the cultivar with positive response to low salinity (IC), in addition of higher dry matter production and broader leaves, membrane integrity was even improved under low salinity (Table 16.6).

Table 16.6 Shoot and root dry weight (mg plant^{-1}), leaf area (cm^{-2} plant^{-1}), leaf K^+ leakage (μg g^{-1} DW min^{-1}) and Na concentration in various fractions of leaf (mg g^{-1} FW) in two cultivars of sugar beet (*Beta vulgaris* L. cvs. 7233 and IC) treated with 50 mM NaCl in the nutrient solution for 2 weeks

	Growth parameters							
	Shoot DW		Root DW		Leaf area		K^+ leakage	
Cultivars	Control	Salinity	Control	Salinity	Control	Salinity	Control	Salinity
IC	675±121^{b}	1228±161^{a}	309±49^{a}	400±65^{a}	379±22^{b}	453±32^{a}	135±27^{a}	70±14^{b}
7233	724±218^{a}	968±201^{a}	320±50^{a}	189±47^{b}	437±11^{a}	348±33^{b}	82±16^{b}	147±29^{a}

	Na concentration and partitioning							
	Total Na concentration		Cell sap Na		Apoplasmic fluid Na		Residual Na	
Cultivars	Control	Salinity	Control	Salinity	Control	Salinity	Control	Salinity
IC	0.94±0.05	8.02±1.09	0.25±0.09 (27)	3.60±0.68 (45)	0.11±0.01 (10)	0.28±0.03 (3)	0.53±0.02 (56)	3.56±0.98 (44)
7233	0.82±0.05	8.85±1.31	0.37±0.09 (45)	3.00±0.32 (34)	0.12±0.02 (15)	0.17±0.06 (2)	0.27±0.02 (33)	5.22±1.33 (59)

[a]Calculated values related to total Na concentration are given in parentheses. Data of each row within each parameters followed by the same letter are not significantly different ($P<0.05$)

[b](Hajibland and Joudmand 2009)

Fractionation of Na in the leaves showed that proportional Na in cell sap (mainly vacuole) was higher in IC, and by contrast, the proportional Na in residual fraction (comprised mainly from cell wall) was lower in this cultivar (Table 16.6). Allocation of more Na to the cell sap may result in facilitating control of water balance of leaf cells and causes an improvement of cell expansion and production of broader leaves (Hajiboland and Joudmand 2009). Finally, salinity stress is known to induce CAM photosynthesis in the facultative CAM species, such as *Mesembryanthemum crystallinum* L., (Aizoaceae) and *Sedum album* (Crassulaceae) (Cushman and Bohnert 2002). Crassulacean acid metabolism (CAM) is a metabolic adaptation of plants to environments with limited availability of water because of restricting CO_2 uptake to the dark hours and reduction of transpirational water loss.

13.4 Potassium Deficiency

The presence of Na in the environment and its uptake by plants can reduce the amount of K required to meet the plants basic metabolic requirements. K functions in plants can be summarized as both biophysical (non-K-specific role as an osmoticum in the vacuole) and biochemical (specific and nonspecific roles in the cytoplasm). The need of monovalent cations in some plant species can also be filled by Na, thus reducing the required critical level of tissue K. In natrophilic species such as sugar beet with a high ability for substitution of K by Na, in old leaves nearly all the K can be replaced by Na that made K available for specific functions in meristematic and expanding tissues. Sodium alleviates K-deficiency symptoms and decreases the critical foliar K concentration at which K-deficiency symptoms appeared (Subbarao et al. 2000).

14 Effects of Silicone Supplementation on Plants Stress Responses

Silicon has been proved to be beneficial for the healthy growth and development of many plant species, particularly graminaceous plants such as rice and sugarcane and some cyperaceous plants (Marschner 1995). The beneficial effects of Si are particularly distinct in plants exposed to abiotic and biotic stresses. Epstein and Bloom (2005) have recently established a new definition for essential elements in higher plants. According to these authors, an element is essential that fulfills either one or both of the following criteria: (1) the element is part of a molecule which is an intrinsic component of the structure or metabolism of the plant, and (2) the plant can be so severely deficient in the element that it exhibits abnormalities in growth, development, or reproduction, i.e., "performance", compared to plants with lower deficiency. Accordingly, Si will be an essential element for higher plants, which is to be generally accepted in the near future. Over last two decades, extensive studies have been performed aiming at understanding of the possible mechanism(s) for Si-enhanced tolerance of higher plants to both abiotic and biotic stresses (Liang et al. 2007). More recently, rapid progress has been also made in Si uptake and transport in higher plants. The uptake of Si was found to be the result of two different transport mechanisms. A low affinity transporter (Lsi1) found on the lateral roots of rice plants is responsible for the uptake of silicic acid from the external solution to the root cortical cells (Ma and Yamaji 2006). The transporter has been localized on the distal cells of exodermis and endodermis. A second transporter has also been identified in rice which is responsible for xylem loading of Si (Mitani and Ma 2005).

In this section we review current knowledge on the roles of Si in conferring tolerance to plants against abiotic stresses. Because of a well-documented role of Si in the plants resistance against biotic stress factors such as pathogens, we also give a brief overview on this effect of Si.

14.1 Drought Tolerance

Optimization of silicon nutrition results in increased mass and volume of roots, giving increased total and adsorbing surfaces (Kudinova 1975). These plants could more efficiently extract water from drying substrate than plants without Si supplementation. Experiment with citrus

(*Citrus* spp.) has demonstrated that with increasing monosilicic acid concentration in irrigation water, the weight of roots increased more than that of shoots (Matichenkov et al. 1999b). The same effect was observed for Bahia grass (*Paspalum notatum* Flügge) (Matichenkov et al. 2000). Greater root–shoot mass ratio provides greater water absorption surface and lower transpiration area leading to a considerable increase in plants tolerance to drought. In the cell walls of xylem elements Si deposition increases resistance of vessels to collapses caused by negative sap pressures particularly under conditions of high transpiration and drought or heat stress. In addition, the silicon-cellulose membrane in epidermal tissue also protects plants against excessive loss of water by cuticular transpiration. This action occurs owing to a reduction in the diameter of stomatal pores and, consequently, a reduction in leaf transpiration (Snyder et al. 2007). In rice plants, Si can alleviate water stress by decreasing transpiration. Rice plants have a thin cuticle and the formation of a cuticle-Si double layer significantly decreases cuticular transpiration. Since water stress causes stomata closure and reduction of photosynthetic rate, Si stimulates the growth and photosynthesis of rice more clearly under water-stresses than nonstressed conditions (Ma et al. 2001). Furthermore, deposition of Si in rice increases the thickness of the culm wall and the size of the vascular bundle preventing lodging. Sterility is related to many factors including excess water loss from the hull. Transpiration from the panicles occurs only from the cuticle of the hull because the hull has no stomata. Silicon deposition on the hull decreases the transpiration from panicles by about 30% at either milky or maturity stage, preventing excess water loss. This is the reason why Si application significantly increases the percentage of ripened grain (Ma et al. 2001).

14.2 Tolerance to Flooding, Lodging and Mutual Shading

Si deposition in the epidermal layer of the leaves is thought to be responsible for improvement of exposure to light, increase of resistance to lodging, and reduction of mutual leaf shading particularly in dense stands of cereals (Marschner 1995). Leaf erectness is an important factor affecting light interception in dense plant stands. The effect of Si on leaf erectness is mainly a function of the Si deposition in the epidermal layers of the leaves and, thus, over a wide range closely related to the Si concentration supplied (Balasta et al. 1989). Also in dicotyledonous species, such as cucumber, Si increases the rigidity of mature leaves, which are held more horizontally, increases their Chl content, and delays their senescence (Adatia and Bestford 1986).

14.3 Tolerance to Frost Damage and Cold

Proper silicon nutrition can increase frost resistance by plants (Matichenkov et al. 1999a). However, this mechanism remains poorly understood. Cool summers (low temperature an insufficient sunlight) usually cause serious damage to rice production. Low temperatures decrease Si uptake by rice and in sufficient sunlight lowers the Si:N ratio, which induces blast. Application of Si under such conditions markedly reduces the incidence of blast in rice (Ohyama 1985). The effect of Si is most evident under low light intensity. The Si effect on rice growth under shaded conditions is larger than that without shading but the mechanism responsible for these phenomena is unknown. It has been hypothesized that Si deposited on the leaf epidermal system might act as a window to facilitate the transmission of light into photosynthetic mesophyll tissue (Agarie et al. 1996). However, evidence supporting this hypothesis could not be obtained at this time.

14.4 Resistance Against Pathogens and Pests

Two possible mechanisms of Si-enhanced plant resistance to pathogens have been proposed. In the first mechanism, polymerized Si can reinforce the cell walls by physically inhibiting fungal germ tube penetration of the epidermis, thereby impeding infections (Hayasaka et al. 2008). Polymerization of monosilicic acid into polysilicic acid and its transform to amorphous silica forms a thickened silicon-cellulose membrane (Aleshin 1988) which can be associated with pectin and Ca ions (Waterkeyn et al. 1982). Such a double-cuticular layer protects plants against attack by fungal pathogens (Yoshida 1975). Silicon forms also some complexes with cell wall components and decreases its sensitivity to enzymes released by the rice blast fungus (*Magnaporthe grisea* M.E. Barr). Indeed, silicon can be associated with lignin–carbohydrate complexes in the cell wall of rice epidermal cells (Inanaga et al. 1995). In the second mechanism, soluble form of Si within plants can induce defense response and Si may act locally as a signal in triggering natural defense response in both dicots and monocots, by stimulating the activity of such enzymes as chitinases, PODs, polyphenol oxidase, and/or by increasing the production of phenolic compounds, phytoalexins, antimicrobial compounds and systemic stress signals (salicylic acid, jasmonic acid, and ethylene) (Ghanmi et al. 2004). Silicone has a similar saturable effect and can significantly change the activity of signaling systems in cells after elicitation, including the mitogen activated protein (MAP) kinases (Fauteux et al. 2005). Only the soluble form of Si within plants can induce defense responses, while the polymerized fraction is almost inert. Therefore, Si-induced plant resistance to pathogens vanishes when Si supply to plants is stopped, even though Si had irreversibly accumulated (Fauteux et al. 2005).

14.5 Silicon-Enhanced Tolerance to Salinity

The mitigative effect of Si on salinity has been examined in rice, mesquite, wheat, barley, cucumber and tomato (Liang et al. 2007). The Na concentration in the shoots of rice and barley was reported to decrease by addition of Si that was attributed to Si-induced reduction in transpiration rate and to the partial blockage of the transpirational bypass flow. Reduction in the uptake and root–shoot transport of Na and increase of that for K has been attributed to Si-induced stimulation of the root plasma membrane H^+ATPase under salt stress (Liang et al. 2006). Added Si decreased the permeability of the plasma membrane of leaf cells, and significantly improved the ultrastructure of chloroplasts which were damaged by the added NaCl with the double membranes disappearing and the grannae being disintegrated in the absence of Si (Liang et al. 2007). Silicon also increases activity of antioxidant enzymes, decreases the malondialdehyde (MDA) concentration, and suppresses membrane leakage in barley under salt stress (Liang et al. 2003) (Fig. 16.6). Silicone supplementation stimulated root H^+ATPase and H^+PP_iase activity in the plasma membranes and tonoplasts and mediated membrane fluidity, suggesting that Si may affect the structure, integrity and functions of plasma membranes by influencing the stress-dependent peroxidation of membrane lipids (Zhu et al. 2004). There are other hypotheses for the ameliorative effect of Si on salt stress include improved photosynthetic activity, enhanced K/Na selectivity ratio, and increased concentration of soluble substances in the xylem, resulting in limited Na adsorption by plants (Snyder et al. 2007).

14.6 Mineral Stress

Mineral stress can be classified into the deficiency of essential elements and the excess of essential and other elements. Many reports have shown the beneficial effects of Si under mineral stress. In this section, the beneficial effects under P deficiency and excess of N are described.

14.6.1 Phosphorus Deficiency

According to the long term field experiment on rice and barley, the effect of Si on plants yield is larger when P is not supplied. Previously, such beneficial effects of Si were explained as an improvement of P availability in soil. However, later experiment showed that Si does not have any effect in P availability in soil and it seems unlikely that interaction between Si and P occurs in the soil. In a solution culture experiment, no

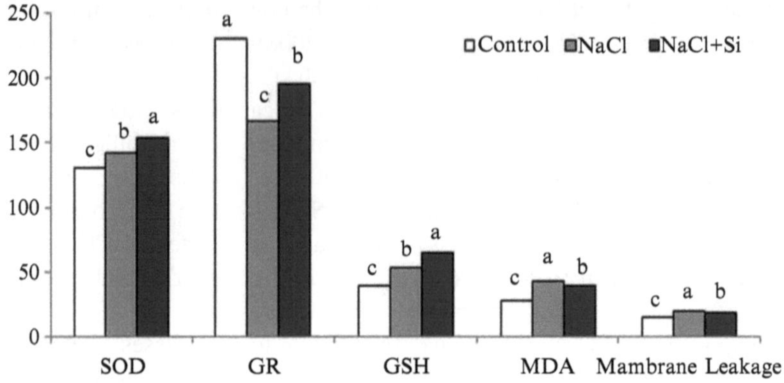

Fig. 16.6 Effect of Si supplementation (1 mM) on the activity of SOD (Unit mg^{-1} protein), GR (NADPH absorption g^{-1} protein) and concentration of GSH (μg g^{-1} FW) and malondialdehyde (MDA) (nmol g^{-1} FW), and electrolytic leakage (%) of the roots of barley plants (*Hordeum vulgare* L. cv. Jian 4) under salt (120 mM) stress (Liang et al. 2003)

significant effect of Si was observed on dry weight of shoot, root, and grain of rice when P was supplied at an adequate level. However, when the P level is decreased, the effects of Si are obvious. Phosphorus uptake is not enhanced by Si in rice, and this implies that Si improves internal P utilization (Ma et al. 2001).

14.6.2 Nitrogen Excess

Application of high levels of N is common for maximum yield in some crop species. Under such cultural conditions, leaf erectness is an important factor affecting light interception. Leaf erectness decreases with increasing N application, but Si application increases leaf erectness, decreasing mutual shading caused by dense planting and high N application. Excess N also increases susceptibility to disease such as blast in rice, but Si decreases the occurrence of blast disease in rice with high N fertilizer applications (Ma et al. 2001).

14.6.3 Mechanisms for Silicon-Mediated Alleviation of Metal Toxicity

Silicon-mediated alleviation of (heavy) metal toxicity in higher plants is widely accepted.

14.6.3.1 Manganese Toxicity

Recently the role of silicon in mitigating Mn toxicity has been investigated extensively in barley, rice, bean, pumpkin, cowpea and cucumber (for a review see Liang et al. 2007). Total Mn in the leaves was unaffected by Si but Si caused Mn to be more evenly distributed instead of being concentrated in discrete necrotic spots. Silicon lowered the apoplastic Mn concentration in cowpea, suggesting that Si may modify the cation binding capacity of the cell wall (Horst et al. 1999). Iwasaki et al. (2002) found that Si supply alleviates Mn toxicity not only by decreasing the concentration of soluble apoplastic Mn through enhanced adsorption of Mn on the cell walls, but also a role of soluble Si in the apoplast in the detoxicification of apoplastic Mn was indicated. Further research suggested that Si may affect the oxidation process of excess Mn mediated by POD through interaction with phenolic substances in the solution phase of the apoplast, maintaining the apoplast in a reduced state, which is thought to be a requirement for improved Mn tolerance of the leaf tissue. Study of Mn toxicity in cucumber clearly showed that plants not treated with Si had higher Mn concentrations in the intercellular washing fluid compared with plants treated with Si despite approximately the same total Mn content in the leaves. The Mn concentration of the intercellular washing fluid was positively correlated with the severity of Mn-toxicity symptoms and negatively correlated with the Si supply (Rogalla and Römheld 2002). Furthermore, in Si-treated plants less Mn was located in the symplast (<10%) and more Mn was bound to the cell wall (>90%) compared with non-Si-treated plants with about 50% in each compartment. Manganese present in Si-treated plants is therefore less available and for this reason less toxic than in plants not treated with Si (Rogalla and Römheld 2002).

Therefore, Si-mediated tolerance of Mn in cucumber is a consequence of stronger binding of Mn to cell walls and a lowering of the Mn concentration within the symplast. In other reports, alleviation of Mn toxicity by Si in cucumber was attributed to a significant reduction in membrane lipid peroxidation caused by excess Mn and to a significant increase in enzymatic (e.g., SOD, APX, and GR) and nonenzymatic antioxidants (e.g., ascorbate and GSH) (Shi et al. 2005a).

14.6.3.2 Aluminum Toxicity

Interactions of Al and Si has been studied in some plant species and alleviating effects of Si on Al toxicity has been reported for sorghum, barley, maize, and soybean (Liang et al. 2007). It has been established that interaction between Si and Al and reduction of the activity of toxic metal ions in the medium, is one possible external mechanism for the detoxification of Al toxicity by Si (Hiradate et al. 1998). The precipitation of subcolloidal, inert hydroxyaluminosilicate species seems to be responsible for the diminished concentration (activity) of phytotoxic Al in solution. The codeposition of Si with Al seems to occur not only in the growth media but also within plants (Liang et al. 2001). It has been reported that added Si increased the shoot Al concentration, which may arise from the formation of hydroxyaluminosilicate complexes in shoots leading to enhanced Al transport from roots to shoots (Birchall 1990). It has been also proposed that low-solubility aluminosilicates or hydroxyaluminosilicates (or both) are formed within the root cell wall (apoplastic) space, thereby reducing the concentration of free, toxic Al^{3+} ions (Cocker et al. 1998). Maize plants exposed to toxic Al concentrations were less inhibited in their growth in the presence of Si that was attributed mainly to the inhibitory effect of Al uptake by the plants because of higher concentrations of malice and formic acids in the presence of added Si, thus a purely internal mechanism related to the physiological processes within plants (Corrales et al. 1997). Silicon may have additional roles in increasing tolerance of Al by mediating the metabolism of phenolic compounds as it has been reported that silicon-treated plants release 15 times more phenolics than untreated maize plants (Kidd et al. 2001). These flavonoid-phenolics (i.e., catechin and quercetin) have a strong Al-chelating ability and may provide metal tolerance in plants.

14.6.3.3 Cadmium Toxicity

It is generally recognized that an external mechanism similar to Si-alleviated toxicity to Al applies to Si-mediated detoxification of Cd in soil/plant systems, i.e., reduction in Cd availability via Cd immobilization arising from a rise in pH. This is true when Na metasilicate, slag or alkaline Si-containing materials such as biosolids are incorporated into Cd-contaminated soils as Si sources (Chen et al. 2000). Moreover, more Cd was found to be in the form of specific adsorbed or Fe-Mn oxides-bound fraction in the Si-amended soil. These results suggest an external interaction between Si and Cd (Liang et al. 2005). Other investigations suggested an internal mechanism for detoxification of Cd by Si occurring within plants. The role of Si in minimizing uptake and root-to-shoot transport of metal ions has recently been confirmed in seedlings of rice grown with toxic Cd (Shi et al. 2005b) and with arsenate (Guo et al. 2005). It has been suggested that (Shi et al. 2005b), the heavy deposition of silica in the vicinity of the endodermis might offer a possible mechanism by which silicon did at least partially physically block the apoplast bypass flow across the roots, and restrained the apoplastic transport of Cd. Cadmium concentration in the xylem exudates was significantly decreased in the Si-amended Cd treatments and Cd was bound to the root cell walls but less to cytosols or symplast in +Si plants than in −Si plants under Cd stress, suggesting a root apoplastic role of Si in detoxification of excess Cd (Liang et al. 2007), a mechanism similar to that responsible for the Si-mediated Mn tolerance in plants (Rogalla and Römheld 2002).

14.6.3.4 Other Heavy Metals

Using electron-energy-loss spectroscopy and other techniques, Neumann and zur Nieden (2001) have reported the occurrence of codeposition of silicon and Zn in heavy metal tolerant *Cardaminopsis halleri*. The formation of

Zn-silicate is part of the mechanism for tolerance to heavy metals and may be responsible for the amelioration of Zn toxicity in *Cardaminopsis halleri*.

15 Effects of Selenium Supplementation on Plants Stress Responses

Selenium is an essential micronutrient for animal and human nutrition because it is an integral part of the enzyme glutathione peroxidase (GPX), a selenoenzyme that prevents oxidative damage to body tissues (Rotruck et al. 1973). According to current thinking, higher plants do not require Se, and genes encoding GPX isolated from several plant species all contained Cys than SeCys in their active sites (Lobanov et al. 2007). Although Se-containing GPX has not been identified in plants, Se supplementation consistently increased GPX activity (Hartikainen et al. 2000).

Growth stimulating effect of trace amounts of Se has been frequently reported in some plant species such as ryegrass (Hartikainen et al. 2000), lettuce (Xue et al. 2001), potato (Seppänen et al. 2003), and different varieties of *Brassica oleracea* (Hajiboland and Amjad 2008). At proper levels, it also delays some of the effects of senescence (Djanaguiraman et al. 2005).

Recent studies have shown that Se at low concentrations can protect plants from several types of abiotic stresses. Selenium strengthens the capacity of plants to counteract the oxidative stress caused by oxygen radicals produced by internal metabolic or external factors. The mechanism of this apparent positive effect of Se on plant growth and stress tolerance may be direct and via higher expression of genes involve in antioxidant activities or indirect, via Se-induced regulation of general stress resistance mechanisms and defense genes of jasmonic acid and salicylic acid pathway (Pilon-Smits et al. 2009).

The growth-promoting response of Se is mainly accompanied with the enhanced antioxidative capacity manifested in decreasing lipid peroxidation, marked increase in GPX activity and a peak concentration of tocopherols, scavengers of lipid peroxide radicals, and singlet oxygen (Xue et al. 2001). Selenium stimulates plants growth even under nonstress conditions. A considerable growth promotion up to 59% has been reported for cabbage plants in response to Se supplementation at 20 μM. This effect was accompanied by increase in the activity of GR and CAT but not SOD, APX, and POD (Hajiboland and Amjad 2007).

15.1 UV Radiation

Damage caused by UV-B radiation has been ascribed to generation of free radicals in the plants (Rozema et al. 1997). In order to survive their deleterious effects, plants need to accumulate molecules, which can quench the free radicals. Selenium protects plants against UV-induced oxidative stress (Pennanen et al. 2002), thereby promoting the growth of plants subjected to high energy radiation (Xue and Hartikainen 2000). In addition, foliar spraying with Se counteracted the effect of ambient UV radiation, resulting in a significant increase in the yield of plants grown in the field (Germ et al. 2005).

15.2 High Temperature

High temperature stress can cause premature leaf senescence, an internally programmed degeneration process that leads to tissue death. High temperature stress directly damages the photosynthetic apparatus and decreases both photosynthetic rate and duration of the assimilate supply and promote accumulation of ROS in the chloroplast, particularly when the antioxidant capacity to detoxify ROS is low (Prasad et al. 2008). Foliar application of Se increased antioxidative function as demonstrated by increased activity of SOD and reduced $O_2^{\bullet-}$ concentration in the high-temperature stressed plants (Djanaguiraman et al. 2010). Selenium application exerted its positive influence under high temperature stress by increasing net CO_2 assimilation and decreasing H_2O_2 content, MDA production, and membrane injury (Fig. 16.7).

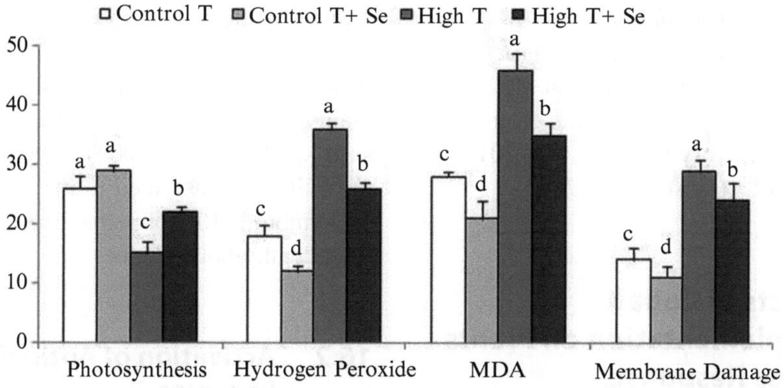

Fig. 16.7 Effect of temperature stress (control temperature, 32/22°C and high temperature, 40/30°C) and selenium (75 ppm foliar spray of sodium selenate) during seed set and seed filling plants on photosynthesis (μmol m^{-2} s^{-1}), hydrogen peroxide (nmol g^{-1} FW) and malondialdehyde (MDA) (nmol g^{-1} FW) content and membrane damage (%) of grain sorghum (*Sorghum bicolor* L. Moench) leaves (Djanaguiraman et al. 2010)

15.3 Chilling Stress

Freezing sensitive plants suffer from significant frost damage in the field even at temperatures slightly below 0°C. Freezing nights are often followed by bright sunlight in the morning and light is an important component of frost injury in the field. Exposure to high intensity light after freezing can significantly increase the severity of freezing injury (Steffen and Palta 1986). In the Se treated plants exposed to high light intensity (600 æmol m ^{2} s ^{1}) at low temperature (4°C), photosynthesis was more protected and Se exerted a synergistic effect on the transcription of CuZn-SOD and GPX (Seppänen et al. 2003).

15.4 Senescence

Senescence is an integral part of plant development and coincides with the production of free oxygen radicals and is regulated by a variety of environmental and autonomous factors. Free radical reactions diminish the value of vegetables on the market and cause postharvest losses. Selenium delays senescence and promote the growth of aging seedlings. The antiaging effect was related to decreased lipid peroxidation and enhanced GPX and SOD activity in Se treated plants (Djanaguiraman et al. 2005). Furthermore, total tocopherols were shown to diminish during plant senescence, but added Se counteracted this process and tended to maintain tocopherol concentration at a higher level than the control plants cultivated without Se (Xue et al. 2001).

15.5 Heavy Metal Toxicity

Published works on the effect of Se in plants response to heavy metal stress all confirm involvement of similar mechanisms for ameliorating effect of Se under other stresses. Selenium supplementation considerably reversed the Cd-induced decrease in fresh mass as well as the changes in lipid unsaturation and peroxidation. Moreover, the presence of Se in medium prevented changes in the DNA methylation pattern triggered in rape seedlings by high Cd concentrations (Filek et al. 2008). Two possible mechanisms for the action of Se were considered, removal of Cd from metabolically active cellular sites, and reduction of oxygen radicals (Filek et al. 2008).

15.6 Se Accumulators

Certain plant species are known to accumulate Se to levels far beyond those observed in other species. The possible functional significance of Se accumulation in these species has been studied in some crops, e.g., *Brassica juncea* (2004) and natural vegetation species, e.g., *Stanleya pinnata* and *Astragalus bisulcatus* (Quinn et al. 2010). It was shown that Se protects plants from fungal infection and herbivory (Quinn et al. 2010) and feeding by aphids (Hanson et al. 2004).

16 Effects of Cobalt Supplementation on Plants Stress Responses

Cobalt has long been known to be a micronutrient for animals and humans, where it is a constituent of vitamin B_{12}. However, a physiological function for this element in higher plants has so far not been established. Vitamin B_{12} is synthesized by soil bacteria, intestinal microbes, and algae, but not in animals and plants. The only physiological role so far definitely attributed to Co in higher plants has been in N fixation by leguminous plants (Marschner 1995). Since Co is essential for mammals, fertilization of crops with Co will have the additional beneficial effect of enhancing its nutritional quality. Similar to other heavy metals, Co causes toxicity to plants at high concentration, and most of the recent literature focuses on the mechanisms through which plants can cope with Co stress (Micó et al. 2008). At low levels, however, Co can have a number of beneficial effects, particularly in leguminous plants.

16.1 Nitrogen Metabolism

Co is a component of cobalamin (vitamin B_{12}), which is required for the activity of several enzymes in N-fixing microorganisms include Rhizobium such as methionine synthase, ribonucleotide reductase, and methylmalonyl-CoA mutase (Marschner 1995). Its importance in N fixation by symbiosis in Leguminosae (Fabaceae) has been established. Soybeans grown with only atmospheric N and no mineral N have rapid N fixation and growth with Co supplementation, but have minimal growth without Co additions (Ahmed and Evans 1960). In pea plants (*Pisum sativum* L.), the application of Co to the soil increased growth, nodule number and weight, plant nutrient levels, as well as seedpod yield and seed quality (Gad 2006). These effects could most likely be ascribed to the essentiality of Co for symbiotic Rhizobia that live in the nodules of these leguminous plants.

16.2 Activation of Antioxidant Enzymes

The high O_2 consumption in nodules for provision of energy also creates a great potential for production of ROS. This is true in particular for leghemoglobin which is also subjected to autooxidation in which $O_2^{\bullet-}$ and H_2O_2 are released. For protection against this toxicity, legume nodules need an efficient defense mechanism (Marschner 1995). Addition of Co to legume plants caused activation of CAT, in parallel with growth improvement and increase in the nodulation and leghemoglobin concentration. Activation of CAT was not observed at higher Co concentration that resulted growth impairment (Jayakumar et al. 2008).

16.3 Resistance to Pathogens

Co stimulated isoquinoline accumulation (an alkaloid) in medicinal plants, through upregulation of the biosynthesis of aromatic amino acid precursors of alkaloids (Palit et al. 1994). This last effect may suggest that Co could indirectly induce biotic stress resistance, but this hypothesis has not been addressed yet. In hyperaccumulators of Co, the high tissue Co levels may also offer direct protection from herbivory or pathogens, as was shown for other hyperaccumulated elements. Alkaloid accumulation in medicinal plants such as *Datura innoxia* Mill., *Atropa caucasica*, *A. belladonna* L., and *Glaucium flavum* Crantz (Talukder and Sharma 2007) is regulated by Co. It also increased rutin (11.6%) and cyanide (67%) levels in different species of buckwheat (*Fagopyrum sagittatum* Gilib., *F. tataricum*

Gaertn., and *F. emargitatum*) (Talukder and Sharma 2007). Cobalt acts as a chelator of salicylidine-o-aminothiophenol and salicylidine-o aminopyridine and exerts biocidal activity against the molds *Aspergillus nidulans* winter and *A. Niger* Tiegh and the yeast *Candida albicans*. Antifungal activities of Co (II) with acetone salicyloyl hydrazone and ethyl methyl ketone salicyloyl hydrazone against *A. Niger* and *A. flavus* have been established (Johari et al. 1987).

16.4 Delay of Senescence

Another beneficial effect reported for Co is retardation of leaf senescence via inhibition of ethylene biosynthesis. The Co (II) ion is an inhibitor of the ethylene biosynthesis pathway, blocking the conversion of 1-amino-cyclopropane-l-carboxylic acid (ACC) (Branden et al. 1987). Senescence in lettuce leaf in the dark is retarded by Co, which acts by arresting the decline of Chl, protein, RNA and, to a lesser extent, DNA. The activities of RNAase and protease, and tissue permeability were decreased, while the activity of CAT increased. Cobalt delays aging and is used for keeping leaves and fruit fresh in vetch (*Vicia* spp.) and apple respectively (Talukder and Sharma 2007). Cobalt inhibits IAA-induced ethylene production in winter wheat and beans, in kiwifruit (*Actinide chinensis* Planch) (Talukder and Sharma 2007) and in wheat seedlings under water stress (Gaal et al. 1988).

Cobalt chloride markedly increases elongation of etiolated pea stems when supplied with indole acetic acid (IAA) and sucrose, but elongation is inhibited by Co acetate. Cobalt in the form of vitamin B_{12} is necessary for the growth of excised tumor tissue from spruce (*Picea glauca* Moench Voss.) cultured in vitro. It increases the apparent rate of synthesis of peroxides and prevents the peroxidative destruction of IAA. It counteracts the inhibition by dinitrophenol (DNP) in oxidative phosphorylation and reduces activity of ATPase and is known to be an activator of plant enzymes such as carboxylases and peptidases (Ahmed and Evans 1960). Cobalt has also been noted to cause repression of developmental distortion such as leaf malformation and accumulation of low-molecular-weight polypeptides in velvet plant (*Gynura aurantiaca* DC) and prevention of 3,6-dichloro-o-anisic acid-induced Chl degradation in tobacco leaves (Talukder and Sharma 2007).

16.5 Drought Resistance

Prevention of auxin-induced stomatal opening in detached leaf epidermis has been observed (Merritt et al. 2001). However, this effect has so far not been studied in intact plants and may cause reduction of water loss and improvement of drought tolerance in plants. Presowing treatment of seeds with Co nitrate increased drought resistance of horse chestnut (*Aesculus hippocastanum* L.) (Tarabrin and Teteneva 1979).

17 Effects of Aluminum Supplementation on Plants Stress Responses

It is well known that high Al concentration in soil solution is the most important factor in restricting plant growth on acid soils (Kochian et al. 2004). No conclusive evidence suggests that Al is an essential nutrient for plants (Marschner 1995).

17.1 Aluminum Accumulators

Relative to Al accumulation, there appears to be two groups of plant species: Al excluders and Al accumulators. Most plant species, particularly crop plants, are Al excluders. Aluminum accumulators are plants with 1,000 mg Al kg^{-1} or greater in leaves (Miyasaka et al. 2007). Aluminum accumulation found frequently among perennial, woody species in tropical rain forests. Tea (*Camellia sinensis* Kuntze) is one crop plant considered to be an Al accumulator, with Al concentrations of 30,700 mg kg^{-1} in mature leaves and 600 mg kg^{-1} in young leaves (Miyasaka et al. 2007). Another well-known Al-accumulating plant is hydrangea (*Hydrangea macrophylla* Ser.), which has blue-colored sepals when the plant is grown in acidic soils and red-colored sepals when grown in alkaline soils. The blue color of hydrangea sepals is due to Al complexing with the anthocyanin, delphinidin 3-glucoside, and the copigment, 3-caffeoylquinic acid (Watanabe and Osaki 2002).

17.2 Beneficial Effects

Low levels of Al (up to 10 μM) sometimes stimulate root growth of nonaccumulators such as turnip and soybean (Miyasaka et al. 2007). In Al accumulators such as tea plants, however, root and shoot growth and leaf numbers and area, respond positively to Al supplementation up to 125 μM Free Al^{3+} activity (300 μM Al concentration). Root axis were growing in length for longer time before elongation ceased, lignification was delayed and relative growth rate of root axis was about two times higher than control plants (Hajiboland et al. 2011). Application of Al on the leaves of tea plant grown in alkaline soil caused the plants to recover from chlorosis. In addition, when seedlings of *Miconia albicans* an Al accumulating species growing in the calcareous soil showed chlorotic leaves, the symptom was completely recovered from after a portion of their root systems were exposed to Al solution (Watanabe and Osaki 2002). These results suggest some physiological role of Al in Al accumulator species. Early explanations for this enhancement in growth include increased Fe solubility and availability, prevention of internal Fe deficiency through displacement of Fe from inactive sites in calcicolous plants, prevention of P toxicity or promotion of P uptake, prevention of Ca depletion, alteration of growth regulators, and protection against Cu/Mn toxicity (Foy et al. 1978). However, hypotheses such as those listed above have only been shown to apply in certain cases. It has been reported that activity of H^+ATPase in plasma-membrane-enriched fraction, which had been treated with Al, showed a 77% increase compared with that in the control (Matsumoto and Yamaya 1986). During growth the activity of H^+ATPase play a critical role for cell wall expansion mediated by auxin. Although studies on the effect of Al on H^+ATPase activity were performed mainly on Al excluder species, this mechanism explains well the considerable stimulatory effect of Al on the elongation of root axis observed in Al accumulator species such as tea.

17.2.1 Alleviation of H^+ Toxicity

There is evidence that the nature of beneficial effects of Al occur through the alleviation of H^+ toxicity by Al^{3+}. Alleviation of H^+ toxicity is a general phenomenon achieved by cations (not solely Al^{3+}), and the effectiveness was dependent upon the charge ($Cat^{3+} > Cat^{2+} > Cat^{1+}$). However, ameliorative effect of Al^{3+} on H^+ toxicity was reported mainly in some crop species such as wheat and maize (Kinraide 1993) and this mechanism is insufficient to explain all the phenomena of Al-induced growth enhancement particularly in plants native to low pH soils. In a study on some Al-accumulators adapted to low pH soils and grow poorly in the absence of Al, other mechanisms such as improved nutrient uptake particularly P has been proposed as mechanism for Al-induced growth improvement.

17.2.2 Alleviation of Boron Deficiency

Low B content because of high leaching losses and high Al^{3+} content are characteristics of acid soils. Inside the plant, Al is likely to be present as $Al(OH)_3$, which is structurally similar to $B(OH)_3$. In the case of sensitive species, Al is assumed to exert its toxic effects in the apoplast through interaction with the negative binding sites of the cell walls, primarily pectin. For B, the predominant function is in the formation of primary cell walls, where it cross-links the pectic polysaccharides (Hu and Brown 1994). Interaction of Al and B was studied mainly in Al sensitive species. Based on the similarities of the molecules and of the symptoms characteristics for Al toxic and B-deficient plants, it has been proposed that Al may exert its toxic effect by inducing B deficiency (Poschenrieder et al. 1995). In tea plants, B deficiency and Al supplementation had marked influence on phenolics metabolism and fractionation in the young and old leaves and roots. A high CO_2 assimilation rate, greater B root–shoot transport and increase in the cell wall bound B fraction are mechanisms for Al-mediated growth amelioration of B-deficient plants. Under these conditions, shoot Al allocated mainly to the old leaves and less Al was retranslocated into young leaves, where most Al was found in the cell wall-bound fraction (Hajiboland and Bastani 2011).

17.2.3 Adaptation to P Deficiency

Al application enhances growth that is accompanied by increased nutrient concentrations, especially P concentrations, in the tissues. In plants adapted to low pH soils in the tropical and temperate regions growth is stimulated by Al application which is assumed to be caused by the stimulation of N, P, and K uptake, although the increase of P content is partly due to Al–P precipitation on the root surface and/or in the Donnan free space. In these species pH decreased around the rhizosphere that may solubilize Al–P precipitates coating the surface of roots (Osaki et al. 1997). In tea plants addition of Al and P, increased P absorption and translocation as well as root and shoot growth (Konishi et al. 1985). Similarly, the Al-accumulating shrub, *Melastoma malabathricum* L., exhibited increased growth of leaf, stem, and roots as well as increased P accumulation when Al was added to culture solutions (Osaki et al. 1997). An increased root length by Al supplementation in tea and other Al accumulator plants native to low pH soils could be an adaptation for these species grow on acid soils with low P availability. Phosphorus acquisition by plants is largely dependent upon spatial availability of P by roots. In addition, greater root length provides more water absorption area and increases considerably drought resistance.

17.2.4 Activation of Antioxidant Enzymes

In the roots of intact plants as well as cultured cells of tea plants higher activity of antioxidant enzymes was observed in the presence of Al (Ghanati et al. 2005). These results indicate that Al-induced increase in the activities of antioxidant enzymes, resulting in increased membrane integrity and delayed lignification and aging, is a possible reason for the stimulatory effects of Al on the growth of tea plants irrespective to the interaction with other micronutrients (Ghanati et al. 2005). In the study on tea plants grown from seeds we observed increase in the activity of antioxidant enzymes and concentration of nonenzymatic antioxidants such as proline in Al-treated plants. Membrane integrity was considerably improved in both leaf and root tissues in the presence of Al (Table 16.7).

Table 16.7 Activity of ascorbate peroxidase (APX), peroxidase (POD), catalase (CAT), superoxide dismutase (SOD), glutathione reductase (GR) and concentration of proline and malondialdehyde (MDA) in the leaves and roots of tea (*Camellia sinensis* L.) plants grown for 6 weeks in the absence (–Al) and presence (+Al) of 300 μM supplemental Al

	Leaves		Roots	
Parameters	–Al	+Al	–Al	+Al
APX (μmol mg^{-1} protein min^{-1})	10.3 ± 1.4^{b}	16.2 ± 1.3^{a}	34.3 ± 3.4^{b}	54.8 ± 5.1^{a}
POD (μmol mg^{-1} protein min^{-1})	14.9 ± 1.5^{a}	18.1 ± 1.3^{a}	19.9 ± 2.2^{b}	28.7 ± 1.5^{a}
CAT (μmol mg^{-1} protein min^{-1})	7.8 ± 1.1^{a}	10.5 ± 1.5^{a}	13.2 ± 2.1^{b}	20.5 ± 3.1^{a}
SOD (Unit mg^{-1} protein)	4.9 ± 0.3^{b}	7.2 ± 0.8^{a}	5.9 ± 0.7^{b}	11.9 ± 0.5^{a}
GR (μmol NADPH mg^{-1} protein min^{-1})	10.5 ± 1.2^{b}	22.8 ± 1.9^{a}	18.8 ± 2.4^{b}	30.5 ± 2.5^{a}
Proline (μg g^{-1} FW)	99 ± 6.1^{b}	224 ± 8^{a}	28 ± 4.5^{b}	70 ± 8.7^{a}
MDA (μg g^{-1} FW)	1.93 ± 0.07^{a}	1.49 ± 0.04^{b}	0.97 ± 0.12^{a}	0.86 ± 0.06^{a}

Data of each enzyme or metabolite within each organ followed by the same letter are not significantly different ($P < 0.05$)

17.2.5 Tolerance to Plant Pathogens

Aluminum can be toxic to pathogenic microorganisms, thus helping plants to avoid disease. Spore germination and vegetative growth of the black root rot pathogen, *Thielaviopsis basicola* (Berk. et Br.) Ferraris, were inhibited by 350 μM Al at pH 5. Similarly, mycelial growth and sporangial germination of potato late blight pathogen, *Phytophthora infestans*, were inhibited by 185 μM Al, and it was speculated that amendment of soils with Al might be used as a means of disease control (Miyasaka et al. 2007).

18 Conclusion and Future Perspective

Under both natural and cultivated ecosystems, plants often experience a combination of various stress factors including drought, high irradiance, UV radiation, chilling, flooding, and salinity. An imbalanced nutrition accentuates effect of stress factors and hampers plant growth and productivity. In this chapter, we try to give evidences on how plants respond to a combination of micronutritional deficiencies and environmental stress factors. Antioxidant defense system is an important cross point between micronutrients and plants stress responses because of changes in the content and activity of its components due to both micronutrients deficiencies and environmental stress factors. However, new evidences on the effects of micronutrients on plants signaling events throw some light on the still poorly known aspects of micronutrients effects on plants interaction with their surrounding environment. Recent evidences on the effect of Mo on ABA signaling pathway and evidences on signaling effect of cell wall-bound B demonstrated that micronutrients may also be involved in plants signal transduction pathways either as components of important enzymes in the signaling or as structural components of a signaling molecule. More investigations are needed on this function of not only micronutrients but also elements that have been defined so far as beneficial elements.

References

Abadía A, Ambard-Bretteville F, Remy R, Trémolières A (1988) Iron-deficiency in pea leaves: effect on lipid composition and synthesis. Physiol Plant 72: 713–717

Abadía J, Morales F, Abadía A (1999) Photosystem II efficiency in low chlorophyll, iron-deficient leaves? Plant Soil 215:183–192

Adams ML, Norvell WA, Philpot WD, Peverly JH (2000) Spectral detection of micronutrient deficiency in Bragg soybean. Agron J 92:261–268

Adatia MH, Bestford RT (1986) The effect of silicone on cucumber plants grown in recirculating nutrient solution. Ann Bot 56:343–351

Agarie S, Agata W, Uchida H, Kubota F, Kaufman PB (1996) Function of silica bodies in the epidermal system of rice (*Oryza sativa* L.): testing the window hypothesis. J Exp Bot 47:655–660

Ahmed S, Evans HJ (1960) Cobalt: a micronutrient for the growth of soybean plants under symbiotic conditions. Soil Sci 90:205–210

Ahn SJ, Im YJ, Chung GC, Cho BH (1999) Inducible expression of plasma membrane H^+ATPase in the roots of fig leaf gourd plants under chilling root temperature. Physiol Plant 106:35–40

Aleshin NE (1988) About the biological role of silicon in rice. Vestnik Agric Sci 10:77–85

Allen DJ, Ort DR (2001) Impacts of chilling temperatures on photosynthesis in warm-climate plants. Trends Plant Sci 6:36–42

Apel K, Hirt H (2004) Reactive oxygen species: metabolism, oxidative stress, and signal transduction. Annu Rev Plant Biol 55:373–399

Apostol KG, Zwiazek JJ (2004) Boron and water uptake in jack pine (*Pinus banksiana*) seedlings. Environ Exp Bot 51:145–153

Arnon DI, Stout PR (1939) The essentiality of certain elements in minute quantity for plants with special reference to copper. Plant Physiol 14:371–375

Aro E-M, Virgin I, Andersson B (1993) Photoinhibition of photosystem II. Inactivation, protein damage and turnover. Biochim Biophys Acta 1143:113–134

Arora A, Sairam RK, Srivastava GC (2002) Oxidative stress and antioxidative system in plants. Curr Sci 82:1227–1238

Asada K (1999) The water-water cycle in chloroplasts: scavenging of active oxygens and dissipation of excess photons. Annu Rev Plant Physiol Plant Mol Biol 50: 601–639

Ashraf M, Foolad MR (2007) Roles of glycine betaine and proline in improving plant abiotic stress resistance. Environ Exp Bot 59:206–216

Ayala MB, Gorgé JL, Lachica M, Sandmann G (1992) Changes in carotenoids and fatty acids in photosystem II of Cu deficient pea plants. Physiol Plant 84:1–5

Bagci SA, Ekiz H, Yilmaz A, Cakmak I (2007) Effect of Zn deficiency and drought on grain yield of field grown wheat cultivars in Central Anatolia. J Agron Crop Sci 193:198–206

Bai C, Reilly CC, Wood BW (2006) Nickel deficiency disrupts metabolism of ureides, amino acids, and organic acids of young pecan foliage. Plant Physiol 140:433–443

Baker NR, Bowyer JR (1994) Photoinhibition of photosynthesis. From molecular mechanisms to the field. BIOS Scientific, Oxford, UK

Balasta MLFC, Perez CM, Juliano BO, Villareal CP, Lott JNA, Roxas DB (1989) Effect of silica level on some properties of *Oryza sativa* straw and hull. Can J Bot 67:2356–2363

Banu MNA, Hoque MAM, Watanabe-Sugimoto M, Matsuoka K, Nakamura Y, Shimoishi Y, Murata Y (2009) Proline and glycinebetaine induce antioxidant defense gene expression and suppress cell death in cultured tobacco cells under salt stress. J Plant Physiol 166:146–156

Bassil E, Hu H, Brown PH (2004) Use of phenylboronic acids to investigate boron function in plants. Possible role of boron in transvascuolar cytoplasmic strands and cell-to-wall adhesion. Plant Physiol 136:3383–3395

Becanne M, Moran JF, Iturbe-Ormaetxe I (1998) Iron dependent oxygen free radical generation in plants subjected to environmental stress: toxicity and antioxidant protection. Plant Soil 201:137–147

Birchall JD (1990) The role of silicon in biology. Chem Britain 26:141–144

Bittner F, Oreb M, Mendel RR (2001) ABA3 is a molybdenum cofactor sulfurase required for activation of aldehyde oxidase and xanthine dehydrogenase in *Arabidopsis thaliana.* J Biol Chem 276:40381–40384

Bloom AJ, Zwieniecki MA, Passioura JB, Randall LB, Holbrook NM, St Clair DA (2004) Water relations under root chilling in a sensitive and tolerant tomato species. Plant Cell Environ 27:971–979

Blumwald E, Aharon GS, Apse MP (2000) Sodium transport in plant cells. Biochem Biophys Acta 1465: 140–151

Bok-Rye L, Kil-Yong K, Woo-Jin J, Jean-Christophe A, Alain O, Tae-Hwan K (2007) Peroxidases and lignification in relation to the intensity of water-deficit stress in white clover (*Trifolium repens* L.). J Exp Bot 58:1271–1279

Boudet AM (1998) A new view of lignifications. Trends Plant Sci 3:67–71

Bowler C, Camp WV, Mantogu MV, Inze D (1994) Superoxide dismutase in plants. Crit Rev Plant Sci 13:199–218

Braconnier S, Bonneau X (1998) Effects of chlorine deficiency in the field on leaf gas exchanges in the PB121 coconut hybrid. Agronomie 18:563–572

Branden R, Jamson K, Nilsson P, Aasa R (1987) Intermediates formed by the Co^{2+} activated ribulose-1-5, biphosphate carboxylase/oxygenase from spinach studied by EPR spectroscopy. Biochem Biophys Acta 916:293–303

Bray TM, Bettger WJ (1990) The physiological role of zinc as an antioxidant. Free Radic Biol Med 8: 281–291

Brown PH (2007) Nickel. In: Barker AV, Pilbeam DJ (eds) Handbook of plant nutrition. CRC Press, Taylor and Francis Group, Boca Raton, FL, pp 395–410

Brown PH, Shelp B (1997) Boron mobility in plants. Plant Soil 193:85–102

Brown PH, Welch RM, Madison JT (1990) Effect of nickel deficiency on soluble anion, amino acid, and nitrogen levels in barley. Plant Soil 125:19–27

Bussler W (1981) Physiological functions and utilization of copper. In: Loneragan JF, Robson AD, Graham RD (eds) Copper in soils and plants. Academic Press, New York, pp 213–234

Cakmak I (2000) Possible role of Zn in protecting plant cells from damage by reactive oxygen species. New Phytol 146:185–205

Cakmak I, Engels C (1999) Role of mineral nutrients in photosynthesis and yield formation. In: Rengel Z (ed) Mineral nutrition of crops. Haworth Press, New York, pp 141–168

Cakmak I, Römheld V (1997) Boron deficiency-induced impairments of cellular functions in plants. Plant Soil 193:71–83

Cakmak I, Yilmaz A, Ekiz H, Torun B, Erenoglu B, Braun HJ (1996) Zinc deficiency as a critical nutritional problem in wheat production in Central Anatolia. Plant Soil 180:165–172

Cakmak I, Ozturk L, Eker S, Torun B, Kalfa HI, Yilmaz A (1997) Concentration of zinc and activity of Cu/Zn superoxide dismutase in leaves of rye and wheat cultivars differing in sensitivity to zinc deficiency. J Plant Physiol 151:91–95

Camacho-Cristóbal JJ, González-Fontes A (2007) Boron deficiency decreases plasmalemma H^{+}ATPase expression and nitrate uptake, and promotes ammonium assimilation into asparagines in tobacco roots. Planta 226:443–451

Cara FA, Sanchez E, Ruiz JM, Romero L (2002) Is phenol oxidation responsible for the short-term effects of boron deficiency on plasma-membrane permeability and function in squash roots? Plant Physiol Biochem 40:853–858

Chaerle L, Leinonen I, Jones HG, Van Der Straeten D (2007) Monitoring and screening plant populations with combined thermal and chlorophyll fluorescence imaging. J Exp Bot 58:773–784

Chen HM, Zheng CR, Tu C, Shen ZG (2000) Chemical methods and phytoremediation of soil contaminated with heavy metals. Chemosphere 41:229e234

Chen W, Yang X, He Z, Feng Y, Hu F (2007) Differential changes in photosynthetic capacity, 77K chlorophyll fluorescence and chloroplast ultrastructure between Zn-efficient and Zn-inefficient rice genotypes (*Oryza sativa* L.) under low Zn stress. Plant Physiol 132: 89–101

Ciurli S (2001) Electronic structure of the nickel ions in the active site of urease. Chemistry (Easton) 2001: 99–100

Cocker KM, Evans DE, Hodson MJ (1998) The amelioration of aluminum toxicity by silicon in higher plants: solution chemistry or an in plants mechanism? Physiol Plant 104:608–614

Cohu CM, Pilon M (2007) Regulation of superoxide dismutase expression by copper availability. Physiol Plant 129:747–755

Corrales I, Poschenrieder C, Barceló J (1997) Influence of silicon pretreatment on aluminum toxicity in maize roots. Plant Soil 190:203–209

Curie C, Briat JF (2003) Iron transport and signaling in plants. Annu Rev Plant Biol 54:183–206

Cushman J, Bohnert HJ (2002) Induction of crassulacean acid metabolism by salinity-molecular aspects. In: Läuchli A, Lüttge U (eds) Salinity: environment-plants-molecules. Kluwer Academic, Dordrecht, pp 361–393

Dalton FN, Maggio A, Piccinni G (2000) Simulation of shoot chloride accumulation: separation of physical and biochemical processes governing plant salt tolerance. Plant Soil 219:1–11

Djanaguiraman M, Durga Devi D, Shanker Arun K, Annie Sheeba J, Bangarusamy U (2005) Selenium – an antioxidative protectant in soybean during senescence. Plant Soil 272:77–86

Djanaguiraman M, Prasad PVV, Seppänen M (2010) Selenium protects sorghum leaves from oxidative damage under high temperature stress by enhancing antioxidant defense system. Plant Physiol Biochem 48:999–1007

Dordas C, Brown PH (2000) Permeability of boric acid across lipid bilayers and factors affecting it. J Membr Biol 175:95–105

Dordas C, Chrispeels MJ, Brown PH (2000) Permeability and channel-mediated transport of boric acid across membrane vesicles isolated from squash roots. Plant Physiol 124:1349–1361

Engel RE, Bruebaker L, Emborg TJ (2001) A chloride deficient leaf spot of durum wheat. Soil Sci Soc Am J 65:1448–1454

Epstein E, Bloom AJ (2005) Mineral nutrition of plants: principles and perspectives, 2nd edn. Sinauer, Sunderland

Fauteux F, Rémus-Borel W, Menzies JG, Bélanger RR (2005) Silicon and plant disease resistance against pathogenic fungi. FEMS Microbiol Lett 249:1–6

Filek M, Keskinen R, Hartikainen H, Szarejko I, Janiak A, Miszalski Z, Golda A (2008) The protective role of selenium in rape seedlings subjected to cadmium stress. J Plant Physiol 165:833–844

Fleischer A, O'Neill MA, Ehwald R (1999) The pore size of non-graminaceous plant cell walls is rapidly decreased by borate ester cross-linking of the pectic polysaccharide rhamnogalactouronan-II. Plant Physiol 121:829–838

Flowers TJ (1988) Chloride as a nutrient and as an osmoticum. In: Tinker B, Läuchli A (eds) Advances in plant nutrition, vol 3. Praeger, New York, pp 55–78

Foy CD, Chaney RL, White MC (1978) The physiology of metal toxicity in plants. Annu Rev Plant Physiol 29:511–566

Gaal I, Ariunaa H, Gyuris M (1988) Influence of various stress on ethylene production in wheat seedlings. Acta Univ Szeged Acta Biol 34:35–44

Gad N (2006) Increasing the efficiency of nitrogen fertilization through cobalt application to pea plant. Re J Agr Biol Sci 2:433–442

Gerendás J, Sattelmacher B (1997) Significance of Ni supply for growth, urease activity and concentrations of urea, amino acids and mineral nutrients of urea-grown plants. Plant Soil 190:153–162

Gerendás J, Polacco J, Freyermuth SK, Sattelmacher B (1999) Significance of nickel for plant growth and metabolism. Z Pflanzenernaehr Bodenkd 162:241–256

Germ M, Kreft I, Osvald J (2005) Influence of UV-B exclusion and selenium treatment on photochemical efficiency of photosystem II, yield and respiratory potential in pumpkins (*Cucurbita pepo* L.). Plant Physiol Biochem 43:445–448

Ghanati F, Morita A, Yokota H (2005) Effects of aluminum on the growth of tea plant and activation of antioxidant system. Plant Soil 276:133–141

Ghanmi D, McNally DJ, Benhamou N, Bélanger RR (2004) Powdery mildew of *Arabidopsis thaliana*: a pathosystem for exploring the role of silicon in plant-microbe interactions. Physiol Mol Plant Pathol 64:189–199

Goldbach HE, Wimmer MA (2007) Boron in plants and animals: is there a role beyond cell-wall structure? J Plant Nutr Soil Sci 170:39–48

Goldbach HE, Wimmer MA, Chaumont F, Matoh T, Volkmann D, Baluska F, Wingender R, Schulz M, Yu Q (2002) Rapid responses of plants to boron deprivation. In: Goldbach HE, Rerkasem B, Wimmer M, Brown PH, Thellier M, Bell RW (eds) Boron in plant and animal nutrition. Kluwer Academic, New York, pp 167–180

Goos RJ, Johnson BE, Holmes BM (1987) Effect of potassium chloride fertilization on two barley cultivars differing in common root rot reaction. Can J Plant Sci 67:395–401

Graham RD, Stangoulis JRC (2005) Molybdenum and disease. In: Datnoff L, Elmer W, Huber D (eds) Mineral nutrition and plant diseases. APS Press, St. Paul, pp 101–118

Grewal HS, Williams R (2000) Zinc nutrition affects alfalfa responses to water stress and excessive moisture. J Plant Nutr 23:949–962

Guo W, Hou YL, Wang SG, Zhu YG (2005) Effect of silicate on the growth and arsenate uptake by rice (*Oryza sativa* L.) seedlings in solution culture. Plant Soil 272:173–181

Hajiboland R (2000) Zinc efficiency in rice plants. PhD Thesis, University of Hohenheim, Verlag Grauer, Stuttgart.

Hajiboland R, Amirazad F (2010a) Growth, photosynthesis and antioxidant defense system in Zn-deficient red cabbage plants. Plant Soil Environ 56:209–217

Hajiboland R, Amirazad F (2010b) Drought tolerance in Zn-deficient red cabbage (*Brassica oleracea* L. var. *capitata* f. *rubra*) plants. Hort Sci 37:88–98

Hajiboland R, Amjad L (2007) Does antioxidant capacity of leaves play a role in growth response to selenium at different sulfur nutritional status? Plant Soil Environ 53:207–215

Hajiboland R, Amjad L (2008) The effects of selenate and sulphate supply on the accumulation and volatilization of Se by cabbage, kohlrabi and alfalfa plants grown hydroponically. Agric Food Sci 17:177–190

Hajiboland R, Bastani S (2011) Aluminum alleviates growth impairment in boron-deficient tea plants. In: Proceeding book of the 11th international conference on the biogeochemistry of trace elements. Florence, Italy, 3–7 July 2011

Hajiboland R, Beiramzadeh N (2008) Growth, gas exchange and function of antioxidant defense system in two contrasting rice genotypes under Zn and Fe deficiency and hypoxia. Acta Biol Szeged 52: 283–294

Hajiboland R, Farhanghi F (2010) Remobilization of boron, photosynthesis, phenolic metabolism and antioxidant defense capacity in boron deficient turnip (*Brassica rapa* L.) plants. Soil Sci Plant Nutr 56:427–437

Hajiboland R, Farhanghi F (2011) Effect of low boron supply in turnip plants under drought stress. Biol Plant doi: 10.1007/s10535-011-0068-9

Hajiboland R, Joudmand A (2009) The K/Na replacement and function of antioxidant defense system in sugar beet (*Beta vulgaris* L.) cultivars. Acta Agric Scand Sect B Soil Plant Sci 59:246–259

Hajiboland R, Joudmand A, Fotouhi K (2009) Mild salinity improves sugar beet (*Beta vulgaris* L.) quality. Acta Agric Scand Sect B Soil Plant Sci 59:295–305

Hajiboland R, Aliasgharzadeh N, Laiegh SF, Poschenrieder C (2010) Colonization with arbuscular mycorrhizal fungi improves salinity tolerance of tomato (*Solanum lycopersicum* L.) plants. Plant Soil 331:313–327

Hamlin RL (2007) Molybdenum. In: Barker AV, Pilbeam DJ (eds) Handbook of plant nutrition. CRC Press, Taylor and Francis Group, Boca Raton, FL, pp 375–394Hajiboland R, Poschenrieder C, Bahrami Rad S, Tolra R (2011) Mechanisms for Al-stimulated growth in tea plants. In: Proceeding book of the 11th international conference on the biogeochemistry of trace elements. Florence, Italy, 3–7 July 2011

Han S, Chen LS, Jiang HX, Smith BR, Yang LT, Xie CY (2008) Boron deficiency decreases growth and photosynthesis, and increases starch and hexoses in leaves of citrus seedlings. J Plant Physiol 165: 1331–1341

Hanson B, Lindblom SD, Loeffler ML, Pilon-Smits EAH (2004) Selenium protects plants from phloem-feeding aphids due to both deterrence and toxicity. New Phytol 162:655–662

Hartikainen H, Xue T, Piironen V (2000) Selenium as an antioxidant and pro-oxidant in ryegrass. Plant Soil 225:193–200

Hasegawa PM, Bressan RA, Zhu JK, Bohnert HJ (2000) Plant cellular and molecular responses to high salinity. Annu Rev Plant Physiol Plant Mol Biol 51:463–499

Hausman JF, Evers D, Thiellement H, Jouve L (2000) Compared responses of poplar cuttings and in vitro raised shoots to short-term chilling treatments. Plant Cell Rep 19:954–960

Hayasaka T, Fujii H, Ishiguro K (2008) The role of silicon in preventing appressorial penetration by the rice blast fungus. Phytopathology 98:1038–1044

Hebbern CA, Laursen KH, Ladegaard AH, Schmidt SB, Dan Bruhn PP, Schjoerring JK, Wulfsohn D, Husted S (2009) Latent manganese deficiency increases transpiration in barley (*Hordeum vulgare*). Physiol Plant 135:307–316

Heckman JR (2007) Chlorine. In: Barker AV, Pilbeam DJ (eds) Handbook of plant nutrition. CRC Press, Taylor and Francis Group, Boca Raton, FL, pp 279–292

Heckman JR, Strick JE (1996) Teaching plant-soil relationships with color images of rhizosphere pH. J Nat Resour Life Sci Educ 25:13–17

Henriques FS (2003) Gas exchange, chlorophyll a fluorescence kinetics and lipid peroxidation of pecan leaves with varying manganese concentrations. Plant Sci 165:239–244

Henzler T, Ye Q, Steudle E (2004) Oxidative gating of water channels (aquaporins) in *Chara* by hydroxyl radicals. Plant Cell Environ 27:1184–1195

Hilal M, Parrado MF, Rosa M, Gallardo M, Orce L, Marta ME, González JA, Prado FE (2004) Epidermal lignin deposition in quinoa cotyledons in response to UV-B radiation. Photochem Photobiol 79:205–210

Hiradate S, Taniguchi S, Sakurai K (1998) Aluminum speciation in aluminum-silica solutions and potassium chloride extracts of acidic soils. Soil Sci Soc Am J 62:630–636

Holmes MG, Keiller DR (2002) Effects of pubescence and waxes on the reflectance of leaves in the ultraviolet and photosynthetic wavebands: a comparison of a range of species. Plant Cell Environ 25:85–93

Horst WJ, Fecht M, Naumann A, Wissemeier AH, Maier P (1999) Physiology of manganese toxicity and tolerance in *Vigna unguiculata* (L.) Walp. J Plant Nutr Soil Sci 162:263–274

Hu H, Brown P (1994) Localization of boron in cell walls of squash and tobacco and its association with pectin (evidence for a structural role of boron in the cell wall). Plant Physiol 105:681–689

Hu H, Brown P (1997) Absorption of boron by plant roots. Plant Soil 193:49–58

Huang LB, Ye ZQ, Bell RW, Dell B (2005) Boron nutrition and chilling tolerance of warm climate crop species. Ann Bot 96:755–767

Huber DM (1989) The role of nutrition in the take-all disease of wheat and other small grains. In: Englehard A (ed) Soil-borne plant pathogens: management of diseases with macro and microelements. American Phytopathological Society, St. Paul, pp 46–74

Humphries JM, Stangoulis JCR, Graham RD (2007) Manganese. In: Barker AV, Pilbeam DJ (eds) Handbook of plant nutrition. CRC Press, Taylor and Francis Group, Boca Raton, FL, pp 351–374

Husted S, Laursen KH, Hebbern CA, Schmidt SB, Pedas P, Haldrup A, Jensen PE (2009) Manganese deficiency leads to genotype-specific changes in fluorescence induction kinetics and state transitions. Plant Physiol 150:825–833

Inanaga S, Okasaka A, Tanaka S (1995) Does silicon exist in association with organic compounds in rice plant? Jpn J Soil Sci Plant Nutr 11:111–117

Iturbe-Ormaetxe I, Moran JF, Arrese-Igor C, Gogorcena Y, Klucas RV, Becana M (1995) Activated oxygen and antioxidant defenses in iron-deficient pea plants. Plant Cell Environ 18:421–429

Iwasaki K, Maier P, Fecht M, Horst WJ (2002) Effects of silicon supply on apoplastic manganese concentrations in leaves and their relation to manganese tolerance in cowpea (*Vigna unguiculata* (L.) Walp). Plant Soil 238:281–288

Jayakumar K, Vijayarengan P, Changxing Z, Gomathinayagam M, Jaleel CA (2008) Soil applied cobalt alters the nodulation, leg-haemoglobin content and antioxidant status of *Glycine max* (L.) Merr. Colloids Surf B Biointerfaces 67:272–275

Jiang C-D, Gao H-Y, Zou Q (2001) Enhanced thermal energy dissipation depending on xanthophyll cycle and D1 protein turnover in iron-deficient maize leaves under high irradiance. Photosynthetica 39:267–271

Jiang C-D, Gao H-Y, Zou Q (2002) Characteristics of photosynthetic apparatus in Mn-starved maize leaves. Photosynthetica 40:209–213

Johari RB, Nagar R, Sharma RC (1987) Studies on copper (II), nickel (II), cobalt (II) and zinc (II) complexes of acetone salicyloyl hydrazone and ethyl methyl ketone salicyloyl hydrazone. Indian J Chem Soc Inorg Phys Theor Anal 26:962–963

Kidd PS, Llugany M, Poschenrieder C, Gunse B, Barcelo J (2001) The role of root exudates in aluminium resistance and silicon-induced amelioration of aluminium toxicity in three varieties of maize (*Zea mays* L.). J Exp Bot 52:1339–1352

Kimura M, Yamamoto YY, Seki M, Sakurai T, Sato M, Abe T, Yoshida S, Manabe K, Shinozaki K, Matsui M (2003) Identification of *Arabidopsis* genes regulated by high light-stress using cDNA microarray. Photochem Photobiol 77:226–233

Kinraide TB (1993) Aluminum enhancement of plant growth in acid rooting media. A case of reciprocal alleviation of toxicity by two toxic cations. Physiol Plant 88:619–625

Kochian LV, Hoekenga OA, Piñeros MA (2004) How do crop plants tolerate acid soils? Mechanisms of aluminum tolerate and phosphorous efficiency. Annu Rev Plant Biol 55:459–493

Konishi W, Miyamoto S, Taki T (1985) Stimulatory effects of aluminum on tea plants grown under low and high phosphorus supply. Soil Sci Plant Nutr 31: 361–368

Kudinova LI (1975) The effect of silicon on growth, size of leaf area and sorbed surface of plant roots. Agrochemistry 10:117–120

Larcher W, Winter A (1981) Frost susceptibility of palms: experimental data and their interpretation. Principes 25:143–152

Liang YC, Yang CG, Shi HH (2001) Effects of silicon on growth and mineral composition of barley grown under toxic levels of aluminum. J Plant Nutr 24:229–243

Liang YC, Chen Q, Liu Q, Zhang W, Ding R (2003) Exogenous silicon (Si) increases antioxidant enzyme activity and reduces lipid peroxidation in roots of salt-stressed barley (*Hordeum vulgare* L.). J Plant Physiol 160:1157–1164

Liang YC, Wong JWC, Wei L (2005) Silicon-mediated enhancement of cadmium tolerance in maize (*Zea mays* L.) grown in cadmium contaminated soil. Chemosphere 58:475–483

Liang YC, Zhang WH, Chen Q, Liu YL, Ding RX (2006) Effect of exogenous silicon (Si) on H-ATPase activity, phospholipids and fluidity of plasma membrane in leaves of salt-stressed barley (*Hordeum vulgare* L.). Environ Exp Bot 57:212–219

Liang YC, Sun W, Zhu Y-G, Christie P (2007) Mechanisms of silicon-mediated alleviation of abiotic stresses in higher plants: a review. Environ Pollut 147:422–428

Lobanov AV, Fomenko DE, Zhang Y, Sengupta A, Hatfield DL, Gladyshev VN (2007) Evolutionary dynamics of eukaryotic selenoproteomes: large selenoproteomes may associate with aquatic life and small with terrestrial life. Genome Biol 8:R198

Ma J-F, Yamaji N (2006) Silicon uptake and accumulation in higher plants. Trends Plant Sci 11:392–397

Ma J-F, Miyake Y, Takahashi E (2001) Silicon as a beneficial element for crop plants. In: Datnoff LE, Snyder GH, Korndörfer GH (eds) Silicon in agriculture. Elsevier Science, Amsterdam, The Netherlands, pp 17–39

Maksymiec W (1997) Effect of copper on cellular processes in higher plants. Photosynthetica 34:321–342

Marschner H (1995) Mineral nutrition of higher plants, 2nd edn. Acedemic Press, London, UK

Marschner H, Cakmak I (1989) High light intensity enhances chlorosis and necrosis in leaves of zinc-, potassium- and magnesium-deficient bean (*Phaseolus vulgaris*) plants. J Plant Physiol 134:308–315

Marschner H, Kirkby EA, Cakmak I (1996) Effect of mineral nutritional status on shoot-root partitioning of photo assimilates and cycling of mineral nutrients. J Exp Bot 47:1255–1263

Matichenkov VV, Bocharnikova EA, Calvert DV, Snyder GH (1999a) Comparison study of soil silicon status in sandy soils of South Florida. Proc Soil Crop Sci Florida 59:132–137

Matichenkov VV, Calvert DV, Snyder GH (1999b) Silicon fertilizers for citrus in Florida. Proc Florida State Hortic Soc 112:5–8

Matichenkov VV, Calvert DV, Snyder GH (2000) Effect of Si fertilization on growth and P nutrition of bahiagrass. Proc Soil Crop Sci Soc Florida 60:30–36

Matsumoto H, Yamaya T (1986) Inhibition of potassium uptake and regulation of membrane-associated

Mg-ATPase activity of pea roots by aluminum. Soil Sci Plant Nutr 32:179–188

Mendel RR, Haensch R (2002) Molybdoenzymes and molybdenum cofactor in plants. J Exp Bot 53: 1689–1698

Merritt F, Kemper A, Tallman G (2001) Inhibitors of ethylene synthesis inhibit auxin-induced stomatal opening in epidermis detached from leaves of *Vicia faba* L. Plant Cell Physiol 42:223–230

Micó C, Li HF, Zhao FJ, McGrath SP (2008) Use of Co speciation and soil properties to explain variation in Co toxicity to root growth of barley (*Hordeum vulgare* L.) in different soils. Environ Pollut 156:883–890

Mitani N, Ma J-F (2005) Uptake system of silicon in different plant species. J Exp Biol 56:1255–1261

Mittler R, Zilinskas BA (1994) Regulation of pea cytosolic ascorbate peroxidase and other antioxidant enzymes during the progression of drought stress and following recovery from drought. Plant J 5:397–405

Miyasaka SC, Hue NV, Dunn MA (2007) Aluminum. In: Barker AV, Pilbeam DJ (eds) Handbook of plant nutrition. CRC Press, Taylor and Francis Group, Boca Raton, FL, pp 439–498

Molinari HBC, Marur CJ, Daros E, De Campos MKF, De Carvalho JFRP, Filho JCB, Pereira LFP, Vieira LGE (2007) Evaluation of the stress-inducible production of proline in transgenic sugarcane (*Saccharum* spp.): osmotic adjustment, chlorophyll fluorescence and oxidative stress. Physiol Plant 130:218–229

Monreal J, Alvarez R, Vidal J, Echevarria C (2003) Characterization of salt stress-enhanced phosphoenolpyruvate carboxylase kinase activity in leaves of *Sorghum vulgare*: independence from osmotic stress, involvement of ion toxicity and significance of dark phosphorylation. Planta 216:648–655

Moore PA Jr, Patrick WH Jr (1988) Effect of zinc deficiency on alcohol dehydrogenase activity and nutrient uptake in rice. Agron J 80:882–885

Morales F, Belkhodja R, Abadía A, Abadía J (2000) Photosystem II efficiency and mechanisms of energy dissipation in iron-deficient, field-grown pear trees (*Pyrus communis* L.). Photosynth Res 63:9–21

Moseley JL, Allinger T, Herzog S, Hoerth P, Wehinger E, Merchant S, Hippler M (2002) Adaptation to Fe-deficiency requires remodeling of the photosynthetic apparatus. EMBO J 21:6709–6720

Möttönen M, Aphalo PJ, Lehto T (2001) Role of boron in drought resistance in Norway spruce (*Picea abies*) seedlings. Tree Physiol 21:673–681

Möttönen M, Lehto T, Aphalo PJ, Kukkola M, Mälkönen E (2003) Response of mature stands of Norway spruce (*Picea abies*) to boron fertilization. Forest Ecol Manage 180:401–412

Möttönen M, Lehto T, Rita H, Aphalo PJ (2005) Recovery of Norway spruce (*Picea abies*) seedlings from repeated drought as affected by boron nutrition. Trees 19:213–223

Moura JCMS, Bonine CAV, Viana JOF, Dornelas MC, Mazzafera P (2010) Abiotic and biotic stresses and changes in the lignin content and composition in plants. J Integr Plant Biol 52:360–376

Müller P, Li XP, Niyogi KK (2001) Non-photochemical quenching. A response to excess light energy. Plant Physiol 125:1558–1566

Mulrooney SB, Hausinger RP (2003) Nickel uptake and utilization by microorganisms. FEMS Microbiol Rev 27:239–261

Murphy TM, Auh CK (1996) The superoxide synthase of plasma membrane preparations from cultured rose cells. Plant Physiol 110:621–629

Neue HU, Quijano C, Senadhira D, Setter T (1998) Strategies for dealing with micronutrient disorders and salinity in lowland rice systems. Field Crop Res 56:139–155

Neumann D, zur Nieden U (2001) Silicon and heavy metal tolerance of higher plants. Phytochemistry 56:685–692

Nicholas DJD (1961) Minor mineral nutrients. Annu Rev Plant Physiol 12:63–90

O'Neill MA, Ishii T, Albersheim P, Darvill AG (2004) Rhamnogalacturonan II: structure and function of a borate cross-linked cell wall pectic polysaccharide. Annu Rev Plant Biol 55:109–139

Ohyama N (1985) Amelioration of cold weather damage of rice by silicate fertilizer application. Agric Hort 60:1385–1389

Osaki M, Watanabe T, Tadano T (1997) Beneficial effect of aluminum on growth of plants adapted to low pH soils. Soil Sci Plant Nutr 43:551–563

Palit S, Sharma A, Talukder G (1994) Effect of cobalt on plants. Bot Rev 60:149–181

Papadakis IE, Bosabalidis AM, Soiropoulos TE, Therios IN (2007) Leaf anatomy and chloroplast ultrastructure of Mn-deficient orange plants. Acta Physiol Plant 29:297–301

Pennanen A, Xue T, Hartikainen H (2002) Protective role of selenium in plant subjected to severe UV irradiation stress. J Appl Bot 76:66–76

Perkasem B, Jamjod S (2004) Boron deficiency in wheat: a review. Field Crop Res 89:173–186

Pilon-Smits EAH, Quinn CF, Tapken W, Malagoli M, Schiavon M (2009) Physiological functions of beneficial elements. Curr Opin Plant Biol 12:267–274

Pinton R, Cakmak I, Marschner H (1994) Zinc deficiency enhanced NAD(P)H-dependent superoxide radical production in plasma membrane vesicles isolated from roots of bean plants. J Exp Bot 45:45–50

Poschenrieder C, Llugany M, Barcelo J (1995) Short term effects of pH and aluminum on maize varieties differing in proton and aluminum tolerance. J Plant Nutr 18:1495–1507

Prasad PVV, Pisipati SR, Mutava RN, Tuinstra MR (2008) Sensitivity of grain sorghum to high temperature stress during reproductive development. Crop Sci 48: 1911–1917

Price AH, Hendry AF (1991) Iron-catalysed oxygen radical formation and its possible contribution to drought damage in nine native grasses and three cereals. Plant Cell Environ 14:477–484

Quartacci MF, Cosi E, Navari - Izzo F (2001) Lipids and NADPH - dependent superoxide production in plasma membrane vesicles from roots of wheat grown under copper deficiency or excess. J Exp Bot 52:77–84

Queiroz CGS, Alonso A, Mares-Guia M, Magalhaes AC (1998) Chilling-induced changes in membrane fluidity and antioxidant enzyme activities in *Coffea arabica* L. roots. Biol Plant 41:403–413

Quinn CF, Freeman JL, Reynolds RJB, Cappa JJ, Fakra SC, Marcus MA, Lindblom SD, Quinn EK, Bennett LF, Pilon-Smits EAH (2010) Selenium hyperaccumulation offers protection from cell disruptor herbivores. BMC Ecol 10:19

Räisänen M, Repo T, Lehto T (2006a) Effect of thawing time, cooling rate and boron nutrition on freezing point of the primordial shoot in Norway spruce buds. Ann Bot 97:593–599

Räisänen M, Repo T, Lehto T (2006b) Does ice crystal formation in buds explain growth disturbances in boron-deficient Norway spruce? Trees 20:441–448

Räisänen M, Repo T, Lehto T (2007) Cold acclimation was partially impaired in boron deficient Norway spruce seedlings. Plant Soil 292:271–282

Räisänen M, Repo T, Lehto T (2009) Cold acclimation of Norway spruce roots and shoots after boron fertilization. Silva Fennica 43:223–233

Ranieri A, Castagna A, Baldan B, Soldatini GF (2001) Iron deficiency differently affects peroxidase isoforms in sunflower. J Exp Bot 52:25–35

Reddy AR, Chaitanya KV, Vivekanandan M (2004) Drought-induced responses of photosynthesis and metabolism in higher plants. J Plant Physiol 161:1189–1202

Redondo-Nieto M, Reguera M, Bonilla I, Bolaños L (2008) Boron dependent membrana glycoproteins in symbiosome development and nodule organogenesis. A model for a common role of boron in organogenesis. Plant Signal Behav 3:298–300

Robinson B, Burne P (2000) Another look at the Merlot problem – could it be molybdenum deficiency? Aust Grapegrow Winemak 427a:21–22, 28th Annual Technical Issue

Rock CD (2000) Pathways to abscisic acid-regulated gene expression. New Phytol 148:357–396

Rogalla H, Römheld V (2002) Role of leaf apoplast in silicon-mediated manganese tolerance of *Cucumis sativus* L. Plant Cell Environ 25:549–555

Rogers LA, Campbel MM (2004) The genetic control of lignin deposition during plant growth and development. New Phytol 164:17–30

Roldan M, Belver A, Rodriguez-Rosales P, Ferrol N, Donaire JD (1992) *In vivo* and *in vitro* effects of boron on the plasma membrane proton pump of sunflower roots. Physiol Plant 84:49–54

Rosenberg RJ, Christensen NW, Jackson TL (1986) Chloride, soil solution osmotic potential, and soil pH effects on nitrification. Soil Sci Soc Am J 50:941–945

Rotruck JT, Pope AL, Ganther HE, Hoekstra WG (1973) Selenium: biochemical role as a component of glutathione peroxidase. Science 179:588–590

Rozema J, Van de Staaij J, Björn LO, Caldwell M (1997) UV-B as an environmental factor in plant life: stress and regulation. Trends Ecol Evol 12:22–28

Sadasivan TS (1965) Effect of mineral nutrients on soil microorganisms and plant disease. In: Baker CF, Snyder WC (eds) Ecology of soil-borne pathogens. Proceedings of the international symposium. University of California Press, Berkeley, pp 460–470

Sagi M, Scazzocchio C, Fluhr R (2002) The absence of molybdenum cofactor sulfuration is the primary cause of the flacca phenotype in tomato plants. Plant J 31:305–317

Sahrawat KL, Mulbah CK, Diatta S, Delaune RD, Patrick WH, Singh BN, Jones MP (1996) The role of tolerant genotypes and plant nutrients in the management of iron toxicity in lowland rice. J Agric Sci 126:143–149

Saito A, Iino T, Sonoike K, Miwa E, Higuchi K (2010) Remodeling of the major light-harvesting antenna protein of PSII protects the young leaves of barley (*Hordeum vulgare* L.) from photoinhibition under prolonged iron deficiency. Plant Cell Physiol 51:2013–2030

Saruyama H, Tanida M (1995) Effect of chilling on activated oxygenscavenging enzymes in low temperature-sensitive and tolerant cultivars of rice (*Oryza sativa* L.). Plant Sci 109:105–113

Schnabl H, Raschke K (1980) Potassium chloride as stomatal osmoticum in *Allium cepa* L., a species devoid of starch in guard cells. Plant Physiol 65:88–93

Schwartz SH, Qin X, Zeevaart JAD (2003) Elucidation of the indirect pathway of abscisic acid biosynthesis by mutants, genes, and enzymes. Plant Physiol 131: 1591–1601

Seppänen M, Turakainen M, Hartikainen H (2003) Selenium effects on oxidative stress in potato. Plant Sci 165:311–319

Sharma PN, Ramchandra T (1990) Water relations and photosynthesis in mustard plants subjected to boron deficiency. Indian J Plant Physiol 33:150–154

Sharma PN, Tripathi A, Bisht SS (1995) Zinc requirement for stomatal opening in cauliflower. Plant Physiol 107:751–756

Shenker M, Plessner OE, Tel-Or E (2004) Manganese nutrition effects on tomato growth, chlorophyll concentration, and superoxide dismutase activity. J Plant Physiol 161:197–202

Shi XH, Zhang CC, Wang H, Zhang FS (2005a) Effect of Si on the distribution of Cd in rice seedlings. Plant Soil 272:53–60

Shi QH, Bao ZY, Zhu ZJ, He Y, Qian QQ, Yu JQ (2005b) Silicon mediated alleviation of Mn toxicity in *Cucumis sativus* in relation to activities of superoxide dismutase and ascorbate peroxidase. Phytochemistry 66: 1551–1559

Singh P, Misra A, Srivastava NK (2001) Influence of Mn deficiency on growth, chlorophyll content, physiology and essential monoterpene oil(s) in genotypes of spearmint (*Mentha spicata* L.). Photosynthetica 39:473–476

Snyder GH, Jones DB, Coale FJ (1990) Occurrence and correction of manganese deficiency in Histosol-grown rice soil. Soil Sci Soc Am J 54:1634–1638

Snyder GH, Matichenkov VV, Datnoff LE (2007) Silicon. In: Barker AV, Pilbeam DJ (eds) Handbook of plant nutrition. CRC Press, Taylor and Francis Group, Boca Raton, FL, pp 551–568

Steffen KL, Palta JP (1986) Effect of light on photosynthetic capacity during cold acclimation in a cold-sensitive and a cold-tolerant potato species. Physiol Plant 66:353–359

Subbarao GV, Wheeler RM, Stutte GW, Levine LH (2000) Low potassium enhances sodium uptake in red-beet under moderate saline conditions. J Plant Nutr 23:1449–1470

Subbarao GV, Ito O, Berry WL, Wheeler RM (2003) Sodium - a functional plant nutrient. Crit Rev Plant Sci 22:391–416

Takano J, Nguchi K, Yasumori M, Kobayashi M, Gajdos Z, Miwa K, Hayashi H, Yoneyama T, Fugiwara T (2002) *Arabidopsis* boron transporter for xylem loading. Nature 420:337–340

Talukder G, Sharma A (2007) Cobalt. In: Barker AV, Pilbeam DJ (eds) Handbook of plant nutrition. CRC Press, Taylor and Francis Group, Boca Raton, FL, pp 499–514

Tanaka K, Masuda R, Sugimoto T, Omasa K, Sakaki T (1990) Water deficiency-induced changes in the contents of defensive substances against active oxygen in spinach leaves. Agric Biol Chem 54:2629–2634

Tao H, Dittert K, Zhang L, Lin S, Römheld V, Sattelmacher B (2007) Effects of soil water content on growth, tillering, and manganese uptake of lowland rice grown in the water-saving ground-cover rice-production system (GCRPS). J Plant Nutr Soil Sci 170:7–13

Tarabrin VP, Teteneva TR (1979) Presowing treatment of seeds and its effect on the resistance of wood plant seedlings against drought. Sov J Ecol 10:204–211

Tewari RK, Kumar P, Sharma PN (2008) Morphology and physiology of zinc-stressed mulberry plants. J Plant Nutr Soil Sci 171:286–294

Thompson DC, Clarke BB, Heckman JR (1995) Nitrogen form and rate of nitrogen and chloride applications for the control of summer patch in Kentucky bluegrass. Plant Dis 79:51–55

Tyerman SD, Niemietz CM, Bramley H (2002) Plant aquaporins: multifunctional water and solute channels with expanding roles. Plant Cell Environ 25:73–194

Upchurch RG (2008) Fatty acid unsaturation, mobilization, and regulation in the response of plants to stress. Biotechnol Lett 30:967–977

Van Breemen N, Castro RU (1980) Zinc deficiency in wetland rice along a toposequence of hydromorphic soils in the Philippines. II. Cropping experiment. Plant Soil 57:215–221

Wang H, Jin JY (2005) Photosynthetic rate, chlorophyll fluorescence parameters, and lipid peroxidation of maize leaves as affected by zinc deficiency. Photosynthetica 43:591–596

Wang ZY, Tang YL, Zhang FS, Wang H (1999) Effect of boron and low temperature on membrane integrity of cucumber leaves. J Plant Nutr 22:543–550

Wang H, Liu RL, Jin JY (2009) Effects of zinc and soil moisture on photosynthetic rate and chlorophyll fluorescence parameters of maize. Biol Plant 53:191–194

Watanabe T, Osaki M (2002) Mechanisms of adaptation to high aluminum condition in native plant species growing in acid soils: a review. Commun Soil Sci Plant Anal 33:1247–1260

Waterkeyn L, Bientait A, Peeters A (1982) Callose et silice epidermiques rapports avec la transpiration culticulaire. La Cellule 73:263–287

Wisnievski M (1995) Deep supercooling in woody plants and the role of cell wall structure. In: Lee RE, Warren GJ, Gusta LV (eds) Biological ice and its applications. American Phytopatological Society Press, St Paul, pp 163–181

Xiong L, Ishitani M, Lee H, Zhu J-K (2001) The *Arabidopsis LOS5/ABA3* locus encodes a molybdenum cofactor sulfurase and modulates cold Stress- and osmotic stress-responsive gene expression. Plant Cell 13:2063–2083

Xiong L, Lee H, Ishitani M, Zhu J-K (2002a) Regulation of osmotic stress-responsive gene expression by the *LOS6/ABA1* locus in *Arabidopsis*. J Biol Chem 277:8588–8596

Xiong L, Schumaker KS, Zhu J-K (2002b) Cell signaling during cold, drought, and salt stress. Plant Cell 14(Suppl):S165–S183

Xue TL, Hartikainen H (2000) Association of antioxidative enzymes with the synergistic effect of selenium and UV irradiation in enhancing plant growth. Agr Food Sci Finland 9:177–186

Xue T, Hartikainen H, Piironen V (2001) Antioxidative and growth–promoting effect of selenium on senescing lettuce. Plant Soil 237:55–61

Ye Z (2005) Effect of low temperature on plant boron nutrition. PhD Thesis, Murdoch University, Perth, Australia

Ye Z, Bell RW, Dell B, Huang L (2000) Response of sunflower to boron supply at low root zone temperature. Commun Soil Sci Plant Anal 31:2379–2392

Yoshida S (1975) The physiology of silicon in rice. Technical bulletin, no. 25. Food and Fertilizer Technology Center, Taipei, Taiwan

Yu Q, Osborne L, Rengel Z (1998) Micronutrient deficiency changes activities of superoxide dismutase and ascorbate peroxidase in tobacco plants. J Plant Nutr 21:1427–1437

Yu Q, Osborne LD, Rengel Z (1999a) Increased tolerance to Mn deficiency in transgenic tobacco overexpressing superoxide dismutase. Ann Bot 84:543–547

Yu Q, Worth C, Rengel Z (1999b) Using capillary electrophoresis to measure Cu/Zn superoxide dismutase concentration in leaves of wheat genotypes differing in tolerance to zinc deficiency. Plant Sci 143:231–239

Zaharieva TB, Abadía J (2003) Iron deficiency enhances the level of ascorbate, glutathione, and related enzymes in sugar beet roots. Protoplasma 221:269–275

Zhu ZJ, Wei GQ, Li J, Qian QQ, Yu JQ (2004) Silicon alleviates salt stress and increases antioxidant enzymes activity in leaves of salt-stressed cucumber (*Cucumis sativus* L.). Plant Sci 167:527–533

[illegible]

Stress-Induced Flowering

17

Kiyotoshi Takeno

Abstract

Plants have a tendency to flower if placed under unsuitable growth conditions. A review of the literature reporting such nonphotoperiodic flowering indicated that most of the factors responsible for flowering could be regarded as stress. Those stress factors include poor nutrition, high or low temperature, high- or low-intensity light, ultraviolet light, and many others. This flowering would be called stress-induced flowering. The plants that were induced to flower by stress reached anthesis, produced fertile seeds, and the progeny developed normally in *Pharbitis nil* and *Perilla frutescens* var. *crispa*. Grafting experiments using two varieties of *P. nil* revealed that a transmissible flowering stimulus is involved in stress-induced flowering. Salicylic acid and/or the flowering gene *FLOWERING LOCUS T* may be involved in the stress-induced flowering of *P. nil*, *P. frutescens*, *Arabidopsis thaliana*, and *Lemna paucicostata*. The stressed plants do not need to wait the arrival of a season when photoperiodic conditions are suitable for flowering, and such precocious flowering might assist in species preservation. Thus, stress-induced flowering might have a biological benefit and should be considered as important as photoperiodic flowering and vernalization.

Keywords

Flowering • *FLOWERING LOCUS T* • *Perilla frutescens* • *Pharbitis nil* • Phenylalanine ammonia-lyase • Salicylic acid • Stress • Transmissible flowering stimulus

K. Takeno (✉)
Department of Biology, Faculty of Science,
Niigata University, Ikarashi, Niigata 950-2181, Japan
e-mail: ktw@bio.sc.niigata-u.ac.jp

P. Ahmad and M.N.V. Prasad (eds.), *Abiotic Stress Responses in Plants: Metabolism, Productivity and Sustainability*, DOI 10.1007/978-1-4614-0634-1_17,

1 Introduction

Flowering is regulated by both endogenous and environmental factors. One endogenous factor is the autonomous pathway of flowering regulation in *Arabidopsis thaliana* (Simpson 2004). Day-neutral plants switch from vegetative to reproductive growth in response to endogenous signals after a certain period of time (McDaniel 1996). Environmental factors that regulate flowering include the duration of the day and night periods in photoperiodic flowering and temperature in vernalization (Thomas and Vince-Prue 1997). Flowering that cannot be classified into these categories has also been sporadically reported. Some plants for which flowering is basically regulated photoperiodically flowered under unsuitable photoperiodic conditions when grown under certain conditions. Experienced flowering physiologists have noticed that plants have a tendency to flower if placed under unsuitable growth conditions. However, such unusual flowering has not been studied systematically. We surveyed the flowering behavior reported in the literature and analyzed the nonphotoperiodic flowering responses in some plant species. We concluded that many cases of nonphotoperiodic flowering are induced by stress. Stress is as important flower-inducing factor as night length and low temperature.

2 History of Stress-Induced Flowering

The short-day plant *Pharbitis nil* (synonym *Ipomoea nil*) can flower under long days when grown in tap water (poor-nutrition conditions), at 12–15°C (low-temperature conditions) or under 15,000–20,000 lux light (high-intensity light conditions) (Shinozaki and Takimoto 1982; Shinozaki et al. 1982, 1988a, b, 1994; Shinozaki 1985; Swe et al. 1985; Hirai et al. 1993, 1994, 1995). These flowering responses were mainly studied by Dr. M. Shinozaki and his coworkers at Kyoto University, Japan, who called such flowering "long-day flowering."

The responses to these conditions differ depending on the cultivar (cv.) (Swe et al. 1985). The most common cultivar, Violet, responded to all these conditions and flowered, whereas cv. Tendan and Kidachi were not induced to flower by poor nutrition or high-intensity light. Kidachi and the white-flowered mutant of Violet responded more sensitively to low temperature than did wild-type Violet (Ishimaru et al. 1996). Chlorogenic acid (CGA) and some other phenylpropanoids were found to accumulate in the cotyledons during the treatments with poor nutrition, low temperature, or high-intensity light (Shinozaki et al. 1988a, b, 1994; Hirai et al. 1993, 1994). Many reports indicated a close correlation between CGA content and flowering response (Ishimaru et al. 1996), in which the number of flower buds increased in parallel with the CGA content. Kidachi that was not induced to flower by poor nutrition or high-intensity light did not accumulate CGA under these conditions. A white-flowered mutant of Violet responded to low temperature more sensitively and accumulated more CGA than did the wild type. Further, the flowering induced by these conditions was accompanied by an increase in phenylalanine ammonia-lyase (PAL) activity (Hirai et al. 1995). Aminooxyacetic acid (AOA) inhibited flowering (Shinozaki et al. 1988a, 1994; Hatayama and Takeno 2003). AOA inhibits the activity of PAL, which catalyzes the conversion of phenylalanine to *t*-cinnamic acid to result in the accumulation of CGA. These findings suggested that endogenous CGA may be involved in long-day flowering. However, exogenous application of CGA could not induce flowering (Shinozaki et al. 1988a, 1994; Ishimaru et al. 1996).

3 Long-Day Flowering May Be Caused by Stress

Although the factors that can induce long-day flowering are not related to each other, PAL is involved in flowering induced by any of these conditions. This suggests that these factors may stimulate flowering through a common signal transduction pathway. Poor nutrition, low temperature, and high-intensity light can be regarded as stress factors, and PAL activity increases under stress conditions (Dixon and

Paiva 1995). Accordingly, we predicted that long-day flowering might be induced by stress.

We found that nonphotoperiodic flowering is not restricted to *P. nil*. We observed that the short-day plant *Lemna paucicostata* flowered under long-day conditions and the long-day plant *Lemna gibba* flowered under short-day conditions when they were grown in tap water, i.e., under poor nutrition conditions (unpublished data). Further, we found that the short-day plant *Perilla frutescens* var. *crispa* flowered under long-day conditions when grown under low-intensity light (Wada et al. 2010b). Low-intensity light can be also regarded as stress factor.

We have noticed that nonphotoperiodic flowering has been sporadically reported. Accordingly, we surveyed past studies and found that many of the conditions under which flowering was induced can be considered to be stress conditions (Wada and Takeno 2010). Some examples are listed in Table 17.1. Flowering has been induced by high- and low-intensity light, ultraviolet-C light, drought, poor nutrition, high and low temperature, and mechanical stimulation. These factors can be regarded as stress, although many of those reports did not mention that stress was responsible for flowering. Papers that clearly mentioned that flowering was induced by stress have appeared only recently (Hatayama and Takeno 2003; Martínez et al. 2004; Kolár and Senková 2008).

Stress can be simply defined as a situation in which the vegetative growth of plants is suppressed. Flowering is the change from vegetative growth to reproductive growth. Therefore, it is quite natural that flowering is accelerated by the suppression of vegetative growth by stress. Plants can modify their development to adapt to stress conditions. Stressed plants may flower to produce the next generation as an emergency response. In this way, plants can preserve the species even in an unfavorable environment. This idea is supported by the recent change in the understanding of shade avoidance responses. The most typical phenotype of the shade avoidance response

Table 17.1 Some cases of stress-induced flowering

Stress factor	Species	Flowering response	References
High-intensity light	*Pharbitis nil*	Induction	Shinozaki et al. (1994)
	Arabidopsis thaliana	Promotion	Marín et al. (2010)
Low-intensity light (shaded by neighbors)	*Lemna perpusilla*	Induction	Takimoto (1973)
	Lemna paucicostata	Induction	Tanaka et al. (1989)
	Perilla frutescens var. *crispa*	Induction	De Zeeuw (1953), Gaillochet et al. (1962), Wada et al. (2010b)
	Arabidopsis thaliana	Promotion	Halliday et al. (1994), Smith and Whitelam (1997), Dorn et al. (2000)
Continuous light	*Arabidopsis thaliana*	Promotion	Marín et al. (2010)
Ultraviolet C	*Arabidopsis thaliana*	Promotion	Martínez et al. (2004)
Ultraviolet	Duckweeds	Induction	Hicks (1932)
Salt	*Mesembryanthemum crystallinum*	Promotion	Adams et al. (1998)
Drought	Douglas-fir	Induction	Ebell (1967)
	Tropical pasture legumes	Induction	Hopkinson (1977)
	Lemon	Induction	Casella (1935), Monselise and Halevy (1964), Nir et al. (1972), Monselise et al. (1981)
	Citrus spp.	Promotion	Monselise (1985)
	Ipomoea batatas	Promotion	Jones (1980)
	Brachypodium distachyon	Promotion	Aronson et al. (1992)
	Lemnaceae	Induction	Krajncic et al. (2006)
Poor nutrition	*Pharbitis nil*	Induction	Shinozaki et al. (1988a), Wada et al. (2010a)
	Macroptilium atropurpureum	Promotion	Imrie (1973)
	Cyclamen persicum	Promotion	Bussler (1969)
	Ipomoea batatas	Promotion	Jones (1980)
	Arabidopsis thaliana	Promotion	Kolár and Senková (2008)

(continued)

Table 17.1 (continued)

Stress factor	Species	Flowering response	References
Poor oxygen	*Pharbitis nil*	Induction	Shinozaki et al. (1982)
High temperature	*Arabidopsis thaliana*	Promotion	Marín et al. (2010)
	Chenopodium polyspermum	Promotion	Chamont et al. (1982)
Low temperature	*Pharbitis nil*	Induction	Hirai et al. (1994), Hatayama and Takeno (2003)
	Chenopodium polyspermum	Promotion	Chamont et al. (1982)
Photochilling	*Arabidopsis thaliana*	Promotion	Marín et al. (2010)
High conc. $GA_{4/7}$	Douglas-fir	Promotion	McMullan (1980)
Crowdedness	*Lemna perpusilla*	Induction	Landolt (1957)
Girdling	Douglas-fir	Induction	Ebell (1971)
Root pruning	*Citrus* sp.	Induction	Iwasaki et al. (1959)
	Pharbitis nil	Induction	Wada (1974)
Mechanical stimulation	*Ananas comosus*	Induction	Metzger (1995)
Suppression of root elongation	*Pharbitis nil*	Induction	Swe et al. (1985)

is rapid stem elongation, but recent articles report that an important component of the shade avoidance syndrome is an acceleration of flowering observable in all shade-avoiding plants (Adams et al. 2009). Accelerated flowering and seed production under unfavorable environments increase the probability of the survival of the individual and therefore of the species. This is true also in flowering induced under stress conditions.

Thus, it is reasonable to assume that stress can induce flowering, and the evidence for this is accumulating (Table 17.1). Therefore, we called such flowering "stress-induced flowering" (Hatayama and Takeno 2003; Wada et al. 2010a, b; Wada and Takeno 2010).

4 Case Studies of Stress-Induced Flowering

4.1 *Pharbitis nil*

P. nil, cv. Violet was induced to flower when grown in a diluted mineral nutrient solution or tap water for 20 days under long-day conditions (Wada et al. 2010a). The vegetative growth of the plants under these poor-nutrition conditions was substantially inhibited. Because the suppression of vegetative growth indicated that the plants were stressed, this flowering can be considered stress-induced flowering. The flowering response was weaker under the weaker stress condition (1/10-strength nutrient solution) than under the stronger stress condition (1/100-strength nutrient solution). The other cultivar, Tendan was not induced to flower even when grown in tap water, although vegetative growth was inhibited. Thus, nutrient stress does not induce flowering in all cultivars. Flowering of the white-flowered mutant of Violet was induced by a low-temperature stress treatment of 13°C for 10 days, whereas the control plants kept at 25°C remained vegetative (Hatayama and Takeno 2003).

The Violet plants induced to flower by poor-nutrition stress produced fertile seeds, and their progeny developed normally (Wada et al. 2010a). Defoliated Violet scions grafted onto rootstocks of Violet or Tendan were induced to flower under poor-nutrition stress conditions. This result indicates that a transmissible flowering stimulus is involved in the induction of flowering by poor-nutrition stress. The poor-nutrition stress-induced flowering and cold-stress-induced flowering were inhibited by AOA, which is an inhibitor of PAL, and this inhibition was almost completely reversed by benzoic acid or salicylic acid (SA) (Hatayama and Takeno 2003; Wada et al. 2010a). However, exogenously applied SA did not induce flowering under nonstress conditions, suggesting that SA may be necessary but not sufficient to induce flowering. *PnFT2*, a *P. nil* ortholog of the flowering gene *FLOWERING LOCUS T* (*FT*) of

A. thaliana, was expressed when the Violet plants were induced to flower by growing in tap water. However, the expression of *PnFT1*, another ortholog of *FT*, was not induced, suggesting a specific involvement of *PnFT2* in stress-induced flowering (Wada et al. 2010a).

4.2 *Perilla frutescens*

The short-day plant *Perilla frutescens* var. *crispa* was induced to flower under long-day conditions when grown under low-intensity light (Wada et al. 2010b). Two forms of *P. frutescens* were planted in vermiculite when the cotyledons had expanded and were grown under long-day conditions with different light intensities. All of the red-leaved plants grown under 30 μmol m^{-2} s^{-1} flowered, whereas the plants grown under 60 or 120 μmol m^{-2} s^{-1} did not. The green-leaved form was also induced to flower under 30 μmol m^{-2} s^{-1}, although the flowering response was lower than that of the red-leaved form. Flowering under low-intensity light accompanied a reduction in stem length. The reduction of vegetative growth results from stress, and therefore the flowering of *P. frutescens* under low-intensity light is another example of stress-induced flowering. *P. frutescens* is an obligatory short-day plant (Jacobs 1982) that does not have a vernalization requirement. Therefore, the flowering under long-day conditions found in this study is independent of photoperiodism and vernalization. Photosynthetic activity may have decreased under low-intensity light conditions. However, it is unlikely that the photosynthetic deficiency induced flowering because the photoassimilate is a flower-inducing factor (Bernier and Périlleux 2005). In fact, sucrose induced flowering of *P. frutescens* cultured in vitro under long-day conditions (Purse 1984). Generally, plants grown under low-intensity light conditions are etiolated and elongated (Lorrain et al. 2008). However, the stem length of *P. frutescens* was shortened, and the leaves became green under low-intensity light. Therefore, the response of *P. frutescens* to low-intensity light was different from the general photomorphogenetic response or shade avoidance response. *P. frutescens* is reportedly induced to flower by poor nutrition (Wada and Totsuka 1982) or low temperature (Zeevaart 1969). Accordingly, the red-leaved *P. frutescens* was treated with several stress factors other than low-intensity light. The plants were grown in tap water or a diluted mineral nutrient solution (poor-nutrition stress), at 5–15°C (low-temperature stress), with 50–400 mM NaCl (salt stress) or poor watering (water stress). None of these factors induced flowering, even though they retarded vegetative growth. This indicates that not all kinds of stress can induce flowering. Although high-intensity light has been well-studied as a stress factor (Chalker-Scott 1999), there are only a few reports on low-intensity light as a stress factor. Red-leaved *P. frutescens* (De Zeeuw 1953; Gaillochet et al. 1962), *Lemna perpusilla* (Takimoto 1973) and *A. thaliana* (Smith and Whitelam 1997) flowered under low-intensity light.

The red-leaved *P. frutescens* that were exposed to low-intensity light when their cotyledons had just expanded were induced to flower by the 3-week treatment, and 100% flowering occurred after the 4-week treatment (Wada et al. 2010b). Flowers were formed even at the cotyledonal nodes. Prolonged treatment for 5 weeks did not increase the number of flowers or inflorescences. The plants could respond to low-intensity light immediately after the cotyledons had expanded. The flowering response decreased with an increase in plant age, and flowering was not induced when the low-intensity light treatment began 2 weeks after the cotyledons had expanded or at any later time. Treatment for at least 3 weeks was required to induce flowering. The low-intensity light stress-induced flowering was inhibited by PAL inhibitors.

4.3 *Arabidopsis thaliana*

Ultraviolet (UV)-C light stress promoted flowering in *A. thaliana* (Martínez et al. 2004). The other stresses of extreme temperature, water deficiency, or high light irradiation did not accelerate flowering. UV-C irradiation accelerated flowering in wild-type Columbia (Col) ecotype in a dose-dependent manner between 0 and 200 mJ cm^{-2}. UV-C irradiation of the same dosage did not promote flowering of the nahG transgenic plants that are expressing bacterial salicylate hydroxylase and are unable to accumulate SA because of the rapid and efficient conversion of SA to catechol. UV-C irradiation increased expression of the SA-responsive *PR1* gene and the gene encoding the SA biosynthetic enzyme in Col. Exogenously applied SA accelerated flowering of Col, but the nahG plants were not responsive to SA treatment. UV-C induced expression of the flowering gene *FT* and moderately induced *CONSTANS* (*CO*) expression in wild type. None of these genes were induced by UV-C in nahG plants. Flowering promoted by SA requires the reduced expression of *FLC* and enhanced expression of *FT*.

Poor-nutrition conditions accelerated the flowering of *A. thaliana* (Kolár and Senková 2008). The authors suggested that this precocious flowering was due to stress. When plants were grown in a full-strength nutrient solution for 3–5 weeks and then transferred to a 1/10- to 1/1,000-strength media, the time to flower was notably shortened. The accelerating effect was stronger when the stress was applied earlier, and the more diluted solution caused greater acceleration. This acceleration was more pronounced in short-day conditions than in long-day conditions. The response was stronger in the ecotype Landsberg *erecta* (L*er*) than in Col. The nutrient-deficient L*er* plants formed normal flowers and fruits with seeds. On the other hand, Marín et al. (2010) reported that flowering of *A. thaliana* was more rapid under low nitrate conditions, although low nitrate did not act via a general stress pathway. They intended to study the specific effect of nitrate on flowering, and therefore added glutamine to the medium as a constitutive nitrogen supply together with nitrate of varied concentrations. The plants that flowered earlier on low nitrate media showed similar growth rates to the plants grown on high nitrate media.

General stress leads to early flowering in *A. thaliana*. High-intensity light of 800 μmol m^{-2} s^{-1}, high temperature of 26°C, photochilling of 800 μmol m^{-2} s^{-1} at 16°C, or continuous light treatments lead to earlier flowering in wild-type L*er*, but not in the *fca1 co-2 ga1-3* triple mutant (Marín et al. 2010).

4.4 *Lemna paucicostata*

When the short-day plant *L. paucicostata* strain 6746 was cultured in tap water under noninductive long-day conditions, the multiplication of fronds decreased and flowering occurred (Shimakawa 2011). Thus, poor-nutrition conditions functioned as a stress factor, and this stress induced flowering. The poor-nutrition stress-induced flowering response was weaker than that induced by short-day treatment. L-2-aminooxy-3-phenylpropionic acid (AOPP), an inhibitor of PAL, inhibited the poor-nutrition stress-induced flowering without preventing the vegetative growth. More SA was detected in the flowered plants than in the control plants. The results suggest the involvement of SA in the stress-induced flowering of *L. paucicostata*. Exogenously applied SA is known to induce flowering in *L. paucicostata* and *L. gibba* (Cleland and Ajami 1974; Cleland and Tanaka 1979; Cleland et al. 1982). However, SA is not considered to be an endogenous flower-regulating factor because the endogenous level of benzoic acid (SA precursor) is not altered by photoperiodic conditions (Fujioka et al. 1983). It is possible that SA is the endogenous flower-regulating factor in stress-induced flowering but not in photoperiodic flowering.

Nitrogen deficiency was reported to induce flowering in *L. paucicostata* in long-day conditions (Tanaka et al. 1988, 1989, 1991). This day length-independent flowering occurred in media supplemented with inorganic salts other than nitrogen, and therefore it is not certain whether this is a kind of poor-nutrition stress-induced flowering. Nitrogen deficiency does not always function as stress as mentioned above for *A. thaliana*.

5 Production of Progeny by Plants that Flowered under Stress Conditions

Stressed plants might flower as an emergency response to produce the next generation. In this way, plants can preserve their species, even in unfavorable environments. In order for this to be a biologically advantageous response, plants induced to flower by stresses must produce fertile seeds and the progeny must develop normally.

P. nil Violet was grown in a 1/10-strength nutrient solution or tap water throughout its life. The plants that were induced to flower by poor-nutrition stress conditions reached anthesis, fruited, and produced seeds (Wada et al. 2010a). The seeds produced by the stressed plants were the same size as or slightly smaller than the control seeds produced by plants that flowered by short-day treatment. All of these seeds germinated, and the progeny developed normally. The progeny responded to short-day treatment and formed floral buds. Furthermore, a normal second generation was produced from the stress progeny.

Red-leaved *P. frutescens* plants were grown under long-day conditions with low-intensity light beginning at the stage in which the cotyledons expanded. Plants were then continuously grown under the same conditions. The plants induced to flower by the low-intensity light stress conditions reached anthesis and formed seeds (Wada et al. 2010b). There were four seeds per flower as in the normal plants. The seeds produced under low-intensity light were heavier than the control seeds produced under usual short-day conditions. The seeds produced under stress conditions germinated, and the progenies grew normally and were induced to flower in response to short-day treatments.

These results in *P. nil* and *P. frutescens* indicate that the stressed plants do not need to await the arrival of a season when photoperiodic conditions are suitable for flowering, and such precocious flowering might assist in species preservation. Therefore, stress-induced flowering might have a biological benefit, and it should be considered as important as photoperiodic flowering and vernalization.

6 Transmissible Flowering Stimulus Produced by Stress

The presence of cotyledons is necessary for the long-day flowering of *P. nil* in response to poor nutrition or low temperature (Shinozaki and Takimoto 1982; Shinozaki 1985). This suggests that a flowering stimulus like florigen, which is involved in photoperiodic flowering, is involved in stress-induced flowering and is produced in cotyledons. If the stress-induced flowering stimulus is transmissible, defoliated scions may flower when grafted onto rootstocks with cotyledons and grown under stress conditions.

P. nil Violet and Tendan were grafted in several combinations, and the grafted plants were grown in tap water under long-day conditions for 20 days (Wada et al. 2010a). The Violet scions grafted onto the Violet rootstocks flowered (Fig. 17.1). The flowering may have been caused by the influence of the rootstocks because all the

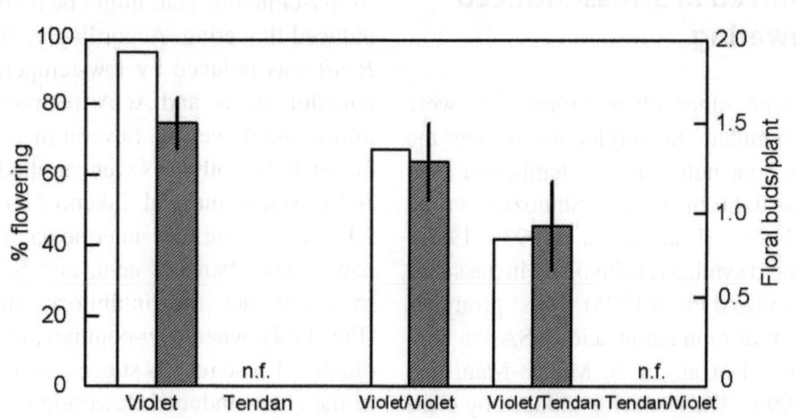

Fig. 17.1 Flowering of grafted plants under poor-nutrition stress conditions in *Pharbitis nil*. Two varieties of *P. nil* were grafted in the combinations as shown (scion/stock) and grown hydroponically in tap water. The nongrafted control plants were grown on vermiculite and fed with tap water. Plants were grown under long-day conditions for 20 days, transferred to nutrient solution, and grown for an additional 2 weeks to score the flowering response; % flowering (*open column*) and number of floral buds/plant (*closed column*). n.f., no flowering occurred. Data adapted from Wada et al. (2010a)

leaves had been removed from the scions and the cotyledons had been maintained on the rootstocks. This suggests that a transmissible flowering stimulus is involved in the stress-induced flowering of *P. nil*.

Violet scions flowered even when grafted onto Tendan rootstocks, although Tendan plants themselves were not induced to flower by the stress treatment. On the other hand, Tendan scions did not flower when grafted onto Violet rootstocks. It was predicted that Tendan would not produce such a flowering stimulus because Tendan did not flower in response to the poor-nutrition stress conditions. If this were the case, Violet would not be expected to flower when grafted onto Tendan rootstocks. However, defoliated Violet scions grafted onto Tendan rootstocks with cotyledons were induced to flower. The difference in flowering response between the scions grafted onto Tendan and those grafted onto Violet was not statistically significant. Therefore, Tendan may produce almost the same amount of the flowering stimulus as does Violet. Conversely, the Tendan scions grafted onto Violet rootstocks were not induced to flower. These results indicate that Tendan produces a transmissible flowering stimulus but does not respond to it.

7 Endogenous Substances Involved in Stress-Induced Flowering

CGA and some other phenylpropanoids were found to accumulate in cotyledons during the treatments by poor nutrition, low temperature, or high-intensity light in *P. nil* (Shinozaki et al. 1988a, b, 1994; Hirai et al. 1993, 1994). Phenylpropanoid synthesis is involved in the stress response (Dixon and Paiva 1995). Stress promotes the metabolism of *t*-cinnamic acid to SA via benzoic acid (Gidrol et al. 1996; Mauch-Mani and Slusarenko 1996). The flowering induced by these conditions is accompanied by an increase in PAL activity (Hirai et al. 1995), and AOA inhibited flowering in *P. nil* (Shinozaki et al. 1988a, 1994; Hatayama and Takeno 2003). Some compound(s) in the metabolic pathway regulated by PAL might act as flowering stimuli. Phenylpropanoids, such as CGA, were prominent candidates for this in earlier studies (Shinozaki et al. 1988a, b, 1994; Hirai et al. 1993, 1994). However, exogenously applied CGA failed to induce flowering (Shinozaki et al. 1988a, 1994; Hatayama and Takeno 2003). No flower-inducing activity was detected in other phenylpropanoids, including 4-*O*-*p*-coumaroylquinic acid, 3-*O*-feruloylquinic acid, dehydrodiconiferylalcohol-13-*O*-β-D-glucoside, and (+)-pinoresinol-β-D-glucoside (unpublished data). Therefore, CGA and related phenylpropanoids are not involved in the stress-induced flowering of *P. nil*. The close positive correlation between CGA content and flowering response was merely coincidence.

In addition to CGA, several compounds, including SA and anthocyanin, are derived from *t*-cinnamic acid of which conversion from phenylalanine is catalyzed by PAL (Dixon and Paiva 1995). Dihydrokaempferol-7-*O*-D-glucoside derived from the pathway from *t*-cinnamic acid to anthocyanin via *p*-coumaric acid has been reported to promote the flowering of *P. nil* (Nakanishi et al. 1995). Furthermore, AOA inhibits 1-aminocyclopropane-1-carboxylic acid (ACC) synthase. ACC synthase catalyzes the conversion of S-adenosylmethionine to ACC, which is converted to ethylene. Such components or other substances in the metabolic pathways derived from *t*-cinnamic acid might be involved in stress-induced flowering. Accordingly, the flowering of *P. nil* was induced by low-temperature or poor-nutrition stress, and AOA treatment was used to inhibit the flowering. Several metabolic intermediates in the pathways were applied together with AOA (Hatayama and Takeno 2003; Wada et al. 2010a). Among the intermediates tested, *t*-cinnamic acid, benzoic acid, and SA were shown to counteract the inhibitory effect of AOA (Fig. 17.2), whereas *p*-coumaric and caffeic acids did not. These results suggest that SA is involved in the stress-induced flowering of *P. nil* and that the pathways to CGA and anthocyanin are not involved. Flowering was completely inhibited in the presence of ACC (Hatayama and Takeno 2003). Thus, the ACC route is not involved. This is consistent with the observation that ethylene

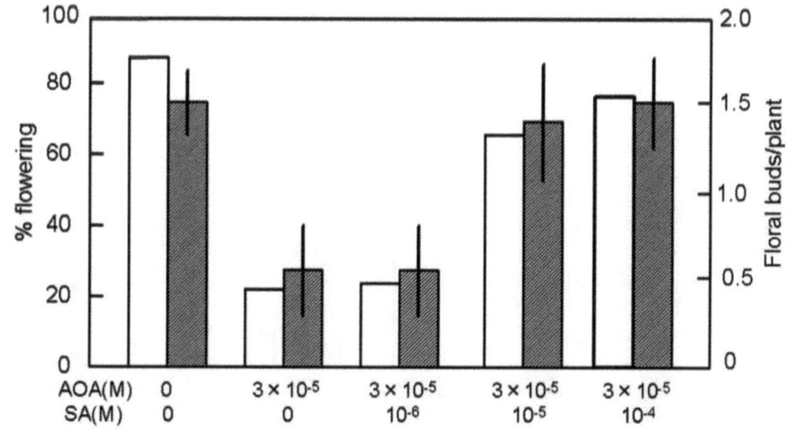

Fig. 17.2 Effects of aminooxyacetic acid (AOA) and salicylic acid (SA) on the poor-nutrition stress-induced flowering of *Pharbitis nil*. The 5-day-old seedlings of *P. nil* cv. Violet were grown in tap water with or without 3×10^{-5} M AOA and SA at various concentrations under long-day conditions for 20 days, transferred to fresh nutrient solution without AOA and SA, and grown for an additional 2 weeks to score the flowering response; % flowering (*open column*) and number of floral buds/plant (*closed column*). Data adapted from Wada et al. (2010a)

derived from ACC inhibits the photoperiodic flowering of *P. nil* (Suge 1972).

The leaves of red-leaved *P. frutescens* were deep green when induced to flower under low-intensity light (Wada et al. 2010b). The greening of the leaves was due to a decrease in anthocyanin content. There was a negative correlation between anthocyanin content and percentage flowering. Therefore, the metabolic pathway related to anthocyanin synthesis may be involved in the regulation of flowering. It is possible that some substances, such as SA, which are synthesized by the common metabolic pathway for anthocyanin synthesis are involved in flowering as mentioned above for *P. nil*. Low-intensity light may induce the flowering of *P. frutescens* by influencing the endogenous level of SA through suppression of PAL activity. However, this conflicts with previous reports. Stress generally increases PAL activity and promotes anthocyanin biosynthesis (Christie et al. 1994; Dixon and Paiva 1995; Chalker-Scott 1999). Actually, PAL activity increases in the stress-induced flowering of *P. nil* as mentioned above. Therefore, it was examined whether the PAL inhibitor could promote or inhibit the low-intensity light stress-induced flowering in *P. frutescens* (Wada et al. 2010b). The PAL inhibitor AOPP did not induce flowering when applied under noninductive normal-intensity light conditions and inhibited flowering in a dose-dependent manner when applied under inductive low-intensity light stress conditions (Fig. 17.3). The treatment with another PAL inhibitor, AOA, gave the same results. These results suggest that the same mechanism is involved in flowering that is induced by low-intensity light in *P. frutescens* and the flowering that is induced by several stress factors in *P. nil*. That PAL inhibitors inhibited stress-induced flowering suggests that the stress increased PAL activity. However, in *P. frutescens*, the decrease in anthocyanin content under low-intensity light suggests that stress limited the activity of PAL. These contradictory results must be explained in future.

8 Involvement of SA in Stress-Induced Flowering

When plants are stressed, they generate stress substances that regulate gene expression to adapt to the stress conditions. The stress substances include reactive oxygen species, nitric acid, jasmonic acid, SA, ethylene, and abscisic acid (Xiong et al. 2002; Moreau et al. 2010; Liu and Zhang 2004; Hey et al. 2010; Jaspers and

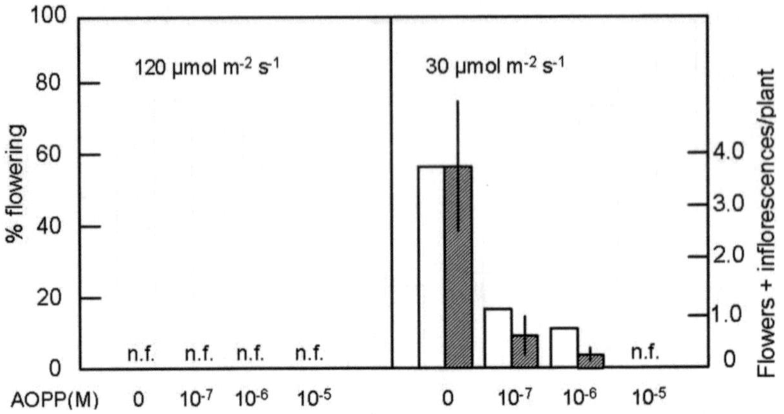

Fig. 17.3 Effects of L-2-aminooxy-3-phenylpropionic acid (AOPP) on flowering in *Perilla frutescens* grown under nonstress light conditions and low-intensity light stress conditions. Red-leaved *P. frutescens* was grown under long-day conditions with a normal light intensity of 120 μmol m^{-2} s^{-1} (*left*) or a low-intensity light of 30 μmol m^{-2} s^{-1} (*right*), and were treated with AOPP for 4 weeks. The treated plants were moved to normal light conditions and grown for an additional 3 (120 μmol m^{-2} s^{-1}) or 5 (30 μmol m^{-2} s^{-1}) weeks to score the flowering response; % flowering (*open column*) and total number of flowers and inflorescences/plant (*closed column*). n.f., no flowering occurred. Data adapted from Wada et al. (2010b)

Kangasjarvi 2010). Among these stress substances, SA and ethylene have been reported to induce flowering. Ethylene induces flowering in the Bromeliaceae, including pineapple. However, this is an exceptional case, and ethylene generally inhibits flowering in many plant species. The most likely stress substance involved in stress-induced flowering may be SA.

UV-C light stress promotes flowering in wild-type *A. thaliana*, but not in SA-deficient nahG transgenic plants (Martínez et al. 2004). UV-C irradiation increased the expression of the SA-responsive *PR1* gene in Col, but not in nahG plants. The transcript of the *SA induction deficient 2/isochorismate synthase 1* (*SID2/ICS1*) gene encoding the SA biosynthetic enzyme increased under UV-C irradiation in Col, but not in nahG plants. These results suggest the involvement of SA in the UV-C stress-induced flowering of *A. thaliana*. Exogenous application of SA at 100 μM accelerated flowering of Col, but the nahG plants were not responsive to the SA treatment (Fig. 17.4). SA also regulates flowering time in nonstressed plants. SA-deficient nahG is late flowering (Martínez et al. 2004). The *siz1* mutant that has elevated SA level is early flowering under short days, and this phenotype is suppressed by expression of *nahG* (Jin et al. 2008).

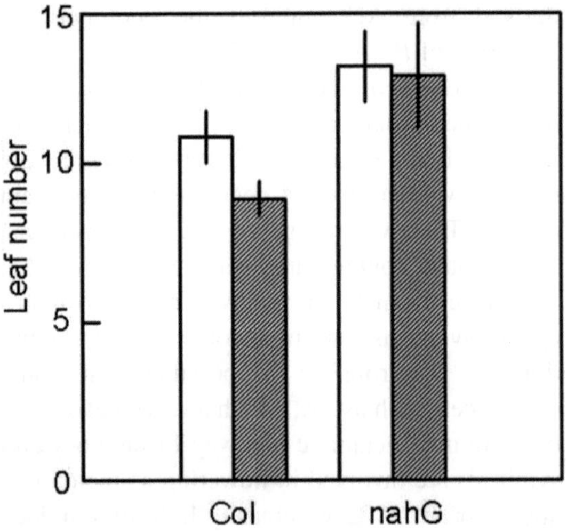

Fig. 17.4 Effect of the exogenous application of salicylic acid (SA) on flowering time in *Arabidopsis thaliana*. Wild-type Col and SA-deficient nahG transgenic plants were treated daily with 100 μM SA solution (*closed column*) or not (*open column*). Total leaf number (rosette plus cauline leaves) was scored when the plants bolted. Data adapted from Martínez et al. (2004)

When *L. paucicostata* 6746 was induced to flower by poor-nutrition stress, a larger amount of SA was detected in the flowered plants than in the control plants (Shimakawa 2011). This result suggests the involvement of SA in the stress-induced flowering of *L. paucicostata*. It is well-known that exogenously applied SA induces flowering in *L. paucicostata*, *L. gibba*, and the other-Lemanceous plants (Cleland and Ajami 1974; Cleland and Tanaka 1979; Cleland et al. 1982). However, SA is not considered to be an endogenous flower-regulating factor in *Lemna* because the endogenous benzoic acid (SA precursor) level is not altered by photoperiodic conditions (Fujioka et al. 1983). SA may be the endogenous flower-regulating factor in stress-induced flowering, but not in the photoperiodic flowering of *Lemna*.

The treatment of *P. nil* with SA and benzoic acid, a precursor of SA, or some benzoic acid derivatives prior to low-temperature treatment enhances the flower-inducing effect of low temperature (Shinozaki 1985; Shinozaki et al. 1982, Shinozaki and Takimoto 1983). In addition to these effects of exogenous application, the flower-inhibiting effects of PAL inhibitors, which may have decreased the endogenous SA level in *P. nil* and *P. frutescens*, provided new evidence to suggest that SA acts as an endogenous regulator of stress-induced flowering (Wada et al. 2010a, b). The flowering response of cultured plumules excised from short-day-treated *P. nil* seedlings was enhanced by benzoic acid (Ishioka et al. 1990). Amagasa et al. (1992) reported that AOA inhibited the photoperiodic flowering of *P. nil*. These observations suggest that SA is also involved in photoperiodic flowering. However, SA did not induce flowering at any concentrations in *P. nil* and *P. frutescens* under non-stress conditions (Wada et al. 2010a, b). SA did not enhance the flowering response under the weak stress conditions. SA may be necessary, but is not sufficient for the induction of flowering. Stress conditions may induce not only SA biosynthesis but also other essential factors to induce flowering.

9 The Genes Involved in Stress-Induced Flowering

Expression of the *CO*, *FT*, and *SOC1* genes that promote flowering was analyzed in *A. thaliana* under UV-C stress conditions (Martínez et al. 2004). UV-C induced expression of *FT*, moderately induced expression of *CO*, and did not induce *SOC1* expression in wild type (Fig. 17.5). Exogenous SA treatment reduced expression levels of the flower-inhibiting gene *FLC*. Thus, flowering promoted by UV-C requires the enhanced expression of *FT* and the reduced expression of *FLC*. SA application induced expression of the sunflower *FT* homolog, *HAFT*, in sunflower (Dezar et al. 2010). The flowering of *A. thaliana*

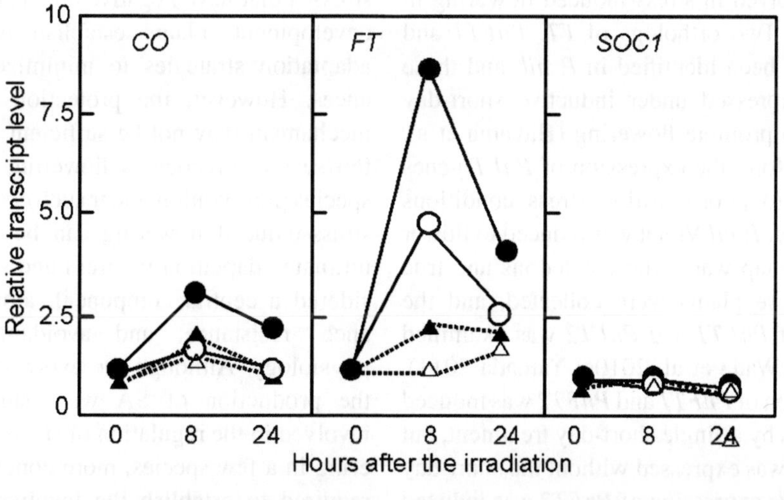

Fig. 17.5 Effect of UV-C irradiation on the expression of flowering genes in *Arabidopsis thaliana* wild-type Col (*circles*) and salicylic acid-deficient nahG transgenic plants (*triangles*) were irradiated with 200 mJ cm^{-2} UV-C light (*closed symbols*) or not (*open symbols*), and the expression of each gene at different hours after the irradiation (as indicated in the abscissa) was quantified by RT-PCR. Data adapted from Martínez et al. (2004)

is induced by long-day conditions, vernalization, autonomous cues, and gibberellins, and these factors operate through a common pathway integrated by *FT* (Boss et al. 2004). It was shown that *FT* is also involved in stress-induced flowering.

Genome-wide analyses of transcriptomes detected the downregulation of *Pathogen and Circadian Controlled 1* (*PCC1*) in SA-deficient plants of *A. thaliana* (Segarra et al. 2010). *PCC1* was initially characterized as a circadian clock-regulated gene that is rapidly upregulated after pathogen inoculation. The expression of *PCC1* was strongly activated by UV-C light irradiation in Col, but not in nahG plants. SA application also activated *PCC1* expression. The activation of *PCC1* expression required *CO*. RNAi transgenic plants contained lower levels of *FT* transcript. The overexpression of *PCC1* did not accelerate flowering, but suppression of its expression by RNAi delayed flowering. UV-C light irradiation of plants accelerates flowering through an SA-dependent process in wild-type but not in RNAi transgenic plants with reduced expression of *PCC1*, suggesting that neither SA nor PCC1 alone is sufficient to accelerate flowering in *A. thaliana*.

The flowering of *A. thaliana* is induced by four previously known factors and stress, and these factors function through the activation of *FT* expression. This suggests that the *FT* homolog could be involved in stress-induced flowering in other plants. Two orthologs of *FT*, *PnFT1* and *PnFT2*, have been identified in *P. nil*, and these genes are expressed under inductive short-day conditions to promote flowering (Hayama et al. 2007). Therefore, the expression of *PnFT* genes in response to poor-nutrition stress conditions was examined. *P. nil* Violet was induced to flower by growth in tap water, the cotyledons and true leaves of these plants were collected, and the expression of *PnFT1* and *PnFT2* was examined by RT-PCR (Wada et al. 2010a; Yamada 2011). The expression of *PnFT1* and *PnFT2* was induced in cotyledons by a single short-day treatment, but neither gene was expressed without the short-day treatment. The expression of *PnFT2* was induced in the cotyledons and true leaves of plants grown under the poor-nutrition conditions for 2 weeks or longer. The level of mRNA expression was closely correlated with the flowering response. Only weak *PnFT2* expression was detected in the true leaves of plants grown under nonstress conditions for 3 weeks. On the other hand, *PnFT1* was not expressed in the cotyledons or true leaves regardless of nutritional conditions. These results suggest that *PnFT2*, but not *PnFT1*, is involved in the poor-nutrition stress-induced flowering of *P. nil*. *PnFT2* is involved in both photoperiodic flowering and stress-induced flowering, whereas *PnFT1* is involved only in photoperiodic flowering. The two *PnFT* genes might have different roles in the regulation of flowering depending on the inductive cue. It is also possible that the essential gene for flowering is *PnFT2* and that *PnFT1* expression is induced only by short-day treatment and redundantly enhances the activity of *PnFT2*. SA might induce the expression of *PnFT2* or the product of *PnFT2* might induce the expression of genes involved in the biosynthesis of, response to, or signal transduction of SA.

10 Conclusion and Future Perspective

It is apparent that plants can flower in response to several stress conditions. Constantly exposed to stresses that have negative effects on growth and development, plants establish protection and adaptation strategies to minimize stress influences. However, the protection or adaptation mechanism may not be sufficient if the stress is too severe. Precocious flowering may assist in species preservation under such conditions. Thus, stress-induced flowering can be considered an ultimate adaptation to stress and should be considered a central component, along with tolerance, resistance, and avoidance, of stress physiology. Although the expression of *FT* and the production of SA were suggested to be involved in the regulation of stress-induced flowering in a few species, more conclusive data are required to establish the involvement of these gene and plant hormone. Further, the interaction between *FT* and SA should be studied in future.

References

Adams S, Allen T, Whitelam GC (2009) Interaction between the light quality and flowering time pathways in Arabidopsis. Plant J 60:257–267

Adams P, Nelson DE, Yamada S, Chmara W, Jensen RG, Bohnert HJ, Griffiths H (1998) Growth and development of *Mesembryanthemum crystallinum* (Aizoaceae). New Phytol 138:171–190

Amagasa T, Ogawa M, Sugai S (1992) Effects of aminooxyacetic acid and its derivatives on flowering in *Pharbitis nil*. Plant Cell Physiol 33:1025–1029

Aronson J, Kigel J, Shimada A, Klein J (1992) Adaptive phenology of desert and Mediterranean populations of annual plants grown with and without water stress. Oecologia 89:17–26

Bernier G, Périlleux C (2005) A physiological overview of the genetics of flowering time control. Plant Biotechnol J 3:3–16

Boss PK, Bastow RM, Mylne JS, Dean C (2004) Multiple pathways in the decision to flower: enabling, promoting, and resetting. Plant Cell 16:S18–S31

Bussler W (1969) Dungungsversuche zu *Cyclamen*. Gartenbauwissenschaft 34:495–510

Casella D (1935) L'agrumicoltura siciliana. Ann R Staz Sper Frutt Agrum Acireale N S 2:165–176

Chalker-Scott L (1999) Environmental significance of anthocyanins in plant stress responses. Photochem Photobiol 70:1–9

Chamont S, Sotta B, Miginiac E (1982) Influence of root thermal treatments on growth, flowering and water status of *Chenopodoum polyspermum* L. Physiol Veg 20:1–10

Christie RJ, Alfenito MR, Walbot V (1994) Impact of low temperature stress on general phenylpropanoid and anthocyanin pathways: Enhancement of transcript abundance and anthocyanin pigmentation in maize seedlings. Planta 194:541–549

Cleland CF, Ajami A (1974) Identification of the flower-inducing factor isolated from aphid honeydew as being salicylic acid. Plant Physiol 54:904–906

Cleland CF, Tanaka O (1979) Effect of daylength on the ability of salicylic acid to induce flowering in the long-day plant *Lemna gibba* G3 and the short-day plant *Lemna paucicostata* 6746. Plant Physiol 64:421–424

Cleland CF, Tanaka O, Feldman LJ (1982) Influence of plant growth substances and salicylic acid on flowering and growth in the Lemnaceae (duckweeds). Aqua Bot 13:3–20

De Zeeuw D (1953) Flower initiation and light intensity in *Perilla*. Proc K Ned Akad Wet C 56:418–422

Dezar CA, Giacomelli J, Manavella PA, Re DA, Alves-Feerreira M, Baldwin IT, Bonaveuture G, Chan RL (2010) HAHB10, a sunflower HD-Zip II transcription factor, participates in the induction of flowering and in the control of phytohormone-mediated responses to biotic stress. J Exp Bot 62:1061–1076

Dixon RA, Paiva L (1995) Stress-induced phenylpropanoid metabolism. Plant Cell 7:1085–1097

Dorn LA, Pyle EH, Schmitt J (2000) Plasticity to light cues and resources in *Arabidopsis thaliana*: testing for adaptive value and costs. Evolution 54:1982–1994

Ebell LF (1967) Bimonthly research note 23. Canadian Department Forestry and Rural Development, Ottawa, pp 26–27

Ebell LF (1971) Girdling: its effect on carbohydrate status and on reproductive bud and cone development of Douglas-fir. Can J Bot 49:453–466

Fujioka S, Yamaguchi I, Murofushi N, Takahashi N, Kaihara S, Takimoto A (1983) Flowering and endogenous levels of benzoic acid in *Lemna* species. Plant Cell Physiol 24:235–239

Gaillochet J, Mathon CC, Stroun M (1962) Nouveau type de reaction et changement du type de reaction au photoperiodisme chez le *Perilla ocymoides* L. C R Acad Sci 255:2501–2503

Gidrol X, Sabelli PA, Fern YS, Kush AK (1996) Annexin-like protein from *Arabidopsis thaliana* rescues *ΔoxyR* mutant of *Escherichia coli* from H_2O_2 stress. Proc Natl Acad Sci USA 93:11268–11273

Halliday KJ, Koornneef M, Whitelam GC (1994) Phytochrome B and at least one other phytochrome mediate the accelerated flowering response of *Arabidopsis thaliana* L. to low red/far-red ratio. Plant Physiol 104:1311–1315

Hatayama T, Takeno K (2003) The metabolic pathway of salicylic acid rather than of chlorogenic acid is involved in the stress-induced flowering of *Pharbitis nil*. J Plant Physiol 160:461–467

Hayama R, Agashea B, Luleya E, King R, Coupland G (2007) A circadian rhythm set by dusk determines the expression of *FT* homologs and the short-day photoperiodic flowering response in *Pharbitis*. Plant Cell 19:2988–3000

Hey SJ, Byrne E, Halford NG (2010) The interface between metabolic and stress signaling. Ann Bot 105:197–203

Hicks LE (1932) Flower production in the Lemnaceae. Ohio J Sci 32:115–131

Hirai N, Kojima Y, Koshimizu K, Shinozaki M, Takimoto A (1993) Accumulation of phenylpropanoids in cotyledons of morning glory (*Pharbitis nil*) seedlings during the induction of flowering by poor nutrition. Plant Cell Physiol 34:1039–1044

Hirai N, Yamamuro M, Koshimizu K, Shinozaki M, Takimoto A (1994) Accumulation of phenylpropanoids in the cotyledons of morning glory (*Pharbitis nil*) seedlings during the induction of flowering by low temperature treatment, and the effect of precedent exposure to high-intensity light. Plant Cell Physiol 35:691–695

Hirai N, Kuwano Y, Kojima Y, Koshimizu K, Shinozaki M, Takimoto A (1995) Increase in the activity of phenylalanine ammonia-lyase during the non-photoperiodic induction of flowering in seedlings of morning glory (*Pharbitis nil*). Plant Cell Physiol 36:291–297

Hopkinson JM (1977) Formative factors – tropical pasture legume seed crops. Proc Aust Seeds Res Conf Tamworth 1:14–23

Imrie BC (1973) Flowering response of *Macroptilium atropurpureum* to daylength, temperature and nitrogen levels. SABRAO Newslett 5:95–100

Ishimaru A, Takeno K, Shinozaki M (1996) Correlation of flowering induced by low temperature and endogenous levels of phenylpropanoids in *Pharbitis nil*: a study with a secondary-metabolism mutant. J Plant Physiol 148:672–676

Ishioka N, Tanimoto S, Harada H (1990) Flower-inducing activity of phloem exudate in cultured apices from *Pharbitis* seedlings. Plant Cell Physiol 31:705–709

Iwasaki T, Owada TA, Iiya Y (1959) Studies on the differentiation and development of the flower bud in citrus I-VI. Bull Tokai Kinki Agric Exp Stn Hortic 5:1–76

Jacobs WP (1982) Comparison of photoperiodic sensitivity of green-leafed and red-leafed *Perilla*. Plant Physiol 70:303–306

Jaspers P, Kangasjarvi J (2010) Reactive oxygen species in abiotic stress signaling. Physiol Plant 138:405–413

Jin JB, Jin YH, Lee J, Miura K, Yoo CY, Kim WY, van Oosten M, Hyun Y, Somers DE, Lee I, Yun DJ, Bressan RA, Hasegawa PM (2008) The SUMO E3 ligase, *AtSIZ1*, regulates flowering by controlling a salicylic acid-mediated floral promotion pathway and through affects on *FLC* chromatin structure. Plant J 53: 530–540

Jones A (1980) Sweet potato. In: Hybridization of Crop Plants. American Society of Agronomy, Crop Science Society of America, Madison, Wis, pp 645–655

Kolár J, Senková J (2008) Reduction of mineral nutrient availability accelerates flowering of *Arabidopsis thaliana*. J Plant Physiol 165:1601–1609

Krajncic B, Kristl J, Janzekovic I (2006) Possible role of jasmonic acid in the regulation of floral induction, evocation and floral differentiation in *Lemna minor* L. Plant Physiol Biochem 44:752–758

Landolt E (1957) Physiologische und ökologische Untersuchungen an Lemnaceen. Ber Schweiz Bot Gez 67:271–410

Liu Y, Zhang S (2004) Phosphorylation of 1-aminocyclopropane-1-carboxylic acid synthase by MPK6, a stress-responsive mitogen-activated protein kinase, induces ethylene biosynthesis in Arabidopsis. Plant Cell 16:3386–3399

Lorrain S, Allen T, Duek PD, Whitelam GC, Fankhauser C (2008) Phytochrome-mediated inhibition of shade avoidance involves degradation of growth-promoting bHLH transcription factors. Plant J 53:312–323

Marín IC, Loef I, Bartetzko L, Searle I, Coupland G, Stitt M, Osuna D (2010) Nitrate regulates floral induction in *Arabidopsis*, acting independently of light, gibberellin and autonomous pathways. Planta 233:539–552

Martínez C, Pons E, Prats G, León J (2004) Salicylic acid regulates flowering time and links defense responses and reproductive development. Plant J 37:209–217

Mauch-Mani B, Slusarenko AJ (1996) Production of salicylic acid precursors is a major function of phenylalanine ammonia-lyase in the resistance of *Arabidopsis* to *Peronospora parasitica*. Plant Cell 8:203–212

McDaniel CN (1996) Developmental physiology of floral initiation in *Nicotiana tabacum* L. J Exp Bot 47: 465–475

McMullan EE (1980) Effect of applied growth regulators on cone production in Douglas-fir and relation of endogenous growth regulators to cone production capacity. Can J For Res 10:405–414

Metzger J (1995) Hormones and reproductive development. In: Davies PJ (ed) Plant Hormones. Kluwer Academic Publishers, Dordrecht, pp 617–648

Monselise SP (1985) *Citrus* and related genera. In: Halevy AH (ed) CRC Handobook of Flowering, vol II, CRC Press. Boca Raton, FL, pp 275–294

Monselise SP, Halevy AH (1964) Chemical inhibition of citrus flower bud induction. Proc Am Soc Hortic Sci 84:141–146

Monselise SP, Goren R, Costo J, Simkhi M (1981) Development of lemon fruits originating at different blossom dates around the year. Sci Hortic 15:23–32

Moreau M, Lindermay C, Durner J, Klessig DF (2010) NO synthesis and signaling in plants – where do we stand? Physiol Plant 138:372–383

Nakanishi F, Kusumi T, Inoue Y, Fujii T (1995) Dihydrokaempferol glucoside from cotyledons promotes flowering in *Pharbitis nil*. Plant Cell Physiol 36:1303–1309

Nir I, Goren R, Leshem B (1972) Effects of water stress, gibberellic acid and 2-chloroethyl trimethylammonium chloride (CCC) on flower differentiation in Eureka lemon trees. J Am Soc Hortic Sci 97:774–778

Purse JG (1984) Phloem exudates of *Perilla crispa* and its effects on flowering of *P. crispa* shoot explants. J Exp Bot 35:227–238

Segarra S, Mir R, Martínez C, León J (2010) Genome-wide analyses of the transcriptomes of salicylic acid-deficient versus wild-type plants uncover Pathogen and Circadian Controlled 1 (PCC1) as a regulator of flowering time in Arabidopsis. Plant Cell Environ 33:11–22

Shimakawa A (2011) Poor-nutrition stress-induced flowering in *Lemna paucicostata*. Dissertation, Niigata University (in Japanese)

Shinozaki M (1985) Organ correlation in long-day flowering of *Pharbitis nil*. Biol Plant 27:382–385

Shinozaki M, Takimoto A (1982) The role of cotyledons in flower initiation of *Pharbitis nil* at low temperatures. Plant Cell Physiol 23:403–408

Shinozaki M, Takimoto A (1983) Effects of some growth regulators and benzoic acid derivatives on flower initiation and root elongation of *Pharbitis nil*, strain Kidachi. Plant Cell Physiol 24:33–39

Shinozaki M, Hikichi M, Yoshida K, Watanabe K, Takimoto A (1982) Effect of high-intensity light given prior to low-temperature treatment on the long-day flowering of *Pharbitis nil*. Plant Cell Physiol 23: 473–477

Shinozaki M, Asada K, Takimoto A (1988a) Correlation between chlorogenic acid content in cotyledons and flowering in *Pharbitis* seedlings under poor nutrition. Plant Cell Physiol 29:605–609

Shinozaki M, Swe KL, Takimoto A (1988b) Varietal difference in the ability to flower in response to poor nutrition and its correlation with chlorogenic acid accumulation in *Pharbitis nil*. Plant Cell Physiol 29: 611–614

Shinozaki M, Hirai N, Kojima Y, Koshimizu K, Takimoto A (1994) Correlation between level of phenylpropanoids in cotyledons and flowering in *Pharbitis* seedlings under high-fluence illumination. Plant Cell Physiol 35:807–810

Simpson GG (2004) The autonomous pathway: epigenetic and post-transcriptional gene regulation in the control of *Arabidopsis* flowering time. Curr Opin Plant Biol 7:570–574

Smith H, Whitelam GC (1997) The shade avoidance syndrome: multiple responses mediated by multiple phytochromes. Plant Cell Environ 20:840–844

Suge H (1972) Inhibition of photoperiodic floral induction in *Pharbitis nil* by ethylene. Plant Cell Physiol 13: 1031–1038

Swe KL, Shinozaki M, Takimoto A (1985) Varietal differences in flowering behavior of *Pharbitis nil* Chois. Mem Coll Agric Kyoto Univ 126:1–20

Takimoto A (1973) Flower initiation of *Lemna perpusilla* under continuous low-intensity light. Plant Cell Physiol 14:1217–1219

Tanaka O, Nagase H, Kitazawa Y, Mizoguchi F, Izawa K (1988) Induction of flowering by nitrogen deficiency in *Lemna paucicostata* 441. Plant Cell Physiol 29: 1379–1384

Tanaka O, Kuwata M, Izawa K, Takeba G (1989) Comparison between induction of flowering by nitrogen deficiency and that by short days or continuous irradiation with blue or low-intensity white light in *Lemna paucicostata* 6746. Plant Cell Physiol 30: 1139–1144

Tanaka O, Yamamoto T, Nakayama Y, Ozaki T, Takeba G (1991) Flowering induced by nitrogen deficiency in *Lemna pauciocostata* 151. Plant Cell Physiol 32:1173–1177

Thomas B, Vince-Prue D (1997) Photoperioidism in Plants 2nd ed. Academic Press, San Diego

Wada K (1974) Floral initiation under continuous light in *Pharbitis nil*, a typical short-day plant. Plant Cell Physiol 15:381–384

Wada K, Totsuka T (1982) Long-day flowering of *Perilla* plants cultured in nitrogen-poor media. Plant Cell Physiol 23:977–985

Wada KC, Takeno K (2010) Wada K, Totsuka (1982) Stress-induced flowering. Plant Signal Behav 5:944–947

Wada KC, Yamada M, Shiraya T, Takeno K (2010a) Salicylic acid and the flowering gene *FLOWERING LOCUS T* homolog are involved in poor-nutrition stress-induced flowering of *Pharbitis nil*. J Plant Physiol 167:447–452

Wada KC, Kondo H, Takeno K (2010b) Obligatory short-day plant, *Perilla frutescens* var. *crispa* can flower in response to low-intensity light stress under long-day conditions. Physiol Plant 138:339–345

Xiong L, Schumaker KS, Zhu JK (2002) Cell signaling during cold, drought, and salt stress. Plant Cell 14(Suppl):S165–S183

Yamada M (2011) The gene regulation of stress-induced flowering in *Pharbitis nil*. Dissertation, Niigata University (in Japanese)

Zeevaart JAD (1969) *Perilla*. In: Evans ET (ed) The induction of flowering. Some case histories. McMillan of Australia, Melbourne, pp 116–155

Postharvest Stress Treatments in Fruits and Vegetables

18

Yoshihiro Imahori

Abstract

Fresh fruits and vegetables are living tissues subject to continuous changes after harvest. While some changes are desirable, most are not. Their commodities are perishable products with active metabolism during the postharvest period. Proper postharvest handling plays an important role in increasing food availability. Postharvest stress treatments have been shown to be generally effective in controlling both insect and fungal pests, reducing physiological disorder or decay, delaying ripening and senescence, and maintaining storage quality in fruits and vegetables. In addition, a moderate stress not only induces the resistance to this kind of severe stress, but also can improve tolerance to other stresses. Postharvest stress treatments can, therefore, be very important to improving shelf life and quality retention during postharvest handling of fruits and vegetables.

Keywords

CA and MA storage • Chilling injury • Ethanol vapor treatment • Heat treatment • Ultraviolet radiation

1 Introduction

Fruits and vegetables are an important source of carbohydrates, proteins, organic acids, vitamins, and minerals for human nutrition. When humans use plants or plant parts, whether for food or for aesthetic purposes, there is always a postharvest component that leads to loss (Fallik 2004). Their losses in quantity and quality affect horticultural crops between harvest and consumption. Thus, to reduce the losses, producers and handlers must understand the biological and environmental factors involved in deterioration (Kader 1992).

Fresh fruits and vegetables are living tissues subject to continuous changes after harvest. While some changes are desirable, most are not (Kader 1992). Their commodities are perishable products with active metabolism and subject to extensive postharvest losses through microbial decay, physical injury, and senescence during the postharvest period. However, postharvest

Y. Imahori (✉)
Graduate School of Life and Environmental Sciences, Osaka Prefecture University, Osaka, Japan
e-mail: imahori@plant.osakafu-u.ac.jp

P. Ahmad and M.N.V. Prasad (eds.), *Abiotic Stress Responses in Plants: Metabolism, Productivity and Sustainability*, DOI 10.1007/978-1-4614-0634-1_18, © Springer Science+Business Media, LLC 2012

changes in horticultural crops cannot be stopped, but they can be slowed within certain limits (Kader 1992). The maintenance or improvement of the postharvest life of fresh fruits and vegetables is becoming increasingly important.

As stress is generally defined as any environmental factor potentially unfavorable to living organisms, with the exception of decay, quality losses in actual postharvest produce can be directly or indirectly attributable to a combination of abiotic stress and stress-induced senescence (Lester 2003). However, it became clear that plant can acquire resistance to abiotic stresses (Capanoglu 2010). The acquired resistance is often associated with enhanced mobilization of defense responses after subsequent exposure of the plants to stress (Capanoglu 2010).

Approaches to modulate or control abiotic stresses in plant tissues can be very important to improving shelf life and quality retention during postharvest handling of fruits and vegetables. In relation to approaches, there have been many strategies using various types of treatments. Temperature treatments, atmospheric treatments, ethanol vapor treatment, and ultraviolet radiation have all shown some potential to maintain and improve the postharvest life of fruits and vegetables

2 Environmental Stress and Postharvest Produce

Changes in the environmental conditions can cause stress to plants (Capanoglu 2010). The effects of stress on metabolism and performance have become a major focus of plant research, especially in postharvest produce. When exposed to unfavorable environments, plants result in some degree of stress and express a fraction of the plants' genetic potential (Cisneros-Zevallos 2003). The plants can be sensitized for more rapid or more intense mobilization of defense responses leading to enhanced resistance to stress, and acquire resistance to abiotic stresses (Capanoglu 2010).

Abiotic stresses, such as drought, salinity, extreme temperatures, chemical toxicity, and oxidative stress, are serious threats to agriculture and result in the primary cause of crop loss worldwide (Wang et al. 2003; Imahori et al. 2008). Abiotic stress leads to a series of morphological, physiological, biochemical, and molecular changes that adversely affect plant growth and productivity (Wang et al. 2001). These stresses affect phytochemical accumulation or loss by inducing an increase or reduction in key enzyme activities of metabolic pathways (Dixon and Paiva 1995; Cisneros-Zevallos 2003; Capanoglu 2010).

These diverse stresses activate similar cell-signaling pathways and cellular responses, such as the production of stress proteins, upregulation of antioxidants, and accumulation of compatible solutes (Wang et al. 2003). Thus, plant stress responses are regulated by multiple signaling pathways that activate gene transcription and its downstream machinery (Wang et al. 2003). The complex plant response to abiotic stress involves many genes and biochemical–molecular mechanisms. The molecular control mechanisms of abiotic stress tolerance, which may result in the use of molecular tools, is based on the expression of specific stress-related genes (Wang et al. 2004). These major tolerance mechanisms include water and ion uptake and transport, such as ion transporter and aquaporins, osmoprotectants, free-radical scavengers, the protection of membranes and proteins, such as heat-shock proteins (HSPs) and chaperones, late embryogenesis-abundant proteins, and factors involved in signaling cascades and transcriptional control (Wang et al. 2003). HSPs are known to be expressed in plants not only when they experience high temperature stress, but also in response to wide range of other abiotic stress, such as water stress, salinity, and osmotic, cold, and oxidative stress (Wang et al. 2004). These play an important role in protecting against stress and in the reestablishment of cellular homeostasis (Wang et al. 2004).

3 Temperature Treatments

3.1 Heat Treatment

Heat treatments have been used for insect disinfestations, decay control, ripening decay, and modification of fruit responses to other stresses and maintenance of quality during storage (Lurie 1998;

Paull and Chen 2000). There are three methods in use to heat commodities; hot water, vapor heat, and hot air. Hot water treatment was originally used for fungal control, and first reported in 1922 to control decay on citrus fruit (Fallik 2004). But its use has been extended to disinfestations of insects. Postharvest heat treatments are often applied for a relatively short time, at temperatures higher than heat treatments designed to kill insect pests located at the interior of a commodity (Lurie 1998; Fallik 2004). Many fruits and vegetables tolerate exposure to water temperatures of 50–60°C for up to 10 min (Lurie 1998). Vapor heat treatment is a method of heating commodity with air saturated with water vapor at temperatures of 50–60°C. This was developed mainly to kill insect eggs and larvae as a quarantine treatment before fresh market shipment (Lurie 1998; Fallik 2004). Hot air treatment can be applied by placing fruits and vegetables in a heated chamber with a ventilating fan or by applying forced hot air, where the speed of air circulation is precisely controlled, and this has been used for both fungal and insect control (Lurie 1998). Heat treatments can also be used to inhibit ripening processes or to induce resistance to chilling injury which is a physiological disorder caused by the exposure of crops to low temperatures above the freezing point, and external skin damage during storage, through changes in gene expression and protein synthesis (Fallik 2004).

3.1.1 Heat Treatment and Fruit Ripening

Heat treatments inhibit biochemical pathways involved in ripening and other processes in many fruits and vegetables (Paull and Chen 2000). The ripening is a readily recognized phenomenon in a fruit by virtue of a series of transformations, including changes in the background color, development of flavor and aroma, increases in respiratory activity and ethylene production, and fruit softening (Imahori et al. 2000a). Exposing to high temperatures attenuates some of these processes while enhancing others (Lurie 1998).

Ethylene synthesis is inhibited by heat treatments. High temperatures of 35–38°C can cause endogenous 1-aminocyclopropane-1-carboxylic acid (ACC) to accumulate in apple and tomato tissue concomitantly with the decrease in ethylene production, as the conversion of ACC to ethylene is highly susceptible to heat damage (Lurie 1998). The rapid loss of ACC oxidase activity occurs in many fruits exposed for short periods to high temperatures (Klein and Lurie 1990; Paull and Chen 1990; Ketsa et al. 1999) due primarily to decrease in ACC oxidase mRNA and cessation of enzyme synthesis (Lurie et al. 1996). ACC synthase is also sensitive to high temperatures, but it is less heat sensitive than ACC oxidase (Atta Aly 1992). Ethylene formation is reversibly inhibited, and when the fruits are removed from heat, they quickly recover their ability for ethylene synthesis (Paull and Chen 2000; Lurie 1998). This recover requires protein synthesis, and both mRNA and protein of ACC oxidase accumulate during recovery from heat treatment (Lurie et al. 1996). Some fruits, such as pears, tomatoes, and bananas, not only inhibit endogenous ethylene production, but also do not respond to exogenous ethylene during heat treatment (Yang et al. 1990; Lurie 1998; Paull and Chen 2000). This loss of sensitivity is associated with inactivation of ethylene receptors or the inability to transfer the signal to the subsequent series of events leading to ripening (Lurie 1998; Paull and Chen 2000).

Respiration rate of ripening is initially enhanced by heat treatment. The response to temperature varies with the physiological age of the plant tissue (Paull and Chen 2000). Heat treatments have an impact on the subsequent climacteric respiratory rise: depending on temperature and length of exposure, they can decrease or increase climacteric respiration peak as well as advance or delay it after treatment (Paull and Chen 2000). The respiration is often lower upon return of heated fruit to ambient temperatures compared to non-heated fruit (Klein and Lurie 1990; Lurie 1998).

Fruit softening is often slowed following exposure to hot air temperatures of 38–40°C (Klein and Lurie 1990; Lurie and Nussinnovitch 1996). The rate of softening increased when heated fruits were returned to ambient temperature, but it was still less than that of non-heated fruits (Lurie 1998). Exposure of apples to 38°C for 4 days resulted in less-soluble pectin and more insoluble pectin compared to non-heated fruits, an indication of inhibition of polyuronide degradation

(Klein et al. 1990; Ben-Shalom et al. 1996). The softening disruption has been ascribed to reduction of cell wall hydrolytic enzymes, such as polygalacturonase and α- and β-galactosidase (Sozzi et al. 1996; Paull and Chen 2000). The disruption is associated with mRNA synthesis and stability or protein synthesis and degradation. In tomato, mRNA polygalacturonase was absent in fruit during a heat treatment of 1–3 days at 38°C and appeared after the fruit was removed from heat (Lurie et al. 1996). Heat disruption of cell wall breakdown has been proposed as the cause of delayed or poor softening (Rose et al. 1998).

Heat treatment can lead to yellowing of cucumbers and zucchini while it delays yellowing of broccoli (Lurie 1998). Above 30°C, papaya fruit fails to ripen normally, with the pulp becoming soft and watery, although color changes in skin or flesh are not affected (An and Paull 1990; Paull and Chen 1990). The difference in responses of different commodities may be an indication of whether new enzymes must be synthesized to effect the color changes or not (Lurie 1998). The inhibition of lycopene in tomatoes is due to the inhibition of transcription of mRNA for lycopene synthase, a key enzyme in the pathway, and this recovers after removal from heat (Lurie et al. 1996; Sozzi et al. 1996).

Flavor and volatile production is affected by a heat treatment. The impact on flavor varies with species, temperature, and duration of the heat treatment (Paull and Chen 2000). Avocado flavor is adversely affected by heat treatment at 43°C for longer than 5 h (Paull and Chen 2000). Apple volatile production, though enhanced during a 38°C treatment, is immediately inhibited after heat treatments and recovers slowly (Fallik et al. 1997). Tomato volatile production also reduced immediately after hot water treatment at 52°C for 15 min (Bai et al. 2011).

3.1.2 Mechanism of Responses to Heat Treatment

The heat shock response is manifested in most living organisms as induction or enhanced synthesis of HSPs. These HSPs are believed to prevent irreversible protein denaturation and breakdown that would be detrimental to the cell and to confer tolerance to heat (Ferguson et al. 2000; Paull and Chen 2000). The heat shock response is primarily regulated at the transcriptional levels (Wang et al. 2003). HSP gene expression and protein synthesis are associated with high temperature exposure of various plant parts. With postharvest heat treatments, HSP transcripts and protein levels have been shown to increase (Ferguson et al. 2000). The lag period for induction for heat shock response is slower than other stress responses, although HSPs are synthesized within 30 min after heat treatment (Paull and Chen 2000). Plant modification for enhanced tolerance is mostly based on the manipulation of genes that protect and maintain the function and structure of cellular components (Wang et al. 2003).

There is a correlation between the development of thermo tolerance and the synthesis of HSP, as well as a correlation between the loss of thermo tolerance and the disappearance of HSP (Lurie 1998). Thus, it also is dependent on the incubation temperature. Temperatures in the range 35–40°C have been found to be effective, depending upon the commodity. At 42°C or higher, however, transcription and translation of HSP are inhibited and HSP synthesis is attenuated (Ferguson et al. 1994). Continued exposure to 42°C, though still allowing heat-shock polypeptide synthesis, is the limit for induction of heat-shock tolerance and enhancement of protein degradation (Ferguson et al. 1994).

The disruption of transcription occurs by mRNA being released from the ribosomes (Stuger et al. 1999). Protein degradation, particularly of rate-limiting enzymes, continues at a higher rate due to the higher temperature (Paull and Chen 2000). During heat treatment, the mRNAs of fruit-ripening genes disappear and those of HSP accumulate (Lurie et al. 1996). Translated polysomal RNA confirmed that new polypeptides are synthesized following heat-shock induction (Paull and Chen 2000). A rapid response of high temperature above 35°C is disassociation of polyribosomes, followed by a reassociation of some ribosomes into polyribosomes which preferentially translate the mRNA of HSP (Ferguson et al. 1994). This response both downregulates normal protein synthesis even without degradation of mRNAs and upregulates HSP synthesis (Lurie 1998).

3.1.3 Tolerance to Chilling Injury

Exposure of plants to a moderate stress not only induces the resistance to this kind of severe stress, but also can improve tolerance to other stresses (Wang et al. 2003). This cross-protection has been shown for different kinds of stress, such as heat pretreatment-induced chilling resistance (Pesis et al. 1997). The heat stress can affect the capacity of biological systems to synthesize proteins, resulting in greater or lesser synthesis of the proteins present and the synthesis of HSPs (Zhang et al. 2005). HSPs function as molecular chaperones and assist in protein folding, assembly and transport, and targeting of damaged proteins for proteolysis; thus, they may also assist in protecting the cells under chilling stress conditions (Sapitnitskaya et al. 2006). Heat stress can condition plants to low temperature. Such high-temperature-conditioning treatment induces an adaptive response in commodities to chilling stress and inhibits the development of chilling injury and external skin damage during cold storage (Lurie 1998; Fallik 2004). This conditioning effect has been demonstrated in tomato (McDonald et al. 2000), avocado (Woolf et al. 1995), mango (Pesis et al. 1997), grapefruit (Porat et al. 2000), and persimmon (Woolf et al. 1997).

The resistance to chilling injury was found to be contingent on the presence of HSP (Lurie 1998). In avocado discs, maximal HSP production was found after 4 h at 38°C and heating provided a significant level of protection from low-temperature injury (Lurie et al. 1996). A role for HSPs has been implicated, arising from persistence of both HSP transcripts and protein at low temperatures after heat treatments (Ferguson et al. 2000). HSP gene expression in tomato fruit was increased by high temperature and expressed when the heated fruits were transferred to low temperature (Kadyrzhanova et al. 1998). Small-molecular-weight HSPs play a specific role in the acquisition of tolerance to chilling stress following heat treatment (Zhang et al. 2005), and protect cellular proteins from thermal aggregation and protein folding activity (Vierling 1997). Thus, heat shock-induced protein aggregation preventive activity is integrated (Lee and Vierling 2000). HSP70 also acts in this way, having essential functions in preventing aggregation and assisting refolding of nonnative proteins under both normal and stress conditions. They are involved in protein import and translocation processes (Wang et al. 2004).

3.2 Low-Temperature Conditioning

Low temperature conditioning is an alternative technique for increasing tolerance to low temperatures. This involves holding cold-sensitive tissue at temperatures just above those at which injury occurs to induce tolerance to these normally damaging low temperatures (Woolf et al. 2003). The crucial factors of this technique are temperature differences between conditioning and storage temperature and the duration of the conditioning treatment (Cai et al. 2006). Typical low-temperature-conditioning treatments effective in reducing chilling injury involve holding fruits, such as zucchini, for 2 days at 15 or 10°C before storage at 5 and 2.5°C (Woolf et al. 2003). Similar low-temperature-conditioning effects have been observed in other chilling-sensitive fruits, such as avocado (Woolf et al. 2003) and grapefruit (Biolatto et al. 2005). As with heat treatments, the low-temperature-conditioning response is time and temperature dependant (Hofman et al. 2003). Longer conditioning treatments were more successful in two grapefruit cultivars, where fruit conditioned for 7 days had significantly less chilling injury after storage at 1°C for 21 days (Cai et al. 2006).

This adaption to lower temperatures is the result of various physiological and biochemical modifications induced by the conditioning treatment (Wang 2010). These modifications include reducing chilling-induced degradation of membrane phospholipids; increasing sugar, starch, and proline content; maintaining high levels of polyamines, squalene, and long-chain aldehydes; and increasing the ratio of unsaturated to saturated fatty acids (Wang 2010).

4 Atmospheric Treatments

4.1 Controlled and Modified Atmosphere Storage

Exposing fresh fruits and vegetables to reduced O_2 and/or elevated CO_2 can either be beneficial or harmful, depending on the concentration of these gases, temperature, exposure duration, and commodity. Controlled atmosphere (CA) or modified atmosphere (MA) storage utilizing reduced O_2 and/or elevated CO_2 are known to maintain quality and consequently extend shelf life of many fresh fruits and vegetables (Kader 1986). The beneficial effects of CA or MA storage include delayed ripening, reduced physiological and pathological disorders, and the possibility for disinfesting fruit (Burdon et al. 2007). The gas composition of CA is monitored and deviations from the set points corrected. MA differs in that it is not actively controlled and the gas composition results from a balance between the plant gas consumption or production and gas diffusion through a permeable membrane (Chervin et al. 1996).

To date, much research has been conducted to evaluate the effects of CA and MA storage on the quality and storability for a large number of fruits and vegetables and specific cultivars of each commodity (Weichmann 1987). However, despite the enormous economic significance of CA or MA storage, accompanying the use of low O_2 or high CO_2 atmospheres for maintaining quality of fresh fruit produce during CA or MA storage is the risk that very low O_2 and/or high CO_2 atmospheres may cause damage to the produce (Burdon et al. 2007). A better understanding of basic biochemical and physiological responses to CA or MA is needed to effectively evaluate storage conditions (Kader 1986).

4.2 Plant Responses to Low O_2 Atmosphere

Exposing fresh fruits and vegetables to low O_2 can be beneficial or harmful, depending on concentrations of these gases, temperature, and exposure duration. Exposing products to stress O_2 levels for long periods can lead to abnormal ripening, browning of tissues, and accumulation of ethanol and acetaldehyde (Imahori et al. 2007a). Oxygen levels as low as 0.2% in the plant cell may result in anaerobic respiration (Kader 1986).

Plant responses to low O_2 concentrations include induction of fermentation pathways, accumulation of fermentation products, and decreases in intracellular pH and ATP levels (Imahori et al. 2003). During fermentation, acetaldehyde which is produced through pyruvate decarboxylation by pyruvate decarboxylase (PDC) is converted to ethanol by alcohol dehydrogenase (ADH) using NADH. On the other hand, lactate is formed in a single step by the reduction of pyruvate by lactate dehydrogenase (LDH) and NADH. Thus, the major function of fermentative metabolism is to use NADH and pyruvate, when electron transport and oxidative phosphorylation are inhibited so that glycolysis can proceed. Both ethanol and lactate are produced to a varying degree by most plants under low O_2. Therefore, many plants have two simultaneous pathways competing for pyruvate and NADH under low O_2 condition (Imahori et al. 2003). The induction of PDC, ADH, and/or LDH is one of the mechanisms for accumulations of anaerobic products. Fermentative metabolism results in the accumulation of anaerobic products by the actions of the enzymes, PDC, ADH, and LDH, under low O_2 concentrations (Imahori et al. 2000b, 2003). However, the activities of ADH and LDH are not necessary the rate-limiting factors for the accumulations of ethanol and lactate in some plant tissues, if the activities of these enzymes are high (Xia and Saglio 1992).

Ke et al. (1995) proposed that fermentative metabolism can be regulated by two mechanisms in avocado fruit: (1) molecular control of PDC, ADH, and LDH and (2) metabolic control of these enzymes in plant tissue under low O_2 stresses. Generally, these increases in activities of enzymes by low O_2 have been found to be largely due to increased transcription and translation, resulting in new mRNA synthesis and de novo synthesis of the corresponding enzyme proteins (Imahori et al. 2003). However, molecular induction of the expression of these enzymes is not the major regulating mechanism, although with limited enzyme level the induction of fermentation enzyme through molecular control (transcription and/or translation) is essential for the accumulation of fermentation products (Ke et al. 1995). Sustained high ADH activities observed in hypoxia-treated pear fruit did not appear to be a function of sustained transcription, but instead may reflect regulated translation of mRNAs or high enzyme stability (Chervin and Truett 1999). The increase in ADH transcript and ADH activity did not correlate with acetaldehyde and ethanol accumulation in bell pepper fruits kept in 0% O_2 (Imahori et al. 2000a). There was no direct correlation between relative levels of gene expression and glycolytic flux, and in many cases mRNA and even enzyme protein reached levels in excess of what would be sufficient to account for the glycolytic flux actually observed (Ricard et al. 1994).

The changes in cytoplasmic pH are considered to be the controlling factor that regulate fermentative metabolism (Imahori et al. 2002b). A self-controlling system for lactate and ethanol production called the pH-stat hypothesis is proposed. This hypothesis suggests that at the onset of anaerobic stress LDH is active at alkaline pH of the cytoplasm and shunts pyruvate and lactate, and that the accumulation of lactate reduces cytoplasmic pH, which, in turn, inhibits LDH and activates PDC leading to ethanol production (Tadege et al. 1999; Imahori et al. 2003).

Concentrations of substrates and cofactors may exert metabolic control on fermentation enzymes. The different K_ms of pyruvate dehydrogenase (PDH) and PDC for pyruvate are the controlling factors that regulate the entry of pyruvate into the TCA cycle or the ethanolic fermentation pathway because the K_m of plant PDHs for pyruvate is in the μM range, whereas that of PDCs is in the mM range (Tadege et al. 1999). However, this would be too low for PDCs, and pyruvate could indeed be the limiting factor. Pyruvate becomes available for the PDC reaction due to a conformation change of the allosteric enzyme through binding to its substrate. The lag phase of ethanol production at the onset of anoxia might not be the result of the need for a drop in cytoplasmic pH, and rather that the lag phase might be required for a buildup of pyruvate (Tadege et al. 1999). Therefore, PDC activity is a key regulator of ethanolic fermentation under conditions of O_2 limitation. Based on the accumulated evidence, ethanolic flux is regulated by a PDH/PDC stat (Imahori et al. 2002a; Imahori et al. 2003).

4.3 Plant Responses to High CO_2 Atmosphere

The responses of fruits and vegetables to elevated CO_2 levels vary considerably within or among species, cultivars, organ types, and developmental stages, and include both undesirable and beneficial physiological and biochemical changes (Beaudry 1999). Moreover, it is well-known that the effect of CO_2 depends on its dosage and environmental conditions, such as temperature (Smith 1992). Carbon dioxide may act both as an inducer and a suppressor of respiration depending on its concentration in situ, duration of exposure, commodity, and temperature (Imahori et al. 2007b). During storage, the physiological effects of elevated CO_2 are a decrease in respiration rate and ethylene production, and retention of chlorophyll content, textural quality, and sensory attributes of horticultural commodities (Herner 1987).

The responses of fruits and vegetables to very high carbon dioxide concentrations include induction of the glycolytic pathway, fermentation pathways, accumulation of succinate and/or alanine, and decreases in pH and ATP levels (Mathooko 1996). During fermentation, acetaldehyde which is produced through pyruvate decarboxylation by PDC is converted to ethanol by ADH using NADH (Imahori et al. 2004a, b). Thus, pyruvate oxidation and NADH use can proceed while electron transport and oxidative phosphorylation are inhibited, and ATP can be produced, albeit at markedly reduced levels by substrate phosphorylation (Imahori et al. 2007a, b). Similarly, an atmosphere enriched with more than 20% CO_2 in the presence of atmospheric oxygen caused ethanol accumulation in lettuce, fig fruit, and strawberry fruit (Mathooko 1996). Elevated CO_2 concentrations, above a level of about 20% or higher, depending on the commodity and the O_2 concentrations, can result in accumulation of ethanol within the tissues

(Kader 1986). The accumulation of ethanol, as a product of fermentative metabolism, indicates that some substrates of energy metabolism are passing through the fermentation pathway (Imahori et al. 2007b).

Studies of the effects of elevated CO_2 on tricarboxylic acid (TCA) cycle intermediates and enzymes have shown accumulation of succinate due to inhibition of succinate dehydrogenase (SDH) activity in apples, pears, and lettuce (Kader 1986). Since SDH catalyzes the conversion of succinate into fumarate in the TCA cycle, SDH appears to be the enzyme most significantly influenced (Imahori et al. 2007b). The inhibition of succinate oxidation to fumarate by CO_2 has been related to the inhibition of SDH, thereby leading to succinate accumulation, a toxicant to plant tissues, and a depletion of malate (Mathooko, 1996). Therefore, the primary action of CO_2 appears to be on the kinetics of reversible reaction within the TCA cycle catalyzed by SDH (Mathooko, 1996). The reduction of the extractable activity of SDH in crisphead lettuce exposed to 20% CO_2 might have been due to a suppression of SDH synthesis, a modification of the enzyme structure, or conformation by the CO_2 treatment, and could also result from depletion of SDH protein due to increased degradation or inactivation of the enzyme in vivo (Ke et al. 1993). The concentrations of CO_2 used in storage of fruits and vegetables may regulate the TCA cycle by an alteration in SDH activity while fermentative metabolism is affected by the activities of ADH, thereby leading to accumulation of ethanol. Thus, the response of a commodity to CO_2-enriched atmosphere treatments includes the molecule's primary action which appears to be based on the kinetics of a reversible reaction within the TCA cycle catalyzed by SDH (Mathooko 1996).

5 Ethanol Vapor Treatment

Ethanol has been found to be beneficial in either counteracting senescent processes or reducing chilling injury (Toivonen 1995). Postharvest ethanol treatment can have beneficial effects on fruit physiology, such as enhancing the sensory quality of apples, reducing astringency of persimmons and bananas, reducing postharvest decay of citrus and stone fruit, and controlling scald in apples (Jamieson et al. 2003).

The application of ethanol to a range of climacteric fruit has been shown to have either a promotory or inhibitory effect on ripening parameters, depending on fruit type (Ritenour et al. 1997). These responses are dependent on a number of factors which include species, cultivar, maturity, applied concentration, mode of application, and duration of exposure (Jamieson et al. 2003). Depending on the maturity of the fruit and the amount of ethanol applied, exposure to ethanol vapors either promotes or inhibits tomato fruit ripening (Beaulieu and Saltiveit 1997). In tomato fruit, ethanol not only reduces ethylene production, but also noncompetitively inhibits ethylene action (Ritenour et al. 1997). In mango discs, low concentrations of ethanol vapor stimulated the production of ethylene (Pesis 2005). It is able to elicit nonenzymatic ethylene production from ACC (Beaulieu et al. 1998).

Exposure to ethanol vapor reduced chilling injury symptoms, which appear as red spots around the lenticels in mangos (Pesis et al. 1997). Exogenous application of ethanol can reduce chilling injury, possibly by altering membrane function (Pesis 2005). In cucumber seeding, ethanol caused changes in membrane-lipid fluidization, although there may be no change in the fatty acid composition (Frenkel and Erez 1996).

6 Ultraviolet Radiation

Ultraviolet (UV) radiation has been used to maintain the postharvest quality and extend the shelf life of several fresh fruits and vegetables (Perkins-Veazie et al. 2008). The UV portion of the electromagnetic spectrum ranges approximately 10–400 nm (Shama 2007). UV radiation has been applied to produce in long- (UV-A: 315–400 nm), medium- (UV-B: 280–315 nm), and short-wave (UV-C: 100–280 nm) dosages. The shortest wavelengths of the UV spectrum are the most energetic ones and more effective biocide for surface sterilization of some food products (Shama 2007; Perkins-Veazie et al. 2008).

Low UV doses induce production of antifungal compounds, ripening delay, and reduction of chilling injury (Pombo et al. 2009). The exposure to UV-C delays fruit softening which is one of the main factors determining fruit postharvest life (Pombo et al. 2009). UV-C decreased the activity of enzymes involved in tomato cell wall degradation and delayed the fruit softening (Pombo et al. 2009; Liu et al. 2011). Treatment with UV-C increases ascorbic acid and total phenolic contents and improves nutritional qualities of tomato fruit (Liu et al. 2011).

UV radiation can affect physiological processes at the genetic level. In parsley, UV-B upregulates genes encoding the flavonoid biosynthetic pathway, such as chalcone synthase and phenylalanine ammonia lyases (PAL), which are key enzymes in anthocyanin formation (Perkins-Veazie et al. 2008). In tomato, this exhibits ethylene production with ripening onset; UV-C treatment has disrupted ethylene production by decreasing the formation of ACC synthase (Perkins-Veazie et al. 2008). Peaches treated with UV-C showed increased activation of genes for β-1,3-glucanase and PAL (Perkins-Veazie et al. 2008).

Hormetic doses of UV-C radiation have been used as a physical treatment to extend postharvest life of several fruits and vegetables (Pombo et al. 2009). Hormesis has been defined as the use of potentially harmful agents at low doses in order to induce a beneficial stress response (Shama and Alderson 2005). Hormetic effects manifest themselves in treated plant tissue through the action of a variety of induced chemical species. They include phytoalexins, such as scoparone in oranges and resveratrol in grapes (Shama 2007). Also induced are enzymes, such as chitinases and glucanases in peaches and PAL in tomatoes (Shama 2007). The deleterious effects of UV light on plant tissues, such as decreased protein synthesis, impaired chloroplast function, and DNA damage, have been shown (Costa et al. 2006). However, low doses of UV could inflict repairable damage to DNA, and this slight trauma would activate repair mechanisms for radiation-induced DNA damage. Sublethal radiation may stimulate vital processes inside the cells and create a positive change in the homeostasis of a plant (Shama and Alderson 2005).

7 Conclusion and Future Perspective

The controlled abiotic stresses would be the basis for designing strategies to develop novel tools that open the possibility of tailoring fresh commodity with enhanced benefit properties for use of the fresh produce and processing industries. Therefore, there is need to understand how different plant tissues and their metabolic pathways respond to different abiotic stresses, applied alone or in combination with others. There is also a need to understand how different stresses trigger the specific enzymes involved in the targeted metabolism, as well as the possible interaction between different stresses and the response of the plant tissue. Such information is invaluable in the development of these treatments for practical commercial use.

References

An RF, Paull RE (1990) Storage temperature and ethylene on the ripening of papaya fruit. J Am Soc Hort Sci 115:949–953

Atta Aly MA (1992) Effects of high temperature on ethylene biosynthesis by tomato fruits. Postharvest Biol Technol 2:19–24

Bai J, Baldwin EA, Imahori Y, Kostenyuk I, Burns J, Brecht J (2011) Chilling and heating may regulate C6 volatile aroma production by different mechanisms in tomato (*Solamum lycopersicum*) fruit. Postharvest Biol Technol 60:111–120

Beaulieu JC, Saltiveit ME (1997) Inhibition or promotion of tomato fruit ripening by acetaldehyde and ethanol is concentration-dependent and varies with initial fruit maturity. J Am Soc Hort Sci 122:392–398

Beaulieu JC, Pesis E, Saltiveit ME (1998) AA or basic pH causes in vitro and nonenzymatic cleavage of ACC to ethylene. J Am Soc Hort Sci 123:675–680

Beaudry RM (1999) Effects of O_2 and CO_2 partial pressure on selected phenomena affecting fruit and vegetable quality. Postharvest Biol Technol 15:293–303

Ben-Shalom N, Hanzon J, Pinto R (1996) Cell wall changes and partial prevention of fruit softening in pre-stotage heat treated 'Anna' apples. J Sci Food Agric 72:231–234

Biolatto A, Cazquez DE, Sancho AM, Carduza FJ, Pensel NA (2005) Effect of commercial conditioning and cold quarantine storage treatments on fruit quality of 'Rouge la Toma' grapefruit (*Citrus paradisi* Macf.). Postharvest Biol Technol 35:167–176

Burdon J, Lallu N, Yearsley C, Burmeister D, Billing D (2007) The kinetics of acetaldehyde and ethanol accumulation in 'Hass' avocado fruit during induction and recovery from low oxygen and high carbon dioxide conditions. Postharvest Biol Technol 43:207–214

Cai C, Xu C, Shan L, Li X, Zhou C, Zhang W, Ferguson I, Chen K (2006) Low temperature conditioning reduces postharvest chilling injury in loquat fruit. Postharvest Biol Technol 41:252–259

Capanoglu E (2010) The potential of priming in food production. Trends Food Sci Technol 21:399–407

Chervin C, Brady CJ, Patterson BD, Faragher JD (1996) Could studies on cell responses to low oxygen levels provide improved options for fruit storage and disinfestations? Postharvest Biol Technol 7:289–299

Chervin C, Truett JK (1999) Alcohol dehydrogenase expression and alcohol production during pear ripening. J Am Soc Hort Sci 124:71–75

Cisneros-Zevallos L (2003) The use of controlled postharvest abiotic stresses as a tool for enhancing the nutraceutical content and adding-value of fresh fruits and vegetables. J Food Sci 68:1560–1565

Costa L, Vicente AR, Civello PM, Chaves AR, Martinez GA (2006) UV-C treatment delays postharvest senescence in broccoli florets. Postharvest Biol Technol 39:204–210

Dixon R, Paiva N (1995) Stress-induced phenylpropanoid metabolism. Plant Cell 7:1085–1097

Fallik E (2004) Prestorage hot water treatments immersion, rinsing and brushing. Postharvest Biol Technol 32:125–134

Fallik E, Archbold AA, Hamilton-Kemp TR, Loughrin JH, Collas RW (1997) Heat treatment temporarily inhibits aroma volatile compound emission from 'Golden Delicious' apples. J Food Chem 45:4038–4041

Ferguson IB, Ben-Yehoshua S, Mitcham EJ, McDonald RE (2000) Postharvest heat treatments: introduction and workshop summary. Postharvest Biol Technol 21:1–6

Ferguson IB, Lurie S, Bowen JH (1994) Protein synthesis and breakdown during heat shock of cultured pear (*Pyrus communis* L.) cells. Plant Physiol 104:1429–1437

Frenkel C, Erez A (1996) Induction of chilling tolerance in cucumber seedlings by endogenous and applied ethanol. Physiol Plant 96:593–600

Herner RC (1987) High CO_2 effects on plant organs. In: Weichmann J (ed) Postharvest physiology of vegetables. Marcel Dekker, Dublin, pp 239–253

Hofman PJ, Stubbings BA, Adkins MF, Corcoran RJ, White A, Woolf AB (2003) Low temperature conditioning before cold disinfestations improves 'Hass' avocado fruit quality. Postharvest Biol Technol 28:123–133

Imahori Y, Kanetune Y, Ueda Y, Chachin K (2000a) Changes in hydrogen peroxide content and antioxidative enzyme activities during maturation of sweet pepper (*Capsicum annuum* L.) fruit. J Jpn Soc Hort Sci 69:690–695

Imahori YK, Kota M, Furukawa H, Ueda Y, Chachin K (2000b) Ethanolic fermentation enzymes, their products and transcription of alcohol dehydrogenase from bell pepper fruit held under various low oxygen atmospheres. J Jpn Soc Hort Sci 69:266–272

Imahori Y, Kota M, Ueda Y, Ishimaru M, Chachin K (2002a) Regulation of ethanolic fermentation in bell pepper fruit under low oxygen stress. Postharvest Biol Technol 25:159–167

Imahori Y, Kishioka I, Uemura K, Yoshioka H, Ueda Y, Ishimaru M, Chachin K (2002b) Physiological responses and quality attributes of Japanese pear 'Kosui' fruit kept in low oxygen atmospheres. J Hort Sci Biotechnol 77:677–682

Imahori Y, Kishioka I, Uemura K, Makita E, Fujiwara H, Nishiyama Y, Ishimaru M, Ueda Y, Chachin K (2007a) Effects of short-term exposure to low oxygen atmospheres on physiological responses of sweetpotato roots. J Jpn Soc Hort Sci 76:258–265

Imahori Y, Matushita K, Kota M, Ueda Y, Ishimaru M, Chachin K (2003) Regulation of fermentative metabolism in tomato fruit under low oxygen stress. J Hort Sci Biotech 78:386–393

Imahori Y, Suzuki Y, Kawagishi M, Ishimaru M, Ueda Y, Chachin K (2007b) Physiological responses and quality attributes of Chinese chive leaves exposed to CO_2-enriched atmospheres. Postharvest Biol Technol 46: 160–166

Imahori Y, Suzuki Y, Uemura K, Kishioka I, Fujiwara H, Ueda Y, Chachin K (2004a) Physiological and quality responses of Chinese chive leaves to low oxygen atmospheres. Postharvest Biol Technol 31:295–303

Imahori Y, Takemura M, Bai J (2008) Chilling-induced oxidative stress and antioxidant responses in mume (*Prunus mume*) fruit during low temperature storage. Postharvest Biol Technol 49:54–60

Imahori Y, Uemura K, Kishioka I, Fujiwara H, Tulio AZ, Ueda Y, Chachin K (2004b) Relationship between low-oxygen injury and ethanol metabolism in various fruits and vegetables. Acta Hort 682:1103–1108

Jamieson LE, Meier X, Smith KJ, Lewthwaite SE, Dentener PP (2003) Effect of ethanol vapor treatments on lightbrown apple moth larval mortality and 'Braeburn' apple fruit characterization. Postharvest Biol Technol 28:391–403

Kader AA (1986) Biochemical and physiological basis for effects of controlled and modified atmospheres on fruit and vegetables. Food Technol 40:99–104

Kader A (1992) Postharvest biology and technology: an overview. In: Kader A (ed) Postharvest technology of horticultural crops. Oakland: Division of Agricultural and Natural Resources, University of California, pp 15–20

Kadyrzhanova DK, Vlachonasios KE, Ververidis P, Dilley DR (1998) Molecular cloning of a novel heat induced chilling tolerance related cDNA in tomato fruit by use of mRNA differential display. Plant Mol Biol 36: 885–895

Ke D, Mateos M, Siriphanich J, Kader AA (1993) Carbon dioxide action on metabolism of organic and amino acids in crisphead lettuce. Postharvest Biol Technol 3:235–247

Ke D, Yahia E, Hess B, Zhou L, Kader AA (1995) Regulation of fermentative metabolism in avocado fruit under oxygen and carbon dioxide stresses. J Am Soc Hort Sci 120:481–490

Ketsa S, Chidtragool S, Klein JD, Lurie S (1999) Ethylene synthesis in mango fruit following heat treatment. Postharvest Biol Technol 15:65–72

Klein JD, Lurie S (1990) Prestorage heat treatment as a means of improving poststorage quality: interaction of time and temperature. Hort Sci 27:326–328

Klein JD, Lurie S, Ben-Arie R (1990) Quality and cell wall components of 'Anna' and 'Granny Smith' apples treated with heat, calcium and ethylene. J Am Soc Hort Sci 115:954–958

Lee GJ, Vierling E (2000) A small heat shock protein cooperates with heat shock protein 70 systems to reactive a heat-denatured protein. Plant Physiol 122:189–197

Lester GE (2003) Oxidative stress affecting fruit senescence. In: Hodges DM (ed) Postharvest oxidative stress in horticultural crops. Haworth, New York, pp 113–129

Liu C, Han X, Cai L, Lu X, Ying T, Jiang Z (2011) Postharvest UV-B irradiation maintains sensory qualities and enhances antioxidant capacity in tomato fruit during storage. Postharvest Biol Technol 59:232–237

Lurie S (1998) Postharvest heat treatments. Postharvest Biol Technol 14:257–269

Lurie S, Handros A, Fallik E, Shapira R (1996) Reversible inhibition of tomato fruit gene expression at high temperature. Plant Physiol 110:1207–1214

Lurie S, Nussinnovitch A (1996) Compression characteristic, firmness and texture perception of heated and unheated apples. Intern J Food Sci Technol 31:1–5

Mathooko FM (1996) Regulation of respiratory metabolism in fruits and vegetables by carbon dioxide. Postharvest Biol Technol 9:247–264

McDonald RE, McCollum TG, Baldwin EA (2000) Temperature of water heat treatments influences tomato fruit quality following low-temperature storage. Postharvest Biol Technol 16:147–155

Paull RE, Chen NJ (1990) Heat shock response in field grown ripening papaya fruit. J Am Soc Hort Sci 115:623–631

Paull RE, Chen NJ (2000) Heat treatment and fruit ripening. Postharvest Biol Technol 21:21–37

Perkins-Veazie P, Collins J, Howard L (2008) Blueberry fruit response to postharvest application of ultraviolet radiation. Postharvest Biol Technol 47:280–285

Pesis E (2005) The role of the anaerobic metabolites, acetaldehyde and ethanol, in fruit ripening, enhancement of fruit quality and fruit deterioration. Postharvest Biol Technol 37:1–19

Pesis E, Faure M, Arie RM (1997) Induction of chilling tolerance in mango by temperature conditioning, heat, low O_2 and ethanol vapours. Acta Hort 455:629–634

Pombo M, Dotto MC, Martniez GA, Civello PM (2009) UV-C irradiation delays strawberry fruit softening and modifies the expression of genes involved in cell wall degradation. Postharvest Biol Technol 51:141–148

Porat R, Pavoncello D, Peretz J, Ben-Yehoshua S, Lurie S (2000) Effects of various heat treatments on the induction of cold tolerance and on the postharvest qualities of 'Star Ruby' grapefruit. Postharvest Biol Technol 18:159–165

Ricard B, Couee I, Raymond P, Saglio PH, Saint-Ges V, Pradet A (1994) Plant metabolism under hypoxia and anoxia. Plant Physiol Biochem 194:791–797

Ritenour MA, Mangrich ME, Beaulieu JC, Rab A, Saltveit ME (1997) Ethanol effects on the ripening of climacteric fruit. 1997. Postharvest Biol Technol 12:35–42

Rose JKC, Hadfield KA, Labavitch JM, Bennet AB (1998) Temporal sequence of cell wall disassembly in rapidly ripening melon fruit. Plant Physiol 117:345–361

Sapitnitskaya M, Maul P, McCollum GT, Guy CL, Weiss B, Samach A, Porat R (2006) Postharvest heat and conditioning treatments activate different molecular responses and reduce chilling injuries in grapefruit. J Exp Bot 57:2943–2953

Shama G (2007) Process challenges in applying low doses of ultraviolet light to fresh produce for eliciting beneficial hermetic responses. Postharvest Biol Technol 44:1–8

Shama G, Alderson P (2005) UV hormesis in fruits: a concept ripe for commercialization. Trends Food Sci Technol 16:128–136

Smith RB (1992) Controlled atmosphere storage of 'Redcoat' strawberry fruit. J Am Soc Hort Sci 117:260–264

Sozzi GO, Cascone O, Fraschina AA (1996) Effects of a high temperature stress on endo-β-mannanase and α- and β-galactosidase activities during tomato fruit ripening. Postharvest Biol Technol 9:49–63

Stuger R, Ranostaj S, Materna T, Forreiter C (1999) Message RNA-binding properties of non-polysomal ribonucleoproteins from heat stressed tomato cells. Plant Physiol 120:23–31

Tadege M, Dupuis I, Kuhlemeier C (1999) Ethanolic fermentation: new functions for an old pathway. Trends Plant Sci 4:320–325

Toivonen PMA (1995) Non-ethylene, non-respiratory volatiles in harvested fruits and vegetables: their occurrence, biological activity and control. Postharvest Biol Technol 12:109–125

Vierling E (1997) The small heat shock proteins in plants are members of an ancient family of heat induced proteins. Acta Physiol Plant 19:539–547

Wang CY (2010) Alleviation of chilling injury in tropical and subtropical fruits. Acta Hort 864:267–274

Wang W, Vinocur B, Altman A (2003) Plant responses to drought, salinity and extreme temperatures: towards genetic engineering for stress tolerance. Planta 218: 1–14

Wang W, Vinocur B, Shoseyov O, Altman A (2001) Biotechnology of plant osmotic stress tolerance: physiological and molecular considerations. Acta Hort 560:285–292

Wang W, Vinocur B, Shoseyov O, Altman A (2004) Role of plant heat-shock protein and molecular chaperones in the abiotic stress response. Trends Plant Sci 9: 244–252

Weichmann J (1987) Low oxygen effects. In: Weichmann J (ed) Postharvest physiology of vegetables. Marcel Dekker, Dublin, pp 231–237

Woolf AB, Watkins CB, Bowen JH, Lay-Lee M (1995) Reducing external chilling injury in stored 'Hass' avocados with dry heat treatments. J Am Soc Hort Sci 120:1050–1056

Woolf AB, Ball S, Spooner KJ, Lay-Yee M, Ferguson IB, Watkins CB, Gunson A, Forbes SK (1997) Reduction of chilling injury in the sweet persimmon 'Fuyu' during storage by dry air heat treatments. Postharvest Biol Technol 11:155–164

Woolf AB, Cox KA, White A, Ferguson IB (2003) Low temperature conditioning treatments reduce external chilling injury of 'Hass' avocados. Postharvest Biol Technol 28:113–122

Xia J, Saglio PH (1992) Lactic acid efflux as a mechanism of hypoxic acclimation of maize root tips to anoxia. Plant Physiol 100:40–46

Yang RF, Cheng MTS, Shewfelt RL (1990) The effect of high temperature and ethylene treatment on the ripening of tomatoes. J Plant Physiol 136:941–942

Zhang J, Hung W, Pan Q, Liu Y (2005) Improvement of chilling tolerance and accumulation of heat shock proteins in grape berries (*Vitis vinifera* cv. Jinxiu) by heat pretreatment. Postharvest Biol Technol 38:80–90

Abscisic Acid Signaling in Plants

19

Radomira Vankova

Abstract

Abscisic acid (ABA) is the key hormone in plant responses to water deficit, caused by abiotic stresses associated with dehydration (drought, salinity, and cold) and during specific developmental stages (e.g., seed maturation). ABA signal transduction pathway is initiated by binding of ABA to its receptors (GTG1, GTG2, Mg protoporphyrin IX chelatase H-subunit, and 14-member family of PYR/PYL/RCAR proteins). PYR/PYL/RCARs directly interact with ABA and the main negative regulators of ABA signaling – type-2C serine threonine protein phosphatases (PP2Cs). Formation of the complex results in derepression of specific protein kinases, especially of SnRK2 family. In cytoplasm, SnRK2.6 (OST1) modulates the activity of ion channels (stimulating anion efflux and inhibiting K^+ influx). By fast stimulation of ion efflux, turgor of stomata cells is reduced, which results in their closure. In the nucleus, ABA signal transduction induces a wide change in the transcriptome not only by stimulation of the expression of defense genes, but also by degradation of specific mRNAs, by change of mRNA stability, splicing, and transport. Elucidation of the mechanism of ABA signal transduction significantly contributes to establishment of a suitable strategy for elevation of plant tolerance to abiotic stresses.

Keywords

Abscisic acid • PYR/PYL/RCAR proteins • PP2C • SnRK2 • Stomatal aperture • Transcriptome • mRNA processing

R. Vankova (✉)
Laboratory of Hormonal Regulations in Plants,
Institute of Experimental Botany AS CR, Prague,
The Czech Republic
e-mail: vankova@ueb.cas.cz

P. Ahmad and M.N.V. Prasad (eds.), *Abiotic Stress Responses in Plants: Metabolism, Productivity and Sustainability*, DOI 10.1007/978-1-4614-0634-1_19, © Springer Science+Business Media, LLC 2012

1 Introduction

Plant hormone abscisic acid (ABA) is a sesquiterpenoid. It is biosynthesized from isopentenyl diphosphate, via β-carotene and *cis*-neoxanthine in plastids. The *cis*-neoxanthine is transported to cytoplasm. Its cleavage represents the rate-limiting step in ABA formation.

ABA consists of a six-member ring with double bond between C2′ and C3′, one methyl group at C2′, keto group at C4′, and two methyl residues at C6′. Six-member side chain has terminal carboxyl group. According to the position of carboxyl, *cis*- and *trans*-isomer can be distinguished, the former one exhibiting the biological activity. Due to the optically active C1′, bearing a hydroxyl group, ABA may occur either as (S)/(+) isomer, which is the most physiologically active, or as much less active (R)/(−) isomer.

2 History and Physiological Functions

ABA was identified independently by several groups in 1960s. It was found in the search for negative regulator of bud growth. As an agent causing sycamore bud dormancy, it was called dormin. The other group searched for the substance which could promote abscission in cotton and called the isolated compound abscisin II. When dormin and absicin II were compared and the same chemical structure was found, the substance was called ABA (Cornforth et al. 1965). The name was maintained, in spite of the fact that ethylene was later recognized as the abscision regulator.

ABA is a plant hormone enhancing the plant (tissue) tolerance to water deficit, both in the response to abiotic stresses associated with dehydration (drought, salinity, and cold) and during the development (e.g., in seed maturation). ABA also downregulates growth, affecting seed germination and postgermination development, as well as the root growth. ABA exhibits both fast effects (modulation of ion flows resulting in stomata closure) and relatively long-term effects on gene expression pattern. ABA-regulated genes represent over 10% of the genome in *Arabidopsis* seedlings (Cutler et al. 2010).

3 Receptors

The most upstream components of ABA signal transduction pathways are ABA receptors. ABA perception by specific receptors represents the primary event which triggers downstream signaling cascades to induce the physiological responses (Shang et al. 2010). Experiments with photolyzable caged ABA indicated the existence of intracellular receptors (Allan et al. 1994). ABA–protein conjugates that were unable to cross the membrane activated both ion channel activity and gene expression (Schultz and Quatrano 1997), which suggested the presence of plasma membrane receptors. Thus, both intracellular and extracellular receptors seem to be involved in ABA signaling.

3.1 FCA: Putative Receptor

In the search for ABA receptors, ABA-binding protein (ABAP1) was identified in barley using anti-idiotypic antibodies (Razem et al. 2004). It exhibited a close homology to *Arabidopsis* protein FCA. Later on, FCA was reported as ABA receptor (Razem et al. 2006). The mode of action for ABA was suggested. ABA was reported to disrupt the complex of FCA and FY, which inhibits the flowering repressor FLC. Derepressed suppressor FLC, then, may inhibit flowering. Unfortunately, the inhibitory effect of ABA on the interaction between FCA and FY could not be reproduced. So the possibility that FCA is ABA receptor was not reiterated (Jang et al. 2008).

3.2 GTG1 and GTG2

As defects in G-protein signaling pathway were found to be associated with disturbance of ABA sensitivity, both in guard cells and during germination and early seedling development

(Wang et al. 2001; Pandey et al. 2006), search for potential ABA receptor among G-protein-coupled receptors (GPCRs) was initiated. G-protein signaling represents an important signal transduction mechanism. The central components of this network are heterotrimeric G-proteins consisting of Gα, Gβ, and Gγ subunits and GPCR (Pandey et al. 2009). Plants have only very limited number of subunits, one Gα (GPA1), one Gβ (AGB1), and two Gγ (AGG1 and AGG2) subunits were found in *Arabidopsis thaliana* (Jones and Assmann 2004). The GPCR homologue (GCR2) was the first reported ABA plasma membrane receptor (Liu et al. 2007), identified as protein interacting with GPA1. The proposed mechanism suggested that ABA binding to GCR2 might disturb heterotrimeric complex which would result in ABA signal transduction. Unfortunately, no independent studies were able to repeat these data, which made the possibility of GCR2 function as ABA receptor controversial (Klingler et al. 2010).

Recently, Pandey et al. (2009) reported two GPCR-type G-proteins (GTG1 and GTG2) which interacted with GPA1 and had intrinsic GTP-binding and GTPase activity. GDP-bound form of GTGs was active. Both proteins specifically bound (S)-ABA with Kd ca 20 nM. The double knockout mutant was impaired in multiple aspects of ABA signaling, e.g., it was hyposensitive to ABA in stomata closure. GTG1 and GTG2 are expressed in all plant organs, which indicate their multiple functions.

3.3 Magnesium Protoporphyrin IX Chelatase H-Subunit

Another ABA-binding protein was isolated from *Vicia faba* leaves (Zhang et al. 2002). It was identified as magnesium protoporphyrin IX chelatase large subunit H (Mg-chelatase H subunit, CHLH). ABA receptor CHLH/ABAR was identified and characterized in chloroplasts in *A. thaliana* (Shen et al. 2006). CHLH/ABAR forms with other subunits (CHLD and CHLI) a complex, which is involved in chlorophyll biosynthesis, catalyzing insertion of Mg^{2+} into protoporphyrin-IX. ABAR seems to be involved in retrograde signaling (Mochizuki et al. 2001; Nott et al. 2006; Wang and Zhang 2008), i.e., signaling between the plastid and nucleus. As many plastid genes are coded by the nucleus, tight regulation of their expression, e.g., under stress conditions, is crucial for coordinated expression in both organelles.

C-terminal part of ABAR molecule was found to play a central role in ABA binding and signaling (Wu et al. 2009). Localization of ABAR was reported to be dependent on the concentration of Mg^{2+}. Receptor was localized in the envelope fraction at high Mg^{2+} (5 mM) or in the stroma fraction, when Mg^{2+} level was low (1 mM). At physiological conditions, ABAR was localized to both inner and outer envelope membranes. Recently, Shang et al. (2010) specified ABAR structure as a transmembrane protein which spans the chloroplast envelope, with N- and C-terminal sequences exposed to cytosol. ABAR C-terminal sequence interacts with ABA in contrast to the N-terminal part, which does not bind ABA, but is functionally required for ABA signaling. C-terminal deletion results in the localization of ABAR predominantly to the stroma. It might be caused by an endocytosis-like mechanism, which involves chloroplast membrane trafficking in response to low Mg^{2+} stress. C-terminal domain is interacting with transcription factor WRKY40, and to lesser extent with WRKY18 and WRKY60. WRKY-mediated signaling is rather complex; WRKY60 seems to be a regulator which may balance the WRKY40/WRKY18 mediated ABA signaling. The three WRKYs as well as ABAR are expressed ubiquitously in different organs/tissues, which may indicate their role at the whole plant level (Wu et al. 2009).

ABAR–WRKY40 interaction is highly stimulated at the presence of ABA. ABA is required for the migration of WRKY40 molecule from the nucleus to the cytosol. This was demonstrated by application of exogenous ABA, which restored WRKY40 cytosolic distribution in ABA-deficient mutant. ABA also downregulates transcript as well as protein level of WRKY40. WRKY40 represses expression of several ABA-responsive genes via binding to their promoter region, namely, to TGAC W-box sequence. These genes involve bZIP transcription factors, ABI5 and ABF4, and Apetala-2 (AP2) domain transcription factor, ABI4, DREB1A, DREB2A, MYB2, and RAB18.

One of the genetically best-characterized and physiologically most important transcription factors is ABI5, which controls seed germination and postgermination growth. In response to high levels of ABA that recruit WRKY40 from the nucleus to cytosol and promote ABAR–WRKY40 interaction, *ABI5* transcription inhibition is relieved. Binding of WRKY40 to *ABI5* promoter region indicates that ABI5 may function directly downstream of WRKY40 in the ABAR–WRKY40-mediated ABA signaling.

Another interesting role of ABAR was recently suggested by Legnaioli et al. (2009), who showed that ABAR could mediate connection between drought ABA signaling and circadian clock.

3.4 PYR/PYL/RCAR Proteins

3.4.1 Identification of PYR/PYL/RCAR Receptors

After a long search for unequivocal ABA receptors, two independent research groups identified the same ABA receptor family using different approaches. In a thorough chemical screening, Cutler's group found compound pyrabactin which was able to inhibit specifically seed germination (Park et al. 2009). Microarray analysis of the ABA and pyrabactin responses in seeds revealed that both compounds induce very similar transcriptional changes, unlike other unrelated germination inhibitors. After isolation of pyrabactin-resistant mutant, PYRABACTIN RESISTANCE 1 (PYR1) gene was identified. In Arabidopsis, 13 related genes were determined, which were named PYL1 – 13 (PYR-like). The other group utilized the fact that PP2C ABI2 is an important negative regulator of ABA signaling. In yeast two-hybrid screening of ABI2 interacting partners, Grill's group found an interacting protein, which was named Regulatory Component of ABA Receptor 1 (RCAR1) (Ma et al. 2009). They identified 13 other family members and called them RCAR2 – 14. Due to the simultaneous announcement of this receptor family by two groups, parallel nomenclature has been used (Table 19.1). In the same year, Rodrigez's group performed yeast two-hybrid screening with another PP2C HAB1 and identified PYL5, PYL6, and PYL8 proteins (Santiago et al. 2009b). Schoeder's group made a search for ABI1 interacting proteins in vivo using affinity purification and MS and identified 9 of 14 PYR/PYL/RCAR proteins (Nishimura et al. 2010).

Table 19.1 Parallel nomenclature of PYR/PYL/RCAR proteins

RCAR1	PYL9
RCAR2	PYL7
RCAR3	PYL8
RCAR4	PYL10
RCAR5	PYL11
RCAR6	PYL12
RCAR7	PY L13
RCAR8	PYL5
RCAR9	PYL6
RCAR10	PYL4
RCAR11	PYR1
RCAR12	PYL1
RCAR13	PYL3
RCAR14	PYL2

3.4.2 Structure of PYR/PYL/RCAR Proteins

PYR/PYL/RCAR proteins are soluble ligand-binding proteins which belong to START domain superfamily, more recently named the Bet v I-fold superfamily (Cutler et al. 2010), because this conserved domain was originally identified in the major pollen allergen of white birch (*Betula verrucosa*, Radauer et al. 2008). A central feature of this family is a seven-stranded bent β-sheet and two small α-helices enfolding a long, carboxy-terminal α-helix, which collectively form a helix-grip fold structure (Klingler et al. 2010), creating a large cavity, that can bind hydrophobic ligands (Iyer et al. 2001; Radauer et al. 2008).

Crystallographic studies revealed that apo receptors contain an open ligand-binding pocket flanked by two loops which form a gate (β3–β4 loop) and latch (β5–β6 loop) (Melcher et al. 2009). Upon ABA binding into the cavity, closure of the SGLPA gate occurs while position of HRL latch does not change substantially. This conformational change enables the receptor to dock into and competitively inhibit the active site of protein phosphatase (PP2C). A conserved tryptophan in the PP2C inserts directly between the gate and latch to interact through water-mediated H-bond with ketone group of ABA cyclohexene ring to stabilize the closed position. In this way, activity of PP2Cs is inhibited. Santiago et al. (2009a) showed in crystallographic studies of PYR1 that receptor is a homodimer with one ABA-binding site occupied by ABA. Upon binding of the second ABA molecule, dissociation occurs and heterodimer is formed with PP2C in the ratio 1:1 (Klingler et al. 2010). Half-maximal inhibition of phosphatase activity was reported after 30 s in the case of RCAR3 and ABI2 complex (Szostkiewicz et al. 2010) while it was 10 s in the case of RCAR1 and ABI2 (Ma et al. 2009).

The individual receptors were studied for their requirement of ABA for the interaction with different PP2Cs. Park et al. (2009) found that interaction among PYR1, PYL1-4, and group A PP2Cs occurs only in the presence of ABA. Point mutation in PYR1 ($PYR1^{P88S}$) severely impaired physical interaction between PYR1 and HAB1 but not ABA binding to PYR1, which demonstrated that ABA binding can be uncoupled from PP2C inhibition (Park et al. 2009). On the other hand, RCAR1(PYL9), RCAR3(PYL8), and RCAR8(PYL5) were shown to interact with ABI1 in the absence of ABA (Ma et al. 2009; Santiago et al. 2009a). The affinity of RCAR1, RCAR3, and RCAR8 did not differ substantially, exhibiting Kd for (S)-ABA of 0.7, 1.0, and 1.1 μM, respectively. Study of (S)-ABA binding to heteromeric receptor complexes revealed more than tenfold lower *K*d values of 64 nM for RCAR1/ABI2 and 38 nM for RCAR8/HAB1 (Raghavendra et al. 2010).

Stereospecificity of different receptors was followed as well. PYR1 and PYL1-4 recognize very well (S)-ABA. PYL2-4 respond to both (S)-ABA and with lower affinity also to (R)-ABA, which makes these proteins candidates for dual stereoreceptors (Park et al. 2009). RCAR1 conferred almost absolute selectivity for (S)-ABA, whereas complexes with RCAR3 responded also to (R)-ABA and *trans*-ABA stereoisomers, even if with more than tenfold lower sensitivity than to (S)-ABA (Szostkiewicz et al. 2010).

Localization of RCAR3/1 and ABI1/2 complex was found both in the cytosol and in the nucleus (Szostkiewicz et al. 2010). The combination of 14 receptors and more than 6 group A PP2Cs (ABI1, ABI2, HAB1, HAB2, PP2CA, AHG1) represents a wide range of different variants which can be regulated in specific way throughout the development and in response to abiotic stresses. Highly differential variation in the expression was found in case of RCAR1, RCAR3, ABI1, and ABI2 at different developmental stages, in different tissues, or under different stress conditions (Szostkiewicz et al. 2010). RCAR1 expression was upregulated in seed coats, flowers, and siliques and RCAR3 expression in xylem (Szostkiewicz et al. 2010) while PYR1 in seeds (Park et al. 2009).

PYR/PYL/RCAR proteins interact not only with group A PP2Cs, but also with other classes of protein phosphatases. Recently, catalytic subunit of PP2A was reported as a negative regulator of ABA signaling as well (Pernas et al. 2007).

PP2Cs have multiple targets in cells; they directly inactivate sucrose non-fermenting-1-related protein kinases class 2 (SnRK2s). PP2Cs interact also with calcium-regulated enzymes CDPKs/CPKs or SnRK3s/CIPKs/PKSs. One of PP2Cs, ABI2, was reported to interact with salt overly sensitive (SOS)2, kinase belonging to SnRK3s, which plays an important role in the salt stress response (Ohta et al. 2003).

4 Regulation of Stomata Aperture

4.1 The Effect of SnRK2.6 on SLAC1 and KAT1

PP2Cs inhibit SnRK2.6 (open stomata 1, OST1) via dephosphorylation of Serine 175, which is in the kinase activation loop (Umezawa et al. 2009; Vlad et al. 2009). SnRK2.6 regulates fast ABA responses resulting in stomata closure. SnRK2.6 directly interacts with slow anion channel-associated 1 (SLAC1), S-type anion channel on the plasma membrane of guard cells. SnRK2.6 phosphorylates the N-terminal, cytoplasmic region of this channel. This phosphorylation is abolished by ABI1 (Geiger et al. 2009). PYR1, in the presence of ABA, is able to remove this inhibition (Fig. 19.1). Derepression of SLAC1 was also achieved with RCAR1(PYL9) (Geiger et al. 2010).

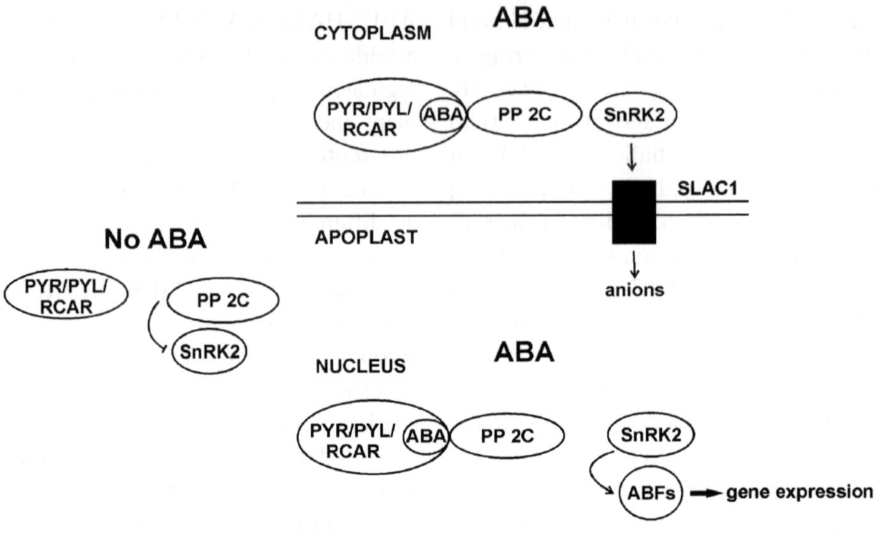

Fig. 19.1 Scheme of ABA derepression of SnRK in the cytoplasm and nucleus

Activation of this channel stimulates anion efflux which causes membrane depolarization. Geiger et al. (2010) reported that SLAC1 is regulated also by Ca-dependent protein kinase CPK23. SLAC1 regulation involves interaction with other Ca-dependent protein kinases, such as CPK3 and CPK6 (Mori et al. 2006).

SnRK2.6 phosphorylates also cation channel KAT1 (inward K^+ channel, Sato et al. 2009). In this case, however, phosphorylation results in inhibition of the activity of channel involved in stomata opening.

The other members of this group are SnRK2.2 and SnRK2.3. They do not seem to be involved in the regulation of stomata aperture. They play a role especially during germination and seedling development. The importance of all SnRK2s as global positive regulators can be deduced from extremely ABA-insensitive phenotype of the triple mutant (Cutler et al. 2010).

4.2 Second Messengers Involved in Stomata Aperture Regulation

SnRK2.6 also interacts and phosphorylates respiratory burst oxidase homolog F (RBOHF), plasma membrane NADPH oxidase, which is involved in the production of ROS, especially of hydrogen peroxide and superoxide. In this way, ABA is able to induce rapid burst of H_2O_2 (Pei et al. 2000). Hydrogen peroxide reversibly inhibits ABI1 through the oxidation of cysteine residues in catalytic center (Meinhard et al. 2002). Thus, hydrogen peroxide generated during the signaling temporarily inactivates negative regulators, which results in amplification of the ABA signal.

Hydrogen peroxide was also found to stimulate activation of calcium channels (Kwak et al. 2003), which in the consequence results in elevation of cytoplasmic calcium level and activation of Ca-dependent protein kinases. Also other second messengers are involved in triggering of Ca^{2+} signal – inositol 1,4,5 triphosphate, produced by phospholipase C, myo-inositol hexakis phosphate, or cyclic ADP-ribose (Schroeder et al. 2001). Another ABA second messenger is phosphatidic acid, produced by phospholipase D (Zhang et al. 2005).

ABA signal transduction involves also nitrogen oxide (NO, Meinhard et al. 2002). Experiments with H_2O_2 scavenger (catalase) and NO synthase inhibitor (L-NAME) showed that catalase was able to abolish production of both ROS and NO while L-NAME prevented only the rise in NO levels (Srivastava et al. 2009). These data indicated that H_2O_2 production preceded NO elevation. NO downregulation was, however, sufficient to suppress stomata closure. NO was reported to exert its effect via cyclic ADP-ribose (Neil et al. 2002).

Another kinase activated by ABA is sphingosine kinase (SPHK1), which produces sphingosine-1-phosphate (Coursol et al. 2003). Overexpression of *SPHK1* led to the increase of stomata sensitivity to ABA, as well as to the decrease of the rate of germination (Worrall et al. 2008). Sphingosine-1-phosphate, thus, functions as another second messenger in ABA signaling. It requires GPA1 for its action.

5 ABA-Induced Changes of the Transcriptome

5.1 Transcription Factors Involved in ABA Signal Transduction

Expression of ABA-response genes is stimulated by transcription factors bZIP, viviparous1 (VP1)/ABI3, and AP2. The bZIP proteins are probably ubiquitously present in nucleus, being activated by phosphorylation. All SnRK2s can in the nucleus directly phosphorylate bZIP transcription factors of ABF/AREB/ABI5 group, which further stimulate the expression of ABA-response genes. Upregulated genes are coding for, e.g., dehydrins, antioxidant enzymes, transporters, PYR/PYL/RCARs, transcription factors, protein kinases, protein phosphatases, and enzymes of phospholipid signaling (Cutler et al. 2010). The bZIP transcription factors interact in dimers with the G-box of ABA-response elements (ABRE) in the promoters of ABA-response genes. The bZIP dimers may interact with 14-3-3 proteins and VP1(ABI3) transcription factor (Himmelbach et al. 2003).

The promoters of ABA-response genes contain also other *cis* elements – dehydration-responsive elements (DREs) and coupling elements (CEs), which both couple AP2 transcription factors (Himmelbach et al. 2003). It is interesting that some DRE-binding factors, e.g., maize DBF1, are positive regulators of ABA-induced transcription while others, e.g., maize DBF2, suppress ABA action (Kizis and Pages 2002). The interaction of the above-mentioned transcription factors and *cis* elements modulates the strength of transcription as well as precisely regulates its initiation, both in time- and tissue-specific manner.

Important role in ABA signaling is also played by MYB and MYC transcription factors (Cutler et al. 2010). These secondary-acting transcription factors are synthesized de novo, as it was demonstrated for *AtMYC2* and *AtMYB2* (Abe et al. 2003). More than 200 transcription factors from at least 20 families have been reported to be involved in ABA signaling (Nemhauser et al. 2006).

The optimal level of transcription factors is regulated also by their degradation with proteasome. When ABI5 is not stabilized by phosphorylation, it is quickly degraded (Lopez-Molina et al. 2001).

Initiation of transcription requires accessible chromatin. ABA can modify access to DNA by inducing the changes in histone modification (Sokol et al. 2007). HAB1 was reported to interact with SWI3B, component of SWI/SNF chromatin-remodeling complexes (Saez et al. 2008). This interaction blocks induction of a subset of ABA-regulated genes. Inhibition of HAB1 by ABA can release SWI3.

5.2 mRNA Processing, Transport, and Degradation

ABA responses are considerably affected by the defects in mRNA processing, transport, and degradation (for review, see Hirayama and Shinozaki 2007). The cap-binding complex (CBC) binds the 5′ terminal cap structure of mRNA, protects it from decapping enzymes, enhances its splicing, and promotes the first-round translation. Cap-binding protein ABA hypersensitive 1 (ABH1) was reported to modulate early ABA transduction (Papp et al. 2004). Hypersensitivity to ABA was reported also upon mutation of SAD1, complex of small nuclear ribonucleoproteins, which affect splicing and export of RNAs (Xiong et al. 2001). Poly(A) tail is processed by ABA hypersensitive at germination 2 (AHG2), poly(A)-specific ribonuclease which destabilizes transcripts induced in response to ABA, abiotic stress, or salicylic acid (Nishimura et al. 2005).

AAPK interacting protein 1 (AKIP1), heterogeneous nuclear ribonucleoprotein, interacts with SnRK2. ABA treatment activates RNA-binding activity of AKIP1 and induces its relocation to nuclear speckles, storage areas for transcription or splicing factors (Li et al. 2002). A DEAD-box RNA helicase, low expression of osmotically responsive genes4 (LOS4), functions in mRNA export from the nucleus (Gong et al. 2005). The link exists between miRNA-mediated gene regulation and ABA signal transduction. Hyponastic leaves 1 (HYL1) is involved in miRNA processing required for ABA response (Lu and Fedoroff 2000; Han et al. 2004). The role of small RNAs in ABA and stress responses was reviewed by Shukla et al. (2008).

6 Conclusion and Future Perspectives

ABA is a plant hormone crucial for plant defense to abiotic stresses associated with dehydration (drought, salinity, and cold). ABA signal transduction allows dynamic regulation of the aperture of stomata, place of more than 90% plant water loss, as well as the adjustment of the transcriptome to the stress conditions. Until recently, rather scattered pieces of knowledge were available on ABA-signaling pathway. The discovery of PYR/PYL/RCAR receptors enabled elucidation of the key regulatory mechanism – derepression of ABA pathway upon inactivation of the repressor (PP2C) by formation of its complex with hormone and receptor. It is interesting that similar mechanism was previously found for other plant hormones – gibberellins, auxin, and jasmonic acid (receptor/repressor: GID1/DELLA proteins, TIR1/AUX/IAA proteins, and COI/JAZ, respectively).

Elucidation of the signaling pathway is a necessary prerequisite for modulation of its activity and/or time of stimulation. Elevation of plant tolerance to abiotic stresses may be achieved by overexpression of individual components of ABA-signaling pathway under tissue- or developmental stage-specific promoter of a suitable strength. Thus, understanding of ABA mode of action may significantly contribute to the establishment of optimal strategy for elevation of plant tolerance to abiotic stresses.

Acknowledgment This work was supported by GA CR, project no. 522/09/2058.

References

Abe H, Urao T, Ito T, Seki M, Shinozaki K, Yamaguchi-Shinozaki K (2003) Arabidopsis AtMYC2 (bHLH) and AtMYB2 (MYB) function as transcriptional activators in abscisic acid signaling. Plant Cell 15:63–78

Allan AC, Fricker MD, Ward JL, Beale MH, Trewavas AJ (1994) Two transduction pathways mediate rapid effects of abscisic acid in *Commelina* guard cells. Plant Cell 6:1319–1328

Cornforth JV, Milborrow BV, Ryback G, Wareing PF (1965) Chemistry and physiology of dormins in sycamore: Identity of sycamore dormin with abscisin II. Nature 205:1269–1270

Coursol S, Fan L, Le Stunff H, Spiegel S, Gilroy S, Assmann SM (2003) Sphingolipid signaling in *Arabidopsis* guard cells involves heterotrimeric G proteins. Nature 423:651–654

Cutler SR, Rodriguez PL, Finkelstein RR, Abrams SR (2010) Abscisic acid: emergence of a core signaling network. Annu Rev Plant Biol 61:651–679

Gong ZZ, Dong CH, Lee H, Zhu JH, Xiong LM, Gong DM, Stevenson B, Zhu JK (2005) A DEAD box RNA helicase is essential for mRNA export and important for development and stress responses in *Arabidopsis*. Plant Cell 17:256–267

Geiger D, Scherzer S, Mumm P, Stange A, Marten I, Bauer H, Ache P, Matschi S, Liese A, Al-Rasheid KAS, Romeis T, Hedrich R (2009) Activity of guard cell anion channel SLAC1 is controlled by drought-stress signalling kinase-phosphatase pair. Proc Natl Acad Sci USA 106:21425–21430

Geiger D, Scherzer S, Mumm P, Marten I, Ache P, Matschi S, Liese A, Wellmann C, Al-Rasheid KAS, Grill E, Romeis T, Hedrich R (2010) Guard cell anion channel SLAC1 is regulated by CDPK protein kinases with distinct Ca^{2+} affinities. Proc Natl Acad Sci USA 107:8023–8028

Han MH, Goud S, Song L, Fedoroff N (2004) The *Arabidopsis* double-stranded RNA-binding protein HYL1 plays a role in microRNA-mediated gene regulation. Proc Natl Acad Sci USA 101:1093–1098

Himmelbach A, Yang Y, Grill E (2003) Relay and control of abscisic acid signaling. Curr Opin Plant Biol 6:470–479

Hirayama T, Shinozaki K (2007) Perception and transduction of abscisic acid signals: keys to the function of the versatile plant hormone ABA. Trends Plant Sci 12: 343–351

Iyer LM, Koonin EV, Aravind L (2001) Adaptations of the helix-grip fold for ligand binding and catalysis in the START domain superfamily. Proteins Struct Funct Genet 43:134–144

Jang YH, Lee JH, Kim J-K (2008) Abscisic acid does not disrupt either the *Arabidopsis* FCAFY interaction or its rice counterpart in vitro. Plant Cell Physiol 49:1898–1901

Jones AM, Assmann SM (2004) Plants: the latest model system for G-protein research. EMBO Rep 5:572–578

Kizis D, Pages M (2002) Maize DRE-binding proteins DBF1 and DBF2 are involved in rab17 regulation through the drought-responsive element in an ABA-dependent pathway. Plant J 30:679–689

Klingler JP, Batelli G, Zhu J-K (2010) ABA receptors: the START of a new paradigm in phytohormone signalling. J Exp Bot 61:3199–3210

Kwak JM, Mori IC, Pei Z-M, Leonhardt N, Torres MA, Dangl JL, Bloom RE, Bode S, Jones JDG, Schroeder JI (2003) NADPH oxidase AtrbohD and AtrbohF genes function in ROS-dependent ABA signaling in *Arabidopsis*. EMBO J 22:2623–2633

Legnaioli T, Cuevas J, Mas P (2009) TOC1 functions as a molecular switch connecting the circadian clock with plant responses to drought. EMBO J 28:3745–3757

Li JX, Kinoshita T, Pandey S, Ng CK, Gygi SP, Shimazaki K, Assmann SM (2002) Modulation of an RNA-binding protein by abscisic acid-activated protein kinase. Nature 418:793–797

Liu X, Yue Y, Li B, Nie Y, Li W, Wu WH, Ma L (2007) A G protein-coupled receptor is a plasma membrane receptor for the plant hormone abscisic acid. Science 315:1712–1716

Lopez-Molina L, Mongrand S, Chua NH (2001) A post-germination developmental arrest checkpoint is mediated by abscisic acid and requires the ABI5 transcription factor in *Arabidopsis*. Proc Natl Acad Sci USA 98: 4782–4787

Lu C, Fedoroff N (2000) A mutation in the *Arabidopsis HYL1* gene encoding a dsRNA binding protein affects responses to abscisic acid, auxin, and cytokinin. Plant Cell 12:2351–2366

Ma Y, Szostkiewicz I, Korte A, Moes D, Yang Y, Christmann A, Grill E (2009) Regulators of PP2C phosphatase activity function as abscisic acid sensors. Science 324:1064–1068

Melcher K, Ng L-M, Zhou XE, Soon F-F, Xu Y, Suino-Powell KM, Park S-Y, Weiner JJ, Fujii H, Chinnusamy V, Kovach A, Li J, Wang Y, Li J, Peterson FC, Jensen DR, Yong E-L, Volkman BF, Cutler SR, Zhu J-K, Xu HR (2009) A gate-latch-lock mechanisms for hormone signalling by abscisic acid receptors. Nature 462:602–608

Meinhard M, Rodriguez PL, Grill E (2002) The sensitivity of ABI2 to hydrogen peroxide links the abscisic acid-response regulator to redox signaling. Planta 214:775–782

Mochizuki N, Brussian JA, Larkin R, Nagatani H, Chory J (2001) Arabidopsis genome uncoupled 5 (GUN5) mutant reveals the involvement of Mg-chelatase H subunit in plastid-to-nucleus signal transduction. Proc Natl Acad Sci USA 98:2053–2058

Mori IC, Murata Y, Yang YZ, Munemasa S, Wang YF, Andreoli S, Tiriac H, Alonso JM, Harper JF, Ecker JR, Kwak JM, Schroeder JI (2006) CDPKs CPK6 and CPK3 function in ABA regulation of guard cell S-type anion- and Ca^{2+}-permeable channels and stomatal closure. PLoS Biol 4:e327

Neil SJ, Desikan R, Clarke A, Hancock JT (2002) Nitric oxide is a novel component of abscisic acid signaling in stomatal guard cells. Plant Physiol 128:13–16

Nemhauser JL, Hong F, Chory J (2006) Different plant hormones regulate similar processes through largely nonoverlapping transcriptional response. Cell 126: 467–475

Nishimura N, Kitahata N, Seki M, Narusaka Y, Narusaka M, Kuromori T, Asami T, Shinozaki K, Hirayama T (2005) Analysis of ABA hypersensitive germination2 revealed the pivotal functions of PARN in stress response in *Arabidopsis*. Plant J 44:972–984

Nishimura N, Sarkeshi A, Nito K, Nito K, Park S-Y, Wang A, Carvalho PC, Lee S, Caddell DF, Cutler SR, Chory J, Yates JR, Schroeder JI (2010) PYR/PYL/RCAR family members are major *in vivo* ABI1 protein phosphatase 2C-interacting proteins in *Arabidopsis*. Plant J 61:290–299

Nott A, Jung HS, Koussevitzky S, Chory J (2006) Plastid-to-nucleus retrograde signalling. Annu Rev Plant Biol 57:739–759

Ohta M, Guo Y, Halfter U, Zhu JK (2003) A novel domain in the protein kinase SOS2 mediates interaction with the protein phosphatase 2C ABI1. Proc Natl Acad Sci USA 100:11771–11776

Pandey S, Chen JG, Jones AM, Assmann SM (2006) G protein complex mutants are hypersensitive to abscisic acid regulation of germination and postgermination development. Plant Physiol 141:243–256

Pandey S, Nelson DC, Assmann SM (2009) Two novel GPCR-type G proteins are abscisic acid receptors in Arabidopsis. Cell 136:136–148

Papp I, Mur LA, Dalmadi A, Dulai S, Koncz C (2004) A mutation in the *Cap Binding Protein 20* gene confers drought tolerance to *Arabidopsis*. Plant Mol Biol 55:679–686

Park S-Y, Fung P, Nishimura N, Jensen DR, Fujii H, Zhao Y, Lumba S, Santiago J, Rodrigues A, Chow TFF, Alfred SE, Bonetta D, Finkelstein R, Provart NJ, Desveaux D, Rodriguez PL, McCourt P, Zhu JK, Schroeder JI, Volkman BF, Cutler SR (2009) PYR/PYL/RCAR family members are major in-vivo ABI1 protein phosphatase 2C-interacting proteins in *Arabidopsis*. Science 324:1068–1071

Pei ZM, Murata Y, Benning G, Tomine S, Klusener B, Allen GJ, Grill E, Schroeder JI (2000) Calcium channels activated by hydrogen peroxide mediate abscisic acid signaling in guard cells. Nature 406:731–734

Pernas M, Garcia-Casado G, Rojo E, Solano R, Sanchez-Solano JJ (2007) A protein phosphatase 2A catalytic subunit is a negative regulator of abscisic acid signaling. Plant J 51:763–778

Radauer C, Lackner P, Breiteneder H (2008) The Bet v1 fold: an ancient, versatile scaffold for binding of large, hydrophobic ligands. BMC Evol Biol 8:286

Raghavendra AS, Gonugunta VK, Christmann A, Grill E (2010) ABA perception and signalling. Trends Plant Sci 15:395–401

Razem FA, Luo M, Liu JH, Abrams SR, Hill RD (2004) Purification and characterization of a barley aleurone abscisic acid-binding protein. J Biol Chem 279: 9922–9929

Razem FA, El-Keremy A, Abrams SR, Hill RD (2006) The RNA-binding protein FCA is an abscisic acid receptor. Nature 439:290–294

Saez A, Rodrigues A, Santiago J, Rubies S, Rodriguez PL (2008) HAB1-SWI3B interaction reveals a link between abscisic acid signaling and putative SWI/SNF chromatin remodeling complexes in *Arabidopsis*. Plant Cell 20:2972–2988

Santiago J, Dupeux F, Round A, Antoni R, Park S-Y, Jamin M, Cutler SR, Rodriguez PL, Marquez JA (2009a) The abscisic acid receptor PYR1 in complex with abscisic acid. Nature 462:665–669

Santiago J, Rodrigues A, Saez A, Rubio S, Antoni R, Dupeux F, Park S-Y, Marquez JA, Cutler SR, Rodriguez PL (2009b) Modulation of drought resistance by the abscisic acid receptor PYL5 through inhibition of clade A PP2Cs. Plant J 60:575–588

Sato A, Sato Y, Fukao Y, Fujiwara M, Umezawa T, Shinozaki K, Hibi T, Taniguchi M, Miyake H, Goto DB, Uozumi N (2009) Threonine at position 306 of the KAT1 potassium channel is essential for channel activity and is a target site for ABA-activated SnRK2/OST1/SnRK2.6 protein kinase. Biochem J 424:439–448

Schroeder JI, Kwak JM, Allen GJ (2001) Guard cell abscisic acid signalling and engineering drought hardiness in plants. Nature 410:711–714

Schultz TF, Quatrano RS (1997) Evidence for surface perception of abscisic acid by rice suspension cells as assayed by Em gene expression. Plant Sci 130:63–71

Shang Y, Yan L, Liu ZQ, Cao Z, Mei C, Xin Q, Wu FQ, Wang XF, Du SY, Jiang T, Zhang XF, Zhao R, Sun HL, Liu R, Yu YT, Zhang DP (2010) The Mg-chelatase H subunit of *Arabidopsis* antagonizes a group of WRKY transcription repressors to relieve ABA-responsive genes of inhibition. Plant Cell 22:1909–1935

Shen YY, Wang XF, Wu FQ, Du SY, Cao Z, Shang Y, Wang XL, Peng CC, Yu XC, Zhu SY, Fan RC, Xu YH, Zhang DP (2006) The Mg-chelatase H subunit as an abscisic acid receptor. Nature 443:823–826

Shukla LI, Chinnusamy V, Sunkar R (2008) The role of microRNAs and other endogenous small RNAs in plant stress responses. Biochim Biophys Acta-Gene Regul Mech 1779:743–748

Sokol A, Kwiatkowska A, Jerzmanowski A, Prymakowska-Bosak M (2007) Up-regulation of stress-inducible genes in tobacco and *Arabidopsis* cells in response to abiotic stresses and ABA treatment correlates with dynamic changes in histone H3 and H4 modifications. Planta 227:245–254

Srivastava N, Gonugunta VK, Mallikarjuna RP, Raghavendra AS (2009) Nitric oxide production occurs downstream of reactive oxygen species in guard cells during stomatal closure induced by chitosan in abaxial epidermis of *Pisum sativum*. Planta 229:757–765

Szostkiewicz I, Richter K, Kepka M, Demmel S, Ma Y, Korte A, Asaad FF, Christmannn A, Grill E (2010) Closely related receptor complexes differ in their ABA selectivity and sensitivity. Plant J 61:25–35

Umezawa T, Sugiyama N, Mizoguchi M, Hayashi S, Myouga F, Yamaguchi-Shinozaki K, Ishihama Y, Hirayama T, Shinozaki K (2009) Type 2C protein phosphatases directly regulate abscisic acid-activated protein kinases in *Arabidopsis*. Proc Natl Acad Sci USA 106:17588–17593

Vlad F, Rubio S, Rodrigues A, Sirichandra C, Belin C, Robert N, Leung J, Rodriguez PL, Lauriere C, Merlot S (2009) Protein phosphatases 2C regulate the activation of the Snf1-related kinase OST1 by abscisic acid in *Arabidopsis*. Plant Cell 21:3170–3184

Wang XQ, Ullah H, Jones AM, Assmann SM (2001) G protein regulation of ion channels and abscisic acid signaling in *Arabidopsis* guard cells. Science 292(5524):2070–2072

Wang X-F, Zhang D-P (2008) Abscisic acid receptors: multiple signal-perception sites. Ann Bot 101:311–317

Worrall D, Liang YK, Alvarez S, Holroyd GH, Spiegel S, Panagopulos M, Gray JE, Hetherington AM (2008) Involvement of sphingosine kinase in plant cell signalling. Plant J 56:64–72

Wu FQ, Xin Q, Cao Z, Liu ZQ, Du SY, Mei C, Zhao CX, Wang XF, Shang Y, Jiang T, Zhang XF, Yan L, Zhao R, Cui ZN, Liu R, Sun HL, Yang XL, Su Z, Zhang DP (2009) The Magnesium-chelatase H subunit binds abscisic acid and functions in abscisic acid signaling: new evidence in *Arabidopsis*. Plant Physiol 150:1940–1954

Xiong LM, Lee BH, Ishitani M, Lee H, Zhang C, Zhu JK (2001) FIERY1 encoding an inositol polyphosphate 1-phosphatase is a negative regulator of abscisic acid and stress signaling in *Arabidopsis*. Genes Dev 15: 1971–1984

Zhang DP, Wu ZY, Li XY, Zhao ZX (2002) Purification and identification of a 42-kilodalton abscisic acid-specific-binding protein from epidermis of broad bean leaves. Plant Physiol 128:714–725

Zhang W, Yu L, Zhang Y, Wang X (2005) Phospholipase D in the signaling networks of plant response to abscisic acid and reactive oxygen species. Biochem Biophys Acta 1736:1–9

Plant Tolerance and Fatty Acid Profile in Responses to Heavy Metals

20

Asiya Hameed, Tabasum N. Qadri, Mahmooduzzafar, and T.O. Siddiqi

Abstract

Heavy metals are more widespread around the world and dangerous for biosphere because they cannot be degraded or destroyed rather tend to be bioaccumulated. Plants can survive even in the extreme environmental conditions, but some environmental factors can affect its various growth aspects and hence the plant productivity. The problem of heavy metal toxicity is further aggravated by the persistence of the metals in the environment. Toxic heavy metals entering the plant tissues inhibit most physiological processes at all levels of metabolism. The extent of inhibition of photosynthesis, ion water uptake, and nitrate assimilation is greatly dependent on the concentration of the metal ions, sensitivity, and tolerance of the plant. There is, therefore, a pressing need to deal with the problem of excess metal already present in the soil and to prevent future contamination.

Keywords

Heavy metals • Toxicity • Contamination • Tolerance • Accumulation • Fatty acids

1 Introduction

Among environmental pollutions, global heavy metal pollution has become a serious environmental concern. Existence of heavy metal in the soil, water, and atmosphere, even at relatively low concentrations, can cause serious risk to all living organisms. Every year, due to an increase of anthropogenic activities, environmental pollution by heavy metal is becoming an important global health concern for human beings as well as for plants. They are toxic to all living organisms even at relatively low concentrations. On the contrary, metals are natural components of all the ecosystems and some of them as micronutrients (e.g., Cu, Fe, Mn, Zn, Mo) are essential for plant growth and yield (Marschner 1983; Purves 1985). Fine particles have high specific area that retains high amounts of metals (Silvia Martinez-Martinez et al. 2010). Elevated amounts of heavy metal contaminants, resulting from industrialization and urbanization, are of serious environmental con-

A. Hameed (✉) • T.N. Qadri • Mahmooduzzafar • T.O. Siddiqi
Department of Botany, Jamia Hamdard, Hamdard Nagar, New Delhi 110062, India
e-mail: hameed901@gmail.com

P. Ahmad and M.N.V. Prasad (eds.), *Abiotic Stress Responses in Plants: Metabolism, Productivity and Sustainability*, DOI 10.1007/978-1-4614-0634-1_20, © Springer Science+Business Media, LLC 2012

cern. Landfill materials and their toxic leachates enrich in heavy metals and other toxic substances (Griffin et al. 1997; Adriano 1986) affect ground water quality and defile soils and aquifers (Menzer and Nelson 1986; Keswick 1984).

Heavy metals are any metallic chemical elements with a relatively high density, i.e., >5–6 $g\,cm^{-3}$ or specific gravity of water (Davies 1980). There are 23 heavy metals. The most common heavy metal pollutants include Cd, Cr, Cu, Pb, Hg, Ni, As, and Zn.

Industrialization has made the environment harsh for plant survival by adding numerous air pollutants into the atmosphere (Oleksyn and Innes 2000) causing stressful conditions for plants. In a broad sense, plant stress can be defined as "any unfavorable conditions or substance that affects or blocks a plant's metabolism, growth or development" (Lichtenthaler 1996). Low stress plant responses can partially be overcome by acclimation and repair while strong or chronic stress effects may cause irreversible damage and cell death (Lichtenthaler 1996). Factors that can cause stress to plants can be classified into seven main classes (Elstner and Osswald 1994) such as light (Low intensity, high intensity), radiation (UV, α, β, γ, X-ray), temperature (high, low, freezing, chilling), hydration (Drought, flooding), chemical (Salts, heavy metals, pH, O_3, SO_2, H_2S, NO_x, NH_3, HF), mechanical (Wind, lightening, fire, snow, cutting, biting, pressing), and biological influence (Flowering, fruit ripening, insects, infections, allelopathy, competition). In most of these stresses including the heavy metal and salinity stress, active oxygen species (AOS) are involved in deleterious processes as central signaling components (Noctor and Foyer 1998).

Today, approximately 60% of the cultivated soils have globally mineral problems, such as metal toxicities along with nutritional and metal deficiencies (Cakmak 2002). Certain heavy metals are natural components of our environment and soils in small quantities and thus necessary for good health, but in excessive level they can be usually poisonous and hazardous to plants, animals, and man. In order to regulate the environmental pollution, the background heavy metal levels of soils and their critical concentrations both in soils and plants are to be set.

Heavy metal contamination results not only from industrial activities such as mining and smelting of metalliferous ores and oil, energy and fuel production, municipal waste products and agricultural chemical application but also from different types of weapon tests or wars. There are several reports showing a severe environmental pollution with metals and other pollutants after or during Second World War, gulf war in 1991, and various other wars (Moniri 2005).

A risk of heavy metal danger is redoubled due to their potential to speed away in thousands of kilometers from their source of release. Heavy metal enter the human bodies via food chain, including herbal drugs, drinking water and air or absorption through the skin as a result of agricultural, industrial and pharmaceutical works or residential activities. These elements are toxic when they are not metabolized by the body and get accumulated and stored in tissues. Heavy metals are carcinogens and mutagens and their toxicity can result in damage or reduction of mental and central nervous function, lower energy levels and damage to blood composition, lungs, kidneys, liver, and other vital organs. Long-term exposure may result in the slowly progressing physical, muscular and neurological degenerative processes that mimic Alzheimer's and Parkinson disease, muscular dystrophy, and sclerosis (Chaney et al. 1999; IOSHIC 1999; Ferner 2001).

Soil is the unconsolidated mineral material on the immediate surface of the earth that serves as a natural medium for plant growth. The composition of the soil at a contaminated site can be extremely diverse and the available heavy metals can exist as components of several different fractions (Salt et al. 1995b). Some conditions or factors which affect plant growth for higher crop production is related to soil's physical, chemical, and biological properties. These factors directly or indirectly affect plant root growth, absorption of water and nutrients, and consequently plant growth and yields.

Microorganisms are generally the first to be affected by discharges of heavy metals into the environment (Wood 1989). These microbial communities are affected by their changes in total viable counts, shifts in the balance of species present or alteration of the metabolic characteristics of the community (Sterritt and Lester 1980). The interaction between heavy metals and humic substrates has been characterized mainly as chelation, complexation and adsorption. Bacteria, cyanobacteria, and fungi can alter the form of occurence of metals through methylation, chelation, complexation, catalysis, or absorption, thus affecting the bioavailability of metals in both aquatic and soil systems and these processes affect the movement of the metal up the food chain. Organisms which produce hydrogen sulfide often exhibit a tolerance to heavy metals because this compound binds with the metal to form insoluble sulfides. This mechanism may also afford protection to non-hydrogen-sulfide-producing microorganisms in the surrounding environment. The production of other extracellular products such as glutathione may also reduce metal toxicity. The production of other extracellular nitrogenous compounds by the heterocystous cyanobacteria reduces the toxicity of mercury (Rath et al. 1986). These play an important role in altering the environmental conditions.

The entry of contaminants into the environment results either from natural processes or human activity. Natural contamination originates from either excessive withering of mineral and metal ions from rocks or from displacement of certain contaminants from the ground water. The heavy metals presence particularly in the agricultural and nonagricultural lands is because of;

- Disposal of industrial effluents
- Sewage sludges
- Deposition of air-borne industrial wastes
- Military operations
- Mining
- Landfill operations
- Industrial solid waste disposal, and
- Use of agricultural chemicals such as pesticides, herbicides, and fertilizers.

Atmospheric pollution from motor vehicles involves the use of leaded petrol for the global dispersion of Pb aerosols. Combustion of fossil fuels results in the dispersion of many elements in the air over a large area. The disposal of ash is a further source of heavy metals. Agricultural fertilizers and pesticides including phosphatic fertilizers, slags from iron manufacture, pesticides, and herbicides contain various combinations of heavy metals either as impurities or active constituents.

Organic manures include pig and poultry manures may contain high concentrations of Cu, as it is fed to improve food conversion efficiency. Sewage sludges usually contains relatively high concentrations of several metals especially those from industrial catchments.

According to the agency for Toxic substances and disease registry (ATSDR) of the U.S. Department of Health and Human Services, mercury is listed as the third-most frequently found (Pb and Ar are first and second) and the most toxic substance in the United States (Anonymous 2001). Annual worldwide emissions of mercury into the atmosphere have been estimated at 2,200 metric tons (Ferrara et al. 2000). One-third of these emissions are estimated to originate from natural sources (volcanic eruptions and decay of mercury-containing sediment) and two thirds from man-made sources. Twenty-five percent of the total worldwide emissions come from fossil fuel combustion.

2 Distribution of Heavy Metals in Soil

Distribution and accessibility of heavy metals to plants is important while assessing the environmental quality of an area. The levels of the heavy metals cadmium (Cd), copper (Cu), lead (Pb), manganese (Mn), nickel (Ni), and zinc (Zn) in the agricultural soils of the Bursa plain have been shown in the Table 20.1 to determine the degree of pollution and was found generally higher than the levels reported in literature for similar soils, suggesting some degree of pollution with heavy metals. The exchangeable forms of the heavy metals, though very low, indicates that the availability of heavy metals to plants is at a minimum (Aydinalp and Marinova 2003).

Table 20.1 Average and range values of total heavy metal content (g kg^{-1}) of all samplings and related background values from different sources

Metal	Values of samplings		Background values from different sources: Bowen (1979)	Shacklette and Boerngen (1984)	Vinogradov (1959)	Rose et al. (1979)	Mitchell (1964)
Cd	Min.	0.1	0.01	–	–	–	–
	Max.	8.7	2.00	–	–	–	–
	Aver.	2	0.35	–	–	–	–
Cr	Min.	42.0	5	1	–	–	5
	Max.	329.2	1,500	2,000	–	–	3,000
	Aver.	124.5	70	54	200	6.3	–
Cu	Min.	14.2	2	1	–	–	10
	Max.	68.9	250	700	–	–	100
	Aver.	40	30	25	20	15	–
Mn	Min.	587.9	20	2	–	–	200
	Max.	3,112	10,000	7,000	–	–	5,000
	Aver.	1,667.1	1,000	550	850	320	–
Ni	Min.	54.6	2	5	–	–	10
	Max.	378.0	750	700	–	–	800
	Aver.	157.8	50	19	40	17	–
Pb	Min.	33.4	2	10	–	–	20
	Max.	163.3	300	700	–	–	8
	Aver.	80.9	35	19	–	17	–
Zn	Min.	187.9	1	5	–	–	–
	Max.	1,087.0	900	2,900	–	–	–
	Aver.	476.7	90	60	50	36	–

Several different soil samples have been investigated for main soil fertility characteristics (pH, humus, available K and P) as well as concentrations of selected heavy metals (As, Cd, Cr, Ni, and Pb).

Soils graded as very acidic cover 46% of the area, which are mainly mountains with acidic parent materials. Content of humus in 41% of soil samples were below 3%. Most of the soils (71%) are weakly supplied available phosphorus, while available potassium in more than 70% is presented in the concentrations enough for good soil quality. As a result, 75% of studied area is characterized with unfavorable soil fertility properties (extremely low soil pH, very low content of available P, about half of the area maintained low soil humus) located under forests, meadows and pastures. Content of heavy metals on studied area in 80% of sampled soils was below maximum allowed concentrations (Saljnikov et al. 2009). However, 19% of samples showed contamination with one or other toxic elements, among which one was the most often pollutant. Contaminated sites are the results of both, geochemical composition of the area and anthropogenic pollution.

Hence the consequences of the status of soil fertility and the level of soil pollution with selected heavy metals builds the foundation for further detailed investigations of effects of higher concentrations of pollutants on plants and other components of biosphere, which in turn would help in finding measures for amelioration and/or prevention of eventual negative consequences.

Agricultural soils contain Hg levels between 0.06 and 0.2 mg/kg, and some edible plants such as carrots, potatoes, and mushrooms have been reported to take up mercury compounds. Some accumulation has been observed in mushrooms, aquatic plants, carrots, and potatoes. Ferrara et al. (1989) observed that the aquatic *Posidonia oceanica* could be a biological indicator for mercury in sediments. Wyttenbach et al. (1989) found that Hg concentrations in spruce needles increased continuously with age.

The metal species commonly found in the soil as a result of the aforementioned human activities include Cu, Pb, Zn, Ni, Co, Hg, and Cd.

Although some of these metals are required in small amounts by living organisms for their normal physiological activities, excessive accumulation is toxic to most life forms

3 Measuring Soil's Heavy Metal Concentrations

The magnitude of heavy metal toxicity is affected by their binding to various ligands (SH- group, carboxylate ion, imidazole, aliphatic amine). Merely a binding to SH-group can lead to multiple disturbances in the metabolism through the reduction of activities of more than hundreds enzymes having a SH-group in their protein molecule (Van Assche and Clijsters 1990; Seregin and Kojevnikova 2006). Differences in an action of metals with the similar properties are defined mainly by the differences in a rate of ligand exchange, but a physiological role of metals directly depends on the physical and chemical properties of their ions (Seregin and Ivanov 2001; Seregin et al. 2003; Ivanov et al. 2003; Seregin and Kojevnikova 2006).

At high concentrations, heavy metals in soil can be phytotoxic to sensitive plants and directly and indirectly affect the plant growth and metabolism, damage various developmental and biochemical parameters, including enzyme activities, inhibit photosynthesis, respiration, and transpiration, change the secondary metabolism, and cause the degeneration of main cell organelles. Their toxic action on plant metabolism is mainly nonspecific (Ernst 1998; Chaney et al. 1999; Ivanov et al. 2003; Gwozdz and Kopyra 2003; Seregin and Kojevnikova 2006). However, different metals have an unequal degree of toxicity to the living organisms. In particular, plants vary in their phytotoxic responses depending on a number of soil factors such as clay and organic matter content, nutritional status, mineral and soil solution composition, pH and redox potential, moisture degree (Ross 1994; Kaschl et al. 2002; Naidu et al. 2003), plant type and their sorptive capacity (Lasat 2000; Mantovi et al. 2003; Al-Najar et al. 2003; Aijen 2004).

To provide a relatively clean environment for growing crop plants is of great importance, because at the end of the food chains are humans. Certain heavy metals, such as Cu, Zn, Mn, Mo are essential for plants, while Fe can be essential and toxic, depending on its amount and charge. The contradiction in the handling of heavy metals is that a part of these are essential, while the other part is toxic: they are sometimes essential for plants, yet toxic to humans.

As growth is a complex process, therefore heavy metals can affect it at several points. The toxic effect of heavy metals is obvious at different levels of cell structures and the function of the plants. Reduction of growth and productivity, changes in membrane structure, enzyme activity, metabolic processes, water and ion uptake, as well as membrane permeability, are the most studied fields (Masarovicova et al. 1999).

In soils, Hg level about 2 mg/kg is toxic. In plants, the Hg content of 1–8 mg/kg vegetative dry matter is associated with yield decrease (Macnicol and Beckett 1985). For animals, the threshold level of Hg toxicity is much lower than that for plants. Hg level about 1 mg/kg dry matter in the diet is toxic to animals.

Cadmium can be transported readily from the soil via the plant root to the upper plant parts. The particular hazard with cadmium is that the plant does not necessarily act as an indicator of levels toxic to humans and animals, because plants tolerate higher levels of cadmium than do animals. The same is true for mercury. Plants can appear healthy but may contain a high concentration of cadmium and mercury, which are completely unacceptable in an animal and human diet (Mengel and Kirkby 2001).

Root growth affects the properties of the rhizospheric soil and stimulates the growth of the microbial consortium. Research has shown that the population of microorganisms in the rhizosphere is several orders of magnitude greater than in the surrounding soil (Anderson 1997). In turn, rhizospheric microorganisms may interact symbiotically with roots to enhance the potential for metal uptake. In addition, some microorganisms may excrete organic compounds which increase bioavailability and facilitate root absorption of essential metals, such as Fe (Crowley et al. 1991) and Mn (Barber and Lee 1974), as well as nonessential metals, such as Cd (Salt et al. 1995a). Soil microorganisms can also directly influence metal solubility by altering their chemical properties. For example, a strain of *Pseudomonas maltophilia* was shown to reduce the mobile and toxic Cr^{6+}, to nontoxic and immobile Cr^{3+}, and also to minimize environmental mobility of other toxic ions such as Hg^{2+}, Pb^{2+}, and Cd^{2+} (Blake et al. 1993; Park et al. 1999). In addition, it has been estimated that microbial reduction of Hg^{2+} generates a significant fraction of global atmospheric Hg^0 emissions (Keating et al. 1997).

4 Effect of Heavy Metals on Plants

The presence of metals in the leaves can hamper plant metabolism and growth even at very low concentrations. The oldest leaves of metal exposed plants exhibit the highest metal content (Ernst 1998; Mills and Scoggins 1998). Under metal stress, plants exhibit a thickening of root tips and a decreased root hair density (Punz and Sieghardt 1992). The reduction in growth and yield depends on the type of metal ions, types of growth media and species, and the stage of plant growth (Woolhouse 1983; Lata 1989a,b). The heavy metal caused inhibition in growth leads to reductions in dry matter (Pahlsson 1989), leaf area and root length (Bhattacharyya and Choudhuri 1995), and yield (Setia et al. 1987). The major effect of Cd^{2+} is on root growth, followed by leaf growth (Di Cagno et al. 1999). It inhibited seed germination and seedling growth in *Triticum aestivum* (Yadav and Yadav 1995). The plant growth is affected not only due to the toxic effect of Cd but also due to a nutritional imbalance created by Cd in the plant (Cordovilla et al. 1996; Sanita di-Toppi and Gabbrielli 1999).

The rate of photosynthesis is often regarded as a major factor regulating the crop productivity. It is the basis of crop yield and provides 90–95% of the plant dry weight, i.e., economic yield (Leuning et al. 1995; Hall and Rao 1999). Leaves are the primary site of photosynthesis and the products of their metabolism fuel the growth of all plant organs. As leaves become older, their photosynthetic capacity decreases (Marschner 1986; Long et al. 1996). Anatomical changes in leaves and structural disorganization of chloroplasts are correlated with the inhibition of photosynthesis (Terashima and Evans 1988). The leaf injuries such as chlorosis, necrosis, browning, and burning adversely affect the physiological functions especially leaf photosynthetic activity (Jafri et al. 1979). The photosynthetic efficiency depends on leaf area, chlorophyll content, and the stomatal response (Pachepsky et al. 1997). Leaves are the highly exposed organs of plants and are affected most by the environmental stresses. The rates of photosynthesis and stomatal conductance are used as early detectors of the potential stress injury to plants. The exact mechanism of the heavy metal action on photosynthesis is, however, still not clearly understood (Krupa et al. 1987; Bhardwaj and Mascarenhas 1989).

Heavy metals decrease leaf expansion, resulting in a more compact leaf structure and increased stomatal resistance (Horvath et al. 1996). They may impair leaf transpiration and CO_2 fixation by decreasing leaf conductance to CO_2 diffusion as a result of stomatal closure (Barcelo et al. 1988). Heavy metals in growth media can function as stressors causing physiological constraints that suppress plant vigor and inhibit plant growth. Heavy metals can inhibit photosynthesis of intact plants at several physiological levels: stomata, pigment synthesis, chloroplast structure and function, and indirectly by affecting various other metabolic pathways (Costa and Spitz 1997). Their treatments inhibit net photosynthesis in various crop plants such as corn and soybean (Bazzaz et al. 1974); tomato (Baszynski et al. 1980), and wheat (Setia et al. 1987). Reduction in chlorophyll content may also be due to the interference of all the metals with chlorophyll synthesis and fat metabolism, inhibiting root shoot growth, photosynthesis, nutrient uptake, leaf area, etc. (Pandey and Tripathi 2011).

Disturbances in plant water relations are widely known as one of the first effect of Cd toxicity (Poschenrieder et al. 1989). In vivo studies have shown that leaves of plants exposed to Cd^{2+} accumulate high amounts of cadmium. Long-term exposure of whole plant to Cd affects chlorophyll with consequences for chloroplast development in young leaves. Cadmium may affect photosynthesis (Greger and Ogren 1991) by altering the chlorophyll content and/or stomatal conductance (Prasad 1995).

Photosynthesis is a very sensitive indication system for toxicity (De Filippis and Pallaghy 1976; Greenfield 1942). Mercury has been shown to interfere with photosynthetic electron transfer mechanisms (Bradeen et al. 1973; Cedeno-Maldonado et al. 1972). Ultimately, mercury causes damage to photosynthetic pigments (Greenfield 1942; Puckett 1976).

Carotenoids are the secondary light absorbing pigments called the accessory pigments. They provide essential photoprotective mechanisms, blocking the formation of ROS (Young and Britton 1990). Carotenoid is less affected (Clijsters and Van Assche 1985) or is generally increased by the heavy metal exposure (Foyer and Harbinson 1994; Ralph and Burchertt 1998). In the green alga *Chlorella vulgaris*, the enzyme protochlorophyllide reductase was found to be inhibited in the presence of sublethal concentration of Hg, which resulted in the reduction of chlorophyll biosynthesis and accumulation of protochlorophyll (De Filippis and Pallaghy 1976).

Proteins are the most abundant molecules in the cell, making up more of the dry weight. They are found in all cellular components forming the basis of the cell structure and function. Each kind of protein is specialized for its biological function. They operate as enzymes, transporting and regulatory proteins and also serve as structure and storage of compounds. Significant alteration in protein metabolism under heavy metal stress has been reported by a number of workers. Some of them observed an increase in protein synthesis (Shah and Dubey 1997) while others observed a decrease (Costa and Spitz 1997). In majority of plant cells, proteins are synthesized in cytoplasmic compartments. Most proteins have a lifetime less than that of cell and therefore are degraded and, if necessary, resynthesized (Nozaki 1986; Nwokolo and Smartt 1996).

Plants appear to contain a diversity of metal binding metallothioneins (MTs) with the potential to perform distinct roles in the metabolism of different metal ions. The change in the biochemical characterization for metal tolerance involved the de novo synthesis of metal binding proteins. Lue-kim and Rauser (1986) reported an induction of Cd-binding protein from crude extracts of roots of tomato with an apparent molecular weight 31,000 Da in high ionic strength and 21,500 Da at low ionic strength. Increase in soluble protein was also reported by Vogeli-Lange and Wagner (1990) in tobacco leaves on Cd exposure at a concentration of 20 μm. Lozano-Rodriguez et al. (1997) observed an increase in the soluble protein content of pea root and shoot upon treatment of Cd at a concentration of 0.05 mM, whereas the same had no effect on maize. Ali et al. (1998) have observed an increase in the protein content of *Bacopa monniera* plantlets on exposure to Cd stress. Hirt et al. (1989) reported stimulation of the protein and RNA synthesis in suspension cells of *Nicotiana tabacum* on exposure to Cd stress at a concentration of 100 μM. They observed that the increase in protein content was probably due to the synthesis of new proteins to detoxify the intracellular Cd by binding with the same and rendering the internal concentration of free Cd low enough to minimize the toxic effect and allow stimulation of RNA synthesis.

Gil et al. (1995) reported that total soluble protein as well as Rubisco decreased with time at Cd concentrations of 15 and 30 mg/L. Rubisco constitutes more than 50% of the leaf soluble protein and is the key enzyme in photosynthesis (Woolhouse 1974); hence, any decline in leaf soluble protein including Rubisco will have an adverse impact on Rubisco activity and ultimately on photosynthesis. Kevreson et al. (1998) observed a decrease in total soluble protein and Rubisco activity in sugar beet plant with decreasing leaf water status and generation of ROS.

Once NO_3 is absorbed by root, it can be assimilated in the root itself, transported to the shoot or stored in vacuoles (Srivasankar and oaks 1996). Assimilation of nitrate reaction takes place in the cytoplasm of cells in both roots and shoots. The uptake and assimilation of nitrate are regulated mainly by an enzyme, nitrate reductase (NR), which plays a rate-limiting role in plant metabolism (Galvan et al. 1992; Khan 1996). This complex enzyme is substrate inducible and predominantly utilizes NADH as a co-factor (Selvaraj et al. 1995). Nitrite (NO^{-2}) is very toxic to plants, but because the activity of nitrite reductase (NIR) is normally higher than that of NR, normally nitrite does not accumulate and is rapidly converted into NH_4^+.

NR was found to be most sensitive cytosolic enzyme, while peroxidase was the most resistant (Ernst 1998). In leaves, NR is activated by photosynthesis, reaching the activation state of 60–80%. In the dark, or after stomatal closure, leaf NR is inactivated down to 20–40% of its maximum activity (Ahmad and Abdin 1999; Kaiser et al. 1999). NR is very sensitive to heavy metal stress and any change in this enzyme affects nitrogen assimilation pathways and thus growth (Hemalatha et al. 1997; Hall and Rao 1999). Reduced NR activity has been reported in many heavy metal treated plants such as *Glycine max*, *Zea mays*, and *Pisum sativum* (Chugh et al. 1992). Percent inhibition in NR activity by toxic metals increased with increasing metal concentrations (Vyas and Puranik 1993). NR shows a considerable variation in response to metal ions, which are species and cultivar specific (Bharti and Singh 1993; Dabas and Singh 1995).

Amino acid catabolism in plants is generally concerned with the production of metabolites for other biosynthetic pathway. They serve as precursor of many kinds of small molecules such as glutathione and proline that have important and diverse biological roles. Increase in amino acid pool was observed by heavy metal stress in maize germinating seed (Nagoor 1999).

Proline is an imino acid with aliphatic side chain, but differs from other members of the set of 20 amino acids in that its side chain is bonded to both the nitrogen and the α-carbon atoms. The resulting cyclic structure markedly influences protein architecture. Usually, glutamate is the precursor of proline. Proline a total free amino-acid accumulated in plants when they experience moisture stress conditions and decrease on release of stress (Pandey and Tripathi 2011). It has been shown to play an important role in ameliorating such conditions as drought, salinity, and heavy metal stress (Andrade et al. 1995). It has been used as a single parameter to measure physiological dryness (Lutts et al. 1999). It oxidizes in turgid tissues rapidly and also gets affected with the duration of stress conditions (Jager and Meyer 1977). Cadmium has a strong and positive relation with proline accumulation. A number of workers reported an increase in proline content under Cd stress (Wu et al. 1995; Nagoor 1999). Wu et al. (1995) studied the impact of Cu^{2+} and Cd^{2+} on intracellular proline level in four species of algae and reported that proline accumulation is the general response of algal cells to Cd stress.

It may be argued that proline acts as a sink for nitrogenous compounds resulting from the degradation of proteins and protects cell metabolism from the harmful nitrogenous compounds (Aspinall and Paleg 1981; Yancey et al. 1982). It is often considered to be involved in stress resistance mechanisms by acting as an osmoprotectant thereby facilitating osmoregulation, protection of enzymes, stabilization of the protein synthesis machinery, and regulation of cytosolic acidity, etc (Alia and Saradhi 1991).

Heavy metal exerted specific influence on the differentiation of various tissues in the root as well as stem. Elevated concentrations of these metals induced drastic anatomical changes. Different heavy metals of supraoptimal concentrations have been shown to inhibit various metabolic processes as in plants resulting in their reduced growth and development (Bala and Setia 1990; Davies 1991). There is paucity of information on the differentiation of tissues in plants in response to heavy metal toxicity.

Differential anatomical changes in the root and stem of *Solanum melongena* have been observed (Mehindirata et al. 1999). The major responses elicited by roots to cadmium treatments appear to be caused by accumulation of these metals in the tissues. These ions seem to attack various cellular components, including cell wall and membranes, resulting in different alterations, which ultimately lead to their disorganizations, as has also been reported by Setia and Bala (1994). Accumulation of Ni and Cd occur in root tissues in toxic amounts in a number of plant species (Woolhouse 1983). The development of large intercellular spaces in root core following heavy metal treatments resembles those resulting from exposure of roots to anaerobiosis due to water logging. The formation of large intercellular spaces (aerenchyma) is an adaptive response to anaerobiosis (Erdmann et al. 1986). With the Cd concentration, root area increases; this might be due to expansion of cells and formation of air spaces in cortical regions. Insufficient supply of essential nutrients and hormones from the root adversely influences the differentiation of tissues in stem (Davies 1991; Setia and Bala 1994). Heavy metals have been shown to affect cells of cortex and pith in root and stem of plants.

A major factor limiting metal uptake into roots is slow transport from soil particles to root surfaces (Nye and Tinker 1977; Barber 1984). With the possible exception of volatile mercury for all other metals, this transport takes place in soil solution. In soil, metal solubility is restricted due to absorption to soil particles. Some of the soil binding sites are not restricted due to adsorption to soil particles. Some of the soil binding-sites are not particularly selective, e.g., they bind Cd as strong as Ca. Non specific binding occurs at clay–cation exchange sites and carboxylic groups associated with soil organic matter. Other sites are more selective and bind Cd stronger than Ca, e.g., most clay particles are covered with a thin layer of hydrous Fe, Mn, and Al oxides. These selective sites maintain the Cd activity in the soil solution at low levels (Chaney 1988).

5 Variations in Heavy Metal Tolerance in Plants

Heavy metal tolerance to plants represents the ability of particular plants to thrive under conditions that are characterized by excess of metal ions, which have toxic effect for other plants (Macnair et al. 2000). The first research on the plants' tolerance toward HM ever made dates back to the beginning of the twentieth century, when it was established that two populations of the species *Silene dioica* have different tolerance toward the excess of Cu (Ernst et al. 1992).

Environmental pollution by heavy metals (HM) represents a real ecological problem. In Bulgaria the soils contaminated by HM cover an area of about 200,000 Da (Grancharov and Popova 2003). The sustainable use of these soils can be achieved by developing various remediation phytotechnologies as well as adaptive agriculture practices (Vassilev et al. 2005).

Tolerance is based on two different strategies (1) to avoid the entry of excess HM into the plants and (2) to achieve effective intracellular detoxification. The main tolerance-related mechanisms are already wellknown, the most important of which include (1) the reduced uptake and/or accelerated excretion of HM by the cells, (2) the metal detoxification and compartmentalization, (3) the control of the metal induced oxidative stress, etc. (Vassilev and Nikolova 2010). The scientific interest with respect to the plant tolerance toward HM has become considerably larger in recent years. On the one hand, this is due to possible usage of tolerant plants for phytoremediation of soils contaminated by HM (Kulakov et al. 2009), and on the other hand, the interest is a result of the possible wider usage of the plants as model objects for ecotoxicological studies (Hock and Elstner 2005). The number of research papers related to the identification of plants that have high tolerance and hyperaccumulative abilities toward HM are constantly increasing (Schulze et al. 2005).

Heavy metal hyperaccumulator plant ecotypes, so-called metallophytes, distinguish from others by their evolving genetically based metal resistance due to their specific detoxification mechanisms. However, there is not a common exact opinion on these mechanisms to date. Several mechanisms of detoxification and inactivation of metals taken up have been proposed (Ernst 2005): compartmentation in physiologically low active compartments of cells and organs (Brooks 1998); binding to the cell wall and sequestration by phytochelatins (Keltjens and Van Beusichem 1998; Seregin and Kojevnikova 2006), metallothioneins (Burdin and Polyakova 1987), intracellular molecules (Seregin and Ivanov, 2001), and low-molecular-weight organic acids (Salt et al. 1999), and their precipitation in the vacuoles (Van Steveninck et al. 1990). The increase of some antioxidant enzyme activities (peroxidase, SOD, catalase) (Schickler and Caspi 1999; Seregin and Ivanov 2001; Guo et al. 2004) and synthesis of osmolytes (Seregin and Kojevnikova 2006) have been proposed to play a role in the resistance mechanisms of tolerant plants. It is significant to note that mechanisms of both a metal toxic action and their detoxification are complex processes and do not become formed by only one mechanism. Plants respond to heavy metal by a number of parallel and/or consecutive processes at molecular, physiological, and morphological levels.

Methods are being employed for the remediation of mercury polluted soils. It involves the use of transgenic plants encoding the bacterial mercury ion reductase (merA) gene. These plants have been shown to grow in and volatilize mercury from soils (Meagher et al. 2000). Microorganisms are manipulated genetically to remove not only mercury but also other toxic elements from the environment (Wood and Wang 1983). It also involves the use of sulfur-containing solutions, as ammonium-thiosulfate, to induce mercury accumulation into aboveground tissues of high-biomass plant species (Moreno et al. 2004). In the latter system, mercury accumulates in the plant causing the plant to die. However, the mercury-laden plant, including roots, can be removed from the soil, thereby allowing mercury removal from the polluted soil. Root mercury accumulation and root area and length are related (Cocking et al. 1995; Heeraman et al. 2001). Plants with large root system, therefore are desirable for removal of mercury in contaminated soils. When the nutrient availability is less than that required for the optimum growth conditions plants suffer nutrient deficiency stress. This results in an inherently low nutrient status of the soil and low mobility of nutrients within the soil. The mobility of nutrients within the soil is governed by a number of factors including mass flow of water, adsorption capacity of the soil, and soil pH. The chemical form of nutrients within the soil also determines the extent of availability. In this way plants have adopted to nutrient deficiency stress, some of which are morphological adaptation to increase the ability of the plant to take up nutrients, such as cluster roots. Plants may also release chemical compounds into the soil environment to increase the efficiency at which nutrients are taken up or increase the number of soil nutrient pools available for uptake. Plants have evolved a mechanism to alleviate nutritional stress by symbiotically associating with microorganisms, such as legumes with Rhizobium and most terrestrial plants with mycorrhizal fungi.

Heavy metals can be removed from polluted sites by phytoextraction, which is a method of phytoremediation and involves the accumulation of pollutants in plant biomass (Zayed et al. 1998). As a result hyperaccumulators (plant species that accumulate extremely high concentrations of heavy metals in their shoots) become particularly useful. In addition, one can genetically engineer these species to improve their metal tolerance and metal-accumulating capacity. A suitable target species for this strategy is Indian mustard (*Brassica juncea*), which has a large biomass production and a relatively high trace element accumulation capacity. Most importantly, it can easily be genetically engineered (Zhu et al. 1999).

Most plants fail to maintain metal homeostasis and develop stress symptoms, when exposed to elevated concentration of micronutrient metals or heavy metals without nutritional functions. Metal-tolerant plants however are adapted to elevated heavy metal concentrations in their growth media. De Vos et al. (1991) observed that metal-tolerant plants do not possess an enhanced tolerance to free radicals (FR) and reactive oxygen species (ROS). By contrast, metal tolerance appears to arise from the prevention of metal-induced oxidative stress through an enhanced capacity and efficiency of metal-ion homeostasis mechanisms (De Vos et al. 1991, 1992, 1993).

Among the metal tolerance mechanisms in plants, metal sequestration has been most extensively documented. Copper has been shown to accumulate in the leaf vacuoles of a Cu-tolerant ecotype of *Armeria maritime* (Lichtenberger and Neumann 1997). An involvement of vacuolar metal sequestration has been proposed in Zn tolerance of *Silene vulgaris* and in metal tolerance of the Zn and Ni hyperaccumulators *Thlaspi caerulescens* and *T. goesingense* (Vazquez et al. 1994). In the Ni-tolerant hyperaccumulator *Alyssum lesbiacum*, and in a heavy metal tolerant ecotype of *Armeria maritime*, metals are accumulated in the epidermis and particularly in leaf trichomes (Neumann et al. 1995; Kramer et al. 1997).

Metal tolerance may be achieved by metal chelation with high-affinity ligands inside and outside the cytoplasm. In *Alyssum lesbiacum*, Ni accumulation was associated with a large and proportional increase in xylem concentrations of the free amino-acid histidine, which was proposed to reflect an increase of intracellular, most probably cytoplasmic histidine concentrations. Furthermore, it was demonstrated that histidine binds Ni in vivo and that exogenous application of histidine reduced Ni toxicity in the nontolerant species *Alyssum montanum* (Kramer et al. 1996). However, the histidine response is unlikely to provide oxidative protection, since histidine complexation of Ni is known to enhance oxidative DNA damage induced by the metal in vitro (Datta et al. 1992, 1993; Misra et al. 1993). This suggests that once the metal has entered the plant cell, high affinity chelation, targeting, and sequestration of the metal chelate are all vital for metal tolerance. Accumulation of Cd in the vacuole is related to the Cd tolerance of plants. Uptake rates depend on the cadmium concentration in the growth medium. Tolerance which is also related to phytochelatins, are a major class of heavy metal chelating peptides that exist in plants (Buchanan et al. 2000). They are low-molecular-weight, enzymatically synthesized cysteine-rich peptides known to bind cadmium and are important for cadmium detoxification (Buchanan et al. 2000; Salt et al. 1995a). A number of metal-binding compounds found in tolerant plants could function as antioxidants, but also have a high affinity for binding metals. So far, evidence has only been provided for metal binding by these compounds, for example phenolic compounds (Lichtenberger and Neumann 1997). Efficient metal binding may be sufficient to prevent FR and ROS formation. However, the possibility of an antioxidant role of such molecules in plants should be investigated.

Higher capacity to accumulate and store HM for a long time have been observed in Lichens because of their morphological and ecological peculiarities and are widely used as plant material to investigate or biomonitor airborne HM. The *Artemisia* species and lichens *(Xantoria parietna, Physcia adscendens)* growing on the different substrates at different distances from a Hydroelectric Power Station were compared for their HM accumulation capacity indicated higher accumulation of HM's in the thalluses of both lichen species than *A. fragrans* situated in the vicinity, which was assumed to be a tolerant/excluder plant. The HM concentrations found in *Xantoria* and *Physcia* considerably exceed the values reported for other species of lichens (Lavrinenko and Lavrinenko 1999) indicating close correlation between the levels of HM in the lichens and atmospheric deposition (Kobayashi et al. 1986). Hence, significantly high HM contents in lichen species consequently provide evidence of a high level of air contamination.

6 Fatty Acids Profile Under Heavy Metal Stress

Adverse environmental conditions such as high and low temperature (Tremolieres et al. 1982; Pleines et al. 1987), salinity (Elenkov et al. 1996; Allakhverdiev et al. 1999), and heavy metals (Fodor et al. 1995; Howlett and Avery 1997; Jemal et al. 2000) change the composition of fatty acids in plants. High temperature has been found to result in a significant increase of $C_{18:1}$ and $C_{18:2}$ desaturation, resulting in a higher $C_{18:3}$ (Tremolieres et al. 1982; Pleines et al. 1987). In a study carried out by Ouarti et al. (1997), cadmium stress has been found to increase the proportion of $C_{16:0}$ and decrease in the $C_{18:2}$ and $C_{18:3}$, in 17 day old tomato seedlings. These results suggest that metal treatment has induced an alteration in the fatty acid desaturation processes. Furthermore, the accumulation of $C_{16:0}$ rather than $C_{18:0}$ indicated an alteration in the ratio of products from the fatty acid synthase. Similarly, Krupa and Baszynski (1989) reported that thylakoids from 4-week-old tomato seedling grown for 14 days in nutrient solutions containing Cd showed a decrease in the content of all individual glycol- and phospholipids to approximately 75% of control. The greatest decrease was in the phosphatidylcholine content. The fatty acid composition of the acyl lipids extracted from the thylakoids was characterized by a significant decrease in the trans-δ-3 hexadecanoic acid component of the phosphatidylglycerol and by a tendency for the linolenic acid content in all lipids to fall.

Kelly-John et al. (2003) reported that metal contamination showed reduction in fatty acids of actinomycetes fungi, etc. Moreover, environmental pollution such as "industrial pollution" also resulted in decrease in the amount of major and minor fatty acids (Patel-Davendra et al. 2004). Abiotic stress including heavy metals cause molecular damage to plant cells and rupture the cell membrane, leading to oxidative stress in plants, i.e., increase in oxidative enzymes (Zhang et al. 2005). This also resulted in production of H_2O_2, which can convert fatty acids to toxic peroxides, thereby destroying biological membrane through lipid peroxidation.

The changes in the composition or molecular arrangement of membranes might also play a role toward heavy metal resistance, either by modifying the permeability of membranes to ions or by altering the membrane-bound enzyme activities (Verkleij and Schat 1990). Cooke and Burden (1990) considered changes in lipid under metal stress as one of the mechanisms most likely involved in the regulation of plant plasma membrane ATPases, which produce the proton electrochemical gradient responsible for primary transport processes in higher plants. This enzyme has already been described as a key regulatory enzyme that controls many important functions including cell division and elongation (Serrano 1989). Zel et al. (1993a, b) reported that heavy metal decreased membrane fluidity in the Al-sensitive fungus *Amanita muscaria*, but an increase in membrane fluidity was observed in Al-resistant fungus *Lactarius piperatus*. Apparently cell decompartmentalization and modification of membrane functions represent the first target for metal toxicity. The changes in Cd can induce disturbance of membrane lipid turnover. Cadmium has been found to enhance lipoxygenase activity (Somashekaraiah et al. 1992) which is responsible for catalyzing lipid peroxidation by using membrane lipids as substrates, particularly unsaturated fatty acids. Likewise, the products of lipoxygenase reaction mainly peroxy, alkoxy, and hydroxyl radicals are themselves reactive and can result in further membrane lipid deterioration and also affect other macromolecules in cells.

7 Conclusion and Future Perspective

Exposure to metals through contaminated air, food, drinking water, beverages, or soil may represent a threat to the existence of many plants and animals. Although some of the major health hazards linked to metals such as mercury and cadmium have been known for centuries, a growing body of evidence is nowadays linking these (and other) metals to adversely affect a number of agricultural crops as well as health effects.

The challenge of responding to the health and environmental impacts of metal emissions from artisanal and small-scale mining operations, abandoned mines and more general occupational exposures will be considered from the perspective of industry and other anthropogenic activities. The arguments presented suggest that the required policy responses are far broader than chemicals management and require an integrated policy approach.

The health and environmental effects associated with metals such as cadmium, lead and mercury are well known and increasingly documented among the scientific community. In the developed world, measures to mitigate adverse exposures have been increasing in number since the middle of nineteenth century. Similarly, traditional end-of-pipe controls and broad-based regulatory policies have led to significant reductions in emissions of metals of concern to the environment and corresponding reductions in exposure and adverse effects within surrounding ecosystems.

References

Adriano DC (1986) Trace elements in the terrestrial environment. Springer, New York, pp 1–533

Ahmad A, Abdin MZ (1999) NADH: nitrate reductase and NAD(P)H: nitrite reductase activities in mustard seedlings. Plant Sci 143:1–8

Aijen 2004 Importance of root growth parameters to Cd and Zn acquisition by non-hyperaccumulator and hyperaccumulator plants. Dissert of Hohenheim University, Verlag Grauer Beuren, Stuttgart

Ali G, Srivastava PS, Iqbal M (1998) Effect of cadmium and copper on growth of *Bacopa monniera* regenerants. Biol Plant 41:635–639

Alia K, Saradhi PP (1991) Proline accumulation under heavy metal stress. J Plant Physiol 138:554–558

Allakhverdiev SI, Nishiyama Y, Suzuki I, Tasaka Y, Murata N (1999) Genetic engineering of the unsaturation of fatty acids in membrane lipids alters the tolerance of synechacystis to salt stress. Proc Natl Acad Sci USA 96:5862–5867

Al-Najar H, Schulz R, Roemheld V (2003) Plant availability of thallium in the rhizosphere of hyperaccumulator plants a key factor for assessment of phytoextraction. Plant Soil 249:97–105

Anderson TA (1997) Development of a phytoremediation handbook: consideration for enhancing microbial degradation in the rhizosphere. Environmental Science and Engineering Fellow, Claudia Sturges, 202–326.

Andrade JL, Larque SA, Trejo CL (1995) Proline accumulation in leaves of four cultivars of *Phaseolus vulgaris* L. with different drought resistance. Phyton Beunas Aires 57:149–155

Anonymous (2001) CERCLA priority list of hazardous substances. ATSDR HazDat Data base information centre: US Department of Health and Human Services

Aspinall D, Paleg LD (1981) Proline accumulation: physiological aspects. In: Paleg LD, Aspinall D (eds) Physiology and biochemistry of drought resistance in plants. Academic, Sydney, pp 215–228

Aydinalp C, Marinova S (2003) Distribution and forms of heavy metals in some agricultural soils. Pol J Environ Stud 12(5):629–633

Bala R, Setia RC (1990) Some aspects of cadmium and lead toxicity in plants. In: Malik CP, Setia RC, Singh P (eds) Advances in frontier areas of plant. Narendra Publishing House, Delhi, India, pp 167–180

Barber SA (1984) Soil nutrient bioavailability: a mechanistic approach. Oxford Claradon Press, Wiley, New York

Barber SA, Lee RB (1974) The effect of micro-organisms on the absorption of manganese by plants. New Phytol 73:97–106

Barcelo J, Vazquez MD, Poschensieder C (1988) Structural and ultrastructural disorders in Cd-treated bush bean plants (*Phaseolus vulgaris* L.). New Phytol 108:37–49

Baszynski T, Wajda L, Krol M, Wolinska D, Krupa Z, Tukendorf A (1980) Photosynthetic activities of cadmium treated tomato plants. Physiol Plant 48:365–370

Bazzaz FA, Rolfe GL, Carlson RW (1974) Effect of Cd on photosynthesis and transpiration of excised leaves of corn and sunflower. Physiol Plant 32:373–376

Bhardwaj R, Mascarenhas C (1989) Cadmium-induced inhibition of photosynthesis in vivo during development of chloroplasts in *Triticum aestivum* L. Plant Physiol Biochem 16:40–48

Bharti N, Singh RP (1993) Growth and nitrate reduction by *Seasum indicum* c.v. PB-1 respond differentially to lead. Phytochemistry 33:531–534

Bhattacharyya M, Choudhuri MA (1995) Heavy metal (Pb^{2+} and Cd^{2+}) stress-induced damage in *Vigna* seedlings and possible involvement of phytochelatin like substances in mitigation of heavy metal stress. Indian J Exp Biol 33:236–238

Blake RC, Choate DM, Bardhan S, Revis N, Barton LL, Zocco TG (1993) Chemical transformation of toxic metals by a *Pseudomonas* strain from a toxic waste site. Environ Toxic Chem 12:1365–1376

Bowen HJM (1979) Environmental chemistry of the elements. Academic, New York

Bradeen DA, Winget GD, Gould JM, Ort DR (1973) Site-specific inhibition of photophosphorylation in isolated spinach chloroplasts by mercuric chloride. Plant Physiol 52:680–682

Brooks RR (1998) Plants that hyperaccumulate heavy metals (Their role in phytoremediation, microbiology, archaeology, mineral exploration and phytomining). CAB International, Wallingford

Buchanan BB, Gruissem W, Jones RL (2000) Biochemistry and molecular biology of plants. Am Soc Plant Physiol, Rockville, MD 1367

Burdin KS, Polyakova EE (1987) Metallothioneins, their structure and functions. Usp Sovern Biol 103:390–400

Cakmak I (2002) Plant nutrition research: priorities to meet human needs for food in sustainable ways. Plant Soil 247:3–24

Cedeno-Maldonado A, Swader JA, Heath RL (1972) The cupric ion an inhibitor of photosynthetic electron transport in isolated chloroplasts. Plant Physiol 50:698–701

Chaney RL (1988) Metal speciation and interactions among elements affecting trace-element transfer in agricultural and environmental food chains. In: Kramer JR, Allen HE (eds) Metal speciation: theory, analysis and applications. Lewis Publishers Inc, Chelsa, MI, pp 218–260

Chaney RL, Li YM, Angel JS, Baker AJM, Reeves RD, Brown SL, Homer FA, Malik M, Chin M (1999) Improving metal hyperaccumulators wild plants to develop commercial phytoextraction systems: approaches and progress. In: Terry Nand Banuelos GS (ed) Phytoremediation of contaminated soil and water. CRC Press, Boca Raton, FL, pp 129–158

Chugh LK, Gupta VK, Sawhney SK (1992) Effect of cadmium on enzymes of nitrogen metabolism in pea seedlings. Phytochemistry 31:395–400

Clijsters H, Van Assche F (1985) Inhibition of photosynthesis by heavy metals. Photosynth Res 7:31–40

Cocking D, Rohrer M, Thomas R, Walker TJ, Ward D (1995) Effect of root morphology and Hg concentration in the soil on uptake by terrestrial vascular plants. Water Air Soil Pollut 80:1113–1116

Cooke DT, Burden RS (1990) Lipid modulation of plasma membrane bound ATPase. Physiol Plant 78:153–159

Cordovilla MP, Ligero F, Lluch C (1996) Growth and nitrogen assimilation in nodules in response to nitrate levels in *Vicia faba* under salt stress. J Exp Bot 47:203–210

Costa G, Spitz E (1997) Infleunce of cadmium on soluble carbohydrates, free amino-acids, protein content of in vitro cultured *Lupinus albus*. Plant Sci 128: 131–140

Crowley DE, Wang YC, Reid CP, Szansiszlo PJ (1991) Mechanism of iron acquisition from siderophores by micro-organisms and plants. Plant Soil 130:179–198

Dabas S, Singh RP (1995) Differential effect of lead on nitrate reductase activity and organic nitcrogen content in mung bean (var. P-105) seedlings. Indian J Plant Physiol 38:155–157

Datta AK, Misra M, North SL, Kasprzak KS (1992) Enhancement by nickel(II) and L-histidine of 2′-deoxyguanosine oxidation with hydrogen peroxide. Carcinogenesis 13:283–287

Datta AK, Shi X, Kasprzak KS (1993) Effect of carnosine, homocarnosine anserine on hydroxylation of the guanine moiety in 2′-deoxyguanosine, DNA and nucleohistone and hydrogen peroxide in the presence of nickel(II). Carcinogenesis 14:417–422

Davies BE (1980) Trace element pollution. In: Davies BE (ed) Applied oil trace elements. Wiley, New York, pp 287–351

Davies MS (1991) Effects of toxic concentration of metals on root growth and development. In: Alkinson D (ed) Plant root growth. Blackwell, London UK, pp 211–227

De Filippis LF, Pallaghy CK (1976) The effect of sublethal concentrations of mercury and zinc on *Chlorella*. In: Growth characteristics and uptake of metals. Z. Pflanzenphysiol. Bd. 78:197–207

De Vos CHR, Schat H, De Waal MAM, Voojis R, Ernst WHO (1991) Increased resistance to copper-induced damage of the root cell plasmalemma in copper-tolerant *Silene cucubalus*. Physiol Plant 82:523–528

De Vos CHR, Ten Bookum W, Vooijis R, Schat H, De Kok LJ (1993) Effect of copper on fatty acid composition and peroxidation of lipids in the roots of copper-tolerant and sensitive *Silene cucubalus*. Plant Physiol Biochem 31:151–158

De Vos CHR, Vonk MJ, Vooijis R, Schat H (1992) Glutathione depletion due to copper-induced phytochelatin synthesis causes oxidative stress in *Silene cucubalus*. Plant Physiol 98:853–858

Di Cagno R, Guidi L, Stefani A, Soldatini GF (1999) Effect of cadmium on growth of *Helianthus annus* seedlings: Physiological aspects. New Phytol 144: 65–71

Elenkov I, Stefanv K, Dimitrova-Konaklieve S, Popov S (1996) Effect of salinity on lipid composition of *Cladophora vagabunda*. Phytochemistry 42:39–44

Elstner EF, Osswald W (1994) Mechanisms of oxygen activation during plant stress. Proc Roy Soc Edinb 102:131–154

Erdmann B, Hoffmann P, Wiedenroth EM (1986) Changes in root system of wheat seedlings following root anaerobiosis. Ann Bot 58:597–605

Ernst WHO (1998) Effects of heavy metals in plants at the cellular and organismic level. In: Schuurmann G, Markert B (eds) Ecotoxicolology. Wiley, Specktrum Akademischer Veriog, New York, pp 587–620

Ernst WHO (2005) Evolution of metal tolerance and the implication on processes from the cellular to the whole plant level. In: Abstract of workshop on metal fluxes and stresses in terrestrial ecosystems. Swiss Federal Institute for Forest, Snow and Landscape Research WSL. Birmensdorf, 31–32

Ernst WHO, Verkeij JAC, Schat H (1992) Metal tolerance in plants. Acta Bot Neerl 41(3):229–248

Ferner DJ (2001) Toxicity, heavy metals. eMed J 2:1

Ferrara R, Maserti BE, Paterno P (1989) Mercury distribution in marine sediment and its correlation with the *Posidonia oceanica* Prairie in a coastal area by a Chloralkali complex. Toxicol Environ Chem 22: 131–134

Ferrara R, Mazzolai B, Lanzillotta E (2000) Temporal trends in gaseous mercury evasion from Mediterranean seawaters. Sci Total Environ 259:183–190

Fodor E, Szabo-Nagy A, Erdei L (1995) The effect of cadmium on the fluidity and H^+ ATPase activity of plasma membrane from sunflower and wheat roots. J Plant Physiol 147:87–92

Foyer CH, Harbinson J (1994) Oxygen metabolism and the regulation of photosynthetic electron transport. In: Foyer CH, Mullineaux PM (eds) Causes of Photooxidative stress and amelioration of defense systems in plants. CRC, London, pp 2–42

Galvan A, Cardenas J, Fernandez E (1992) Nitrate reductase regulates expression of nitrite uptake and nitrite reductase activities in *Chlamydomonas reinhardtii*. Plant Physiol 98:422–426

Gil J, Morel R, Gomez I, Novarro-Pedrerio J, Matrix J (1995) Effect of cadmium on physiological and nutritional aspects in tomato plant. II. Soluble and rubisco proteins and nutrient evolution. Fresenius Environ Bull 4:436–440

Grancharov I, Popova S (2003) Heavy metals pollution around the metallurgy plants in some regions in Bulgaria. In: Proceedings of the workshop "Bulgarian Priorities in Chemical Risk Assessment and Management", held on 12 September 2003 Sofia, 38–47

Greenfield SS (1942) Inhibitory effects of inorganic compounds on photosynthesis in *Chorella*. Am J Bot 29:121–131

Greger M, Ogren E (1991) Direct and indirect effects of cadmium on photosynthesis in sugarbeet (*Beta vulgaris*) Physiol. Plant 83:129–135

Griffin M, Antikainen M, Hon WC, Pihakaski-Maunsbach K, Yu XM, Chun JU, Yang DSC (1997) Antifreeze proteins in winter rye. Plant Physiol 100:327–332

Guo T, Zhang G, Zhou M, Wu F, Chen J (2004) Effects of aluminium and cadmium toxicity on growth and antioxidant activities of two barley genotypes with different Al resistance. Plant Soil 258:241–248

Gwozdz EA, Kopyra M (2003) Plant cell responses to heavy metals-biotechnological aspects. Biotechnologia 62:107–123

Hall DO, Rao KK (1999) Photosynthesis 6th Ed. Cambridge University Press, UK

Heeraman DA, Claassen VP, Zasoki RJ (2001) Interaction of lime, organic matter and fertilizer on growth and uptake of arsenic and mercury by Zorrofescue (*Vulpia myuros* L.). Plant Soil 234:215–231

Hemalatha S, Anburaj A, Francis K (1997) Effect of heavy metals on certain biochemical constituents and nitrate reductase activity in *Oryza sativa* L. seedlings. J Environ Biol 18:313–319

Hirt H, Casari G, Barta A (1989) Cadmium enhanced gene expression in suspension culture cells of tobacco. Planta 179:414–420

Hock B, Elstner E (2005) Preface. In: Hock B, Elstner E (eds) Plant toxicology, 4th edn. Marcel Dekker, New York

Horvath G, Droppa M, Oraveez A, Raskin VI, Marder JB (1996) Formation of the photosynthetic apparatus during greening and cadmium poisoning barley leaves. Planta 199:238–243

Howlett NG, Avery SV (1997) Relationship between cadmium sensitivity and degree of plasma membrane fatty acid unsaturation in *Saccharomyces cerevisiae*. Appl Microbiol Biotechnol 48:539–546

International Occupational Safety and Health Information Centre (1999) Basics of chemical safety. International Labour Organization, Geneva

Ivanov VB, Bistrova EI, Seregin IV (2003) Comparison of influence of heavy metals on root growth in connection with problems of specificity and selectivity of their action. Plant Physiol 50:445–454 (in Russian)

Jafri S, Srivastava K, Ahmad KJ (1979) Environmental pollution and epidermal structure in *Syzygium cumini* L. Skeel Indian J Air pollut Control 2:74–77

Jager HG, Meyer HR (1977) Effect of water stress on growth and proline oxidation by water stress. Plant Physiol 59:930–932

Jemal F, Zarrouk M, Ghorbal MH (2000) Effects of Cd on lipid composition of pepper. Biochem Soc Trans 28:907–912

Kaiser WM, Weiner H, Huber SC (1999) Nitrate reductase in higher plants: a case study for transduction of environmental stimuli into control of catalytic activity. Physiol Plant 105:385–390

Kaschl A, Romheld V, Chen Y (2002) The influence of soluble organic matter from municipal solid waste compost on trace metal leaching in calcareous soils. Sci Total Environ 291:45–57

Keating MH, Mahaffey KR, Schoney R, Rice GE, Bullock OR, Ambrose RB, Swartout J, Nichols JW (1997) Mercury Study Report to congress. Washington DC, 3(1):6–7

Kelly-John J, Haggblom Max M, Tate-Robert L (2003) Effects of heavy metal contamination and remediation on soil microbial communities in the vicinity of a zinc smelter as indicated by analysis of microbial community phospholipids fatty acid profiles. Biol Fertil Soils 38:65–71

Keltjens WG, Van Beusichem ML (1998) Phytochelatins as biomarkers for heavy metal stress in maize (*Zea mays* L.) and wheat (*Triticum aestivum* L.): combined effects of copper and cadmium. Plant Soil 203: 119–126

Keswick BH (1984) Sources of ground water pollution. In: Bitton G, Gerba CP (eds) Ground water pollution Microbiology. John Wiley, New York, USA, pp 39–69

Kevreson S, Ksristic B, Popovic M, Kovacev L, Pajevic S, Kandrac J, Malenic D (1998) Biochemical changes in sugar beet lines as dependence on soil moisture. Biol Plant 40:245–250

Khan MG (1996) Nitrate and nitrite reductase activities in soybean plants raised with saline water. Indian J Plant Physiol 1:128–129

Kobayashi T, Nakagawa Y, Mitsugi H, Watanabi H (1986) Estimation of atmosphere pollution by mercury by means of epiphyte lichens. Tayki osen gakkaysi, J Jap Soc Air Pollut 2:151–155 (in Japanese)

Kramer U, Cotter-Howells JD, Charnock JM, Baker AJM, Smith JAC (1996) Free histidine as a metal chelator in plants that accumulate nickel. Nature 379:635–638

Kramer U, Grime GW, Smith JAC, Hawes CR, Baker AJM (1997) Micro-PIXE as a technique for studying nickel localization in leaves of the hyperaccumulator plant *Alyssum lesbiacum*. Nucl Instrum Methods Phys Res B 130:346–350

Krupa Z, Baszynski T (1989) Acyl lipid composition of thylakoid membranes of cadmium treated tomato plants. Acta Physiol Plant 11:111–116

Krupa Z, Skorzynska F, Maksymiec W, Baszynski T (1987) Effect of Cd^{2+} treatment on the photosynthetic apparatus and its photochemical activities in greening radish seedlings. Photosynthetica 21:156–164

Kulakov P, Zeeb K, Teritze I, Atanasov A, Vassilev R, Donkova, 2009. Textbook on phytoremediation of contaminated soils and sites. Academic Press, Agricultural University – Plovdiv. 104 (In Bulgarian)

Lasat MM (2000) Phytoextraction of metals from contaminated soil: a review of plant/soil/metal interaction and assessment of pertinent agronomic issues. J Hazard Subs Res 2:5–25

Lata S (1989a) Effects of cadmium on seedling growth, mobilization of food reserves and activity of hydrolytic enzymes in *Phaseolus aureus* cv. T-44 seeds. Acta Bot Indica 17:290–293

Lata S (1989b) Phasic pretreatment effect cadmium on seedling growth and activity of certain hydrolytic enzymes of *Phaseolus aureus* cv. T-44 seeds. Acta Bot Indica 17:222–224

Lavrinenko OV, Lavrinenko IA (1999) Heavy metal accumulation by lichens from Peltigora genus. Mater. of Intern. Conf. "Plant Physiology – Sci. of III millenium", Moscow, Russia

Leuning R, Kelliher FM, De-Pury DGG, Schulze ED (1995) Leaf nitrogen, photosynthesis, conductance and transpiration: scaling from leaves to canopies. Plant Cell Environ 18:1183–1200

Lichtenberger O, Neumann D (1997) Analytical electron microscopy as a powerful tool in plant cell biology:

Examples using electron energy loss spectroscopy and X-ray microanalysis. Eur J Cell Biol 73:378–386

Lichtenthaler HK (1996) An introduction to the stress concept in plants. J Plant Physiol 148:4–14

Long SP, Farage PK, Garcia RL (1996) Measurement of leaf and canopy photosynthetic CO_2 exchange in field. J Exp Bot 47:1629–1642

Lozano-Rodriguez E, Hernandez LE, Bonay P, Carpena-Ruiz RO (1997) Distribution of cadmium in root tissues of maize and pea plants: physiological disturbances. J Exp Bot 48:123–128

Lue-Kim H, Rauser WE (1986) Partial characterization of cadmium binding proteins from roots of tomato. Plant Physiol 81:896–900

Lutts S, Majerus V, Kinet JM (1999) NaCl effects on proline metabolism in rice (*Oryza sativa*) seedlings. Physiol Plant 105:450–458

Macnair MR, Tilstone GH, Smith SE (2000) The genetics of metal tolerance and accumulation in higher plants. In: Terry N, Banuelos G (eds) Phytoremediation of contaminated soils and water. CRC Press LLC, Boca Raton, pp 235–250

Macnicol RD, Beckett PHT (1985) Critical tissue concentrations of potentially toxic elements. Plant Soil 85: 107–129

Mantovi P, Bonassi G, Maestri E, Marmirolli N (2003) Accumulation of copper and zinc from liquid manure in agricultural soils and crop plants. Plant Soil 250: 249–257

Marschner H (1983) Heavy metals. In: Lauchli A, Bieleski RL (eds) Inorganic plant nutrition encyclopedia of plant physiol, vol 15A. Springer, Berlin, pp 39–49

Marschner H (1986) Mineral nutrition of higher plants. Academic, London, 254

Martínez-Martínez S, Faz Á, Acosta JA, Carmona DM, Zornoza R, BüYükkiliç A, Kabas S (2010) Heavy metals distribution in soil particle size fractions from a mining area in the southeast of Spain. In: 19th World Congress of Soil Science, Soil Solutions for a Changing World 1–6 August 2010, Brisbane, Australia. Published on DVD

Masarovicova E, Cicak A, Stefancik I (1999) Plant response to air pollution and heavy metal stresses. In: Pessarakli M (ed) Handbook of plant and crop Stresses, 2nd edn. Marcel Dekker, Basel, AG, pp 569–599

Meagher RB, Rugh CL, Kandasamy MK, Gragson G, Wang NJ (2000) Engineered phytoremediation of mercury pollution in soil and water using bacterial genes. In: Terry N, Banuelos G (eds) Phytoremediation of contaminated soil and water. Lewis, Boca Raton, FL, p 201

Mehindirata S, Ali ST, Siddiqi MTO, Iqbal M (1999) Cadmium-induced changes in foliar responses of *Solanum melongena* L. Phytomorphology 49:295–302

Mengel K, Kirkby EA (2001) Principles of plant nutrition, 5th edn. Kluwer Academic, Dordrecht, The Netherlands, 849

Menzer RE, Nelson JO (1986) Water and soil pollutants. In: Kaassen CD, Andur MO, Doull J (eds) The basic science of poisons. Lasarett and Doull's toxicology. Macmillan, NY, USA, pp 825–856

Mills HA, Scoggins HL (1998) Nutritional levels for anthurium: young versus mature leaves. J Plant Nutr 21:199–203

Misra M, Olinski R, Dizdaroglu M, Kasprzak KS (1993) Enhancement by l-histidine of nickel(II)-induced DNA-protein cross-linking and oxidative DNA base damage in the rat kidney. Chem Res Toxicol 6:33–37

Mitchell RL (1964) Trace elements in soils. In: Bear FE (ed) Chemistry of the soil. Reinhold, New York, pp 320–368

Moniri HM (2005) Influence of pollutants on vegetation in Southwest Iran resulting from the burning oil wells of Kuwait in 1991. In: Siddiqi TO, Ahmad A (eds.) Book of a satellite Session of ICPEP-3 on "Herbal Drugs and Environmental Pollution". Absr. Depart. Of Botany, Faculty of Sci, Jamia Hamdard Univ, New Delhi, pp 73–74

Moreno FN, Anderson CWN, Stewart RB, Robinson BH (2004) Phytoremediation of mercury contaminated mine tailing by induced plant-mercury accumulation. Environ Pract 6:165–175

Nagoor S (1999) Physiological and biochemical responses of heavy metals to graded levels of heavy metal. II. Effect on protein metabolism in maize seedlings. Adv Plant Sci 12:424–433

Naidu R, Oliver D, Mc Cornell S (2003) Heavy metal phytotoxicity in soils. In: Langley A, Gilbey M, Kennedy B (eds) Proceedings of the 5th National Workshop on the Assessment of site contamination. NEPC Service Corp, Adelaide, pp 235–241

Neumann D, Nieden U, Lichtenberger O, Leopold I (1995) How does *Armeria maritama* tolerate high heavy metal concentrations? J Plant Physiol 146:704–717

Noctor G, Foyer CH (1998) Ascorbate and glutathione: keeping active oxygen under control. Ann Rev Plant Physiol Plant Mol Biol 49:249–279

Nozaki JO (1986) Determination of the concentration of protein by dry wt: a comparison with spectrophotometric methods. Arch Biochem Biophys 249:437–446

Nwokolo E, Smartt J (1996) Food and feed from legumes and oil seeds. Chapman and Hall, New York, USA

Nye PH, Tinker TB (1977) Solute movement in the soil-root system. University of California Press, Berkeley, CA, 342

Oleksyn J, Innes JL (2000) Air pollution and forests in heavily industrialized regions: An introduction. In: Innes JL, Oleksyn J (eds) Forest dynamics in heavily polluted regions, 1st edn. CABI Publishing, Oxon, pp 1–7

Ouarti O, Bussama N, Zarrouk M, Cherif A, Ghorbal MH (1997) Cadmium and copper induced changes in tomato membrane lipids. Phytochemistry 45:1343–1350

Pachepsky LB, Acock B, Hoffman-Benning S, Willmitzer L, Fisahn J (1997) Estimation of the anatomical, stomatal and biochemical components of differences in photosynthesis and transpiration of wild type and transgenic (expressing yeast-derived invertase targeted to the vacuole) tobacco leaves. Plant Cell Environ 20: 1070–1078

Pahlsson AMB (1989) Toxicity of heavy metals (Zn, Cu, Cd and Pb) to vascular plants. Water Air Soil Pollut 47:287–319

Park CH, Keyhan M, Matin A (1999) Purification and characterization of chromate reductase in *Pseudomanas putida*. Abs Gen Meet Am Soc Microbiol 99:536

Patel-Davendra K, Kumar R, Prasad S (2004) Variation in the chemical constituents of soybean due to industrial pollution. J Serb Chem Soc 69:635–640

Pleines S, Marquard R, Friedt W (1987) Recurrent selection for modified polyenoic fatty acid composition in rapeseed (*Brassica napus* L.) 7th Rapesees Congress Poznan (Poland) 140–145

Poschenrieder C, Gunse B, Barcelo J (1989) Influence of cadmium on water relations, stomatal resistance and abscisic acid content in expanding bean leaves. Plant Physiol 90:1365–1371

Prasad MNV (1995) Cd Toxicity and tolerance in vascular plants. Environ Exp Bot 35:525–545

Pandey P, Tripathi AK (2011) Effect of heavy metals on morphological and biochemical characteristics of *Albizia procera* (Roxb.) Benth. seedlings. Int J Environ Sci 5(1):1009–1018

Puckett KJ (1976) The effect of heavy metals on some aspects of lichen physiology. Can J Bot 54: 2695–2703

Punz WF, Sieghardt H (1992) The response of roots of herbaceous plant species to heavy metals. Environ Exp Bot 33:85–98

Purves D (1985) Trace element contamination of the environment. Elsevier, Amsterdam

Ralph PJ, Burchertt MD (1998) Photosynthetic response of *Halophila ovalis* to heavy metal stress. Environ Pollut 103:91–101

Rath P, Panigrahi AK, Misra BN (1986) Effects of both inorganic and organic Mercury on the ATPase activity of *Westiellopsis prolifica* Janet. Environ Pollut (Series A) 42:143–149

Rose AW, Hawkes FE, Webb JS (1979) Geochemistry in mineral exploration. Academic, London, England

Ross SM (1994) Toxic metals in soil-plant systems. Wiley, Chichester

Saljnikov E, Mrvic V, Cakmak D, Nikoloski M, Perovic V, Kostic L, Brebanovic B (2009) Spatial distribution of selected heavy metals and soil fertility status in South eastern Serbia. Geophys Res Abs 11:1376

Salt DE, Blaylock M, Kumar NPBA, Vratcheslav D, Ensley BD (1995a) Phytoremediation: a novel strategy for the removal of toxic metals from the environment using plants. Biotechnology 13:468–474

Salt DE, Prince RC, Baker AJM, Raskin I, Pickering IJ (1999) Zinc ligands in the metal hyperaccumulator *Thlaspi caerulescens* as determined using X-ray absorption spectroscopy. Environ Sci Tech 33: 713–717

Salt DE, Prince RC, Pickering IJ, Raskin I (1995b) Mechanisms of cadmium mobility and accumulation in Indian mustard. Plant Physiol 109:1427–1433

Sanita di-Toppi L, Gabbrielli R (1999) Response to cadmium in higher plants. Environ Exp Bot 41: 105–130

Schickler H, Caspi H (1999) Response of antioxidant enzymes to nickel and cadmium stress in hyperaccumulator plants of the genus *Alyssum*. Physiol Plant 105:39–44

Schulze E, Beck E, Muller-Hohenstein K (2005) Plant Ecol. Springer, Berlin, Heidelberg, 702

Selvaraj K, Hemalatha C, Gandhi SJ (1995) Differential nitrate reductase activity in NO_3^- grown cluster bean (*Cyamopsis tetragonoloba* L.) and green gram (*Vigna radiate* L.): inhibitory effects of ammonia, amino acids and urea. Indian J Exp Biol 33:456–458

Seregin IV, Ivanov VB (2001) Physiological aspects of cadmium and lead toxic action on higher plants. Plant Physiol 48:606–630 (in Russian)

Seregin IV, Kojevnikova AD (2006) Physiological role of nickel and its toxic action on higher plants. Plant Physiol 53:285–308 (in Russian)

Seregin IV, Kojevnikova AD, Kozyumina EM, Ivanov VB (2003) Toxic action and distribution of nickel in maize roots. Plant Physiol 50:793–800 (in Russian)

Serrano R (1989) Structure and function of plasma membrane ATPase. Ann Rev Plant Physiol Plant Mol Biol 40:60–94

Setia RC, Bala R (1994) Anatomical changes in root and stem of wheat (*Triticum aestivum* L.) in response to different heavy metals. Phytomorphology 44:95–104

Setia RC, Setia N, Malik CP (1987) The pod wall structure and function in relation to seed development in some legumes. Phyton 27:205–220

Shacklette HT, Boerngen JG (1984) Element concentrations in soils and other surface materials of the contiguous United States. USGS professional paper 1270. U.S. Government Printing Office, Washington, DC

Shah K, Dubey RS (1997) Effect of cadmium on protein, aminoacids and protease, amino peptidase and carboxypeptidases in rice seedlings. Plant Physiol Biochem 24:89–95

Somashekaraiah SV, Padmaja K, Prasad ARK (1992) Phytotoxicity of cadmium ions on germinating seedlings of mung bean (*Phaseolus vulgaris*): involvement of lipid peroxidation in chlorophyll degradation. Physiol Plant 85:85–89

Srivasankar S, Oaks A (1996) Nitrate assimilation in higher plants: the effect of metabolites and light. Plant Physiol Biochem 34:609–620

Sterritt RM, Lester JN (1980) Interaction of heavy metals with bacteria. Sci Total Environ 14:5–17

Terashima I, Evans JR (1988) Effects of light and nitrogen nutrition on the organization of the photosynthetic apparatus in spinach. Plant Cell Physiol 29:143–155

Tremolieres A, Dubacq JP, Drapier D (1982) Unsaturated fatty acids in maturing seeds of sunflower and rape: regulation by temperature and light intensity. Phytochemistry 21:41–45

Van Assche F, Clijsters H (1990) Effects of metals on enzyme activity in plants. Plant Cell Environ 13: 195–206

Van Steveninck RFM, Van Steveninck ME, Wells AJ, Fernando DR (1990) Zinc tolerance and the binding of zinc as zinc phytate in *Lemna minor.* X-ray microanalytical evidence. J Plant Physiol 137:140–146

Vassilev A, Berova M, Stoeva N, Zlatev Z (2005) Phytotechnologies for sustainable use management of metal contaminated soils: short review. Manage Sustain Dev 3(4):90–96 (In Bulgarian)

Vassilev A, Nikolova A (2010) Mechanisms of plant metal tolerance. Ecology and Future Vol IX 2:3–13 (In Bulgarian)

Vazquez MD, Poschenrieder CH, Barcelo J, Baker AJM, Hatton P, Cope GH (1994) Compartmentation of zinc in roots and leaves of the zinc hyperaccumulator *Thlaspi carulescens*. Bot Acta 107:243–250

Verkleij JAC, Schat H (1990) Mechanism of metal tolerance in higher plants. In: Shaw AJ (ed) Heavy metal tolerance in plants: evolutionary aspects. CRC, Boca Raton¸FL, pp 179–193

Vinogradov AP (1959) The geochemistry of rare and dispersed chemical elements in soils. Consultants Bureau. Inc., New York, NY

Vogeli-Lange R, Wagner GJ (1990) Subcellular localization of cadmium and cadmium binding peptides in tobacco leaves. Plant Physiol 92:1086–1093

Vyas J, Puranik RM (1993) Inhibition of nitrate reductase activity by mercury in bean leaf segments. Indian J Plant Physiol 1:57–60

Wood JM (1989) Transport, bioaccumulation and toxicity of elements in microorganisms under environmental stress. In: Proceedings International Conference on Heavy Metals in the Environment. Geneva CEP Consultants Ltd. Edinburgh 1:1–12

Wood JM, Wang HK (1983) Microbial resistance to heavy metals. Environ Sci Tech 17:582–590

Woolhouse HW (1974) Cell ageing and cell death. In: Davies I, Sigee DC (eds) Cambridge University Press, London 123–153

Woolhouse HW (1983) Toxicity and tolerance in the response of plants to metals. In: Lange OL, Nobel PS, Osmond CB, Zeigler H (eds) Physiological plant ecology III, vol 12. Springer, New york, pp 245–300

Wu JT, Chand SJ, Chou TL (1995) Intracellular proline accumulation in some algae exposed to copper and cadmium. Bot Bull Acad Sin 36:89–93

Wyttenbach A, Tobler L, Bajo S (1989) Na, Cl and Br in needles of Norway spruce and in the aerosol adhering to the needles. Toxicol Environ Chem 18:249–256

Yadav VK, Yadav N (1995) Influence of cadmium on germination, seedling growth and biochemical traits of wheat. Plant Physiol Biochem 22:74–77

Yancey PH, Clarke ME, Hand SC, Bowlus RD, Somero GN (1982) Living with water stress: evolution of complete system. Science 217:1214–1222

Young AJ, Britton G (1990) Carotenoids and antioxidative stress. Curr Res Photosynth 4:587–590

Zayed AM, Gowthaman S, Terry N (1998) Phytoaccumulation of trace elements by wetland plants: I. Duckweed. J Environ Qual 27:715–721

Zel J, Sehara M, Svetek J, Nemee M (1993a) Influence of Al on the membrane mycorrhizal fungi. Water Air Soil Pollut 71:101–109

Zel J, Svetek J, Crne H, Sahara M (1993b) Influence of Al on the membrane fluidity of mycorhizal fungus *Anamita muscaria*. Physiol Plant 89:172–176

Zhang H, Jiang Y, He Z, Ma M (2005) Cadmium accumulation and oxidative burst in garlic (*Allium sativum*). J Plant Physiol 162:977–984

Zhu YL, Pilon-Smits EAH, Jouanin L, Terry N (1999) Cadmium tolerance and accumulation in Indian mustard is enhanced by overexpressing γ–glutamylcysteine synthase. Plant Physiol 121:1169–1177

Cd Accumulation and Subcellular Distribution in Plants and Their Relevance to the Trophic Transfer of Cd

21

M.S. Monteiro and A.M.V.M. Soares

Abstract

Cadmium (Cd) is an elemental substance that occurs naturally in the earth's crust, and is variously taken up and incorporated into plant biological systems. Environmental contamination with this non-essential metal presents a challenge to plant species because they are often not equipped to regulate internal concentrations of Cd or to employ proper detoxification mechanisms. This review examines the movement of Cd from soil to plants, how different plant species handle Cd stress by evolutive acquisition of different mechanisms of tolerance and accumulation patterns. The consequences and toxic effects of Cd (hyper)accumulation in plants and other organisms to animal consumers in terrestrial ecosystems are highlighted. Understanding Cd uptake of plants, how they handle Cd contamination and how they made Cd bioavailable in trophic food chains is critical to the long-term safety and conservation of agricultural resources and ecosystems services and functions.

Keywords

Cadmium • Metal compartmentalization • Metal trophic transfer • Plant hyperaccumulation • Ecotoxicology

1 Introduction

The contamination of soils with some metals, especially in agricultural soils, has reached levels that can threaten food safety via plant uptake (Amini et al. 2005; Peris et al. 2007). Compared to other metals, cadmium (Cd) is of great concern due to its high toxicity to animals and humans. Cadmium in soils is easily concentrated in particular edible parts of plants, which can threaten human health via food chains.

Cadmium is a naturally occurring element, and its presence has been detected in more than 1,000 species of aquatic and terrestrial flora and fauna (Eisler 1985). With one known exception, there is no evidence that Cd is biologically essential or beneficial; on the contrary, it has been

M.S. Monteiro (✉) • A.M.V.M. Soares
CESAM and Department of Biology,
University of Aveiro, 3810-193 Aveiro, Portugal
e-mail: mmonteiro@ua.pt

P. Ahmad and M.N.V. Prasad (eds.), *Abiotic Stress Responses in Plants: Metabolism, Productivity and Sustainability*, DOI 10.1007/978-1-4614-0634-1_21, © Springer Science+Business Media, LLC 2012

implicated in several human diseases and various deleterious effects in wildlife (Eisler 1985). The exception is a Cd-dependent carbonic anhydrase found in the marine diatom *Thalassiosira weissflogii* (Lane and Morel 2000; Lane et al. 2005); a similar role has been postulated for the metal hyperaccumulating plant *Thlaspi caerulescens* (Liu et al. 2008). These authors have suggested that Cd may play a physiological role by enhancing the activities of carbonic anhydrase, a typical Zn-requiring enzyme. In all life forms, including microorganisms, higher plants and animals (in particular humans), Cd is toxic when present in sufficient concentrations (Eisler 1985).

Pollution of the biosphere with this toxic metal has accelerated dramatically since the beginning of the industrial revolution (Nriagu 1996) and Cd accumulation in soil and water now poses a major environmental and human health problem. Cadmium is now included in the list of priority pollutants established by European Community (2455/2001/EC 2001) and also by EPA (Environmental Protection Agency, USA).

Historically, the study of metal uptake by plants has focused on micronutrient metals important in agricultural production, whereas non-essential metals, such as Cd, Hg and Pb, have generally received less attention. However, over the last three decades Cd has been the subject of several investigations in plant research mainly because of its potential for bioaccumulation through soil–plant–animal food chain. For example, the consequences of soil contamination by Cd, through application of treated sewage sludge (biosolids) (McLaughlin et al. 2006) and Cd-enriched phosphate fertilizers (e.g. He and Singh 1994a, b) to soils have been extensively studied. However, the driving force of this research area has been the concern for the risk to human health, not for the state of the plant itself.

Most of the research on Cd pollution focused on the processes involved in Cd accumulation in crop plants and on the consequences of this accumulation on human health (Wagner 1993). Cadmium phytotoxicity is, however, a relevant problem, especially in some highly metal-polluted regions, where a decrease in agricultural crop productivity has been observed (Vassilev and Yordanov 1997). More recently, metal hyperaccumulator plant species have been used to re-examine mechanisms of metal uptake by plants in light of the potential for phytoremediation of metal-contaminated soils (Chaney et al. 1997; Padmavathiamma and Li 2007; Pilon-Smits 2005; Shao et al. 2010). On the other hand, Cd uptake by plants has great impact and relevance not only to plants but also to the ecosystem, in which plants form an integral component. Therefore, understanding Cd uptake of plants and its trophic transfer is critical to the long-term safety and conservation of agricultural resources and ecosystems services and functions.

An important issue concerning Cd accumulation in plants centres on the fact that Cd could pose a risk to the animal consumer health, even if plant tissue concentrations are not generally phytotoxic (McLaughlin 2002). Indeed, the ability of some plant species to uptake and hyperaccumulate Cd in edible parts increases the risk of Cd assimilation by animal consumers through trophic transfer. Because there is very limited knowledge regarding the trophic transfer of metallic contaminants between plants and consumers of plants, this chapter presents an examination of how plants handle Cd uptake, how it is compartmentalized at the subcellular level within plant leaves, and subsequently the significance of that compartmentalization on the bioavailability of Cd for trophic transfer.

2 Soil Cd Contamination

The release of Cd into the environment constitutes a significant pollution problem. The release of Cd from anthropogenic activities is estimated to be about 4,000 to 13,000 tons per year, with major contributions from mining activities, and burning of fossil fuels (ATSDR 1999). The Cd-yellow oil colours used by landscape painters, including Claude Monet (Fig. 21.1a) is just one of the many valuable uses of Cd, contrasting with the pernicious effects that Cd can cause in plants (Fig. 21.1b). Other important applications of Cd are in metallurgical industry and in the manufacture of nickel–cadmium batteries,

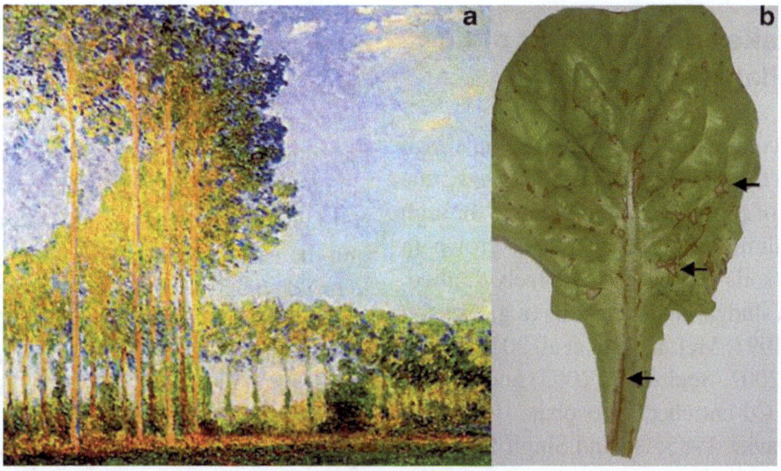

Fig. 21.1 Cadmium, a beautiful toxic colour. (**a**) *Poplars in the sun* by Claude Monet, 1891 (www.monet-on-canvas.com/prod197.htm), showing the powerful use of Cd yellow pigments and (**b**) a lettuce leaf reflecting the toxic effects of Cd exposure (*arrow*: Cd-induced necrosis)

pigments, plastic stabilizers and anti-corrosive products, phosphors for television sets, scintillation counters and X-ray screens, semiconductors and ceramic glazes (Robards and Worsfold 1991). As a consequence of this widespread and diverse usage, large quantities of Cd end up in sewage.

Treated sewage sludge ("biosolids") and phosphate fertilizers (He and Singh 1994a, b; McLaughlin et al. 2006; Singh and Agrawal 2007; Speir et al. 2003) are important sources of Cd contamination in agricultural soils. The usage of Cd in developed countries has, however, begun to decline because of its toxicity. For instance, Cd is one of the six substances banned by the European Union's Restriction on Hazardous Substances directive, which bans carcinogens in computers (2002/95/EC 2002).

Wagner (1993) estimated that non-polluted soil solutions contain Cd concentrations ranging from 0.04 to 0.32 μM. Cadmium concentrations in non-polluted soils are, however, highly variable, depending on sources of minerals and organic material. For instance, Eisler (1985) reported Cd concentrations of 0.01–1.00 mg/kg in soils of nonvolcanic origin and up to 4.50 mg/kg in soils of volcanic origin. Soil solutions that have a Cd concentration varying from 0.32 to about 1 μM are considered as moderately polluted (Sanitá di Toppi and Gabbrielli 1999). Topsoil concentrations are often more than twice as high as subsoil levels as the result of atmospheric fallout and contamination (Pierce et al. 1982). Cadmium levels up to 800 mg/kg have been reported for soils in polluted areas (IARC 1993). Jung and Thornton (1996) have found Cd concentrations up to 40 mg/kg in surface soils taken from a mining area in Korea; and more recently, Cd contaminated river water (65–240 μg/l, 0.58–2.13 μM) downstream from a mining area in Bolivia has increased the soil concentration of Cd to 20 mg/kg and the concentration of Cd in soil solutions to 27 μg/l (0.24 μM) (Oporto et al. 2007).

Contamination of topsoil is likely the most important route for human exposure to Cd, mediated through uptake of soil Cd into edible plants (IARC 1993). Cadmium concentrations of 0.5 mg/kg or more have been found in rice grown in Cd-polluted areas of Japan (Nogawa et al. 1989) and China (Cai et al. 1990). Furthermore, in a recent field study in Europe performed by Peris et al. (2007) the Cd content in edible parts of vegetables such as lettuce was found to be above the maximum levels established by the Commission Regulation no. 466/2001 for horticultural crops (466/2001/EC 2001).

3 Uptake and Transport of Cd by Plants

Cadmium accumulation by higher plants can occur through foliar or root uptake. However, the primary point of entry for Cd into plants is through the roots. Cadmium uptake by plants grown in contaminated soils has been extensively studied, particularly in sludge-amended soils (e.g. Jackson and Alloway 1991; McLaughlin et al. 2006; Singh and Agrawal 2007; Speir et al. 2003) and in soils treated with Cd-enriched phosphate fertilizers (Crews and Davies 1985; He and Singh 1994a, b; Huang et al. 2003). In general, metals have to be in an available form to be taken up by plants. Alternatively plants must have mechanisms to make the metals available. The degree to which higher plants are able to take up Cd depends on its concentration in the soil and its bioavailability. Cadmium bioavailablity in soils is modulated by the presence of organic matter, pH, redox potential, temperature, light intensity, cation exchange capacity and concentrations of other elements (Greger 1999; He and Singh 1993; Sanitá di Toppi and Gabbrielli 1999). In particular, Cd ions seem to compete with other micro and macronutrients such as calcium and zinc for the same transmembrane carriers (Sanitá di Toppi and Gabbrielli 1999), which might lead to plant nutrient deficiencies (Krupa et al. 2002). As is the case for other metals, Cd uptake tends to be reduced at low pHs because of competition with H^+ ions at root uptake sites; however, Cd bioavailability increases with decreasing pH in soil (Greger 1999). The presence of colloids from which there is a release of metals at low pH, increases the metal concentration in pore water and thus also in the roots (Greger 1999). For instance, acid rain and the resulting acidification of soils and surface waters are known to increase the geochemical mobility of Cd (Campbell 2006). Cadmium uptake also appears to be decreased in the presence of dissolved organic matter because ligands on the organic matter effectively bind Cd ions (He and Singh 1993; Prasad 1995). Chloride levels would also be expected to affect Cd availability as soil sodium chloride has an antagonistic effect on metal toxicity (Bhartia and Singh 1994).

Cadmium is believed to enter the root through the cortical tissue till the stele either by apoplastic and/or a symplastic pathway (Sanitá di Toppi and Gabbrielli 1999). The apoplast continuum of the root epidermis and cortex is readily permeable to solutes. The cell walls of the endodermal cell layer act as a barrier for apoplastic diffusion into the vascular system. In general, solutes have to be taken up into the root symplasm before they can enter the xylem (McLaughlin 2002). The cell membrane plays a key role in metal homeostasis, preventing or reducing entry into the cell. However, examples of exclusion or reduced uptake mechanisms in higher plants are limited (Benavides et al. 2005). The mechanism for metal transport across the plasma membrane to the stele is still not completely understood (McLaughlin 2002). For all cationic metals, such as Cd^{2+}, the main route for uptake across the plasma membrane is the large negative electrochemical potential produced as a result of the membrane H^+ translocating adenosine triphospatase (ATPases) (Krämer 2010; McLaughlin 2002). For instance, Costa and Morel (1994) reported that in lettuce grown in hydroponic solution with Cd concentrations from 0.05 μM to 5 μM, high amounts of Cd in roots were correlated with high contributions from H^+-ATPase in the active process of Cd uptake. Other authors contend, however, that the main route for uptake of divalent metals is via ion channels, such as Cd^{2+} and Mg^{2+} channels (McLaughlin 2002 and references therein). Subsequent to metal uptake into the root symplasm, three processes govern the movement of metals from the root into the xylem: sequestration of metals inside root cells, symplastic transport into the stele and release into the xylem (Clemens et al. 2002).

During their transport through the plant, metals become bound to cell walls, which can explain why normally Cd^{2+} ions are mainly retained in the roots, and only small amounts are translocated to the shoots (Cataldo et al. 1983; Greger 1999). But once loaded in the xylem sap, Cd is translocated to the aerial parts of plants through the transpiration stream, where they might be present as a divalent ion (Greger 1999) or complexed by several ligands, such as amino acids, organic acids and/or phytochelatins (Briat and

Lebrun 1999; Gong et al. 2003; Salt et al. 1995; Sanitá di Toppi and Gabbrielli 1999).

4 Mechanisms of Tolerance

4.1 Metal-Binding Ligands

Plants, like all living organisms, have evolved a suite of mechanisms that control and respond to the uptake and accumulation of both essential and non-essential metals. These mechanisms include the chelation and sequestration of metals by particular ligands and, in some cases, the subsequent compartmentalization of the ligand–metal complex in vacuoles.

The vacuole of plant cells plays an important role in the homeostasis of the cell (Barkla and Pantoja 1996). In most plant cells the vacuole comprises more than 80–90% of the cell volume and acts as a central storage compartment for ions, amino acids, sugars and CO_2 in the form of malate and also play a key role in the sequestration of toxic ions and xenobiotics (Barkla and Pantoja 1996; Briat and Lebrun 1999). The vacuolar membrane, named tonoplast, functions as an effective and selective metal diffusion barrier (Briat and Lebrun 1999). Vacuolar compartmentalization prevents the free circulation of Cd ions in the cytosol and forces them into a limited area (Sanitá di Toppi and Gabbrielli 1999). Several studies have shown that the vacuole is the site of accumulation of a number of metals including Cd (Hall 2002; Salt and Wagner 1993). One example is the accumulation of Cd and phytochelatins (PCs) in the vacuole involving an ATP-binding cassette (ABC) transporter (Hall 2002). Oat root tonoplast vesicles were found to accumulate Cd^{2+} by a $2H^{+}$/ion antiport mechanism (Salt and Wagner 1993).

Several metal-binding ligands have now been recognized in plants and include organic acids, amino acids, peptides and polypeptides (Rauser 1999). Among the metal-binding ligands in plant cells, the PCs and metallothioneins (MTs) are the best characterized. MTs are cysteine-rich polypeptides encoded by a family of genes whereas PCs are a family of enzymatically synthesized cysteine-rich peptides (Cobbett and Goldsbrough 2002).

In plants, PC–Cd complexes are sequestered in the vacuole (Cobbett and Goldsbrough 2002). In mesophyll protoplasts derived from tobacco plants exposed to Cd, almost all of both the Cd and PCs accumulated were confined to the vacuole (Vogeli-Lange and Wagner 1990). For instance, the non-accumulator field pennycress *Thlaspi arvense* also possesses detoxification mechanisms in which PCs play an important role (Ebbs et al. 2002; Maier et al. 2003), whereas the alpine pennycress *T. caerulescens* was found to mainly store Cd^{2+} in electron-dense granules inside vacuoles by means of complexation with malate (Ma et al. 2005; Ueno et al. 2005).

4.2 Plant Metal Accumulation and Hyperaccumulation

Plants respond to high concentrations of environmental metals in three main ways: (1) a metal excluder plant pursues the metal tolerance strategy of restricting metal movement into shoots and maintains low and constant metal concentration in their shoots up to a critical soil value; (2) indicator plants have internal metal concentrations that reflect the external metal levels; (3) metal accumulators have high accumulation of metal at very low external metal concentration (Greger 1999; Krämer 2010). The term hyperaccumulator describes a plant accumulating metals primarily in the shoots and maintaining low metal concentrations in the roots (Krämer 2010). Hyperaccumulator plants are found in metalliferous soils, such as calamine (with high levels of Zn, Pb and Cd) and serpentine soils (with high levels of Ni, Cr and Co) (Greger 1999).

Although Cd is not an essential or beneficial element for plants (with the exception mentioned before), they generally exhibit measurable Cd concentrations, particularly in roots, but also in leaves, most probably as a result of inadvertent uptake and translocation (Assunção et al. 2003). A Cd foliar concentration above 100 μg/g DW (0.01%) is considered exceptional and it is used as a threshold value for Cd hyperaccumulation (100 mg/Kg DW) (see Table 21.1, Krämer 2010; Reeves and Baker 2000). The metal

Table 21.1 Normal range, minimum values for status as Cd accumulator and hyperaccumulator in terrestrial plants

Normal range[a]	0.1–3
Critical toxicity level[a] ($\mu g\ g^{-1}$)[b]	6–10
Accumulator threshold[a]	20
Hyperaccumulation concentration criterion ($\mu g\ g^{-1}$)[b]	>100
Hyperaccumulator taxa (no.)[b]	5
Hyperaccumulator families (no.)[b]	2

[a] Reported by Reeves and Baker (2000)
[b] Reported by Krämer (2010)

hyperaccumulation characteristic is not common in higher terrestrial plants and less than 0.2% of all angiosperms have been identified as metal hyperaccumulators (Reeves and Baker 2000). Metal hyperaccumulation has evolved multiple times and the Brassicaceae plant family is well represented among the reported hyperaccumulators (Krämer 2010). Cadmium hyperaccumulator species are distributed along five taxa of two different plants (Krämer 2010).

T. caerulescens, of the Brassicaceae family began to attract increasing attention in the early 1990s and its now one of the best known hyperaccumulator plants with a capacity to hyperaccumulate Zn, Cd and Ni (Assunção et al. 2003). *T. caerulescens* plants have been found by Reeves and Baker (2000) to contain more than 100 mg/Kg Cd frequently, and more than 1,000 mg/Kg Cd occasionally, with very large variations between sites and populations, and considerable intrasite variability. Several studies have shown that *T. caerulescens* ecotype from metalliferous soils of a Zn/Pb mine spoil in the southern France (Ganges ecotype) is far superior in Cd accumulation to other ecotypes (e.g. Prayon from Belgium); in hydroponic conditions it was able to accumulate >10,000 mg/kg Cd in the shoots without showing any symptoms of phytotoxicity (Lombi et al. 2000).

T. caerulescens and *Arabidopsis halleri* have been the most successful models used in advancing our molecular understanding of metal hyperaccumulation and hypertolerance in plants (e.g. Hanikenne et al. 2008; Küpper and Kochian 2010; Schmidt and Bancroft 2011; Shahzad et al. 2010). The main driving force on the research on metal (hyper)accumulators has been the intention to use these plants, or the molecular mechanisms operating therein, for the development of phytoremediation, phytomining, bio-fortification technologies, or for the improvement of crop nutrient efficiency (Clemens 2001; Clemens et al. 2002; Krämer 2010; Shao et al. 2010).

Metal hypertolerance is an example of an extreme abiotic stress resistance trait, that, most likely, has evolved as an elemental defence against herbivory and/or pathogen attack (Boyd 2007). The hyperaccumulation trait has been hypothesized to perform several ecological functions in hyperaccumulator plants: (1) metal tolerance/disposal; (2) interference with other plants (elemental allelopathy); (3) drought resistance and (4) defence against some herbivores and pathogens (Boyd 2007). The elemental defence hypothesis predicts that metal hyperaccumulation will negatively affect at least some herbivores/pathogens in a given habitat (Boyd 2004). Vesk and Reichman (2009) have suggested that in general, high metal diets deterred insects but not gastropods. This suggests that the evolution of hyperaccumulation may have differing selective pressures depending upon the suite of herbivores the plants are naturally exposed to. Defence can be achieved by two known mechanisms: (1) by acute toxicity of high metal-containing plant tissue, in which ingestion results in mortality of the animal; via deterrence of herbivory, where if a choice is provided, high-metal containing material is ingested in lesser extent than low-metal containing tissue (Boyd 2007). In the other hand, some species of herbivores have evolved to utilise metals bioaccumulated from ingested plant biomass as a defence against subsequent predation. The implication here is that some animal consumers/predators are able to detect and selectively ingest or avoid metals in prey/food items (Boyd 2007).

5 Trophic Ecotoxicology of Cd Accumulation

Apart from the question and research around of ecological function and evolutionary value of hyperaccumulation, the consequences of this trait for other species in hyperaccumulator habitats is also raising questions and concern in ecotoxicology (e.g. Gonçalves et al. 2007; Monteiro et al. 2008). High metal levels in plant tissue will have consequences for other organisms that share those habitats. In the absence of metal avoidance behaviours, trophic biomagnification of metals, including Cd might be expected. Specialist herbivores that consume metal hyperaccumulators without harm are also of great concern, since elevated metal levels in adapted herbivores may have consequences for organisms at other trophic levels in the food webs.

For instance, the effects of Ni hyperaccumulation in the Brassicaceae endemic to serpentine soils in NE Portugal *Alyssum pintodasilvae* to the isopod *Porcellio dilatatus* were studied by Gonçalves et al. (2007). These authors suggested that the effects of hyperaccumulator litter on the activity of this important detritivore species may be significantly impaired with potential consequences on the decomposition processes.

In another study, Notten et al. (2005) examined an area with elevated soil metal concentrations, where the most dominant plant species was the stinging nettle *Urtica dioica*. This species contained very low metal concentrations, far below the maximum values found in plants from non-polluted sites. Nevertheless, the herbivore snail feeding on these plants, *Cepaea nemoralis*, contained high metal concentrations (Notten et al. 2005). Cadmium in particular was accumulated to very high levels, with consequent negative effects on reproduction (Notten et al. 2005, 2006). Dietary accumulation of Cd has also been demonstrated in aphids. In an examination of the trophic movement of Cd and zinc (Zn) between wheat grown on Cd-contaminated soils and aphids (*Rhopalosiphum padi* and *Sitobion avenae*), aphids were demonstrated to bioaccumulate both Cd and Zn up to ten times the concentrations in wheat (Merrington et al. 1997a, b).

Trophic transfer of metals within terrestrial food chains, especially in what concerns terrestrial plants, has been less studied. Therefore, hereafter, several studies concerning aquatic organisms will be presented as examples, as well as few known studies on terrestrial plants.

5.1 Factors Affecting Trophic Transfer of Metals

The bioaccumulation of metals is known to differ among species and metals because of differences in uptake and loss rates, exposure pathways and influences of environmental parameters (Fisher and Reinfelder 1995; Wang and Fisher 1999). However, less is known about the influence of these factors in the internal storage and detoxification of accumulated metal and subsequent impacts on trophic transfer. Since the ingestion of metal-contaminated food can serve as a source of metals to consumers and can result in sub-lethal toxicity (e.g. Fisher and Hook 2002), understanding the mechanisms that influence metal trophic transfer is a critical step in the management of metal-contaminated ecosystems. In general, to completely understand metal cycling through trophic levels, several factors which control the bioavailability of tissue-bound metals to consumers must be considered and understood (e.g. tissue metal distributions and concentrations, duration of exposure, nutritional status and exposure history of predator). Different species will accumulate and partition metals in varying ways depending on the detoxification mechanisms employed. The subsequent bioavailability of those partitioned metals to a consumer will be dictated by digestive and assimilative mechanisms of its digestive tract and gut passage time (Rainbow and Smith 2010; Wang and Fisher 1999). Added to this complexity is the varying ability of consumers to discriminate between different foods and contaminants, their nutritional status at the time of consumption, the degree of exposure, their excretion capacity and the exposure history for the metal in question, all of which can influence the degree of metal assimilation (Wang and Fisher 1999).

5.2 Metal Assimilation and Assimilation Efficiency

One critical parameter in understanding the trophic transfer and accumulation of a metal is its assimilation efficiency (AE) in animals from the ingested food (Wang and Fisher 1999). Assimilation efficiency has been defined as the fraction of ingested metal that is assimilated across the gut lining into the body tissue (Wang and Fisher 1999). Assimilation efficiency measurements are difficult to make and often yield variable results (Fisher and Reinfelder 1995); they can be determined by the mass balance method in which ingested and egested masses are compared to each other or to the mass retained in the animal after an appropriate gut clearance period (Fisher and Reinfelder 1995).

Determination of AEs is an important endpoint when addressing contaminant bioavailability, it is considered a first-order physiological parameter that can be quantitatively compared among different chemicals, species, and food particles under various environmental conditions (Wang and Fisher 1999). Furthermore, AE for metals has been shown to be directly proportional to metal bioaccumulation, which highlights the significance of AE in understanding and predicting metal bioaccumulation (Fisher et al. 1996).

The various factors that affect metal assimilation are reflected in the wide variety of Cd AEs that have been reported in organisms of different food chains fed biologically contaminated food. For instance AEs ranging from 1% have been reported in rats fed snail viscera (Hispard et al. 2008), to 4.7% in the lizard *Podarcis carbonelli* fed crickets (Mann et al. 2006), to 52% in the isopod *P. dilatatus* fed lettuce (Calhôa et al. 2006), and up to 76.2–94.2% for whelk *Thais clavigera* fed five different species of prey (Cheung and Wang 2005). High body-burdens among herbivorous/detritivorous invertebrates occur because of high dietary-Cd assimilation efficiencies (up to 100%, Calhôa et al. 2006; Laskowski and Hopkin 1996; Zidar et al. 2003) and low rates of elimination (e.g. Witzel 2000).

5.3 Subcellular Partition of Metals

The internal distribution and detoxification of metals within an organism, including plants, can be used to explain trophic transfer of metals but also to predict metal toxicity for the organism itself. The internal metal sequestration strategies of different species are complex and variable and the determination of the metal concentrations in different compartments can be used to understand the complex relationship between metal accumulation and toxicity (Rainbow 2002, 2007).

Over the past decades, chemistry-orientated models have been developed to predict the bioavailability and toxicity of metals focusing on identifying which metal forms are present in the aquatic environment, and investigating their interaction with the biological site of action (Paquin et al. 2002). The free ion activity model (FIAM) relied on the free metal ion activity and assumed that uptake from solution was determined by the availability of free metal ions, whereas the biotic ligand model (BLM), which is an extension of the FIAM, assumes that the effect is proportional to the concentration of metal bound to the target site (biotic ligand) and that this site is in direct contact with the external environment. These models perform well in the prediction of metal bioavailability in waterborne exposures of aquatic organisms, but also for plants (Antunes et al. 2006). When considering the contribution of the dietary route of metal exposure the gut/instestine can also act as a biotic ligand (Hogstrand et al. 2002) and metal speciation and/or dietary form is likely to be an important factor for metal assimilation.

Metals can be present in various chemical forms in an organism (e.g. plants), including the following: (a) free ionic form or complexed ion species (e.g., $CdCl_2$, $CdCl^+$, $CdCl^{3-}$); (b) bound in the active centre of functional proteins and enzymes; (c) bound to low molecular weight organic acids (e.g., citrate, malate); (d) bound to sequestration proteins (MTs and PCs); (e) bound in vesicles of the lysosomal system, as intracellular granules; (f) precipitated in extracellular granules, mineral deposits, residual bodies and exoskeletons; (g) bound to cellular constituents potentially causing dysfunction (e.g. DNA) (Vijver et al. 2004).

The various internal metal fractions all have their own binding capacity for metals, which has implications for food chain transfer to higher trophic levels. A study on the relationship between

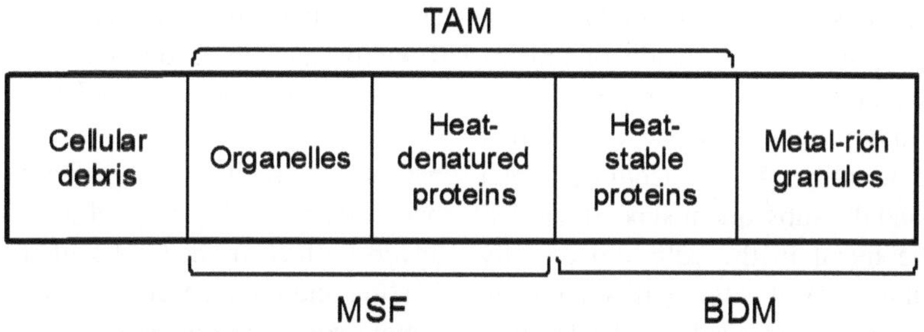

Fig. 21.2 The five subcellular fractions of metals identified in aquatic organisms and their biological significance as attributed by Wallace et al. (2003): TAM – trophically available metal; MSF – metal sensitive fraction; and BDM – Biologically detoxified metal

subcellular Cd distribution in an oligochaete and its trophic transfer to a predatory shrimp showed that only the metal present in the soluble fraction (organelles and protein fraction) of the prey is available for the predator (Wallace et al. 1998). Factors influencing the subcellular distribution in the prey will directly alter trophic transfer to predators. Wallace et al. (1998) showed that differences in subcellular distribution of Cd between resistant and non-resistant worms directly affected Cd availability for the predatory shrimp. When fed resistant worms, shrimp absorbed about four times less Cd than when fed non-resistant worms (Wallace et al. 1998). Similar conclusions were found in a study using bivalves as prey (Wallace et al. 2003), where the metal partitioning to organelles, denaturated proteins, and MTs comprise a subcellular compartment that was considered as trophically available metal (TAM) to predators (see Fig. 21.2).

A subcellular fractionation procedure (Wallace et al. 2003; Wallace and Luoma 2003) has been successfully applied in several studies of dietary accumulation of metals, particularly in marine food chains, with the purpose of explaining the variability observed in metal accumulation across the different species and food chains. This method has been considered by other authors to be a simple and pragmatic approach in the prediction of trophic transfer of metals and a first step towards a practical tool that could explain most of the variability observed in metal accumulation in organisms (Vijver et al. 2004).

Wallace and Luoma (2003) building on previous studies (Wallace and Lopez 1997; Wallace et al. 1998), postulated that Cd associated with the subcellular fractions, organelles, heat-denatured proteins (HDP) and heat-stable proteins (HSP) of prey was TAM (Fig. 21.2) and was assimilated at an efficiency of approximately 100% by the predator, while Cd bound to metal-rich granules was less bioavailable to predators. Using this procedure the accumulated metals associated with different subcellular compartments were separated into five different fractions by differential centrifugation: cellular debris, granules, organelles, HDPs and HSPs (MTs and PCs). Such subcellular partitioning is dynamic in response to metal exposure and other environmental conditions and is metal- and organism-specific. The different metal pools are not equally bioavailable to predators/consumers; thus, the determination of the metal concentration in the different subcellular compartments and the differences in its assimilation by consumers can be a useful tool to understand metal transfer to higher trophic levels. A similar procedure of subcellular fractionation to the one developed by Wallace and co-workers (Wallace et al. 2003; Wallace and Luoma 2003) was adopted by Monteiro et al. (2008) as a tool to explain the variability observed in Cd assimilation by isopods fed plants with different patterns of Cd accumulation. More recently, Calhôa et al. (2011) examined the influence of Cd speciation on metal bioavailability to *P. dilatatus* fed with food amended with either cadmium cysteinate or cadmium nitrate followed by an examination of the sub-cellular distribution in isopods using the same centrifugal fractionation protocol described above.

Other applications of this approach have been proposed. Recent studies in aquatic organisms have revealed that the subcellular partitioning model (SPM) may provide an improved method to predict Cd toxicity. As intracellular metal accumulation and the subsequent subcellular distribution of the metal in the cells are directly related to metal toxicity, it is likely that the metal concentration in a particular subcellular fraction will serve as a better toxicity predictor than the activity of the free metal ion in bulk solution (Wang and Rainbow 2006). In this approach different combinations of the five subcellular fractions have been proposed to represent a metal-sensitive fraction (organelles and HDP) and a biologically detoxified metal fraction (HSP and granules) (Wallace et al. 2003). Rainbow (2002) proposed that when accumulated metal destined for storage in a detoxified form (e.g. by MTs and granules) exceeds the detoxified binding capacity, the metals are subsequently bound with other (metabolically available) forms, with the potential to cause toxicity to the organism. The significance of the subcellular distribution of accumulated metals in toxicity assessments is now receiving increasing attention among aquatic (e.g. Cheung et al. 2006; Goto and Wallace 2010; Perceval et al. 2006; Steen Redeker et al. 2007) and terrestrial organisms (Gimbert et al. 2008; Vijver et al. 2006, 2007).

The internal metal sequestration strategies of different plant species are complex and variable and the determination of the metal concentrations in specific cellular compartments can be used to understand the complex relationship between metal accumulation and toxicity. In this way, several recent studies in aquatic organisms have revealed that the SPM may provide an improved method to predict metal toxicity (Giguère et al. 2003; Perceval et al. 2006; Rainbow 2002). To our knowledge, no similar approaches with terrestrial plants have been performed. Non-essential metals such as Cd, when associated with HSP and MRG fractions are usually considered to have been detoxified. It is generally assumed that when accumulated metal destined for storage in a detoxified form exceeds the binding capacity of detoxification systems, the metals spill over into the metabolically available pool, with the potential to cause toxicity to the organism (Rainbow 2002). Toxicity can then be related to a threshold concentration of metabolically available metal that has a specific biological significance. However, in nature, such relations are difficult to demonstrate, particularly in the case of toxicity. A lack of adverse effects in the field could be due to detoxification and compensation strategies that counter metal toxicity within cells.

5.4 Trophic Bioavailability of Plant Accumulated Cd

In 1955 in Japan, Cd toxicity was found to be the cause of Itai–itai disease. For the first time, Cd pollution was shown to have severe consequences on human health. Cadmium contaminations were attributed to the effluents from a zinc mine located in the upper reaches of Jinzu river and profoundly affected the health of the human population living in that area (Inaba et al. 2005). More recently, Kobayashi et al. (2009) demonstrated that eating Cd-polluted rice has a greater impact on the occurrence of Itai–itai disease as compared to drinking and/or cooking with Jinzu River water. This suggests that biological incorporated Cd might be more assimilated and be a more relevant form and/or route of exposure in trophic chains.

Preliminary studies in our laboratory concerning Cd trophic transfer from plants examined the role of biological metal sequestration in the assimilation efficiency of Cd in a terrestrial isopod (Calhôa et al. 2006). The results obtained in this study highlighted differences in Cd assimilation among isopods when they were provided with a plant-based food with either Cd biologically incorporated into plant tissue or superficially amended with ionic Cd^{2+} (Calhôa et al. 2006). In more detail, Monteiro et al. (2008) examined how subcellular partitioning of Cd in plants with different strategies to store and detoxify Cd would affect trophic transfer of Cd to *P. dilatatus*. After a 14-day feeding trial, net assimilation of Cd in isopods following consumption of *T. caerulescens* and *T. arvense* leaves reached 16.0 ± 2.33 and 21.9 ± 1.94 g/g animal, respectively. Cadmium assimilation efficiencies were significantly lower in isopods fed *T. caerulescens* ($10.0 \pm 0.92\%$)

than in those fed *T. arvense* (15.0±1.03%). In additional experiments by the same authors, Cd assimilation efficiencies were determined among isopods provided with purified subcellular fractions of the three plants (Monteiro et al. 2008). On the basis of their results, Cd bound to HSPs (metallothionein-like proteins) was the least bioavailable to isopods (14.4–19.6%), while Cd bound to HDPs was the most trophically available to isopods (34.4–52.8%). This work highlighted that the different strategies of plants to cope with metal stress and to internally compartmentalize them might have direct impact and a relevant role in the assimilation of metal by the animal consumer in detriment of the level of metal content accumulated in the plant.

6 Conclusion and Future Perspective

Cadmium is one of the most studied trace pollutants, and its toxic effects on humans, animals and plants are well known. The interaction of Cd with plants and its toxicological effects at the plant physiological level, as well as factors conditioning its uptake by plants, are well described. In particular, studies concerning Cd concentration in soil and its bioavailability that is modulated by the presence of organic matter, pH, redox potential, temperature and concentrations of other elements.

One of the main concerns when considering Cd pollution is its bioaccumulation and biomagnification capacity. Accumulator and tolerant species have evolved mechanisms to respond to metal uptake and accumulation, which include the chelation and sequestration of metals by particular ligands and, in some cases, the subsequent compartmentalization of the ligand–metal complex in vacuoles, achieving cellular metal homeostasis and consequent detoxification. Accumulator species generally exhibit high Cd concentrations in aboveground tissues, such as edible leaves. Plant hyperaccumulators are widely used model species in fundamental research on functional pathways involved in metal uptake-transport-accumulation processes, but they also provide valuable model systems to study the mechanisms of Cd bioaccumulation and risk to higher trophic levels.

Among the various aspects of metal ecotoxicology, the trophic transfer of metals along food chains has only recently generated a great deal of interest and research, particularly in regard to the implications for consumers of metals sequestered within prey species (Goto and Wallace 2010; Rainbow 2007; Rainbow and Smith 2010; Wang and Rainbow 2006), including some studies concerning cadmium biologically incorporated in plant species (Calhôa et al. 2006; Monteiro et al. 2008). Traditionally, dietary toxicity studies have added contaminants directly to the food source. However, contaminants that are biologically incorporated into live prey are likely to be sequestered into various sub-cellular compartments and are likely to be bound within various chemical complexes. These different compartmentalized chemical species may have very different bioavailabilities when fed to predator species. In addition, the degree to which metals are transferred within a food chain is not easily predictable, because both the metal-binding properties of the plant/prey species and subsequent bioavailability to the consumer/predator are likely to be highly variable. Future investigation on the factors influencing the internal storage and detoxification of the accumulated metal and subsequent bioavailability and assimilation of Cd through trophic food chains will improve the knowledge of Cd bioaccumulation in food webs. Both the understanding of Cd uptake of plants and the ability to predict the risk to animal consumers/predators are critical tools for the long-term safety and conservation of agricultural resources and ecosystems services and functions.

Acknowledgements The Portuguese Foundation for Science & Technology (FCT) supported the postdoctoral fellowship of M.S. Monteiro (SFRH/BPD/45911/2008).

References

Amini M, Afyuni M, Fathianpour N, Khademi H, Flühler H (2005) Continuous soil pollution mapping using fuzzy logic and spatial interpolation. Geoderma 124(3–4):223–233

Anonymous (2001a) Decision no. 2455/2001/EC of the European Parliament and of the Council of 20 November 2001 establishing the list of priority substances

Anonymous (2001b) European Commission Regulation (EC) no. 466/2001 of 8 March 2001 setting maximum levels for certain contaminants in foodstuffs

Anonymous (2002) Directive on the restriction of the use of certain hazardous substances in electrical and electronic equipment (RoHS). European community

Antunes PMC, Berkelaar EJ, Boyle D, Hale BA, Hendershot W, Voigt A (2006) The biotic ligand model for plants and metals: technical chalanges for field application. Environ Toxicol Chem 25(3):875–882

Assunção AGL, Schat H, Aarts MGM (2003) *Thlaspi caerulescens*, an attractive model species to study heavy metal hyperaccumulation in plants. New Phytol 159(2):351–360

ATSDR (1999) Toxicological profile for cadmium. Agency for Toxic Substances and Disease Registry, Draft for Public Comment. Public Health Service, U.S. Department of Health and Human Services, Atlanta, GA

Barkla BJ, Pantoja O (1996) Physiology of ion transport across the tonoplast of higher plants. Annu Rev Plant Physiol Plant Mol Biol 47(1):159–184

Benavides MP, Gallego SM, Tomaro ML (2005) Cadmium toxicity in plants. Braz J Plant Physiol 17(1):21–34

Bhartia N, Singh RP (1994) Antagonistic effect of sodium chloride to differential heavy metal toxicity regarding biomass accumulation and nitrate assimilation in *Sesamum indicum* seedlings. Phytochemistry 35(5): 1157–1161

Boyd RS (2004) Ecology of metal hyperaccumulation. New Phytol 162(3):563–567

Boyd R (2007) The defense hypothesis of elemental hyperaccumulation: status, challenges and new directions. Plant Soil 293(1):153–176

Briat J-F, Lebrun M (1999) Plant responses to metal toxicity. Compt Rendus Acad Sci III Sci Vie 322(1):43–54

Cai SW, Yue L, Hu ZN, Zhong XZ, Ye ZL, Xu HD, Liu YR, Ji RD, Zhang WH, Zhang FY (1990) Cadmium exposure and health effects among residents in an irrigation area with ore dressing wastewater. Sci Total Environ 90:67–73

Calhôa CF, Soares AMVM, Mann RM (2006) Cadmium assimilation in the terrestrial isopod, *Porcellio dilatatus* – is trophic transfer important? Sci Total Environ 371(1–3):206–213

Calhôa CF, Monteiro MS, Soares AMVM, Mann RM (2011) The influence of metal speciation on the bioavailability and sub-cellular distribution of cadmium to the terrestrial isopod, *Porcellio dilatatus*. Chemosphere 83(4):531–537

Campbell PGC (2006) Cadmium – a priority pollutant. Environ Chem 3(6):387–388

Cataldo DA, Garland TR, Wildung RE (1983) Cadmium uptake kinetics in intact soybean plants. Plant Physiol 73(3):844–848

Chaney RL, Malik M, Li YM, Brown SL, Brewer EP, Angle JS, Baker AJM (1997) Phytoremediation of soil metals. Curr Opin Biotechnol 8(3):279–284

Cheung M-S, Wang W-X (2005) Influence of metal compartmentalization in different prey on the tranfer of metals to a predatory gastropod. Mar Ecol Prog Ser 286:155–166

Cheung M-S, Fok EMW, Ng TY-T, Yen Y-F, Wang W-X (2006) Subcellular cadmium distribution, accumulation, and toxicity in a predatory gastropod, *Thais clavigera*, fed different prey. Environ Toxicol Chem 25(1):174–181

Clemens S (2001) Molecular mechanisms of plant metal tolerance and homeostasis. Planta 212(4):475–486

Clemens S, Palmgren MG, Kramer U (2002) A long way ahead: understanding and engineering plant metal accumulation. Trends Plant Sci 7(7):309–315

Cobbett C, Goldsbrough P (2002) Phytocelatins and metallothioneins: roles in heavy metal detoxification and homeostasis. Annu Rev Plant Biol 53(1):159–182

Costa G, Morel JL (1994) Efficiency of H+–Atpase activity on cadmium uptake by 4 cultivars of lettuce. J Plant Nutr 17(4):627–637

Crews HM, Davies BE (1985) Heavy metal uptake from contaminated soils by 6 varieties of lettuce (*Lactuca sativa* L). J Agric Sci 105:591–595

Ebbs S, Lau I, Ahner B, Kochian L (2002) Phytochelatin synthesis is not responsible for Cd tolerance in the Zn/Cd hyperaccumulator *Thlaspi caerulescens* (J. & C. Presl). Planta 214(4):635–640

Eisler R (1985) Cadmium hazards to fish, wildçilfe and invertebrates (Contaminant Hazard Reviews). Patuxent Wildlife Research Center, U.S. Fish and Wildlife Service, Laurel, MD

Fisher NS, Hook SE (2002) Toxicology tests with aquatic animals need to consider the trophic transfer of metals. Toxicology 181:531–536

Fisher NS, Reinfelder JR (1995) The trophic transfer of metals in marine systems. In: Tessier A, Turner DR (eds) Metal speciation and bioavailability in aquatic systems. Wiley, London, pp 363–406

Fisher NS, Teyssie JL, Fowler SW, Wang WX (1996) Accumulation and retention of metals in mussels from food and water: a comparison under field and laboratory conditions. Environ Sci Technol 30(11):3232–3242

Giguère A, Couillard Y, Campbell PGC, Perceval O, Hare L, Pinel-Alloul B, Pellerin J (2003) Steady-state distribution of metals among metallothionein and other cytosolic ligands and links to cytotoxicity in bivalves living along a polymetallic gradient. Aquat Toxicol 64(2):185–200

Gimbert F, Vijver MG, Coeurdassier M, Scheifler R, Peijnenburg WJGM, Badot P-M, Vaufleury AD (2008) How subcellular partitioning can help to understand heavy metal accumulation and elimination kinetics in snails. Environ Toxicol Chem 27(6):1284–1292

Gonçalves M, Gonçalves S, Portugal A, Silva S, Sousa J, Freitas H (2007) Effects of nickel hyperaccumulation in *Alyssum pintodasilvae* on model arthropods representatives of two trophic levels. Plant Soil 293(1): 177–188

Gong J-M, Lee DA, Schroeder JI (2003) Long-distance root-to-shoot transport of phytochelatins and cadmium in *Arabidopsis*. Proc Natl Acad Sci USA 100(17): 10118–10123

Goto D, Wallace WG (2010) Metal intracellular partitioning as a detoxification mechanism for mummichogs (*Fundulus heteroclitus*) living in metal-polluted salt marshes. Mar Environ Res 69(3):163–171

Greger M (1999) Metal availability and bioconcentration in plants. In: Prasad MNV, Hagemeyer J (eds) Heavy metal stress in plants – from molecules to ecosystems. Springer, Berlin, Germany, pp 1–27

Hall JL (2002) Cellular mechanisms for heavy metal detoxification and tolerance. J Exp Bot 53(366):1–11

Hanikenne M, Talke IN, Haydon MJ, Lanz C, Nolte A, Motte P, Kroymann J, Weigel D, Kramer U (2008) Evolution of metal hyperaccumulation required cis-regulatory changes and triplication of HMA4. Nature 453(7193):391–395

He QB, Singh BR (1993) Effect of organic-matter on the distribution, extractability and uptake of cadmium in soils. J Soil Sci 44(4):641–650

He QB, Singh BR (1994a) Crop uptake of cadmium from phosphorus fertilizers: I. Yield and cadmium content. Water Air Soil Pollut 74(3–4):251–265

He QB, Singh BR (1994b) Crop uptake of cadmium from phosphorus fertilizers: II. Relationship with extractable soil cadmium. Water Air Soil Pollut 74(3–4):267–280

Hispard F, de Vaufleury A, Cosson RP, Devaux S, Scheifler R, Coeurdassier M, Gimbert F, Martin H, Richert L, Berthelot A, Badot PM (2008) Comparison of transfer and effects of Cd on rats exposed in a short experimental snail-rat food chain or to CdCl2 dosed food. Environ Int 34(3):381–389

Hogstrand C, Wood CM, Bury NR, Wilson RW, Rankin JC, Grosell M (2002) Binding and movement of silver in the intestinal epithelium of a marine teleost fish, the European flounder (*Platichthys flesus*). Comp Biochem Physiol C Toxicol Pharmacol 133(1–2):125–135

Huang B, Kuo S, Bembenek R (2003) Cadmium uptake by lettuce from soil amended with phosphorus and trace element fertilizers. Water Air Soil Pollut 147(1–4):109–127

IARC (1993) Beryllium, cadmium, mercury, and exposures in the glass manufacturing industry (IARC monographs on the evaluation of carcinogenic risks to humans). Lyon, France

Inaba T, Kobayashi E, Suwazono Y, Uetani M, Oishi M, Nakagawa H, Nogawa K (2005) Estimation of cumulative cadmium intake causing Itai-itai disease. Toxicol Lett 159(2):192–201

Jackson AP, Alloway BJ (1991) The bioavailability of cadmium to lettuce and cabbage in soils previously treated with sewage sludges. Plant Soil 132(2):179–186

Jung MC, Thornton I (1996) Heavy metal contamination of soils and plants in the vicinity of a lead-zinc mine, Korea. Appl Geochem 11(1–2):53–59

Kobayashi E, Suwazono Y, Dochi M, Honda R, Kido T (2009) Influence of consumption of cadmium-polluted rice or Jinzu river water on occurrence of renal tubular dysfunction and/or Itai-itai disease. Biol Trace Elem Res 127(3):257–268

Krämer U (2010) Metal hyperaccumulation in plants. Annu Rev Plant Biol 61(1):517–534

Krupa Z, Siedlecka A, Skórzynska-Polit E, Maksymiek W (2002) Heavy metal interactions with plant nutrients. In: Prasad MNV, Strzalka K (eds) Physiology and biochemistry of metal toxicity and tolerance in plants. Kluwer, Dordrecht, Netherlands, pp 287–301

Küpper H, Kochian LV (2010) Transcriptional regulation of metal transport genes and mineral nutrition during acclimatization to cadmium and zinc in the Cd/Zn hyperaccumulator, *Thlaspi caerulescens* (Ganges population). New Phytol 185(1):114–129

Lane TW, Morel FOMM (2000) A biological function for cadmium in marine diatoms. Proc Natl Acad Sci USA 97(9):4627–4631

Lane TW, Saito MA, George GN, Pickering IJ, Prince RC, Morel FMM (2005) Biochemistry: a cadmium enzyme from a marine diatom. Nature 435(7038):42

Laskowski R, Hopkin SP (1996) Accumulation of Zn, Cu, Pb and Cd in the garden snail (*Helix aspersa*): implications for predators. Environ Pollut 91(3):289–297

Liu M-Q, Yanai J, Jiang R-F, Zhang F, McGrath SP, Zhao F-J (2008) Does cadmium play a physiological role in the hyperaccumulator *Thlaspi caerulescens*? Chemosphere 71(7):1276–1283

Lombi E, Zhao FJ, Dunham SJ, McGrath SP (2000) Cadmium accumulation in populations of *Thlaspi caerulescens* and *Thlaspi goesingense*. New Phytol 145(1):11–20

Ma JF, Ueno D, Zhao FJ, McGrath SP (2005) Subcellular localisation of Cd and Zn in the leaves of a Cd-hyperaccumulating ecotype of *Thlaspi caerulescens*. Planta 220(5):731–736

Maier EA, Matthews RD, McDowell JA, Walden RR, Ahner BA (2003) Environmental cadmium levels increase phytochelatin and glutathione in lettuce grown in a chelator-buffered nutrient solution. J Environ Qual 32(4):1356–1364

Mann RM, Serra EA, Soares AMVM (2006) Asimilation od cadmium in a european lacertid lizard: is trophic transfer important? Environ Toxicol Chem 25(12):3199–3203

McLaughlin MJ (2002) Bioavailability of metals to terrestrial plants. In: Allen HE (ed) Bioavailability of metals in terrestrial ecosystems: importance of partitioning for bioavailability to invertebrates, microbes, and plants. Society of Environmental Toxicology and Chemistry, Lawrence, MA, pp 39–68

McLaughlin MJ, Whatmuff M, Warne M, Heemsbergen D, Barry G, Bell M, Nash D, Pritchard D (2006) A field investigation of solubility and food chain accumulation of biosolid-cadmium across diverse soil types. Environ Chem 3(6):428–432

Merrington G, Winder L, Green I (1997a) The uptake of cadmium and zinc by the bird-cherry oat aphid *Rhopalosiphum padi* (Homoptera:Aphididae) feeding on wheat grown on sewage sludge amended agricultural soil. Environ Pollut 96(1):111–114

Merrington G, Winder L, Green I (1997b) The bioavailability of Cd and Zn from soils amended with sewage sludge to winter wheat and subsequently to the grain aphid *Sitobion avenae*. Sci Total Environ 205(2–3): 245–254

Monteiro M, Santos C, Soares AMVM, Mann RM (2008) Does subcellular distribution in plants dictate the trophic bioavailability of cadmium to *Porcellio dilatatus* (Crustacea, Isopoda)? Environ Toxicol Chem 27(12):2548–2556

Nogawa K, Honda R, Kido T, Tsuritani I, Yamada Y, Ishizaki M, Yamaya H (1989) A dose-response analysis of cadmium in the general environment with special reference to total cadmium intake limit. Environ Res 48(1):7–16

Notten MJM, Oosthoek AJP, Rozema J, Aerts R (2005) Heavy metal concentrations in a soil-plant-snail food chain along a terrestrial soil pollution gradient. Environ Pollut 138(1):178–190

Notten MJM, Oosthoek AJP, Rozema J, Aerts R (2006) Heavy metal pollution affects consumption and reproduction of the landsnail *Cepaea nemoralis* fed on naturally polluted *Urtica dioica* leaves. Ecotoxicology 15(3):295–304

Nriagu JO (1996) A history of global metal pollution. Science 272(5259):223–224

Oporto C, Vandecasteele C, Smolders E (2007) Elevated cadmium concentrations in potato tubers due to irrigation with river water contaminated by mining in Potosi, Bolivia. J Environ Qual 36(4):1181–1186

Padmavathiamma PK, Li LY (2007) Phytoremediation technology: hyper-accumulation metals in plants. Water Air Soil Pollut 184(1–4):105–126

Paquin PR, Gorsuch JW, Apte S, Batley GE, Bowles KC, Campbell PGC, Delos CG, Di Toro DM, Dwyer RL, Galvez F (2002) The biotic ligand model: a historical overview. Comp Biochem Physiol C Toxicol Pharmcol 133(1–2):3–35

Perceval O, Couillard Y, Pinel-Alloul B, Campbell PGC (2006) Linking changes in subcellular cadmium distribution to growth and mortality rates in transplanted freshwater bivalves (*Pyganodon grandis*). Aquat Toxicol 79(1):87–98

Peris M, Mico C, Recatala L, Sanchez R, Sanchez J (2007) Heavy metal contents in horticultural crops of a representative area of the European Mediterranean region. Sci Total Environ 378(1–2):42–48

Pierce FJ, Dowdy RH, Grigal DF (1982) Concentrations of six trace metals in some major Minnesota soil series. J Environ Qual 11(3):416–422

Pilon-Smits E (2005) Phytoremediation. Annu Rev Plant Biol 56(1):15–39

Prasad MNV (1995) Cadmium toxicity and tolerance in vascular plants. Environ Exp Bot 35(4):525–545

Rainbow PS (2002) Trace metal concentrations in aquatic invertebrates: why and so what? Environ Pollut 120(3):497–507

Rainbow PS (2007) Trace metal bioaccumulation: models, metabolic availability and toxicity. Environ Int 33(4):576–582

Rainbow PS, Smith BD (2010) Trophic transfer of trace metals: subcellular compartmentalisation in bivalve prey and comparative assimilation efficiencies of two invertebrate predators. J Exp Mar Biol Ecol 390(2): 143–148

Rauser W (1999) Structure and function of metal chelators produced by plants. Cell Biochem Biophys 31(1): 19–48

Reeves RD, Baker AJM (2000) Metal-accumulating plants. In: Raskin I, Ensley BD (eds) Phytoremediation of toxic metals: using plants to clean up the environment. Wiley, New York, pp 193–229

Robards K, Worsfold P (1991) Cadmium: toxicology and analysis – a review. Analyst 116:549–568

Salt DE, Wagner GJ (1993) Cadmium transport across tonoplast of vesicles from oat roots. Evidence for a Cd^{2+}/H^{+} antiport activity. J Biol Chem 268(17): 12297–12302

Salt DE, Blaylock M, Kumar NPBA, Dushenkov V, Ensley BD, Chet I, Raskin I (1995) Phytoremediation: a novel strategy for the removal of toxic metals from the environment using plants. Nat Biotechnol 13(5): 468–474

Sanitá di Toppi L, Gabbrielli R (1999) Response to cadmium in higher plants. Environ Exp Bot 41(2): 105–130

Schmidt R, Bancroft I (2011) Perspectives on genetics and genomics of the Brassicaceae. In: Schmidt R, Bancroft I (eds) Genetics and genomics of the Brassicaceae, vol 9. Springer, New York, pp 617–632

Shahzad Z, Gosti F, Frérot H, Lacombe E, Roosens N, Saumitou-Laprade P, Berthomieu P (2010) The five *AhMTP1* zinc transporters undergo different evolutionary fates towards adaptive evolution to zinc tolerance in *Arabidopsis halleri*. PLoS Genet 6(4):e1000911

Shao H-B, Chu L-Y, Ni F-T, Guo D-G, Li H, Li W-X (2010) Perspective on phytoremediation for improving heavy metal-contaminated soils. In: Ashraf M, Ozturk M, Ahmad MSA (eds) Plant adaptation and phytoremediation. Springer, Netherlands, pp 227–244

Singh RP, Agrawal M (2007) Effects of sewage sludge amendment on heavy metal accumulation and consequent responses of *Beta vulgaris* plants. Chemosphere 67(11):2229–2240

Speir TW, Van Schaik AP, Percival HJ, Close ME, Pang LP (2003) Heavy metals in soil, plants and groundwater following high-rate sewage sludge application to land. Water Air Soil Pollut 150(1–4):319–358

Steen Redeker E, van Campenhout K, Bervoets L, Reijnders H, Blust R (2007) Subcellular distribution of Cd in the aquatic oligochaete *Tubifex tubifex*, implications for trophic availability and toxicity. Environ Pollut 148(1):166–175

Ueno D, Ma JF, Iwashita T, Zhao FJ, McGrath SP (2005) Identification of the form of Cd in the leaves of a superior Cd-accumulating ecotype of *Thlaspi caerulescens* using Cd-113-NMR. Planta 221(6):928–936

Vassilev A, Yordanov I (1997) Reductive analysis of factors limiting growth of cadmium-treated plants: a review. Bulg J Plant Physiol 23:114–133

Vesk P, Reichman S (2009) Hyperaccumulators and herbivores – a bayesian meta-analysis of feeding choice trials. J Chem Ecol 35(3):289–296

Vijver MG, van Gestel CAM, Lanno RP, van Straalen NM, Peijnenburg WJGM (2004) Internal metal sequestration and its ecotoxicological relevance: a review. Environ Sci Technol 38(18):4705–4712

Vijver MG, van Gestel CAM, van Straalen NM, Lanno RP, Peijnenburg WJGM (2006) Biological significance of metals partitioned to subcellular fractions within earthworms (*Aporrectodea caliginosa*). Environ Toxicol Chem 25:807–814

Vijver MG, Koster M, Peijnenburg WJGM (2007) Impact of pH on Cu accumulation kinetics in earthworm cytosol. Environ Sci Technol 41(7):2255–2260

Vogeli-Lange R, Wagner GJ (1990) Subcellular localization of cadmium and cadmium-binding peptides in tobacco leaves: implication of a transport function for cadmium-binding peptides. Plant Physiol 92(4):1086–1093

Wagner GJ (1993) Accumulation of cadmium in crop plants and its consequences to human health. Adv Agron 51:173–212

Wallace WG, Lopez GR (1997) Bioavailability of biologically sequestered cadmium and the implications of metal detoxification. Mar Ecol Prog Ser 147(1–3):149–157

Wallace WG, Luoma SN (2003) Subcellular compartmentalization of Cd and Zn in two bivalves. II. Significance of trophically available metal (TAM). Mar Ecol Prog Ser 257:125–137

Wallace WG, Lopez GR, Levinton JS (1998) Cadmium resistance in an oligochaete and its effect on cadmium trophic transfer to an omnivorous shrimp. Mar Ecol Prog Ser 172:225–237

Wallace WG, Lee BG, Luoma SN (2003) Subcellular compartmentalization of Cd and Zn in two bivalves. I. Significance of metal-sensitive fractions (MSF) and biologically detoxified metal (BDM). Mar Ecol Prog Ser 249:183–197

Wang W-X, Fisher NS (1999) Assimilation efficiencies of chemical contaminants in aquatic invertebrates: a synthesis. Environ Toxicol Chem 18(9):2034–2045

Wang WX, Rainbow PS (2006) Subcellular partitioning and the prediction of cadmium toxicity to aquatic organisms. Environ Chem 3(6):395–399

Witzel B (2000) The influence of zinc on the uptake and loss of cadmium and lead in the woodlouse, *Porcellio scaber* (Isopoda, Oniscidea). Ecotoxicol Environ Saf 47(1):43–53

Zidar P, Drobne D, Štrus J, Blejec A (2003) Intake and assimilation of zinc, copper, and cadmium in the terrestrial isopod *Porcellio scaber* Latr. (Crustacea, Isopoda). Bull Environ Contam Toxicol 70(5): 1028–1035

The Role of Soil Organic Matter in Trace Element Bioavailability and Toxicity

22

Gabrijel Ondrasek and Zed Rengel

Abstract

Soil organic matter (SOM) is a highly important pedospheric variable for agricultural practice and ecological functions. A decline in SOM content during the last two centuries has affected agricultural areas in many countries. Among multifunctional roles of SOM, one of the most crucial is that, due to its huge reactive interfaces, SOM strongly competes with other soil matrix constitutes/ligands in trace element (TE) adsorption/chemosorption. Thus, declining SOM content may cause soil degradation, particularly from the standpoint of disturbing soil's capacity to retain potentially toxic TEs and therefore increase a risk of their migration in the environment. Using geochemical modelling with realistic natural conditions, we highlighted the importance and complexity of SOM in the rhizosphere interactions with some of the widespread and potentially toxic TEs. It was shown that biogeochemistry of Cd, Zn and Cu may vary distinctly in relatively similar environmental conditions (e.g. narrow pH range, same SOM content and temperature), thus influencing their mobility and bioavailability, i.e. toxicity in the soil–plant–animal continuum.

Keywords

Soil organic matter • Metals • Rhizosphere • Geochemical modelling

G. Ondrasek (✉)
Faculty of Agriculture, University of Zagreb, Zagreb, Croatia
e-mail: gondrasek@agr.hr

Z. Rengel
School of Earth and Environment, University of Western Australia, Crawley, WA, Australia
e-mail: zed.rengel@uwa.edu.au

P. Ahmad and M.N.V. Prasad (eds.), *Abiotic Stress Responses in Plants: Metabolism, Productivity and Sustainability*, DOI 10.1007/978-1-4614-0634-1_22, © Springer Science+Business Media, LLC 2012

1 Introduction

During the past two centuries of intensive agricultural management (e.g. conventional tillage, intensive cultivation, reduction in animal manure but increased application of inorganic fertilisers) and other human activities (e.g. urbanisation, deforestation), negative influences on soil organic matter (SOM) content (i.e. declining), accompanied with land overloading by different contaminants. Many traditionally agricultural areas in developing as well as developed countries are seriously depleted in SOM content (i.e. chemically degraded). According to recent estimates, mostly as a consequence of reduced SOM, around 20% of agricultural land over the world can be considered chemically degraded, whereas around 90% of European soils are in the category from low to medium SOM content (i.e. 0–6% w/w organic carbon; Ondrasek 2008 and references therein). Chemically degraded land areas with declining SOM content in conjunction with areas contaminated by inorganic pollutants (e.g. trace elements; TEs) are identified among the most important hazards to (1) natural ecosystems, (2) food/feed production, (3) human health and (4) economy of European countries (e.g. European Commission 2006; Ondrasek et al. submitted). According to the same sources, total costs to European economy caused by soil contamination/SOM depletion are estimated in the range of US$8–32 billion annually, and do not include the damage to the soil ecological functions as these are almost impossible to quantify.

Over millennia of agricultural practice, SOM was recognised as a highly desirable soil property given its positive influence on many pedological (fertility, water/air capacity, topsoil worming, biological functioning) and crop (yield high/quality, earlier fructification) characteristics. Increasing the SOM content is a relatively slow process that takes several decades, while its substantial portion may be mineralised in only several years of cultivation. Among SOM multifunctionalities, one of the most crucial is that it has huge electro-potentially reactive interfaces, which enables SOM to strongly compete with other soil matrix constitutes (clays, hydroxides) in TEs adsorption/chemosorption. Depending on its origin (plant/animal) and characteristics (e.g. complex humics or simple root-exuded organics), SOM simultaneously represents an important source and active surface (buffer) of relatively bioavailable fractions of TEs. Declining of the SOM pool may further exacerbate soil degradation and disturb soil's buffering capacity to retain potentially toxic TEs, thus elevating a risk of exposing natural resources and food crops to their stronger influence.

Among a wide range of TE metal(loid)s, some (under specific conditions) such as Cu and Cd have a strong potential for organo-complexation, whereas Zn exhibits strong interaction with inorganic ligands (e.g. chlorides; Ondrasek et al. submitted). A wide range of metallic ligands (humics, organic acids, chlorides, carbonates, sulphates) exist in cultivated soil surface horizons, and influence presumably phytoavailable (to a certain extent) TE fractions such as free cationic metal forms. However, in which direction and to what extent chelation of particular TE will be pronounced, depends on many variables, with the two master ones being SOM (Rengel 2007) and pH (Adriano et al. 2004). In the next section, we discuss SOM (origin, dynamics in terrestrial ecosystems, characterisation) and the importance of its fractions in TEs (Cu, Zn, Cd) biogeochemistry in the rhizosphere.

2 The Origin of Soil Organic Matter

The origin of soil organic matter (SOM) varies considerably depending of the particular ecosystem. For instance, in the natural forest or grassland systems, SOM is mostly in situ produced from biomass sources (Fig. 22.1). The largest contribution to maintenance and/or rising (accumulation) of SOM comes from above-ground plant biomass production such as leaf and woody litter in addition to belowground root necromass and exuded organic substances (Kalbitz et al. 2000). For instance, leaf contribution in overall litterfall biomass production in forest/woodland

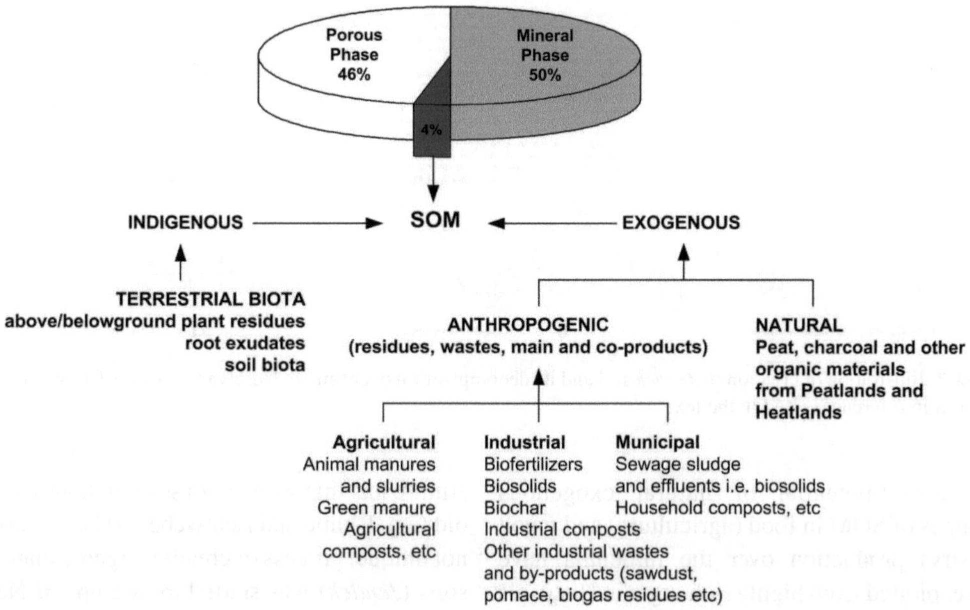

Fig. 22.1 Possible sources of SOM for commonly cultivated mineral soil (adapted from Ondrasek 2008 and references therein)

ecosystems is 70–75%, in wooded grasslands 50–60% and in sclerophyllous Mediterranean forests 54% (Matthews 1997).

Contrary, in today's modern and intensive agro-ecosystems, with possible several crop rotations over the short period, such as associated rain-fed cereal and irrigated horticultural production, the main part of above-ground biomass and litterfall (e.g. yield, straw) is exported from the system Ondrasek et al. (in press). To prevent depletion and retain its multifunctional role, SOM must be replenished, i.e. introduced into the arable topsoil layer through (1) specific land use and management practices and/or (2) other natural and/or anthropogenic exogenous resources (Fig. 22.1).

Anthropogenic exogenous sources of SOM are different sorts of organic wastes and/or co-products from urban (e.g. biosolids from sewage sludge) and peri-urban or rural activities such as agriculture (e.g. animal manures/slurries) and industry (e.g. biosolids from industrial wastewaters) (Fig. 22.1). In the last several decades there have been increasing environmental regulations that have resulted in more animal wastes treatment options, and thus affecting characteristics of residues that are subsequently applied to land. According to recent estimations (Van-Camp et al. 2004) around 1.65 billion tonnes of exogenous SOM are produced in the European Union (EU) each year, from which 61% represent animal wastes, 25% crop residues, 7% industrial wastes and 7% urban and municipal wastes (e.g. sewage sludge, biowastes and green wastes). The predominant application of exogenous SOM on arable land areas in EU is (on a weight basis) in the form of animal manure and slurries (97%), whereas almost a negligible portion comes from industrial wastes (2%) and sewage sludge (1%) application.

Some of mentioned organic wastes (precursors of organic soil amendments) before application must be stabilised, i.e. degraded and decomposed to a certain degree through composting or similar processes, to final product such as biosolids. Data about biosolids production and management practices across Europe and the USA (e.g. Epstein 2003; Ondrasek 2008 and references therein) show that in EU countries, from the total biosolids production (cca 7.4 million dry tones per year) 42% is disposed of in landfills, 36% is used in agriculture and 11% is incinerated, whereas in the USA, from total production (cca 4.1 million dry tones per year) 60% of biosolids is beneficially used (e.g. land application), 17% is disposed and 22% is incinerated.

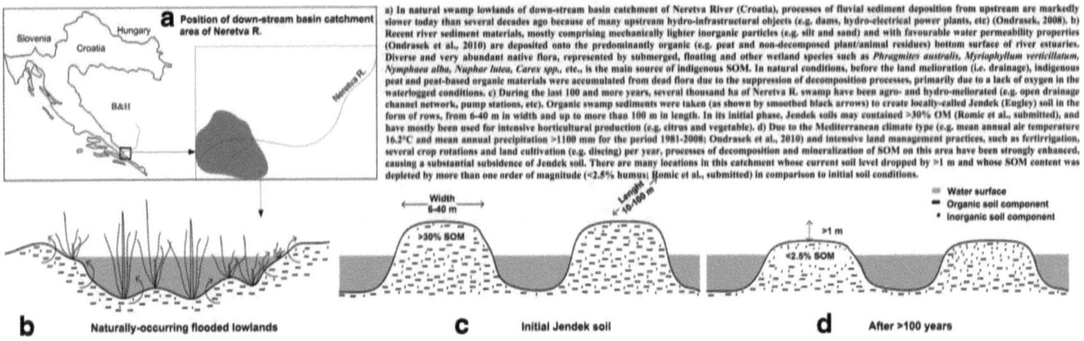

Fig. 22.2 Illustration of creation of *Jendek* soil and its decomposition over time in Neretva R. swamp, Croatia according to data from references cited in the text

A great potential of natural exogenous resources of SOM in food (agriculture) and wood (forestry) production over the millennia have been exploited from highly rich organic Histosols of peatlands and heathlands. Both systems were used as indigenous (after drainage/drying) or exogenous (as soil amendments) SOM sources for different purposes in crop production (nutrients source) and animal (bedding material, feed supplements) food, as well for wood (forestation) production. Heathlands are unique ecosystems found over the world in temperate upland regions, characterised by low-growing vegetation and organically enriched soils, which develop because environmental factors such as waterlogging and acidity constrain biomass decomposition (Holden et al. 2007). Peatlands are distributed just on around 3% of the global land surface (Strak 2008), but comprise 20–30% of the global, and ~50% of the UK's soil carbon stock (Yallop and Clutterbuck 2009). At the European scale, the current area of peatlands is estimated at 340,000 km^2 (Byrne et al. 2004), of which almost 50% has already been artificially drained for forestry (90,000 km^2), agriculture (65,000 km^2) and peat extraction (2,300 km^2).

One of the oldest examples of exogenous SOM application (peat, charcoal, animal residues, etc.) to naturally occurring nutrient poorly Oxisol was found in more than 10,000-year-old man-made Terra Preta Anthrosols of Amazonia (Woods et al. 2006). Similar examples of improvement of naturally low-fertile Arenosols with different organic amendments (peat, manures, forest litter) can be recognised in European Plaggic and Terric Anthrosols that were up to several thousand years old (e.g. Blume and Leinweber 2004). Unusual, if not unique, process of creating organic man-made soils (*Jendek*) was started in swamps of Neretva River valley (Croatia) more than 100 years ago (Fig. 22.2). Over decades that initially organic soil (>30% OM) was totally changed (up to 1-m depth) in most physical, chemical and biological properties, and became mineral soil (<2% OM) due to intensive land management practices and favourable natural conditions (i.e. Mediterranean climate) for decomposition and mineralization of indigenous OM (Fig. 22.2).

Land use and management practices may also be a powerful tool for SOM conservation and/or even improvement, when an application of exogenous organic materials is restricted or non-achievable (e.g. small livestock production) Ondrasek et al. in press). Many case studies on long-term changes of SOM in natural and agricultural systems worldwide confirmed that native (vs. arable) systems have greater SOM, i.e. carbon (C) storage, and that conversion from native to arable cropping significantly influence soil C (Soussana et al. 2004). Conversion from grassland and forest to cropland usually results in a decrease in SOM content, and the opposite conversion increases SOM content (Ondrasek 2008 and references therein). Higher SOM content in grassland and forest vs. cropland may be due to many factors: greater return of plant residues, return of dung during grazing, absence of soil disturbance and therefore restricted aeration causing decreased mineralisation.

Increased concerns for healthy food production and environment protection, i.e. increased emphasis on sustaining the productive capacity of natural resources, have raised interest in the maintenance and increasing of SOM for various land uses and management practices. For example, conservation tillage systems in which ≥30% of the crop residues remains on the soil surface after planting (IPCC 2000), mainly with the aim to restrict soil erosion, result in reduced soil compaction, disturbance and energy consumption, i.e. they conserve plant-available water and SOM. Land management practices from (1) reduced (minimum) tillage, where ploughing replaces surface tillage and/or strip tillage, to (2) complete absence (zero) tillage are increasingly used worldwide on >70 millions ha (e.g. Cerri et al. 2004). There are also other agricultural practices (crop selection, green manure, (fert)irrigation, etc.) which may improve/conserve SOM in arable topsoils.

3 Classification, Composition and Structure of SOM

The SOM content is one of the most varying pedosphere properties, ranging from <0.2% in salt-affected sandy soils of arid and semi-arid climate zone (Khoshgoftarmanesh et al. 2006) up to >99% in highly organic Histosoils from uplands in humid and boreal climates of northern hemisphere (Ondrasek 2008). In cultivated topsoils, SOM represents a fairly small fraction (1–4%), but it is one of the most complex, dynamic and multifunctional soil components. Although its importance was recognised from different scientific perspectives, a consensual definition of SOM is still missing, mostly because in disparities such as the following (Ondrasek 2008 and references therein): (1) inclusion/exclusion of living biomass, litter, fragmentation and humification layers and (2) threshold degree of decomposition. Disregarding this, SOM can be defined as soil biota and non-living biomass, i.e. a wide range of organic residues and effluents (Fig. 22.3).

Non-living SOM can be present in soil as fresh matter (litterfall) that is biologically and chemically altered i.e. decomposed and/or polymerised (re-synthesised) and stabilised to certain extent by soil micro-organisms (Fig. 22.3). A typical fresh plant biomass is composed of (hemi)cellulose (~65%) and lignin (~20%), however, its quantity/quality vary among natural ecosystems, influencing the rate of decomposition and the balance between mineralisation/immobilisation.

As a portion of the solid SOM, there is a pool of potentially dissolved OM (DOM), which is during last few decades typically quantified as dissolved organic carbon (DOC) i.e. dissolved OM fraction able to pass through a filter 0.4–0.7 μm (Peichl et al. 2007; Ondrasek 2008). The concentration of DOC in the bulk soil and/or rhizosphere solution may differ considerably (Fig. 22.3) and is controlled by the following: (1) total SOM content (Ondrasek et al. 2009a), pH and ionic strength (Kalbitz et al. 2000), soil watering status (Yallop and Clutterbuck 2009), crop species/genotype (Rengel 2002), rhizosphere salinity (Ondrasek et al. submitted) or trace elements (TEs) contamination (e.g. Chiang et al. 2011). Usual DOC concentrations in mineral soil solutions may be up to one order of magnitude lower (e.g. <4 mg/L from mineral soil; Peichl et al. 2007) than in drainage solution from organic (peat) soils (20–110 mg/L; Kalbitz et al. 2000; Glatzel et al. 2003; Ondrasek et al. submitted). (Fig. 22.3).

The solid/dissolved SOM fractions may contain labile low-molecular-weight (LMW; <10%) and more stable high-molecular-weight (HMW; >90%) substances (Fig. 22.3). LMW materials refer to carbohydrates, small proteins/peptides, organic acids (OAs), amino acids, fats, siderophores, phenolics, vitamins, hormones, etc. (e.g. Neumann and Romheld 2000; Rengel 2002). Besides preferentially originating from decomposed biomass, LMW compounds are also released as root exudates and/or microbial metabolites (Rengel and Marschner 2005). Recently, Chiang et al. (2011) have observed a significant increment in concentration of volatile (acetic, propionic and butyric acids) and non-volatile (lactic, maleic, succinic and malic) LMW-OAs in the millet rhizosphere contaminated by Cd. Another example is secretion of LMW siderophores by

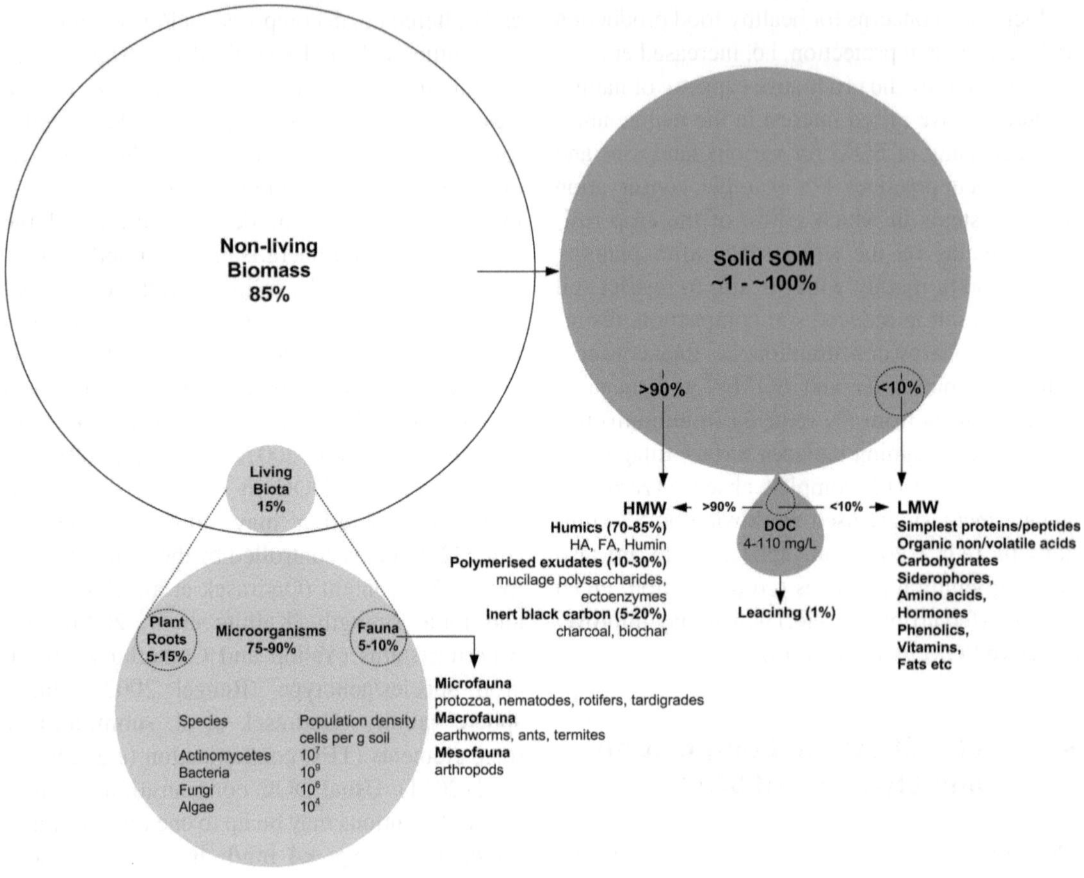

Fig. 22.3 A conceptual classification and composition of SOM according to adapted data from references cited in the text (% inside the circles are expressed on a dry weight basis)

soil microbes and phytosiderophores by plant roots that can chelate some TEs (Fe, Zn) and increase their bioavailability (Rengel 2002; Rengel and Marschner 2005).

Bulk of DOC-LMW compounds from the soil solution are rapidly decomposed/mineralised by microorganisms and are typically maintained at low concentrations (<50 μM; Van Hees et al. 2005) representing a labile (easily degradable) organic pool. Notwithstanding that the labile LMW pool has one of the shortest residence times (1–10 h), its presence in the soil (rhizosphere) solution appears to be: (1) stable, being replenished continuously by rhizosphere exudation and from fresh litterfall (Van Hees et al. 2005), and (2) much lower in comparison to a stable HMW pool. In agricultural/forest lands, Sachse et al. (2005) found that DOC-HMW compounds (humics + polysaccharides) dominated in all samples (55–77%), whereas DOC-LMW (e.g. acids) portion was only several %s; however, not all organic fractions were accounted for (Fig. 22.3); thus, contribution of DOC-LMW/-HMW pools could be higher. For example, after a 90-day incubation, Kalbitz et al. (2003) found for fresh and less humified organic material (straw, forest floor litter) that DOC was mineralised 61–93% and the labile fraction comprised 60–90% of DOC, whereas for humified peats and Oa forest layers the mineralised and labile pools represented only 4–9% and 3–6% of DOC, respectively. Therefore, for humified sources the predominant portion (>90%) of OM in dissolved phase appears to be stable DOC-HMW substances, which is similar to a solid SOM fraction (Fig. 22.3).

A negligible portion of DOC or ~1% may be lost from pedosphere by leaching (Peichl et al. 2007), whereas the rest of SOM is stored in a complex and poorly characterised HMW pool comprising the following: (1) humics that represent the majority or 70–85% of HMW pool, (2) highly polymerised exudates from plant root (e.g. mucilage polysaccharides and secretory proteins or ectoenzymes; Neumann and Romheld 2000) or soil microbes (e.g. glomalin or Fe-containing glycoprotein from arbuscular mycorrhizal fungi) representing 10–30% of HMW and (3) an inert organic pool, i.e. black carbon (C), which is the most recalcitrant SOM fraction, contributing 5–20% of HMW (Ondrasek 2008 and references therein) (Fig. 22.3).

Even though humus is still not fully defined, there are many hypothetical chemical structures (Schnitzer 1978) from which is possible to get insight about humus complexity, heterogeneity and continuous transformation (e.g. over H-bonds or Van der Waals forces). It is known that humics are chemically highly polymerised organics containing both aromatic and aliphatic monomers. Therefore, the crucial role in humics characterisation is played by their particular constitutes i.e. functional groups; carboxyl, hydroxyl, aldehyde, ketone, ester, amino, nitro, thiol, etc. Oxygen (O) containing groups (e.g. -COOH, -OH) are much more abundant than nitrogen (N) and sulphur (S) groups (e.g. $-NH_2$, –SH) (Essington 2004), but all may act as acid/base and serve as proton (H^+) donor/acceptor (i.e. TEs acceptors) under certain pHs depending on their acidity constants (p*K*a). Given that in the most abundant carboxyl groups of humus, the p*K*a values range between 4 and 6, the average p*K*a of humics is 4–4.5 (Tan 2003). Certain humus fractions may be extracted from soil after suspending in NaOH and filtering (e.g. XAD-8 resin column). The compounds absorbed onto the resin include humic (HA) and fulvic (FA) acids that are further fractionated by acidification. In pH<2, precipitated fraction represents HAs, a major component of natural humus materials (Schnitzer 1978), whereas FAs remain in the solution after acidification and are soluble at all pHs. According to Shintzer's (1978) "model" HA and FA data, i.e. average values compiled from different geographical (from Artic to tropical climatic zone) and pedological (acid, neutral soils) conditions, it may be concluded that the portion of C and N is higher and that of H and O lower in HAs compared to FAs.

In the soil HMW pool there are many other organics with complex and even more recalcitrant chemical structure than those elaborated for humics. One of them is organic charred material (biochar, charcoal), i.e. black C, which is during last few decades under intensive observation, principally for its influences on properties, processes and functioning of natural ecosystems (Hockaday et al. 2006). The contribution of black C as a percentage of SOM-C may be substantial (~20–50%) in specific soil type such as terric Anthrosol (e.g. Hockaday et al. 2006 and references therein). Similarly to humics, some of biochar active radicals act as proton donors/acceptors, resulting in coexisting areas whose properties can range from acidic to basic and from hydrophilic to hydrophobic. Due to high enrichment with polyaromatic rings, biochar is persistent even up to nano-scale fraction. Ultrafine biochar dust fraction, formed by condensed aromatic fullerene-like structures, is presumed to be the most recalcitrant OM in the nature, whose residence times is 10 to 1,000-fold longer than most of SOM (Lehmann et al. 2009 and references therein). Through natural weathering or oxidative depolymerization, soil charcoal slowly degrades to relatively less recalcitrant forms of humic substances (Hockaday et al. 2006).

4 Dynamics of SOM in Soil–Microbial–Plant Continuum

A bulk of SOM predominantly comprises different % of C (40–60), O (35–40) and H (4–6) (e.g. $C_{19}H_{17}O_{10}$ or calculated empirical formula of Suwanne River-FA), which are ultimately mineralised mostly to gaseous C (CO_2) and H_2O (Ondrasek 2008 and references therein). In spite of the fact that in general, dominant part of arable land areas in the topsoil horizons (usually the most SOM-enriched) contain only several %s of SOM, i.e. C, at the global scale, soil has a crucial function in overall C budget, either by sequestering (over photosynthetic assimilation) or by emitting C (over plant/soil respiration).

Van Hees et al. (2005) distinguished three respiration processes in the soil–microbial–plant continuum: (1) autotrophic or release of photosynthetically assimilated organics through plant shoot/root and associated symbiotic mycorrhizal fungi, (2) heterotrophic or release of C from free-living decomposers biota (e.g. microbes, micro/meso fauna) during SOM degradation and (3) rhizosphere or coupled autotrophic/heterotrophic respiration in the root zone. Via respiration processes, SOM flows among different pools to be finally released into the atmosphere. There are several critical factors that may significantly influence SOM transformation/degradation, such as nutrients (NO_3^-, NH_4^+), CO_2 and H_2O. Some of them are (1) elemental composition of SOM (Kalbitz et al. 2000) and (2) environmental conditions such as pH, oxidation state, temperature and population of below-ground biota (e.g. Rengel and Marschner 2005).

Organic rhizodeposition could be tightly coupled with photosynthetic activity. From net fixed C, 30–60% in annual (e.g. cereals up to 60% and vegetables up to 44%) and >70% in perennial plants (*Bouteloua gracilis*) is transferred to soil, of which 40–90% may be released into the rhizosphere in the form of organic exudates/secretions and respiration gases by roots and associated micro-organisms (Lynch and Whipps 1990). In the majority of reviewed crops (8/11) the root-translocated C was dominantly (25–76%) released as CO_2 and the rest (4–29%) was exudate-C. This is in good agreement with later findings by Liljeroth et al. (1994) who obtained similar results investigating the shoot–root translocation of net assimilated C from maize (26–34%) and wheat (40–58%). In the same study, the contribution of rhizodeposited C (exuded+respiratory) varied between 41 and 45% (maize), namely, 43–57% (wheat) and in both crops dominated C realised by respiration; in wheat 74–85% and in maize 78–86%. Recently, analysing 43 tracer studies dealing with C efflux from shoots to roots and further to rhizosphere, Nguyen (2003) have summarised that on average, shoots export ~50% of the net fixed C to roots, half of which stay incorporated in root tissues, ~33% is respired by root+rhizomicrobs and the rest >15% is released as organic exudates.

5 Trace Elements and Their Dynamics in Soil–Plant Continuum

European Commission (2006) stated that contamination by certain trace elements (TEs) and nutrients (Cd, Cu, Pb, Zn, Hg, As, Ni, Cr, N and P), and persistent organic pollutants are recognised as one of the main threats to environmental resources and human health in EU. The same source estimates that around 3.5 million sites in EU may be considered contaminated, whereas annual costs to European economy caused by soil contamination (without the damage to soil ecological functions) are estimated to be in the range of US$3.5–25 billion.

TEs are heavy metals and metalloids that include essential nutrients as well as some of the most toxic elements for living organisms. The total content of TEs in solutions of non-contaminated surface horizons of mineral soils range from 1 to 100 mg/kg (e.g. Kabata-Pendias 2004), whereas in contaminated soils these values may be several orders of magnitude higher. Although TEs in plant tissues usually positively correlate with the content in the surrounding environment, that is not always the case because of variable bioavailability. A usual decreasing order of TEs bioavailable forms is: water soluble free metal (e.g. Cd^{2+}) > specifically complexed (e.g. $CdHCO_3^-$, $CdCl^+$) > organically complexed (e.g. FA–Cd) > Fe/Mn/Al (hydr)oxide-complexed (e.g. $\equiv FeOCd^+$) > associated with secondary clay minerals i.e. residual form (e.g. Kabata-Pendias 2004; Ondrasek et al. 2009a).

Various metals fractions in the soil solid/dissolved phase may be quantified by sequential extraction, isotopic dilution, electro-analytical methods, etc., or predicted by computational approach (see Sect. 6). Generally, bulk and rhizosphere soil/solutions considerably differ in most of physical (porosity), chemical (pH, SOM content/

composition, elements speciation) and biological parameters (microbial activity) (e.g. Hinsinger et al. 2005). For instance, in the rhizosphere (vs. bulk soil) due to rhizodeposition and microbial secretion of different inorganic/organic compounds (H^+/HCO_3^-, LMW-OAs, etc.) (Rengel and Marschner 2005) it may be expected that a relatively high proportion of TEs is in the relatively mobile free form (e.g. Zn and Mn; Rengel 2004; Rengel and Marschner 2005) and/or in their simple complexed forms (e.g. Cd-LMW-OAs, $CdCl^+$, $ZnCl^+$; Chiang et al. 2011; Ondrasek et al. submitted). Therefore, it may be concluded that solubility (mobility) and bioavailability (toxicity) of TEs, which are dominantly governed by mentioned properties, are markedly different over relatively small distances, i.e. bulk soil vs. rhizosphere. In the next sections, the focus is on the properties of the rhizosphere soil solution because it represents the interface between TE sources and root cells, i.e. soil–plant interface.

5.1 Uptake, Translocation and Deposition of TEs Within the Plant

The majority of plant species have a weak ability to selectively extract essential TEs from the rhizosphere (Marschner 1995); therefore, they take up non-essential (Si, Co) and even highly toxic ones (Hg, Pb, Cd). Some of potentially toxic metals (Cr, Ag, Sn) due to lower solubility in the most soil circumstances are practically phytounavailable. In contrast, Cd is one of the most soluble TEs with relatively high bioavailability (Adriano et al. 2004; Clemens 2006). The preferred metal forms that are taken up by roots are soluble free cations (Cd^{2+}, Zn^{2+}, Cu^{2+}), but also complexes with organic ligands (e.g. Zn, Broadley et al. 2007). Free metal forms create (bioavailable?) complexes with in/organic natural and/or synthetic ligands dissolved in the rhizosphere. However, confirmation of metallo-complex acquisition by roots is relatively scarce and based mostly on assumptions.

Increased uptake and phytoaccumulation of Cd, Zn and/or Cu in the presence of excessive inorganic ligands (e.g. Cl^-) in the soil/rhizosphere have been showed in a wide range of experimental conditions (Smolders and McLaughlin 1996; Weggler et al. 2004; Ondrasek et al. 2009a, b). Also, Chiang et al. (2011) confirmed that LMW-OAs concentration in root exudates was positively correlated with the amount of Cd accumulated in millet shoots/roots, whereas Hoffland et al. (2006) found that Zn uptake efficiency in rice correlates with exudation rates of LMW organic anions. Similar results were obtained with giant alga *Chara* whereby Al^{3+} complexation with sulphate and citrate anions increased Al uptake by up to greater than twofold (Taylor et al. 2000). Likewise to LMW, positive influence of HMW organic ligands (e.g. HAs) on phytoextraction of TEs (Cd, Cu, Zn, etc.) was confirmed in tobacco (Evangelou et al. 2004), forage (Neunhauserer et al. 2001) and recently in fodder radish (Bandiera et al. 2009) (see Sect. 6).

Undoubtedly, metal chelation and/or complexation with in/organic ligands enhance desorption from soil solids, and having mobilised TEs into a dissolved phase, improve their root uptake/phytoaccumulation, although underlying mechanisms are unclear. Uptake of metals from the rhizosphere is mediated by a specific (Cu, Fe, Zn) and/or nonspecific (e.g. Cd) transporter proteins embedded into the plasma membrane of root cells (e.g. see review by Clemens 2006). Due to similar physical (ionic radius) and chemical (redox-activity, Lewis acidity) properties among TEs, non-essential and highly toxic ones (e.g. Cd) most probably enter roots via highly specific routes for essential nutrients (e.g. Zn), especially under stressful conditions (e.g. Ondrasek et al. 2009a).

The TEs may be adsorbed/sequestered by numerous reactive groups and constituents in the apoplast (cell wall) and finally after crossing the plasma membrane in the symplast. In most cultured species, the largest portion of taken TEs is retained in the below-ground tissues (e.g. >70% Cd, Chiang et al. 2011; 65% Mn, Pearson and Rengel 1995) and the remaining portion is translocated to aerial parts in the following decreasing order: stem > old leaves > young leaves > fruits > seed. Information on the speciation of metals

Table 22.1 Possible intake of TEs via the foodstuffs consumption [adapted from ANZFA (1997) and Ondrasek (2008a)]

	Cd Intake	
Foodstuff	µg per day	%
Potato	4.26	46.7
Wheat	1.48	16.3
Cocoa and related products	1.14	12.5
Meat of mammals	0.96	10.5
Seafood	0.55	6.0
Peanuts	0.28	3.1
Rooty vegetables	0.23	2.5
Leafy vegetables	0.17	1.9
Rice	0.05	0.5
Total	9.12	100
µg per 100 g fresh tissue weight		
Radish hypocotyls[a]	Cd 2.7(42); Zn 195(167); Cu 9.8(9.9)	
Strawberry fruit[a]	Cd 1.3(14); Zn 188(167); Cu 36(33)	
Lettuce leaf[a]	Cd 3.4(54); Zn 382(348); Cu 24(22)	
Melon pulp[a]	Cd 0.55(3.50); Zn 97.6(87); Cu 14(11.6)	
Melon leaf[a]	Cd 15(220); Zn 510(470); Cu 64(57)	

[a]Soil background TEs concentrations were (mg/kg): Cd 0.33; Zn 14.7; Cu 12.9 whereas in parenthesis are values obtained under Cd-contaminated soil conditions (mg/kg): ~10 (in trial with melon) and ~5 (in all other trials)

along the long-distance transport is deficient. Given alkaline conditions (pH ~8) and high concentration of LMW/HMW organic and inorganic (e.g. Cl^-) ligands in the phloem, polyvalent metal cations are probably more complexed there than in the xylem (Ondrasek 2008 and references therein).

Although TEs concentration in edible crop tissues may be many times lower than in non-edible ones (Table 22.1), consumption of food/feed crops grown in soils containing high concentration of TEs represents one of the main routes of metals for humans/animals. Metal intake via food, drink and/or medicaments (e.g. Al intake in tea and medications; Rengel 2004) for the majority of human population is unavoidable, even though the amount of intake would depend on the selection of foodstuff (Table 22.1). According to ANZFA (1997), for the Australian and N. Zealand population, consumption of plant foodstuffs represent a dominant route (84%) for Cd intake, with more 47% of the total Cd taken in originating from potato (Table 22.1).

Many TEs have a long biological half-life (Cd > 10 and Pb ≈ 20 years) and their accumulation in human body gradually increases with time, causing a number of health problems/diseases that are the main reasons for mandatory reductions of annual emissions of certain TEs in many countries (e.g. Lado et al. 2008).

6 SOM and TEs Interaction: Influence on Bioavailability and Toxicity of TEs

During the last decades, Cd, Zn and Cu have been intensively studied from environmental, agricultural, medical, economical and/or social science perspectives. Only a few TEs, including Cu and Zn, are phyto-essential in small amounts, whereas at excessive concentrations they can become toxic. In contrast, Cd is one of the most biotoxic (cancerogenic, teratogenic, etc.) metals (Ondrasek 2008 and references therein).

As reviewed in previous sections, one of the most important roles of SOM is that its huge reactive, mostly negatively charged interfaces strongly compete with inorganic ligands in complexing TEs and therefore affect availability and phyto-extraction of metals. In the next part, using a geochemical modelling approach, we discuss the influence of different forms of dissolved OM substances on Cu, Zn and Cd interactions, bioavailability and toxicity for relatively realistic rhizosphere conditions.

Table 22.2 Chemical composition of "generalised" rhizosphere solution

Mineral components		*LMW organic components*	
NO_3 (mM)	1.8	Citrate (μM)[a]	70
PO_4 (mM)	0.5	Malate (μM)[b]	1,472
SO_4 (mM)	1.7	Formate (μM)[c]	563
NH_4 (mM)	0.1	Oxalate (μM)[c]	100
K (mM)	2.2	Lactate (μM)[d]	48(82[f])
Ca (mM)	4.8	Succinate (μM)[d]	–(124[f])
Mg (mM)	3.9	Acetate (μM)[d]	851(4,018[f])
Na (mM)	1.9	Propionate (μM)[d]	173(1,932[f])
Cl (mM)	1.6	Butyrate (μM)[d]	271(2,052[f])
CO_3 (mM)	4.7	*HMW organic components*	
Mn (μM)	0.55(5.5[e])	DOC (mg/L)	10
Zn (μM)	0.75(7.5[e])	FA1-(6) (mg/L)	0.082
Cu (μM)	0.42(4.2[e])	FA2-(6) (mg/L)	0.026
Cd (μM)	0.05(0.5[e])	*pH conditions*	3.5–9.0

[a]Grierson (1992)
[b]Bolan et al. (1994)
[c]Baziramakenga et al. (1995)
[d]Chiang et al. (2011)
[e]Concentrations for contaminated rhizosphere conditions (increased by one order of magnitude vs. non-contaminated)
[f]Concentrations from contaminated rhizosphere solution with 300 mg Cd/kg soil

6.1 A Model Overview

Geochemical modelling may be a powerful tool for studying various types of anthropogenic/natural pollutants, their presence/mobility in natural systems, and for environmental risk assessment purposes. Among the four main groups of geochemical models (speciation-solubility, reaction path, inverse mass balance and coupled mass transport), the first one based on a thermodynamic equilibrium conceptual approach is suitable for explaining OM–TEs interactions and assess bioavailability and toxicity of metals (e.g. Zhu and Anderson 2002).

A chemical composition of "generalised" rhizosphere solution was compiled from different soils databases, and can be considered appropriate for mineral soils (Table 22.2). Modelling on a Visual MINTEQ interface (Gustafsson 2006) was performed for two levels of TEs (non-contaminated and contaminated system) and in a pH range 3.5–9.0 (covering the most naturally occurring conditions in different soil ecosystems; e.g. Tipping 2005). However, in extreme (acid/alkaline) conditions (e.g. pH <3.5 and >10.5) it is almost impossible to characterise in full the proton affinity distribution because of a lack of reliable data (Kinniburgh et al. 1999). Input data for mineral components in non-contaminated conditions were obtained from the database of saturated soil paste extracts of arable alluvial, mostly mineral soils (Ondrasek et al. unpublished), whereas total concentration of TEs in contaminated conditions were increased (vs. non-contaminated) by one order of magnitude in accordance to Kabata-Pendias (2004).

Due to continuous CO_2 release in the rhizosphere as a consequence of the root/microbial respiration, its presence is usually markedly higher than in the atmosphere (Ondrasek 2008); hence, in all models partial CO_2 pressure was increased 15-fold compared to atmospheric. TEs speciation was calculated using a thermodynamic database from the Visual MINTEQ ver. 3.0, and all component activities were calculated with the Davies equation at 22°C (suitable for moderate soil climate conditions). An oxidising state was assumed in all models (i.e. pH + pe = 15). Mineral dissolution/precipitation, describing possible TEs immobilisation as a consequence of oversaturation under certain chemical circumstances, was assessed by calculated mineral saturation indices (SI) without allowing solids to precipitate during the modelling. In a case of $SI = 0$, the certain mineral is at thermodynamic equilibrium in solution with its dissolved products (ions). When $SI > 0$, solution is oversaturated (i.e. mineral is precipitated), and if $SI < 0$ solution is undersaturated (i.e. mineral is dissolved) (Zhu and Anderson 2002). The LMW organic pool is strongly dependent on non/contaminated conditions (e.g. Chiang et al. 2011), i.e. may differ by more than an order of magnitude depending on metal contamination (Table 22.2), and may have very short residence time (Sect. 3). Those are the main reasons that the LMW pool was varied, whereas the HMW pool, which is mostly stable over the short/mid term period (Sect. 3), remained constant (Table 22.2).

Organo-complexation/chemosorption between TEs and HMW organics was assessed by using the Non Ideal Competitive Adsorption (NICA)-Donnan isotherm model (Kinniburgh et al. 1999) as one of the most advanced models for competitive metal ion binding to humic substances (Weng et al. 2001). The input variables required by this sub-model were presumed (e.g. DOC 10 mg/L for mineral soil solutions) or retained as default settings. For instance, whereas all of the DOC may not be reactive with respect to proton and metal binding, the ratio of active dissolved organic matter (DOM) to DOC was by default set to 1.4., which is close to the average conversion factor of 1.36 suggested by Bryan et al. (2002) and has been found to work well for modelling soil systems (Apul et al. 2010). Also, it was assumed that all active DOM is comprised from FAs, what is in a good agreement for a majority of natural soil environments and tested pHs in which FAs are soluble compared to mostly insoluble HAs (Sect. 3). However, in the case that there was not enough total FAs to make up the total active DOM, the model adjusted this parameter so that HAs may contribute to the DOM.

6.2 Cadmium

According to modelling results, the uncomplexed free Cd^{2+} pool was the predominant (50–74%) in acid ($pH \leq 5.0$) conditions, but its activity linearly decreased with pH increasing, it was almost fully depleted (<1%) at pH 9.0 (Fig. 22.4). In the very narrow range of slight acidity (pH 6.0–6.5) Cd-organocomplexation was the most pronounced (44–45%), especially in the LMW pool (>35%), whereas inorganic complexation of Cd dominated (38–97%) in basic ($pH \geq 8$) conditions (Fig. 22.4). In contaminated rhizosphere conditions (vs. non-contaminated), distribution in Cd^{2+} and Cd-inorganic pools stayed rather stable, whereas in other pools Cd was redistributed from the HMW to the LMW fraction. The main cause of such Cd redistribution inside organic pools is that in contaminated conditions certain LMW organic acids were increased in the rhizosphere many times (Table 22.2) as a consequence of pronounced rhizodeposition (see the next).

Although Cd-inorganic complexation started to increase from pH 5.5 (Fig. 22.4a), $CdCl^+$ activity from the Cd-inorganic pool continued decreasing throughout the pH range (Fig. 22.4b). The main reason is that Cd sorption with other ligands (HCO_3^-, CO_3^{2-}, HPO_4^{2-}, etc.) dominated over Cd-chlorocomplexation (data not shown). The most dominant LMW- and HMW-Cd forms were Cd-malate and FA1-Cd(6) (i.e. Cd bound to FAs via carboxylic groups) over the whole pH range (data not shown), with Cd-malate concentrations being the highest in slightly acidic pH (5.5–6.0), and those of FA1-Cd(6) in neutral to slightly basic pH (7–7.5) (Fig. 22.4b). In contaminated conditions, the curves showing activity of Cd forms retained the same shape but the activities were an order of magnitude higher than in non-contaminated model (data not shown).

Calculated saturation indices for over 200 possible mineral phases included in the Visual Minteq database were checked, and several of observed TE minerals were found to be oversaturated (i.e. $SI > 0$), but only in contaminated and basic conditions: at pH 8 malachite (Cu mineral), hydrozincite and smithsonite (Zn minerals) and otavite (Cd mineral) and at pH 9 all Zn and Cd minerals as at pH 8 plus zincite (Zn mineral) (data not shown).

The mobility and uptake of Cd, Zn and Cu is strongly dependent on chemical speciation/distribution and concentrations (activities) of TEs in

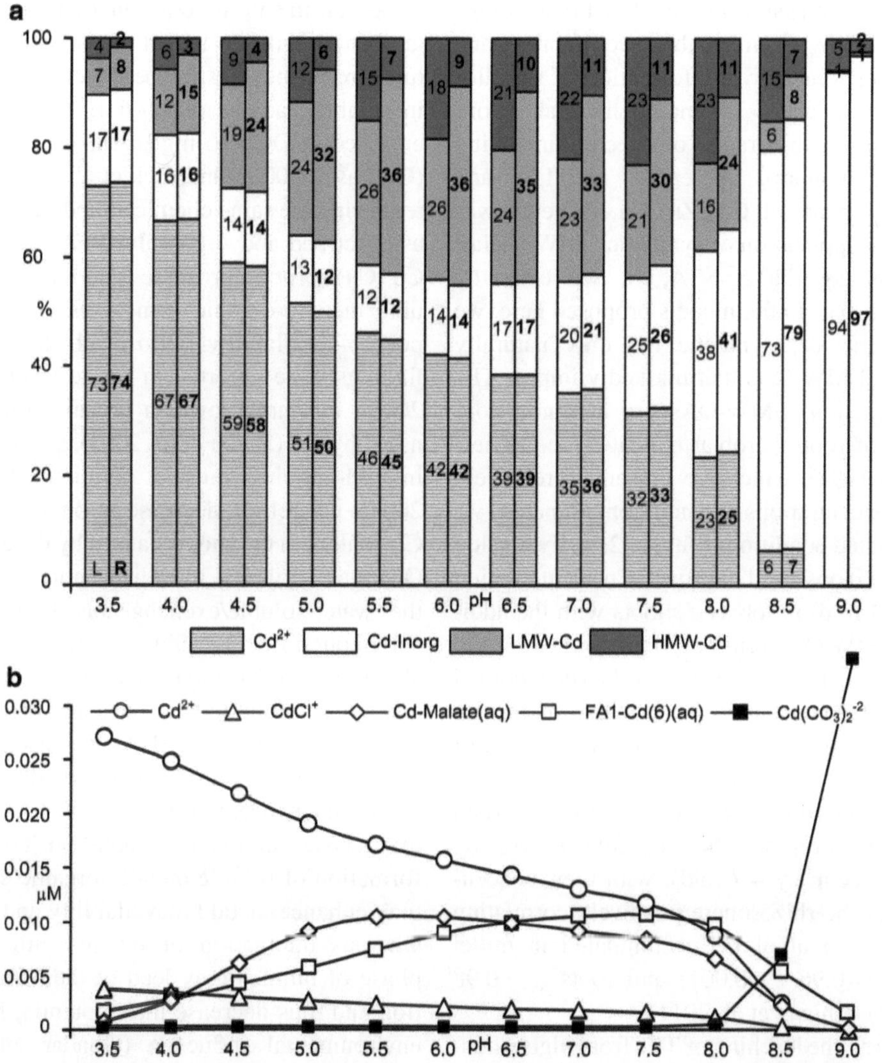

Fig. 22.4 Distribution of Cd species (**a**) among four pools [non-contaminated conditions = L (left bars); contaminated conditions = R (right bars with *bold* %)] and activities of some of the most abundant Cd species (**b**) inside each pool (non-contaminated conditions only)

the rhizosphere solution. Although it is believed that only uncomplexed i.e. free cationic metal form may be taken up by roots, there is increasing evidence that Cd (Zn, Cu) can be mobilised in soil/nutrient solution and then taken up/phytoaccumulated complexed with inorganic ligands such as chlorides (Smolders and McLaughlin 1996; Khoshgoftar et al. 2004; Weggler et al. 2004; Khoshgoftarmanesh et al. 2006; Ondrasek et al. 2009a) or sulphates (McLaughlin et al. 1998b). Indeed, Weggler et al. (2004) observed that shoot Cd concentrations of wheat grown in a biosolid-amended soil were most closely correlated with $CdCl^+$ activity in soil solution, whereas the correlation with the Cd^{2+} activity was weak. McLaughlin et al. (1998a) observed that Cd shoot/root concentrations in Swish chard were unaffected by additions of sulphate to nutrient solution despite Cd^{2+} activities decreasing markedly in the rhizosphere. The above studies suggested that Cd-inorganic complexes ($CdSO_4$, $CdCl^+$) could be phytoavailable and enter root plasma membrane either directly as a metal-complex and/or dissociating in the apoplast and entering cells as the free metal cation (Smolders and McLaughlin 1996).

With pH decreasing, the acid functional groups of LMW/HMW organic substances deprotonate, influencing solubility and formation of metallo-organo complexes. In many field/laboratory experiments with a range of species (including hyperaccumulators), increased mobility and improved uptake of Cd (Zn, Cu, Pb, etc.) was elicited by application of synthetic LMW chelating agents (e.g. EDTA, NTA, etc. See review by Schmidt 2003). In the models proposed here, we did not consider synthetic but only naturally occurring LMW-OAs that markedly impact TEs phytoextraction. LMW-OAs are common constituents of root/microbial exudates (see review by Jones 1998), which are present in relatively lower concentrations in non-contaminated vs. contaminated conditions (Table 22.2). Evangelou et al. (2006) observed improved Cu accumulation (up to 2.3-fold) in tobacco shoots with the addition of LMW-OAs (citric, tartaric and oxalic) to soil, similarly to Nigam et al. (2001) who recorded enhanced Cd concentrations in maize shoots (up to greater than twofold) after soil application of citric and malic acid. Rhizosphere contamination with Cd induced higher root exudation of certain LMW-OAs (propionic by 11.2-fold, butyric by 7.6-fold, acetic by 4.7-fold), with their concentrations in the rhizosphere positively correlating with the amount of Cd accumulated in millet shoots ($r^2=0.96$ $P<0.001$) and roots ($r^2=0.98$ $P<0.001$) (Chiang et al. 2011).

Studying the leaching of TEs from highly contaminated soils, Fischer et al. (1998) depleted total Cd by 75% (Cu by 54% and Zn by 56%) with the application of acidic (pH 4.4) grass silage effluent. Such metal removal efficiency was attributed to effluent's main component i.e. lactic acid, which under acid pHs acts as a proton-donor and strong metallo-complexing agent. According to the same authors, there are several crucial mechanisms that explain TE mobilisation by LMW-OAs: (1) chelation, (2) proton-initiated solubilisation, (3) surface complexation, (4) ion exchange and (5) reductive dissolution of metal binding substrates i.e. hydr/oxides.

Complexation of free metals ions (the most bioavailable and potentially the most toxic forms) with certain HMW SOM substances could be important in safe food production on metal-contaminated soils. In recent studies with horticultural crops (melon, strawberry, lettuce and radish) in slightly acidic pH (~6.0) and organically enriched (DOC > 76 mg/L) rhizosphere solution (Ondrasek 2008; Ondrasek et al. 2009a, submitted) using the same chemical modelling approach as described above, dissolved FAs dominated in Cd (Cu) sorption processes and formation of soluble metallo-organic complexes with relatively poor bioavailability (Otto et al. 2001). Similar findings were reported in maize (Shuman et al. 2002), mustard grown in organic soil contaminated by Cd (Bolan et al. 2003), sorghum grown in OM-enriched nutrient solution (Pinto et al. 2004), whereby a decrease in phytotoxicity and Cd content in the shoots caused by the addition of OM was attributed to redistribution of Cd from the water-soluble/exchangeable to the organically bound fraction. Similarly to FAs, HAs may also influence TE bioavailability given they also contain acid functional groups. However, in most natural soil conditions, HAs are insoluble compared to naturally more soluble FAs.

However, when added to soil, FAs and HAs may cause contrasting effects on TEs, i.e. the formation of soluble metallo-organic complexes may enhance metal bioavailability and leaching, whereas interaction of metals with the solid phase of humics may lead to their immobilisation and thus decrease their potential hazardous environmental influence (Gondar and Bernal 2009). The phytoavailable forms of Cd (Zn, Cu, Pb) in soils were increased by adding HAs to two metal contaminated mineral soils (Halim et al. 2003). In contrast, application of HAs to Cd-contaminated mineral soil did not change phytoavailable soil Cd, but phytoaccumulation of Cd in tobacco increased by up to 65% accompanied by toxicity.

6.3 Zinc

Although total Zn concentration in the modelled rhizosphere solution was more than an order of magnitude higher compared to Cd (Table 22.2), modelling results show that a free Zn^{2+} pool was

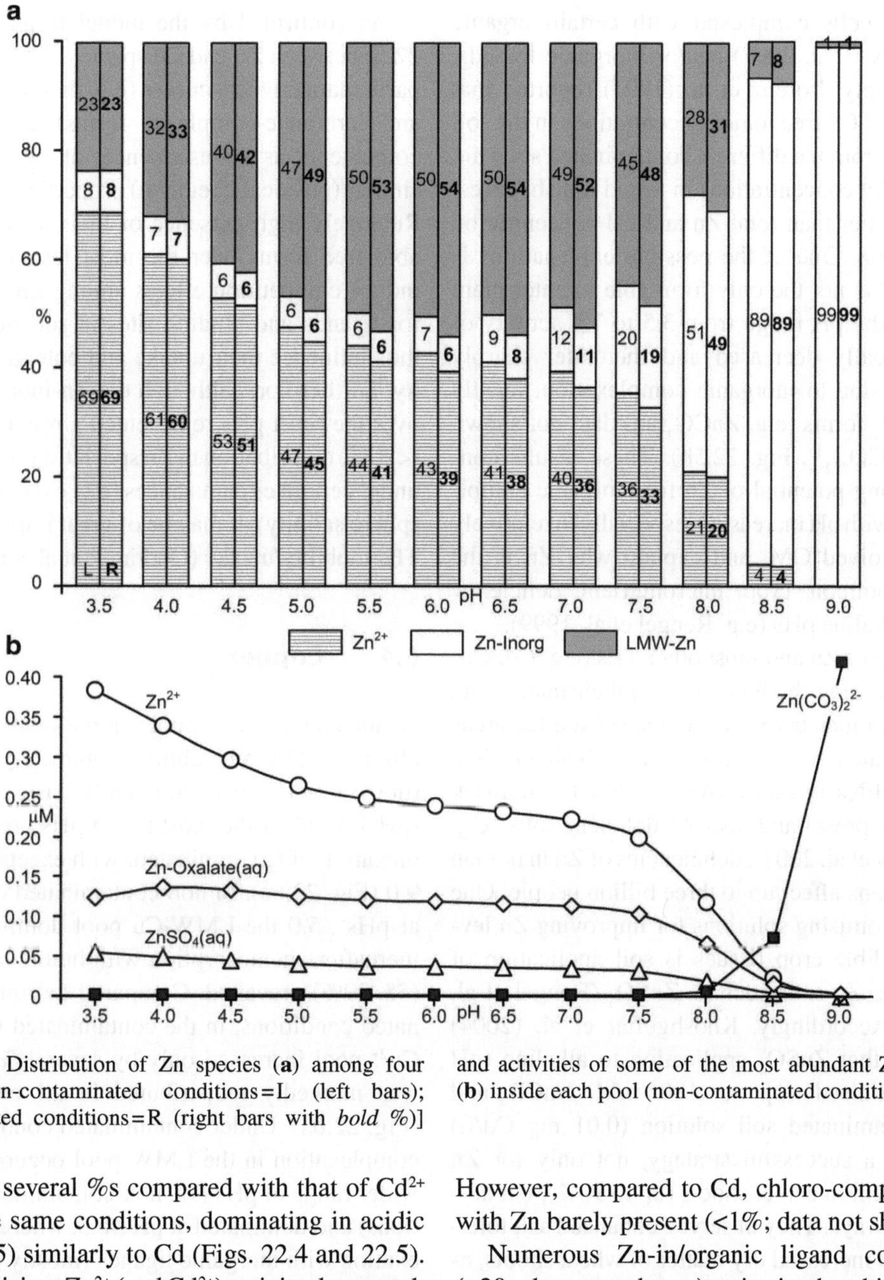

Fig. 22.5 Distribution of Zn species (**a**) among four pools [non-contaminated conditions = L (left bars); contaminated conditions = R (right bars with *bold %*)] and activities of some of the most abundant Zn species (**b**) inside each pool (non-contaminated conditions only)

lower by several %s compared with that of Cd^{2+} under the same conditions, dominating in acidic pHs (≤4.5) similarly to Cd (Figs. 22.4 and 22.5). With pH rising, Zn^{2+} (and Cd^{2+}) activity decreased, whereas the activities of $ZnSO_4$ and Zn-oxalate stayed rather stable up to pH 7.5, and thereafter got depleted (Fig. 22.5b). Inorganic complexation of Zn, mostly with SO_4^{2-} under low pHs, and CO_3^{2-} under high pHs, dominated in basic conditions (pH ≥ 8.0). LMW organo-complexation prevailed in slightly acidic to neutral pH (5.0–7.5) and that with HMW was negligible over the whole tested conditions (Fig. 22.5a). However, compared to Cd, chloro-complexation with Zn barely present (<1%; data not shown).

Numerous Zn-in/organic ligand complexes (>20; data not shown) exist in the rhizosphere solution and are strongly pH dependent (Fig. 22.5). In slightly acid pHs (5.5–6.5), even in Zn-contaminated conditions (i.e. 7.5 μM Zn) Zn^{2+} typically accounts for around 40% of the soluble Zn fraction (Fig. 22.5a). Nutrient solution studies commonly suggest that free ionic metal form is the one most easily absorbed by plants, and Zn^{2+} is recognised as dominant plant-available Zn fraction, although there is a possibility it can enter the root cells complexed with certain organic (Broadley et al. 2007) and/or inorganic ligands. Accordingly, Lorenz et al. (1997) reported that for Zn and Cd free ionic concentrations in the soil solution from ten different contaminated soils did not predict concentrations in tested radish leaves/tubers better than total Zn and Cd concentration in solution. One of the possible explanations is that Zn^{2+} is not the only form able to enter plant root. In the pH range from 3.5 to 7.5 activity of Zn^{2+} lineally decreased and thereafter sharply dropped due to inorganic complexation, mostly with CO_3 forms (e.g. $ZnCO_3$(aq) data not shown and $Zn(CO_3)_2^{2-}$, Fig. 22.5b). These results confirm strong potential of Zn for inorganic complexation with pH increasing, especially in relatively low dissolved OM, and explain why Zn is the most common crop micronutrient deficiency under alkaline pHs (e.g. Rengel et al. 1999).

Intake of Zn and most other TEs (e.g. Cd, Cu) via consuming the food crops is their main route into the human body. Given that substantial areas of cultivated land worldwide are alkaline/saline (high pH/Na in saturated soil extract; Ondrasek et al. (in press) and also Zn-deficient soils (e.g. Broadley et al. 2007), deficiencies of Zn in human populations affect up to three billion people. One of the promising solutions for improving Zn levels in edible crop tissues is soil application of inorganic Zn salts such as $ZnSO_4$ (Rengel et al. 1999). Accordingly, Khoshgoftar et al. (2004) showed that $ZnSO_4$ application to alkaline (pH ~8), salinised (up to 180 mM NaCl) and Cd-contaminated soil solution (0.01 mg Cd/L) may be a successful strategy, not only for Zn enrichment in cereal shoots (up to 90%), but also in other ways. They observed enhanced salt tolerance and increased dry matter of wheat shoots, as well as reduced shoot Cd accumulation (down to less than 50%) with Zn salt application. Similar antagonistic relationship between Zn (or Cu) and Cd in soil solution and their shoot phytoaccumulation under salinity has been recently confirmed in wheat genotypes by Khoshgoftarmanesh et al. (2006) and in radish by Ondrasek et al. submitted), what authors mainly contribute to the Zn–Cd competition for ligands as well as soil adsorption and root uptake sites.

As confirmed by the model (Figs. 22.4 and 22.5) between Zn and Cd species distribution and particular activities curves (e.g. their free cationic and carbonate-complexed forms), exist a quite coincidence as a consequences of their explained similar (physical/chemical) properties (Sect. 5.1). Relatively high presence of their most bioavailable free forms over the most tested pHs may induce competition effects among Zn^{2+} and Cd^{2+} for ligands and binding sites in soil matrix, and thus influence their uptake and potentially toxicity. Furthermore, although the Zn-inorganic pool over the most pHs represented a relatively small (<20%) contribution in Zn speciation (Fig. 22.5a), under certain circumstances (e.g. excessive rhizosphere salinity), it may be of great importance for TEs mobility/uptake (Ondrasek et al. submitted).

6.4 Copper

An uncomplexed Cu pool was the smallest among observed TEs and almost completely depleted (down to ~1%) even at slightly basic conditions (pH 7.5). Over the most tested pHs organo-complexation of Cu dominated, with exception at pH 9.0 (Fig. 22.6a). In non-contaminated conditions at pHs ≤5.0 the LMW-Cu pool dominated, and thereafter chemosorption with humic substances (55–99%) prevailed. Compared to non-contaminated conditions, in the contaminated model the Cu^{2+} pool increased only by several %s, but Cu was markedly redistributed in all other pools (Fig. 22.6a). Under contaminated conditions, Cu complexation in the LMW pool occurred over a wider range of pHs (vs. non-contaminated conditions) and dominated at pH ≤6.5, whereas complexation with inorganic ligands (mostly with CO_3; data not shown) started to increase from the neutral pH, and dominated (63%) in the basic (9.0) pH (Fig. 22.6a).

As confirmed by the model, Cu possesses the highest affinity to OM among observed TEs. In the model were included organics differing in functional, mostly acidic radicals, with strong potential to complex Cu. Soluble metallo-organo complexes undergo microbial degradation and thereafter (1) a portion of metal is able to enter

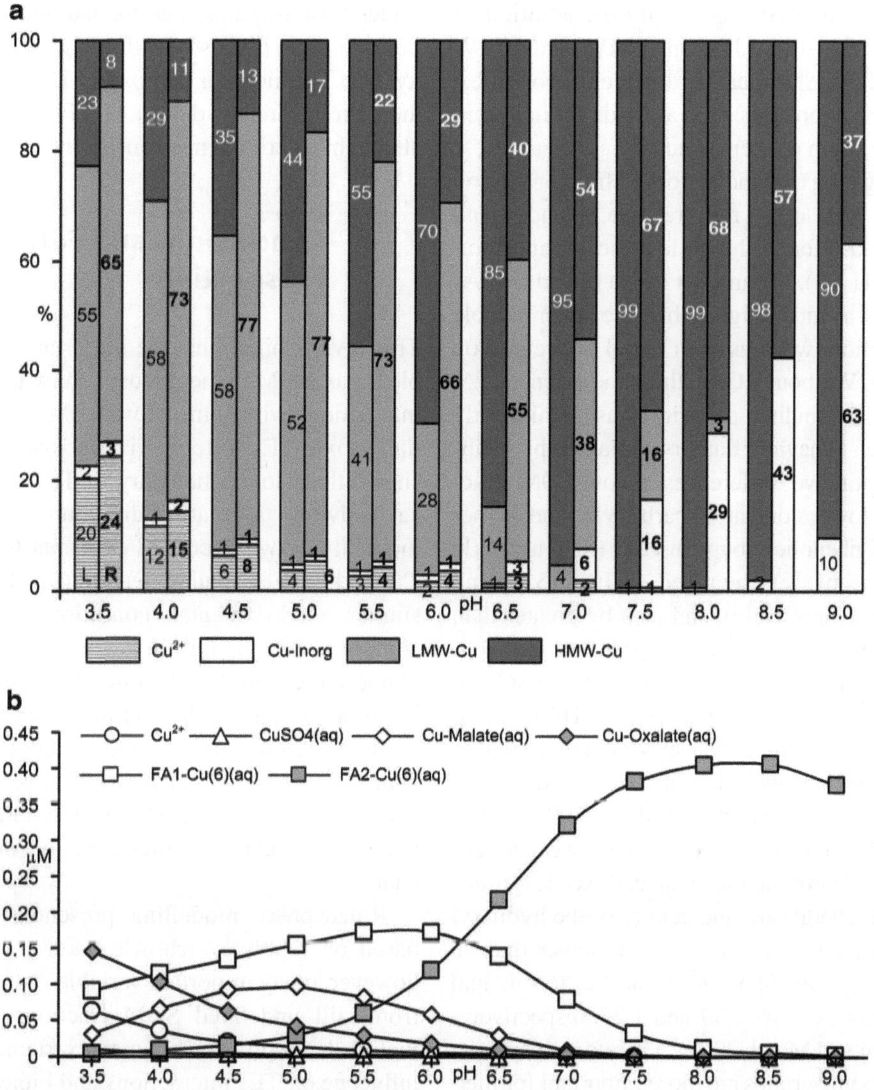

Fig. 22.6 Distribution of Cu species (**a**) among four pools [non-contaminated conditions = L (left bars); contaminated conditions = R (right bars with *bold* %)] and activities of some of the most abundant Cu species (**b**) inside each pool (non-contaminated conditions only)

plants roots as complex fragments (Evangelou et al. 2004), whereas the remaining portion may be (2) leached to deeper soil profiles/groundwater and/or (3) re-adsorbed to soil matrix (e.g. Gondar and Bernal 2009). To what extent a certain process would prevail depends on many physical (soil temperature, moisture), chemical (thermodynamic stability of particular metallo-complexes, salinity, DOC) and biological (microbial activity) conditions in the rhizosphere. Recently, Ondrasek et al. (submitted) working with salty (0–60 mM NaCl) and Cd-contaminated (4.9–39 μg/L) rhizosphere solution have confirmed a significant decrease in DOC (probably under diminished photosynthetical and microbial activity) and an increase in Cu (Zn, Cd) concentrations, either in the rhizosphere or in leaf/fruit tissues of radish. In the same study, under slightly acidic rhizosphere (pH ~6) organically complexed Cu predominated (>99%) in all tested conditions. Evangelou et al. (2006) observed improved Cu uptake and accumulation (up to

2.3-fold) in tobacco shoots with the addition of LMW-OAs to soil. Although LMW and HMW organics have significantly different properties, particular components such as acidic radicals are inherent to both of them (Sect. 3).

Investigating Cu binding to OM fractions in soil amended with olive mill residue, Gondar and Bernal (2009) found that at low metal concentrations (at pH 5.3), the amount of Cu bound by FAs and HAs was much higher than the water soluble (WS) fraction, whereas with metal concentration increasing, WS bound Cu to the same extent as FA, but the HA binding capacity was significantly lower. Such behaviour authors explained by qualitative/quantitative differences among OM fractions, i.e. lower content of carboxylic and higher content of phenolic groups in HAs (0.96 and 2.06 me/g, respectively) compared to FAs (5.32 and 1.26 me/g, respectively), and also by greater than twofold lower metal binding capacity of HAs (vs. FAs). Also, FAs are known to form more soluble and mobile metallo-complexes than HAs due to their higher acidic functional group content, smaller molecular weight and solubility over a wider range of pHs (2–12) (e.g. Smith and March 2007).

Phenolic functional groups have weaker tendency for deprotonation than carboxylic groups under acid conditions, and acidity of the hydroxyl groups in phenols (p*K*a 10–12) is lower than in carboxylic groups (e.g. p*K*a of acetic, formic and oxalic acids are 4.76, 3.77 and 1.27, respectively) (e.g. Smith and March 2007). Therefore, it appears that carboxylic groups are more important for metals binding under lower pHs, whereas in higher pHs other functional groups such as phenolic could be more involved in metallo-complexation. This hypothesis has been confirmed by the proposed model (Fig. 22.6b) where organo-complexation of Cu with carboxylate anions from LMW (e.g. oxalate and malate) and HMW [i.e. FA1-Cu(6)] dominated in acidic pHs (<6.0), whereas a continuous increase in organo-complexation with an increase in pH was due to enhanced deprotonation of LMW phenolic groups and Cu chemosorption with phenolate anions [i.e. FA2-Cu(6)].

With their huge reactive interfaces, predominantly polycarboxylic (FAs) and polyphenolic (HAs) (e.g. Gondar and Bernal 2009), HMW ensure a plenty of sorption sites for Cu and other TEs. Such sorption is mostly with a higher metal affinity than for mono/di carboxylic LMW-OAs, and therefore indirectly decreases TE bioavailability, although underlying mechanisms remain unclear.

7 Conclusions and Future Perspective

This review highlights the importance and complexity of SOM in the rhizosphere with respect to interactions with some of widespread and potentially toxic TEs in terrestrial ecosystems. It was shown that biogeochemistry of TEs (e.g. chemical activities, speciation distribution) may vary markedly between certain elements (e.g. Zn vs. Cu) and for a particular element (e.g. Cu) in very similar environmental conditions (e.g. from slightly acidic to slightly basic pHs at the same temperature and SOM content). It is of great importance that such environmental variations exist on a micro-scale and are inherent to the root zone (e.g. plant root vs. rhizosphere solution) and may therefore be relevant for TEs mobility, i.e. toxicity to plants and thus entering into a food chain.

Rhizosphere modelling presented here was based on relatively realistic natural conditions; however, many important variables, such as those from still undefined SOM fractions, were not included, but presumably may exert considerably influence on TEs interactions and bioavailability. The recent intensive scientific work and significant improvements in instrumental methodology (e.g. solid-state CP/MAS ^{13}C NMR spectroscopy, FT-ICR mass spectrometry) have ensured characterisation of various highly complex and recalcitrant SOM fractions (humics, black C) that should further help in elucidating interactions between SOM and TEs. Therefore, some of the future research perspectives, more focused on (in)organic complexation of TEs, are likely going to uncover novel mechanisms for uptake of certain TE forms that currently have not featured in the uptake studies (e.g. metal-HAs) or re-evaluate some that have been presumed to be taken up (e.g. metal-chloride complexes).

Acknowledgements This work was supported by the Croatian Science Foundation (Con. O-3510-2010) and Croatian Ministry of Science, Education and Sports (Pr. No: 0178039).

References

Adriano DC, Wenzel WW, Vangronsveld J, Bolan NS (2004) Role of assisted natural remediation in environmental cleanup. Geoderma 122:121–142

ANZFA (1997) Australia New Zealand Food Agency. www.foodstandards.gov.au

Apul DS, Diaz ME, Gustafsson JP, Hundal LS (2010) Environ Eng Sci 27:743–755

Bandiera M, Moscaa G, Vamerali T (2009) Humic acids affect root characteristics of fodder radish (*Raphanus sativus* L. var. *oleiformis* Pers.) in metal-polluted wastes. Desalination 246:78–91

Baziramakenga R, Simard RR, Leroux GD (1995) Determination of organic acids in soil extracts by ion chromatography. Soil Biol Biochem 27:349–356

Blume HP, Leinweber P (2004) Plaggen soils: Landscape history, properties, and classification. J Plant Nutr Soil Sci 167:319–327

Bolan NS, Naidu R, Mahimairaja S, Baskaran S (1994) Influence of low-molecular-weight organic-acids on the solubilization of phosphates. Biol Fertil Soils 18: 311–319

Bolan NS, Adriano DC, Duraisamy P, Mani A (2003) Immobilization and phytoavailability of cadmium in variable charge soils. III. Effect of biosolid compost addition. Plant Soil 256:231–241

Broadley MR, White PJ, Hammond JP, Zelko I, Lux A (2007) Zinc in plants. New Phytologist 173:677–702

Bryan SE, Tipping E, Hamilton-Taylor J (2002) Comparison of measured and modeled copper binding by natural organic matter in freshwaters. Comp Biochem Physiol C133:37–49

Byrne KA, Chonjicki B, Christensen TR, Drosler M, Freibauer A, Friborg T, Frolking S, Lindroth A, Mailhammer J, Malmer N, Selin P, Turunen J, Valentini R, Zetterberg L (2004) EU peatlands: current carbon stocks and trace gas fluxes. In: Christensen TR, Friborg T (eds) Carbo-Europe Report

Cerri CC, Bernoux M, Cerri CEP, Feller C (2004) Carbon cycling and sequestration opportunities in South America: the case of Brazil. Soil Use Manag 20: 248–254

Chiang PN, Chiu CY, Wang MK, Chen B-T (2011) Low-molecular-weight organic acids exuded by Millet (*Setaria italica* (L.) Beauv.) roots and their effect on the remediation of cadmium-contaminated soil. Soil Sci 176:33–38

Clemens S (2006) Toxic metal accumulation, responses to exposure and mechanisms of tolerance in plants. Biochimie 88:1707–1719

Epstein E (2003) Land application of sewage sludge and biosolids. Lewis, Boca Raton

Essington ME (2004) Soil and water chemistry: an integrative approach. CRC, Boca Raton, FL

European Commission (2006) Communication from the Commission to the Council, the European Parliament, the European Economic and Social Committee and the Committee of the Regions, Thematic Strategy for Soil Protection, COM (2006) 231

Evangelou MWH, Dagan H, Schaeffer A (2004) The influence of humic acids on the phytoextraction of cadmium from soil. Chemosphere 57:207–213

Evangelou MWH, Ebel M, Schaefferet A (2006) Evaluation of the effect of small organic acids on phytoextraction of Cu and Pb from soil with tobacco *Nicotiana tabacum*. Chemosphere 63:996–1004

Fischer K, Bipp H-P, Riemschnider P, Leidmann P, Bieniek D, Kettrup A (1998) Utilization of biomass residues for the remediation of metal-polluted soils. Environ Sci Technol 32:2154–2161

Glatzel S, Kalbitz K, Dalva M, Moore T (2003) Dissolved organic matter properties and their relationship to carbon dioxide efflux from restored peat bogs. Geoderma 113:397–411

Gondar D, Bernal MP (2009) Copper binding by olive mill solid waste and its organic matter fractions. Geoderma 149:272–279

Grierson PF (1992) Organic-acids in the rhizosphere of *Banksia integrifolia*. L. Plant Soil 144:259–265

Gustafsson JP (2006) Arsenate adsorption to soils: Modelling the competition from humic substances. Geoderma 136:320–330

Halim M, Conte P, Piccolo A (2003) Potential availability of heavy metals to phytoextraction from contaminated soils induced by exogenous humic substances. Chemosphere 52:265–275

Hinsinger P, Gobran GR, Gregory PJ, Wenzel WW (2005) Rhizosphere geometry and heterogeneity arising from rootmediated physical and chemical processes. New Phytol 168:293–303

Hockaday WC, Grannas AM, Kim S, Hatcher PG (2006) Direct molecular evidence for the degradation and mobility of black carbon in soils from ultrahigh-resolution mass spectral analysis of dissolved organic matter from a fire-impacted forest soil. Org Geochem 37:501–510

Hoffland E, Wei C, Wissuwa M (2006) Organic anion exudation by lowland rice (*Oryza sativa* L.) at zinc and phosphorus deficiency. Plant Soil 283:155–162

Holden J, Shotbolt L, Bonn A, Burt TP, Chapman PJ, Dougill AJ, Fraser EDG, Hubacek K, Irvine B, Kirkby MJ, Reed MS, Prell C, Stagl S, Stringer LC, Turner A, Worrall F (2007) Environmental change in moorland landscapes. Earth Sci Rev 82:75–100

IPCC (2000) Land use, land-use change and forestry. Special report of the intergovernmental panel on climate change. Cambridge University Press, Cambridge

Jones DL (1998) Organic acids in the rhizosphere – a critical review. Plant Soil 205:25–44

Kabata-Pendias A (2004) Soil-plant transfer of trace elements – an environmental issue. Geoderma 122: 143–149

Kalbitz K, Solinger S, Park JH, Michalzik B, Matzner E (2000) Controls on the dynamics of dissolved organic matter in soils. Soil Sci 165:277–304

Kalbitz K, Schmerwitz J, Schwesig D, Matzner E (2003) Biodegradation of soil-derived dissolved organic matter as related to its properties. Geoderma 113:273–291

Khoshgoftar AH, Shariatmadari H, Karimian N, Kalbasi M, van der Zee SEATM, Parker DR (2004) Salinity and Zn application effects on phytoavailability of Cd and Zn. Soil Sci Soc Am J 68:1885–1889

Khoshgoftarmanesh AH, Shariatmadari H, Karimian N, Kalbasi M, van der Zee SEATM (2006) Cadmium and zinc in saline soil solutions and their concentrations in wheat. Soil Sci Soc Am J 70:582–589

Kinniburgh DG, van Riemsdijk WH, Koopal LK, Borkovec M, Benedetti MF, Avena MJ (1999) Ion binding to natural organic matter: competition, heterogeneity, stoichiometry and thermodynamic consistency. Colloid Surface A151:147–166

Lado LL, Hengl T, Reuter HI (2008) Heavy metals in European soils: a geostatistical analysis of the FOREGS geochemical database. Geoderma 148: 189–199

Lehmann J, Czimczik C, Laird D, Sohi S (2009) Stability of biochar in the soil. In: Lehmann J, Joseph S (eds) Biochar for environmental management: science and technology. Earthscan

Liljeroth E, Kuikman P, Van Veen JA (1994) Carbon translocation to the rhizosphere of maize and wheat and influence on the turnover of native soil organic matter at different soil nitrogen levels. Plant Soil 161: 233–240

Lorenz SE, Hamon RE, Holm PE, Domingues HC, Sequeira EM, Christensen TH, McGrath SP (1997) Cadmium and zinc in plants and soil solutions from contaminated soils. Plant Soil 189:21–31

Lynch JM, Whipps JM (1990) Substrate flow in the rhizosphere. Plant Soil 129:1–10

Marschner H (1995) Mineral nutrition of higher plants, 2nd edn. Academic, London

Matthews E (1997) Global litter production, pools, and turnover times: estimates from measurement data and regression models. J Geophys Res 102:18771–18800

McLaughlin MJ, Andrews SJ, Smart MK, Smolders E (1998a) Effects of sulfate on cadmium uptake by Swiss chard. I. Effects of complexation and calcium competition in nutrient solutions. Plant Soil 202: 211–216

McLaughlin MJ, Lambrechts RM, SmoldersE SMK (1998b) Effects of sulfate on cadmium uptake by Swiss chard: II. Effects due to sulfate addition to soil. Plant Soil 202:217–222

Neumann G, Romheld V (2000) The release of root exudates as affected by the plant's physiological status. In: Pinton R, Varanini Z, Nannipieri P (eds) The rhizosphere: biochemistry and organic substances at the soil-plant interface. Marcel Dekker, New York, NY, pp 41–93

Neunhauserer C, Berreck M, Insam H (2001) Remediation of soils contaminated with molybdenum using soil amendments and phytoremediation. Water Air Soil Pollut 128:85–96

Nguyen C (2003) Rhizodeposition of organic C by plants: mechanisms and controls. Agronomie 23:375–396

Nigam R, Srivastava S, Prakash S, Srivastava MM (2001) Cadmium mobilisation and plant availability – the impact of natural organic acids commonly exuded from roots. Plant Soil 230:107–113

Ondrasek G (2008) Salt stress and rhizosphere contamination by cadmium: phytoaccumulation of nutrients and toxic metal. PhD Thesis, University of Zagreb, Croatia

Ondrasek G, Rengel Z, Romic D, Savic R. Salinity decreases dissolved organic carbon in the rhizosphere and increases trace elements phytoaccumulation (submitted)

Ondrasek G, Romic D, Rengel Z, Romic M, Zovko M (2009a) Cadmium accumulation by muskmelon under salt stress in contaminated organic soil. Sci Total Environ 407:2175–2182

Ondrasek G, Rengel Z, Romic D, Poljak M, Romic M (2009b) Accumulation of non/essential elements in radish plants grown in salt-affected and cadmium-contaminated environment. Cereal Res Commun 37:9–12

Ondrasek G, Rengel Z, Romic D, Savic R (2010) Environmental salinisation processes in agro-ecosystem of Neretva River estuary. Novenytermeles 59:223–226

Ondrasek G, Rengel Z, Veres S. Soil salinisation and salt stress in crop production. In: Shanker AK, Venkateswarlu B (eds) Abiotic Stress in Plants - Mechanisms and Adaptations. ISBN 978-953-307-394–1. InTech (in press)

Otto WH, Carper WR, Larive CK (2001) Measurement of cadmium(II) and calcium(II) complexation by fulvic acids using Cd-113 NMR. Environ Sci Technol 35: 1463–1468

Pearson JN, Rengel Z (1995) Uptake and distribution of ^{65}Zn and ^{54}Mn in wheat grown at sufficient and deficient levels of Zn and Mn I. During vegetative growth. J Exp Bot 46:833–839

Peichl M, Moore TR, Arain MA, Dalva M, Brodkey D, McLaren J (2007) Concentrations and fluxes of dissolved organic carbon in an age-sequence of white pine forests in Southern Ontario, Canada. Biogeochemistry 86:1–17

Pinto AP, Mota AM, de Varennes A, Pinto FC (2004) Influence of organic matter on the uptake of cadmium, zinc, copper and iron by sorghum plants. Sci Total Environ 326:239–247

Rengel Z (2002) Genetic control of root exudation. Plant Soil 245:59–70

Rengel Z (2004) Aluminium cycling in the soil-plant-animal-human continuum. Biometals 17:669–689

Rengel Z (2007) Cycling of micronutrients in terrestrial ecosystem. In: Marschner P, Rengel Z (eds) Nutrient cycling in terestrial ecosystem. Springer, Berlin, Germany, pp 93–113

Rengel Z, Marschner P (2005) Nutrient availability and management in the rhizosphere: exploiting genotypic differences. New Phytol 168:305–312

Rengel Z, Batten GD, Crowley DE (1999) Agronomic approaches for improving the micronutrient density in edible portions of feld crops. Field Crops Res 60:27–40

Romic D, Romic M, Rengel Z, Zovko M, Bakic H, Ondrasek G. Trace metals in the coastal soils developed from estuarine floodplain sediments in the Croatian Mediterranean region (submitted)

Sachse A, Henrion R, Gelbrecht J, Steinberg CEW (2005) Classification of dissolved organic carbon (DOC) in river systems: Influence of catchment characteristics and autochthonous processes. Org Geochem 36: 923–935

Schmidt U (2003) Enhancing phytoextraction: the effect of chemical soil manipulation on mobility, plant accumulation, and leaching of heavy metals. J Environ Qual 32:1939–1954

Schnitzer M (1978) Humic substances: chemistry and reactions. In: Schnitzer M, Khan SU (eds) Soil organic matter. Elsevier, Amsterdam, pp 1–58

Shuman LM, Dudka S, Das K (2002) Cadmium forms and plant availability in compost-amended soil. Commun Soil Sci Plant Anal 33:737–748

Smith MB, March J (2007) Advanced organic chemistry: reactions, mechanisms, and structure, 6th edn. Wiley, New York

Smolders E, McLaughlin MJ (1996) Effect of Cl and Cd uptake by Swiss chard in nutrient solution. Plant Soil 179:57–64

Soussana JF, Loiseau P, Viuchard N, Ceschia E, Balesdent J, Chevallier T, Arrouays D (2004) Carbon cycling and sequestration opportunities in temperate grasslands. Soil Use Manag 20:219–230

Strak M (2008) Peatlands and climate change. International Peat Society, Vapaudenkatu 12, 40100 Jyväskylä, Finland. ISBN 978-952-99401-1-0

Tan KH (2003) Humic matter in soil and the environment, principles and controversies. Marcel Dekker, Inc., New York

Taylor GJ, McDonald-Stephens JL, Hunter DB, Bertsch PM, Elmore D, Rengel Z, Reid RJ (2000) Direct measurement of aluminum uptake and distribution in single cells of *Chara corallina*. Plant Physiol 123: 987–996

Tipping E (2005) Modelling Al competition for heavy metal binding by dissolved organic matter in soil and surface waters of acid and neutral pH. Geoderma 127:293–304

Van Hees PAW, Jones DL, Finlay R, Godbold DL, Lundstrom US (2005) The carbon we do not see - the impact of low molecular weight compounds on carbon dynamics and respiration in forest soils: a review. Soil Biol Biochem 37:1–13

Van-Camp L, Bujarrabal B, Gentile A-R, Jones RJA, Montanarella L, Olazabal C, Selvaradjou S-K (2004) Reports of the technical working groups established under the thematic strategy for soil protection. EUR 21319 EN/3. Office for Official Publications of the European Communities, Luxembourg, p 872

Weggler K, McLaughlin MJ, Graham RD (2004) Effect of chloride in soil solution on the plant availability of biosolid-borne cadmium. J Envirom Qual 33: 496–504

Weng L, Temminghoff EJM, Van Riemsdijk WH (2001) Contribution of individual sorbents to the control of heavy metal activity in sandy soil. Environ Sci Technol 35:4436–4443

Woods WI, Falcao NPS, Teixeira WG (2006) Biochar trials aim to enrich soil for smallholders. Nature 443:144

Yallop AR, Clutterbuck B (2009) Land management as a factor controlling dissolved organic carbon release from upland peat soils 1: spatial variation in DOC productivity. Sci Total Environ 407:3803–3813

Zhu C, Anderson G (2002) Environmental applications of geochemical modeling. Cambridge University Press, Cambridge, UK

Oxidative Stress and Phytoremediation

23

Kinga Drzewiecka, Mirosław Mleczek, Agnieszka Waśkiewicz, and Piotr Goliński

Abstract

Numerous environmental factors such as air and soil pollutants, soil conditions including heavy metal ions, temperature shock, drought, UV radiation, paraquats, pathogens and their secondary metabolites can cause the phenomenon of oxidative stress. Nowadays, the understanding of oxidative stress and plant resistance mechanisms seems to be of great importance. Plants sensitive to selected stressors have been employed as active or passive bioindicators of environmental pollution worldwide. By contrast, plants with enhanced tolerance versus pollutants are a promising tool in efficient bioremediation of areas contaminated with heavy metals and xenobiotics. Phytoremediation is a rapidly developing technique of soil and water cleanup in the case of both metals/metalloids and xenobiotics pollution. A lot of studies investigate the diverse aspects of phytoremediation including the mechanism of oxidative stress of plants and its impact on the process efficiency. The use of green plants capable of sequestering heavy metals in their aerial organs combined with uninterrupted high biomass production allows to achieve high efficiency of cleaning process. In our review, we discuss environmental factors causing oxidative stress in plants effecting the efficiency of remediation.

Keywords

Heavy metals • Mycotoxins • Oxidative stress • Phytoremediation • Tropospheric ozone • Salicylic acid

K. Drzewiecka • M. Mleczek • A. Waśkiewicz
• P. Goliński (✉)
Department of Chemistry, University of Life Sciences,
ul. Wojska Polskiego 75, 60-625 Poznań, Poland
e-mail: piotrg@up.poznan.pl

P. Ahmad and M.N.V. Prasad (eds.), *Abiotic Stress Responses in Plants: Metabolism, Productivity and Sustainability*, DOI 10.1007/978-1-4614-0634-1_23, © Springer Science+Business Media, LLC 2012

1 Introduction

Increasing anthropogenic contamination causes degradation of the surface, biologically active layer of the lithosphere. Degraded areas, particularly as a result of intensive industrial activity, not only constitute an ecological problem but also cease to be commercially attractive. Management of such areas is a complicated problem which involves issues related to social and engineering sciences, particularly planning and urban development. Management of degraded areas should be undertaken in a comprehensive manner, i.e. taking into consideration interdependencies between elements of the environment (soil, water, air) and the economy (land use, land value, reclamation costs). Post-industrial areas are defined as "degraded, unused or underused areas originally allocated to economic activity, which has since then ceased to be run". These are areas degraded to a degree limiting the potential for development and/or restoration of their economic functions, excluding areas used for military and agricultural purposes. We also need to consider health risks and relationships between the degraded site and the economy. These criteria may be applied both to the areas in which economic activity is no longer run, and to those which continue to be used for industrial purposes and as such may be further degraded. They do not cover the specific character of post-industrial areas and may be applied in the management of such areas only to a limited degree. For this reason it is necessary to conduct a survey of post-industrial areas, on the basis of which it will be possible: (1) to develop an information system for degraded areas, based on databases co-operating with geographical information systems and facilitating the application of mathematical modeling; (2) to identify priorities and ordering of reclamation and renewal tasks, on the basis of previously established uniform criteria; and thus also (3) to plan and implement an effective earth surface protection policy.

The management system for post-industrial areas should contain information on development planning of areas included in urban renewal projects. Efficient management also requires the introduction of certain legal changes, concerning first of all the scope and manner of soil and land quality analyses and principles of preparing land records, and possibly closing down of industrial plants, mines, burial grounds, hazardous waste disposal sites, etc., as entities having a potential considerable environmental impact, to create financial incentives for prospective investors in those areas.

We may distinguish the following post-industrial areas: (1) chemically degraded – requiring treatment, (2) physically degraded – requiring reclamation, and (3) not degraded physically/chemically, but no longer serving their previous economic functions. Chemically degraded areas need to be remediated, since chemical substances may migrate from the earth surface (from contamination sources) to the receptor (e.g. the aquifer). The following phenomena occur most frequently in post-industrial areas: (1) changes in water relations (surface waters and groundwater), (2) changes in overconsolidation of non-cohesive soils, resulting from fluctuating loads and (3) penetration of pollutants (particularly chemicals) from sewage and waste to the subsoil.

The concept of post-industrial areas is connected with the manner of earth surface use and denotes areas serving industrial functions: economic, social and spatial. Economic functions are connected with the extraction and acquisition of mineral resources and their transformation into means of production and means of consumption, as well as stimulation of development in other branches of the economy. Social functions are connected with the generation of new jobs, improvement of living conditions and elevation of the level of education. In turn, spatial functions result in the transformation of the natural environment (e.g. mines, industrial plants, urbanization).

The basic division comprises three types of post-industrial areas: (1) type I – areas allocated to production processes (delivery or acquisition of raw materials, energy and workforce, production of goods and dispatch of products, processing/management/disposal of wastes; (2) type II – areas serving auxiliary functions (e.g. administrative, research and design centres, company cultural, sports and health-care facilities, vocational training facilities, housing facilities for employees, freight depots, logistics and communication centres, water intakes and sewage treatment plants, isolation green belts and protection zones); and (3) type III – areas of physical impact (contamination) and economic impact (the involvement of industry in the generation of the national product). Industrial wastes to a considerable degree contribute to the formation of degraded areas, major contributors being coal mining, extraction of mineral raw materials, power engineering and metallurgy. Hazardous wastes, due to their origin, chemical and biological composition, as well as other properties and circumstances, constitute a particular threat to human life and health or to the environment.

Since plants are settled permanently in the environment, they have developed efficient mechanisms of adaptation and stress avoidance to survive fluctuations and adverse conditions caused by either human activities or natural sources, i.e. air pollution with organic and inorganic compounds (O_3, SO_2, NO_x, xenobiotics), soil conditions including salinity and heavy metal ions, temperature shock, drought, UV radiation, nutrient limitation, paraquats, toxic secondary metabolites (mycotoxins) and attack by pathogens and herbivores. However, these factors can cause the phenomenon of oxidative stress, with induction and development pathways still not fully understood. Nowadays, the understanding of oxidative stress and plant resistance mechanisms seems to be of great importance. Plants sensitive to selected stressors have been employed as active or passive bioindicators of environmental pollution worldwide, e.g. *Lemna minor*, *Nicotiana tabacum*, *Trifolium repens*, naturally growing tree and shrub species. By contrast, plants with enhanced tolerance versus pollutants (*Salix* and *Populus* spp.) are a promising tool in efficient bioremediation of areas contaminated with heavy metals and xenobiotics.

A common feature of various stress factors is overproduction of reactive oxygen species (ROS) such as superoxide anion, hydroxyl radical and hydrogen peroxide, as well as organic free radicals and peroxides resulting in direct peroxidation of cell continuants including lipids, proteins and DNA, membrane breakdown and eventually in cell collapse (necrosis), or in controlled signaling and in turn induction of defence mechanisms, which may be accompanied by suicidal cell death (apoptosis). Plants synthesize a wide range of enzymatic and non-enzymatic antioxidants able to scavenge toxic oxygen derivatives and limiting oxidative damage, i.e. superoxide dismutase, ascorbate peroxidase, glutathione reductase, catalase, ascorbic acid, glutathione, α-tocopherol, carotenoids and phenolics. Nevertheless, the antioxidative potential of the above compounds is limited. Thus, oxidative stress occurs as a result of an imbalance between ROS generation and the plant's abilities to detoxify them with its antioxidant system or due to the plant's reaction to a stress factor and a controlled oxidative burst within the cell interior (Arora et al. 2002; Greene 2002).

2 Signaling Compounds in Plant Reaction to Stress

Salicylic acid (SA) (*ortho*-hydroxybenzoic acid) is one of the phenolic compounds commonly present in higher plants. The highest contents of salicylic acid in the free form as well as in the form of its glucoside have been found and reported in plants of *Salicaceae*, *Betulaceae* and *Ericaceae* families. Volatile methyl salicylate (MeSA) is one of the main ingredients of essential oils of *Apiaceae*, *Violaceae* and *Rosaceae* families, attracting insects and aiding plant reproduction this way (Shlaev et al. 1997). Salicin, an active extract from *Salix alba* L. bark, had been used for centuries as a pain-relieving and fever-reducing remedy of which the active ingredient was first isolated as a crystalline compound in 1828 and further converted into a sugar and salicylic acid (Pierpoint 1997). According to some authors, salicylic acid belongs to a group of plant hormones according to the function it exhibits in plant growth and development, but its content in plants is substantially higher than other phytohormones, e.g. jasmonic acid or ethylene. Salicylic acid level in plants is species- and tissue-specific in a concentration range from a few nanograms up to 75 $\mu g\ g^{-1}$ of fresh weight. The highest contents have been documented for thermogenic plants of *Araceae and Nymphaeaceae* families, the compound's function and role lying in an alternative form of respiration to generate heat, spreading their scent, and plants challenged

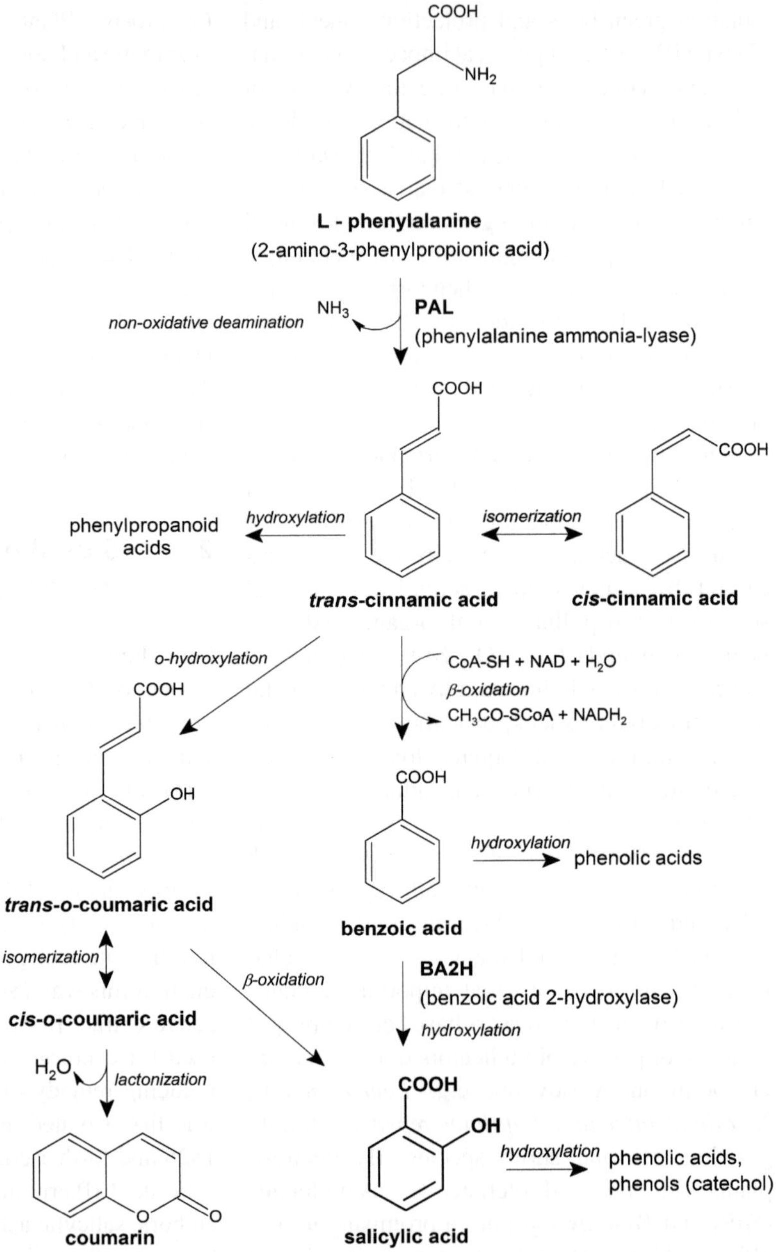

Fig. 23.1 Salicylic acid biosynthesis in plant cells [according to Czerpak and Bajguz (1998)]

with incompatible pathogen infection (Pierpoint 1997; O'Donnell et al. 2001). Salicylic acid influences several physiological and metabolic processes, regulates seed germination, growth of the root system and leaves, chlorophyll biosynthesis, as well as flowering and thermogenesis (Pancheva et al. 1996; Raskin 1992). In plant cytosol, amino acid phenylalanine (2-amino-3-phenylpropionic acid) is a precursor of salicylic acid biosynthesis in the phenylpropanoid pathway via benzoic or coumaric acids as alternative substrates (Metroux 2002) (Fig. 23.1). Salicylic acid synthesis is often accompanied by the generation of other compounds playing a role in resistance mechanisms

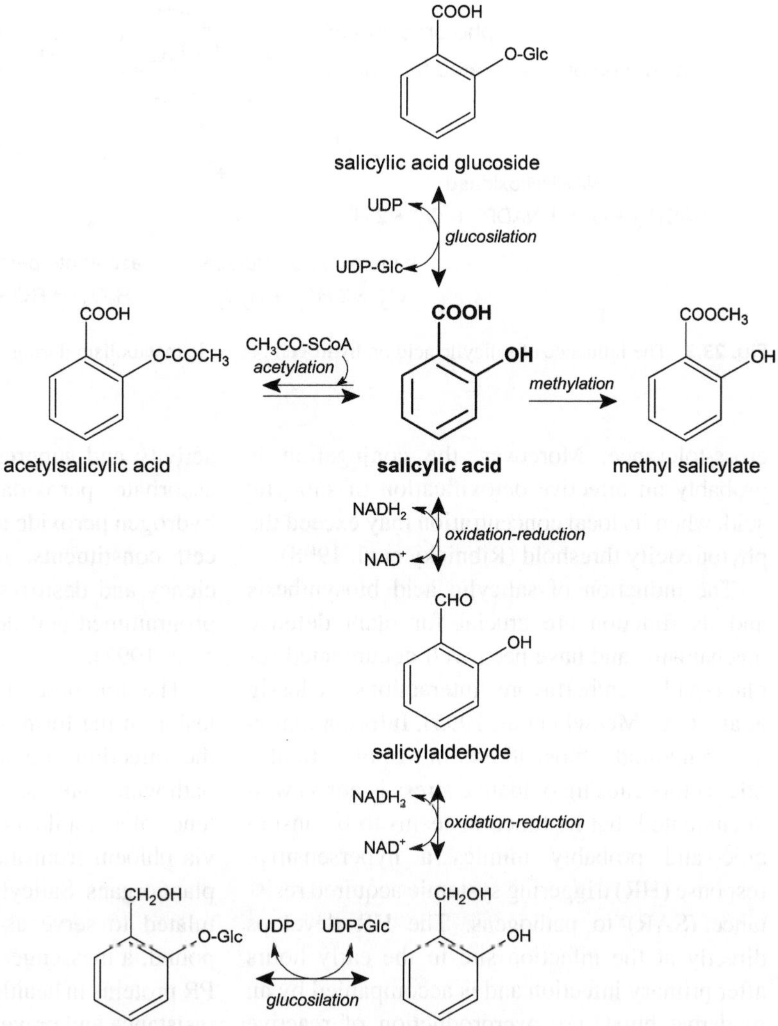

Fig. 23.2 Conjugates of salicylic acid in plant cells [according to Czerpak and Bajguz (1998)]

of plant cells, including phenolic acids (cinnamic, benzoic, coumaric, caffeic, ferulic, etc.) and flavonoids of antioxidant and chelating properties, phytoalexins and lignin (Chong et al. 2001; Metroux 2002). Salicylic acid can be further conjugated into methyl salicylate, a volatile intra- and interplant signal transducer, alerting plant organs as well as neighbouring plants of adverse conditions in the environment (Fig. 23.2). The biosynthesis of methyl salicylate was found for *Nicotiana tabacum* L. infected with tobacco mosaic virus (TMV). Furthermore, exogenous MeSA induced the biosynthesis of salicylic acid and pathogenesis-related (PR) proteins in tobacco leaves (Lee et al. 1995; Shlaev et al. 1997). Among the conjugated forms, salicylic acid *ortho*-β-D-glucoside was observed at the highest concentration level in plant cells. Salicylic acid self-regulates the biosynthesis of the glucoside by activating a specific cytoplasmic UDPglucose:SA glucosyltransferase (GTase) catalyzing its conjugation with a glucose molecule (Fig. 23.2) (Lee et al. 1995), which confirms an active storage mechanism of a locally and systemically active free salicylic acid, probably for the purpose of further exposure to stress and

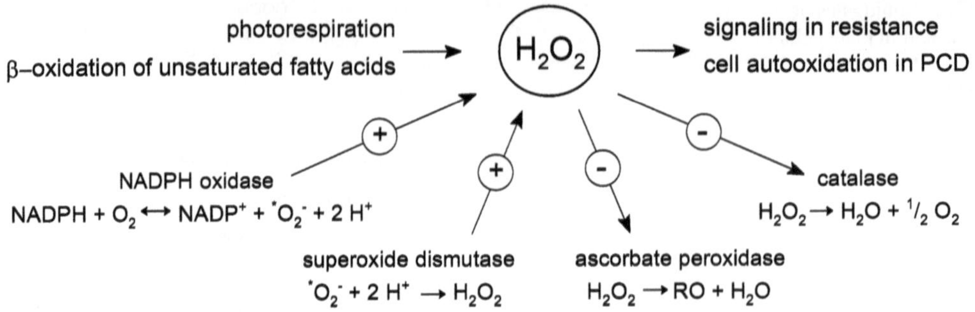

Fig. 23.3 The influence of salicylic acid on hydrogen peroxide metabolism during oxidative stress

cross-tolerance. Moreover, the conjugation is probably an effective detoxification of salicylic acid, when its local concentration may exceed the phytotoxicity threshold (Ribnicky et al. 1998).

The induction of salicylic acid biosynthesis and its function are crucial for plant defence mechanisms and have been well documented for plant–pathogen/herbivore interactions (Klessig et al. 2000; Meuwly et al. 1995). Information on the compound's biosynthesis in response to abiotic factors causing oxidative stress is not as well documented, but its function seems to be unspecific and probably mimics a hypersensitive response (HR) triggering systemic acquired resistance (SAR) to pathogens. The HR develops directly at the infection site in the early hours after primary infection and is accompanied by an oxidative burst, i.e. overproduction of reactive oxygen species (ROS), enhanced biosynthesis of phenolic compounds (including salicylic acid), cell wall lignification and induction of PR proteins at the infection site and in surrounding cells (Klessig et al. 2000; Wojtaszek 1997). Salicylic acid is also a non-specific regulator of plant resistance to pathogens and probably to other environmental factors responsible for oxidative stress, i.e. tropospheric ozone, xenobiotics, heavy metals, toxic secondary metabolites (mycotoxins), UV radiation, salt and drought stress, etc. (Koch et al. 2000; Pal et al. 2002; Pasqualini et al. 2002; Zhu 2002). During the hypersensitive response, salicylic acid alters hydrogen peroxide metabolism within plant cells by the induction of superoxide dismutase (SOD), NADPH oxidase activity and suppression of catalase (CAT) and ascorbate peroxidase (APx), responsible for hydrogen peroxide accumulation, which oxidizes cell constituents, reduces photosynthesis efficiency and destroys cell membranes, leading to programmed cell death (PCD) (Fig. 23.3) (Rao et al. 1997).

The function of the apoptosis is to form a lesion in the form of a ring of dead cells around the infection site to prevent the spread of the pathogen. Later developed SAR requires the existence of a mobile molecule for signal transduction via phloem from the infection site to uninfected plant organs. Salicylic acid in the free form is postulated to serve as a signal transduction compound, a messenger inducing the biosynthesis of PR proteins in healthy plant parts to enhance plant resistance and prevent future infections.

In our earlier studies ((Drzewiecka et al. 2012), we investigated the impact of ambient ozone on the biosynthesis of salicylic acid in leaves of two tobacco cultivars showing diverse sensitivity to ozone (Bel-W3 – sensitive, and Bel-B – resistant). The main aim of the study was to generate information on the possibility of the compound's application as a biomarker of ozone-caused oxidative stress informing of the possible negative influence of tropospheric ozone on trees, crops and plants used in phytoremediation. Tropospheric (ground level) ozone has been known to be a highly oxidizing molecule for plant tissue since the first observations of ozone-caused injury of crops in the 1950s in the United States. Proliferation and constantly increasing concentrations make ozone one of the constituents of the ambient air causing considerable devastation among wild-growing plants, reducing annual tree increments and negatively influencing species biodiversity. In addition, ozone reduces the yield of sensitive crops and decreases their commercial value (Black et al. 2000). Ozone concentration in the troposphere has increased fourfold since the beginning of the industrial era and its peak values in the most industrialized countries achieved 100–400 ppb (Kley et al. 1999). Mean ozone concentrations in Europe and North America during summer months continue to grow (by 0.2–1% annually) and are sufficient to cause damage in ozone sensitive plants (Drzewiecka et al. 2012; Vingarzan 2004).

Tobacco plants (*Nicotiana tabacum* L.) of both cultivars were exposed to ambient air of the city of Poznan (a city located in west-central Poland) and surrounding rural areas according to the standard VDI methodology (2000) applied in the biomonitoring of air contamination with ozone in European countries in years 1999–2002 (Klumpp et al. 2006). The exposure of the Bel-W3 plants to ozone caused a significant increase, on average fourfold, in the content of free salicylic acid (SA) and a nearly 20-fold increase in the total salicylic acid (TSA – free and in the form of glucoside) in leaves showing ozone-caused injuries (compared with control plants). The exposure of the ozone-tolerant Bel-B cultivar resulted in a slight, insignificant increase in the salicylic acid content only. SA observed for the tobacco leaves of the Bel-W3 cultivar exhibiting almost 100% injuries was at the concentration level close to nearly 20 $\mu g\ g^{-1}$ FW. This was several times lower than the salicylic acid content in tobacco leaves inoculated with TMV 6 hours after inoculation (approximately 75 $\mu g\ g^{-1}$ FW) (Enyedi et al. 1992). It is possible that distinct threshold concentrations of salicylic acid are required to induce PCD in plants challenged with different stress factors (Thulke and Conrath 1998). The exposure of the Bel-W3 tobacco plants reduced the proportion of SA and TSA contents in leaves from about 30% before the exposure to about 12% after it, which confirms a regulatory function of salicylic acid in the biosynthesis of its glucoside. A strong direct proportional relation between BA2H activity and TSA content and a slightly weaker one with SA were found, proving that the hydroxylation of benzoic acid on the *ortho* position is catalyzed by benzoic acid 2-hydroxylase as a response to ozone presence. A weaker correlation of BA2H activity with the content of free salicylic acid may indicate a relatively rapid transformation of this compound into the glucoside followed by salicylic acid release from the glucoside (Fig. 23.4).

Enhanced biosynthesis of salicylic acid in leaves of the Bel-W3 plants confirms its involvement in PCD via the impact on hydrogen peroxide metabolism in response to tropospheric ozone. According to Durner et al. (1997), salicylic acid influences the cellular redox state triggering lesion formation. By contrast, plants unable to accumulate salicylic acid (e.g. NahG plants) or having reduced ability to perceive SA (e.g. hybrid poplar clone NE-388) show weaker induction of antioxidant enzymes and a lack of ozone-induced PCD (Koch et al. 2000; Samuel et al. 2000). Our results supplement earlier reports on two-stage accumulation of ROS, mainly H_2O_2, in the leaves of both tobacco cultivars exposed to ozone in fumigation chambers (Pellinen et al. 1999; Schraudner et al. 1998; Wohlgemuth et al. 2002). In the case of the Bel-B cultivar the first phase during the exposure to ozone only as a result of ozone reactions in the apoplast was observed. In the case of the Bel-W3 cultivar, also the second phase was observed after the exposure, as the result of the induction of defence pathways and a controlled intracellular response to ozone, including biosynthesis of salicylic acid. Furthermore, a highly significant positive linear correlation between the content of salicylic acid and ozone-induced injuries of tobacco leaves was observed. However, in the case of moderate injuries, SA was strongly correlated with the number of lesions observed on the leaf surface (Fig. 23.5). This confirms the highest rate of salicylic acid biosynthesis as the plants' earliest response to stress.

Fig. 23.4 Linear correlation between the activity of benzoic acid 2-hydroxylase (BA2H) and salicylic acid content in tobacco Bel-W3 leaves after exposure to tropospheric ozone (**a**) free salicylic acid (SA); (**b**) salicylic acid released from glucoside (SAG); (**c**) total salicylic acid (TSA)

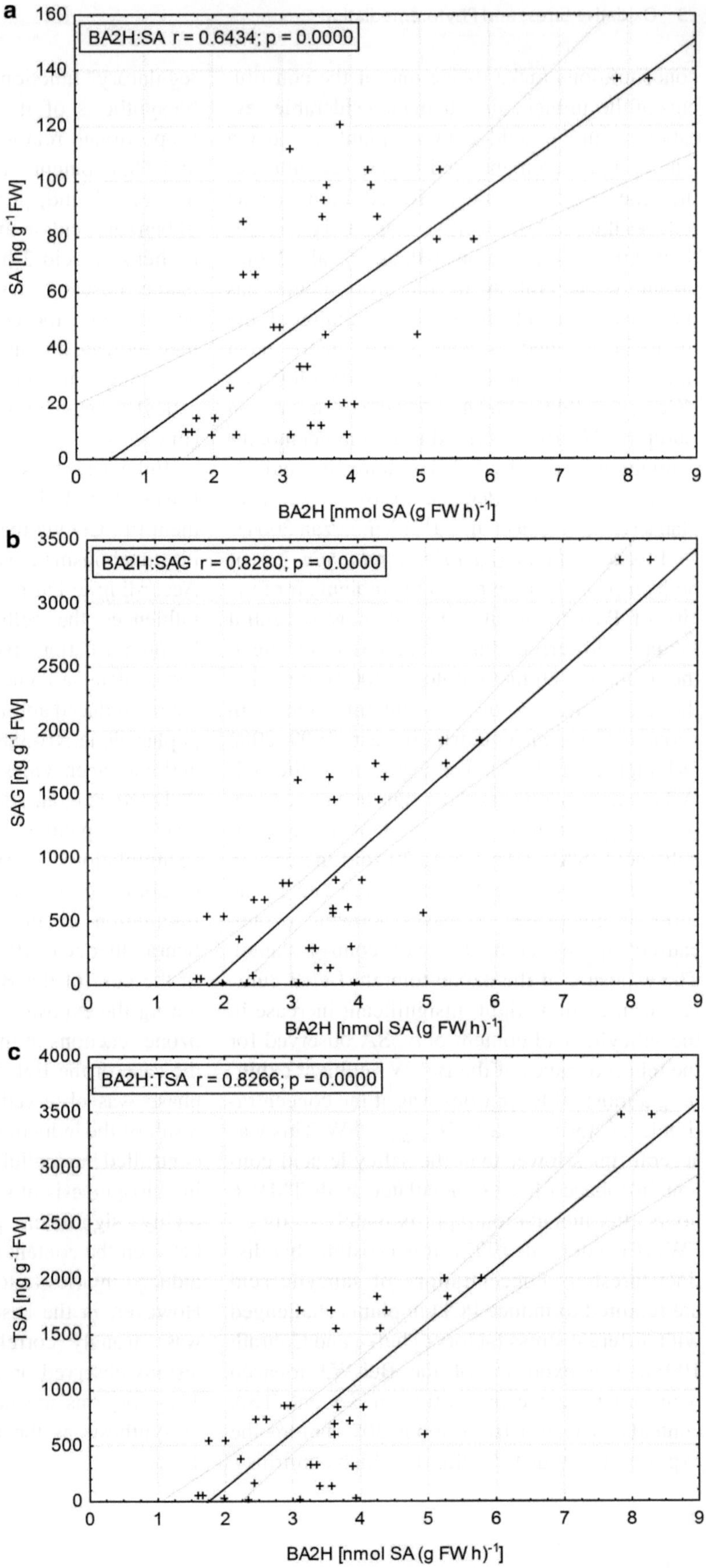

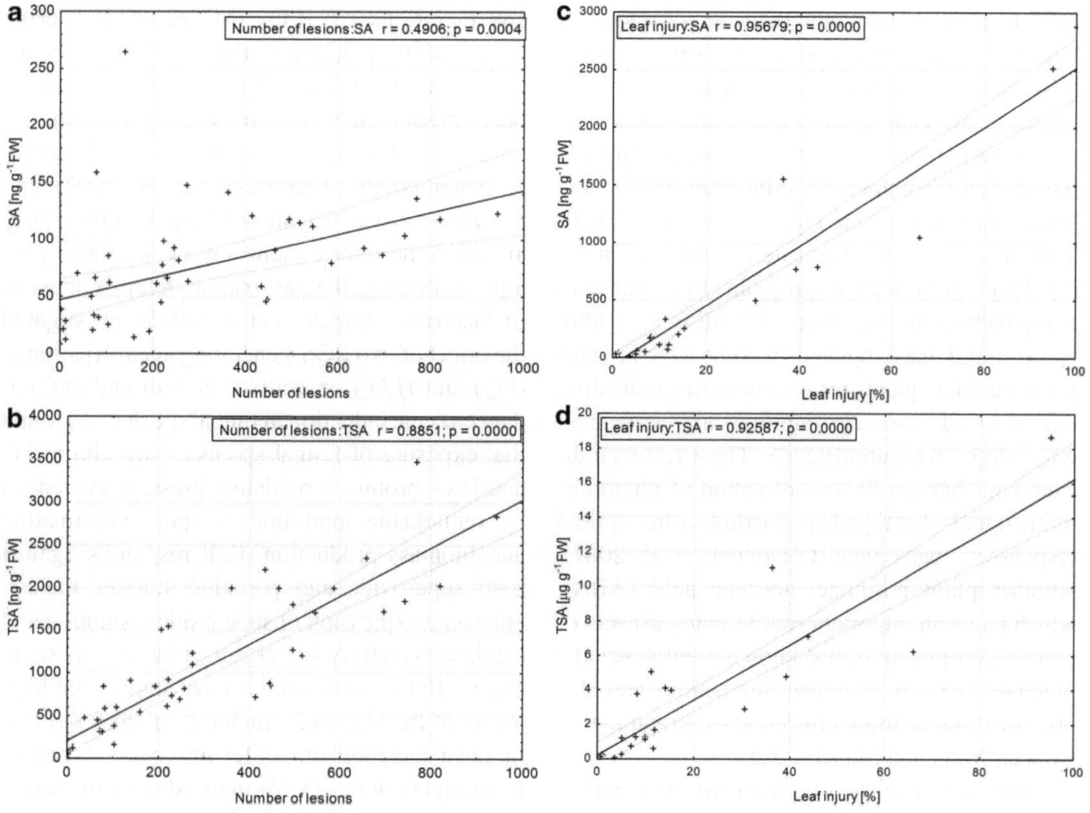

Fig. 23.5 Linear correlations between number of lesions (**a**, **b**) or leaf injury (**c**, **d**) and salicylic acid (SA – free salicylic acid; TSA – total salicylic acid, i.e. free and in the form of glucoside) content in tobacco Bel-W3 leaves after exposure to tropospheric ozone (for medium and full-scale injuries, respectively)

3 Toxic Secondary Metabolites of Fungi (Mycotoxins) in Oxidative Stress

Plants encounter many different pathogens during their lifetime, and are armed and ready to defend themselves with a combination of pre-formed and inducible defence mechanisms (Thatcher et al. 2005). Interactions of plants and their pathogenic fungi are now an interesting and rapidly developing field in plant science, with a significant impact in new strategies for plant protection. The plant response to infection is determined by the genetic background of the host as well as the pathogen (Molodchenkova et al. 2002). Pathogen attack triggers complex signaling cascades regulated by signaling molecules such as salicylic acid, jasmonic acid (JA) and ethylene (ET), resulting in the expression of defence-related genes such as those encoding pathogenesis-related proteins, and the production of antimicrobial secondary metabolites (Glazebrook 2001). Often, the type of induced response that is effective against a given pathogen varies, depending on the lifestyle of the pathogen (Thatcher et al. 2009).

Fungi are aerobic organisms dependent on oxygen for survival, so they have to cope with the consequences of its presence, i.e. the formation of reactive oxygen species (ROS). These reactive compounds are formed during metabolic processes, such as glucose respiration and fatty acid metabolism, by the activity of NADPH oxidases and other oxygenases.

Biosynthesis of salicylic acid – as a key molecule involved in the immune response in plants – accompanies oxidative stress as confirmed in plants of many species in the case of pathogen attack (O'Donnell et al. 2001; Vasyukova and Ozeretskovskaya 2007). Constitutive levels of salicylic acid vary not only between plant species but also between cultivars of the same species (Vasyukova and Ozeretskovskaya 2007) and organs of the same plant. SA accumulates at sites where the pathogen invaded the plant and then is transported via phloem to uninfected parts of the plant (Vasyukova and Ozeretskovskaya 2007). There is a very important role for ethylene in the development of disease symptoms in response to many pathogenic organisms (Ciardi et al. 2000). It was found that exogenous feeding of SA to ethylene-deficient plants restores necrosis, indicating that reduced disease symptoms are associated with failure to accumulate SA. These results indicate a mechanism for co-ordination of phytohormone signals that together constitute a susceptible response to pathogens (O'Donnell et al. 2001). Another plant hormone, abscisic acid (ABA), which plays an important role in many aspects of plant development, is also known to influence the outcome of plant–pathogen interactions and abiotic stress has a strong effects on its accumulation (Mauch-Mani and Mauch 2005).

Increased activity of oxidative metabolism was frequently found in localized infection of foliar tissues by obligate biotrophic (which derive nutrients from living host cells) or necrotrophic (which derive nutrients from dead cells) pathogens associated with a rapid development of HR (Beckers and Spoel 2006). By contrast, very little is known about oxidative metabolism in plant resistance to pathogens that do not induce HR, such as the fungi that invade the plant vascular system (Gracia-Limones et al. 2002). Cells can tolerate a small to moderate amount of oxidative stress, which usually stimulates increased synthesis of antioxidants to restore a correct balance between the formation of ROS and the protective antioxidant mechanisms of cells (catalase, superoxide dismutases, peroxidases, etc.) as well as non-enzymatic protective molecules such as glutathione and thioredoxin (Pitzschke et al. 2006).

For instance, catalase is thought to play a key role as an antioxidant, protecting aerobic organisms from the toxic effects of hydrogen peroxide – the filamentous penicillin-producing fungus *Penicillium chrysogenum* is especially resistant to the oxidative stress caused by high concentrations of H_2O_2 (Mayer et al. 2001). Information regarding the *Fusarium* genus remains fragmentary and only a few species have been investigated. The most extensive studies focused on *F. oxysporum* (Angelova et al. 2005), *F. acuminatum* (Ayar-Kayali et al. 2002) and *F. equiseti* (Ayar-Kayali and Tarhan 2004). They all led to the same conclusion, i.e. catalases play a pivotal role in the antioxidant defence network of *Fusarium*. Angelova et al. (2005) investigated the effect of two ROS-generating agents (paraquat (PQ) and H_2O_2) on cellular growth and antioxidant enzyme induction in fungal species and found that exposure of fungal spores or mycelia to PQ and H_2O_2 promoted oxidative stress, as evidenced by remarkable inhibition of spore germination and biomass production. Cell responses against both superoxide and peroxide stresses include enhanced expression of superoxide dismutase and catalase, which are key enzymes for directly scavenging ROS. Other studies have shown that high levels of hexokinase, a major regulatory enzyme in sugar metabolism, confer improved resistance to methyl viologen (MV)-induced oxidative stress and pathogen infection (Sarowar et al. 2008). A strong increase in extracellular ROS content has a directly inhibitory effect on the development of pathogenic fungi and, by activating the antioxidant system, promotes lignification of cell walls (Troshina et al. 2007).

The response of filamentous fungi to oxidative stress is not only based on antioxidant stimulation. Low levels of ROS, produced at specific points during the life cycle, play a crucial role in the fungal cell. For instance, the transition from conidia to the germ tube, the establishment of apical dominance and the onset of secondary metabolism are each under the control of reactive oxygen species (Semighini and Harris 2008).

Mycotoxins are harmful and often carcinogenic secondary metabolites produced by a range of widespread fungi belonging in the main to *Fusarium*, *Aspergillus* and *Penicillium* genera (Goliński et al. 2009). In general, they are low-molecular-weight compounds synthesized by filamentous fungi and are capable of causing disease and death in plants, animals and humans (Bennett and Klich 2003). While in the literature there are many reports indicating high toxicity of mycotoxins, little is known about their role in plant–pathogen interactions. The relationship between the decrease in cell proliferation, the presence of oxidative stress generated by the enhancement of intracellular ROS production and ROS-induced lipid peroxidation by mycotoxins is of a priority (Ferrer et al. 2009). Mycotoxins currently considered of importance from the toxicological point of view include patulin, aflatoxins, ochratoxin A, zearalenone as well as trichothecenes and fumonisins, and their occurrence is now regulated by legal limits in all the developed countries (Goliński et al. 2009).

Several secondary metabolites are synthesized by fungi during morphological and metabolic transitions, when the accumulation of ROS occurs (e.g. *Aspergillus parasiticus*) (Reverberi et al. 2008). *Aspergillus*, oxidative stress and aflatoxin production are closely linked (Fanelli et al. 2004; Huang et al. 2009). Such antioxidants as butylated hydroxytoluene (BHT), trihydroxybutyrophenone (THB), propylparaben (PP) and butylated hydroxyanisole (BHA) can directly inhibit aflatoxin biosynthesis in aspergilli (Passone et al. 2005). Under in vivo experimental conditions, BHA, PP and BHA/PP mixtures had a negative impact on the microflora and aflatoxin accumulation in maize grains (Nesci et al. 2003). In stored maize it was observed that BHA and PP at a concentration of 20 nmol l^{-1} affected the mycoflora and *Aspergillus* section *Flavi* populations (Nesci et al. 2008). Plant compounds involved in plant-fungi interactions are able to interfere with mycotoxin biosynthesis in host tissues (Boutigny et al. 2008). Moreover, a comparative study between toxigenic and non-toxigenic strains of *Aspergillus* suggested that oxidative stress is a "prerequisite" for aflatoxin production (Jayashree and Subramanyam 2000). Reverberi et al. (2007) demonstrated an association between oxidative stress and aflatoxin biosynthesis during *Aspergillus parasiticus* growth in maize seeds.

Anti-aflatoxigenic activity of certain chemicals such as eugenol (Jayashree and Subramanyam 1999) and hydrolysable tannins (Mahoney and Molyneux 2004) as well as some plant components (Joseph et al. 2005) is due to their antioxidant capacities. Epoxides, which can lead to lipid peroxidation of fungal cells, stimulated aflatoxin biosynthesis. It seems reasonable to assume that ethylene inhibited aflatoxin biosynthesis is a result of a cascade of events involving a decrease of ROS formation, lipid peroxidation alleviation, glutathione redox state reduction and ultimately down-expression of aflatoxin biosynthesis genes and a decrease of aflatoxin production (Huang et al. 2009). Other studies show that the application of ethylene is effective in reducing the level of contamination of peanuts with aflatoxin during their long-term storage (Gunterus et al. 2007) and that this compound modulates toxin biosynthesis in *Aspergillus* at least in part at the transcriptional level, and an ethylene sensor-mediated signaling pathway probably participated in the process (Roze et al. 2004). Molyneux et al. (2007) claim that the ability of a structurally diverse suite of phenolic antioxidants to suppress aflatoxin production is confirmatory evidence for the hypothesis that aflatoxin biosynthesis is a response of the fungus to oxidative stress. Interaction of a pathogen with the cell wall of a host plant induces a cascade of reactions in the plasma membrane ultimately resulting in the formation of hydrogen peroxide and lipid peroxides (Lamb and Dixon 1997). These products result in direct effects on the pathogen, sealing of the wound against further damage and induced gene regulation resulting in defensive reactions such as formation of phytoalexins. In response, the pathogen (in this case *Aspergillus flavus*) initiates its own stress response to combat the defensive attack of the plant (Molyneux et al. 2007).

Studies in recent years have shown that ochratoxin A (OA) from *Aspergillus* and *Penicillium* strains on *Arabidopsis thaliana* induced an evident oxidative burst in the leaves with an increase of reactive oxygen species and concomitant down-regulation of antioxidant defence enzymes and up-regulation of lipid peroxidation (Peng et al. 2010). These results suggested that OA damage might result from reactive oxygen species pathways.

It is apparent that environmental oxidative stress cannot only concern aflatoxin and ochratoxin biosynthesis but also other mycotoxins in other fungi. *Fusarium* spp. are ubiquitous fungi found in soil worldwide as both pathogenic and non-pathogenic strains. The signals leading to disease or the absence of disease are poorly understood (Bouizgarne et al. 2006a). *Fusarium* triggered transient H_2O_2 production, calcium influx and alkalinization of the extracellular medium (Olivain et al. 2003). Later defence responses were also observed, such as increased activities of peroxidase and phenylalanine ammonia-lyase (PAL) (He et al. 2002), reinforcement of the cell wall (Benhamou and Garand 2001; He et al. 2002; Salerno et al. 2000) and accumulation of potential antimicrobial compounds, such as phytoalexins (Cachinero et al. 2002; Daayf et al. 2003).

The potential role for salicylic acid in induced resistance of asparagus to *F. oxysporum* f. sp. *asparagi* (*Foa*) was described by He and Wolyn (2005). They reported that exogenous SA activated peroxidase (POD) and phenylalanine ammonia-lyase, as well as lignifications upon *Foa* attack. Mandal et al. (2009) demonstrated that the exogenous application of 200 μM salicylic acid through root feeding or foliar spray could induce resistance against *F. oxysporum* f. sp. *lycopersici* (*Fol*) in tomato. The activities of PAL and POD were about five times higher than for the control plants after 168 h of salicylic acid feeding through the roots, and almost four times higher after treatments through foliar spray. The salicylic acid-treated tomato plants challenged with *Fol* exhibited significantly reduced vascular browning and leaf yellowing wilting. So far, little is known about changes in endogenous salicylic acid content in plants challenged with *Fusarium* infection and the induction of intracellular mechanisms of resistance. In our studies the main goal was the elucidation of the plant–pathogen interaction with the impact of the infection by *Fusarium proliferatum* and *F. oxysporum* on the level of salicylic acid and mycotoxins (moniliformin and fumonisin B_1) concentrations in asparagus tissues (Waśkiewicz et al. 2011). In addition, the electron paramagnetic resonance (EPR) spectroscopy was used to evaluate the impact of the infection by both *Fusarium* species on the level of free radicals and other paramagnetic species in asparagus (Dobosz et al. 2011). According to Edgar et al. (2006), exogenous salicylic acid treatment prior to inoculation activated *PR1* and *BGL2* defence gene expression in leaves and provided increased *F. oxysporum* systemic resistance as evidenced by reduced foliar necrosis and plant death. Exogenous SA treatment of the foliar tissue did not activate defence gene expression in the roots of plants. This suggests that salicylate-dependent defences may function in foliar tissue to reduce the development of pathogen-induced wilting and necrosis. Molodchenkova et al. (2002) suggested that exogenous salicylic acid was involved in the induction of trypsin and lectin inhibitors that are important in the formation of defences against *F. moniliforme* in maize sprouts.

In the case of *Fusarium* toxins, the addition of hydrogen peroxide and diamine – which acts as an oxidant mainly towards thiols such as glutathione – to the culture media induces the biosynthesis of type B trichothecenes, deoxynivalenol and 15-acetyl-deoxynivalenol (Ponts et al. 2006). Ye et al. (2006) found that *Fusarium* infection also resulted in increased activities of antioxidant enzymes together with increased levels of ROS and lipid peroxidation in cucumber roots and that apparently did not lead to a localized resistant response. Recently, beauvericin (BEA), fusarenon X (FX), nivalenol (NIV), deoxynivalenol (DON), diacetoxyscirpenol (DAS), neosolaniol (NEO), HT-2 and T-2 toxin have been demonstrated to induce cell death and alteration to the ascorbate metabolism in tomato protoplasts (Paciolla et al. 2004; Reverberi et al. 2010). BEA is reported to be phytotoxic for the plant system and to induce oxidative stress at micromolar concentrations (Logrieco et al. 2002). In turn, in tomato plant–pathogen interactions in which the phytopathogens are T-2 producing species of the *Fusarium* genus, one of the roles of this mycotoxin could be to bind to lignin and so to destroy it, thus facilitating the passage of the pathogen into the host cell (Paciolla et al. 2008). Other studies, conducted to clarify the impact of oxidative stress on type B trichothecene (TR) production by *F. graminearum* and *F. culmorum* and show the higher adaptation to oxidative stress developed by nivalenol isolates, are consistent with the higher virulence of these *Fusarium* strains on maize compared with DON isolates (Carter et al. 2002) and also suggested that the effect of H_2O_2 might depend on the species and/or the chemotype of the considered *Fusarium* strain (Maier et al. 2006). Ponts et al. (2009) reported a strong effect of H_2O_2 stress on NIV+FX biosynthesis by *F. graminearum* and *F. culmorum* and significant inhibition of NIV+FX production by H_2O_2. Actually, isolates of the DON and NIV chemotypes were reported to be equally pathogenic to wheat (Carter et al. 2002). The host/virulence interactions could be ascribed to different host H_2O_2-production patterns in response to *F. graminearum* and *F. culmorum* infection (Ponts et al. 2009). Nishiuchi et al. (2006) reported that the T-2 toxin and another TR produced by *F. sporotrichioides* induces H_2O_2 production, inhibits protein synthesis and simulates cell death in the non-host plant *A. thaliana* whereas fumonisin B_1 (FB_1) inhibits ceramide synthase, impairing the sphingolipid metabolism and causing cell death in *Arabidopsis* (Asai et al. 2000).

A nonspecific toxin produced by most *Fusarium* spp. is fusaric acid (FA), which could induce typical early defence responses such as ROS generation, increase in cytosolic calcium, modifications of ion fluxes, and delayed response such as phytoalexin synthesis (Bouizgarne et al. 2006b). It has recently been observed that FA could elicit various plant defence responses at 100 nM without toxic effects, and the effect of this toxin on root and root hairs, the probable first site of contact between the fungi and the host, was investigated (Bouizgarne et al. 2004). Other studies demonstrated that FA at 10^{-7} M failed to induce salicylic acid- and jasmonic acid/ethylene-dependent defence-related genes but inhibited the germination of the angiosperm parasite *Orobanche ramosa* in contact of FA-pretreated *Arabidopsis thaliana* seedlings. These data suggest that FA at nontoxic concentrations could activate signal transduction components necessary for plant defence responses that could contribute to biocontrol activity of *Fusarium* spp. (Bouizgarne et al. 2006a).

Other studies have shown that cucumber plants have autotoxic potential by releasing autotoxic substances such as cinnamic and benzoic acids (Yu et al. 2000). Both direct and indirect effects of cinnamic acid might be an important mechanism in soil sickness of cucumber. Autotoxic cinnamic acid could indirectly exert detrimental effects by triggering oxidative stress, then predisposing cucumber plants to infection by *Fusarium* pathogens and finally to colonization of the vascular bundle system in roots (Ye et al. 2006).

The expression of mycotoxin biosynthesis genes is highly regulated by temperature, pH, humidity and host genomic background and conditions that impose stress on the fungus (Schmidt-Heydt et al. 2008). The effect of external factors on mycotoxin biosynthesis is exerted at the level of transcription. It was demonstrated that the *fum1* gene of *F. verticillioides* is activated under stress conditions.

4 Heavy Metals in Different Environmental Matrices

The problem of such heavy metals and metalloids as cadmium, mercury, nickel, lead or chromium accumulation in various ecosystems is not new and is increasing significantly because of the improvement of the level of life, population expansion and heavy industry development. High natural levels of the pollutants in environmental matrices are the result of agricultural and semi-industrial activities, energy supply, mining or waste disposal. Heavy metal ions are ubiquitous in water (ground or surface water), plants, animals, soil as sediments and sewage sludge. They are present in all environmental matrices but in diverse concentration levels, depending on the metal, the matrix and the distance from pollutant sources, which – considering their toxicity and easy translocation in living organisms – causes the real threats.

Heavy metal ions are able to modify many vital mechanisms, to inhibit enzymatic processes by binding to bio-molecules and in consequence to destroy organisms (animals, plants). Recognition of the heavy metal concentration level in environmental matrices is of particular importance due to their significant effects on vegetation followed by human health risk (entering and accumulating in the food chain – fruits, vegetables, crops, plant tissue, etc). Uptake of heavy metal ions by plants and/or animals is one of the main causes of their presence in food. In environmental matrices there are several heavy metals essential in animal nutrition (As, Co, Cr, Cu, Fe, Mn, Mo, Ni, Se, Sn, V and Zn) or in plant growth (B, Cu, Fe, Mo, Mn, Ni, Zn), although a few of them (Cd, Hg and Pb) are not recognized as essential in living organisms. Owing to the acute toxicity of the above-listed elements, they and also Cu, Ni and Zn are included in the US Environmental Protection Agency's (EPA) list as priority pollutants and also in the lists of the American Agency for Toxic Substances and Disease Registry (ATSDR).

The presence of metals and metalloids in matrices influences almost the whole ecosystem, which necessitates environmental monitoring followed by application of proper, continually improved and modernized remediation technology. The heavy metal removal efficiency depends on the method, concentration level and kind of pollutants in the matrix and the presence of other pollutants (interaction, change of redox potential and pH). In contrast to organic compounds (possible to be decomposed or mineralized with final CO_2 and H_2O products), heavy metals are not decomposable to a simpler form and will never be completely removed from polluted matrices but can be immobilized (technical methods) or taken up by plants (phytoremediation) only.

The uptake of heavy metals by plants and their accumulation in tissue has been of great interest recently, not just because of their negative impact on ecosystems and human health. The improved resistance of some plant species to toxic elements has been found as a promising tool in phytoremediation techniques, with efficient extraction of heavy metals from contaminated soil, sewage and water followed by their accumulation in plant tissue (Kumar et al. 1995). Plants have developed numerous mechanisms to compensate for toxicity of heavy metals including avoidance, detoxification and non-specific resistance (Zenk 1996). However, when their concentration in the environment exceeds the species-specific phytotoxicity threshold, the excessive accumulation of heavy metals within cells leads to numerous toxic effects, such as rapid inhibition of overall growth, leaf chlorosis and premature leaf senescence due to the reduction of photosynthesis rate and transpiration, reduction of root elongation and inhibition of seed germination (Borowiak et al. 2011; Gąsecka et al. 2011; Obrouchova et al. 1998). Accumulation of heavy metals in leaves via roots and/or stomata causes a reduction in the size of guard cells as the effect of enhanced biosynthesis of abscisic acid and further decline in the size of the photosynthetic area. Heavy metals interfere with active sites of many enzymes (binding directly with thiol groups of the active centre or with carboxyl groups stabilizing the secondary structure of the enzyme protein), including phosphatase, ATPase and enzymatic antioxidants (catalase, glutathione reductase, ascorbate peroxidase, superoxide dismutase), which results in inhibition of enzyme activity (van Assche and Clijsters 1990; Verma and Dubey 2003). Heavy metals also inhibit the biosynthesis of chlorophyll, cause structural changes in chloroplasts depressing the photosynthesis rate, bind to nucleic acids inducing the aggregation and condensation of chromatin and inhibit replication and transcription, reduce cellular respiration, and lead to macro- and microelement deficits (iron, potassium, magnesium and chlorine) (Borowiak et al. 2011; Seregin et al. 2004; Stroiński and Kozłowska 1997).

Rhizofiltration by symbiotic fungi and bacteria associated with the root system and the exudation of chelating agents into the rhizosphere, i.e. hydrocarbons, organic acids, amino acids and glycoproteins, are the first mechanisms of heavy metal avoidance (Marchner et al. 1996). In the root apoplast, heavy metal transport up to aerial plant organs is restricted due to their binding with carboxyl groups of galacturonic and glucuronic acids present in the structure of the cell wall. Simultaneously, root endodermis prevents radial transport of heavy metals into vascular tissue and further to upper parts of the plant. Among heavy metals, lead is a hardly mobile element, and 70–95% of its content in plants is associated within roots (Piechalak et al. 2002). Heavy metals transported through vascular tissue to upper plant parts accumulate in leaves. However, atmospheric deposition of particulate matter – taken up through stomata or adsorbed by epicuticular wax – is a significant source of toxic elements in photosynthetic tissue (Woźny 1995). At low concentrations, heavy metals are immobilized within the apoplastic space, bound to cell wall polysaccharides. However, higher concentrations perpetuate their transport via the cell membrane to the cell interior (passive – simple or facilitated diffusion; active – with *trans*-membrane proteins, and endocytosis) (Samardakiewicz and Woźny 2000). Plant resistance to heavy metals is mostly dependent on the efficiency of the detoxification mechanisms in intercellular liquid, among which chelation and further subcellular compartmentation in vacuoles is the most abundant one (Clemens 2001; Piechalak et al. 2002). Phytochelatins (PCs), a family of cysteine-rich peptides (glutathione-derived oligomers) are metal-chelating molecules that are crucial in plant tolerance versus metal ions (Cobbett and Goldsbrough 2002; Yruela 2005). PC synthesis via glutathione transpeptidation is catalyzed by phytochelatin synthase induced by heavy metal ions bound to the thiol group of glutathione (Chen et al. 1997). Metal ions complexed with PCs are further transported to the vacuole, where after being released by hydrogen ions they react with organic acids, polyphenols and glucosides, and as a result form less or non-toxic complexes (Cobbett 2000; Zenk 1996). However, Landberg and Greger (2004) did not detect PCs in willow – a plant considered in application of phytoextraction – during long-term treatment of different *Salix* genotypes with a wide range of heavy metals. The mechanism of willow tolerance may be determined by other factors, perhaps metal complexation with low-molecular carboxylic organic acids in the cytoplasm (Clemens 2001; Gąsecka et al. 2011).

Furthermore, plants employ non-specific resistance mechanisms such as enhanced amino acid biosynthesis, mainly proline which chelates heavy metal ions and serves as a free radical scavenger, thus preventing cell membrane lipids from being oxidized (Alia et al. 2001; Mehta and Gaur 1999). Heavy metals induce chemical changes in cell wall composition, i.e. enhanced formation of callose and suberin, accumulation of reactive oxygen species, induction of the antioxidant system in the apoplast and cell interior, as well as biosynthesis of signaling compounds, phytohormones and other regulatory compounds in a controlled response to heavy metal stress (Małecka et al. 2001; Zenk 1996).

5 Biological Methods of Environment Decontamination: Significance and Improvement of Phytoremediation

There is no doubt that increasing consumption to some extent is a sign of the times indicating an increasing and improved level of life, which in consequence results in deterioration of the environment by pollution (rubbish, refuse and debris). The large and growing number and range of pollutants and xenobiotics present in the soil, sediments, sewage and surface water makes it necessary to develop new, more efficient methods of environment cleaning. Elaborated methods are divided into different groups depending on the pollutant specificity, site and cost of the method used. The economic dimension is especially significant when reclamation is performed on a large polluted area. The new environment cleaning methods, including application of nanomaterials to environment protection (mesoporous and siliceous materials usually used for water but also for polluted soil), are still too expensive for the industrial scale.

Most frequently, technical (chemical and physical, such as excavation and burial of soil, reverse osmosis, ion exchange, microfiltration or fixation/inactivation) and biological (bioremediation and phytoremediation) methods are applied (Juwarkar et al. 2010; Jabeen et al. 2009). Technical methods cause a lot of irreversible changes in properties of polluted matrices (e.g. soil destruction) and the final product is often completely deprived of the valuable elements necessary for plants' development. For this reason, biological methods are of prime concern, since they are almost non-invasive to the environment. In these methods plants and microorganisms active in natural processes and present in soil or water ecosystems are usually used.

Phytoremediation ("*phyto*" in Latin means plant and "*remedium*" means restore) is defined as a biological method using selected species/varieties of plants for the effective accumulation of inorganic pollutants or degradation of organic ones (Vamerali et al. 2010). An important restriction is the long time required to achieve satisfactory effects of the process, but it also has a lot of benefits, which means that this effective method may support technical methods or may be used as an individual method of deteriorated environment cleaning. Additionally, the methods are gradually replacing technical methods, especially in the case of areas temporarily excluded from practical use or production.

The development of phytoremediation in the last 20 years is focused on the selection of new plants and/or modification of the process conditions to increase its effectiveness (Lone et al. 2008). The weak point – at the beginning – was the lack of information on complex mechanisms of plant–soil interactions and their influence on transport (translocation) of pollutants to the above-ground plant tissues. Recently, a wide range of studies on the mechanisms of processes in physiology, biochemistry and botany, and with the support of genetic engineering, with emphasis on rapid development of phytoremediation, have been performed. In recent years, phytoremediation as an interdisciplinary technology has been examined in terms of its practical effectiveness in cleaning post-industrial areas, characterized by high contents of various pollutants (organic and inorganic). The most spectacular progress is observed in phytoextraction and phytostabilization studies. The uptake, chelation, translocation or volatilization of heavy metals needs to be developed, since the expected results are very important in elucidation of pollutant transport pathways in the ecosystem, estimation of the risk of phytoremediation for natural processes in the environment, and also identification and introduction of genes in transgenic plants.

Changes in phytoremediation are observed mostly in terms of plant features. Hyperaccumulators able to accumulate above average amounts of pollutants are of prime concern in the area (Memon and Schröder 2009). As the next step, a combination of phytoaccumulation with high and fast efficiency in plant biomass production is important to improve the process rationalization (simpler harvest of plants), and to increase the accumulation with the perspective of wood utilization in energy production. Results presented recently underline physiological and biochemical aspects of phytoremediation with the influence of environmental factors (plant–fungus or plant–soil interactions) with possibilities of more efficient implementation of this process. An interesting prospect in this scope may be the exploration of combined techniques (e.g. selected plants with phytoremediation abilities with such technical or semi-technical methods as electrokinetic remediation under constant voltage across the soil).

Heavy metals belong to an ecologically significant and toxicologically unique class of toxicants, because they are spread everywhere, particularly in industrialized areas. Effective phytoextraction requires the regular (not significantly inhibited) growth of plants in polluted areas followed by the activation of defence mechanisms. Plants with phytoremediation abilities have to meet several fundamental criteria: high effectiveness of phytoaccumulation/phytodegradation, high biomass increase, well-developed root system, high resistance to pollutants, easy adaptation to different environmental conditions and simple environmental requirements (Vangronsveld et al. 2009). Plants used in phytoremediation should exhibit no or very small risk of metals' transport to higher trophic levels (reduced possibility to contaminate the food chain).

To date the following topics have been elucidated and are well known: removal of pollutants by aerial plant organs, transport of metals through the plasmalemma, and detoxification of pollutants in the cell (chaperones, phytochelatins, metallothioneins, low-molecular-weight organic acids – LMWOAs) (Pal and Rai 2010). The above possibilities allow plants to defend themselves against viruses, microbes and fungi with effective phytoaccumulation/phytodegradation of pollutants present in the environment (Rascio and Navari-Izzo 2011).

6 Ecological (Mycorrhizal) and Chemical (Ca/Mg Ratio, LMWOAs) Factors Influencing Phytoremediation Efficiency

Phytoremediation efficiency depends on a complex sequence of factors influencing the number of interactions, one of the most important being the species/variety of used plant. In different studies significant diversity in the phytoextraction/phytodegradation effectiveness was confirmed, not only within the species but also among varieties of the same plant species (Mleczek et al. 2010). The most important differences of plant traits are as follows: structure, size, immunity and individual traits such as environmental and climatic requirements (water, nutrients or temperature).

Natural bioremediation with selected bacterial strains and fungi is an interesting solution in decontamination of areas polluted with organic compounds (Juwarkar et al. 2010). Like other methods, bioremediation is also limited (by interaction of microbes with existing microorganisms, presence of toxic substances inhibiting microbial development or low bioavailability of xenobiotics). Hence, co-operation of both methods has a significant role in increase of phytoremediation efficiency. Along with many strategies focused on plants' accommodation to unfriendly pollutants, a symbiosis with mycorrhizal fungi seems to be very helpful (Vamerali et al. 2010). Fungi in the rhizosphere are a significant factor in phytoaccumulation/phytodegradation efficiency increase, to stimulate plant growth as well as to increase the resistance of plants to concentrations of pollutants found in the environment. The synergism is especially significant in the case of hyperaccumulators. According to the literature data mentioned above, more than 400 plant species are documented as hyperaccumulators and they belong to the following families: *Asteraceae*, *Brassicaceae*, *Caryophyllaceae*, *Cyperaceae*, *Cununiaceae*, *Fabaceae*, *Flacourtiaceae*, *Lamiaceae*, *Poaceae*, *Violaceae* and *Euphorbiaceae*.

The presence of microbial clusters (different genes in rhizoremediation) may decrease levels of plant stress hormone. Especially significant is a combination of plant and plant growth promoting rhizobacteria (PGPR). Inoculation of selected plant species (including hyperaccumulators) with endophytic bacteria, e.g. *Achromobacter xylosoxidans*, *Bacillus pumilus*, *Corynebacterium flavescens*, protects plants against the phytotoxic effects of heavy metals and/or different xenobiotics (Glick 2010). In this case a significant role of plants able to exude pollutant-degrading enzymes into the rhizosphere can be found in plants, fungi, endophytic bacteria and root-colonizing bacteria (e.g. root-specific laccase (LAC1), peroxidases, haloalkane dehydrogenase (DhaA), P450 monooxygenases, phosphatases, nitrilases). These enzymes are able to transform pollutants without their uptake (Dowling and Doty 2009; Gerhardt et al. 2009).

The efficiency of selected heavy metals' phytoaccumulation depends on the mutual relations of macroelements important in plant growth (nutrition) and development in the polluted matrix. When macroelements are present in excess or deficiency, oxidative stress begins but also different ratios of them are important in the phytoremediation process. As an example, when compared to the physiological Ca/Mg ratio (4:1), an increase of calcium ion concentration in relation to magnesium ions in studied soil (Mleczek et al. 2011) resulted in decreased cadmium and lead phytoextraction efficiency by *Salix viminalis* 'Cinamomea'. Additionally, *Salix* growth was restrained under 1:10 Ca/Mg ratio while it was stimulated under 20:1 ratio, which is opposite to cadmium and lead sorption.

Additionally, a change of Ca/Mg ratio influences the amount and kind of low molecular weight organic acids (LMWOAs) exuded into the rhizosphere. A model experiment where the efficiency of formation of selected LMWOAs depending on cadmium, copper, lead and zinc concentration was tested and indicated selective exudation of acids depending on the concentration and the kind of heavy metal. In physiological 4:1 Ca/Mg ratio the following acids formed complexes with particular heavy metal ions: citric, lactic, maleic and succinic acids with Zn^{2+}, and malonic acid with Pb^{2+} and Zn^{2+}. A change of Ca/Mg ratio to 1:10 caused that citric (Cd^{2+}, Zn^{2+} complexation), maleic and succinic (Cd^{2+}, Cu^{2+}, Pb^{2+}, Zn^{2+}) acids were observed in the rhizosphere (Magdziak et al. 2011).

The rhizosphere, as the space in the immediate vicinity of roots, is permanently influenced by their exudates. Moreover, it differs – in relation to other soil fractions – in the composition and large amounts of bacterial cells (the phenomenon of bacteriolysis) and fungi (mycorrhiza), with diversified decay of plant roots, soil structure, composition of organic matter, pH, humidity and microorganism activity. The properties often change in a particular site, demonstrating frequently dynamic changes in time (Macek et al. 2007). All the above-listed factors influence the solubility and uptake of pollutants, both indirectly, through the change of their microbiological activity and the root growth dynamics, and also directly through changes of soil reactions, chelation, precipitation of deposits and oxidation–reduction reactions.

In the broad spectrum of organic compounds present in the rhizosphere, particular attention is focused now on LMWOAs. Organic acids such as: malic, oxalic, acetic or citric are recognized as the most significant ones in many different processes in the rhizosphere. Depending on their degree of dissociation (efficiency), and the amount of carboxylic groups in the molecule, acids can appear in the form of differently charged anions, which in consequence results in the possibility of metal cations' complexation and relocation from the soil. This is the reason that acids are reported as components of the soil environment which in the rhizosphere take part in many processes, e.g. in dissolving and uptake of nutrients (e.g. P and Fe) by plants and microorganisms, decrease of stress associated with anaerobic conditions, dissolving soil minerals leading to pedogenesis, and detoxification of heavy metals by plants (e.g. Al).

LMWOAs exuded by plant roots play a significant role in bacterial microflora composition in relation to nutrients and amounts of available forms of elements in the soil (Magdziak et al. 2011). Moreover, acids influence decomposition of organic matter, and structural formation with particular physical–chemical soil properties. Heavy metals are present in polluted soil in a form insoluble in water, because afore-mentioned water-soluble LMWOAs, as rhizosphere components excluded by the plant root system, change the rhizosphere features, which results in heavy metals' complexation to insoluble forms in soil. LMWOAs usually appear as anions. This allows for instantaneous reaction with metal ions, in the water phase, soil solution and in constant phases, which makes it an important element in the phytoremediation process. It is worth underlining that interaction of organic acids with metals and other elements closely depends on the kind of soil. For example, for a nutrient such as phosphorus, dissolving and mobilization of ions of this element by selected LMWOAs (oxalic and citric acids) is closely related to the soil type. Similar relations exist for other exuded LMWOAs playing a significant role in macroelement mobility increase (e.g. Cu, Cd, Zn) and mechanisms of heavy metal immobilization (Al, Cd, Ni).

The problems presented above and associated with the type of soil and its physical-chemical properties indicate ambiguous information about LMWOAs' function in the rhizosphere. Some data inform about mobilization and elution from the soil of heavy metals after soluble complex formation with the acids, but this information is fragmentary and insufficient to answer the following questions: (1) do organic acids released by roots influence the mobilization and uptake of heavy metals by plants from the rhizosphere or (2) is their amount dependent on the concentration and chemical character of the metal or (3) does the amount of acids indicate activation of the plant defence mechanism against stress? Such information can also elucidate the role of plant genetic factors in increase of heavy metals' availability and uptake from soil followed by improved effectiveness of heavy metal accumulation. For that reason more detailed studies are needed to fully answer the above questions.

7 Conclusions and Future Perspectives in Phytoremediation

To conclude, phytoremediation – being one of the most interesting methods of environment cleaning – seeks for an alternative solution to hyperaccumulators or high biomass green plants. Symbiosis between plants and bacteria or fungi together with genetically modified plants are of the greatest chance for phytoremediation development. The major concern of the process efficiency is plants' resistance to environmental stress factors represented by the pollutant its self (heavy metal, xenobiotic), as well as by accompanying stressors (air pollution at contaminated site, fungal infections, etc.). Thus, the understanding of oxidative stress and defence mechanisms of plants used in phytoremediation is of a great importance. In our studies, tropospheric ozone induced the biosynthesis of salicylic acid in ozone-sensitive tobacco plants which was strongly correlated with the level of injuries observed on leaves after exposure in ambient air conditions. We could assume, that ozone has a strong negative impact on plants causing the ozone-induced oxidative burst.

Several secondary metabolites are synthesized by fungi during morphological and metabolic transitions, when the accumulation of ROS occurs (e.g. *Aspergillus parasiticus*) what effects in oxidative stress and mycotoxins (e.g. aflatoxin) biosynthesis. In conclusion, we believe that oxidative stress promotes secondary metabolism and mycotoxins (secondary metabolites) are part of differentiation process in fungi. On the contrary, plant compounds involved in plant–fungi interactions are able to interfere with mycotoxin biosynthesis in host tissues.

An ideal plant with all traits important in phytoremediation is not available in the environment, which indicates the need for the introduction of new genetically modified organisms (GMO) and their application in phytoremediation. It is believed that through modification of the plant genome, by implanting DNA of one plant in another organism, and thus obtaining a new plant with significantly improved phytoremediation abilities, it will be possible to introduce into practice more valuable plant material. To construct the perfect hyperaccumulator, it is necessary to elucidate the complex of mechanisms in the field, to meet all the basics of phytoremediation. Depending on the goals, selection of important genes followed by model experiments (hydroponic, aeroponic) should be the first step in studies on phytoextraction/phytodegradation efficiency.

The literature in recent years indicates that genes coding heavy metal ion transporters and associated ligands are of prime concern. Generally, it is possible to underline a few fundamental types of genetic plant modifications including increase of resistance to herbicides (two enzyme systems have a significant role: cytochrome P450 monooxygenases (P450s) and glutathione S-transferases (GSTs)), insects, diseases caused by fungi and viruses, as well as unfavourable environmental conditions (Dowling and Doty 2009). Probably all these modifications will be used to improve plants' phytoremediation abilities required depending on the site and conditions of the process (Kawahigashi 2009).

A significant influence on the heavy metal phytoaccumulation efficiency and plant resistance in the case of high levels of heavy metals and other pollutants in the environment is exerted by protein origin heavy metal chelators such as phytochelatins (PCs) and metallothioneins (MT). Gene modifications responsible for phytochelatins or glutathione (GSH, c-L-glutamyl-L-cysteinylglycine) synthesis next to metallothionein genes will probably exemplify one of the most important ways of preparing transgenic plants in the near future (Dowling and Doty 2009). Transport of heavy metal ions from the root system to shoots or leaves requires the presence of specific transport proteins. Additionally, some enzymes essential in biotransformation are able to catalyze the oxidation of toxic heavy metal ions (e.g. As^{3+}, Cr^{6+}, Sb^{3+}), as shown by the new results of various studies (published recently worldwide) indicating new heavy metal ion transporters.

So far, several transporter families have been studied: ABC (ATP-Binding Cassette) – *YCF1*, ATPases type P (P_{1B}) – *AtHMA4*, CDF (Cation Diffusion Facilitator) – cdf1, cdf2, Nramp (Natural resistance associated macrophage protein), YSL (Yellow Stripe Like) – *YS1*, ZIP (Zinc-regulated transporter (ZRT), Iron-regulated transporter (IRT)-like Protein) – *AtZIP1* (Memon and Schröder 2009). The transporters are selective to pollutants but on the contrary being located in the cell membrane, or in membranes of the endoplasmic reticulum and vacuole, also transport essential elements, necessary in normal growth and development of the plant (Ca^{2+}, K^{+}, Mg^{2+}, N, P and S), or microelements (Cu^{2+}, Fe^{2+}, Mn^{2+}, Mo^{2+}, Ni^{2+}, Zn^{2+}) including in this group toxic heavy metal ions (Cd^{2+}, Co^{2+}, Cr^{3+}, Cr^{6+} or Pb^{2+}).

An interesting and ambitious challenge in the area of genetic engineering for phytoremediation is the creation of systems consisting of transgenic plants with bacterial genes. Single-celled prokaryote microorganisms have a significant potential in genes responsible for mechanisms of heavy metal detoxification. Additionally, studies on the role and influence of autochthonic bacteria and mycorrhizal fungus in phytoremediation efficiency will help to enrich and improve the knowledge concerning mechanisms and basics of natural remediation (Yadav et al. 2010).

Probably controversies concerning genetically modified organisms in the case of GM plants applied in phytoremediation will not take place. Since plant hyperaccumulators usually are not in the food chain, it is believed that because of the lack of direct contact, proteins formed during transgene expression will not modify cell metabolism and cause harmful effects to human beings. It is also worth pointing out the dynamic increase of genetic engineering known in farm products (transgenic rice (*Oryza sativa*) – human cytochrome 450 gene *CYP1A1* or human genes encoding human *CYP1A1*, *CYP2B6* and/or *CYP2C19* in rice (Abhilash et al. 2009) through agrobacterium-mediated transformation or transgenic tobacco – *Enterobacter cloacae* as a gene source with pentaerythritol tetranitrate reductase enzyme) introduced in phytoremediation studies (*Brassica juncea*, selected taxa of *Populus*, *Arabidopsis thaliana* or *Phragmites australis*) (James and Strand 2009). Another example is that of transgenic plants in phytodegradation of explosives (Jabeen et al. 2009) by expression of bacterial nitroreductases and cytochrome P450s (e.g. glycerol trinitrate hexahydro-1,3,5-trinitro-1,3,5-triazine and 2,4,6-trinitrotoluene) (Eapen et al. 2007). The above genes, especially in hyperaccumulators (*Thlaspi*, *Brassica*) in combination with significant (high) biomass production, will probably be very interesting trends in phytoremediation studies. The future in the field of genetic engineering in phytoremediation will undoubtedly necessitate a complex approach to this issue, but not only introduction of single genes/traits. In the case of plant–fungi or plant–microbe interactions, attention should be focused on mutual relations between microorganisms and rhizosphere components, including plant exudates. Playing a significant role in dynamics of phytoremediation development will be co-operation between many disciplines (novel gene and enzyme identification with metagenomics and genomic sequencing projects).

Considering the development of phytoremediation, the problem of polluted biomass should be considered. Plants in polluted areas usually exhibit concentrated pollutants, when compared with the relatively low level of pollutants in the soil. In the case of plants characterized by a significant biomass increase in combination with high phytoextraction efficiency, heavily polluted biomass is the final product. Such biomass is periodically collected; so the amount of it is significant and requires particular treatment before further utilization.

The amount of pollutants can be reduced by ocean dumping, deep well injection or approved secure landfills, but biomass volume can be reduced by physical, chemical and thermal methods and with the use of selected microbial cultures. Phytoremediation as a green technology should not generate toxic substances and should improve the quality of the environment. Taking into consideration this important fact, it seems promising to apply thermo-chemical methods to utilize polluted biomass (pyrolysis, still works, combustion or gasification).

Considering that the phytoremediation process is characterized by significant efficiency of pollutants removed from the matrix and great biomass increase, this technology can generate additional financial benefits. The product of biomass thermal disintegration is ash, which has been an interesting subject of studies for many years with additional possibilities of phytoremediation application (phytomining – bio-ore). The biomass obtained after the phytoaccumulation process, including contaminants (heavy metals), can be the substrate in the steel industry, since there is a chance for their permanent recovery as valuable products (recovery of the heavy metals in pure form or in alloys).

Acknowledgements We would like to acknowledge the financial support of the Polish Ministry of Science and Higher Education (grants no.: NN310301934, NN 305045836 and NN305372538) in chapter preparation.

References

Abhilash PC, Jamil S, Singh N (2009) Transgenic plants for enhanced biodegradation and phytoremediation of organic xenobiotics. Biotechnol Adv 27:474–488

Alia, Mohanty P, Matysik J (2001) Effect of proline on the production of singlet oxygen. Amino Acids 21: 195–200

Angelova MB, Pashova SB, Spasova BK, Vassilev SV, Slokoska LS (2005) Oxidative stress response of filamentous fungi induced by hydrogen peroxide and paarquat. Mycol Res 109:150–158

Arora A, Sairam RK, Srivastava GC (2002) Oxidative stress and antioxidative system in plants. Curr Sci 82:1227–1238

Asai T, Stone JM, Heard JE, Kovtun Y, Yorgey P, Sheen J, Ausubel FM (2000) Fumonisin B_1-induced cell death in Arabidopsis protoplasts requires jasmonate-, ethylene-, salicylate-dependent signaling pathways. Plant Cell 12:1823–1836

Ayar-Kayali H, Tarhan L (2004) The effect of glucose and maltose concentrations on pyruvate and ascorbate production, antioxidant enzyme activities and LPO levels in *Fusarium equiseti*. Process Biochem 39: 1823–1828

Ayar-Kayali H, Ozer N, Tarhan L (2002) Intracellular superoxide dismutase, catalase, and glutathione peroxidase and membrane lipid peroxide levels in *Fusarium acuminatum* upon environmental changes in a defined medium. Arch Biochem Biophys 400: 265–272

Beckers GJM, Spoel SH (2006) Fine-tuning plant defence signalling: salicylate versus jasmonate. Plant Biol 8:1–10

Benhamou N, Garand C (2001) Cytological analysis of defense related mechanisms in pea root tissues in response to colonization by nonpathogenic *Fusarium oxysporum* Fo47. Phytopathology 91:730–740

Bennett JW, Klich M (2003) Mycotoxins. Clin Microbiol Rev 16:497–516

Black V, Black C, Roberts J, Stewart C (2000) Impact of ozone on the reproductive development of plants. New Physiol 147:421–447

Borowiak K, Drzewiecka K, Magdziak Z, Gasecka M, Mleczek M (2011) Effect of Ca/Mg ratio on copper uptake, photosynthesis activity and growth of Cu (II) – treated *Salix viminalis* L. "Cannabina". Photosynthetica – manuscript in revision

Bouizgarne B, Brault M, Pennarun AM, Rona JP, Ouhdouch Y, Hadrami I, Bouteau F (2004) Electrophysiological responses to fusaric acid of root hairs from seedlings of date palm susceptible and resistant to *Fusarium oxysporum* f. sp. *Albedinis*. J Phytopathol 152:321–324

Bouizgarne B, El-Maarouf-Bouteau H, Madiona K, Biligui B, Monestiez M, Pennarun AM, Amiar Z, Rona JP, Ouhdouch Y, Hadrami IE, Bouteau F (2006a) A putative role for fusaric acid in biocontrol parasitic angiosperm *Orobanche ramosa*. Mol Plant Microbe Interact 19:550–556

Bouizgarne B, El-Maarouf-Bouteau H, Frankart C, Reboutier D, Madiona K, Pennarun AM, Monestiez M, Trouverie J, Amiar Z, Briand J, Brault M, Rona JP, Ouhdouch Y, Hadrami EI, Bouteau F (2006b) Early physiological responses of *Arabidopsis thaliana* cells to fusaric acid: toxic and signaling effects. New Phytol 169:209–218

Boutigny AL, Richard-Forget F, Barreau C (2008) Natural mechanisms for cereal resistance to *Fusarium* mycotoxins accumulation. Review. Eur J Plant Pathol 121: 411–423

Cachinero JM, Hervas A, Jimenez-Diaz RM, Tena M (2002) Plant defense reactions against fusarium wilt in chickpea induced by incompatible race 0 of *Fusarium oxysporum* f. sp. *ciceris* and non host isolates of *F. oxysporum*. Plant Pathol 15:765–776

Carter JP, Rezanoor HN, Holden D, Desjardins AE, Plattner RD, Nicholson P (2002) Variation in pathogenicity associated with the genetic diversity of *Fusarium graminearum*. Eur J Plant Pathol 108:573–583

Chen J, Zhou J, Goldsbroubh PB (1997) Characterization of phytochelatin synthase from tomato. Physiol Plant 101:165–172

Chong J, Pierrel M-A, Atanassova R, Werck-Reichhart D, Fritig B, Saindrenan P (2001) Free and conjugated benzoic acid in tobacco plants and cell cultures. Induced accumulation upon elicitation of defense responses and role as salicylic acid precursors. Plant Physiol 125:318–328

Ciardi JA, Tieman DM, Lund ST, Jones JB, Stall RE, Klee HJ (2000) Response to *Xanthomonas campestris* pv. *vesicatoria* in tomato involves regulation of ethylene receptor gene expression. Plant Physiol 123:81–92

Clemens S (2001) Molecular mechanisms of plant metal tolerance and homeostasis. Planta 212:475–486

Cobbett CS (2000) Phytochelatin biosynthesis and function in heavy-metal detoxification. Curr Opin Plant Biol 3:211–216

Cobbett CS, Goldsbrough P (2002) Phytochelatins and metallothioneins: roles in heavy metal detoxification and homeostasis. Annu Rev Plant Biol 53:159–182

Czerpak R, Bajguz A (1998) Physiological and metabolic activity of salicylic acid in plants. Kosmos 47:83–93 (in polish)

Daayf F, El Bellaj M, El Hassni M, J'Aiti F, El Hadrami I (2003) Elicitation of soluble phenolics in date palm

(*Phoenix dactylifera*) callus by *Fusarium oxysporum* f. sp. *albedinis* culture medium. Environ Exp Bot 49:41–47

Dobosz B, Drzewiecka K, Waśkiewicz A, Irzykowska L, Bocianowski J, Karolewski Z, Kostecki M, Kruczyński Z, Krzyminiewski R, Weber Z, Goliński P (2011) Free radicals, salicylic acid and mycotoxins in asparagus after inoculation with *Fusarium proliferatum* and *F. oxysporum*. Appl Magn Reson 41:19–30

Dowling DN, Doty SL (2009) Improving phytoremediation through biotechnology. Curr Opin Biotech 20:204–206

Drzewiecka K, Borowiak, K, Bandurska H, Goliński P (2012) Salicylic acid – a potential biomarker of oxidative stress and ozone – induced cell heath caused by tropospheric ozone under ambient conditions. Acta Biol Hung 63:00–00 (article in press)

Durner J, Shah J, Klessig D (1997) Salicylic acid and disease resistance in plants. Trends Plant Sci 2:266–274

Eapen S, Singh S, D'Souza SF (2007) Advances in development of transgenic plants for remediation of xenobiotic pollutants. Biotechnol Adv 25:442–451

Edgar CI, McGrath KC, Dombrecht B, Manners JM, Maclean DC, Schenk PM, Kazan K (2006) Salicylic acid mediates resistance to the vascular wilt pathogen *Fusarium oxysporum* in the model host *Arabidosis thaliana*. Aust Plant Pathol 35:581–591

Enyedi AJ, Yalpani N, Silverman P, Raskin I (1992) Localization, conjugation and function of salicylic acid in tobacco during the hypersensitive reaction to tobacco mosaic virus. Proc Natl Acad Sci USA 89:2480–2484

Fanelli C, Ricelli A, Reverberi M, Fabbri AA (2004) Aflatoxins and ochrotoxins in cereal grains: an open challenge. Recent Res Dev Crop Sci 1:295–317

Ferrer E, Juan-Garcia A, Font G, Ruiz MJ (2009) Reactive oxygen species induced by beauvericin, patulin and zearalenone in CHO-K1 cells. Toxicol In Vitro 23: 1504–1509

Gąsecka M, Mleczek M, Drzewiecka K, Magdziak Z, Chadzinikolau T, Rissmann I, Goliński P (2011) Physiological and morphological changes in *Salix viminalis* as a result of plant exposure to copper. J Environ Sci Heal A (article accepted for publication)

Gerhardt KE, Huang X-D, Glick BR, Greenberg BM (2009) Phytoremediation and rhizoremediation of organic soil contaminants; potential and challenges. Plant Sci 176:20–30

Glazebrook J (2001) Genes controlling expression of defence responses in *Arabidopsis* – 2001 status. Curr Opin Plant Biol 4:301–308

Glick BR (2010) Using soil bacteria to facilitate phytoremediation. Biotechnol Adv 28:367–374

Goliński P, Waśkiewicz A, Gromadzka K (2009) Mycotoxins and mycotoxicoses under climatic conditions of Poland. Pol J Vet Sci 12:581–588

Gracia-Limones C, Hervas A, Navas-Cortes JA, Jimenez-Diaz RM, Tena M (2002) Induction of an antioxidant enzyme system and other oxidative stress markers associated with compatible and incompatible interactions between chickpea (*Cicer arietinum* L.) and *Fusarium oxysporum* f. sp. *ciceris*. Physiol Mol Plant Pathol 61:325–337

Green R (2002) Oxidative stress and acclimation mechanisms in plants. The Arabidopsis Book: Vol. 49, No. 1 pp. 1–20, BioOne Publishers, Washington, USA

Gunterus A, Roze LV, Beaudry R, Linz JE (2007) Ethylene inhibits aflatoxin biosynthesis in *Aspergillus parasiticus* grown on peanuts. Food Microbiol 24:658–663

He CY, Wolyn DJ (2005) Potential role for salicylic acid in induced resistance of asparagus roots to *Fusarium oxysporum* f. sp. *asparagi*. Plant Pathol 54:227–232

He CY, Hasiang T, Wolyn DJ (2002) Induction of systemic disease resistance and pathogen defence responses in *Asparagus officinalis* by nonpathogenic strains of *Fusarium oxysporum*. Plant Pathol 51: 225–230

Huang J-Q, Jiang H-F, Zhou Y-Q, Lei Y, Wang S-Y, Liao B-S (2009) Ethylene inhibited aflatoxin biosynthesis is due to oxidative stress alleviation and related to glutathione redox state changes in *Aspergillus flavus*. Int J Food Microbiol 130:17–21

Jabeen R, Ahmad A, Iqbal M (2009) Phytoremediation of heavy metals: physiological and molecular mechanisms. Bot Rev 75:339–364

James CA, Strand SE (2009) Phytoremediation of small organic contaminants using transgenic plants. Curr Opin Biotechnol 20:237–241

Jayashree T, Subramanyam C (1999) Antiaflatoxigenic activity of eugenol is due to onhibition of lipid peroxidation. Lett Appl Microbiol 28:179–183

Jayashree T, Subramanyam C (2000) Oxidative stress is a prerequisite for aflatoxin production by *Aspergillus parasiticus*. Free Radic Biol Med 10:981–985

Joseph GS, Jayaprakasha GK, Selvi AT, Jena BS, Sakariah KK (2005) Antiaflatoxigenic and antioxidant activities of Garcinia extracts. Int J Food Microbiol 101: 153–160

Juwarkar AA, Singh SK, Mudhoo A (2010) A comprehensive overview of elements in bioremediation. Rev Environ Sci Biotechnol 9:215–288

Kawahigashi H (2009) Transgenic plants for phytoremediation of herbicides. Curr Opin Biotechnol 20:225–230

Klessig DF, Durner J, Noad R, Navarre DA, Wendehenne D, Kumar D (2000) Nitric oxide and salicylic acid signaling in plant defense – colloquium. Proc Natl Acad Sci USA 97:8849–8855

Kley D, Kleinman M, Sandermann H, Krupa S (1999) Photochemical oxidants: state of the science. Environ Poll 100:19–42

Klumpp A, Ansel W, Klumpp G, Calatayud V, Garrec JP, He S, Peñuelas J, Ribas À, Ro-Poulsen H, Rasmussen S, Sanz MJ, Vergne P (2006) Ozone pollution and ozone biomonitoring in European cities. Part I: Ozone concentrations and cumulative exposure indices at urban and suburban sites. Atmos Environ 40: 7963–7974

Koch JR, Creelman RA, Eshita SM, Seskar M, Mullet JE, Davis KR (2000) Ozone sensitivity in hybrid poplar correlates with intensity of both salicylic acid and jasmonic acid. The role of programmed cell death in lesion formation. Plant Physiol 123:487–496

Kumar NP, Dushenkov V, Motto H, Raskin I (1995) Phytoextraction: the use of plants to remove heavy metals from soils. Environ Sci Technol 29:1232–1238

Lamb C, Dixon RA (1997) The oxidative burst in plant disease resistance. Annu Rev Plant Physiol Plant Mol Biol 48:251–275

Landberg T, Greger M (2004) No phytochelatins (PC2 and PC3) detected in *Salix viminalis*. Physiol Plant 121:481–487

Lee H-J, Leon J, Raskin I (1995) Biosythesis and metabolism of salicylic acid. Proc Natl Acad Sci USA 92:4076–4079

Logrieco A, Moretti A, Ritieni A, Caiaffa MF, Macchia L (2002) Beauvericin: chemistry, biology and significance. In: Upadhyay R (ed) Advances in microbial toxin research and its biotechnological exploitation. Kluwer Academic, New York

Lone MI, He Z-L, Stoffella PJ, Yang X-E (2008) Phytoremediation of heavy metal polluted soils and water: progresses and perspectives. J Zhejiang Univ Sci B 9:210–220

Macek T, Kotrba P, Svatos A, Novakova M, Demnerova K, Mackova M (2007) Novel roles for genetically modified plants in environmental protection. Trends Biotechnol 26:146–152

Magdziak Z, Kozlowska M, Kaczmarek Z, Mleczek M, Chadzinikolau T, Goliński P, Drzewiecka K (2011) Influence of Ca/Mg ratio on phytoextraction properties of *Salix viminalis*. II. Secretion of low molecular weight organic acids to the rhizosphere. Ecotox Environ Safe 74:33–40

Mahoney N, Molyneux RJ (2004) Phytochemical inhibition of aflatoxigenicity in *Aspergillus flavus* by constituents of walnut (*Juglans regia*). J Agric Food Chem 52:1882–1889

Maier FJ, Miedaner T, Hadeler B, Felk A, Salomon S, Lemmens M, Kassner H, Schäfer W (2006) Involvement of trichothecenes in fusariose of wheat, barley and maize evaluated by gene disruption of the trichodiene synthase (Tri 5) gene in three field isolates of different chemotype and virulence. Mol Plant Pathol 7:449–461

Małecka A, Jarmuszkiewicz W, Tomaszewska B (2001) Antioxidative defence to lead stress In subcellular compartments of pea root cells. Acta Biochim Pol 48:687–698

Mandal S, Mallick N, Mitra A (2009) Salicilic acid-induced resistance to *Fusarium oxysporum* f. sp. *lycopersici* in tomato. Plant Physiol Biochem 47:642–649

Marchner P, Godbold DL, Jutschhe G (1996) Dynamics of lead accumulation in mycorrhizal and non-mycorrhizal Norway spruce (*Picea abies* L. karst.). Plant Soil 178:239–245

Mauch-Mani B, Mauch F (2005) The role of abscisic acid in plant-pathogen interactions. Curr Opin Plant Biol 8:409–414

Mayer AM, Staples RC, Gil-ad NL (2001) Mechanisms of survival of necrotrophic fungal plant pathogens in hosts expressing the hypersensitive response. Phytochemistry 58:33–41

Mehta SK, Gaur JP (1999) Heavy metal induced proline accumulation and its role in ameliorating metal toxicity in *Chlorella vulgaris*. New Physiol 143:253–259

Memon AR, Schröder P (2009) Implications of metal accumulation mechanisms to phytoremediation. Environ Sci Pollut Res Int 16:162–175

Metroux J-P (2002) Recent breakthroughs in the study of salicylic acid biosynthesis. Trends Plant Sci 2:1–3

Meuwly P, Molders W, Buchala A, Metroux J-P (1995) Local and systemic biosynthesis of salicylic acid in infected cucumber plants. Plant Physiol 109:1107–1114

Mleczek M, Rutkowski P, Rissmann I, Kaczmarek Z, Goliński P, Szentner K, Strażyńska K, Stachowiak A (2010) Biomass productivity and phytoremediation potential of *Salix alba* and *Salix viminalis*. Biomass Bioenerg 34:1410–1418

Mleczek M, Kozłowska M, Kaczmarek Z, Magdziak Z, Goliński P (2011) Cadmium and lead uptake by *Salix viminalis* under modified Ca/Mg ratio. Ecotoxicology 20:158–165

Molodchenkova OO, Adamovskaya VG, Yu AL, Gontarenko OV, Sokolov VM (2002) Maize response to salicylic acid and *Fusarium moniliforme*. Appl Biochem Microbiol 38:381–385

Molyneux RJ, Mahoney N, Kim JH, Campbell BC (2007) Mycotoxins in edible tree nuts. Int J Food Microbiol 119:72–78

Nesci A, Rodriguez M, Etcheverry MG (2003) Control of *Aspergillus* growth and aflatoxin production using antioxidants at different conditions of water activity and pH. J Appl Microbiol 95:279–287

Nesci A, Ferrari L, Etcheverry M (2008) Effect of synthetic antioxidants on stored maize grain mycoflora *in situ*. J Sci Food Agric 88:797–804

Nishiuchi T, Masuda D, Nakashita H, Ichimura K, Shinozaki K, Yoshida S (2006) *Fusarium* phytotoxin trichothecenes have an elicitor-like activity in *Arabidopsis thaliana*, but the activity differed significantly among their molecular species. Mol Plant Microbiol Interact 19:512–520

O'Donnell PJ, Jones JB, Antonine FR, Ciardi J, Klee HJ (2001) Ethylene-dependent salicylic acid regulates an expanded cell death response to a plant pathogen. Plant J 25:315–323

Obrouchova NV, Bystrova EI, Ivanov VB, Anupova O, Seregin I (1998) Root growth responses to lead in young maize seedlings. Plant Soil 200:55–61

Olivain C, Trouvelot S, Binet MN, Cordier C, Pugin A, Alabouvette C (2003) Colonization of flax roots and early physiological responses of flax cells inoculated with pathogenic and nonpathogenic strains of *Fusarium oxysporum*. Appl Environ Microbiol 69:5453–5462

Paciolla C, Dipierro N, Mule G, Logrieco A, Dipierro S (2004) The mycotoxins beauvericin and T-2 induce cell death and alteration to the ascorbate metabolism in tomato protoplasts. Physiol Mol Plant Pathol 65:49–56

Paciolla C, Ippolito MP, Logrieco A, Dipierro N, Mule G, Dipierro S (2008) A different trend of antioxidant defence responses makes tomato plants less susceptible to beauvericin than to T-2 mycotoxin phytotoxicity. Physiol Mol Plant Pathol 72:3–9

Pal R, Rai JPN (2010) Phytochelatins: peptides involved in heavy metal detoxification. Appl Biochem Biotechnol 160:945–963

Pal M, Szalai G, Horvath E, Janda T, Paldi E (2002) Effect of salicylic acid during heavy metal stress. Acta Biol Szeged 46:119–120

Pancheva T, Popova L, Uzunova A (1996) Effect of salicylic acid on growth and photosynthesis in barley plants. J Plant Physiol 149:57–63

Pasqualini S, Torre G, Ferranti F, Ederli L, Piccioni C, Reale L (2002) Salicylic acid modulates ozone-induced hypersensitive cell death in tobacco plant. Physiol Plant 115:204–212

Passone MA, Resnik SL, Etcheverry MG (2005) In vitro effect of phenolic antioxidants on germination, growth and aflatoxin B_1 accumulation by peanut *Aspergillus* section *Flavi*. J Appl Microbiol 99:682–691

Pellinen R, Palva T, Kangasjrvi J (1999) Subcellular localization of ozone-induced hydrogen peroxide production in birch (*Betula pendula*) leaf cells. Plant J 20:349–356

Peng XL, Xu WT, Wang Y, Huang KL, Liang ZH, Zhao WW, Luo YB (2010) Mycotoxin ochratoxin A-induced cell death and changes in oxidative metabolism of *Arabidopsis thaliana*. Plant Cell Rep 29:153–161

Piechalak A, Tomaszewska B, Barałkiewicz D, Malecka A (2002) Accumulation and detoxification of lead in legumes. Phytochemistry 60:153–162

Pierpoint WS (1997) The natural history of salicylic acid; plant product and mammalian medicine. Interdiscipl Sci Rev 22:45–52

Pitzschke A, Forzani C, Hirt H (2006) Reactive oxygen species signaling in plants. Antioxid Redox Signal 8:1757–1764

Ponts N, Pinson-Gadais L, Verdal-Bonnin MN, Barreau C, Richard-Forget F (2006) Accumulation of deoxynivalenol and its 15-acetylated form is significantly modulated by oxidative stress in liquid cultures of *Fusarium graminearum*. FEMS Microbiol Lett 258:102–107

Ponts N, Couedelo L, Pinson-Gadais L, Verdal-Bonnin M-N, Barreau C, Richard-Forget F (2009) *Fusarium* response to oxidative stress by H_2O_2 is trichothecene chemotype-dependent. FEMS Microbiol Lett 293:255–262

Rao MV, Paliyath G, Ormrod DP, Murr DP, Watkins CB (1997) Influence of salicylic acid on H_2O_2 production, oxidative burst, and H_2O_2-metabolizing enzymes. Plant Physiol 115:137–149

Rascio N, Navari-Izzo F (2011) Heavy metal hyperaccumulating plants: how and why do they do it? And what makes them so interesting? Plant Sci 180:169–181

Raskin I (1992) Role of salicylic acid in plants. Annu Rev Plant Physiol Plant Mol Biol 43:439–462

Reverberi M, Zjalic S, Punelli F, Ricelli A, Fabbri A, Fanelli C (2007) *Apyap1* affects aflatoxin biosynthesis during *Aspergillus parasiticus* growth in maize seeds. Food Addit Contam 24:1070–1075

Reverberi M, Zjalic S, Ricelli A, Punelli F, Camera E, Fabbri C, Picardo M, Fanelli C, Fabbri A (2008) Modulation of antioxidant defence in *Aspergillus parasiticus* is involved in aflatoxin biosynthesis: a role for the Ap*yapA* gene. Eukaryot Cell 7:988–1000

Reverberi M, Ricelli A, Zjalic S, Fabbri AA, Fanelli C (2010) Natural functions of mycotoxins and control of their biosynthesis in fungi. Appl Microbiol Biotechnol 87:899–911

Ribnicky DM, Shulaev V, Raskin I (1998) Intermediates of salicylic acid biosynthesis in tobacco. Plant Physiol 118:565–572

Roze LV, Calvo AM, Gunterus A, Beaudry R, Kall M, Linz JE (2004) Ethylene modulates development and toxin biosynthesis in *Aspergillus* possibly via an ethylene sensor-mediated signaling pathway. J Food Prot 67:438–447

Salerno MI, Gianinazzi S, Gianinazzi-Pearson M (2000) Effects on growth and comparison of root tissue colonization patterns of *Eucalyptus viminalis* by pathogenic or non pathogenic strains of *Fusarium oxysporum*. New Phytol 146:317–324

Samardakiewicz S, Woźny A (2000) The distribution of lead in duckweed (*Lemna minor* L.) root tip. Plant Soil 226:107–111

Samuel M, Miles G, Ellis B (2000) Ozone treatment rapidly activates MAP kinase signaling in plants. Plant J 22:367–376

Sarowar S, Lee J-Y, Ahn E-R, Pai H-S (2008) A role of hexokinases in plant resistance to oxidative stress and pathogen infection. J Plant Biol 51:341–346

Schmidt-Heydt M, Magan N, Geisen R (2008) Stress induction of mycotoxins biosynthesis genes by abiotic factors. FEMS Microbiol Lett 284:142–149

Schraudner M, Moder W, Wiese C, Camp WV, Inze D, Langebartels C, Sandermann H (1998) Ozone-induced oxidative burst in the ozone biomonitor plant, tobacco Bel-W3. Plant J 16:235–245

Semighini CP, Harris SD (2008) Regulation of apical dominance in *Aspergillus nidulans* hyphae by reactive oxygen species. Genetics 179:1919–1932

Seregin I, Shpigun L, Ivanov V (2004) Distribution and toxic effects of cadmium and lead on maize roots. Russ J Plant Physiol 51:525–533

Shlaev V, Silverman P, Raskin I (1997) Airborne signaling by methyl salicylate in plant pathogen resistance. Nature 385:718–721

Stroiński A, Kozłowska M (1997) Cadmium-induced oxidative stress in potato tuber. Acta Soc Bot Pol 66:189–195

Thatcher LF, Anderson JP, Singh KB (2005) Plant defence responses: what have learnt from Arabidopsis? Funct Plant Biol 32:1–19

Thatcher LF, Manners JM, Kazan K (2009) *Fusarium oxysporum* hijacks COI1-mediated jasmonate signaling to promote disease development in Arabidopsis. Plant J 58:927–939

Thulke OU, Conrath U (1998) Salicylic acid has a dual role in the activation of defence-related genes in parsley. Plant J 14:35–43

Troshina NB, Yarullina LG, Valeec AS, Maksimov IV (2007) Salicylic acid induces resistance to septoria nodorum Berk in wheat. Biol Bull 34:451–456

Vamerali T, Bandiera M, Mosca G (2010) Field crops for phytoremediation of metal-contaminated land. A review. Environ Chem Lett 8:1–17

Van Assche F, Clijsters H (1990) Effect of metal on enzyme activity in plants. Plant Cell Environ 13:195–206

Vangronsveld J, Herzig R, Weyens N, Boulet J, Adriaensen K, Ruttens A, Thewys T, Vassilev A, Meers E, Nehnevajova E, van der Lelie D, Mench M (2009) Phytoremediation of contaminated soils and groundwater: lessons from the field. Environ Sci Pollut Res Int 16:765–794

Vasyukova NI, Ozeretskovskaya OL (2007) Induced plant resistance and salicylic acid: a review. Appl Biochem Microbiol 43:367–373

Verma S, Dubey R (2003) Lead toxicity induces lipid peroxidation and alters the activities of antioxidant enzymes in growing rice plants. Plant Sci 164:645–655

Vingarzan R (2004) A review of surface ozone background levels and trends. Atmos Environ 38:3431–3442

Waśkiewicz A, Drzewiecka K, Bocianowski J, Karolewski Z, Irzykowska L, Dobosz B, Kostecki M, Weber Z, Krzyminiewski R, Goliński P (2011) Plant-pathogen interactions after asparagus inoculation with *Fusarium proliferatum* and *Fusarium oxysporum* – a multidisciplinary approach. Int J Food Microbiol – manuscript under review

Wohlgemuth H, Mittelstrass K, Kschieschan S, Bender J, Weigel H, Overmyer K, Kangasjarvi J, Sandermann H, Langebartels C (2002) Activation of an oxidative burst is a general feature of sensitive plants exposed to the air pollutant ozone. Plant Cell Environ 25:717–726

Wojtaszek P (1997) Oxidative burst: an early plant response to pathogen infection. Biochem J 322:681–692

Woźny A (1995) Lead in plant cells; uptake, reactions, resistance. Sorus, Poznan, Poland (in polish)

Yadav R, Arora P, Kumar S, Chaudhury A (2010) Perspectives for genetic engineering of poplars for enhanced phytoremediation abilities. Ecotoxicology 19:1574–1588

Ye SF, Zhou YH, Sun Y, Zou LY, Yu JQ (2006) Cinnamic acid causes oxidative stress in cucumber roots, and promoted incidence of *Fusarium* wilt. Environ Exp Bot 56:255–262

Yruela I (2005) Toxic metals in plants: copper in plants. Braz J Plant Physiol 171:145–156

Yu JQ, Shou SY, Qian YR, Hu WH (2000) Autotoxic potential in cucurbit crops. Plant Soil 223:147–151

Zenk M (1996) Heavy metal deteoxification. Curr Opin Plant Biol 3:211–216

Zhu JK (2002) Salt and drought stress signal transduction in plants. Annu Rev Plant Biol 53:247–273

Phytoremediation of Low Levels of Heavy Metals Using Duckweed (*Lemna minor*)

24

Lué-Merú Marcó Parra, Gosmyr Torres, Adolfo David Arenas, Erick Sánchez, and Korina Rodríguez

Abstract

The treatment of waste waters by bioremediation is nowadays an interesting approach due to low cost, simplicity, and the quality of being more environment friendly as compared to other remediation techniques. Certain plants as *Lemna minor* are able to remove heavy metals from contaminated waters. The removal efficiency of this plant was evaluated with different experiments for the contaminants arsenic, mercury, lead, chromium, copper, and zinc. The elements were monitored in water as function of time during a period of 15–25 days, for the experiments using plants, in contaminated waters and controls, with and without plants. The foliar tissues were also analyzed to determine the mass of contaminant accumulated by the specie. The removal efficiency varied from 3 to 30% depending on the element. The contaminants did not affect significantly the agronomical behavior of the *Lemna minor* at the levels used in the experiments (low levels). The plant is well-suited for the phytoremediation in particular mercury phytoremediation.

Keywords

Bioremediation • Mercury • Arsenic • Lead • Chromium • Zinc • Copper • *Lemna minor*

L.-M.M. Parra (✉) • G. Torres • A.D. Arenas • E. Sánchez • K. Rodríguez
Dpto. de Química y Suelos, Universidad Centroccidental
Lisandro Alvarado, Cabudare, Venezuela
e-mail: mparra@ucla.edu.ve

P. Ahmad and M.N.V. Prasad (eds.), *Abiotic Stress Responses in Plants: Metabolism, Productivity and Sustainability*, DOI 10.1007/978-1-4614-0634-1_24,

1 Introduction

The treatment of waste water for the removal of toxic elements requires in general high costs, but becomes a necessary procedure to preserve the water resource and the public health. The toxic elements come into the food chain when the crops are irrigated with high levels of heavy metals waste water (Das et al. 2004). The bioaccumulation on plant and animal tissues could occur (Vidya and Chandrasekaran 2010).

Conventional methods for removing low heavy metal levels in contaminated waters are ionic exchange, adsorption and separation through membranes (Daus et al. 2004), adsorption–coprecipitation using aluminum and iron salts (Song et al. 2006), adsorption onto activated alumina (Lin and Wu 2001), activated coal or bauxite (Daus et al. 2004), inverse osmosis, ionic interchange (Kim and Benjamin 2004), and nanofiltration among others (Pena et al. 2005; Maity et al. 2005). These methods have high costs and difficult implementation and maintenance. Biosorption technology based on the ability of biomass for capture metallic species has been received special attention due to its potentiality for polluted water treatment (Özer et al. 2000; Leusch and Volesky 1995; Upendra and Bandyopadhyay 2006). As an example, the dried roots of water hyacinth (*Eichornia crassipes*) have been used to remove As from water (Shaban et al. 2005). Netzahuatl-Muñoz et al. (2008) evaluated the hexavalent chromium [Cr(VI)] and total chromium removal by the external layer of the Haden variety mango seed. The external layer of the mango seed was able to decrease the concentration of Cr(VI) significantly, from 102 to 1.56 mg L^{-1} in 120 h. It was determined that the external layer of the mango seed is capable of removing chromium from aqueous solutions by two mechanisms: chemical reduction and biosorption. The dead tissue of macrophytes allows the simultaneous heavy metal removal (Miretzky et al. 2006). Among natural sorbents, Chitosan has a high potentiality and low cost. The availability is guaranteed since it is prepared from the exoskeleton of crabs and arthropods (a waste in many cases). The chitosan has been successfully tested for the adsorption and removal of Cr^{+6} (Owlad et al. 2009).

The bioremediation of soils and waters (Juwarkar et al. 2006; Cheng and Wong 2002) is a useful tool to face the troubleshoot of high costs and complexity with the use of plants (Esteve-Turrillas et al. 2005; Zhuang et al. 2007) and other living individuals (Wen et al. 2004, 2006). It is more environment friendly as other remediation techniques. Toxic elements are tolerated by some plants (Tripathi et al. 2007). Some aquatic plants have a high capability to accumulate heavy elements or toxic ones by different mechanisms, and then allow the purification of high-contaminated waters, due to industrial or agrochemical discharges (Maine et al. 2001; Chua 1998; So et al. 2003). Zayed et al. (1998a) evaluated the potential of the *Lemna minor* for the removal of Cd, Cr, Cu, Ni, Pb, and Se. The results showed that at experimental laboratory conditions the plant is a good accumulator of Cd, Se, and Cu, moderate of Cr, and poor of Ni and Pb. Then, the plant has a capacity for the adsorption of heavy metals as could be also mercury, and becomes an alternative for the remediation of this toxic element from lakes and surface waters (Ávila et al. 2007). The species water hyacinth (*Eichhornia crassipes*) and Lesser Duckweed (*Lemna minor*) were previously used for decontamination or reduction of As levels in water (Alvarado et al. 2008). The behavior of these species under As contamination was not well-documented in the literature. The arsenic concentration was determined as function of time in water and foliar tissue at a level of contamination of 0.15 mg L^{-1} and in control groups. The agronomical parameters percentage of dried weight, growth rate, and plant density were evaluated to determine the reliability of the plants for the bioremediation of water contaminated with the element. The importance of the monitoring of water without plants to determine the effect of the element kinetic on the whole process was demonstrated.

In a recent review about the phytoremediation of arsenic by aquatic macrophytes, it was detailed the state of the art (Azizur and Hasegawa 2011). The *Lemna minor* is established as one of the promising species for the arsenic phytoremediation.

Kara and Kara (2005) examined the ability of *Lemna trisulca* to remove soluble cadmium from water. After four days of span, the plant removed about 75–85% of the element from 100 mL of waste water at a level of contamination between 3 and 7 mg L^{-1}.

Oporto et al. (2001) evaluated the potential of *Lemna minor* for Cr(VI) removal from waste water. The methodology included three phases: laboratory tests in batch reactors. Then, the mathematical model to describe the Cr(VI) removal process was developed, and finally the proposed mathematical model was validated in a pilot duckweed system. Chua (1998) determined the bioaccumulation of rare earth elements by *Eichhornia crassipes*. The use of plants for bioremediation has the lack of phytotoxicity (Posada and Arroyave 2006; Prieto et al. 2009) that restricts the application to low levels of contamination with heavy metals.

Posada and Arroyave (2006) performed a description of the mercury effects on aquatic tropical plants and recommended a protocol for the toxicity test using *Lemna minor* and different concentrations of mercury salts at laboratory conditions. They found that the bioavailability of the element depends on the factors influencing the solubility of the mercury in water. Some plants act as biomonitors taking selectively from medium. Other plants act as bioaccumulators, tolerating the element and giving chance to the accumulation through the trophic chain. In the case of *Lemna minor*, the mercury concentrations for toxicity test must be in the range of 0.01–10.0 mg L^{-1}.

Leblebici and Aksoy (2011) evaluated the growth and lead accumulation capacity of *Lemna minor* and the interactions with the nutrient enrichment. They found that the relative growth rates and photosynthetic pigment levels were negatively correlated with metal exposure. The nutrient addition was found to suppress this effect and to diminish the chlorophyll decrease. The study concluded that nutrient enrichment increases the tolerance of *Lemna minor* to metals. The plant is suitable for the bioremediation of low concentrations of lead.

In this work, the evaluation of the aquatic plant *Lemna minor* for the bioremediation of Cr^{+3}, Cr^{+6}, As, Hg, Cu, Zn, and Pb from contaminated water at low levels of these elements is performed focusing also on the behavior of the elements in water without plants at the natural condition of pH 7. The removal percents were calculated on the basis of the amount of elements retained in the plant tissue.

2 Material and Methods

2.1 Bioremediation with Aquatic Plants

2.1.1 Assay Location

The experiments were carried out in a greenhouse at the Estación Experimental "Miguel Luna Lugo" of the Decanato de Agronomía, Universidad Centroccidental "Lisandro Alvarado," Tarabana, Municipio Palavecino, Edo. Lara, Venezuela. The area is characterized as a very dry tropical forest, with a height of 510 m.o.s.l, with a mean of 658.3 mm of rain, potential evapotranspiration of 2048.1 mm $year^{-1}$, mean temperature of 25.1°C, a mean photoperiod of 7.9 h, and relative humidity of 70%. The latitude is 10°1′25″N and the longitude 69°17′W.

2.1.2 Experimental Design

The experimental design was random with three treatments and five repetitions for each element or elemental specie: seven experimental groups, seven control groups, and an additional seven controls of water without plants. The treatments are described in Table 24.1. The experimental units were identical plastic vessels MANAPLAST® of 7-L capacity and surface area of water lamina 0.27 m^2, deep of 13.3 cm, with plants and treatments (as shown in Table 24.1). The initial plant biomass was 100 g.

Table 24.1 Treatments used for the experiment

Treatment	Description
T1i	Control: Water without I element, *Lemna minor* plants
T2i	Water with I element, *Lemna minor* plants
T3i	Control: Water with I element, without plants

Elements (i): As, Hg, Pb, Cr^{+3}, Cr^{+6}, Cu, Zn

2.1.3 Plants

Young plants of Lesser Duckweed (*Lemna minor*), collected in field, were taken in big plastic bags with the water to reduce the stress. Then, the plants were carefully selected and reproduced under the assay conditions to ensure the adaptation. During the adaptation period, the associated species as algae and other aquatic plants were eliminated. A commercial nutrient solution was added to the vessels. After the adaptation period, the plants were again selected for the assay according to the criteria of leaves number and size and taken to the vessels. This procedure is the same as followed by Alvarado et al. (2008).

2.1.4 Experimental Setup

The vessels were random placed in the greenhouse to avoid the preferential effect of temperature gradient or air flow in the groups that could allow different evaporation rates. To each vessel was added the nutrient solution (0.4 $g L^{-1}$ of fertilizer SOLUB® in a volume of 20 L) with a stock composition of SOLUB® as follows: $N–NH_3$ 8%, $N–NO_3$ 10%, P_2O_5 18%, K_2O 18%, MgO 1%, S 1% , B 0.01%, Cu–EDTA 0.019%, Fe–EDTA 0.04%, Mn–EDTA 0.05%, Mo 0.001%, Zn–EDTA 0.019%, carbonate free, potassium sources free of Cl and Na. The solution was prepared with distilled water. The water volume was kept constant during the whole experiment. The water sampling procedure was performed always after water reposition. The element solutions were prepared from stock 1,000 $mg L^{-1}$ solutions Titrisol (Merck, Darmstadt, Germany) by dilution to a nominal concentration in the treatments for mercury (0.1 $mg L^{-1}$), copper (10 $mg L^{-1}$), zinc (0.6 $mg L^{-1}$), and arsenic (0.15 $mg L^{-1}$). The solution for Pb contamination was prepared using $Pb(NO_3)_2$ supplied by Merck in a final concentration in water of 0.7 $mg L^{-1}$. The Cr^{+3} was added as $Cr(NO_3)_3$ and Cr^{+6} as $K_2Cr_2O_7$ both by Merck in a final nominal concentration of 1.4 $mg L^{-1}$. The plant densities were 0.3 $kg m^{-2}$ for *Lemna minor*.

2.1.5 Sampling Procedure

Water Samples

The water samples were taken as function of time, with plastic syringes, after the water reposition, to plastic, 100-mL closed vessels. The first milliliters of sample were used for vessel washing. The water samples were taken with a 20-mL syringe from different deeps and points of the vessel. The samples were collected interdaily for a period of 21 days. They were preserved at pH less than 2, following the methodology described in section 1060C of the Standard Methods for the Examination for Water and Wastewater (Eaton et al. 1995).

Foliar Tissue

The whole plant tissue was collected from each vessel at the 21st day and washed with tap water and then with distilled water. Samples were stored after drying in sealed plastic bags at –20°C.

2.1.6 Arsenic, Lead, Chromium, Zinc, Copper, and Mercury Determination

Lead, chromium, copper, and zinc in water and foliar tissue were determined by Flame Atomic Absorption spectrometry (FAAS) in a Perkin-Elmer® Spectrometer 3110, Waltham, MA, USA. The total arsenic and mercury were determined in water and foliar tissue samples by hydride generation (HG) FAAS with a hydride generation module with quartz cell and a single hollow cathode lamp.

Analysis of Water Samples

The elements Pb and Cr were determined directly in the water samples. For the Hg and As determinations, subsamples of 0.1 or 0.2 mL were aphorized to 10 mL with HCl 0.5 M after a prereduction step. The analysis was carried out according to the Standard Methods for Water and Wastewater Examination (Eaton et al. 1995).

Analysis of Plant Tissue

Subsamples of 0.5 or 1.5 g of dry tissue were digested using the method 3030F of the Standard Methods for Water and Wastewater Examination (Eaton et al. 1995) with a mixture of HCl/HNO_3; final digested sample volume was 50 mL. Then, the analyses by HG-FAAS and FAAS were performed in a similar way as for the water samples.

2.1.7 Agronomical Variables

The variables fresh and dry weight were determined at the beginning of the experiment and at the end of the experiment (day 21). Samples were taken, weighed, and dried in oven at 70°C for 48 h. After the drying procedure, the samples were weighed again to determine the percent of dry weight and the relationship among fresh and dry weight, the estimated total dry weight in each vessel, and total biomass.

2.1.8 Statistical Analysis

The statistical analysis of the results was performed with the program STATISTIX 8 (Analytical Software 2003). The normality of the analyzed variables (elements in water, elements in tissue, in treatments, dry weight %, biomass production, and other parameters) was tested. For variables with normality in error (Wilk–Shapiro test), variance homogeneity (Barlett's test), a parametric test (ANOVA), was applied. For those variables that do not accomplish these criteria, the nonparametric test of Kruskal–Wallis was applied.

3 Results and Discussion

3.1 Bioremediation of Toxic Elements with Duckweed

The concentration of each element in water decreased as function of time in the experimental treatments (with an added amount of the element) (see Fig. 24.1). This is a consequence of the bioabsorption of the elements by the plants, but it must be taken into account that each element, depending on the pH and chemical properties, independently, experiments change in water as function of time. The concentration of the elements decreases in a different way, and depends also on the chemical specie, as it is the case of chromium. The concentration of the elements in water without plants was then monitored. The initial concentration did not remain constant in the treatments without plants, the element concentration changed in some cases below the detection limits, and the availability of the elements could be different, affecting the absorption process (see Fig. 24.2).

Fig. 24.1 Concentration of Cr^{+3} (*unfilled circle*), Pb (*filled circle*), Hg (*filled triangle*), and As (*filled square*) in water as function of time in the presence of Lesser Duckweed (*Lemna minor*)

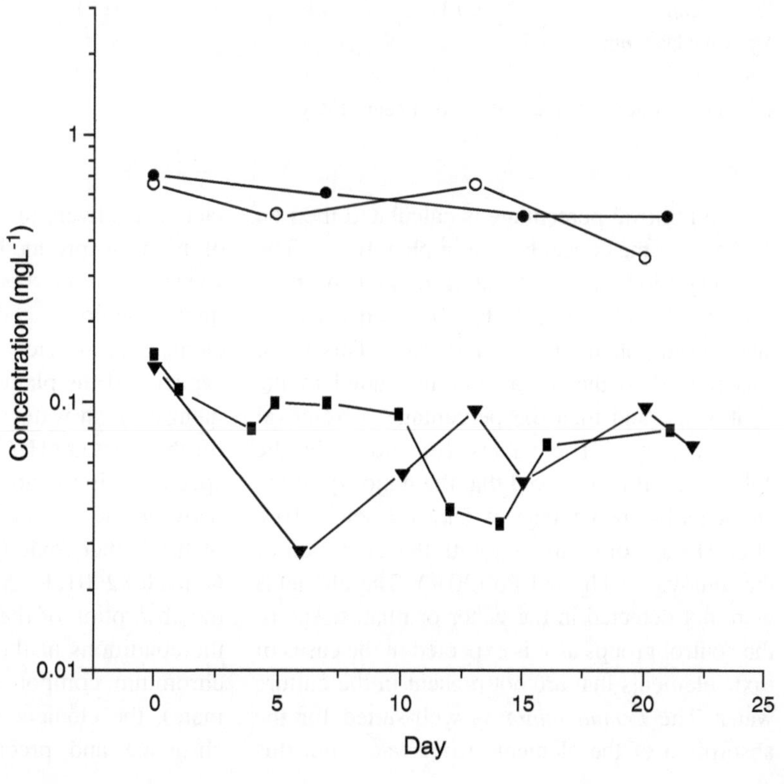

Fig. 24.2 Concentration of Cr^{+3} (*filled square*), Pb (*unfilled circle*), Hg (*filled circle*), and As (*filled triangle*) in water without plants as function of time

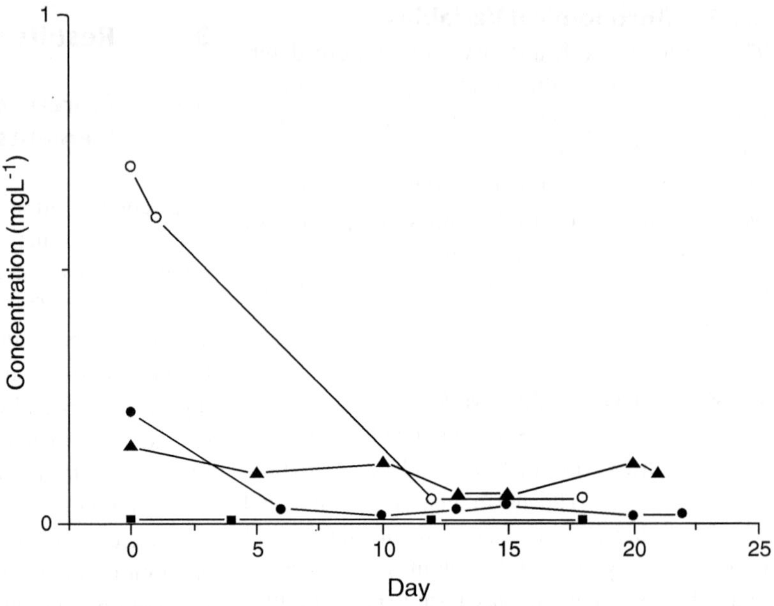

Table 24.2 Percent of removal, final concentration in tissue, and added amount of toxic elements for bioremediation assays with *Lemna minor*

Specie	As	Hg	Cr(III)	Cr(VI)	Pb
% of removal					
Lemna minor	5	30	6,5	3.8	30
Final concentration in plant tissue (mg kg^{-1}) on dry basis					
Lemna minor	2.5±0.3	40±13	100±21	50±23	76±8
Amount added (mg)					
Lemna minor	1.9	1	10	10	2.5

pH value in water 7, period of the experiment 21 days

The removal percentage is calculated then on the basis of the concentration in plant tissue. The total dry biomass in all the cases and the total amount of each element absorbed and accumulated by the plants were determined. This value was related to the initial amount added in the treatments, and then the percentage of removal was calculated. The values are shown in the Table 24.2. It is observed that the majority of the elements has percentage of removal below 10% after 21 days of treatment, with the exception of the removal of Hg and Pb (30%). The elements were not detected in the water or plant tissue of the control groups as it is expected in the cases of toxic elements that are not present in the culture water. The *Lemna minor* is well-suited for the absorption of the elements from water, but this fact is negatively influenced by the lesser amount of biomass production when compared to the water hyacinth (Alvarado et al. 2008). The experiments demonstrated that the removal of each element is affected by the biomass production capacity of the plant and the kinetic of each element itself in water that changes the availability for the plant at pH 7. Regarding to the chromium species, significant differences were found between the Cr^{+3} and Cr^{+6}, and are consequence of the higher toxicity of the Cr^{+6} (Aportela and González 2001; Rodríguez et al. 2002) that affects the absorption of the element from water. Under the conditions of the experiment and due to the chromium compound used (potassium dichromate), the element was mainly in the form of chromate and precipitated (Netzahuatl-Muñoz et al. 2008; Rodríguez et al. 2002). The Cr in the Cr^{+3} treatment also precipitated, as shown in Fig. 24.2, but in a minor level, since it was added as the very soluble nitrate. It is important to point that the assays were carried out using low concentrations of the elements in water to avoid the effect of the toxicity (Zayed et al. 1998b). All concentrations were below 4 mg L^{-1}. For mercury and *Lemna* 0.113 mg L^{-1}, arsenic 0.150 mg L^{-1}, chromium 0.5 mg L^{-1}, and 1.5 and 0.7 mg L^{-1} for Pb. Higher concentrations could be deleterious for the species viability and the process efficiency (Posada and Arroyave 2006).

3.2 Bioremediation of Arsenic and Agronomical Parameters

The fresh and dry biomass were significantly reduced due to the climatic conditions of the experiment, which also negatively influenced the growth rate (Alvarado et al. 2008). This effect was more pronounced in the dry biomass of the control (without As). In wet basis (fresh biomass), nonsignificant differences were found between treatments as consequence of the higher humidity percentage in plants of the control and also due to a more slow generational change in the presence of the arsenic in the experimental treatment. This fact is confirmed with the percentage of dry matter at the end of the experiment. There are significant differences between treatments and a higher percentage of dry matter was observed in the experimental treatment at the end of the experiment, when compared to the control (see Table 24.3).

The lower concentration of the element in water corresponded to the 14th day of the experiment and it is in agreement with the maximum concentration in tissue. After the 14th day, there was a release of the element to water as a consequence of tissue death (Alvarado et al. 2008).

The mass of arsenic removed and the removal rate for *Lemna minor* were low with respect to the initial arsenic level, but the bioaccumulation factor was 16. The harvest of plants must be performed in longer periods for the efficient removal of the contaminant or higher plant densities must be used.

3.3 Bioremediation of Mercury and Agronomical Behavior

The agronomical behavior of the plant was tested with the variables fresh and percentage of dry weight. There are significant differences between the percent of dry biomass at the beginning with respect to the end of the experiment (21 days). $P=0.05$ for both groups (control and experimental). No significant differences were found between the groups regarding to the percentage of dry matter at the end of the experiment.

In the control group (*Lemna* in water without mercury), an increment in the fresh biomass of 64.46 g was found at the end of the experiment (Table 24.3).

Table 24.3 Agronomical variables of *Lemna minor* in contaminated water and controls

Group	Fresh initial biomass (g)	Initial dry weight (g)	Initial dry weight %	Fresh biomass (g)	Final dry weight (g)	Final % of dry weight
Mercury						
Control	100.00	5.3±0.4	5.3±0.4	164±36	10±3	6.7±0.3
Experimental	100.00	5.3±0.4	5.3±0.4	162±24	11±2	6.9±0.8
Arsenic						
Control	100.00	9±1	9±1	75±10	5.9±0.8	8±1
Experimental	100.00	9±1	9±1	60±5	7±1	12±2
Chromium (III)						
Control	100.00	5.3±0.4	5.3±0,4	109±5	7.9±0,6	6.5±0.3
Experimental	100.00	5.3±0.4	5.3±0,4	121±4	9±2	9±1
Chromium (VI)						
Control	100.00	5.3±0.4	5.3±0.4	109±5	9±2	9±1
Experimental	100.00	5.3±0.4	5.3±0.4	116±5	9±2	8±1

In the experimental group (*Lemna* with mercury), the biomass increment was 62.18 and had no significant differences with respect to the control at the end of the experiment; $P = 0.05$. In general, an increment in fresh and dry biomass was observed with respect to the beginning of the experiment. No significant differences were found also in the dry biomass at the end of the experiment between the control and experimental groups; $P = 0.05$.

These facts demonstrated that the concentration of the element in water did not had a hard toxic effect on the plants, evidencing the tolerance for the biological development at the level of contamination used in the experiment (0.133 mg L^{-1}) in concordance to Posada and Arroyave (2006). In that work, in the range of 0.1–1 mg L^{-1}, the plant adapted and observed a fast recovery in the growth.

An important diminution of the mercury concentration was observed in the control of mercury without plants after 6 days of the experiment (from 0.22 at the day 0 till 0.031 mg L^{-1} at the 6th day). After this period, the concentration did not change significantly. This fact implies that the kinetic of the element in water must be taken into account. Part of the added amount of Hg possibly precipitated or was adsorbed in the vessel. It must be taken into account also that the element is volatile and the climatic conditions as temperature and evapotranspiration could also activate the loss of mercury. In the experimental group of mercury with plants, the initial concentration was 0.135 mg L^{-1} and fell to 0.027 mg L^{-1} at the 6th day. At the end of the experiment, the mercury foliar concentrations were 40 μg g^{-1} dry weight, demonstrating that a significant part of the mercury was absorbed by the plant. The removal capacity is then 30% as mentioned before and it is shown in Table 24.2.

The bioremediation is a two-step process according to Wang et al. (1996), and the metal accumulation by aquatic individuals consists in a first step of the fast absorption in the biological surface. The second step is the slow and irreversible transport into the cells (bioaccumulation) by diffusion of the metallic ion through the cell membrane or by the active transport using proteins (Metcalf & Eddy, Inc. 1995; Miretzky et al. 2006). As a surface phenomenon, there is saturation according to the monocape model (Oporto et al. 2001). After the 6th day, part of the retained mercury is released to the water, making it available. Then, the concentration in water raised. At the day 13, the results are according to Burke and Weis (2000). They established that aquatic plants accumulate heavy metals in tissue, which are further released to the water. Posada and Arroyave (2006) observed that *Lemna minor*, at the concentration range of 0.10–1 mg L^{-1}, experimented a decrease in growth till the 4th day. After that period, *Lemna minor* recovered and grew up constantly. According to this experience of Posada and Arroyave (2006), a percentage of the plants died, releasing the absorbed element till the day 13. After this day, the reabsorption occurs again within a new recovery period. As a consequence, the concentration in water decreases till day 20. The adaptation to the contaminated medium was observed. A process of absorption, release, and reabsorption occurs in a cyclic fashion, as happened with the arsenic (Alvarado et al. 2008). The mass balance shows that the decrease of the element concentration in water could be the result not only of the bioabsorption by plants, but also a consequence of different processes as precipitation, adsorption in vessels, and even phytovolatilization. 300 μg of Hg were found in the whole *Lemna minor* tissue and the final mass in water was 520 μg (60); the remaining 180 μg of the added 1,000 μg could be not only precipitated, but also volatized as showed by Wollwnberg and Peters, 2006. The transpiration of Hg(0) was demonstrated by the plants. Carvalho (2001) also reported this behavior of the macrophytes with the selenium.

3.4 Bioremediation of Chromium and Agronomical Variables

The fresh biomass increased in the three treatments as function of time. There are not significant differences among treatments when compared at the end of the experiment with respect to the beginning. The higher fresh biomass increment was found in the treatment with Cr^{+3}. At the end of the experiment, there are no significant differences between the treatments Cr^{+3} and Cr^{+6} ($P=0.05$) and the treatment Cr^{+6} and control. There were found significant differences between the treatment Cr^{+3} and the control regarding to fresh biomass after 21 days. The comparison of the parameter dry matter indicated that significant differences did not exist among the three treatments at the end of the experiment, but there are significant differences and higher values regarding to the beginning of the experiment. The higher percentage of dry matter corresponded to the control treatment with significant differences with respect to the treatments with Cr^{+3} and Cr^{+6}. No differences were found between the treatments with Cr. The presence of the element independently of the chemical specie affects the percentage of dry matter. This is explained by Kabata-Pendias (2000) by the adaptation of the plants reducing the biomass production.

The absorption of the element occurs in the two media with the Cr species. The higher removal percentage corresponded to the treatment with Cr^{+3} due to the toxicity effect of the Cr^{+6} and to the fact that in this experiment the availability of the element was lower for this treatment as shown in Fig. 24.2. The element was not detected in water and tissue of the control treatment. In the treatment without plants, it was observed that the specie Cr^{+6} is not available in the solution even at the beginning of the experiment while the Cr in the Cr^{+3} treatment is more available, although part of the element also precipitated (Fig. 24.2). The initial concentration was 1.25 mg L^{-1} and the final one in water was 0.35 mg L^{-1} with a 72% of removal, taking into account the reduction of the concentration in water. But the results of concentration in the tissue plant showed only 6.5% of Cr removal in the Cr^{+3} treatment and 3.8 in the Cr^{+6} treatment. It is observed that Cr concentration is significantly higher (twice) in the Cr^{+3} treatment than in the Cr^{+6} treatment ($P=0.05$). This is a consequence of the precipitation of the Cr in the Cr^{+6} treatment and the toxicity of this specie (Aportela and González 2001; Panda and Choudhoury 2005). The Cr^{+6}, at neutral pH, could be converted into chromate and then to Cr^{+3} and precipitated as an unavailable specie or fraction, with the restriction of the absorption. The specie of Cr^{+3} added as chloride was more soluble than Cr^{+6} at pH 7.

3.5 Bioremediation of Copper and Zinc and Agronomical Behavior

These elements are essential for the plants and the bioremediation process could differ from that of the toxic elements. The fresh biomass and the gain of the control and the treatment with Zn at the end of the experiment (18 days) were not significantly different. On the contrary, a higher fresh biomass and gain were achieved in the treatment with Cu. This is in agreement with the observation of the behavior of the elements in water. The total amount of Cu was available during the entire experiment and the contribution from nutrient solution was low. Zinc was added in similar amounts in the control and experimental groups since the contribution from the nutrient solution was higher than the initial amount added to the experimental treatment.

As it was observed with other elements in this work, the climatic conditions affected the plant growth. Although Velázquez (1994) indicates an ideal temperature of 25°C for normal development of *Lemna minor* and a range of tolerance till 35–40°C, Ponce et al. (2005) established that the optimal production is observed at pH 6.5–7 and below 27°C. The temperature of all the assays was above 27°C inside the greenhouse.

Regarding to the dry weight, no significant differences were found among treatments at the end of the experiment. A significant increment of this parameter was observed with respect to the initial value of dry weight or biomass (5.3 ± 0.4 g). Zn and Cu affected the percentage of dry weight at the end of the experiment, as it is observed in Table 24.4. This parameter is significantly higher in the control treatment with respect to Zn and Cu treatments at the end of the experiment. The increment in the value with respect to the beginning was observed only in the control, without

Table 24.4 Agronomical parameters for Cu and Zn and control and concentration of Zn and Cu

Treatment	Fresh weight (g)	Dry weight (g)	% of Fresh biomass gain	% of dry weight	μg g^{-1} (Zn)	μg g^{-1} (Cu)
Zinc	116±1a	6.5±0.4a	16±1a	5.9±0.3a	330±50a	–
Control	113±2a	7.0±0.5a	13±2a	6.6 ±0.3b	260±4a	Less than 4a
Copper	134±.9b	7.0±0.6a	34±8b	5.4 ±0.4a	–	15±2b

Values with the same letter in a column mean no significant differences

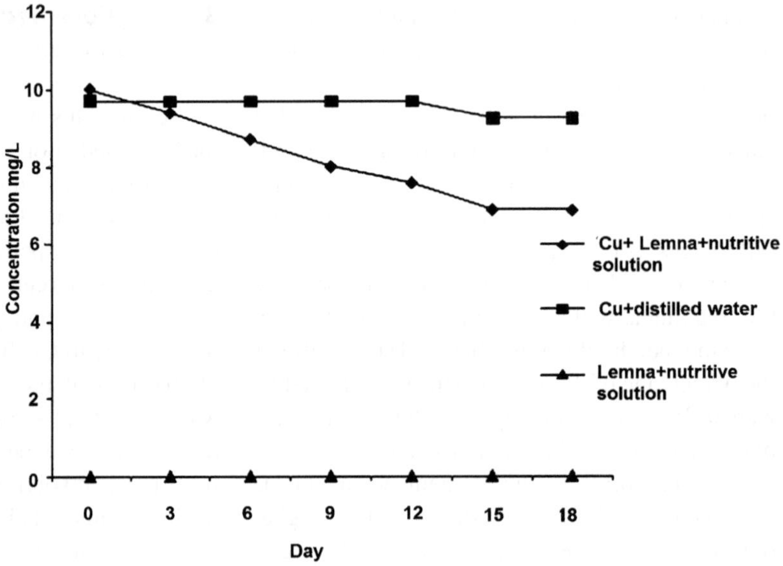

Fig. 24.3 Concentration of copper in water as function of time

changes in the experimental treatments at this respect.

The final zinc concentration of the *Lemna* foliar tissue in the treatment with Zn was not significantly different from that of the control, but the copper concentration was significantly different. With respect to the initial concentration values of these essential elements, an increment was observed: from 24 to 330 mg kg^{-1} for Zn and from a value less than 4 to 15 mg kg^{-1} for Cu. Considering that the concentration in water was significantly higher than the concentration of Zn, the higher bioaccumulation capability of Zn of the plant *Lemna minor* is evident with respect to the Cu.

The concentration of copper in the water decreased within the time in the experimental treatment (*Lemna*+Cu+nutritive solution), evidencing that the element was absorbed by the plant and it was accumulated and used for the physiological functions (see Fig. 24.3).

The removal percentage from water was 31.24% after 18 days. The element was available in water during the entire experiment, as shown the Fig. 24.3. For the water without plants, initial concentration was 9.71±0.27 mg L^{-1} and final 9.25±0.12 mg L^{-1}. Although the nutritive solution contained Cu as essential element, the concentration in water of the control group was undetectable (less than 0.05 mg L^{-1}) and the concentration in the plant tissue controls was below 4 mg kg^{-1} (quantification limit of the analytical method employed in this work). These results are in agreement with Zayed et al. (1998b) who demonstrated the high potential of the plant for the accumulation of Cu from waste waters. Nevertheless, Boniardi et al. (1999), on the contrary, expressed that concentrations higher than 1 mg L^{-1} could be deleterious. Kabata-Pendias (2000) cited, regarding this fact, that a passive absorption in the toxic range could occur. The Cu mobility inside the plant tissue depended directly of the concentration in the substrate. This is observed in the experiment, since the plant progressively reduced the Cu concentration in the

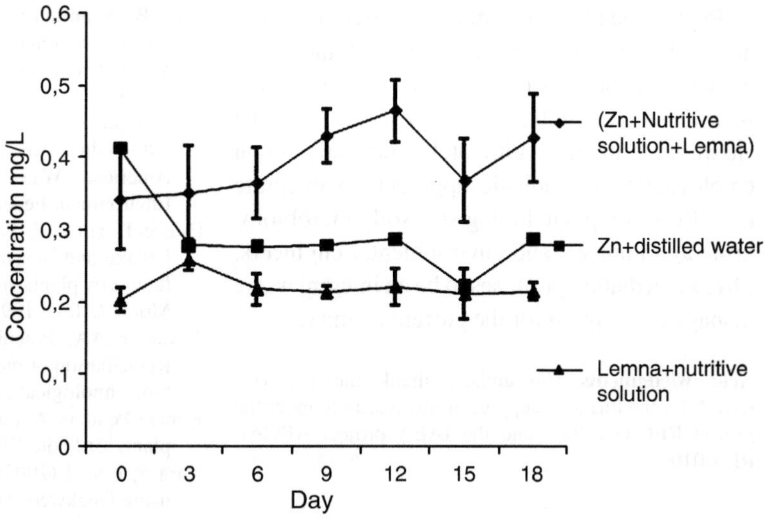

Fig. 24.4 Concentration of zinc in water as function of time

water of the experimental treatment, absorbing and accumulating the element.

This element was added in two ways: (a) an initial amount in the order of 0.5 mg L^{-1} in the experimental group and (b) within the nutrient solution during the entire experiment, as showed the composition of the nutrient solution. As occurs in the case of Cu, the element was present in the control group but in a lesser amount. The concentration of Zn in the water (see Fig. 24.4) of the treatment zinc + *Lemna* + nutritive solution increased 35.29% till day 12 (from 0.34 ± 0.07 to 0.46 ± 0.04 mg L^{-1}) and after that decreased 21.73% till day 15 (0.36 ± 0.06 mg L^{-1}). The plant consumption is low during the first 12 days and the element amounts added within the nutritive solution accumulated in the water. In the second period (day 12–15), a major demand and absorption of the element by the plants occurs. After 15 days, it is observed that the senescence and the concentration of the element in water is consequence of the release to water as well the addition within the nutritive solution.

In the control treatment (*Lemna* + nutritive solution), the concentration in water is constant during the entire experiment. The plant absorbed the essential element from water, but the concentration in the medium reached the equilibrium. The concentration in the treatment without plants decreased about 50%, showing that the element in water also precipitated or was adsorbed in the vessels. The effect is suppressed when nutritive solution is added due to the presence of the chelating agent EDTA.

4 Conclusion and Future Perspective

The plant Lesser Duckweed (*Lemna minor*) is well-suited for the remediation of mercury, arsenic, chromium, zinc, and lead at low levels of contamination (less than 0.5 mg L^{-1}). The harvest of the plant must be performed in the period from 14 to 21 days. The kinetic of the elements and chemical species must be taken into account, since the decrease of the concentration in water is not totally related to the bioaccumulation of the element by the plant. The lower removal percent of the Cr^{+6} specie with respect to the Cr^{+3} is related to the chemical properties in water at pH 7 and to the higher toxicity. The removal of As is determined by the low biomass production of the plant. The plant did not remove so efficiently the specie Cr^{+6}, As, and Cu as mercury. *Lemna minor* has a high capability for the bioaccumulation of zinc at low levels in the medium. In general, both kind of elements, toxics, and essentials did not affect significantly the agronomical behavior of the plant at the conditions used in this work. The main effect on the biomass production is related to the climatic conditions temperature and evapotranspiration.

Phytoremediation technology is used for remediating heavy metal-contaminated soil and water bodies. This process is environment friendly and is still under research as many questions has to be answered. Success ultimately depends upon employment of a holistic approach to integrate the efforts of plant biologists, soil microbiologists, agronomists, and environmental engineers. Phytoremediation promises to be an integral waste management option for the current century.

Acknowledgments The authors thank the CDCHT-UCLA for the financial support of this research under the project REG-001-2007 and the IAEA project ARCAL RLA 010.

References

Alvarado S, Guédez M, Marcó L, Graterol N, Anzalone A, Arroyo J, Záray G (2008) Arsenic removal from waters by bioremediation with the aquatic plants Water Hyacinth (*Eichhornia crassipes*) and Lesser Duckweed (*Lemna minor*). Bioresour Technol 19:8436–8440

Analytical Software (2003) STATISTIX 8 for Windows. Analytical Software, Tallahassee, FL

Aportela P, González Y (2001) Evaluación toxicológica del dicromato de potasio en plantas de lechuga (*Lactuca sativa*). Anuario Toxicol 1:98–103

Arroyave MP (2004) La lenteja de agua (*Lemna minor* L): una planta acuática promisoria. Rev EIA 1:33–38

Ávila H, Soto A, Torres J, Araujo M, Gutiérrez E, Pirela R (2007) Metales pesados en la lenteja acuática (*Lemna* spp.) de la zona costera del Lago de Maracaibo. Bol Centr Invest Biol 41:114–122

Azizur MR, Hasegawa H (2011) Aquatic arsenic: phytoremediation using floating macrophytes. Chemosphere 83:633–646

Boniardi N, Rota R, Nano G (1999) Effect of dissolved metals on the organic load removal efficiency of *Lemna gibba*. Water Res 33(2):530–538

Burke D, Weis J (2000) Release of metals by the leaves of the salt marsh grasses *Spartina alterniflora* and *Phragmites australis*. Estuar Coast Shelf Sci 51:153–159

Carvalho K (2001) Removal of aqueous selenium by four aquatic plants. J Aquat Plant Manag 39:33–36

Cheng J, Wong MH (2002) Effects of earthworms on Zn fractionation in soils. Biol Fertil Soils 36:72–78

Chua H (1998) Bioaccumulation of environmental residues of rare earth elements in aquatic flora *Eichhornia crassipes* (Mart.) Solms in Guangdong Province of China. Sci Total Environ 214:79–85

Das HK, Mitra AK, Sengupta PK, Hossain A, Islam F, Rabbani GH (2004) Arsenic concentrations in rice, vegetables, and fish in Bangladesh: a preliminary study. Environ Int 30:383–387

Daus B, Wennrich R, Weiss H (2004) Sorption materials for arsenic removal from water: comparative study. Water Res 38:2948–2954

Eaton AD, Clesceri LS, Greenberg AE (1995) Standard methods for the examination of water and wastewater, 19th edn. American Public Health Association, American Water Works Association, and Water Environment Federation, Washington, DC

Esteve-Turrillas F, Scott W, Pastor A, Dean J (2005) Uptake and bioavailability of persistent organic pollutants by plants grown in contaminated soil. J Environ Monit 7:1093–1098

Juwarkar AA, Prachi J, Singh SK, Rayalu S (2006) Remediation of metal contaminated soil using a novel biotechnological approach. Environ Sci 1:4–6

Kabata-Pendias A (2000) Trace elements in soils and plants, 3rd edn. CRC, Boca Raton, 413 pp

Kara Y, Kara I (2005) Removal of cadmium from water using Duckweed (*Lemna trisulca* L.). Int J Agric Biol 7:660–662

Kim J, Benjamin MM (2004) Modeling a novel ion exchange process for arsenic and nitrate removal. Water Res 38:2053–2062

Leblebici Z, Aksoy A (2011) Growth and lead accumulation capacity of *Lemna minor* and *Spirodela polyrhiza* (Lemnaceae): interactions with nutrient enrichment. Water Air Soil Pollut 214:175–184

Leusch A, Volesky B (1995) The influence of film diffusion on cadmium biosorption by marine biomass. J Biotechnol 43:1–10

Lin TF, Wu JK (2001) Adsorption of arsenite and arsenate within activated alumina grains: equilibrium and kinetics. Water Res 35:2049–2057

Maine MA, Duarte MV, Sune NL (2001) Cadmium uptake by floating macrophytes. Water Res 35:2629–2634

Maity S, Chakravarty S, And BS, Roy BC (2005) A study on arsenic adsorption on polymetallic sea nodule in aqueous medium. Water Res 39:2579–2590

Metcalf & Eddy, Inc. (1995) Ingeniería de aguas residuales. Tratamiento, vertido y reutilización. McGraw-Hill Interamericana de España S.A., Madrid, 1485 pp

Miretzky P, Saralegui A, Fernandez C (2006) Simultaneous heavy metal removal mechanism by dead macrophytes. Chemosphere 62:247–54

Netzahuatl-Muñoz AR, Cristiani-Urbina MC, Cristiani-Urbina E (2008) Uso de la semilla de mango variedad Haden para la remoción de cromo hexavalente de soluciones acuosas. Rev Cub Quím 20:91–98

Oporto C, Arce O, De Pauw N, Van den Broeck E (2001) Evaluación del potencial de *Lemna minor* para la remoción de Cr(VI) de aguas residuales. Rev Boliv Ecol 10:17–27

Owlad M, Aroua MK, Daud WAW, Baroutian S (2009) Removal of hexavalent chromium-contaminated water and wastewater: a review. Water Air Soil Pollut 200: 59–77

Özer D, Özer A, Dursun G (2000) Investigation of zinc(II) adsorption on *Cladophora crispata* in a two-staged reactor. J Chem Technol Biotechnol 75:410–416

Panda S, Choudhoury S (2005) Chromium stress in plants. Braz J Plant Physiol 17:95–102

Pena ME, Korfiatis GP, Patel M, Lippincott L, And MX (2005) Adsorption of As(V) and As(III) by nanocrystalline titanium dioxide. Water Res 39:2327–2337

Ponce J, Febrero J, Gonzalez R, Romero O, Estrada O (2005) Perspectiva de la *Lemna* sp. para la alimentacion de peces. REDVET (revista electrónica de veterinaria) IV(3). http://www.veterinaria.org/revista/redvet/n030305.html. January 10th 2011

Posada M, Arroyave M (2006) Efectos del mercurio sobre algunas plantas acuáticas tropicales. Rev EIA 6: 57–67

Prieto J, González C, Román A, Prieto F (2009) Plant contamination and phytotoxicity due to heavy metals from soil and water. Trop Subtrop Agroecosyst 10: 29–44

Rodríguez C, Gómez H, Scilia R (2002) Estudio de la Sensibilidad espectro fotométrica del cromo III y del cromo VI. Rev Per Quím Ing Quím 5:29–36

Shaban W, Rmalli A, Harrington CF, Ayub M, Parvez I, Haris I (2005) A biomaterial based approach for arsenic removal from water. J Environ Monit 7: 279–282

So LM, Chu LM, Wong PK (2003) Microbial enhancement of Cu^{2+} removal capacity of *Eichhornia crassipes* (Mart.). Chemosphere 52:1499–1503

Song S, López-Valdivieso A, Hernández Campos DJ, Peng C, Monroy-Fernández MG, Razo-Soto I (2006) Arsenic removal from high arsenic water by enhanced coagulation with ferric ions and coarse calcite. Water Res 40:364–372

Tripathi RD, Srivastava S, Mishra S, Singh N, Tuli R, Gupta DK, Maathuis FJM (2007) Arsenic hazards: strategies for tolerance and remediation by plants. Trends Biotechnol 25:158–165

Upendra K, Bandyopadhyay M (2006) Fixed bed column study for Cd(II) removal from wastewater using treated rice husk. J Hazard Mater 129:253–259

Velázquez J (1994) Plantas acuáticas de Venezuela. Universidad Central de Venezuela, CDCH, Caracas, pp 830–839

Vidya R, Chandrasekaran N (2010) Ecotoxicological studies on the bioaccumulation of the heavy metals in the Vellore population, Tamil Nadu, India. Sci Technol 2:60–65

Wang TC, Weissman JC, Ramesh G, Benemann JR (1996) *Parameters for removal of toxic heavy metals by water milfoil (Myriophyllum spicatum).* Bulletin of Environmental Contamination and Toxicology 57:779–786

Wen B, Hu X, Liu Y, Wang W, Feng M, Shan X (2004) The role of earthworms (*Eisenia fetida*) in influencing bioavailability of heavy metals in soils. Biol Fertil Soils 40:181–187

Wen B, Liu Y, Hu X, Shan X (2006) Effect of earthworms (*Eisenia fetida*) on the fractionation and bioavailability of rare earth elements in nine Chinese soils. Chemosphere 63:1179–1186

Wollenberg JL, Peters SC (2009) Diminished mercury emission from waters with duckweed cover. J Geophys Res 114:10–29

Zayed A, Mel Lytle C, Quin J-H, Terry N (1998a) Chromium accumulation translocation and chemical speciation in vegetable crops. Planta 206: 293–299

Zayed A, Gowtheman S, Terry N (1998b) Phytoaccumulation of trace elements by wetland plants: I. Duckweed. J Environ Qual 27:715–721

Zhuang X, Chen J, Shim H, Bai Z (2007) New advances in plant growth-promoting rhizobacteria for bioremediation. Environ Int 33:406–413

Index

P. Ahmad and M.N.V. Prasad (eds.), *Abiotic Stress Responses in Plants: Metabolism, Productivity and Sustainability*, DOI 10.1007/978-1-4614-0634-1, © Springer Science+Business Media, LLC 2012

B

C

MIX
Papier aus verantwortungsvollen Quellen
Paper from responsible sources
FSC® C105338

If you have any concerns about our products,
you can contact us on
ProductSafety@springernature.com

In case Publisher is established outside the EU,
the EU authorized representative is:
Springer Nature Customer Service Center GmbH
Europaplatz 3, 69115 Heidelberg, Germany

Printed by Libri Plureos GmbH
in Hamburg, Germany